The Science of Grapevines

The Science of Grapevines

Third Edition

Markus Keller
Irrigated Agriculture Research and Extension Center,
Washington State University, Prosser,
WA, United States

Academic Press is an imprint of Elsevier
125 London Wall, London EC2Y 5AS, United Kingdom
525 B Street, Suite 1650, San Diego, CA 92101, United States
50 Hampshire Street, 5th Floor, Cambridge, MA 02139, United States
The Boulevard, Langford Lane, Kidlington, Oxford OX5 1GB, United Kingdom

© 2020, 2015, 2010 Markus Keller. Published by Elsevier Inc. All rights reserved.

No part of this publication may be reproduced or transmitted in any form or by any means, electronic or mechanical, including photocopying, recording, or any information storage and retrieval system, without permission in writing from the publisher. Details on how to seek permission, further information about the Publisher's permissions policies and our arrangements with organizations such as the Copyright Clearance Center and the Copyright Licensing Agency, can be found at our website: www.elsevier.com/permissions.

This book and the individual contributions contained in it are protected under copyright by the Publisher (other than as may be noted herein).

Notices
Knowledge and best practice in this field are constantly changing. As new research and experience broaden our understanding, changes in research methods, professional practices, or medical treatment may become necessary.

Practitioners and researchers must always rely on their own experience and knowledge in evaluating and using any information, methods, compounds, or experiments described herein. In using such information or methods they should be mindful of their own safety and the safety of others, including parties for whom they have a professional responsibility.

To the fullest extent of the law, neither the Publisher nor the authors, contributors, or editors, assume any liability for any injury and/or damage to persons or property as a matter of products liability, negligence or otherwise, or from any use or operation of any methods, products, instructions, or ideas contained in the material herein.

Library of Congress Cataloging-in-Publication Data
A catalog record for this book is available from the Library of Congress

British Library Cataloguing-in-Publication Data
A catalogue record for this book is available from the British Library

ISBN 978-0-12-816365-8

For information on all Academic Press publications
visit our website at https://www.elsevier.com/books-and-journals

Publisher: Charlotte Cockle
Acquisition Editor: Nancy Maragioglio
Editorial Project Manager: Kelsey Connors
Production Project Manager: Vignesh Tamil
Cover Designer: Miles Hitchen

Typeset by SPi Global, India

Contents

About the author ix
Preface to the third edition xi

CHAPTER 1 Taxonomy and anatomy 1
   1.1 Botanical classification and geographical distribution 1
          Domain Eukaryota 3
          Kingdom plantae 3
          Division (synonym phylum) Angiospermae (synonym Magnoliophyta) 4
          Class dicotyledoneae (synonym Magnoliopsida) 4
          Order vitales (formerly Rhamnales) 4
          Family Vitaceae 5
          Genus Muscadinia 5
          Genus *Vitis* 6
          American group 8
          Eurasian group 10
   1.2 Cultivars, clones, and rootstocks 11
          1.2.1 Variety versus cultivar 11
          1.2.2 Cultivar classification 20
          1.2.3 Clones 22
          1.2.4 Rootstocks 24
   1.3 Morphology and anatomy 26
          1.3.1 Roots 28
          1.3.2 Trunks and shoots 36
          1.3.3 Nodes and buds 44
          1.3.4 Leaves 47
          1.3.5 Tendrils and clusters 53
          1.3.6 Flowers and grape berries 56

CHAPTER 2 Phenology and growth cycle 61
   2.1 Seasons and day length 61
   2.2 Vegetative cycle 65
          2.2.1 Bleeding 65
          2.2.2 Budbreak 66
          2.2.3 Shoot growth 69
          2.2.4 Growth cessation and senescence 76
          2.2.5 Bud development and dormancy 78
          2.2.6 Root growth 82

2.3 Reproductive cycle 86
      2.3.1 Inflorescence and flower formation 86
      2.3.2 Pollination and fertilization 89
      2.3.3 Fruit set 94
      2.3.4 Berry and seed development 97

CHAPTER 3 Water relations and nutrient uptake 105
   3.1 Osmosis, water potential, and cell expansion 105
   3.2 Transpiration and stomatal action 110
   3.3 Water and nutrient uptake and transport 114
      3.3.1 Driving forces and resistances 114
      3.3.2 Root water uptake 119
      3.3.3 Symplast and apoplast 120
      3.3.4 Channels and transporters 122
      3.3.5 Transpiration stream and growth 124

CHAPTER 4 Photosynthesis and respiration 129
   4.1 Light absorption and energy capture 129
   4.2 Carbon uptake and assimilation 134
      4.2.1 Gas exchange 134
      4.2.2 Carbon assimilation 135
      4.2.3 Carbohydrate production 137
   4.3 Photorespiration 140
   4.4 Respiration 142
   4.5 From cells to plants 145

CHAPTER 5 Partitioning of assimilates 149
   5.1 Photosynthate transport and distribution 149
      5.1.1 Sources and sinks 149
      5.1.2 Assimilate transport 152
      5.1.3 Loading, transport, and unloading 156
      5.1.4 Allocation and partitioning 161
   5.2 Canopy–environment interactions 168
      5.2.1 Light 170
      5.2.2 Temperature 179
      5.2.3 Wind 183
      5.2.4 Humidity 184
      5.2.5 The "ideal" canopy 185
   5.3 Nitrogen assimilation and interaction with carbon metabolism 188
      5.3.1 Nitrate uptake and reduction 188
      5.3.2 Ammonium assimilation 190
      5.3.3 From cells to plants 192

CHAPTER 6 Developmental physiology 199
6.1 Yield formation 199
6.1.1 Yield components and compensation 199
6.1.2 Yield potential and its realization 203
6.2 Grape composition and fruit quality 211
6.2.1 Cell walls and membranes 211
6.2.2 Water 215
6.2.3 Sugars 219
6.2.4 Acids 223
6.2.5 Nitrogenous compounds and mineral nutrients 226
6.2.6 Phenolics 230
6.2.7 Lipids and volatiles 245
6.3 Sources of variation in fruit composition 251
6.3.1 Fruit maturity 254
6.3.2 Light 255
6.3.3 Temperature 259
6.3.4 Water status 264
6.3.5 Nutrient status 268
6.3.6 Crop load 272
6.3.7 Rootstock 275

CHAPTER 7 Environmental constraints and stress physiology 279
7.1 Responses to abiotic stress 279
7.2 Water: Too much or too little 284
7.2.1 Water surplus 285
7.2.2 Water deficit 287
7.3 Mineral nutrients: Deficiency and excess 301
7.3.1 Macronutrients 307
7.3.2 Transition metals and micronutrients 322
7.3.3 Salinity 330
7.4 Temperature: Too hot or too cold 335
7.4.1 Heat acclimation and damage 335
7.4.2 Chilling stress 340
7.4.3 Cold acclimation and damage 342

CHAPTER 8 Living with other organisms 357
8.1 Biotic stress and evolutionary arms races 357
8.1.1 Competitors, herbivores, and pathogens 358
8.1.2 Defense strategies 361
8.2 Pathogens: Defense and damage 366
8.2.1 Bunch rot 367
8.2.2 Powdery mildew 371

8.2.3 Downy mildew 374
 8.2.4 Bacteria 376
 8.2.5 Viruses 378

Glossary 383
Abbreviations and symbols 393
Bibliography 395
Internet resources 519
Index 521

About the author

Markus Keller is the Chateau Ste. Michelle Distinguished Professor of Viticulture at Washington State University's Irrigated Agriculture Research and Extension Center in Prosser. He received his master's degree in agronomy (plant science) in 1989 and a doctorate in natural sciences in 1995 from the Swiss Federal Institute of Technology in Zürich. Having grown up on a diversified farm that included wine grape production among other crops and livestock, he began his research and teaching career in viticulture and grapevine physiology at the Federal Research Station for Fruit-Growing, Viticulture and Horticulture in Wädenswil (now Agroscope Changins-Wädenswil), Switzerland. He then moved to Cornell University in Geneva, New York, and from there to Charles Sturt University in Wagga Wagga, Australia, before coming to eastern Washington. He also has been a regular guest lecturer at the Universidad Nacional de Cuyo in Mendoza, Argentina. He was awarded the Swiss Agro-Prize for innovative contributions to Switzerland's agriculture industry. His research focuses on developmental and environmental factors, as well as vineyard management practices, as they influence crop physiology and production of wine and juice grapes. He has had a long involvement with the American Society for Enology and Viticulture and currently serves as the science editor for the society's journals.

Preface to the third edition

The Science of Grapevines was first published 10 years ago. Science (Latin *scientia* = knowledge), or the pursuit of knowledge, does not stand still, of course, which poses a challenge for any printed textbook. For the third edition, all chapters have been thoroughly revised and updated. I have reviewed and integrated novel material to present the latest information available from the scientific literature. Additionally, I have closed or at least narrowed several gaps by revisiting some of the older, classic literature. The task has not been trivial; the bibliography of the third edition contains more than 2700 literature references.

Grapes were among the first fruit species to be domesticated and remain the world's most economically important fruit crop, mainly due to the value that is added during postharvest processing. According to the International Organisation of Vine and Wine, grapevines were grown in 2018 on about 7.4 million hectares of vineyard land producing more than 73 million metric tons of fruit. The Food and Agriculture Organization of the United Nations estimates the global production value of grapes to be close to 70 billion US dollars. While more than 70% of the world's total grape crop was long used to make wine, this proportion has now declined to less than 50% owing to the rapid increase in table grape production, especially in China, since the start of this millennium. A small portion of wine is further distilled to brandy. In 2015, 36% of the global grape production was consumed as fresh fruit (table grapes), 8% was consumed as dried fruit (raisins), 6% was processed to grape juice, and less than 3% was transformed into vinegar, jam, jelly, or grape seed oil or extract.

The Science of Grapevines explores the state of knowledge of the construction and life of this economically and socially vital plant species. The book is an introduction to the physical structure of the grapevine, its various organs and tissues, their functions, their interactions with one another, and their responses to the environment. The study of an organism's external form and structure is termed morphology, its internal structure is studied in anatomy, and the functions of its organs are explored in the field of physiology. Following a summary of the morphology and anatomy of grapevines, this book focuses essentially on the physical and biological functions of whole plants and organs rather than the metabolism and molecular biology of individual cells. It is nonetheless necessary to review some fundamental processes at the cell and tissue levels in order to build up an appreciation of whole-plant function. The book covers those elements of anatomy and physiology that will enhance our understanding of grapevine function and their implications for practical vineyard management. Most physiological processes (water movement through the vine's hydraulic system and evaporation through plant surfaces may be exceptions) are rooted in biochemistry. They are driven or at least facilitated by enzymes which, in turn, are built based on blueprints provided by genes. I have therefore taken it for granted that it is understood that a developmental process or change in chemical composition implies a change in enzyme activity, which in turn implies a change in the activity, or expression, of one or more genes. This does not imply, as used to be thought, that "one gene makes one enzyme," or that "one enzyme makes one chemical," but means merely that all enzymatic processes are rooted in the dynamic expression of certain genes.

Many biochemical and biophysical processes apply to many or even all plants. Perhaps no process is truly unique to grapevines. Chances are if grapes employ a solution to a survival issue, then some or many other species do the same thing, because they share a common ancestor that invented the trick a long time ago. For example, microbes hit upon photosynthesis and respiration long before these discoveries enabled some of them to join forces and evolve into plants. Consequently, although this book is about grapevines, and primarily about the species *Vitis vinifera* to which most wine and table grape cultivars belong, I have borrowed heavily from knowledge gained from other plant species. These include wild and cultivated, perennial and annual, as well as woody and herbaceous forms, including that "queen of weeds"—at least in the fast-paced world of modern molecular biology—*Arabidopsis thaliana*, the otherwise inconspicuous thale cress. I have even taken the liberty of borrowing insights gained using microorganisms such as the yeast *Saccharomyces cerevisiae*, which gives us the wine, beer, and bread that help us to think about these issues.

This book aims to be global in scale. It covers topics ranging from the physiological aspects of tropical viticulture near the equator all the way to those that pertain to the production of ice wine at the temperate latitudinal margins of grape growing. It moves from vineyards at sea level to vineyards at high altitude. It considers the humid conditions of cool, maritime climates, the moist winters and dry summers of Mediterranean climates, as well as the arid environment, often combined with hot summers and cold winters, typical of continental climates in the rain shadows of massive mountain ranges. Yet a book of this nature is necessarily incomplete, and so is the selection of published information included in the text. No one can read everything that has been and is being published, even in the admittedly relatively narrow field of grapevine anatomy and physiology. The magnitude of the task of reviewing as much of the pertinent literature as possible often forced me to rely on review papers where they were available. I apologize to those friends and colleagues whose work I did not cite or cited incompletely or, worse, incorrectly. Science—and scientists—can only ever hope to approximate the truth. This and the simple fact of "*errare humanum est*" will guarantee a number of errors throughout the text. These are entirely my responsibility, and I would be grateful for any feedback that might help improve this book and further our understanding of the world's most important and arguably most malleable fruit crop. After all, the full quote from Seneca the Younger, who was a contemporary of Columella, the Roman author of agriculture and viticulture textbooks, reads "*errare humanum est, sed in perseverare diabolicum*" ("to err is human, but to persevere is devilish").

As with the first and second editions, many people have contributed to bringing this third edition to fruition. I am indebted to all these generous individuals. My wife, Sandra Wran, has supported this project throughout and waited countless weekends for me to read yet another paper or revise another section. Lynn Mills helped with data collection and some of the most recalcitrant illustrations. Feedback from colleagues and students has helped to eliminate numerous mistakes, some of them more glaring than others. I am also grateful for the encouragement from Nancy Maragioglio, Tasha Frank, and Kelsey Connors at Elsevier, who were always quick to answer my questions *du jour*.

CHAPTER 1

Taxonomy and anatomy

Chapter outline

1.1 Botanical classification and geographical distribution 1
    Domain Eukaryota 3
    Kingdom plantae 3
    Division (synonym phylum) Angiospermae (synonym Magnoliophyta) 4
    Class dicotyledoneae (synonym Magnoliopsida) 4
    Order vitales (formerly Rhamnales) 4
    Family Vitaceae 5
    Genus Muscadinia 5
    Genus *Vitis* 6
    American group 8
    Eurasian group 10
1.2 Cultivars, clones, and rootstocks 11
    1.2.1 Variety versus cultivar 11
    1.2.2 Cultivar classification 20
    1.2.3 Clones 22
    1.2.4 Rootstocks 24
1.3 Morphology and anatomy 26
    1.3.1 Roots 28
    1.3.2 Trunks and shoots 36
    1.3.3 Nodes and buds 44
    1.3.4 Leaves 47
    1.3.5 Tendrils and clusters 53
    1.3.6 Flowers and grape berries 56

1.1 Botanical classification and geographical distribution

The basic unit of biological classification is the species. According to the "biological species concept," a species is defined as a community of individuals—that is, a population or group of populations whose members can interbreed freely with one another under natural conditions but not with members of other populations (Mayr, 2001; Soltis and Soltis, 2009). In other words, such communities are reproductively isolated. Although each individual of a sexual population is genetically unique, each species is a closed gene pool, an assemblage of organisms that do not normally exchange genes with other species. Their genes compel the individuals belonging to a species to perpetuate themselves over many generations. Yet all life forms on Earth are interrelated; they all ultimately descended from a common ancestor and "dance" to the same genetic code, whereby different combinations of three consecutive nucleotides of each organism's deoxyribonucleic acid (DNA) specify different amino acids that can be assembled into proteins. Because they are thus interrelated, organisms can be grouped according to the degree of their genetic similarity, external appearance, and behavior. In the classification hierarchy, closely related species are grouped into a genus, related genera into a family, allied families into an order, associated orders into a class, similar classes into a division (plants) or a phylum (animals), related divisions or phyla into a kingdom, and, finally, allied kingdoms into a domain. The "evolutionary species concept" recognizes this ancestor–offspring connection among populations that may follow distinct evolutionary paths to occupy separate ecological niches but may continue to interbreed for some time (Soltis and Soltis, 2009). For example, although they have been geographically isolated for over 20 million years, Eurasian and North American *Vitis* species are still able to interbreed readily.

As is the case with many plants, the species of the genus *Vitis* are not very well defined because of the extreme morphological variation among and within populations of wild vines (Currle et al., 1983; Hardie, 2000; Mullins et al., 1992). This implies the following: (i) all *Vitis* species are close relatives that share a relatively recent common ancestor and (ii) evolution is still at work, throwing up new variants all the time (see Section 2.3). Many vine species are actually semispecies—that is, populations that partially interbreed and form hybrids under natural conditions, which is in fact common among plants and may be an important avenue for the evolution of new species (Aradhya et al., 2013; Soltis and Soltis, 2009). Despite some hybridization where their natural habitats overlap, however, the various *Vitis* gene pools usually stay apart so that the populations remain recognizably different. Nonetheless, species that occur in close proximity are more similar than distant species in similar habitats. Grapevines are a good example of the limits of taxonomic systems. They demonstrate that there is a continuum of differentiation rather than a set of discrete, sexually incompatible units.

As early as 1822, the Rev. William Herbert asserted that "botanical species are only a higher and more permanent class of varieties," and in 1825 the geologist Leopold von Buch postulated that "varieties slowly become changed into permanent species, which are no longer capable of intercrossing" (both cited in Darwin, 1859). A few decades later, Charles Darwin expressed it clearly: "Wherever many closely allied yet distinct species occur, many doubtful forms and varieties of the same species likewise occur" and, furthermore, "there is no fundamental distinction between species and varieties," and, finally, "varieties are species in the process of formation" (Darwin, 1859). Indeed, >150 years after the first publication of Darwin's revolutionary insights, modern genetic research has confirmed that the various *Vitis* species evolved within the last 18 million years from a common ancestor (Aradhya et al., 2013; Péros et al., 2011; Wan et al., 2013; Zecca et al., 2012). They have not yet developed the complete reproductive isolation that normally characterizes biological species. Thus, *Vitis* species are defined as populations of vines that can be easily distinguished by morphological traits, such as the shape and size of their leaves, flowers, and berries, and that are isolated from one another by geographical, ecological, or phenological barriers; such species are termed *ecospecies* (Hardie, 2000; Levadoux, 1956; Mullins et al., 1992). The following is a brief overview of the botanical classification of grapevines, starting with the domain and finishing with a selection of species.

Domain Eukaryota

All living beings, making up the Earth's biological diversity or biodiversity, are currently divided into three great domains of life: the Bacteria, the Archaea, and the Eukaryota. The Eukaryota (eukaryotes; Greek *eu* = true, *karyon* = nucleus) include all terrestrial, sexually reproducing "higher" organisms with relatively large cells (10–100 μm) containing a true cell nucleus, in which the DNA-carrying chromosomes are enclosed inside a nuclear membrane, and cell organelles such as mitochondria and plastids that are enclosed in their own membranes (Mayr, 2001). The eukaryotes evolved following injections of oxygen into the atmosphere caused by abiotic (i.e., nonbiological) factors such as plate tectonics and glaciation (Lane, 2002).

The vast majority of life and the bulk of the world's biomass—the small (1–10 μm), single-celled prokaryotes (Greek *pro* = before) with cell walls composed of protein–polysaccharides called peptidoglycans—is grouped into the two other domains. However, both the photosynthetic organelles and the "power plants" of eukaryotic cells have descended from symbiotic bacteria that invaded other single-celled organisms, or were "swallowed" by them, over 1 billion years ago. The chloroplasts, which house the photosynthetic machinery, are derived from cyanobacteria, and the mitochondria, which generate most of the eukaryotes' energy, originate from proteobacteria. These organelles still retain some of their own DNA (i.e., genes), although >95% of their original genes have since been and are still being lost or transferred to their host's nucleus (Timmis et al., 2004; Green, 2011). The mitochondria, in turn, have acquired a sizeable fraction (>40%) of genes from the chloroplasts, which may have returned the favor by incorporating some mitochondrial genes (Goremykin et al., 2009). Genetic modification or transformation, resulting in transgenic organisms, is evidently a natural process.

Some bacteria cause diseases of grapevines; for example, crown gall is caused by *Agrobacterium vitis* and Pierce's disease by *Xylella fastidiosa* (see Section 8.2).

Kingdom plantae

The Eukaryota comprise at least five major lineages that are termed supergroups or kingdoms (Green, 2011); the number and associations change as the relationships among organisms become better known. The Plantae have a haplo-diploid life cycle, in which a haploid form (with n chromosomes) alternates with a diploid form (with $2n$ chromosomes; see Section 2.3), and cell walls composed of cellulose. This supergroup is studied in the field of Botany and includes the plants, the green algae, and some other algae. There are approximately 500,000 plant species, which are classified into 12 phyla or divisions based largely on reproductive characteristics. The higher or vascular (Latin *vasculum* = small vessel) plants, to which grapevines belong on account of their water conduits, form the subkingdom Tracheobionta (Greek *trachea* = windpipe). The Animalia, which are the domain of Zoology, comprise the multicellular, diploid animals whose cells lack cell walls. The Fungi (singular fungus) include the haploid mushrooms, molds, and other fungi, with cell walls composed of glucans and chitin; they are studied in Mycology. Because they are more closely related to each other than to any other eukaryotes, the animals and fungi are now grouped together into the Opisthokonta (Green, 2011). The number and names of the other lineages and the relationships among them remain contested; they include all other "higher-order" organisms, from single-celled microbes or microorganisms (Greek *mikros* = small), including some that resemble unicellular fungi, plants (some algae), and animals (protozoans), to large, multicellular seaweeds (marine algae such as kelps).

Whereas one group of fungi, the yeasts (especially *Saccharomyces cerevisiae*), turns grapes into wine through fermentation, other fungi cause diseases of grapevines; for example, gray rot is caused by *Botrytis cinerea* and powdery mildew by *Erysiphe necator* (see Section 8.2). Animals also can be important pests of grapevines, especially certain insects (e.g., phylloxera, *Daktulosphaira vitifoliae*), mites, and nematodes.

Division (synonym phylum) Angiospermae (synonym Magnoliophyta)

The angiosperms or, in new terminology, the magnoliophytes are one of about 14 divisions or phyla of the kingdom Plantae. Angiospermae or Magnoliophyta are the two names given to the group of the flowering, or seed-bearing, plants. This group includes at least 200,000 and perhaps as many as 400,000 species, all of which are believed to have evolved from a common ancestor that lived approximately 160 million years ago during the late Jurassic period. The angiosperms make up the most evolutionarily successful group of plants. They are the plants with the most complex reproductive system: They grow their seeds inside an ovary that in turn is embedded in a flower (Greek *angeion* = pot, vessel, *spérma* = seed). After the flower has been fertilized, the other flower parts fall away, and the ovary swells to become a fruit, such as a grape berry (see Sections 1.3 and 2.3). Indeed, the production of fruits is what defines the angiosperms and sets them apart from the gymnosperms (Greek *gymnos* = naked) that include the conifers and their relatives and with whom they are classed in the superdivision Spermatophyta, or seed plants.

Class dicotyledoneae (synonym Magnoliopsida)

The angiosperms have long been divided into the dicotyledons and the monocotyledons (Greek *kotúlē* = cup) based on the number of leaves that their embroys form in the seed. Like all members of the Dicotyledoneae, which are often called dicot plants, grapevines start their life cycle with two seed leaves called cotyledons. The class of the dicodyledons is large and very diverse. Most plants (~200,000 species), including most trees, shrubs, vines, and flowers, and most fruits, vegetables, and legumes, belong to this group. However, the Angiosperm Phylogeny Group no longer recognizes the dicots as a systematic unit, because modern molecular genetic analysis has shown that its members are not all descended from a common ancestor (Angiosperm Phylogeny Group, APG IV, 2016). Those almost 75% of dicot species that are descended from a common ancestor, and which therefore form a so-called clade (Greek *klados* = branch), are now grouped together as the Eudicotidae or eudicots (Greek *eû* = good). Further subdivisions then lead to another clade, the core eudicots, which, among others contains an even smaller clade comprising over 25% of all seed plants, the superrosids. The great majority of species (~70,000) within the superrosids in turn are grouped together as the rosids which can be split into a large and diverse group, the eurosids (true rosids), and a small group, the Vitales.

Order vitales (formerly Rhamnales)

Grapevines belong to the order Vitales, which gets its name from the genus *Vitis*. The Vitales make up one of perhaps 17 orders within the clade of rosids, and this order is evolutionarily distant from the other 16 orders which together form the eurosids. This means that apples and oranges are more closely related to each other, and even to their fellow eurosids cucumbers and cabbages, than any of them is to grapes. And yet, all these species are genetically closer to one another than they are to the classic model fruit crop, the tomato. The order Vitales has a single family, the Vitaceae (APG IV, 2016; Jansen et al., 2006).

Family Vitaceae

The members of this family are collectively termed vines. All of them are descended from a common ancestor that probably lived in North America about 18 million years ago, some of whose descendants slowly spread first to eastern Asia and eventually to western Eurasia (Wan et al., 2013; Zecca et al., 2012). The family contains approximately 850 species assigned to 14 genera that are typically shrubs or woody lianas and can be divided into two subgroups, the Leeoideae (synonym Leeaceae) and the Vitoideae (APG IV, 2016). Although morphologically distinct, plants of the group Leeoideae, which comprises a single genus, the *Leea*, are visibly related to grapevines, being shrubs or trees with flowers aggregated in inflorescences, black berries, and seeds that resemble grape seeds, also named pips. Unlike the Leeoideae, members of the group Vitoideae (Latin *viere* = to attach) climb by means of their leaf-opposed tendrils (Gerrath et al., 2015). Most species of the Vitaceae family reside in the tropics or subtropics, and yet a single species from the temperate zones has become one of the world's leading fruit crops grown in about 90 countries for wine and juice production or as fresh table grapes or dried grapes (a.k.a. raisins). Vitaceae roots are generally fibrous and well branched, and they can grow to several meters in length. The leaves are alternate, except during the juvenile phase of plants grown from seeds, and can be simple or composite. The fruits are usually fleshy berries with one to four seeds. The Vitaceae comprise several genera containing species that have become ornamental plants, such as Virginia creeper and Boston ivy of the genus *Parthenocissus*, or kangaroo vine and grape ivy of the genus *Cissus*. However, all grapevines cultivated for their grapes belong to either the genus *Muscadinia* ($2n = 40$ chromosomes) or the genus *Vitis* ($2n = 38$ chromosomes). Although most species of the Vitaceae have perfect flowers, with functional male and female organs in the same flower, the wild *Muscadinia* and *Vitis* species have imperfect male or female flowers on different plants (Gerrath et al., 2015). Unlike their other family members, therefore, these species have male and female plants and are thus said to be dioecious (Greek dis = double, $oikos$ = house). The former classification of *Muscadinia* and *Euvitis* as either subgenera or sections of the genus *Vitis* has fallen out of favor among taxonomists (Mullins et al., 1992). Because of the different numbers of chromosomes, crosses between these two genera rarely produce fertile hybrids. Key morphological characteristics of the two genera include the following:

- simple leaves
- simple or forked tendrils
- generally unisexual flowers—that is, either male (staminate) or female (pistillate)
- fused flower petals that form a calyptra or "cap" and separate at the base
- soft and pulpy berry fruits.

Genus Muscadinia

Members of the genus *Muscadinia* usually have glabrous (hairless) mature leaves, unbranched tendrils, nonshredding bark with conspicuous white lenticels, nodes without diaphragms, and hard wood

(Currle et al., 1983; Gerrath et al., 2015; Mullins et al., 1992; Olmo, 1986). Because they do not root from dormant cuttings, they are usually propagated by layering, although they do root easily from green cuttings. The "homeland" of this genus extends from the southeastern United States to Mexico. The genus has only three species, which are all very similar and may not even deserve to be classed as separate species (Currle et al., 1983; Mullins et al., 1992; Olien, 1990).

- *Muscadinia rotundifolia* Small (formerly *Vitis rotundifolia* Michaux): The "muscadine grapes" are native of the southeastern United States, occupying warm and wet forest habitats (Callen et al., 2016). The female-flowered Scuppernong variety has unusual greenish ("white") berries and may have been cultivated by Native Americans before European contact. Breeding has since yielded a few perfect-flowered cultivars, such as Noble, Carlos, or Magnolia that are regionally grown as table, jelly, or wine grapes (Olmo, 1986). Due to the small cluster size of two to eight berries and their uneven ripening, the berries are often harvested individually rather than as whole clusters. However, the strong musky flavor and thick skins of the fruit can be unattractive. The species has coevolved with and therefore resists or tolerates, albeit to varying degrees, the grapevine diseases and pests native to North America, including the fungi powdery mildew and black rot (*Guignardia bidwellii*), the slime mold downy mildew (*Plasmopara viticola*), the bacterium causing Pierce's disease, the aphid phylloxera, and the dagger nematode *Xiphinema index* (which transmits the grapevine fanleaf virus from plant to plant), but is sensitive to winter freeze and lime-induced chlorosis (Alleweldt and Possingham, 1988; Olien, 1990; Olmo, 1986; Ruel and Walker, 2006). Although usually incompatible in both flowering and grafting with *Vitis* species, it occasionally does produce fertile hybrids with *Vitis rupestris*, which allows it to be used in modern breeding programs, especially breeding efforts aimed at producing nematode-resistant rootstocks (e.g., Ferris et al., 2012).
- *Muscadinia munsoniana* Small (Simpson): The "Munson's grape" is native to Florida and the Bahamas, growing in floodplain forests and low pine woods. It is gathered locally in the wild for fresh consumption or jelly or wine production but is not cultivated, even though it has larger fruit clusters and better flavor and skin characteristics than *M. rotundifolia*.
- *Muscadinia popenoei* Fennell: The "Totoloche grape" is native to southern Mexico and Guatemala but has remained rather obscure.

Genus *Vitis*

The genus *Vitis* occurs predominantly in the temperate and subtropical climate zones of the Northern Hemisphere (Mullins et al., 1992; Wan et al., 2008a). All members of this genus are perennial vines or shrubs with tendril-bearing shoots. This genus probably comprises 60–70 extant species, plus up to 30 fossil species and 15 doubtful species, spread mostly throughout Asia (~40 species) and North America (~20 species) (Alleweldt and Possingham, 1988; Gerrath et al., 2015; Wan et al., 2008b, c). The Eurasian species *Vitis vinifera* L. gave rise to the overwhelming majority of grape varieties cultivated today. Plants that belong to this genus have hairy leaves with five main veins, forked tendrils, bark that shreds when mature, nodes with diaphragms, and soft secondary wood. They all can form adventitious roots, a trait that permits propagation by cuttings, yet only *V. vinifera*, *V. riparia*, and *V. rupestris* root easily from dormant cuttings. Although the ancestor of all *Vitis* species may have had perfect flowers (McGovern, 2003), the extant wild species are dioecious, whereas the cultivated varieties of *V. vinifera* again have perfect or, in a few cases, physiologically female flowers (Boursiquot et al., 1995; Levadoux, 1956; Negrul, 1936; Pratt, 1971; see also Fig. 1.1). Members of this genus are very diverse in both habitat and form. Nevertheless, all species within the genus can readily interbreed to form fertile interspecific crosses called hybrids, which implies that they had a relatively recent common ancestor (Gerrath et al., 2015). Moreover, all *Vitis* species can be grafted onto each other. The genus is often divided into two major groups: the American and the Eurasian groups. The dominant species of the two groups differ greatly in their useful agronomic traits (Table 1.1), which makes them attractive breeding partners (Alleweldt and Possingham, 1988; This et al., 2006). Unfortunately, none of the many attempts and thousands of crosses that have been tested to date have truly fulfilled the breeders' hopes of combining the positive attributes while eliminating the negative ones contained in the natural genetic variation of the two groups. Perhaps the genes conferring disease resistance are coupled to those responsible for undesirable fruit composition. Indeed, hybrids have often been banned in European wine-producing countries because of their perceived poor fruit (and resulting wine) quality. Confident predictions of the future availability and spread of newly bred cultivars in the New World (Olmo, 1952) have been wrecked by the reality of taste conservatism of producers and consumers alike. The only unequivocal success story thus far has been the grafting of phylloxera-susceptible European wine grape cultivars to rootstocks that are usually hybrids of tolerant American *Vitis* species (see Section 1.2).

FIG. 1.1

Flower types in the genus *Vitis*: perfect flower (left), female flower (center), and male flower (right).

Table 1.1 Broad viticultural traits of American and Eurasian grapevine species.

Trait	Eurasian species	American species
Fruitfulness	Good	Poor or highly variable
Fruit quality	Good	Poor
Usefulness	Highly diverse products	Niche products, rootstocks
Propagation capacity	Good	Variable
Lime tolerance	Good	Highly variable
Phylloxera tolerance	Poor	Good or variable
Disease resistance	Poor	Good or variable

American group

Depending on the taxonomist, this group contains between 8 and 34 species, of which several have become economically important as wine or juice grapes. Because of their varying resistance to the North American grapevine diseases and pests, members of this group are also being used as rootstocks (see Section 1.2) or crossing partners in breeding programs (Alleweldt and Possingham, 1988; This et al., 2006). As an aside, crosses are always listed as maternal parent × paternal parent, that is, the mother's name comes first. The species of this group generally have thinner shoots with longer internodes and less prominent nodes than the Eurasian species. They also have smaller buds, and the leaves have very shallow indentations termed sinuses between their lobes and often a glossy surface. All grape species native to North America are strictly dioecious (i.e., none of them has perfect flowers), and most of them grow near a permanent source of water, such as a river, stream, or spring (Callen et al., 2016; Gerrath et al., 2015; Kevan et al., 1985, 1988; Morano and Walker, 1995; Fig. 1.2). Following is an incomplete list and brief description of some of the more important species. Much of this information is derived from the Plants Database (plants.sc.egov.usda.gov) of the United States Department of Agriculture Natural Resources Conservation Service:

- *Vitis labrusca* L.: The "northern fox grape" climbs vigorously and is native to southeastern Canada and the northeastern and eastern United States to Georgia, with Indiana as its western limit (Callen et al., 2016). Unlike its other family members, this species has continuous tendrils, that is, a tendril at every node of its shoots. Some of its cultivars (e.g., Catawba, Concord, Niagara) are commercially grown in the United States for juice, jam, jelly, and wine production. However, unlike wild *V. labrusca*, but like most cultivated *V. vinifera*, these cultivars have perfect flowers and are examples of fertile interspecific crosses. The *V. vinifera* cultivar Sémillon is the likely father of Catawba and grandfather of Concord; in both cases the mother was a wild *V. labrusca* or hybrid plant (Huber et al., 2016). Because these hermaphroditic cultivars probably arose through natural

FIG. 1.2

The "bank grape," *Vitis riparia*, growing in a forest in upstate New York (left) and the "canyon grape," *Vitis arizonica* Engelmann, growing up a riverside tree in Utah's Zion National Park (right). Note the large size of the wild vines and the long "trailing trunks" at the bottom right.

Photos by M. Keller.

hybridization, they have also been classed as *Vitis* × *labruscana* L. Bailey (Cahoon, 1986; Mitani et al., 2009; Pratt, 1973; Sawler et al., 2013). Another such hybrid, Kyoho, which however was bred intentionally as a table grape in Japan, is now the world's most widely planted grape cultivar, mainly due to its recent and rapid rise in China. The distinct foxy flavor caused by methyl anthranilate that characterizes this species is popular in the United States and eastern Asia but strange to Europeans. The species is cold tolerant, resistant to powdery mildew and crown gall, and tolerant of phylloxera. It is, however, susceptible to downy mildew, black rot, and Pierce's disease and has poor lime tolerance, preferring acid soils. Hybrids of *V. labrusca* were exported to Europe at the beginning of the nineteenth century. Some of these plants carried powdery mildew, downy mildew, black rot, and phylloxera, which drove most populations of wild vines extinct and brought the European wine industry to the verge of destruction.

- *Vitis aestivalis* Michaux: The "summer grape" is a vigorous climber native to eastern North America, from Quebec to Texas, and closely related to *V. labrusca*, growing in dry upland forests and bluffs (Callen et al., 2016; Gerrath et al., 2015; Miller et al., 2013; Wan et al., 2013). It is very cold hardy (to approximately −30°C) and drought tolerant, tolerates wet and humid summers as well, and is resistant to phylloxera, powdery and downy mildew, and Pierce's disease. The species is very difficult to propagate from cuttings. Its fruits are used to make grape jelly, and the cultivars Norton and Cynthiana are commercially grown as wine grapes in the southern and midwestern United States (Tarara and Hellman, 1991). It seems likely, however, that the two names are synonyms for the same cultivar and that it is a hybrid between *V. aestivalis* and *V. labrusca* and/or *V. vinifera* (Reisch et al., 1993; Sawler et al., 2013).
- *Vitis riparia* Michaux: The "bank grape" is widespread in North America from Canada to Texas and from the Atlantic Ocean to the Rocky Mountains (Callen et al., 2016). This species climbs in trees and shrubs along riverbanks and prefers deep alluvial soils, but it struggles in calcareous soils (i.e., prefers acid soils), and its shallow roots make it susceptible to drought—a trait it also confers on the rootstocks derived from its crosses with other species. It is the earliest of all the American species to break buds and ripen its dark-fleshed fruit, matures its shoots early, is very cold hardy (to approximately −36°C), tolerant of phylloxera, and resistant to fungal diseases but susceptible to Pierce's disease.
- *Vitis rupestris* Scheele: The "rock grape" is native to the southeastern United States from Texas to Tennessee. It is closely related to *V. riparia* (Callen et al., 2016; Miller et al., 2013; Zecca et al., 2012) but is now almost extinct. It is found in rocky creek beds with permanent water, and despite its vigorous growth, it is shrub-like and rarely climbs. It has deep roots for anchorage but is not very drought tolerant on shallow soils, and its lime tolerance is variable. The species tolerates phylloxera and is resistant to powdery mildew, downy mildew, and black rot, but susceptible to anthracnose (*Elsinoë ampelina* (de Bary) Shear).
- *Vitis berlandieri* Planchon: The "fall grape" is native to central Texas and eastern Mexico. It is now regarded as a subspecies or variety of *V. cinerea* but is one of very few American *Vitis* species that have good lime tolerance and thus grows well on high-pH soils. It climbs on trees on deeper limestone soils between ridges. Its deep root system makes it relatively drought tolerant, but it is very susceptible to waterlogging. The species breaks buds and flowers much later than other species and is the latest ripening of the American group with very late shoot maturation. It is somewhat tolerant of phylloxera and resistant to fungal diseases and Pierce's disease, but it is very difficult to propagate and to graft (Mullins et al., 1992).

- *Vitis cinerea* Engelmann: The "winter grape" is a sprawling, vigorous, but relatively low climber native to much of the eastern and southeastern United States through Texas (Callen et al., 2016). It thrives in moist woodlands and near streams and prefers relatively acid soils. The species has distinctly heart-shaped leaves and is resistant to powdery mildew.
- *Vitis candicans* Engelmann (synonym *Vitis mustangensis* Buckley): The "mustang grape" is a very vigorous climber native to the southern United States and northern Mexico. It is drought tolerant, relatively tolerant of lime and phylloxera, and resistant to powdery and downy mildew and Pierce's disease, but it is difficult to propagate. Its berries are very bitter and acidic but are used regionally for juice and jelly and, in Texas, to make "mustang wine." Other southern species, such as *V. champinii* Planchon and *V. longii* Prince, are probably natural hybrids of *V. candicans*, *V. rupestris*, and other native species (Péros et al., 2011; Pongrácz, 1983). They are highly resistant to nematodes (Ferris et al., 2012).

Eurasian group

There are approximately 40 known species in this group, most of them confined to eastern Asia. Chinese species are particularly diverse, growing in the dry southwest, the northern and southern foothills of the Himalayas, the very cold northeast, and the hot and humid southeast. Most of these species are little known, and there may be additional species that have not yet been described. Most Eurasian species are not resistant to the North American grapevine diseases to which they have had no exposure until after the discovery of the New World. Nevertheless, some species have at least some level of resistance and may tolerate high humidity (Li et al., 2008; Wan et al., 2007, 2008a). One species from Eurasia has come to dominate the grape and wine industries throughout the world.

- *Vitis vinifera* L.: The "wine grape" is native to western Asia and Europe at latitudes from 30°N to 50°N. During the Ice Ages it was temporarily confined, in more or less isolated populations, to the humid and forested to arid, volcanic mountain ranges of the southern Caucasus between the Black Sea and the Caspian Sea and to the Mediterranean region (Grassi et al., 2006; Hardie, 2000; Levadoux, 1956; Maghradze et al., 2012; Zohary and Hopf, 2001). This is the most well-known species of the Eurasian group, as it gave rise to most of the cultivated grapes grown today. The species is highly tolerant of lime, even more so than *V. berlandieri*, and drought. Because almost all cultivated grapevines have hermaphroditic flowers, they are often grouped into a subspecies variously named *V. vinifera* ssp. *sativa* or *V. vinifera* ssp. *vinifera*. But according to many taxonomists, *V. vinifera sativa* is merely the domesticated form of *V. vinifera* ssp. *sylvestris*. According to this view, the differences between the two forms are the result of the domestication process (Levadoux, 1956; This et al., 2006), that is, *V. vinifera sativa* arose through human rather than natural selection. Indeed, the two forms began to diverge genetically only between about 22,000 and 30,000 years ago, which is at least 10,000 years before the advent of grain farming (Zhou et al., 2017). However, the concept of subspecies, which are also called geographical races, as taxonomically identifiable populations below the species level ought to be abandoned, as it is highly subjective and inefficient and, therefore, biologically meaningless (Wilson and Brown, 1953).
- *Vitis sylvestris* (or *silvestris*) (Gmelin) Hegi: The "forest grape" is native to an area spanning central Asia to the Mediterranean region. It comprises the dioecious wild vines—also called lambrusca vines—of Asia, Europe, and northern Africa, growing mainly in damp woodlands on alluvial soils of river valleys and hillsides (Ghaffari et al., 2013; Maghradze et al., 2012). In their home range of

the Caucasus mountains between 40°N and 43°N they may climb to an elevation of up to 1200 m. Taxonomists debate whether this group of vines deserves species status or whether it is a subspecies of *V. vinifera* (ssp. *sylvestris*, but see above) because, apart from their flowers, the two look so similar and interbreed readily (Mullins et al., 1992). Although Levadoux (1956) concluded that they belong to the same species, namely *V. vinifera*, the wild form is genetically somewhat distinct from *V. vinifera sativa* (De Andrés et al., 2012; Laucou et al., 2011; Miller et al., 2013). It has almost disappeared in Europe and is considered an endangered species in many countries, mainly because of the destruction of its natural habitats and its susceptibility to phylloxera and the mildews introduced from North America (Arnold et al., 1998; Levadoux, 1956). The high humidity and periodic flooding in wooded river valleys may have protected the remaining populations from destruction by phylloxera. These wild grapes are thought to be resistant to leafroll and fanleaf viruses (Arnold et al., 1998).

- *Vitis amurensis* Ruprecht: The genetically diverse "Amur grape" is native of northeastern China and Russian Siberia. Although it may be the most winter-hardy of all *Vitis* species, budbreak occurs about a month earlier than in *V. vinifera*, making it vulnerable to spring frosts (Alleweldt and Possingham, 1988; Wan et al., 2008a). Bloom and fruit maturity are also advanced, which seems to be true for many other Chinese species as well (Wan et al., 2008c). This species is resistant to crown gall, downy mildew, anthracnose, and *Botrytis* but susceptible to phylloxera (Du et al., 2009; Li et al., 2008; Szegedi et al., 1984). Several cultivars with female or, rarely, hermaphroditic flowers and small clusters of small and very dark-colored berries are grown in northeastern China for wine production.
- *Vitis davidii* (Romanet du Caillaud) Foëx: One of >35 Chinese species, the "bramble grape" is native to subtropical regions of southern China. It is a vigorous climber that grows in river valleys and adjacent hillsides. It has large, heart-shaped leaves and is adapted to warm, humid conditions but has poor cold hardiness. It is the only *Vitis* species that produces small thorn-like structures termed prickles on its shoots and petioles (Ma et al., 2016). The species is rather resistant to powdery mildew and *Botrytis* but susceptible to downy mildew (Liang et al., 2013). Some cultivars are grown for wine production.
- *Vitis romanetii* Romanet du Caillaud: This species is native to southeastern China. It climbs on trees in forests and shrubland, and on hillsides, and is resistant to powdery mildew and anthracnose (Li et al., 2008). The fruit is used regionally to make wine.
- *Vitis coignetiae* Pulliat: The "crimson glory vine" is native to Japan and Korea and climbs vigorously in forest trees. It develops bright red leaves in fall but otherwise resembles the American *V. labrusca*, although the two species do not seem to be closely related genetically (Kimura et al., 1997; Mullins et al., 1992; Wan et al., 2013). It is grown as an ornamental plant and used locally for jam production and some winemaking.

1.2 Cultivars, clones, and rootstocks

1.2.1 Variety versus cultivar

The study of the botanical description, identification, and classification of plants belonging to the genus *Vitis* and of their usefulness for viticulture, the cultivation of grapevines, is termed *ampelography* (Greek *ampelos* = vine, *graphein* = to write). The descriptors have traditionally included visual traits such as the shape, hairs, and pigmentation of the shoot tips, the shape of the leaf blade with its lobes, sinuses, and serrations, the size and shape of the fruit clusters, and the size, shape, and pigmentation of the berries (Galet, 1985, 1998; Viala and Vermorel, 1901–1909). Not surprisingly, for example, leaf shape in grapevines is rather highly heritable (Chitwood et al., 2014). The advent of DNA fingerprinting has led to the adoption of this technique for the identification of selections in grape collections and is increasingly being used to test ampelography-based classifications of species and cultivars and to uncover the historical origins and genetic relationships of grapevines.

Cultivated grapevines of sufficiently similar vegetative and reproductive appearance are usually called "grape varieties" by growers and "cultivars" by botanists. Botanically speaking, a variety includes individuals of a wild population that can interbreed freely. A cultivar, however, is "an assemblage of plants that (a) has been selected for a particular character or combination of characters, (b) is distinct, uniform, and stable in these characters, and (c) when propagated by appropriate means, retains those characters" according to the international code of nomenclature for cultivated plants (Brickell et al., 2009). Plants that retain all their characters when propagated are called "true to type" and are assumed to be genetically identical, or at least nearly so. In principle, a cultivar can be produced both sexually (i.e., from a seedling) and asexually (i.e., as a vegetative clone), but in grapevines only the latter method will give descendants that are true to type. Seedlings of the so-called grape varieties are not identical copies of their mother plant, because they inherit half of their DNA from their father's pollen. At any given chromosome position, called locus, a new seedling may have two slightly different versions of the same gene, one from the mother and one from the father (see Section 2.3). The seedling is thus called heterozygous (Greek *heteros*=other, different, *zugōtós*=yoked) at that locus; it is called homozygous (Greek *homós*=same) at a locus that carries two identical copies of a gene. Because grapevines are heterozygous across many loci in their genomes, each seed may give rise to a cultivar with distinct characteristics (Bacilieri et al., 2013; Laucou et al., 2011; Mullins et al., 1992; Pelsy et al., 2010; Thomas and Scott, 1993; see also Section 2.3).

Grapes were among the first fruit crops to be domesticated, along with olives, figs, and dates (Maghradze et al., 2012; Zohary and Hopf, 2001; Zohary and Spiegel-Roy, 1975). Although all four wild fruits had ranges stretching far beyond the eastern Mediterranean region, the deliberate cultivation of grapes for winemaking as well as fresh fruit and raisin production probably started about 7000–8000 years ago in the Caucasus region (now mostly Georgia, Armenia, and Azerbaijan) and the northern part of the Fertile Crescent that spans the modern-day countries of the Near East: Egypt, Israel, Lebanon, Jordan, Syria, eastern Turkey, Iraq, and western Iran (McGovern et al., 1996, 2017). This is roughly the same time that people began producing the first clay jars and metal tools, and that farmers invented irrigation; it precedes by approximately 2000 years the earliest civilization (Latin *civitas*= city, state) with a state government and writing, that of Sumer in southern Mesopotamia. It seems fitting that the "tree of life" of many ancient civilizations is the grapevine. These early societies regarded wine as "nectar of the gods," and wine drinking was considered to be a hallmark of civilization; "savages" or "barbarians" drank no wine. A simple explanation for this phenomenon might be that winemaking called for a settled lifestyle; nomadic hunter-gatherers had neither the time nor the infrastructure required for this technological process. The biblical legend, modeled on the world's oldest surviving myth, the Sumerian Gilgamesh legend (McGovern, 2003), of Noah's ark stranding near volcanic Mount Ararat on the Turkish/Iranian/Armenian border and his subsequent planting of the first vineyard also reflects the probable geographic origin of viticulture. Indeed, the oldest winery known as yet, dating to about 6000 years ago, was discovered in southeastern Armenia at an elevation of almost 1000 m

(Barnard et al., 2011). Even today, this arid region is home to wild forms of *V. vinifera* (or *V. sylvestris*) with a greater genetic diversity than is found anywhere else (Grassi et al., 2006; Maghradze et al., 2012). Moreover, many cultivated varieties of *V. vinifera* are more closely related to *V. sylvestris* from the Near East than to their wild relatives from Western Europe (Myles et al., 2011). Nevertheless, archeological discoveries suggest that wine might have been made from wild grapes in China as far back as 9000 years ago (McGovern et al., 2004). In Europe, grapes were grown at least 4000 years ago (Rivera Núñez and Walker, 1989). Clearly, humans quickly learned how to use fermentation as one of the most important food preservation technologies; they also learned to preserve grapes as raisins by drying them. Domestication of *V. vinifera* was also accompanied by a change from dioecious to hermaphroditic reproduction and an increase in seed, berry, and cluster size (Bouby et al., 2013; This et al., 2006; Zohary and Spiegel-Roy, 1975). Perhaps early viticulturists selected not only vines that produced more fruit than others but also plants that were self-pollinated, thus eliminating the need for fruitless pollen donors.

Today, an estimated 10,000 or so grape cultivars are being grown commercially, although DNA fingerprinting suggests that a more accurate figure may be approximately 5000 (This et al., 2006). Many cultivated grapes are closely related; in fact, many of them are siblings or cousins (Cipriani et al., 2010; Di Vecchi Staraz et al., 2007; Robinson et al., 2012). Moreover, many cultivars are known by several or even many synonyms (different names for the same cultivar) or homonyms (identical name for different cultivars). A good example of the latter is Malvasia, a name that has been applied to dozens of related and unrelated cultivars, of which many have several synonyms, with a range of berry colors and flavor profiles (Lacombe et al., 2007).

The vast majority of cultivars belong to the species *V. vinifera*, and comparisons of the DNA contained in chloroplasts suggest that they might have originated from at least two geographically distinct populations of *V. sylvestris*: one in the Near and Middle East and the other in a region comprising the Iberian Peninsula, Central Europe, and Northern Africa (Arroyo-García et al., 2006). Most cultivated grapevines have a very large effective population size and evidently did not experience a genetic bottleneck during domestication, which means that domestication and subsequent breeding did little to reduce the genetic diversity in grapes (Fournier-Level et al., 2010; Laucou et al., 2011; Myles et al., 2010, 2011, Zhou et al., 2017). Such reduction of genetic diversity did, however, occur when phylloxera and mildew pathogens introduced from North America in the second half of the nineteenth century wreaked havoc across European vineyards (Cipriani et al., 2010). Homogenization of consumer preferences have since contributed to a further focus of the global wine industry on fewer cultivars. By 2010, the top 35 cultivars accounted for two-thirds of the world's wine grape area, and three out of four cultivars grown globally had their origins in just three countries: 36% from France, 26% from Spain, and 13% from Italy (Anderson, 2014; see also Table 1.2).

Most current cultivars are not products of deliberate breeding efforts. Rather, they are the results of continuous selection of chance seedlings and vegetative (i.e., clonal) propagation over many centuries of grapevines that were spontaneously generated by mutations (Latin *mutatio*=change). Seedlings may have been selected that resulted from sexual reproduction following either self-pollination within the flowers of a single mother vine or cross-pollination involving different vines (see Section 2.3). The pronounced inbreeding depression in grapevines, which is a consequence of the accumulation over time of deleterious, but usually recessive, mutations due to prolonged vegetative propagation, has ensured that only about 2% of the current cultivars are the result of self-pollination (Lacombe et al., 2013; Zhou et al., 2017). Intraspecific hybridization due to cross-pollination was facilitated by the ancient

Table 1.2 Country of origin, skin color, and planted area of the world's top 35 wine grape cultivars in 2010.

Cultivar	Origin	Skin color	Area (ha)
Cabernet Sauvignon	France	Red	290,091
Merlot	France	Red	267,169
Airén	Spain	White	252,364
Tempranillo	Spain	Red	232,561
Chardonnay	France	White	198,793
Syrah	France	Red	185,568
Garnacha	Spain	Red	184,735
Sauvignon blanc	France	White	110,138
Trebbiano	Italy	White	109,772
Pinot noir	France	Red	86,662
Carignan	Spain	Red	80,178
Bobal	Spain	Red	80,120
Sangiovese	Italy	Red	77,709
Monastrell	Spain	Red	69,850
Graševina	Croatia	White	61,200
Rkatsiteli	Georgia	White	58,641
Cabernet franc	France	Red	53,599
Riesling	Germany	White	50,060
Pinot gris	France	White	43,563
Macabeo	Spain	White	41,046
Malbec	France	Red	40,688
Cayetana blanca	Spain	White	39,741
Alicante Bouschet	France	Red	38,985
Aligoté	France	White	36,119
Cinsaut	France	Red	36,040
Chenin blanc	France	White	35,164
Montepulciano	Italy	Red	34,947
Catarratto bianco	Italy	White	34,863
Tribidrag	Croatia	Red	32,745
Gamay noir	France	Red	32,671
Isabella	USA	Red	32,494
Colombard	France	White	32,076
Muscat blanc	Greece	White	31,112
Cereza	Argentina	White	29,189
Muscat of Alexandria	Egypt	White	26,336

Modified from Anderson, K., 2014. Changing varietal distinctiveness of the world's wine regions: evidence from a new global database. J. Wine Econ. 9, 249–272.

practice of growing multiple varieties in the same vineyard, a custom that was widespread until the nineteenth century (Cipriani et al., 2010). Somatic (Greek *soma*=body) mutations are mutations that occur in the dividing cells of the shoot apical meristem (Greek *merizein*=to divide; see Section 1.3). Because somatic mutations typically happen during bud formation (see Section 2.2), clones that result from propagation of affected shoots are also called bud sports. Moreover, since plants, unlike animals, do not separate their somatic cells from their germ line cells (Walbot and Evans, 2003), and because somatic mutations are often nonlethal, such somatic mutations will also be carried forward into the new seeds derived from mutated cell lines (see Section 2.3).

Mutations in the grape genome arise through rare, chance mistakes in DNA replication during the process of cell division. Whenever a cell divides, it first must double (i.e., copy) its DNA so that each daughter cell gets a complete set of chromosomes. Millions of nucleotides are regularly copied with amazing accuracy, and sophisticated "spell-checking" and repair mechanisms attempt to fix most mistakes that do occur. However, every so often, a nucleotide is inadvertently switched with another one, or a whole group of nucleotides (termed a DNA sequence) is inserted, repeated, omitted, or moved to a different location on the chromosome. Many DNA sequences, called "jumping genes" (a.k.a. transposons, transposable elements, or mobile elements), even move around a chromosome on their own, some of them employing "cut-and-paste" or "copy-and-paste" strategies (Benjak et al., 2008; Cardone et al., 2016; Carrier et al., 2012; Godinho et al., 2012; Lisch, 2009). Some of these transposons might be remnants of old virus infections following the loss of a virus's ability to move from cell to cell. Most of these repetitive DNA sequences and transposons are silenced, meaning their activity is suppressed by the addition of methyl groups (Pecinka et al., 2013). Such DNA methylation hinders transcription, or the copying of the DNA into RNA, which in turn prohibits gene expression, that is, the manufacture of RNA and protein from the segment of DNA making up a gene. Nonetheless, new genes quite often arise from duplication of an existing gene and subsequent mutation of one of the two copies (Díaz-Riquelme et al., 2009; Firn and Jones, 2009; Guo et al., 2014). Many mutations are also caused by damage to the DNA by so-called reactive oxygen species that are produced during oxidative stress (Halliwell, 2006; Møller et al., 2007; see also Section 7.1).

Mutations, if they are not lethal or silenced, can give rise to slightly different forms of the same gene that are termed alleles (Greek *allelos*=each other), just as does the recombination that occurs during sexual reproduction (see Section 2.3). The number of these gene variants at any chromosome locus of grapevines varies from as little as 3 to >30 (Bacilieri et al., 2013; De Andrés et al., 2012; Laucou et al., 2011). The *Vitis* genome is approximately 500 million DNA base pairs long and comprises an estimated 30,000 or so genes (Di Genova et al., 2014; Jaillon et al., 2007; Velasco et al., 2007). By encoding (i.e., prescribing) the amino acid makeup of proteins these genes form both a vine's "building plan," including its flexible architectural and engineering design, and its "operating manual." For comparison, we seemingly far more complex humans are thought to have fewer than 25,000 genes. Like in humans, though, most of the genetic diversity occurs *within* local populations of wild grapevines, rather than *among* them (De Andrés et al., 2012).

Because there are genes that influence growth habit, leaf shape, disease resistance, cluster architecture, berry size and color, seedlessness, flavor, and other quality attributes, some of the mutations also affect these traits (Bessis, 2007; Bönisch et al., 2014a; Cardone et al., 2016; Fernandez et al., 2006, 2010; Pelsy, 2010). For instance, dark-skinned (i.e., anthocyanin-accumulating) fruit is the "default" version in the Vitaceae. It appears that virtually all so-called white *V. vinifera* cultivars with green-yellow fruit have a single common ancestor that arose from mutations of two neighboring genes of an original dark-fruited grapevine (Boss and Davies, 2009; Cadle-Davidson and Owens, 2008; This et al., 2007; Walker et al., 2007). At least one of these mutations was probably caused by insertion of a jumping gene into the DNA sequence that codes for a transcription factor named *MYB* (short for myeloblastosis gene, which was first discovered in humans), and another one resulted from the loss of that entire section of DNA. Transcription factors are proteins whose function it is to infer information from a DNA sequence in response to developmental and environmental cues and regulate expression of certain genes by binding to a stretch of DNA in the so-called promoter region of those genes. This protein binding then helps to either activate (i.e., upregulate) or repress (i.e., downregulate) the target genes, akin to an "on/off" switch. The mutations that compromised the *MYB* transcription factor render grapes unable to "switch on" the anthocyanin assembly line (Hichri et al., 2011; Kobayashi et al., 2004). The chances of these mutations occurring together are extremely small; this might have been a one-time event in the history of grapevines that is thought to have occurred only *after* grapevines were first domesticated (Fournier-Level et al., 2010; Mitani et al., 2009). Although this implies that most white-fruited cultivars are likely to be genetically more closely related to each other than are cultivars with dark fruit, roughly half of all current grape cultivars have fruit with "white" skin. Light berry skin color must have been a sought-after trait that was heavily selected for during the history of grape cultivation. Instability of at least one of these mutations may account for the occasional appearance of dark-skinned variants of white cultivars (Boss and Davies, 2009; Lijavetzky et al., 2006).

The same jumping gene that caused the appearance of white grapes, moreover, is also responsible for much of the berry color variation among dark-skinned cultivars (Fournier-Level et al., 2009). In a dose-dependent, cultivar-specific manner, the presence of this *MYB* gene, in concert with a few others, permits more or less of the enzymes that produce anthocyanins to be synthesized, so that cultivars vary greatly in their ability to accumulate anthocyanin pigments. For example, Pinot noir has one intact and one mutated *MYB* allele in each of its two meristematic cell layers (see Section 1.3), whereas Pinot gris lacks the intact version in its inner (unpigmented) cell layer that, moreover, has partly displaced cells from the outer (pigmented) layer, and Pinot blanc carries only the mutated version in both cell layers (Vezzulli et al., 2012; see also Fig. 1.3).

The *MYB* story does not end here, however. Because there are slightly different versions of such transcription factors in different organs and tissues, they regulate gene expression in an organ- and tissue-specific manner. Consequently, just because a white grape cultivar cannot produce anthocyanins in its berries, this does not mean that it is also unable to produce these red pigments in other organs. Indeed, the presence of leaf-specific *MYB* genes that were unaffected by the mutations described earlier enables the leaves of both red and white cultivars to accumulate anthocyanins, for example when young leaves unfold under bright sunlight (Matus et al., 2017). But courtesy of the different *MYB* transcription factors, malvidin-based pigments tend to dominate the anthocyanin profile in red berries (see Section 6.2), while cyanidin and peonidin derivatives are much more abundant in red leaves (Kobayashi et al., 2009; Matus et al., 2017; Wenzel et al., 1987; Keller, unpublished data).

Another small mutation of great historic and economic consequences gave rise to almost all seedless table grapes around the world. Such grapes are said to be stenospermocarpic and do contain small seed traces (see Section 2.3). The mutation resulted in the substitution of the amino acid leucine for arginine in a single location on another transcription factor, this one named *AGL11* (short for agamous-like). Because *AGL11* is crucial in orchestrating the normal development of seed coats, the mutation renders seeds unable to form, which leads to stenospermocarpy (Royo et al., 2018). This mutation was selected, probably millennia ago, in the Middle Eastern grape variety Kishmish

FIG. 1.3

Pinot noir (left) growing next to Pinot gris, some of whose berries resemble Pinot blanc due to displacement of the pigmented epidermis cells by underlying nonpigmented cells (right).

Photo by M. Keller.

(synonyms Sultanina, Sultana, Thompson Seedless). Whereas almost all wine and juice grapes contain seeds, many raisin and table grape cultivars are seedless because they are direct descendants of Kishmish and thus have a very narrow genetic base (Adam-Blondon et al., 2001; Boursiquot et al., 1995; Ibáñez et al., 2009).

In addition to generating new cultivars, mutations also add to the genetic variation within existing cultivars, and because mutations accumulate over time, the variation also increases. Jumping genes, of which over 100 are known in *V. vinifera*, are responsible for the majority of mutations that lead to the diversity among grapevine clones (Carrier et al., 2012). In addition, some clonal differences in gene expression may also arise from variation in gene silencing, whereby genetically identical regions may have a different DNA methylation pattern—such differences are, therefore, epigenetic (Pecinka et al., 2013). In other cases, a change in a few of the several hundred amino acids that compose a protein may render that protein ineffective. For example, substitution of only 3 out of the approximately 480 amino acids in a monoterpene glucosyltransferase (an enzyme that attaches a sugar molecule to a monoterpene; see Section 6.2) accounts for the difference between two Gewürztraminer clones, one aromatic and the other nonaromatic (Bönisch et al., 2014a). However, although ancient cultivars (e.g., Grenache, Pinot noir) can be quite heterogeneous and are planted as many different clones, the genetic similarity of these clones typically is still on the order of 95–99% (Bessis, 2007; Blaich et al., 2007; Cabezas et al., 2003; Levadoux, 1956; Pelsy, 2010; Wegscheider et al., 2009).

One drawback of prolonged vegetative propagation is that the descendant clones tend to accumulate deleterious mutations and systemic pathogens such as viruses and bacteria, both of which will eventually curtail the vines' agronomic performance (McKey et al., 2010). But because the frequency at which mutations occur tends to increase as plants age, and because some of these mutations may increase the plant's adaptive fitness (Thomas, 2013), one might argue that clonal selection programs could gain from focusing on the oldest available vineyards.

In addition to mutation, hybridization has also added to the diversity among cultivated grapevines. Hybridization occurs by cross-pollination and fertilization of flowers from different plants that are genetically distinct and thus gives rise to a genetically novel individual arising from a seed (see Section 2.3). Both deliberate interbreeding and natural hybridization have occurred many times in the history of viticulture. Many cultivars were originally selected from domesticated local wild grapes (*V. sylvestris* or *V. vinifera* ssp. *sylvestris*) and further developed by selection of seedlings that arose from intentional or accidental interbreeding with other wild grapes, or by the introduction of exotic varieties (Arroyo-García et al., 2006; Bacilieri et al., 2013; Bouby et al., 2013; Grassi et al., 2003; Levadoux, 1956; Maghradze et al., 2012; Myles et al., 2011; Salmasco et al., 2008; Santana et al., 2010; Sefc et al., 2003). Some hybridization also occurs in the opposite direction, whereby pollen from cultivated grapevines finds its way onto the flowers of nearby wild plants (De Andrés et al., 2012; Di Vecchi-Staraz et al., 2009). Intriguingly, although the majority of the French wine grape cultivars are genetically closely related to each other and to the wild grapes of western Europe, most Italian and Iberian cultivars are different from these and are also more distinct among themselves, while most table grapes are in a different "genetic league" altogether. It appears that the Italian and Iberian peninsulas have long served as "melting pots" and exchange sites for cultivars arriving by sea or land (Bacilieri et al., 2013).

Many of the numerous cultivars with "muscat" aroma, used to produce raisin or table grapes as well as various wine styles, and many of them with several synonyms, are direct descendants of the ancient Muscat à petits grains (synonym Muscat blanc) or its offspring Muscat of Alexandria (Cipriani et al., 2010; Crespan and Milani, 2001; Lacombe et al., 2013). Muscat of Alexandria in turn is one parent of the Argentinian Torrontés cultivars—which are unrelated to several different Spanish cultivars named Torrontés (Agüero et al., 2003; Gago et al., 2009). Nevertheless, many "noble" grape lineages are as interwoven as those of their human selectors.

For instance, it is thought that Cabernet franc and Savagnin (synonym Traminer) may have been selected from wild grapes, whereas Cabernet Sauvignon resulted from a natural cross between Cabernet franc and Sauvignon blanc (Bowers and Meredith, 1997; Cipriani et al., 2010; Levadoux, 1956). Cabernet franc, which is closely related to Petit Verdot, also fathered Merlot and Carmenère (Boursiquot et al., 2009; Lacombe et al., 2013; Salmasco et al., 2008). Mutations within Savagnin over time gave rise to what are now considered three distinct cultivars: the original Savagnin blanc with white berries, Savagnin rose with pink berries, and Savagnin rose aromatique (synonym Gewürztraminer) with pink and highly aromatic berries (Pelsy et al., 2010). Similarly, Pinot blanc and Pinot gris are somatic berry-pigment mutants of Pinot noir that arose independently of each other (Vezzulli et al., 2012; see also Fig. 1.3). Such berry-pigment mutants exist for Garnacha (synonym Grenache) as well (Cabezas et al., 2003). Indeed, variation in berry color seems to be a fairly common outcome of somatic mutation (Bessis, 2007; Furiya et al., 2009; Müller-Stoll, 1950; Regner et al., 2000).

The Pinot family itself is thought to descend from the aforementioned Savagnin, which also features as a parent of Sauvignon blanc, Sylvaner, and many other cultivars (Myles et al., 2011; Regner et al., 2000; Salmasco et al., 2008; Sefc et al., 1998). Savagnin and Sauvignon blanc, in turn, may be the parents of Chenin blanc (Cipriani et al., 2010). Pinot appears to be one of the more distant ancestors of Syrah (synonym Shiraz), but Syrah more directly is the offspring of a cross between Dureza and Mondeuse blanche (Meredith et al., 1999; Vouillamoz and Grando, 2006). Syrah crossed with Peloursin, a relative of Malbec and Marsanne, later gave rise to Durif (synonym Petite Sirah).

Chardonnay, Gamay noir, Aligoté, Auxerrois, Melon, and other French cultivars all originated from the same two parents, Pinot and Gouais blanc (Bowers et al., 1999; Regner et al., 2000). This not only means that these cultivars are full siblings, but also that they are rather inbred, since Gouais blanc (synonym White Heunisch) in turn has been proposed to be an offspring of Pinot with an unknown partner (Roach et al., 2018). Gouais blanc, crossed with a Savagnin offspring, probably gave rise to Riesling, and with Chenin blanc it yielded Colombard and other cultivars. Even Lemberger (synonym Blaufränkisch), which is genetically similar to many southeastern European wine grapes, may have been the result of a Gouais blanc cross with an unknown partner. In fact, based on genetic evidence, the ancient Croatian cultivar Gouais blanc has thus far been proposed to feature in the pedigree of >60 European cultivars (Lacombe et al., 2013). Its prominent position as a parent of many premium wine grape cultivars is somewhat ironic because its fruit quality has long been considered inferior. Yet its vigorous growth, high fruitfulness, and low risk of spring frost due to late budbreak made it a favorite with growers. For centuries they grew it alongside selections from wild vines, such as Savagnin whose yields varied widely, so that it became widespread throughout Europe in the Middle Ages. Moreover, because of their sour berries, Gouais vines were often planted as buffers around vineyards to deter potential grape thieves. Another old cultivar from the Adriatic coast has also attained some popularity in both the Old and the New World, albeit under different names: Tribidrag or Crljenak kaštelanski in Croatia, Primitivo in southern Italy, and Zinfandel in California (Fanizza et al., 2005; Maletić et al., 2004).

The case of Gouais blanc demonstrates that obscurity is not unusual for parents of current noble grape cultivars. Further examples of widely grown cultivars, one of whose parents, when finally determined by DNA fingerprinting, turned out to have fallen into oblivion, include Merlot and Tempranillo (Boursiquot et al., 2009; Ibáñez et al., 2012; Salmasco et al., 2008). Modern DNA fingerprinting has further uncovered that what breeders get is not always what they set out to develop. Despite the breeders' best efforts to exclude nondesired pollen from their crosses, the DNA-derived pedigrees of about one-third of all tested crosses turned out to be at variance with those reported by the breeders (Bautista et al., 2008; Ibáñez et al., 2009; Lacombe et al., 2013). The most well-known example is the German wine grape cultivar Müller-Thurgau, which has long been disseminated as a deliberate cross of Riesling flowers with Sylvaner pollen—hence the synonym Riesling × Sylvaner (Becker, 1976). >100 years after its introduction, however, genetic analysis exposed Madeleine Royale, another likely progeny of Pinot noir, as the illegitimate father (Dettweiler et al., 2000). Similarly, Cardinal, a table grape cultivar with worldwide distribution, may be derived from Alphonse Lavallée (a possible progeny of Gros Colman and Muscat Hamburg) and Königin der Weingärten (probably a descendant of Pearl of Csaba) instead of, as has been assumed, Flame Tokay (Cipriani et al., 2010; Ibáñez et al., 2009).

Natural hybridization can also cross species boundaries, as exemplified by the North American juice grape cultivar Concord, which was selected from a wild *V. labrusca* seedling that was pollinated by Catawba, which in turn was the chance offspring of a wild seedling pollinated with the *V. vinifera* cultivar Sémillon (Huber et al., 2016; Mitani et al., 2009; Sawler et al., 2013). Misidentification of planting material imported to new grape growing regions is another common problem. For instance, the so-called Bonarda in Argentina and Charbono in California are not related to the various Italian cultivars of the same names but are in fact both identical to the old French cultivar Corbeau (Martínez et al., 2008).

Although each cultivar originally began as a single vine that grew from a seedling, most major cultivars grown today have been propagated vegetatively for a long time, some for many centuries and perhaps even millennia (Bessis, 2007; Di Vecchi Staraz et al., 2007; Hardie, 2000; Pelsy et al., 2010). Propagation by cuttings was undoubtedly in use by Roman times 2000 years ago (Columella, ca. 70 ad; Pliny the Elder, ca. 70 ad). For instance, Gouais blanc is thought to have been brought from Dalmatia (now Croatia) to Gallia (now France) along with Viognier around ad 280 by the Roman Emperor Probus, who encouraged vineyard development to enhance economic stability. Indeed, the Croatian cultivar Vugava bijela seems to be identical to Viognier and a close relative of the Italian Barbera and the Swiss Arvine. Viticulture also expanded greatly as a result of the state-sponsored settlement of and vineyard establishment by Roman army veterans (Bouby et al., 2013). In fact, if a map of the distribution of viticulture in Europe, the Near East, and North Africa were to be overlaid with a map of the Roman Empire at its greatest extent, there would be an almost exact geographical overlap.

A long gap in cultivar description and identification followed the demise of the Roman Empire. The cultivar name Traminer was first mentioned in 1349, Pinot gris (synonym Ruländer) in 1375, Pinot in 1394, Riesling in 1435, Chasselas (synonym Gutedel) in 1523, and Sangiovese in 1590, although it is not certain that the same name was consistently applied to the same cultivar (von Bassermann-Jordan, 1923). Traditional European vineyards were—and in some areas still are—composed of a population of heterogeneous vines. Sometimes several cultivars were planted together in the same block. This practice would have favored Gouais blanc-type cross pollination with the result that many cultivated varieties have no defined origin. The identification and cultivation of "pure" cultivars is quite recent. Nonetheless, the genetic diversity, as well as similarity, among extant *V. vinifera* cultivars agrees well with the idea that viticulture originated in the Near East and spread to Europe via routes around the northern and southern shores of and across the Mediterranean Sea (Bacilieri et al., 2013). More details about the origins, genetic relationships, current distribution, as well as viticultural and enological characteristics, of 1368 extant cultivars can be found in the comprehensive review by Robinson et al. (2012).

1.2.2 Cultivar classification

Following several millennia of cultivation and repeated selection of spontaneous mutants and natural as well as man-made intraspecific and interspecific crosses in many different regions, there is a vast range of cultivated forms and types of grapevines. Because thousands of grape cultivars are being grown commercially, various attempts have been made to group them into families or tribes. Unlike botanical classifications, these groupings are not always based on phenotypic or genotypic differences, whereby the genotype is the sum of the genetic material of an organism, and the phenotype is its visible physical properties that arise from the interaction of the genotype with the environment during development (Mayr, 2001). The most common methods to categorize grape cultivars involve classifications on the basis of place or climate of origin, viticultural characteristics, final use, or winemaking characteristics.

Arguably the first cultivar classification was attempted by Plinius (ca. ad 50) in Rome, who created three groups according to berry color and yield:

1. *Anemic cultivars*: Varieties with small, white grapes
2. *Nomentanic cultivars*: Low-yielding varieties with red grapes
3. *Apianic cultivars*: High-yielding varieties of poor quality

During the Middle Ages, European wine grapes were simply divided into two groups according to perceived wine quality (von Bassermann-Jordan, 1923):

1. *Vinum francicum*: "Frentsch" grapes, low-yielding varieties of high quality (e.g., Traminer)
2. *Vinum hunicum*: "Huntsch" grapes, high-yielding varieties of poor quality (e.g., Heunisch)

The division proposed by the Russian botanist Negrul (cited in Levadoux, 1956, and Maghradze et al., 2012) distinguishes three main ecological or ecogeographical groups of varieties, called "proles" (Latin *proles* = offspring), based on their region of origin:

1. *Proles pontica*: Fruitful varieties with small to medium-sized, round berries with seeds ranging from very small to very large that originated from the shores of the Aegean and Black Seas and spread throughout eastern and southern Europe. Examples include Furmint, Clairette, Black Corinth, and Rkatziteli.
2. *Proles occidentalis*: Highly fruitful wine grape varieties of central and western Europe with small clusters of small berries with small seeds. Examples include Riesling, Chardonnay, Sémillon, Sauvignon blanc, Gewürztraminer, the Pinots, and the Cabernets.
3. *Proles orientalis*: Mostly table grape varieties with low bud fruitfulness and large clusters of large, elongate berries with either large seeds or partly to wholly absent seeds (the so-called seedless cultivars) originating from the Near East, Iran, Afghanistan, and central Asia. Examples include Thompson Seedless, the Muscats, and Cinsaut.

Modern genetic research has mostly borne out Negrul's grouping (Aradhya et al., 2003; Bacilieri et al., 2013). Such research also confirmed that many of the *proles occidentalis* cultivars are closely related to the wild *V. sylvestris* (or *V. vinifera* ssp. *sylvestris*), that almost all the *proles pontica* cultivars are closely related to each other, and that the *proles orientalis* group is genetically much more diverse than and distinct from the other groups. Yet Negrul's classification is not clear-cut and contains some ampelographic errors. In addition, all these classifications suffer from the fact that grape cuttings have forever been carried to distant locations and used for deliberate or natural interbreeding with locally domesticated and selected variants. Therefore, cultivars have also been grouped according to their viticultural characteristics, although these can be modified by local soil and climatic conditions and cultural practices:

- Time of maturity (Table 1.3): The early maturing Chasselas is used as a basis or standard
- Vigor: Rate of shoot growth
- Productivity: Yielding ability

Classification is also possible in terms of what the grapes will be used for (Galet, 2000), although some cultivars are used for various purposes:

- Table grapes: Large, fleshy or juicy grapes, often seedless, some of them with muscat or foxy aromas. Examples include Cardinal, Cinsaut, Chasselas, and Muscat of Alexandria.
- Raisin grapes: Predominantly seedless grapes. Examples include Thompson Seedless, Flame Seedless, Black Corinth (synonym Zante Currant), and Delight.
- Juice grapes: Some highly aromatic grapes, especially in the United States. Examples include Concord and Niagara.

Table 1.3 Classification of some grape cultivars based on their relative heat requirement to reach acceptable fruit maturity, subjectively defined as 18–20°Brix.

Group	Red-wine cultivars	White-wine cultivars
1 (early)		Madeleine Angevine, Pearl of Csaba
2	Blue Portuguese	Chasselas, Müller-Thurgau
3	Gamay, Dolcetto	Pinot gris, Pinot blanc, Aligoté, Gewürztraminer
4	Pinot noir	Chardonnay, Sauvignon blanc, Sylvaner
5	Cabernet franc, Lemberger	Riesling
6	Merlot, Malbec, Tribidrag, Tempranillo, Cinsaut, Barbera, Sangiovese	Sémillon, Muscadelle, Chenin blanc, Marsanne, Roussanne, Viognier
7	Cabernet Sauvignon, Syrah, Nebbiolo	Colombard, Palomino
8	Petit Verdot, Aramon, Carignan, Garnacha, Monastrell	Muscat of Alexandria, Trebbiano
9 (late)	Tarrango, Terret noir	Clairette, Garnacha blanca

Modified from Viala, P., Vermorel, V., 1901–1909. Ampélographie. Tomes I-VII, Masson, Paris; Gladstones, J., 1992. Viticulture and Environment. Winetitles, Adelaide, Australia; Huglin, P., Schneider, C., 1998. Biologie et Ecologie de la Vigne, second ed. Lavoisier, Paris.

- Wine grapes: Very sweet, juicy grapes, often low yielding. Examples include Riesling, Chardonnay, Sémillon, Sauvignon blanc, Gewürztraminer, the Pinots and Cabernets, Merlot, Tempranillo (synonyms Aragonez, Cencibel, Tinta Roriz), and Nebbiolo.
- Brandy (distillation) grapes: Generally white grapes producing bland, acidic wines with low alcohol content. Examples include Ugni blanc (synonym Trebbiano), Colombard, Folle blanche, and Baco blanc.

Of course, cultivars can also be classified based on winemaking characteristics, but this method is strongly influenced by the market environment and changing consumer demand:

- Grape composition: Basic characteristics (sugar, acid, pH, tannins, flavors, and aroma) important for winemaking.
- Varietal aroma: White grapes can be aromatic (e.g., Riesling, Gewürztraminer, and Muscats) or nonaromatic (e.g., Chardonnay, Sémillon, and Sylvaner).
- Production costs: Value for winemaking reflected in wine and grape pricing structure and appellation systems.

1.2.3 Clones

In viticultural parlance, a clone (Greek *klon*=twig) is a group of grapevines of a uniform type that have been vegetatively propagated, usually by cuttings, from an original mother vine that would normally have been selected for one or more desired traits. Such traits may include low vigor, high yield, loose clusters, large or small berries, seedlessness, different or more intense fruit pigmentation or other perceived quality attributes, or disease resistance. Due to vegetative propagation, such clonal traits arise from somatic mutations (see Section 1.3) rather than during sexual reproduction (see Section 2.3). Nonetheless, somatic mutations may ultimately be propagated both vegetatively and by seeds and result in individual plants of the same cultivar having slightly different genotypes and sometimes phenotypes (Franks et al., 2002; Riaz et al., 2002; Roach et al., 2018; This et al., 2006). This genotypic diversity, which accumulates over time, is termed clonal variation. Even "clonemates" and clones of relatively recent origin are unlikely to be strictly genetically identical, because the rules of statistics make somatic mutations of at least some of grapevines' 500 million DNA bases a probable event during cell division in the vine's various meristems (Lushai and Loxdale, 2002). Most of these mutations, however, occur on the very long stretches of DNA that do not encode any genes. Most mutations are therefore "invisible," which means they have no phenotype. Nonetheless, mutations in the DNA's coding regions also tend to accumulate over time. Because their cumulative effects may vary among different parts of the plant, grapevines may evolve over their lifetime, and clonal descendants, even when propagated from a single mother plant, can be genetically heterogeneous (McKey et al., 2010; Roach et al., 2018). Many small changes over a long period may result in equally large phenotypic changes as does a major change due to a single large mutation event. If such phenotypic changes are sufficiently distinct, clones may come to be called cultivars, as exemplified by the fruit skin color mutants of Pinot: Pinot noir, Pinot gris, and Pinot blanc (Bessis, 2007; Regner et al., 2000). In fact, many clones of the Pinot family are chimeras, which are defined as plants with more than one genetically different cell population that arose from a mutation in only one of the two functionally distinct cell layers or lineages of the shoot apical meristem (Riaz et al., 2002; Thompson and Olmo, 1963; Vezzulli et al., 2012). This includes another old Pinot clone (or Pinot ancestor), Pinot Meunier, whose leaves are densely coated with white hair (Franks et al., 2002; Hocquigny et al., 2004). Chimeras also exist among clones of Chardonnay and Cabernet Sauvignon, which suggests that layer-specific mutations in the apical meristem may be an important source of clonal and varietal variation in grapevines (Moncada et al., 2006; Pelsy, 2010). Therefore, as in the case of species discussed in Section 1.1, there is no clear-cut distinction between clones and cultivars, a fact recognized by Charles Darwin while he developed his revolutionary theory of evolution by random mutation and directional natural selection (Darwin, 1859).

Although clonal vines were selected and vegetatively propagated for fruitfulness, fruit flavor, and wine quality in Roman times (Columella, ca. 70 ad), organized and methodical clonal selection only began in Germany in the late nineteenth century and did not begin in France until the 1960s (Pelsy, 2010). Selection is usually based on the absence of symptoms of virus diseases (e.g., leafroll and fanleaf), healthy growth, and on performance indicators such as consistent yields and high wine quality, but criteria can also include specific viticultural traits such as fruit set, disease resistance, or drought tolerance. Nevertheless, many so-called clones are actually phenotypes caused by various combinations and degrees of virus infections rather than genuine genetic differences among the plants themselves. Today, clones exist for most major grape cultivars, and their success is based on the ability of some clones to perform differently in diverse environments. Differences among clones, and thus within a cultivar, offer viticulturally relevant diversity and potentially better planting material. For example, some clones of Cabernet franc have less compact fruit clusters with smaller berries and juice with a lower pH than others, or may be less vigorous, while others may be less prone to some fungal diseases (van Leeuwen et al., 2013). Nonetheless, the genetic differences among clones of a cultivar are typically very small. For example, all of the more than three dozen Sangiovese clones analyzed using DNA assays were found to have been derived from the same mother vine (Di Vecchi Staraz et al., 2007; Filippetti et al., 2005). Similarly, although there are Pinot noir, Grenache, and Tempranillo clones with compact clusters and others with loose clusters, no consistent differences in gene expression among such clones have yet been found (Grimplet et al., 2019).

It is currently fashionable to plant clonal selections in vineyards, although choices are often based on the "fallacy of the perfect clone"—that is, on the false assumption that the "best" clone found on one site will also perform best in a new location. Yet the performance or suitability of a clone on a particular site is strongly modified by the site's environment (soil, climatic conditions, cultural practices, etc.) and also depends on the desired end use of the grapes produced on that site and even regulatory circumstances, such as legal yield restrictions. This makes it nearly impossible to predict clonal performance in new environments. Nevertheless, in some cultivars, certain clonal traits, such as yield, cluster architecture, grape sugar, or acid content, can be quite stable across relatively diverse environments (Huglin and Schneider, 1998). This may be particularly true for old cultivars with very high clonal diversity; examples include the Pinots, Savagnins, and Cabernet franc. Planting several clones of a cultivar in the same vineyard block not only enables growers to conduct their own evaluation at a particular site but also provides some insurance against fluctuations in yield, fruit quality, and disease susceptibility that come with the absence of genetic diversity of a single clone (see also Section 8.1).

1.2.4 Rootstocks

Rootstocks are specialized stock material to which grape cultivars with desirable fruit properties are grafted. The shoot portion of the two grafting partners is termed the scion, whereas the rootstock provides the root system to the fused combination of genotypes. The callus cells that form at the graft union initially show an immune response following detection of the nonself nature of the grafting partners (Cookson et al., 2014). For the grafting operation to be successful, the vascular cambiums (Latin *cambium* = change), responsible for cell division (see Section 1.3), of the two grafting partners must make contact with each other so that they can "build" a connection between their separate "plumbing" systems for water and nutrient supply. Grafting of grapevines was common in ancient times; in his *De Re Rustica*, the Roman writer Columella described techniques known to his ancestors. In the late nineteenth century, the use of rootstocks derived from North American *Vitis* species saved grape growing in Europe from extinction due to the introduction, on imported planting material, of the aphid-like insect phylloxera (see Section 8.1). On the downside, rootstocks or their progeny that escaped, for example, from abandoned vineyards have also become feral across Europe, and their interbreeding populations are in some areas behaving as invasive species (Arrigo and Arnold, 2007; Laguna, 2003).

Commonly used rootstocks are either individual *Vitis* species or crosses of two or more species (Table 1.4). Due to the dioecious nature of their parents, they are either male or female plants. Examples of male rootstocks include Teleki 5C and SO4 and Riparia gloire de Montpellier, whereas female rootstocks include Kober 5BB, 101-14 Millardet et de Grasset, and Fercal (Meneghetti et al., 2006). Most rootstocks in use today are hybrids of three species: *V. riparia, V. rupestris*, and *V. berlandieri* (Galet, 1998). These three species are considered the most tolerant of phylloxera, whereas some other North American species, such as *V. labrusca*, are susceptible, and *V. vinifera* is highly susceptible. However, the genetic basis of the world's rootstocks is extremely narrow: As many as 90% of all *V. vinifera* vines are grafted to fewer than 10 different rootstock varieties. Most rootstocks have remained unchanged following initial breeding efforts before 1900 (Galet, 1998; Huglin and Schneider, 1998). As discussed previously for scion cultivars, this strategy puts vineyards at risk from mutant strains of soil pests, including phylloxera, with potentially devastating effects if a resistance breaks through. An example is the failure due to phylloxera in the 1980s, after decades of widespread

1.2 Cultivars, clones, and rootstocks

Table 1.4 Agronomic characteristics of important grapevine rootstocks.

Rootstock	Parent species	Phylloxera	Root knot nematode	Dagger nematode	Crown gall	Phytophthora	Drought	Flooding	Lime	Salinity	Acid soil	Clay soil	Sandy soil	Mg deficiency[b]	K deficiency[b]	Grafted scion vigor	Fruit maturation[c]	Ease of grafting	Ease of rooting
Riparia Gloire	V. riparia	●	●	●			◐	●	◐	●		●	◐	☒		●	◐	◐	○
Rupestris St. George	V. rupestris	◐	◐	◐	●	◐	◐	◐	◐	●	●	◐	●		☒	●	◐	◐	○
Rupestris du Lot	V. rupestris			○			○	◐	◐	●		●	◐				●	◐	◐	
420A Millardet et de Grasset	V. berlandieri × V. riparia	◐	◐	◐		●	◐	◐	◐	○		○	◐		☑	◐	◐	◐	●
5BB Kober	V. berlandieri × V. riparia	◐	◐				●	◐	●	○		◐	●	☑	☑	◐	◐	◐	◐
SO4	V. berlandieri × V. riparia	◐	◐	◐	●	●	◐	●	◐	○	◐	○	●	☑	☒	○	◐	◐	◐
8B	V. berlandieri × V. riparia	◐	◐				◐		◐							○			●
5C Teleki	V. berlandieri × V. riparia	◐	◐	◐			◐	◐	◐			○	●			●	◐	◐	◐
161-49 Couderc	V. berlandieri × V. riparia	●	◐				●	◐	◐	●		◐	◐			○		◐	●
99 Richter	V. berlandieri × V. rupestris	◐	◐		●		◐	○	◐	●		◐	◐	☑	☑	●	◐	◐	◐
110 Richter	V. berlandieri × V. rupestris	◐	○				●	○	●	◐	○	◐	◐	☑	☑	●	◐	◐	◐
140 Ruggeri	V. berlandieri × V. rupestris	◐	○				●	◐	●	●	○	◐	●	☒	☑	●	◐	◐	◐
3309 Couderc	V. riparia × V. rupestris	◐	●	○	◐	●	◐	●	◐	●		◐	●	☒		◐	◐	◐	○
101-14 Millardet et de Grasset	V. riparia × V. rupestris	◐	◐	◐	◐	●	◐	●	●	○	○	◐	●			○	◐	◐	◐
Schwarzmann	V. riparia × V. rupestris	◐			●		◐	●	◐	○		◐				○		◐	●
Gravesac	V. berl. × V. rip. × V. rup.	◐	●				○	○	◐	●						◐		○	◐
44-53 Malègue	V. rip. × V. cordif. × V. rup.	◐	◐	◐			●	●	●		○	○	●	☑	☒	○	◐	●	◐
1103 Paulsen	V. solonis × V. riparia	◐	◐	◐	◐	◐	●	○	●	●	◐	○	○	☒	☑	●	◐	◐	○
1616 Couderc	V. solonis × V. riparia	○	○	○		●	○	●	◐	●		●	●			◐	◐	●	●
Ramsey	V. champinii	◐	●	◐		◐	◐		◐			●	●			◐		●	●
Dog Ridge	V. champinii	◐	○	◐		●	◐		◐			●	●			◐		●	●
Harmony	Complex[a]	◐	○	○	●		◐		◐			●	●			◐		○	◐
Freedom	Complex	◐	●	○	●		◐		●			●	○			◐		○	○

● ◐ ○ ◐ ●
Excellent/high .... Poor/low

[a] V. champinii × V. longii × V. vinifera × V. riparia × V. labrusca   [b] ☑: Yes; ☒: No   [c] ◐: Advanced; ◐: Delayed

Modified from Currle, O., Bauer, O., Hofäcker, W., Schumann, F., Frisch, W., 1983, Biologie der Rebe. Neustadt an der Weinstrasse. Meininger, Germany; Dry, P.R., Coombe, B.G., 2004. Viticulture. Volume 1—Resources, second ed. Winetitles, Adelaide, Australia; Ferris, H., Zheng, L., Walker, M.A., 2012. Resistance of grape rootstocks to plant-parasitic nematodes. J. Nematol. 44, 377–386; Galet, P., 1998. Grape Varieties and Rootstock Varieties (J. Smith, Trans.). Oenoplurimédia, Chaintré, France; Pongrácz, D.P., 1983. Rootstocks for Grape-Vines. David Philip, Cape Town, South Africa.

use, of the *V. vinifera × V. rupestris* rootstock Aramon × *rupestris* Ganzin 1 (A × R1) in California. Moreover, except for those derived from *M. rotundifolia*, the so-called resistant rootstocks are not immune to phylloxera or nematodes. They just tolerate the feeding of these insects and microscopic roundworms better than do *V. vinifera* cultivars and therefore suffer less from infestation (Grzegorczyk and Walker, 1998; Huglin and Schneider, 1998; Mullins et al., 1992). Tolerant species or cultivars may grow quite well in the presence of these pests, but they do not prohibit pest numbers from building up over time. They can also be symptomless carriers of virus diseases. Infested propagation material is therefore a potent source of inoculum (infectious material) for new vineyards and new viticultural regions (see also Section 8.1).

The technique of grafting combines the tolerance of soil-borne pests of American *Vitis* species with the winemaking quality (i.e., consumer acceptance) of *V. vinifera*. Today, grape rootstock varieties are used not only for their tolerance of or resistance to root parasites, such as phylloxera and nematodes, but also for their ability to influence crop maturity or their tolerance of adverse soil conditions such as drought, waterlogging, excess lime, acidity, or salinity (May, 1994; Pongrácz, 1983; see also

Table 1.4). Rootstocks can also contribute to the management of vine vigor and of grape maturity and composition (Currle et al., 1983; Dry and Coombe, 2004; Galet, 1998). Although the influence on scion vigor is an important consideration in rootstock choice, there are no truly dwarfing grape rootstocks, unlike in the case of, for instance, apple. On the contrary, many interspecific *Vitis* hybrids grow more vigorously than their parents, perhaps as a result of the effect termed "hybrid vigor" or heterosis and/or of reduced investment in sexual reproduction, which may free up resources for reproductive growth (McKey et al., 2010).

The influence of rootstocks on the composition of the fruit produced by the scion cultivar is discussed in Section 6.3. Although it seems possible for a rootstock and a scion to exchange DNA pieces at the graft union (Stegemann and Bock, 2009), the rootstock effect on fruit composition is indirect and relatively minor. Some of the lesser known eastern Asian species, such as *V. amurensis*, are used increasingly in modern rootstock breeding programs for their cold tolerance (Alleweldt and Possingham, 1988), even though this diminishes the phylloxera resistance of the resulting crosses. Similarly, *M. rotundifolia* is used in rootstock breeding programs for its broad resistance to nematodes, although it roots and grafts inconsistently and does not tolerate cold winters (Ferris et al., 2012).

1.3 Morphology and anatomy

Many grapevine species are very vigorous, woody climbers named lianas, while others have a more brush-type growth habit (see Section 1.1). All are perennial (Latin *per* = through; *annus* = year), polycarpic (Greek *polus* = many, much; *karpos* = fruit, grain), and deciduous (Latin *decido* = to fall down). These terms are used to describe the vines' typical life form: they live longer than 2 years, they flower and reproduce repeatedly during their life, and they shed their leaves every year. With the aid of their coiling tendrils and flexible, slender trunks, wild vines climb on trees to a height of 30 m or more, spreading their foliage over the tree canopy (see Fig. 1.2). Unlike trees, vines can maximize their leaf area and shoot growth rate while minimizing their investment in a supporting stem structure (Ewers and Fisher, 1991; Ichihashi and Tateno, 2015; Wyka et al., 2013). As "structural parasites" the vines use the trees for support, seizing them as a natural trellis system. This enables them to grow to tremendous size; the canopy of a single vine may cover a surface area of dozens of square meters.

Grapevines can live to several hundred years of age: Famous examples include the "Great Vine" in London's Hampton Court Palace, which is thought to have been planted before 1770 and continues to produce 200–300 kg of fruit each year, the "Old Vine" of Maribor in Slovenia, and the "Mother Vine" on North Carolina's Roanoke Island, both of which are believed to be over 400 years old (Vršič et al., 2011). Vegetative propagation, whether naturally by layering or artificially from cuttings, bud grafts, tissue culture, or other means, extends a vine's life span virtually indefinitely. In fact, both rooted cuttings and grafted plants show clear signs of rejuvenation compared to their "mother" plants from which the cuttings were taken or to which they were grafted (Munné-Bosch, 2008). Growth rates, leaf photosynthetic rates, and fruiting of recently propagated vines are like those of young plants and are independent of the age of the vines from which the propagation material was taken.

Like other higher plants, grapevines comprise vegetative organs and reproductive or generative organs or fruiting structures. The vegetative organs include the roots, trunks, shoots, leaves, and tendrils (Fig. 1.4), while the reproductive organs include the clusters with flowers or berries. A vine's vegetative part can be divided into a belowground portion (roots) and an aboveground portion (trunks and shoots). The aboveground vegetative organs together with the reproductive organs is termed

1.3 Morphology and anatomy

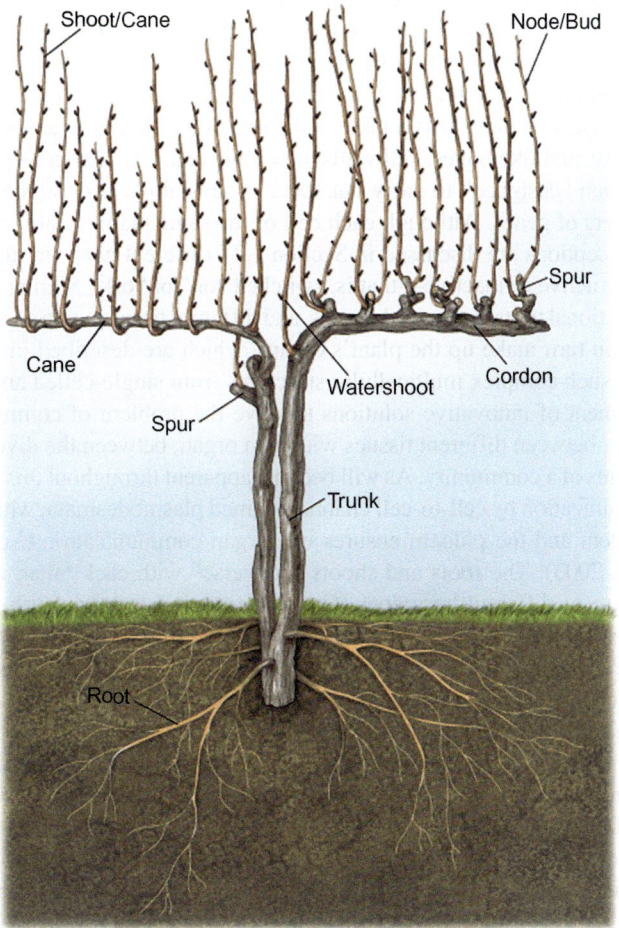

FIG. 1.4

Own-rooted (ungrafted) grapevine body with vegetative organs, but without leaves and tendrils. The left half is head-trained and cane-pruned, the right half is cordon-trained and spur-pruned.

© *Elsevier Inc. Illustration by L. O'Keefe after M. Keller.*

the vine's canopy. The roots, the trunks, and their man-made extensions called cordons, together form the permanent structure of the vine, which makes up 50–75% of the biomass (dry matter) of cultivated grapevines. The share of the permanent structure is much lower than 50% in young plants and increases with plant age and size. It is lower in heavily cropped vines than in lightly cropped vines, and it is lower in humid climates than in dry climates.

The entire plant body is rigid and strong yet flexible and adaptive enough to explore soil resources, intercept sunlight, and develop seeds for propagation. This combination of strength and flexibility is possible because plants employ a "bricks-and-mortar" strategy for growth. The vine's basic building blocks, the cells, are able to divide and expand while being joined together by an adhesive matrix, the cell walls.

Thus the cells function as bricks, while their walls serve as mortar. More precisely, the two cell walls of adjoining cells are "glued" together by the so-called middle lamella, which prevents sliding between cells. Growing cells secrete a thin (<0.1 µm), flexible primary wall, which is reinforced in mature cells that have ceased expanding with a thick (>1 µm), lignified secondary wall for extra mechanical strength (Doblin et al., 2010). The cell walls partly take on the function of bones in animals: They give the plant body form and rigidity, and like bones, cell walls are rich in calcium. The cells themselves differentiate into various types, each "designed" to carry out specific physiological or structural tasks under the direction of different sets of genes. Although each cell of the same plant usually contains the same complement of genes (exceptions are discussed in Section 1.2), different genes in different locations and at different times can be active or inactive—that is, switched "on" or "off." Similar and specialized groups of cells that form functional units are termed tissues, and different tissues, such as epidermis, parenchyma, phloem, and xylem, in turn make up the plant's organs, which are described in the following sections.

The evolution of such complex multicellular structures from single-celled ancestors required the simultaneous development of innovative solutions to solve the problem of communication between the cells forming a tissue, between different tissues within an organ, between the diverse organs of the plant body, and among plants of a community. As will become apparent throughout this book, plants master the art of intercell communication by cell-to-cell channels named plasmodesmata, whereas a vascular system composed of the xylem and the phloem ensures interorgan communication (Lough and Lucas, 2006; Roberts and Oparka, 2003). The roots and shoots "converse" with each other, at least in part, via the production and release of different hormones (Greek *hormaein* = to stimulate); cytokinin and abscisic acid (ABA) are the main root hormones, whereas auxin and gibberellin are the main shoot hormones. Hormones are defined as extracellular signaling molecules that act on target cells distant from their site of production. Perhaps even more fascinatingly, information exchange between neighboring plants is achieved by emitting and receiving volatile chemicals transmitted through the air and by releasing soluble chemicals into the soil and recruiting symbiotic microorganisms that infect and thereby interconnect the roots of different plants. Of course, successful communication requires not only the production and export of appropriate signaling molecules from transmitting cells (output signals) but also the presence of a sophisticated network of signal perception, decoding, and transformation in the receiving cells (input signals). Such signals integrate genetic programs that perceive and respond to both developmental and environmental cues to direct cell, tissue, organ, and plant growth and physiology (Lough and Lucas, 2006). The following sections give an overview of the external and internal form and structure, that is, of the morphology and anatomy, of the principal organs and tissues of grapevines, thereby laying a foundation for the discussion of their functions, or physiology, in subsequent Chapters.

1.3.1 Roots

The root system is the interface between a grapevine and the soil. It provides physical support for the vine in the soil and is responsible for water and nutrient uptake. The roots also serve as storage organs for carbohydrates and other nutrients, which support the initial growth of shoots and roots in spring, and for water. In addition, they are a source of plant hormones, especially cytokinins and ABA, which modify shoot physiology.

True roots develop from the stem named hypocotyl (Greek *hypo* = below) of the embryo (Greek *embryon* = unborn child) contained within a seed. In vines grown from seeds, seed germination begins with water absorption, called imbibition, which enables the embryonic root, or radicle

(Latin *radicula* = little root), that has been preformed in the seed, to grow and rupture the seed coat, forming a primary root or tap root. Multiple secondary roots, which are also called branch roots or lateral roots, then grow from the primary root. These lateral roots in turn develop higher-order lateral roots. This type of branched root morphology is typical of dicotyledons and is called an allorhizic root system (Bellini et al., 2014; Osmont et al., 2007).

Grapevine seedlings, which arise from sexual reproduction (see Section 2.3), are rarely planted in vineyards; they are mostly grown in breeding programs to develop and test new cultivars. The vast majority of vineyards across the world are planted with vegetatively propagated grafted vines or rooted cuttings that are very often produced in specialized nurseries. In vegetatively propagated grapevines, the roots originate from or near the cambium layer of woody cuttings (Pratt, 1974). Such stem-derived roots are termed adventitious roots (Bellini et al., 2014). The growth hormone auxin promotes the initiation of root growth from the cambium, whereas other hormones like cytokinin and strigolactone suppress root outgrowth (Rasmussen et al., 2013). Most of these adventitious roots form near the nodes of a cutting, but they also grow throughout the internodes. The roots that emerge directly from a cutting are often called main roots, and they branch off into secondary, tertiary, and higher-order lateral roots to form a branched root system wose complicated architecture resembles the secondary homorhizic root system of seedlings (Osmont et al., 2007). In summary, primary roots develop from seeds, main roots develop from cuttings, and lateral roots develop from other roots, regardless of whether those roots were derived from seeds or from cuttings (Bellini et al., 2014).

Unlike the strict arrangement of shoot lateral organs, the number and placement of lateral roots are not predetermined but depend on the availability of water and nutrients in the soil (Malamy and Benfey, 1997). The capacity to generate a highly flexible root architecture is important, because grapevines cannot move away from poor soils. Instead they rely on their ability to detect these environmental cues and extend lateral roots to take advantage of favorable soil regions. In contrast to many other plant species, lateral root initiation in grapevines is not restricted to the unbranched apical zone. Grapevines are also able to grow new lateral roots on older parts of the roots that have already developed a vascular cambium. Under conditions of very high humidity and high temperature, moreover, trunks and other aboveground parts of the vine can form aerial roots. Whereas this phenomenon is mostly restricted in the genus *Vitis* to greenhouse conditions (and is rare even there), it is quite common in the natural habitat of the *Muscadinia* (Viala and Vermorel, 1901–1909). Nonetheless, *V. sylvestris* vines growing in humid forest habitats readily form aerial roots, too (Levadoux, 1956).

The root system of established grapevines comprises a far-reaching, highly branched structure with a surface area far exceeding that of the leaf canopy it supports. A mature, cultivated grapevine can have >100 km of total root length with a surface area >100 m^2, whereas its leaf area is usually <10 m^2. It has been estimated that a vineyard may contain >30 t ha^{-1} of roots (Huglin and Schneider, 1998). The woody roots, whose diameter rarely exceeds 3 or 4 cm, serve to anchor the vine and transport and store soil-derived and leaf-derived nutrients. It is the small absorbing roots of 0.1–1 mm in diameter, which are called fine roots, that are responsible for most of the acquisition from the soil of resources such as water and nutrients. The woody roots of mature vines are widely distributed, with horizontal roots exploring the soil for distances of up to 10 m from the trunk (Huglin and Schneider, 1998; Smart et al., 2006; Fig. 1.5). Although most roots, especially the fine roots, are normally concentrated in the nutrient-rich and oxygen-rich top 0.5–1 m of the soil, roots can grow to a depth of >30 m when they encounter no impermeable barriers (Galet, 2000; Lehnart et al., 2008; Morlat and Jacquet, 1993; Pourtchev, 2003; Viala and Vermorel, 1901–1909). Indeed, grapevines are among the most deep-rooted plants, and their root biomass

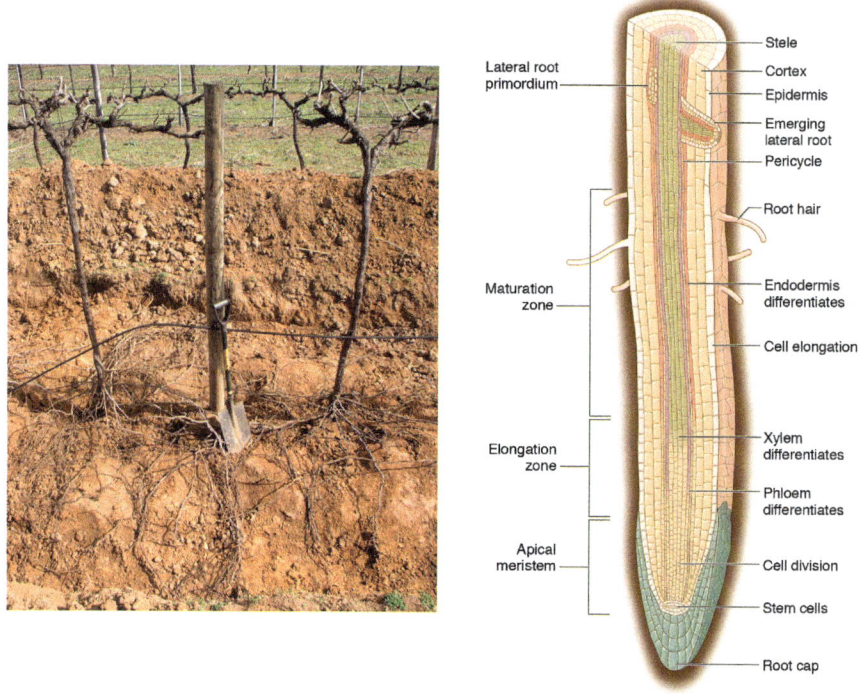

FIG. 1.5

Distribution of Merlot root system in a drip-irrigated vineyard (left) and diagrammatic longitudinal section of the apical region of the root (right).

Left: photo by M. Keller.

can range from 5 to 40 t ha^{-1}, which may reflect the competition for water and nutrients during the vines' coevolution with their "trellis" trees (Huglin and Schneider, 1998; Smart et al., 2006).

The growing root tip or apex is covered by a slimy root cap whose starch-containing central portion is named the columella. The cap protects the root meristem from abrasion, facilitates penetration of the soil, and contains gravity sensors, which guide the root downward through the soil (see Section 2.2). As the root advances and encounters new regions of moisture and nutrients, the cap is continuously sloughed off and replenished from the inside, whereas the polysaccharide-rich slime, called mucigel, that the cap secretes covers the surface of the maturing root. The mucigel harbors microorganisms and probably aids in the establishment of symbiotic mycorrhiza (Greek *mykes* = fungus, *rhíza* = root) in the root tips, which form a network of fungal mycelium that connects plants belowground. This network helps with the uptake of nutrients and water in exchange for carbohydrates supplied to the fungi by the plants (Bais et al., 2006; Marschner, 1995; Smith et al., 2001; Smith and Smith, 2011). In addition, the mucigel and the dead cells lubricate the growing root to reduce the friction between the root surface and the surrounding soil (Jin et al., 2013). The apical meristem lies just behind the ephemeral root cap and is the site of active cell division (Fig. 1.5). The apical meristem contains "stem cells" supplying "founder cells" or "initials" that, following several cycles of cell division, differentiate into the various root tissues according to their position.

Cell division is stimulated by the growth-promoting hormone auxin (indole-3-acetic acid and related compounds), which is produced locally in addition to being delivered from the unfolding leaves near the shoot tips and is then recycled, or refluxed, within the epidermis of the root tip (Aloni, 2013; Friml, 2003; Kramer and Bennett, 2006; Petrášek and Friml, 2009; Zhao, 2018). Cell elongation and differentiation, on the other hand, are promoted by a group of growth hormones termed gibberellins that are produced, upon stimulation by auxin, close to or at their sites of action—that is, in tissues with rapidly expanding cells (Ross et al., 2011; Yamaguchi, 2008). The activity of the meristem requires high concentrations of the "cell-division hormone" hormone cytokinin (*trans*-zeatin and related compounds), produced in the root tips, and low concentrations of the "cell-expansion hormone" gibberellin in the endodermis, whereas cell elongation requires the opposite (Wang and Li, 2008; Weiss and Ori, 2007). By countering the effect of auxin on cell division, cytokinin controls the rate of cell differentiation in the transition zone, so that the length of the meristem is determined by the balance between the division signal auxin and the differentiation signal cytokinin (Del Bianco et al., 2013). The growth-suppressing hormones ABA and the gaseous ethylene (C_2H_4) inhibit differentiation of the stem cells and the initials; they thereby help sustain the meristem during periods of stress that interfere with root growth (Zhang et al., 2010b).

As the newly produced cells begin to differentiate, the first phloem (Greek *phloos* = bark) cells appear behind and to the outside of the apical meristem (Fig. 1.5). Approximately 0.7–1.5 mm behind the root tip the meristematic region transitions into the elongation zone. The location of the transition zone between these two growth regions appears to be determined by a gradient of competing reactive oxygen species: $O_2^{\cdot-}$ toward the tip and H_2O_2 toward the base (Chandler, 2011). The elongation zone is a short region of cell expansion and further differentiation; the first xylem (Greek *ksúlon* = wood) cells appear in this zone. Since the phloem and xylem cells differentiate behind the root tip, hormones as well as the water and nutrients feeding the dividing cells must move across several layers of cells after leaving the vascular tissues.

The cells produced in the meristem form a central ring called the endodermis, which divides the root into two regions—the cortex toward the outside and the stele toward the inside (Fig. 1.5). The cortex is responsible for nutrient uptake from the soil, whereas the stele is responsible for nutrient transport up and down the plant; both sections also store starch and other nutrients (see Section 3.3). The stele differentiates into the pericycle, primary phloem, and primary xylem and is thus also called the vascular cylinder, whereas the cortex develops the exodermis and epidermis, which form the root's dermal tissues or "skin" (Evert, 2006; Galet, 2000). The epidermis, however, is short-lived, and the underlying exodermis (the hypodermis of the root) forms the boundary between root and soil even a few millimeters behind the root tip (Gambetta et al., 2013; Storey et al., 2003a). The endodermis is the innermost layer of the cortex and consists of a single layer of cells with thickened radial and transverse cell walls, named Casparian strips or Casparian bands, that are impregnated with lignin (Geldner, 2013; see also Section 3.3). In addition, behind the elongation zone, both the exodermis and the endodermis also incorporate suberin in their cell walls (Gambetta et al., 2013). The waxy biopolyester suberin is hydrophobic (water-repellent) but not impermeable, and the suberization pattern is interrupted by occasional unsuberized cells called passage cells (Geldner, 2013; Vandeleur et al., 2009).

The pericycle beneath the endodermis, rather than the apical meristem, gives rise to the lateral roots by renewed cell division, following dedifferentiation of pericycle founder cells, where the meristem transitions into the elongation zone (Mapfumo et al., 1994a; Nibau et al., 2008; Osmont et al., 2007; Viala and Vermorel, 1901–1909). Lateral root growth is initiated by shoot-derived auxin and aided by locally produced growth hormones termed brassinosteroids (which resemble animal steroid hormones) and small amounts of ethylene, which is produced by and released from the differentiating xylem. Apparently, the incoming auxin induces this ethylene production, which in turn blocks further auxin movement. The resulting local auxin accumulation in some pericycle cells adjoining the primary xylem then induces lateral root growth (Aloni et al., 2006b; Del Bianco et al., 2013). Higher concentrations of ethylene, however, along with cytokinin moving up from the root tips inhibit lateral root formation (Ivanchenko et al., 2008; Osmont et al., 2007). By blocking the production of auxin transport proteins, cytokinin disrupts auxin accumulation at the appropriate locations, so that lateral roots cannot be initiated (Santner et al., 2009). The fact that cytokinin concentration is highest closest to the site of its production ensures that lateral roots are not formed too close to the tip, which would interfere with its own continued growth (Aloni et al., 2006b). Moreover, ABA seems to control the activation of the new lateral root meristem after emergence of the lateral root from its parent root: A high ABA concentration (e.g., due to water stress or high nitrogen availability) inhibits meristem activation and keeps the lateral root in a dormant state (De Smet et al., 2006; Malamy, 2005). Once activated, auxin usually maintains the cell-producing activity of the meristem. Thus, the meristem and elongation zones together form the zone of active root growth in length, and a signaling network consisting of interactions among auxin, ethylene, cytokinin, brassinosteroid, and ABA regulates root architecture in response to environmental and developmental influences (Osmont et al., 2007).

A few millimeters behind the root's growth region follows the absorption zone, which is densely covered by root hairs. The colorless root hairs are long and thin protrusions of epidermal cells (epidermis cells that form root hairs are termed trichoblasts) and are mostly responsible for water and nutrient uptake (Gilroy and Jones, 2000; Pratt, 1974). Their small diameter (10–15 µm) and ability to flatten permit them to squeeze through soil pores of <2 µm. Root hairs can make up >60% of the root's surface area, which greatly increases the contact surface between root and soil, or the plant–soil interface, and the exploited soil volume (Clarkson, 1985; Sondergaard et al., 2004; Watt et al., 2006). The initiation and elongation of root hairs is stimulated by auxin and ethylene; the two hormones simultaneously slow down the elongation of the root's epidermis cells (Muday et al., 2012; Strader et al., 2010). As the root grows, new root hairs constantly grow behind the elongation zone, while the older ones are worn away along with the epidermis, the original cortex, and the endodermis. Hence, the absorption zone advances along with the root tip, leaving behind the conducting zone that continues into the trunk and all aboveground organs of the vine.

Procambial cells, which are derived from cells of the ground tissue named parenchyma (Greek *pará* = beside, *enchyma* = infusion) and serve as vascular stem cells, generate both the primary phloem and the primary xylem. The xylem cells closest to the root tip are termed protoxylem, its conducting cells deposit helical thickenings in their secondary cell walls, which enable these cells to stretch as the root elongates (Schuetz et al., 2013). Behind the elongation zone, the xylem cells transition into metaxylem, whose conducting cells are larger than those of the protoxylem and grow thick, pitted secondary cell walls that make them strong yet rigid. At about the same time or slightly later, about 10 cm behind the root tip, the procambium develops into the vascular cambium during the transition from primary to secondary growth (Gambetta et al., 2013; Schuetz et al., 2013). The vascular cambium serves as a lateral meristem and forms a continuous sleeve of several cell layers within the root–shoot axis.

Once elongation growth has ceased, the cambium forms new, secondary phloem cells toward the outside and secondary xylem cells, which collectively form the wood, toward the inside (Viala and Vermorel, 1901–1909; Rathgeber et al., 2016). Thus, if the apical meristem in the root tip is responsible for primary growth, the cambium is responsible for secondary or radial growth of a root and moves outward as the root grows thicker. Just like the cells within the apical meristem, the cambium cells can also be regarded as stem cells. Their division and subsequent differentiation are stimulated by transcellular flow of auxin from the shoot tip toward the roots in addition to locally produced strigolactone, cytokinin produced in the root tips, and physical pressure (Agusti et al., 2011; Aloni, 2001; Aloni et al., 2006b; Dengler, 2001; Matsumoto-Kitano et al., 2008; Ye, 2002). As auxin moves actively from cell to cell across the cell membranes and cell walls and mostly down the cambium "sleeve" and xylem parenchyma, its concentration remains highest in these dividing and differentiating cells (Schuetz et al., 2013).

A new meristem, the cork cambium or bark cambium, forms in the pericycle and generates cork toward the outside and secondary cortex toward the inside. This cambium is also called the phellogen, the cork it produces is termed the phellem, and the new cortex cells are termed the phelloderm; together these secondary protective tissues form the periderm (Evert, 2006; Pratt, 1974). Because the outermost cell layers, including the exodermis, cortex, and endodermis, rupture and are sloughed off, the mature, woody root consists of periderm on the outside and phloem, vascular cambium, and xylem toward the center (Viala and Vermorel, 1901–1909; Gambetta et al., 2013). Contrasting with the wood, which is mostly xylem, to the inside of the vascular cambium, the tissue layers to the outside of this cambium are collectively referred to as the bark (Cuneo et al., 2018; Evert, 2006). The deposition within the colorless suberin that accumulates in the cork cell walls of oxidized phenolic compounds released from dying cells give the mature root a brownish color.

The phloem, vascular cambium, and xylem together are called the vascular tissues, or vascular system, which forms discrete vertical strands of conducting tissues named vascular bundles, separated by rays consisting of at least one layer of compact parenchyma cells (Currle et al., 1983; Stafford, 1988; Viala and Vermorel, 1901–1909). The phloem and xylem are placed in parallel within each vascular bundle—a pattern termed collateral vascular bundle—and are interconnected by the parenchyma rays (Fig. 1.6), which allows for the radial transfer of water and nutrients between the two "circulatory systems." In other words, the phloem and the xylem are hydraulically interconnected (Pfautsch et al., 2015).

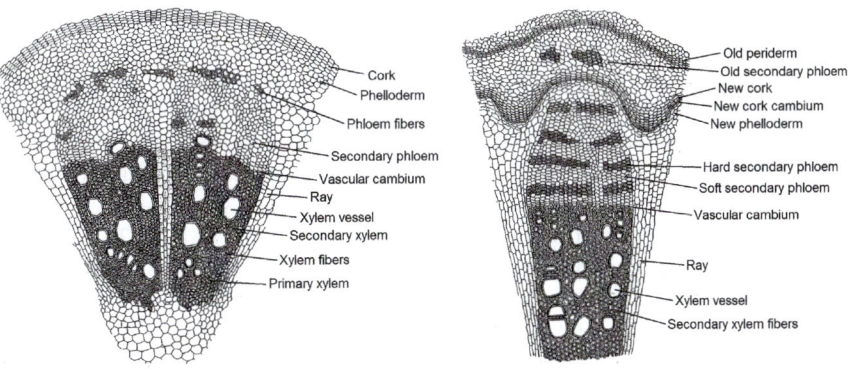

FIG. 1.6

Cross sections of one-year-old root (left) and three-year-old root of *V. vinifera* (right). Cork + Cork cambium + Phelloderm = Periderm.

Modified after Viala, P., Vermorel, V., 1901–1909. Ampélographie. Tomes I-VII, Masson, Paris.

The phloem is composed of sieve elements, companion cells, and phloem parenchyma. Before the sieve elements become functional, they resorb their nucleus and vacuole and retain only their cell membrane, a thin layer of cytoplasm, plastids, and a few enlarged mitochondria, tied together so they are not dragged along with the phloem flow (van Bel, 2003). In a unique division of labor, the hollow and almost "clinically" dead sieve elements become the conduits for sugar and other products, whereas the companion cells, which are crammed with mitochondria, carry out the metabolic functions abandoned by the sieve elements and act as transfer cells, loading sugar and other materials from and to the surrounding parenchyma cells into and out of the sieve elements (see Section 5.1). To this end, pairs of companion cells and sieve elements are interconnected through numerous "lifelines" termed pore/plasmodesma units, and the companion cells and parenchyma cells are connected through membrane-lined channels called plasmodesmata (Roberts and Oparka, 2003; Turnbull and Lopez-Cobollo, 2013). The sieve elements, with a length of 100–400 μm and a diameter of 10–40 μm, have thicker cell walls than the adjacent parenchyma cells. They are stacked end-to-end and interconnected across their end walls to form long sieve tubes. The specialized end walls are called sieve plates because they are perforated by 20–200 large sieve pores derived from plasmodesmata (Crespo-Martínez et al., 2019; Knoblauch et al., 2016; Tyree et al., 1974; van Bel, 2003; Viala and Vermorel, 1901–1909). The sieve pores have a diameter of 1–2 μm and are lined with callose, which enhances their rigidity and prevents collapse under pressure (Barratt et al., 2011; Bondada, 2014; Knoblauch and Oparka, 2012). In addition, extra callose may be deposited in the cell walls surrounding the pores, which restricts passage of molecules through the opening; this callose deposition is reversible in order to regulate plasmodesmal traffic (Burch-Smith and Zambryski, 2012; Wu et al., 2018). Circulation of auxin in the sieve tubes, where the hormone is deposited by the leaves, serves to ensure that the amount of callose remains just right (Aloni, 2013). Unlike the water-conducting xylem cells, the sieve elements retain their plasma membrane, which is necessary to generate osmotic gradients for sugar transport in a direction opposite to the transpiration stream (see Section 5.1).

The xylem consists of vessel elements, tracheids, fibers, and xylem parenchyma. The vessel elements and tracheids together are called the tracheary elements. Once they are fully grown, they form strong, thick secondary cell walls, and both their primary and secondary cell walls become lignified by the incorporation of lignin (Latin *lignum* = wood), which is mostly composed of the phenolic building blocks coniferyl alcohol and sinapyl alcohol (Rathgeber et al., 2016). The tracheary elements are arranged longitudinally and embedded in the fibers and parenchyma, the latter forming radial rays (a.k.a. medullary rays) that separate the bundles of tracheary elements (Fig. 1.6). The tracheary elements are the actual xylem conduits or water pipes of the vine. Their differentiation is induced by the movement of auxin, which arrives from the shoot tips, in the growing cells that are to become the tracheary elements, whereas the formation of their secondary cell walls is induced by brassinosteroids, and lignification may be induced by ethylene (Aloni, 2013; Rathgeber et al., 2016). Before they can assume their function, however, these cells must die; their organelles and cytoplasm are disassembled, digested, and removed in an orderly manner termed programmed cell death, and the dead cells become the wood (Jones and Dangl, 1996; Plomion et al., 2001; Schuetz et al., 2013; Turner et al., 2007). Calcium influx, which is stimulated by brassinosteroids, serves as the trigger for the rapid cell death program (Rathgeber et al., 2016). The tracheary elements' thickened secondary cell walls provide mechanical strength to prevent the pipes from collapsing under the negative pressure generated by transpiration (Ye, 2002; see also Section 3.3). The lignification also renders the cell walls relatively waterproof, although lignin is not completely impermeable to water. Because much of the lignification happens postmortem, that is, after the death of the tracheary elements, the lignin building blocks are manufactured by the neighboring parenchyma cells, attached to glucose, and exuded to the cell walls, where the glucose is removed and the resulting monolignols oxidized and assembled into the lignin polymer (Barros et al., 2015; Schuetz et al., 2013).

Some fiber cells, whose differentiation is triggered by gibberellin arriving via the phloem from the leaves, also commit suicide in a tightly controlled manner after the death of the tracheary elements (Aloni, 2013; Courtois-Moreau et al., 2009). The very long (~0.5mm) and thin (~20μm) fibers reinforce and seal their secondary cell walls completely with lignin before they die, and their massive (~4μm) cell walls give the root mechanical support. Other lignified fiber cells may remain alive for some time (Knipfer et al., 2016). The remaining living cells in the mature xylem of woody roots are those of the parenchyma, but even these cells die after a few years (Cuneo et al., 2018; Plomion et al., 2001). Parenchyma cells have thin primary walls but no well-defined secondary walls, serving as "ground" tissue and storage sites for water and nutrients such as carbohydrates, proteins, and amino acids (Schuetz et al., 2013; Zapata et al., 2004). They also contain tannins and raphides, which are needle-shaped crystals composed of calcium-oxalate that help deter herbivores, and they are involved in the defense against pathogen penetration and can be recruited to repair wounds.

Multiple vessel elements are stacked end-to-end and interconnected by perforation plates in the end walls to form long hollow columns termed vessels. Perforation plates are holes created by targeted removal of the cell walls and membranes; they can be simple or scalariform. Simple perforation plates have a single hole, while scalariform perforation plates have several elongated holes that make the remaining cell wall appear like rungs of a ladder. The much rarer tracheids, which look much like vessels without perforation plates and, because they are single cells, are much shorter (1–10mm) than the vessels; many vessels in grapevines are >100mm long, and some exceed 1m. Both vessels and tracheids are radially interconnected (although not with each other) through lined-up pairs of porous cell wall depressions named pits. Whereas the connections between individual vessel elements within a vessel are provided by either pits or perforation plates, those between one vessel and another or between neighboring tracheids are provided by pits (Bondada, 2014; Brodersen et al., 2011; De Boer and Volkov, 2003; Sun et al., 2006). The perforation plates within small vessels are generally scalariform, especially in the primary xylem, whereas those within larger vessels are usually simple and often surrounded by bordered pits. Bordered pits are pores with overarching secondary cell walls that have a narrower aperture or pit channel opening into a wider pit chamber (Brodersen et al., 2018). Unlike the pits connecting vessel elements within a vessel, the pits between adjoining vessels or tracheids contain inside their chamber a water-permeable "membrane" consisting of the middle lamella and primary cell walls. Although not a true cell membrane because it does not contain a lipid bilayer, the fine pore size of 5–20nm in diameter of these specialized cell wall modifications termed pit membranes tends to slow water flow but also limits the spread of air bubbles (see Section 3.3) and pathogens (Pérez-Donoso et al., 2010). Because of their more open construction, vessels can conduct water more efficiently than can tracheids. In addition to its longitudinal movement up the plant body, water also moves laterally through pits, both between parallel vessels or tracheids, and from and to the surrounding parenchyma cells, but not to fibers (Pfautsch et al., 2015). Due to their length (~0.5mm) and large diameter (10–150μm), individual vessel elements can be surrounded by and connected with many dozens of the much smaller parenchyma cells (Sun et al., 2006; Viala and Vermorel, 1901–1909).

1.3.2 Trunks and shoots

The aboveground or aerial axis of the grapevine, which comprises the shoots, cordon arms, and trunk, is called the stem by botanists. It provides support for the growing vine and a scaffold for the leaves, and is responsible for the transport of water, sugar, and other nutrients. Like the roots, the flexible stem also serves as a storage organ for carbohydrates and other nutrients, which support early growth in spring and during stress periods, and for water. Because of their liana nature, cultivated grapevines typically require a trellis system for support, unless they are trained very close to the ground, such as the freestanding so-called bush vines. The trunk is often extended along a horizontal wire to form one or more permanent arms, or cordons, that support the 1-year-old wood, which in turn gives rise to the fruiting shoots (Fig. 1.4). Such vines are said to be cordon-trained, whereas vines without cordons are said to be head-trained.

The shoots, which are called canes after they have matured and the leaves have fallen off, are generally pruned back in winter to canes of varying length; canes with less than four buds are called spurs (Fig. 1.4). Spur pruning retains one to three buds per spur, whereas cane pruning typically retains eight or more buds per cane. Both spur and cane pruning are usually done manually, which permits maximum control over both bud number and position on the vine. Cordon-trained grapevines may also be pruned by machine, leaving a range of shorter and longer spurs centered on the cordon; this is also called a box cut (Poni et al., 2016). Although specialized modern equipment can prune very close to the cordon, no mechanical pruning system can as yet control the number and position of the spurs left on the vines like manual pruning does. Consequently, machine pruning is sometimes done as a prepruning operation only, followed by manual touch-up to control bud numbers. Alternatively, very light mechanical pruning, which trims off only the ends of the shoots or canes, may be done in summer and/or winter, a method termed minimal pruning (Clingeleffer, 1984; Poni et al., 2016; Possingham, 1994). This practice relies on the vines' self-pruning ability, in which the immature apical portion of the shoots dies back and falls off in winter (see Section 5.1).

In addition to pruning, an assortment of combinations of training systems and other cultural practices are employed in commercial viticulture. They all have as their common goals to constrain the natural vigor of grapevines and maintain them at a small, manageable size; to sustain their shape and productivity over the several decades of a vineyard's lifespan; to optimize fruit production and quality depending on the intended end use of the grapes; and to facilitate labor and permit various degrees of labor-reducing mechanization.

The shoots, of course, are the carriers of grapevines' main photosynthetic lateral organs, the leaves. The points at which these lateral organs connect to the shoot are called nodes, and the nodes are separated by internodes (Fig. 1.11). The pattern of lateral organs on the shoot and especially the regular arrangement of leaves in space is described by phyllotaxy, or phyllotaxis (Greek *phyllon* = leaf, *taxis* = order). Vines grown from seeds go through a juvenile phase during which the embryonic shoot, named epicotyl, emerges from the embryo, forming a primary shoot without tendrils. On this shoot the leaves are arranged in a spiral fashion in which they are offset by the so-called golden angle of 137.5°, displaying 2/5 (sometimes 3/7) phyllotaxy (Galet, 2000; Viala and Vermorel, 1901–1909). A 2/5 phyllotaxy means that, moving up the shoot from any one leaf, two revolutions around the shoot must be made to find a leaf that is located directly above the reference leaf, which is five leaves below. The juvenile phase of seedlings generally ends when 6–10 leaves have developed and the vine transitions to adult development (Mullins et al., 1992). The shoot now forms the first tendrils, and all new leaves are being produced alternately—that is, on two opposite sides of the shoot with a single leaf at each node. This design with 180° angles between successive leaves is called distichous or 1/2 phyllotaxy.

1.3 Morphology and anatomy

In vegetatively propagated grapevines or adult seedlings, primary shoots usually originate from the buds at nodes of woody canes (Fig. 1.11), and their leaves always display 1/2 phyllotaxy. Shoot growth from buds is sometimes divided into "fixed" growth and "free" growth. Fixed growth occurs from leaf primordia and compressed internodes preformed in the bud during the previous growing season and overwintering in the dormant bud. Fixed growth is responsible for the rapid growth of the first 6–12 (occasionally up to 18) leaves in spring; the number increases with increasing temperature during bud development (Buttrose, 1970a; Greer and Weston, 2010a; Morrison, 1991). Free growth, on the other hand, occurs later in the season from the production of new leaf primordia and internodes by the shoot's apical meristem concurrently with internode elongation.

The apical meristem is the site of cell division, where all lateral organs of the shoot are initiated and where the highly variable pattern of the shoot system—and hence of the grapevine canopy—is established. This contrasts with the situation in the roots, where lateral roots arise from the pericycle rather than the root apical meristem; the shoot, however, does not have a pericycle. The shoot apical meristem consists of three domains or subpopulations of cells: the central zone, the rib zone or rib meristem, and the peripheral zone (Clark, 1997; Evert, 2006; Kerstetter and Hake, 1997). The central zone has very small, relatively slowly dividing cells with small vacuoles and is located at the very apex of the apical meristem; it produces and maintains stem cells that serve as a source of cells for the other two zones, which are involved in cell differentiation. The rib zone at the base of the apical meristem forms the vascular tissues and other tissues in the central part of the shoot. The peripheral, or morphogenetic, zone has small, rapidly dividing cells with small vacuoles; it surrounds the central zone on the flanks of the apical meristem and produces the outer shoot tissues and the lateral meristems. The lateral meristems are produced from a few founder cells that lose their meristem identity and acquire a new organ identity to give rise to the primordia (Latin *primus* = first, *ordiri* = to begin to weave) which, in turn, develop into the lateral organs (Sassi and Vernoux, 2013; see Fig. 1.7). In the great majority of species of the family Vitaceae, including all species of economic importance, there are two types of lateral meristems: One is responsible for leaf formation and the other for inflorescence and tendril formation (Carmona et al., 2007; Gerrath and Posluszny, 2007).

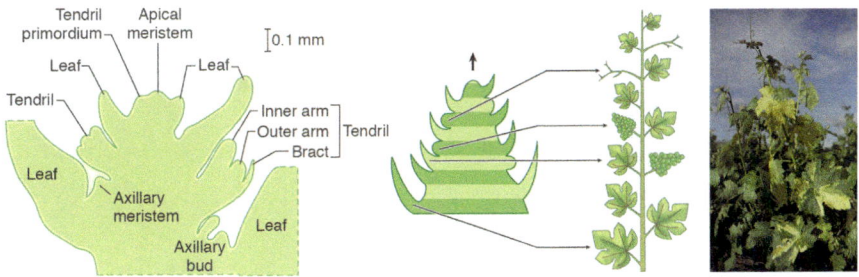

FIG. 1.7

Diagrammatic longitudinal section of Concord shoot tip (left); lateral organs arising from various positions in the dormant bud, illustrating the repeating three-node pattern unit of the shoots of many *Vitis* species (center); and chimeric Cabernet Sauvignon shoot (right).

Left: © *Elsevier Inc., illustration after Viala, P., Vermorel, V., 1901–1909. Ampélographie. Tomes I-VII, Masson, Paris; center: © Elsevier Inc., illustration after Carmona, M.J., Cubas, P., Martínez-Zapater, J.M., 2002. VFL, the grapevine FLORICAULA/LEAFY ortholog, is expressed in meristematic regions independently of their fate. Plant Physiol. 130, 68–77; right: photo by M. Keller.*

The three meristematic zones overlap with another structural feature of the apical meristem: The horizontal layering of cells divided into the tunica on the outside and the corpus on the inside (Evert, 2006; Thompson and Olmo, 1963). The tunica usually consists of two parallel cell layers, the outermost of which generates the epidermis. Whereas the cells of the corpus divide in various directions, those of the tunica divide perpendicularly to the surface. Because in this so-called anticlinal pattern of cell division the daughter cells remain in the same layer as their parent cells, the tunica's two layers perpetuate separate cell lineages (Clark, 1997; Jenik and Irish, 2000; Kerstetter and Hake, 1997; Sassi and Vernoux, 2013). Mutations occurring in only one of the two layers can therefore give rise to chimeric plants (see Fig. 1.7) and to clonal variation within grape cultivars (see Section 1.2).

The meristem cells do not contain chlorophyll, so they cannot perform photosynthesis and must import all carbohydrates and other nutrients to sustain their proliferation. The construction of the photosynthetic machinery begins as soon as the cells are released from the meristem, unless those cells remain in darkness. The latter occurs, for example, in a trunk sucker arising from below the soil surface and leads to the formation of colorless, etiolated shoots that elongate rapidly under the influence of gibberellin in search of light (Hartweck, 2008). Once the photosynthetic machinery has been "installed" in the chloroplasts of the differentiating cells, their chlorophyll content is responsible for the green appearance of growing shoots (Kriedemann and Buttrose, 1971).

The initial cell from which all subsequent cells of a lateral organ are derived is termed a founder cell (Chandler, 2011). The founder cell divides into a set of cells called preprimordium, or "anlage" (German *Anlage* = conception, layout, investment), whose cells then proliferate into a primordium. When these primordia cells differentiate, they give rise to lateral organs (Fig. 1.7). The lateral organs include leaves, tendrils, fruiting clusters, and axillary buds. Axillary buds in turn contain meristem primordia that are called secondary, axillary, or lateral meristems and form several compressed shoot segments before becoming dormant; they can be reactivated later to produce lateral shoots, whose leaves form at right angles to those of the main shoot (Gerrath and Posluszny, 1988a; McSteen and Leyser, 2005; Pratt, 1974; Fig. 1.8). Likewise, the inflorescence primordia later form their own flower meristems that produce the grape flowers with their different floral organs, which are themselves modified leaves—a concept originally proposed by the German writer and philosopher Johann Wolfgang von Goethe (1790). Contrary to their influence in the root tip, cytokinins stimulate cell division in the shoot

FIG. 1.8

Repeating three-node pattern of a Syrah shoot (left); mistakes do happen in nature—three consecutive tendrils on a *V. vinifera* shoot (center); and dormant bud and lateral shoot in a leaf axil (juncture between petiole and shoot) of a Malbec main shoot (right).

Photos by M. Keller.

apical meristem and are necessary for the transition from undifferentiated cells to differentiated primordia (Dewitte et al., 1999; Mok and Mok, 2001; Werner et al., 2003).

Just how the vine "knows" when and where to initiate the various lateral organs is still somewhat of a mystery. Whatever their identity, however, primordia cannot form without auxin, which is produced mostly from the amino acid tryptophan (Gallavotti, 2013; Wang and Jiao, 2018; Zhao, 2018). Many genes respond within minutes to auxin upon being triggered by a transcription factor named ARF (auxin response factor). When auxin accumulates in a cell, it activates ARF by binding to a receptor protein named TIR (short for transport inhibitor response) in the cell nucleus, and the resulting complex in turn binds to, and induces degradation of, another protein that normally keeps ARF inactive (Fukui and Hayashi 2018). Local accumulation of auxin in the peripheral zone, perhaps supported by cytokinin, seems to be the signal for organ formation by activating the cell wall-loosening protein expansin (see Section 3.1) in a founder cell, and the identity of each lateral organ may be determined by hormone gradients (Chandler, 2011; Petrášek and Friml, 2009; Pien et al., 2001; Reinhardt et al., 2000; Wang and Li, 2008). The incipient primordia are sinks for auxin, but once initiated, the primordia themselves also begin to produce and export auxin back toward the apical meristem (Wang and Jiao, 2018). This polarity of auxin transport depletes auxin in the surrounding cells and thereby prevents their growth, which ensures the regular, phyllotactic positioning of lateral organs (Benjamins and Scheres, 2008; Berleth et al., 2007; Kuhlemeier, 2007; Sassi and Vernoux, 2013). The local auxin minima, followed by a brief cytokinin pulse, also prescribe the location of axillary meristems that will become the axillary buds in the future leaf axils (Wang and Jiao, 2018).

Meristems and leaves are in constant communication and clearly influence each other; preexisting leaves are required for the correct positioning of a new leaf by defining the routes of auxin transport to the meristem (Petrášek and Friml, 2009; Piazza et al., 2005; Reinhardt et al., 2000). The older leaves even influence specific traits of newly developing leaves, such as leaf size and the density and size of stomata, so that the unfolding leaves are adapted to the environment into which they are "born" (Chater et al., 2014; Lough and Lucas, 2006). Unlike the new leaves, newly formed axillary buds are not connected to the shoot's vascular bundles; this connection is established at the time of bud outgrowth. It is thought that the buds release auxin as they begin to grow by renewed cell division and cell expansion, and this auxin induces differentiation of vascular tissues that connect the buds—and hence the growing lateral organs—to the preexisting vascular network of the shoot (Aloni, 2001). Thus, all active apical meristems export auxin, which moves away from the meristem via active, polar cell-to-cell transport (Leyser, 2010).

Tendrils and fruiting clusters are formed at the same position as the leaves but on the opposite side of the shoot (Fig. 1.7). The production of leaf-opposed tendrils and clusters appears to be unique to the Vitaceae family and is typically discontinuous; that is, two of every three nodes bear a tendril (Gerrath and Posluszny, 2007; Pratt, 1974). One notable exception is *V. labrusca*, which has a continuous pattern; that is, a tendril at every node. Why the other members of the family leave a "blank" at every third node and how they keep count is still mysterious. The unusual position of clusters and tendrils opposite leaves is a result of the lack of elongation of the internodes separating nodes that bear only rudimentary leaves, termed bracts, from those that bear true leaves (Morrison, 1991; Tucker and Hoefert, 1968). The first two or three nodes of each shoot usually carry only leaves; the next two nodes are generally the ones with clusters, followed by a node with a leaf only and then a succession of repeating mirror images of three-node units with the clusters replaced by tendrils (Figs. 1.7 and 1.8). The repeating units, or shoot segments, are called phytomers, and each phytomer consists of a node, an internode, a leaf attached to the node, and an axillary bud containing an axillary meristem for branching (McSteen and Leyser, 2005).

Some of the cells produced by the apical meristem do not become primordia for lateral organs. They instead develop into the internodes of the shoot axis. During primary shoot growth, leading to elongation of the internodes, these cells differentiate into the epidermis, cortex, endodermis, phloem, procambium, and xylem (Viala and Vermorel, 1901–1909; Esau, 1948; Galet, 2000). The outermost cell layer, the epidermis, develops a waxy cuticle (Latin *cutis* = skin) on its thick outer cell walls. The cuticle serves as a protective layer of all aboveground organs except woody stems that develop a periderm to replace their lost cuticle (Fich et al., 2016; see also below). The epidermis also has stomata, various types of hairs called trichomes (Greek *trichos* = hair) and, on young organs, often small and short-lived translucent structures variously named pearls, pearl glands, pearl bodies, or sap balls (Gerrath et al., 2015; Ma et al., 2016; Pratt, 1974; Fig. 1.9). Pearls develop on shoots, tendrils, inflorescences, petioles, and the abaxial surface of leaves, and they are especially frequent under warm, humid conditions and high nutrient availability promoting vigorous growth (Paiva et al., 2009; Viala and Vermorel, 1901–1909). Although the pearls remain covered by a thin epidermis that sometimes develops a stoma, they are formed of highly vacuolated, thin-walled subepidermal cells that accumulate sugars, oils and, perhaps, proteins. Pearls have no glandular activity but instead are also termed food bodies because they ostensibly attract and feed ants (Fig. 1.9), which in turn help defend the vulnerable young plant organs against herbivorous insects (Paiva et al., 2009).

The cortex consists mostly of parenchyma cells containing chlorophyll, starch, calcium oxalate crystals, and phenolic compounds that include red-pigmented anthocyanins on the sun-exposed side of the

FIG. 1.9

Ant feeding on pearls on a Chardonnay shoot.

Photo by M. Keller.

shoots of many cultivars (Fig. 1.8). Some parenchyma cells about one to four layers below the epidermis differentiate into strands of small collenchyma (Greek *kolla*=glue) cells with thick primary walls, strengthening the growing shoot (Esau, 1948). Shoots that grow on windy sites tend to develop much thicker-walled collenchyma than do shoots at sheltered sites, which is probably why the latter are more easily broken off by passing machinery. The innermost layer of the cortex, located just outside the phloem, forms the endodermis. The endodermis is also called the starch sheath because of its high concentration of amyloplasts, which are specialized plastids responsible for starch production from imported sucrose and its storage in starch grains or granules (Emes and Neuhaus, 1997; Martin and Smith, 1995; Pratt, 1974). But other parenchyma cells also accumulate starch, and they also store protein reserves inside specialized protein storage vacuoles. In spring the starch is broken down into sugar and the proteins are disassembled into amino acids for remobilization and export to the new shoots.

The vascular tissues to the inside of the endodermis comprise yet more parenchyma cells that form the primary rays and alternate radially with the vascular bundles like the spokes of a wheel (Fig. 1.10). The vascular bundles, bounded on the outside by caps of long, narrow cells with thick, lignified cell walls, called primary phloem fibers, contain the primary phloem, procambium, and xylem (Bondada, 2012; Esau, 1948). To the inside of the ring of vascular bundles, there is a core of unlignified and rather soft parenchyma cells called the pith or medulla.

The differentiation of the vascular bundles is induced by active, acropetal (toward the tip) movement of auxin at a rate of $\sim 1\,\mathrm{cm\,h^{-1}}$ from developing leaf primordia and young leaves (Benjamins and Scheres, 2008; Berleth et al., 2007; Petrášek and Friml, 2009; Woodward and Bartel, 2005). In addition, gibberellin imported via the phloem and/or produced by the growing xylem cells may regulate xylem cell differentiation and elongation (Israelsson et al., 2005). As in the root tip, the first phloem cells appear before the first xylem cells, providing a conduit for the import of carbohydrates and other material at the earliest stages of shoot and leaf development. Each new leaf generated by the apical meristem is connected with preexisting vascular bundles in the shoot by, normally, five divergent bundles called leaf traces. Some leaf traces may arise up to three nodes below their point of departure to a leaf,

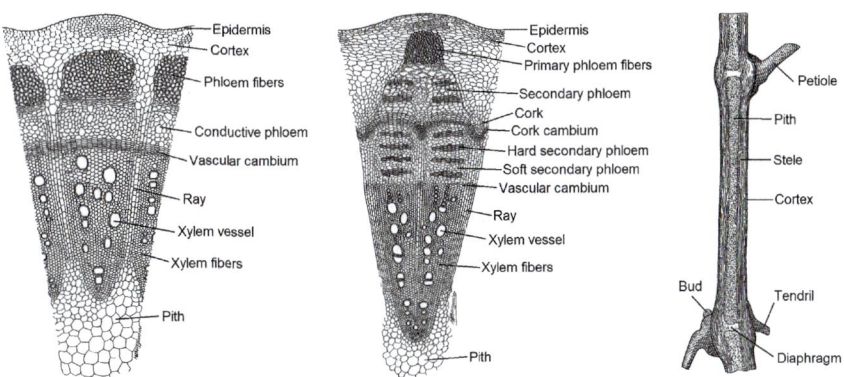

FIG. 1.10

Cross section of *V. vinifera* shoot during secondary growth but prior to periderm formation (left) and after periderm formation that has isolated two layers of secondary phloem (center), and longitudinal section through *Vitis* shoot (right).

Modified after Viala, P., Vermorel, V., 1901–1909. Ampélographie. Tomes I-VII, Masson, Paris.

whereas tendril and axillary bud traces usually begin one node below their point of departure (Chatelet et al., 2006; Gerrath et al., 2001). The shoot vascular bundles that continue their course through the next internode and ensure the continuity of the vine's vascular system are termed sympodial bundles. Their xylem vessels are connected with the vessels of the leaf traces via intervessel pits, whereas the leaf traces ensure continuous vessels between the shoot and the leaf blade. The vessels terminate in the leaf at approximately 50–60% of the leaf length; the xylem conduits beyond this limit are tracheids rather than vessels (Chatelet et al., 2006).

As primary growth ceases in the internodes that stop elongating, the procambium develops into the vascular cambium or, more specifically, the so-called intrafascicular cambium. During the ensuing secondary growth, the intrafascicular cambium produces secondary phloem toward the outside and secondary xylem toward the inside. Soon after the emergence of the intrafascicular cambium, some parenchyma cells between the vascular bundles develop the interfascicular cambium that produces more parenchyma cells (Campbell and Turner, 2017). The latter become the secondary rays separating the vascular bundles. The rays' thick-walled but heavily pitted parenchyma cells can accumulate large amounts of storage starch in their amyloplasts (Bondada, 2014; Buttrose, 1969b; Earles et al., 2018; Kriedemann and Buttrose, 1971). Within the secondary phloem, tangential bands of conducting tissue composed of sieve tubes, companion cells, and parenchyma cells, collectively termed the soft phloem, alternate with bands of fibers, termed the hard phloem, whose thick secondary walls add mechanical strength to the shoot (Bondada, 2012; Esau, 1948; see also Fig. 1.10).

As in the root, the vascular cambium is responsible for radial growth of the shoot, cordon, and trunk. Shoot tip-derived auxin, supported by cytokinin and strigolactone, stimulates this cambial activity by promoting cell division (Campbell and Turner, 2017; Nieminen et al., 2008; Sanchez et al., 2012; Werner et al., 2003). The faster the rate of auxin transport down the vascular cambium and xylem parenchyma, the more active is cell division in the cambium. Because removal of the shoot tip temporarily eliminates the crucial auxin source, viticultural practices such as shoot topping, trimming, or hedging tend to reduce secondary growth. The "translation" of the auxin signal into secondary growth involves the local production of strigolactone from carotenoids (Agusti et al., 2011). In addition, cytokinins are produced from the purine base adenine in the root tips and unfolding leaves and are transported upward (mostly as zeatin-type cytokinins) in the xylem and downward (mostly as isopentenyladenine-type cytokinins) in the phloem (Aloni et al., 2005; Ha et al., 2012; Matsumoto-Kitano et al., 2008; Nordström et al., 2004).

The vascular cambium becomes dormant in late summer, at approximately the time the grapes begin to ripen, and is reactivated in spring by root-derived cytokinin in response to auxin released by the swelling buds (Aloni, 2001). As the cambium moves outward with the growing trunk, it leaves only one or two, but occasionally up to four, rings of phloem at the end of each growing season, only one of which is reactivated in spring (Davis and Evert, 1970; Esau, 1948). Consequently, the seasonality of cambium activity results in annual rings of secondary xylem; when we count annual rings in a trunk or cordon cross-section, we are counting rings of xylem. The annual rings of grapevines are called diffuse-porous because the xylem cells produced in spring are similar in size to those produced in summer (Mullins et al., 1992; Sun et al., 2006). Since annual rings are composed almost entirely of xylem cells, xylem comprises the bulk of cordons and trunks. The xylem remains functional for several years; older, dysfunctional xylem then becomes the heartwood.

Some xylem parenchyma cells form outgrowths into the lumen of tracheary elements, entering via the interconnecting pits; such outgrowths are termed tyloses (Greek *tylos* = knob, knot). They serve mainly to seal injured xylem in the shoots, for instance due to summer pruning, and perhaps to prevent entry of pathogens (Evert, 2006; Sun et al., 2006; Viala and Vermorel, 1901–1909). Unlike growing shoots, mature canes respond to injury, for instance due to winter pruning, by secreting a pectin-rich gel (a.k.a. gum) mixed with antimicrobial phenolic compounds to seal the xylem to a depth of at least 1 cm at the site of injury (Sun et al., 2008; Price et al., 2015).

Some time before secondary growth due to the activity of the vascular cambium ceases, some secondary phloem cells form a new, additional cambium termed the cork cambium or phellogen. The cork cambium produces cork (phellem) toward the outside and secondary cortex (phelloderm) toward the inside and later cuts off the primary and some secondary phloem, as well as the cortex and the epidermis from internal tissues; the exterior tissues gradually die (Bondada, 2012; Esau, 1948; Evert, 2006; Galet, 2000; Pratt, 1974). Colletively, the phellem, phellogen, and phelloderm are called the periderm. Its multilayer combination of dead but elastic and heavily suberized cork cells and waxes protects shoots and older perennial stems from wounding, water loss, and nutrient loss. The periderm not only takes over the function of the epidermis/cuticle it replaces but also provides some insulation against temperature fluctuations (Franke and Schreiber, 2007; Lendzian, 2006). In fact, suberin is chemically almost identical to cutin, but unlike cutin, it is deposited on the inner side of the cell walls (Fich et al., 2016). The cork layer is interspersed with multicellular, raised pores called lenticels that are connected to underlying intercellular air spaces. The lenticels partly assume the role of the vanished stomata, permitting continued gas exchange between the interior tissues and the atmosphere—albeit without the stomata's ability to control the rate of gas exchange (Lendzian, 2006; Swanepoel et al., 1984). Additional protection, especially against fungal infection, arises from the incorporation in the parenchyma of phenolic compounds such as resveratrol and ε-viniferin that belong to the group of stilbenes, which is one of the reasons woody tissues decay very slowly (Pawlus et al., 2013; see also Section 8.2).

All tissues outside the vascular cambium are collectively referred to as the bark; these include epidermis, cortex, periderm, and phloem. The periderm, along with the cortex and the outermost layers of secondary phloem, dies starting at the base and moving up toward the tip (Esau, 1948). Dehydration of these dead outer tissues leads to shrinkage of the shoot diameter (Van de Wal et al., 2018). At the same time the shoot turns brown from oxidation of cellular components leaked from the dying cells. Thus, shoot browning is caused by the death of the green bark, rather than by lignification, as is often mistakenly believed. The invisible process of lignification instead occurs in all secondary cell walls, but especially in those of the xylem cells, independently of and starting long before the beginning of browning. Because lignification is a slow process, however, it may continue in the late-formed xylem cells even after shoot browning (Rathgeber et al., 2016). Browning is also called shoot "maturation" (French *aoûtement*, from *aoûté* = mature) and is accompanied by deposition of starch in the xylem and phloem parenchyma cells (Eifert et al., 1961; Plank and Wolkinger, 1976). It ends with the death and abscission (Latin *abscissio* = separation) of the apical meristem and the leaves. Leaf abscission is followed within days by the sealing of the sieve pores and plasmodesmata inside the sieve tubes with callose for overwintering (Davis and Evert, 1970; Esau, 1948; Pouget, 1963; Viala and Vermorel, 1901–1909). Like cellulose (β-1,4-glucan) and starch (α-1,4- and α-1,6-glucan), callose (β-1,3-glucan with some β-1,6 branches) consists of long chains of glucose molecules. It is thought that the "dormancy" hormone abscisic acid triggers this callose deposition. In spring the callose is degraded under the influence of auxin and gibberrelin, which restores the surviving phloem's transport function for another growing season (Aloni, 2013; Aloni et al., 1991; Esau, 1948; Pouget, 1963; Wu et al., 2018).

Phloem reactivation is very fast and occurs almost concurrently in the vine's canes and trunk. The reactivated phloem from the previous season is important during the early growth period, before it is replaced with newly formed phloem and discarded around midseason, concurrent with renewed periderm formation (see Fig. 1.10). Depending on the species, the vascular cambium resumes its function from approximately 2 weeks (*V. vinifera*) to 2 months (*V. riparia*) after the phloem has been reactivated, while a new cork cambium forms within the remaining inner cortex or the secondary phloem in midsummer (Davis and Evert, 1970; Esau, 1948). In contrast to phloem reactivation, cambium reactivation is spatially and temporally linked to the swelling and breaking buds. Reactivation begins beneath those buds and progresses downward, so that the vascular cambium of the trunk, and finally that of the roots, resumes its function weeks after the first cell divisions occurred in the canes.

The new cork cambium isolates most of the secondary phloem produced in the previous growing season. This phloem and surrounding cells soon die and dehydrate, causing the trunk diameter to shrink somewhat. The dead phloem, together with the old periderm, is eventually sloughed off as strips of dead bark (Davis and Evert, 1970; Esau, 1948; Van de Wal et al., 2018). The dead, fibrous tissue layers that are eliminated annually from woody organs such as trunks and cordons are collectively termed the rhytidome (Greek *rhytidùma* = wrinkle) or outer bark or ring bark, which contrasts with the living inner bark that is separated from the outer bark by a layer of dead cork cells and is comprised of secondary phloem (Evert, 2006; Viala and Vermorel, 1901–1909).

1.3.3 Nodes and buds

The lateral organs of the shoot, including the leaves, tendrils, clusters, and buds, are attached at nodes, which can be distinguished from the rest of the shoot or cane by their characteristic swelling due to a thicker pith and cortex (Figs. 1.10 and 1.11). In all *Vitis* species, but not in *Muscadinia* species, a

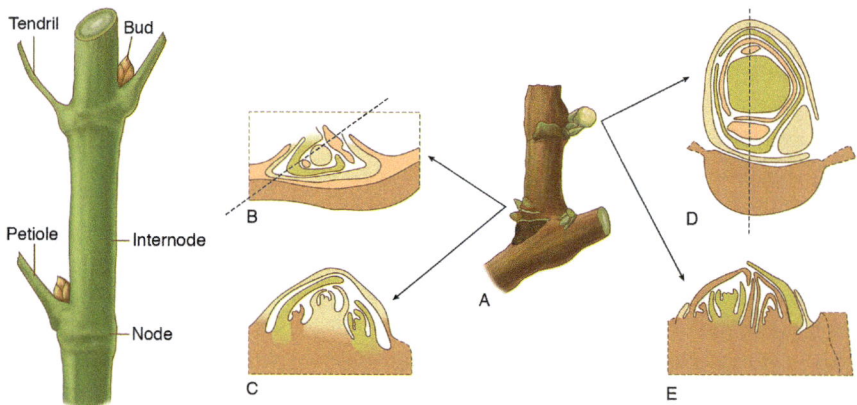

FIG. 1.11

Location of the main features of a *Vitis* shoot (left) and one-node Concord spur with one count node and three basal buds (right: A, spur with buds; B, cross-section of basal bud; C, longitudinal section of basal bud; D, cross-section of compound bud; E, longitudinal section of compound bud).

Left: © Elsevier Inc.; right: © Elsevier Inc., illustration after Pool, R.M., Pratt, H.D., 1978. Hubbard, Structure of base buds in relation to yield of grapes. Am. J. Enol. Vitic. 29, 36–41.

so-called diaphragm consisting of small, hard, thickened pith cells with sclerified (Greek *skleros*=dry, hard) cell walls divides the pith at the node (Gerrath et al., 2015). It is thought that the diaphragm aids in directing vascular bundles, and hence the flow of water and nutrients, into the leaves. The leaf and bud are inserted just above the node, and each leaf is connected to the shoot via its own vascular bundles inside a petiole. When the petiole falls off at the end of the growing season, a leaf scar remains on the node of the cane.

Buds are young, compressed shoots whose internodes have not yet elongated (Fig. 1.12) and are enclosed in scales or bracts. They are green in spring and summer and turn brown during shoot maturation. The initiation of bracts is the first sign of bud formation, while their browning is associated with the entry of the buds into dormancy (van der Schoot and Rinne, 2011; van der Schoot et al., 2014). These leaflike structures without petiole or lamina protect the bud from desiccation and freezing. In addition, long, woolly epidermal hairs growing on the inside of the bracts form a down that cushions the bud and gives it a woolly appearance when it opens during budbreak. Grapevine buds are always axillary; that is, they arise in the axil (upper angle) between a shoot and a leaf, directly above the petiole insertion point (Galet, 2000; Viala and Vermorel, 1901–1909; Fig. 1.11). All buds are formed by the shoot apical meristem. There are three types of buds: prompt buds, dormant buds, and latent buds.

Prompt buds are the only true axillary buds of the primary or main shoot and are also called lateral buds (Gerrath and Posluszny, 2007; Pratt, 1974). They develop from the earliest formed lateral meristem and can remain dormant or break in the current growing season and give rise to secondary shoots called lateral shoots (Fig. 1.8). Many prompt buds form several leaf primordia with their own, smaller lateral buds in their axils, which again may remain dormant or grow out (Gerrath et al., 2015). Species such as *V. rupestris, Vitis arizonica*, or *Vitis monticola* often grow many long laterals, which in turn can give rise to tertiary shoots (Viala and Vermorel, 1901–1909). By contrast, tertiary shoots are rare in *V. vinifera*, and laterals form mainly if the apical dominance of the primary shoot is broken (Alleweldt

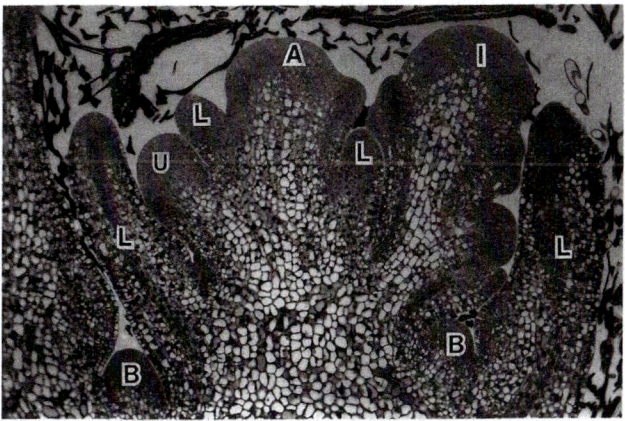

FIG. 1.12

Longitudinal section of dormant Cabernet Sauvignon bud. A, shoot apical meristem; B, lateral bud primordium; I, inflorescence primordium; L, leaf primordium; U, uncommitted primordium.

Reproduced with permission of University of Chicago Press from Morrison, J.C., 1991. Bud development in Vitis vinifera *L. Bot. Gaz. 152, 305–315; permission conveyed through Copyright Clearance Center, Inc.*

and Istar, 1969), for example, by removal of the shoot tip as part of canopy management practices or due to hail damage (see Section 2.2). Although prompt buds are usually not fruitful, some clusters, referred to as "second crop," can grow on lateral shoots. Removal of the shoot tip appears to promote the development of such second-crop clusters (May, 2000). These clusters are typically small, and their development lags well behind that of the clusters on the main shoot, so they rarely compete with the latter for resources. Many lateral shoots fail to form a proper periderm and are abscised after having been killed by the first hard frost in late fall or early winter.

A dormant bud could be called a secondary lateral bud, because it is formed later than the prompt bud by its own shoot apical meristem in the axil of the basal leaf (which is reduced to a small, scalelike leaf called a prophyll) of a lateral shoot, irrespective of whether or not that lateral shoot emerges from the prompt bud (Gerrath and Posluszny, 1988a, 2007; Gerrath et al., 2015; Pratt, 1974; Viala and Vermorel, 1901–1909). An imaginary longitudinal cut down a shoot between prompt buds and dormant buds reveals that all prompt buds are located on the same side and all dormant buds on the other; the prompt-bud side is called the dorsal side, and the dormant-bud side is called the ventral side (Huglin and Schneider, 1998). Some of these dormant buds will give rise to next year's fruit crop, but they remain dormant until the following spring, which is why they are also called "winter buds." A dormant bud is in fact an overwintering compound bud, also called an "eye," because it generally contains three separate buds—a larger, primary bud in the center flanked by two smaller, secondary buds (Fig. 1.11). The latter are also termed accessory buds and are often separately named a secondary bud and a tertiary bud because they develop in the axils of the first and second prophyll, respectively, of the shoot primordium that forms the primary bud (Morrison, 1991; Pratt, 1974). Higher-order buds have been reported, but these usually remain rudimentary (Zelleke and Düring, 1994).

Just like the shoot apical meristem, the apical meristem of the primary bud generates two types of lateral meristems: One is responsible for leaf production and the other for inflorescence and tendril production (Carmona et al., 2007). These meristems form as many as 18 leaf primordia (8–10 on average) and their associated prompt buds, up to several inflorescence primordia, and several tendril primordia before they become dormant (see Section 2.2). Initially, three or four leaf primordia are produced, each of them flanked by a pair of scales and wrapped in abundant hair (Morrison, 1991; Snyder, 1933; Srinivasan and Mullins, 1981). The bud's apical meristem then starts to produce additional lateral meristems opposite the subsequent leaf primordia. The first two or three of these will normally give rise to inflorescences, and the remainder will become tendrils. In most cultivars, only the buds produced from a shoot's fixed growth (i.e., at the first 6–10 nodes) are able to initiate inflorescences; all buds formed during the shoot's subsequent free-growth phase only initiate tendrils (Morrison, 1991; Sánchez and Dokoozlian, 2005). The inflorescence meristems in turn produce several branch meristems in spiral phyllotaxy, each of them subtended by a bract (Gerrath and Posluszny, 1988b; May, 2004; Srinivasan and Mullins, 1981).

Secondary buds generally remain small and are less fruitful, that is, they produce fewer and smaller inflorescences, than the primary buds, and tertiary buds are rarely fruitful (Noyce et al., 2016). Secondary or tertiary buds often grow out when the primary bud has been damaged, for example, by spring frost or insects, but also when vines have been pruned severely in winter, so that their shoot system is out of balance with the capacity of the root system (Friend et al., 2011).

Latent buds are buds that remain dormant or hidden for several years and often become part of the permanent structure of the vine. Even decades-old cordons and trunks still carry viable latent buds that can give rise to so-called watershoots or suckers, especially following frost events or on vines that have been pruned too severely (Galet, 2000; Lavee and May, 1997; Sartorius, 1968). Such shoots are usually not fruitful, although exceptions do occur.

Basal buds, located at the base of a shoot or cane (see Fig. 1.11), are formed in the axils of the shoot's prophylls, and thus can be prompt buds, dormant buds, or latent buds (Pratt, 1978). Because the basal internodes elongate only insignificantly, these buds appear to be situated in a whorl or "crown" around the shoot base (Pratt, 1974; Viala and Vermorel, 1901–1909). They are therefore also called "crown buds" and can be quite numerous, though they are typically inconspicuous and do not have leaves associated with them. Although basal buds are usually less fruitful than the other buds, this varies by cultivar: Gamay, Sangiovese, Zinfandel, and Muscat of Alexandria have particularly fruitful basal buds, whereas fruitfulness is very low in the basal buds of Gewürztraminer, Viognier, and Thompson Seedless. Other cultivars, such as Pinot noir, have fruitful basal buds, but the resulting clusters are smaller and often fewer in number than those from other buds (Skinkis and Gregory, 2017). Moreover, the fruitfulness of basal buds is much higher in many interspecific hybrids than in *V. vinifera* cultivars (Huglin and Schneider, 1998; Pool et al., 1978). These differences have implications for the type of winter pruning: cultivars with low basal-bud fruitfulness are better suited for cane pruning or pruning to longer spurs, whereas those with high fruitfulness can be pruned to very short spurs.

1.3.4 Leaves

A leaf differentiates along with a corresponding node from a leaf primordium produced by the apical meristem in the shoot tip (Fig. 1.7). The time, or thermal time, between the formation of two successive leaf primordia is termed a plastochron (Greek *plastos* = formed, *chronos* = time). We distinguish four types of leaves according to their position on the shoot: cotyledons, scales, bracts, and foliage leaves. Cotyledons are also called embryonic leaves, because they are preformed in the embryo and are the first two leaves to emerge from the embryo during seed germination. They are short lived and fall off soon after germination. The scales form around buds and protect them from water loss and mechanical injury. Bracts are small, scalelike leaves at branch points on the stem of inflorescences and tendrils. The first two leaves of a shoot grown from a bud ordinarily also develop as bracts and are separated by very short internodes. Subsequent leaves develop into mature leaves termed foliage leaves, which are separated from each other by elongated internodes. A foliage leaf consists of the leaf blade called lamina, the leaf stalk called petiole and, at the base of the petiole, a pair of stipules, which are short-lived leaflike structures or sheaths nearly surrounding the shoot's node (Viala and Vermorel, 1901–1909; see also Fig. 1.9).

The petiole (Latin *petiolus* = little foot) is the leaf stem that connects the lamina to the shoot and typically contains 12–14 vascular bundles serving the lamina (Viala and Vermorel, 1901–1909). Several vascular bundles are often derived from the same shoot vascular bundle (i.e., leaf trace); this branching is called anastomosis. Depending on species and cultivar, the petiole can be between 2 and 12 cm long and grows toward the light to position the leaf for optimal sunlight interception (Fig. 1.13). Where the petiole joins the lamina, it divides to form the lamina's five main veins. An abscission layer forms at each end of the petiole toward the end of the growing season and leads to leaf fall.

The basic function of the lamina (Latin *lamina* = leaf) is to capture sunlight for energy production (in the form of ATP) and carbon dioxide (CO_2) for carbohydrate production to support the vine's metabolism and growth (see Chapter 4). To maximize light absorption, leaves must be as wide as possible, whereas to maximize gas exchange, they must be as flat as possible; both of these properties, however, make leaves inherently vulnerable to overheating and dehydration (Tsukaya, 2006). The predominant

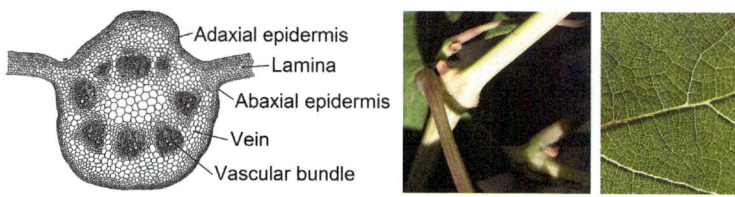

FIG. 1.13

Cross section of a *V. vinifera* leaf with prominent main vein (left), Zinfandel petiole bending toward the light (center), and pattern of Concord leaf venation (right).

Left: Modified after Viala, P., Vermorel, V., 1901–1909. Ampélographie. Tomes I-VII, Masson, Paris; right: photos by M. Keller.

FIG. 1.14

Chardonnay leaf (left) and Merlot leaf (right).

leaf form of the genus *Vitis* is palmate, in which the five main or major veins serve the leaf's five lobes (Fig. 1.14). In most *V. vinifera* cultivars the five lobes are partly separated from each other by more or less deep gaps, termed sinuses (Mullins et al., 1992; Viala and Vermorel, 1901–1909; see Fig. 1.14). Different species and cultivars of grapevine differ considerably in their leaf shape; consequently, leaf morphology forms the main basis of ampelography (see Section 1.1). The main veins, which originate from procambial cells, divide into an ever finer, elaborate network whose branches, termed minor veins, are interconnected or eventually (at the fifth order) end blindly between the leaf mesophyll cells (Fig. 1.13). Interconnected veins form enclosed areas or islets termed areoles, approximately 1 or 2 mm across, and 10 μm of minor veins services up to 10 mesophyll cells (Wardlaw, 1990).

The vein network, which can make up several kilometers per square meter of leaf area, functions as an "irrigation/collection system," with the major veins acting as a rapid supply network and the minor veins forming a slow distribution and collection network (Canny, 1993; Carvalho et al., 2017). This system delivers water and mineral nutrients from the roots, and it collects photosynthetically produced sugar, relatively evenly to and from all over the lamina. Because much more water flows in the xylem, to satisfy the leaf's transpiration demand, than in the phloem, the conductive area of the xylem is about

10 times that of the phloem across all vein orders (Carvalho et al., 2017). The veins are enclosed by a set of thickened parenchyma cells termed the bundle sheaths. Grapevine leaves are classified as heterobaric (Greek *baros* = weight), because their prominent bundle sheaths extend all the way to the epidermis, so that they are visible as almost transparent, projecting "ribs," especially on the abaxial side of the leaf (Liakoura et al., 2009; Pratt, 1974; see Fig. 1.13). The walls of the bundle sheath cells are waterproofed by deposition of suberin (Sack and Holbrook, 2006). The bundle sheaths function to control the flow of water away from the xylem to the mesophyll cells and serve as "windows," transferring light to the photosynthetic mesophyll cells, in addition to providing mechanical support to protect the lamina against collapse during severe dehydration and other stresses.

The leaf margins are serrated, and both the number of the so-called teeth and their sharpness, or sinus depth, vary greatly among *Vitis* species and cultivars (Fig. 1.14). Across plant species, it is thought that the sinus depth, but not the tooth number, is inversely related to the rate and/or duration of cell division during leaf formation (Tsukaya, 2006). This means that leaves with shallower sinuses should have greater cell numbers, but this association has not been investigated in grapevines. The teeth "funnel" the tracheids of the terminal veins into water pores called hydathodes, which can discharge droplets of xylem sap (Evert, 2006; Pratt, 1974). This process is called guttation (Latin *gutta* = drop) and should not be confused with dew, which is the formation of droplets of atmospheric water that condense on cool surfaces. Guttation is especially pronounced during warm nights with high relative humidity when transpiration is minimal and root pressure may force water up the vine (see Sections 2.2 and 3.3). Scavenging cells that surround the tracheids near their endings actively and selectively collect valuable xylem solutes, such as nutrient ions and hormones, and recycle them back to the phloem for reexport (Sakakibara, 2006). Conversely, the hydathodes excrete undesirable solutes, such as excess calcium. Thus, the hydathodes may function as overflow valves that can rid the leaves of excess water and potentially toxic solutes.

The leaf's shape and the basic architecture of its vascular system are highly correlated and are determined by genetic and hormonal controls during the initial stages of leaf formation (Chitwood et al., 2014). Since auxin stimulates cell division at high concentration and cell expansion at low concentration, the hormone may exert its growth-promoting influence via differences in its content among cells. Such differences, or auxin gradients, arise from active cell-to-cell transport away from the auxin source and determine the extent and direction of expansion of these cells by means of sugar and water import (see Section 3.1). Accumulation of auxin begins in a single epidermal cell at the leaf margin and leads to the initiation and outgrowth of a tooth by promoting local cell division (Kawamura et al., 2010). In addition, a hydathode differentiates at the site of the auxin maximum. Further auxin production in the cells around the hydathode of leaf primordia and unfolding leaves is, at least in part, responsible for the formation of the characteristic leaf shapes, a process termed patterning. It also leads to the differentiation of vascular bundles, whereby cells that experience rapid auxin flow will differentiate into veins (Aloni, 2001; Aloni et al., 2003, 2006b; Benjamins and Scheres, 2008; Berleth et al., 2007; Dengler, 2001). The formation of teeth in simple leaves is similar to the development of compound leaves and of leaf primordia in general (Kawamura et al., 2010). The herbicide 2,4-D (2,4-dichlorophenoxyacetic acid) is a synthetic, nontransportable auxin that stimulates cell division but inhibits cell expansion and differentiation. In the highly sensitive grapevines 2,4-D interferes strongly with patterning, leading to severely distorted leaves (Bondada, 2011).

Newly formed leaf cells require approximately 2 weeks to expand to their full size—unlike root cells, which finish expanding within a few hours. Their relatively slow growth gives leaves enough developmental flexibility to adjust their final size to environmental conditions prevailing during leaf formation. For example, leaf size and shape can be markedly altered by gradients of water availability across the leaf during the period of leaf expansion. The final size of a leaf is usually closely tied to its cell number, although greater cell expansion often partially compensates for lower cell number (Gonzalez et al., 2012; Tsukaya, 2006). Because both cell division and cell expansion are dependent on sugar supply, leaves developing when a vine is photosynthetically "challenged" will produce few and small cells (van Volkenburgh, 1999). However, while cell division is highly sensitive to sugar availability, cell expansion is more responsive to water availability. Consequently, the earliest stages of leaf development are mostly under the control of carbon supply, whereas later stages are predominantly controlled by water supply (Pantin et al., 2012). Therefore, leaves can grow to a large size if resources are abundant during their expansion, but they will remain small if they form while environmental conditions are unfavorable.

The lamina grows by cell division and expansion in basal and intercalar (Latin *intercalare* = to insert) meristems and consists mainly of primary tissue; there is no secondary growth to speak of (Evert, 2006). Cell division ceases first in the leaf tips and later toward the base (Gonzalez et al., 2012). Division generally stops when the leaf is only half or less of its final size so that the remainder of leaf growth is caused solely by the expansion of preformed cells, which occurs mainly due to enlargement of their vacuoles (see Section 3.1). In addition to the influence of auxin, both cell division and cell expansion are stimulated by cytokinins produced by the dividing leaf cells as well as by cytokinins imported via the xylem from the root tips. Moreover, cell expansion is also promoted by gibberellins that stimulate loosening of the cell walls by inducing the activity of expansin and other wall-relaxing enzymes (Claeys et al., 2014). Gibberellin production by expanding cells just behind the shoot apical meristem and by the young, rapidly expanding leaves and petioles near elongating internodes is induced by auxin (Hartweck, 2008; Ross et al., 2011).

As the rate of cell division decreases, the expanding leaf cells also build up their photosynthetic machinery, which is housed in the chlorophyll-rich chloroplasts (see Section 4.1). By the time the various tissues of the leaf blade begin to differentiate, the leaf is generally only six cell layers thick (Pratt, 1974). The outer surface layers differentiate into the adaxial (upper, top, and sun-exposed) epidermis and the abaxial (lower, bottom, and shaded) epidermis (Fig. 1.13). The epidermis does not contain chloroplasts but can have various types of hairs called trichomes, particularly along the veins on the abaxial side (Ma et al., 2016; Viala and Vermorel, 1901–1909). The number and length of trichomes varies with species and cultivar and gives the underside of some leaves (e.g., *V. candicans*; Clairette, Meunier) a matte whitish appearance, whereas others (e.g., *V. monticola*; Grenache) are glabrous (Latin *glaber* = hairless, bald). Trichome density is especially high in young leaves, but it decreases after unfolding due to gradual shedding of trichomes, especially on the adaxial side (Karabourniotis et al., 1999). Trichomes can be simple, twisted, or glandular. The former two are outgrowths of a single epidermis cell, whereas glandular trichomes, which occur only in Asian *Vitis* species, may comprise many cells, including cells from subepidermal tissues, and they also contain a single stoma at their tip (Ma et al., 2016). Trichomes have several functions: They help reduce water loss by regulating the leaf surface temperature and increasing the boundary layer resistance (see Section 3.2); they deter insects as mechanical barriers; they provide protection from damaging ultraviolet radiation because their phenolic compounds absorb and scatter radiation (see Section 5.2); and they may even provide some insulation.

The outside walls of epidermis cells are much thicker than the cell walls of interior tissues (Kutschera, 2008a, b). They contain cutin, a strong, extremely durable, elastic biopolyester composed of polymerized, long-chain hydroxy-fatty acids with 16 (e.g., palmitic acid) or 18 (e.g., stearic acid) carbon atoms, and are covered with a thin, continuous, extracellular membrane termed the cuticular membrane or cuticle (Bargel et al., 2006; Domínguez et al., 2011; Fich et al., 2016; Kerstiens, 1996; Schreiber, 2010; Yeats and Rose, 2013). As a specialized modification and extension of the cell wall, the cuticle consists of cutin, polysaccharides that belong to and anchor it in the cell wall, phenolics (especially hydroxycinnamic acids and flavonoids), and lipids that make up the intracuticular wax embedded in the outer parts of the polymer matrix. It is covered by a layer of overlapping solid, semicrystalline lipid platelets—the epicuticular wax. The intra- and epicuticular waxes are collectively referred to as cuticular wax, which seals the plant surface. Although waxes are complex mixtures of different lipids, they are essentially (>90%) made up of very long chains of 20–34 methylene (CH_2) groups that are assembled within expanding epidermis cells and then secreted to the surface along with cutin monomers as the leaf unfolds and expands (Domínguez et al., 2011; Samuels et al., 2008; Schönherr, 2006; Yeats and Rose, 2013). The self-cleaning properties of this water-repellent, or hydrophobic (Greek *hydros* = water, *phobos* = fear), surface film keep the leaves clean by greatly reducing the adhesion of water and particles (Bargel et al., 2006). Together with the thick outer cell walls of the epidermis to which it is "glued" by cell wall pectins, the strong and stiff cuticle provides mechanical support that maintains the integrity of leaves and other plant organs, defines the boundaries of these organs by preventing fusion of adjacent organs, and determines their growth rate and extent of growth (Fich et al., 2016; Javelle et al., 2011). The epidermis/cuticle skin also forms the first mechanical barrier to invading pathogens as well as physical or chemical injury by repelling fungal spores and dust particles thanks to its self-cleaning surface, reduces nutrient leaching by rainfall, and provides some insulation against temperature extremes (Bargel et al., 2006; Domínguez et al., 2011; Kerstiens, 1996; Yeats and Rose, 2013). Its most important function, however, is the protection of the plant from desiccation, a trait that early plants evolving from cuticleless algae ancestors probably acquired when they began to colonize the land about 450 million years ago (Fich et al., 2016).

The wax deposited in and on the cuticle greatly restricts gas diffusion so that most water vapor and other gases must pass through openings called stomata (Greek *stoma* = mouth; Fig. 1.15). These small pores are formed by degradation of the cell wall joining two specialized cells, the kidney-shaped guard cells, which, in contrast to the surrounding epidermis cells from which they are derived, contain a few chloroplasts (Lawson, 2008). *Vitis* leaves are said to be hypostomatous because the stomata are located almost exclusively in the abaxial epidermis. The near-absence of stomata on the adaxial leaf surface is characteristic of plant species that evolved in warm and dry climates and helps to reduce water loss by transpiration.

Grapevine leaves typically have between 50 and 400 stomata per square millimeter, with *V. vinifera* and most rootstocks at the lower end ($<250 \, mm^{-2}$) and *V. labrusca* and *V. cinerea* at the higher end ($>300 \, mm^{-2}$) of this range (Düring, 1980; Liu et al., 1978; Scienza and Boselli, 1981). The stomatal density is determined by the environmental conditions not only of a developing leaf itself but also by those that act on the mature leaves (Chater et al., 2014; Engineer et al., 2016). Mature leaves detect the light and CO_2 environment around them and send a (still unknown) signal to the newly developing leaves to adjust the number of stomata that will form on those leaves. For example, when the mature leaves experience shade or high CO_2, the stomatal density of the newly developing leaves is reduced (Chater et al., 2014; Engineer et al., 2016).

Stomata act as valves that are responsible for regulating the exchange of gases between a leaf and the surrounding atmosphere. Considering that these gases include mainly CO_2 for photosynthesis,

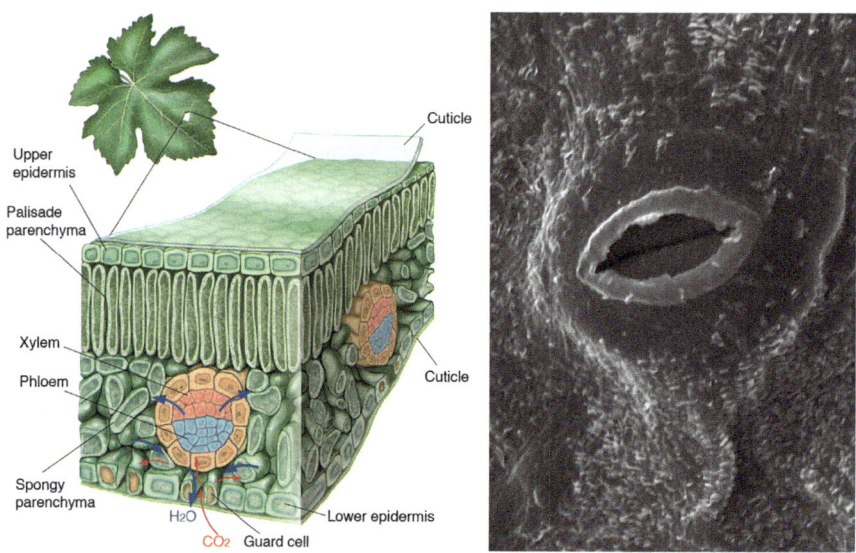

FIG. 1.15

Cross section of a leaf (left) and stoma on a Chardonnay leaf, surrounded by crystalline platelets of epicuticular wax (right).

Left: © Elsevier Inc. illustration by L. O'Keefe; right: photo by M. Keller.

oxygen (O_2) for respiration, and water vapor (H_2O) from transpiration, the stomata's major function is to balance the uptake of CO_2 with the loss of water (see Section 3.2). Stomata open when the guard cells take up water and swell, which more than doubles their volume; because the two guard cells are attached to each other at both ends, they cannot increase in length and therefore bend outward. Given a length of 15–40 µm, the aperture, or pore width, of open stomata is approximately 8 µm, whereas it is almost zero when the stomata are closed, for example, at night or during severe water stress.

The tissue between a leaf's two epidermal layers is called the mesophyll (Fig. 1.15). As the leaf expands, the mesophyll cells stop growing before the epidermis cells do. This pulls the mesophyll cells apart and leads to the formation of a large interconnected network of intercellular spaces that facilitate the diffusion of gases such as water vapor, O_2, and CO_2 (van Volkenburgh, 1999). This gas-distribution network is very important for both photosynthesis and respiration, because O_2 and CO_2 diffuse through air about four orders of magnitude more rapidly than they diffuse through water. The mesophyll consists of a single layer of elongated cells termed palisade parenchyma and four to six layers of irregularly shaped, loosely packed cells termed spongy parenchyma (Galet, 2000; Pratt, 1974; Viala and Vermorel, 1901–1909). Both cell types contain large numbers of chloroplasts (Greek *chloros* = light green, yellow), which are the actual centers of photosynthesis and assimilation.

The palisade parenchyma lies below the adaxial epidermis and has small intercellular spaces. It also contains crystals of calcium oxalate, which are arranged inside large cells (typically 50–100 µm long and 20–25 µm wide) in bundles of needlelike crystals called raphides and, in older leaves inside very small cells (~10 µm in diameter) lining the veins, in aggregates of octahedral (star-shaped) crystals called druses (Evert, 2006; Viala and Vermorel, 1901–1909). By scattering the incoming light on their irregular surfaces, the druses may help distribute light evenly to the chloroplasts to maximize the leaf's photosynthetic efficiency (He et al., 2014). The spongy parenchyma lies toward the abaxial side and has large intercellular spaces, so that the internal leaf surface is approximately 10–40 times the surface area of the leaf. Many of the mesophyll cell walls are thus exposed to the air inside the leaf, that is, to the leaf's internal atmosphere. Water evaporates from the epidermal and mesophyll cell walls into these intercellular spaces, which are continuous with the outside air when the stomata are open so that water vapor is discharged into the atmosphere in the process of transpiration (Fig. 1.15; see also Section 3.2). The CO_2 needed for photosynthesis follows the reverse diffusion route into the leaf (see Section 4.2).

1.3.5 Tendrils and clusters

The grapevine's climbing organs, the tendrils, and its fruiting structures, the inflorescences or flower clusters, are generally considered homologous based on anatomical, morphological, and physiological similarities. Darwin (1875) concluded from observations of grapevines growing in his backyard that "there can be no doubt that the tendril is a modified flower-peduncle." Indeed, studies of gene expression also suggest that tendrils are modified reproductive organs that have been adapted during evolution as climbing organs (Calonje et al., 2004; Díaz-Riquelme et al., 2009). As sterile reproductive structures, they are prevented from completing floral development by the cell-elongation hormone gibberellin (Alleweldt, 1961; Boss and Thomas, 2002; Boss et al., 2003; Srinivasan and Mullins, 1978, 1980, 1981). The absence of gibberellin production or suppression of its activity at the site of organ initiation gives rise to an inflorescence, whereas the presence of gibberellin results in a tendril. The cell-division hormone cytokinin, in contrast, promotes inflorescence formation over tendril formation (Crane et al., 2012; Srinivasan and Mullins, 1978, 1980, 1981). It seems likely that the response to these hormones is quantitative because there is a continuum of transitional structures that are partly tendril-like and partly inflorescence-like. Moreover, the degree of differentiation depends largely on the cultivar and environmental conditions during the time of cluster initiation and differentiation (see Section 2.3). Both tendrils and inflorescences, in turn, are modified shoots, and once more there is a variety of intermediary forms between inflorescence, tendril, and shoot (Fig. 1.16). The individual flowers on an inflorescence, again, are modified shoots, with the separate floral organs having evolved from modified leaves.

Like leaves, tendrils grow in intercalary fashion in addition to growth in their own apical meristem (Tucker and Hoefert, 1968). They also differentiate an epidermis with numerous stomata, a spongy

FIG. 1.16

Intermediary forms of Syrah inflorescence/tendril/shoot (left; inset: inflorescence that would rather be a tendril) and structure of a grape cluster with berries removed (right).

Left: photos by M. Keller.

FIG. 1.17

Vitis shoot tip showing tendrils with two and three tips (left) and tips of a tendril coiling around a trellis wire (right).

Photos by M. Keller.

parenchyma, and a large but short-lived hydathode at each of their tips, which soon become suberized (Gerrath and Posluszny, 1988a; Tucker and Hoefert, 1968). Although two tips are most common, vigorously growing shoots often form tendrils with three or more tips (Fig. 1.17). Tendrils enable wild vines to access sunlight at the top of tree canopies with a relatively small investment in shoot biomass per unit height gain. A tendril's tips search for surrounding objects by making sweeping, rotating movements during growth (Darwin, 1875; Galet, 2000; Viala and Vermorel, 1901–1909); this oscillatory growth pattern is termed circumnutation. When one of the tips detects a support via its contact-sensitive epidermis cells, the arms rapidly coil around the support in opposing directions (Fig. 1.17). Following this so-called thigmotropic (Greek *thigma* = to touch, *trope* = turn) movement, the entire tendril lignifies and stiffens to prevent unwinding (Braam, 2005). Tendrils that fail to find a support are of no use to the vine; they die and are abscised at the point of attachment to the shoot.

The fruiting structures of the grapevine are panicles and are known as inflorescences. Like tendrils, they are always inserted opposite a leaf and are initially protected by bracts covered with trichomes. After fruit set, when the flowers have turned into fruit, the inflorescence is called a cluster or bunch. Its main stem consists of the hypoclade or peduncle, which attaches the inflorescence or cluster to the shoot, and the central axis termed rachis (Greek *rhákhi* = spine). The rachis is in turn made up of a main axis, which is also called the inner arm, and a lateral wing or shoulder, which is also called the outer arm. Both arms carry multiple secondary branches, and many of these in turn carry tertiary branches, some of which carry quaternary branches. The shoulder can vary in size or may be missing altogether, or it may be a tendril or, occasionally, a shoot. Just as in tendrils, each of the arms and branches is subtended by a bract.

The rachis forms the entire branched inflorescence axis apart from the pedicels of the individual flowers or, following fruit set, the grape berries. Each flower or berry is attached to the rachis via a pedicel that widens into the receptacle carrying the flower or berry (Figs. 1.18 and 1.20). The flowers usually occur in groups of three or five; these basic floral units are termed a triad, a dichasium, or a

1.3 Morphology and anatomy 55

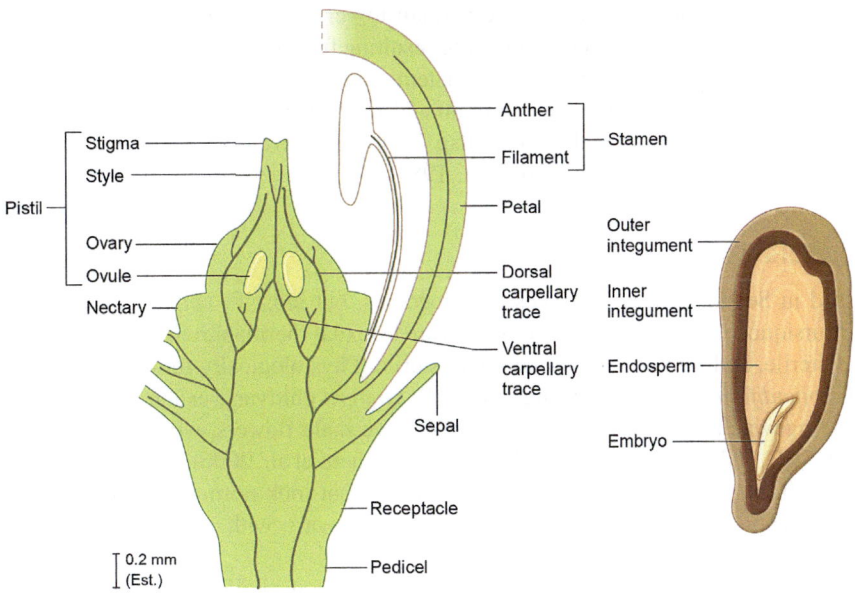

FIG. 1.18

Diagrammatic longitudinal section of a *Vitis* flower (left), longitudinal section of a *Vitis* seed (top right).
Left: © Elsevier Inc., illustration after Rafei, M.S., 1941. Anatomical studies in Vitis and allied genera. I. Development of the fruit. II. Floral anatomy. PhD thesis, Oregon State College; right: © Elsevier Inc.

cyme (Gerrath and Posluszny, 1988b; Gerrath et al., 2015; May, 2004; Mullins et al., 1992). In each triad two, often slightly smaller, flowers flank a central, terminal flower. A single inflorescence may carry between less than a handful and well over a thousand flowers; this variation contributes to large variations in yield in cultivated grapes (see Section 6.1).

The hypoclade, which varies in length from 2 to 10 cm depending on the cultivar, contains around 30 vascular bundles that are separated by rays (Viala and Vermorel, 1901–1909). The vascular bundles run the entire length of the cluster framework, then branch in the receptacles to serve the individual berries (Bondada and Keller, 2012; Hall et al., 2011). The parenchyma cells of the rays and to the inside of the vascular bundles accumulate starch and may serve as a transient nutrient storage compartment (Bondada, 2014; Bondada et al., 2017; Gourieroux et al., 2016). The epidermis on the exterior contains numerous stomata and is covered with a cuticle. The epidermis, however, does not last. Like the shoot, the peduncle exhibits both primary and secondary growth and starts to develop a brown periderm approximately by the time the berries begin to ripen (Bondada et al., 2017; Gourieroux et al., 2016; Hall et al., 2011). Consequently, the stomata, especially those of the pedicels and receptacles, are converted into suberized lenticels for continued gas exchange; additional lenticels can also form in stomata-free areas (Bondada and Keller, 2012; Theiler, 1970). The lenticels may serve as an important avenue for oxygen supply to grape berries, whose skin is rather impermeable to gases such as O_2 and CO_2 (Xiao et al., 2018).

Fruit clusters vary widely in length from 3 to 5 cm in wild grape species to >50 cm in some table grape cultivars. Variation in length is accompanied by varying degrees of branching, which can result in clusters that range in appearance from very small to very large and from very tight to very loose (Tello and Ibáñez, 2018). The cluster architecture resulting from this branching is probably determined by auxin released from developing flower meristems. Indeed, the development of the inflorescence requires active transport of auxin in the cambium and xylem parenchyma down the inflorescence axis (Morris, 2000). Loss of flowers therefore also limits the development of the subtending rachis (Gourieroux et al., 2016; Theiler and Coombe, 1985).

1.3.6 Flowers and grape berries

As discussed in Section 1.1, wild grapes are dioecious, but most *V. vinifera* cultivars have perfect flowers (Boursiquot et al., 1995). Perfect flowers are bisexual (hermaphroditic), whereas male flowers are female sterile, and female flowers are male sterile. Physiologically female flowers are common among *M. rotundifolia* cultivars but rare among *V. vinifera* cultivars; examples of the latter include Madeleine Angevine and Picolit. Many rootstocks have male flowers, whereas some (e.g., Ramsey, 101-14 Mgt, 5BB, and 41B) have female flowers (Meneghetti et al., 2006). Female flowers are difficult to distinguish from perfect flowers: they have stamens that look normal but produce sterile pollen, whereas in male flowers the pistil is reduced to a tiny ovary with undeveloped style and stigma (Caporali et al., 2003; see also Fig. 1.1).

Regardless of their sex, grapevine flowers are green, inconspicuous, and small; their size varies, depending on species, from about 2 mm in *V. berlandieri* to 6–7 mm in *V. labrusca*. A network of vascular bundles, or traces, supplies each of the floral organs (Fig. 1.18). The floral organs comprise a receptacle, which is sometimes called a torus, that carries five sepals, five petals, five stamens, and the pistil (Gerrath et al., 2015; May, 2004; Meneghetti et al., 2006; Srinivasan and Mullins, 1981; Swanepoel and Archer, 1988). The sepals are fused and collectively form the calyx, and the petals have interlocking bordering cells to form a protective cap termed a calyptra or corolla (Fig. 1.19). Both the sepals and petals are specialized, modified leaves; together they form the sterile flower organs, collectively termed the perianth.

The stamens are the male reproductive organs (Fig. 1.18), and the set of stamens of a flower are collectively termed the androecium. A stamen comprises a stem, termed filament, that carries the bi-lobed anther. Each of the anther's two lobes contains two pollen sacs with >1000 pollen grains. A pollen grain, or simply a pollen, is composed of a large vegetative cell and two sperm cells, the male gametes, which are enclosed in an extremely tough and durable outer wall composed of a mixture of fatty acids, phenolics, and carotenoids.

The pistil is the female reproductive organ; it comprises a superior ovary surmounted by a short style and a papillate stigma (Considine and Knox, 1979a; Gerrath et al., 2015; Hardie et al., 1996b; Pratt, 1971; Fig. 1.18). The female floral parts together are named the gynoecium (Greek $gune$ = female, $oikos$ = house). The gynoecium of grape flowers generally consists of two seed pockets that are derived from modified leaves and are called carpels. The two carpels are fused to form the pistil. Like the calyptra, the pistil develops a cuticle-covered epidermis with 10–20 stomata per square-millimeter of surface area (Fig. 1.19). The stomata enable the flowers to carry out their own gas exchange (Blanke and Leyhe, 1987, 1988, 1989a). The pistil tissues contain chloroplasts and are rich in calcium oxalate crystals, starch, and phenolic compounds, although the starch disappears during berry development (Considine and Knox, 1979a; Hardie et al., 1996a; Sartorius, 1926). The base of the ovary is encircled by a whorl of odor glands termed osmophors but sometimes referred to as

FIG. 1.19

Tip of Chardonnay flower cap with interlocking cells and protruding stomata.

Photo by O. Viret.

nectaries. The central region in the ovary, where the two inner carpel walls meet, is termed the septum. It divides the ovary, and later the fruit, into two, rarely three, cavities termed locules that serve as seed chambers, each normally containing two seed primordia, called ovules (Nitsch et al., 1960). An ovule is formed by a placental wall to which it remains attached via a stalk termed funiculus (Latin *funiculus* = thin rope) that serves the same function as the umbilical cord in mammals. The point where the funiculus joins the ovule later remains visible as a dark bulge, called chalaza, on the upper, dorsal portion of the seed that develops from the ovule (Gerrath et al., 2015).

The ovules, with their fertilized egg (zygote), can develop into heart- or pear-shaped seeds whose main role is to protect and nurture the developing embryo (Roberts et al., 2002). Thus, a grape berry with two locules can have a maximum of four seeds, and one with three locules can have six seeds. In practice, the seed number is usually one or two in most *V. vinifera* cultivars but two or three in many Chinese *Vitis* species (Wan et al., 2008c). Depending on cultivar and seed number, the seeds make up between 2% and 7% of the fresh weight of mature grapes (Viala and Vermorel, 1901–1909). After fertilization, the pistil develops into the fruit, with the ovary wall becoming the pericarp—that is, the flesh and skin (Fig. 1.20). Consequently, the fruit can be viewed as a swollen or mature ovary (Gerrath et al., 2015; Gillaspy et al., 1993). Grapes are classified as a berry fruit because the thick pericarp encloses the seeds.

The ovule is made up of the outer and inner integuments, the nucellus, and the embryo sac (Pratt, 1971). The latter is also called the female gametophyte or megagametophyte and is composed of seven cells: the egg cell flanked by two synergid cells, the central cell, and the three antipodals. After fertilization, the integuments will form the seed coat, termed testa, and the nucellus will surround the embryo, which develops from the egg cell, and the endosperm, which develops from the central cell (Fig. 1.18). The developing seed's thick outer integument consists of several layers of palisade parenchyma cells and the protoderm, which soon becomes the epidermis and forms a cuticle on the outside

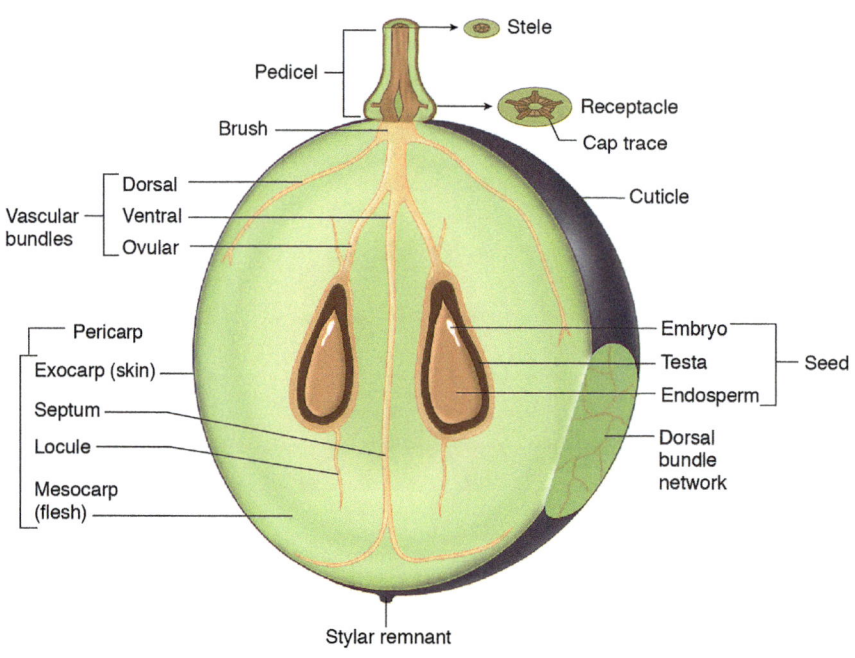

FIG. 1.20

Structure of a ripe grape berry.

© *Elsevier Inc., after Coombe, B.G., 1987. Distribution of solutes within the developing grape berry in relation to its morphology. Am. J. Enol. Vitic. 38, 120–127.*

that is impregnated with suberin (Javelle et al., 2011; Schreiber, 2010). The outer integument is sometimes further divided into a soft outer integument and a hard middle integument; the outer and inner integuments, as well as the epidermis, accumulate tannins, whereas the middle integument instead becomes lignified as the seed matures (Cadot et al., 2006). The protective testa contains inclusions of calcium oxalate crystals and is rich in phenolic compounds such as tannins that protect the seed against premature "consumption" by microorganisms or insects (Dixon et al., 2005). The apparent folding of the integuments leaves a small opening or pore, termed micropyle, through which the pollen tube enters during pollination (Sartorius, 1926; see also Section 2.3); during seed germination, water will enter and the embryonic root will exit through the micropyle.

During early seed development, the endosperm progressively expands at the expense of the nucellus, which may serve as a food supply for the embryo and endosperm (Cadot et al., 2006; Nitsch et al., 1960). The embryo produces an embryonic shoot termed epicotyl, two embryonic or seed leaves termed cotyledons, an embryonic shoot base termed hypocotyl, and an embryonic root termed radicle, complete with their vascular systems. The epicotyl contains at its tip the shoot apical meristem, and the radicle contains at its tip the root apical meristem, both of which will continue to provide stem cells to build plant organs throughout the life of the vine that will grow from the embryo. This epicotyl–radicle arrangement also defines the apical–basal axis of the embryo and future seedling.

The vascular bundles that previously served the ovary give rise to a complex network of vascular traces that supply the seeds and the pericarp. The vascular traces serving the seeds are called ovular bundles, and those serving the pericarp are divided into the ventral and dorsal bundles (Figs. 1.18 and 1.20). The ventral bundles are also called the central or axial vascular bundles; they run from the pedicel end of a berry to its stylar end, where they interconnect with the dorsal bundles. The ventral bundles and their associated parenchyma cells are termed the "brush" and often remain partly attached to the pedicel when a ripe berry is plucked from the cluster. The dorsal, or peripheral, vascular bundles of most *V. vinifera* cultivars lie within the berry flesh, approximately 200–300 µm beneath the surface of a mature berry; in *V. labrusca* these bundles are approximately twice as far from the surface (Alleweldt et al., 1981). Like the minor veins in a leaf, the berry's dorsal bundles are interconnected by many anastomoses, which can give some berries a "chicken-wire" appearance (Fig. 1.20). The tracheary elements in the berry are mostly vessel elements with some tracheids; they are approximately 250–300 µm long with an average diameter of 3.5–5 µm (Chatelet et al., 2008a). Although existing tracheary elements are stretched at the beginning of ripening, most of them remain intact, and new elements are added throughout berry development.

The pericarp consists of three anatomically distinct tissues: the exocarp, mesocarp, and endocarp (Galet, 2000; Viala and Vermorel, 1901–1909). The exocarp is sometimes called the epicarp; it forms the grape berry's dermal system or "skin." Depending on skin thickness and berry size, both of which vary by species and cultivar, the exocarp makes up between 5% and 18% of the fresh weight of mature berries. It is composed of the outer epidermis covered with a waxy cuticle and the underlying outer hypodermis (Bessis, 1972a; Considine and Knox, 1979b; Fougère-Rifot et al., 1995; Hardie et al., 1996b). Epicuticular wax, often referred to as "bloom," covers the surface of the cuticle as overlapping platelets forming a strongly hydrophobic layer that protects the berry from water loss and forms the first barrier against pathogen invasion (Grncarevic and Radler, 1971; Possingham et al., 1967; Rosenquist and Morrison, 1988). The structure of the epicuticular wax changes as the berry develops. Although the wax is normally crystalline, the crystals are gradually degraded, fusing into amorphous masses and eventually into a more or less continuous layer on top of the normal, uninterrupted wax layer (Casado and Heredia, 2001; Rogiers et al., 2004a; Fig. 1.21). Sun-exposed berries have considerably more amorphous wax than do shaded berries (Rogiers et al., 2004a). This is probably a consequence of solar heating of the berry surface; the transition from crystalline to amorphous wax occurs at

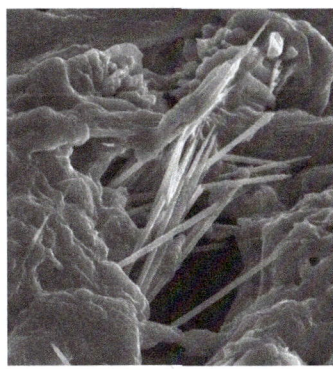 

FIG. 1.21

Scanning electron micrograph detail of ruptured Syrah berry surface with epicuticular wax and raphide crystals of calcium oxalate (left), and stylar end of Syrah berry covered with epicuticular wax (right).

Left: photo by S. Rogiers; right: photo by M. Keller.

approximately 50°C (Khanal et al., 2013). The wax material is rather soft and can be altered or removed by the impact of rain, by abrasion from windblown particles, or by contact with other berries, leaves, or other adjacent objects (Shepherd and Griffiths, 2006). However, because of its vital importance as a desiccation protectant, a damaged or removed wax layer is quickly regenerated, especially in young berries.

The cuticle is 1.5–4 μm thick, and the epidermis consists of a single cell layer that is 6–10 μm thick (Alleweldt et al., 1981; Bessis, 1972a; Hardie et al., 1996b). Unlike the hypodermis and mesocarp cells, the cells of the epidermis divide in a strictly anticlinal manner (i.e., normal to the surface), thus maintaining their cell "lineage" (Considine and Knox, 1981). The initially single-layered outer hypodermis increases to 6–10 layers of thick-walled cells after fruit set and then transiently even to 15–20 layers, reaching 100–250 μm in thickness. The hypodermal cells contain chloroplasts and tannin-rich vacuoles, interspersed with amyloplasts, which form within a few days after fruit set but disappear during berry development, and the so-called idioblast (Greek *idios* = unique, *blastos* = germ, sprout) cells rich in bundles of needlelike calcium oxalate crystals called raphides (Evert, 2006; Fougère-Rifot et al., 1995; Fig. 1.21). As an aside, a cell may contain more than one kind of vacuole, so it may be possible for different components to be stored in specialized vacuoles within the same cell and for neighboring cells to differ in the content of their vacuoles (Marty, 1999; Müntz, 2007). The outermost hypodermal cells become stretched and radially compressed, and their cell walls thicken approximately 10-fold as the berry expands (Considine and Knox, 1979b; Hardie et al., 1996b). The innermost cells expand and seem to be progressively converted to mesocarp cells. Thus, the skin thins to four or five cell layers in the ripening berry and the vascular bundles, which initially marked the "boundary" between hypodermis and mesocarp, are now situated within the mesocarp (Chatelet et al., 2008a; Fougère-Rifot et al., 1995). From fruit set to maturity, the relative thickness of the skin declines from approximately one-eighth to one-hundredth of the total berry diameter.

The mesocarp, which is commonly called the "flesh" or "pulp" of the grape berry, forms soon after fruit set and consists of 25–30 layers of large, thin-walled, and highly vacuolated storage parenchyma cells that come in a wide variety of sizes (50–400 μm) and shapes (Gray et al., 1999). The diverse vacuoles also vary greatly in size, ranging from 1 to 50 μm in diameter, and can make up as much as 99% of the cell volume in ripe grape berries (Diakou and Carde, 2001; Fontes et al., 2011b; Storey, 1987). They serve as an internal reservoir storing sugars, organic acids, and other nutrients (see Section 6.2). Mesocarp cells outside the network of peripheral vascular bundles of the pericarp are termed the outer mesocarp; those toward the inside are the inner mesocarp, which, at maturity, makes up almost two-thirds of the berry volume. By this time the mesocarp cells are about 75 times larger than the skin cells, and their radial width is around 10 times that of the skin cells (Hardie et al., 1996b). Nevertheless, the cell walls in the skin are almost 20 times thicker than those in the mesocarp. It is no wonder, then, that the skin sets the upper limit to which a berry can expand.

The innermost tissue of the pericarp, the endocarp, is often difficult to distinguish from the mesocarp in grapes (Mullins et al., 1992). It surrounds the locular space around the seeds and consists of the inner epidermis and the inner hypodermis with cells rich in star-shaped calcium oxalate crystals called druses (Fougère-Rifot et al., 1995; Hardie et al., 1996b). The inner epidermis, like the stigma and the integuments, is derived from the same cell layer in the apical meristem that also forms the outer epidermis (Jenik and Irish, 2000).

CHAPTER 2

Phenology and growth cycle

Chapter outline

- 2.1 Seasons and day length 61
- 2.2 Vegetative cycle 65
  - 2.2.1 Bleeding 65
  - 2.2.2 Budbreak 66
  - 2.2.3 Shoot growth 69
  - 2.2.4 Growth cessation and senescence 76
  - 2.2.5 Bud development and dormancy 78
  - 2.2.6 Root growth 82
- 2.3 Reproductive cycle 86
  - 2.3.1 Inflorescence and flower formation 86
  - 2.3.2 Pollination and fertilization 89
  - 2.3.3 Fruit set 94
  - 2.3.4 Berry and seed development 97

2.1 Seasons and day length

Phenology (Greek *phainein* = to show, *logos* = word, knowledge) is the study of natural phenomena that recur periodically in plants and animals and of the relationship of these phenomena to seasonal changes and climate. In other words, phenology is the study of the annual sequence of plant development. Its aim is to describe the causes of variation in timing of developmental events by seeking correlations between weather indices and the dates of particular growth events and the intervals between them. Phenology thus investigates a plant's reaction to the environment and attempts to predict its behavior in new environments. In viticulture, phenology is mainly concerned with the timing of specific stages of growth and development in the annual cycle. Such knowledge can be used for site and cultivar selection, vineyard design, planning of labor and equipment requirements, and timing of cultural practices as part of vineyard management (Dry and Coombe, 2004).

Grapevines, like other plants, monitor the seasons by means of an endogenous "clock" (Greek *endon* = inside, *genes* = causing) in order to prevent damage by unfavorable environments. Every plant cell contains a clock (perhaps even several clocks), and the clocks of different cells, tissues, and organs act autonomously (McClung, 2001, 2008). The clock is driven by a self-sustaining oscillator consisting of proteins that oscillate with a 24-h rhythm in response to light detected by phytochromes (see Section 5.2) and other light-sensitive proteins that translate the light signal into a clock input that then uses this time-encoded information to regulate physiological functions (Fankhauser and Staiger, 2002; Spalding and Folta, 2005). Additional input is provided through feedback loops that arise from various compounds produced by the plant's metabolism (McClung, 2008). Just how plants manage to integrate and transform this environmental and metabolic information into daily and seasonal functions and how they avoid being "fooled" by bright moonlight is still mysterious, but it could involve fluctuations in calcium (Ca^{2+}) concentration in the cytosol. Cytosolic Ca^{2+} also oscillates with a 24-h period, with a peak before dusk, and it participates in the "translation" of many signals, both internal and external (Hotta et al., 2007; McClung, 2001; Salomé and McClung, 2005; see also Section 7.3).

In contrast to the temperature dependence of most biochemical processes, the period of the rhythm is temperature compensated so that it remains constant at 24 h over a wide range of temperatures. The gradual shift in the time of sunrise and sunset as the seasons progress serves to reset the phase of the clock every day and enables plants to "know" when the sun will rise even before dawn (Fankhauser and Staiger, 2002). This synchronization of the circadian (Latin circa = about, *dies* = day) clock to the external environment is termed entrainment. It translates day length, or photoperiod, into a vine's internal time as an estimate of both the time of day and the time of year, enabling it to anticipate and prepare for daily and seasonal fluctuations in light and temperature. For example, the genes responsible for "building" enzymes involved in the production of ultraviolet (UV)-protecting phenolics are most active before dawn, opening and closing of the stomata can anticipate dawn and dusk, photosynthesis-related genes peak during the day, and genes associated with stress are induced in the afternoon (Fankhauser and Staiger, 2002; Hotta et al., 2007).

On a seasonal scale, flowering, onset of fruit ripening, bud dormancy, and leaf senescence (Latin *senescere* = to grow old) are typical responses to day length, although the precise timing of each of these developmental processes is modulated by temperature, and some can be altered by stress factors such as drought (see Section 7.2), nutrient deficiency or excess (see Section 7.3), or infection by pathogens (see Section 8.2). In general, higher temperatures accelerate plant development and advance grapevine phenology (Keller and Tarara, 2010; Parker et al., 2011; Sadras and Moran, 2013). For example, the time of budbreak, bloom, or onset of fruit ripening of a given *V. vinifera* cultivar grown in a specific location can vary by >1 month from one year to another (Boursiquot et al., 1995; Duchêne et al., 2010). Even extremely high temperatures, such as daily means near 35°C, still shorten the time from budbreak to bloom compared with lower temperatures (Buttrose and Hale, 1973). Therefore, one of the most conspicuous consequences of climate change associated with the current and future increase in temperature due to the man-made rise of atmospheric CO_2 is a shift of phenological stages to earlier times during the growing season (Chuine et al., 2004; Duchêne et al., 2010; Wolfe et al., 2005).

Because day length during the growing season increases with distance from the equator, the effects of photoperiod and latitude on the internal clock are very similar (Salomé and McClung, 2005). In tropical climates, with little or no change in day length (and temperature at lower elevations) throughout the year, grapevines behave as evergreens with continuous growth and strong apical dominance, and often continuous fruit production, throughout the year (Dry and Coombe, 2004; Lavee and May, 1997; Mullins et al., 1992). With appropriate pruning strategies and sometimes deliberate defoliation, often in combination with imposed water deficit and sprays of ethephon or other chemicals to induce

2.1 Seasons and day length

near-dormancy, followed by applications of hydrogen cyanamide to induce uniform budbreak, two crops can often be harvested each year in the tropics (Bammi and Randhawa, 1968; Favero et al., 2011; Lin et al., 1985). In temperate climates (and at high elevations in the tropics), where cold winter temperatures prevent the survival of leaves and reproductive organs, grapevines have a discontinuous cycle with alternating periods of growth and dormancy. Under these conditions, the time of active growth generally occurs from March/April through October/November in the Northern Hemisphere, and from September/October through April/May in the Southern Hemisphere. Earlier spring warming and later autumn cooling in warmer regions permits longer growing seasons than in cooler regions.

Several distinct developmental stages or key events have been identified in grapevines (Figs. 2.1 and 2.2) and have been given a variety of names. These stages include dormancy (rest),

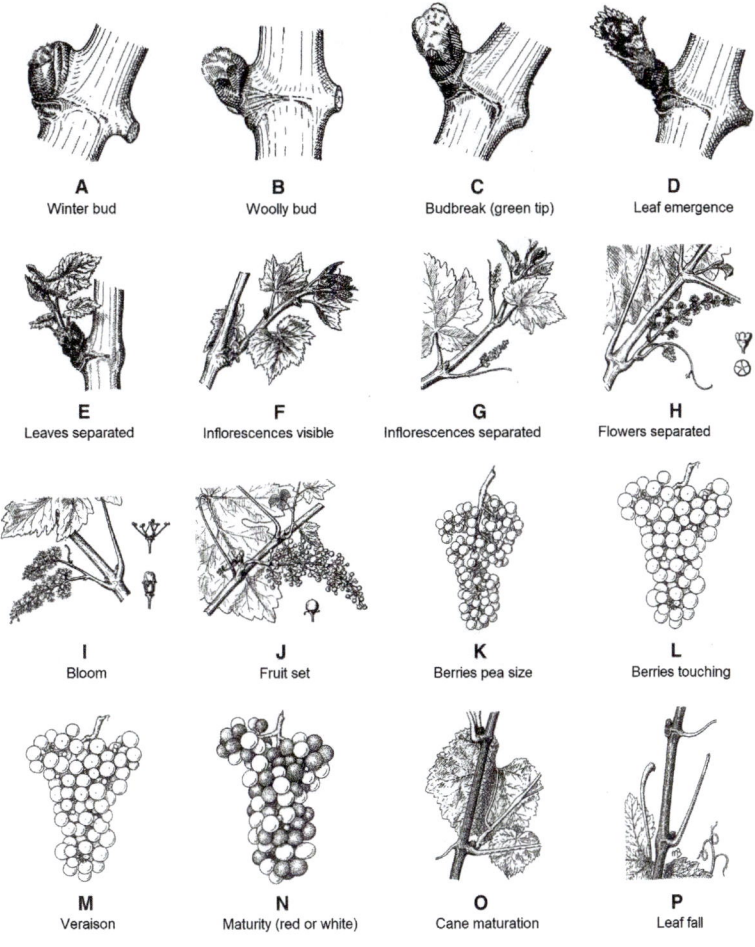

FIG. 2.1

Grapevine growth stages according to Baillod and Baggiolini (1993).

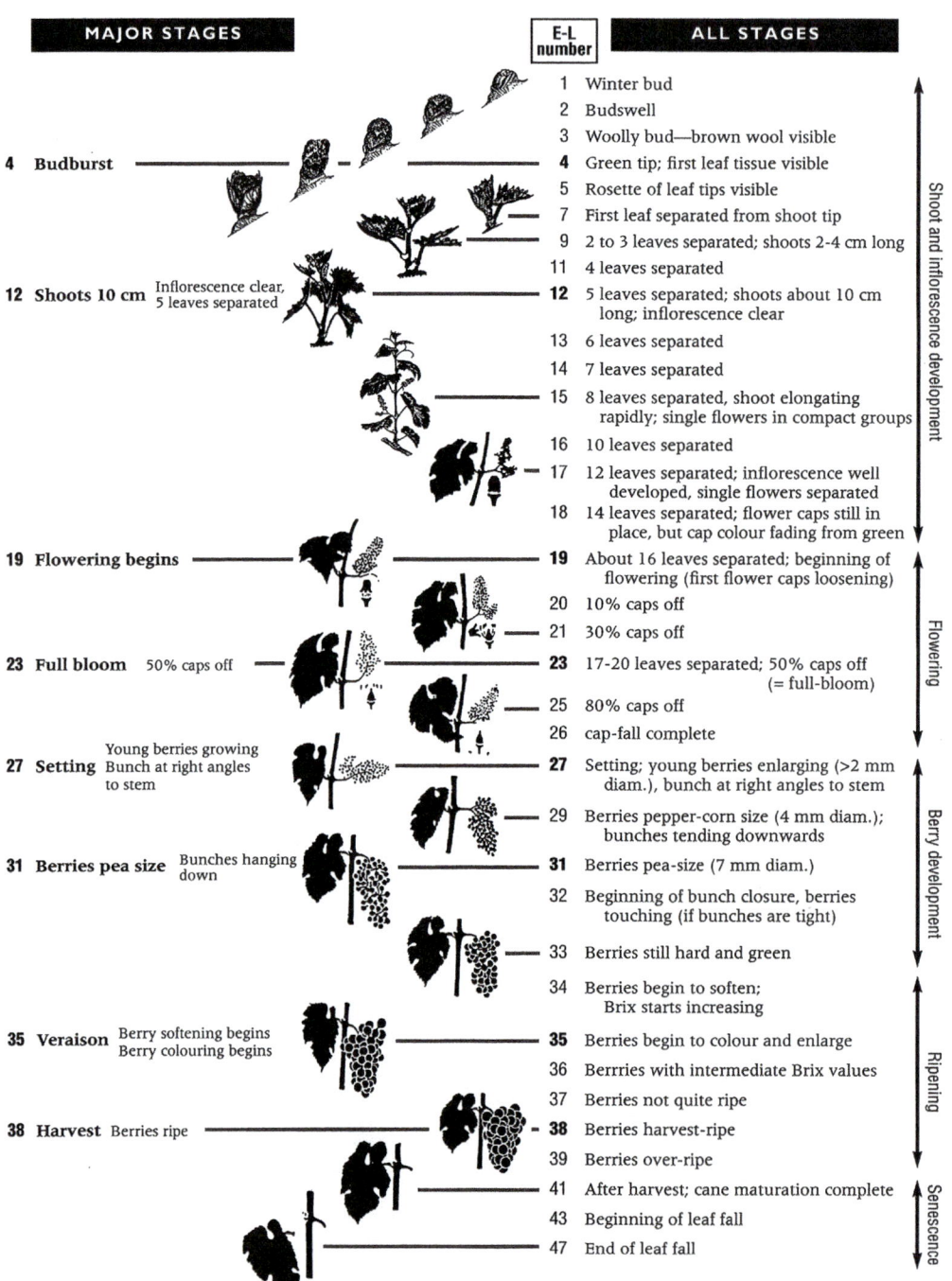

FIG. 2.2

Grapevine growth stages according to Eichhorn and Lorenz.

Reproduced with permission of Wiley-Blackwell from Coombe, B.G., 1995. Adoption of a system for identifying grapevine growth stages. Aust. J. Grape Wine Res. 1, 104–110; permission conveyed through Copyright Clearance Center, Inc.

budbreak (budburst), bloom (anthesis, flowering), fruit set (berry set, setting), veraison (French *vérer* = to change; color change, onset of ripening), harvest (ripeness, maturity), and leaf senescence and subsequent leaf fall (abscission). Although no visible growth occurs during dormancy, metabolism does not rest completely, but high concentrations of the dormancy hormone abscisic acid (ABA) keep it at a minimum necessary for survival of the buds and woody tissues. Whereas the division of meristematic cells in the buds and cambium is blocked, chromosome duplication and protein synthesis resume during the later stages of dormancy in preparation for the activation of growth when temperature and soil moisture become favorable in spring.

The annual growth cycle of mature, fruiting grapevines is often divided into a vegetative and a reproductive cycle.

2.2 Vegetative cycle

2.2.1 Bleeding

In late winter or early spring, grapevines often exude xylem sap from pruning surfaces and other wounds that have not yet been suberized (Fig. 2.3). Such sap flow, called "bleeding" by viticulturists, marks the impending transition from dormancy to active growth. Initiation of bleeding is related to the restoration of metabolic activity in the roots. It is influenced by soil temperature, soil moisture, and rootstock, but on average it begins when the soil temperature rises above approximately 7 °C (Alleweldt, 1965). Indeed, root respiration, as a proxy for metabolic activity in the roots, is closely coupled to both soil temperature and soil moisture (Franck et al., 2011; Hernández-Montes et al., 2017). Bleeding can last for a few days or several weeks, depending on whether or not air temperatures are conducive to budbreak; it can also be a stop-and-go process that fluctuates with changes in soil temperature (Andersen and Brodbeck, 1989b; Reuther and Reichardt, 1963). A vine can exude bleeding sap at rates of <0.1 L to >1 L per day, with the highest rates occurring on warm and moist soils (Alleweldt, 1965; Currle et al., 1983; Houdaille and Guillon, 1895).

Bleeding is caused by root pressure (Fischer et al., 1997; Priestley and Wormall, 1925; Sperry et al., 1987). Its existence in grapevines was first demonstrated by the English cleric Stephen Hales (1727). Root pressure arises from the remobilization of nutrient reserves from proteins and starch and pumping

FIG. 2.3

Bleeding grapevine cane (left) and swelling, woolly bud just before budbreak (right).

Photos by M. Keller.

of amino acids and sugars into the xylem conduits (Reuther and Reichardt, 1963; Roubelakis-Angelakis and Kliewer, 1979). The concentration of sugars apparently increases exponentially as the temperature declines close to and below 0 °C; this response occurs within a few hours and is equally rapidly reversible when the temperature rises again (Moreau and Vinet, 1923; Reuther and Reichardt, 1963). Sugars such as glucose and fructose, however, are only a minor component of the bleeding sap. Organic acids such as malate, tartrate, and citrate; amino acids such as glutamine; and mineral nutrient ions such as potassium and calcium together contribute >90% of the total solutes present (Andersen and Brodbeck, 1989a; Glad et al., 1992; Marangoni et al., 1986; Priestley and Wormall, 1925). The resulting increase in the osmotic pressure of the xylem sap provides the driving force for water uptake by the roots from the soil. Such water uptake in turn generates a positive hydrostatic pressure in the xylem of 0.1–0.4 MPa at the trunk base that declines at a rate of 0.01 $MPa\,m^{-1}$ due to gravity as the sap rises (Knipfer et al., 2015; Scholander et al., 1955; Sperry et al., 1987). Since grapevines have not yet produced new water-absorbing fine roots by this time, water uptake occurs across the bark of mature, woody roots (Cuneo et al., 2018). Because a pressure of 0.1 MPa can support a 10 m high water colum, root pressure lifts water up the vine and to the buds of even tall, wild grapevines (see also Section 3.3).

Root pressure serves to dissolve and push out air bubbles that have formed during the winter in the xylem, thereby restoring xylem function (Dixon and Joly, 1895; Scholander et al., 1955; Sperry et al., 1987). This is necessary because when xylem sap freezes, the gas dissolved in the sap is forced out of solution because gases are practically insoluble in ice. This physical process can lead to cavitation (see Section 3.3) upon thawing if the bubbles expand instead of going back into solution. Given that bleeding typically starts before the buds begin to swell, root pressure may also be necessary to rehydrate the buds and reactivate them with cytokinin that is delivered from the roots (Field et al., 2009; Lilov and Andonova, 1976). In fact, the buds' water content quickly rises from <50% to approximately 80% during the budswell phase (Fuller and Telli, 1999; Lavee and May, 1997; Pouget, 1963). The delivery of sugars in the xylem before phloem flow resumes may enable the buds to resume cell division in their apical meristem and production of auxin—or its release from storage pools. Though root pressure–driven water ascent may enable bud swelling and budbreak in spring, the emerging shoots quickly become dependent on the delivery in the reactivated phloem of sugar and other nutrients remobilized from the permanent vine organs.

The maximum rate of bleeding occurs a few days before budbreak and generally stops 10–14 days later, although root pressure may occasionally remain active for >1 month after budbreak (Alleweldt, 1965; Reuther and Reichardt, 1963). However, both bleeding and root respiration decline as soil moisture decreases, and leakage of root membranes increases sharply if the soil becomes too dry (Currle et al., 1983; Huang et al., 2005b). Therefore, bleeding is minimal or even absent in vines growing in dry soil, for example due to insufficient winter precipitation. This can cause stunted shoot growth and abortion of fruit clusters when the shoots begin to grow with stored water from the trunk and root parenchyma in the absence of sufficient soil water to supply the unfolding leaves as they begin to transpire.

2.2.2 Budbreak

Budreak marks the onset of vegetative growth in spring, but cell division and auxin production in the buds starts 1–3 weeks earlier, beginning with the basal (closest to the base) leaf primordia of the distal (closest to the tip) buds of a cane. At the same time, ABA declines strongly in the buds, relieving the growth inhibition that keeps the vine dormant while low temperatures would threaten its survival. The release of auxin from the buds and its movement from the apex toward the base in the cambial region begins suddenly and rushes like a continuous wave down the vine, stimulating the cambium cells to resume division and differentiation into new phloem and xylem cells (Aloni, 2001). The cambium cells become sensitive to auxin when they are exposed to cytokinin, which is supplied from the roots via the xylem as a result of root pressure. Although it remains to be demonstrated whether the root tips are capable of cytokinin production this early in spring, cytokinins have indeed been found in the bleeding sap of grapevines and, in addition to being influenced by the rootstock, higher root temperatures favor cytokinin export from the roots (Field et al., 2009; Skene and Antcliff, 1972; Zelleke and Kliewer, 1981). As an aside, the same interaction of bud-derived auxin and root-derived cytokinin probably also induces the cambium on either side of new grafts to produce undifferentiated cells termed callus and to form a graft union, complete with new vascular tissues, between scion and rootstock (Cookson et al., 2013). Before budbreak, however, differentiation of cells produced in the previous growing season to form the first xylem vessels of the new shoot begins even before the cambium starts producing new cells, although these vessels lignify and thus mature only after budbreak. Warmer temperatures during early vessel differentiation, perhaps by enhancing the supply of remobilized nutrient reserves to the cambium, result in wider vessels (Fonti et al., 2007). Because wide vessels are more effective than narrow ones at supplying water to the developing shoots later in the season, this early-season temperature effect could be an important driving variable for seasonal shoot vigor. Indeed, the greater number and wider diameter of vessels in vigorous shoots support more rapid water flow to the leaves than in weaker shoots (Keller et al., 2015a; Pagay et al., 2016).

Since cambium reactivation occurs well before the new leaves are mature enough to supply sugar from photosynthesis, the energy and carbon requirements of a young shoot's dividing cells must be met by remobilization of stored starch and export of sugar from the permanent structure of the vine. The necessary starch breakdown may be induced by gibberellin, whose production is stimulated by the auxin coming from the buds and shoot tips (Frigerio et al., 2006; Yamaguchi, 2008).

Although warm temperatures in spring induce budbreak in grapevines, the buds are generally unresponsive to this warming effect unless they have previously experienced a period of cool temperatures. The "chilling requirement," which will be discussed later, is often defined as the period of low temperature that is necessary to permit 100% of the buds to break under favorable temperatures (Cooke et al., 2012). Longer duration of chilling and lower temperatures down to about −3 °C typically accelerate the rate of budbreak and enhance the uniformity of budbreak once warmer temperatures return (Cragin et al., 2017; Dokoozlian, 1999; Koussa et al., 1994; Lavee and May, 1997; Pouget, 1963). Consequently, budbreak occurs earlier, more rapidly, and more uniformly on vines that have experienced a period of cool winter weather (Antcliff and May, 1961; Currle et al., 1983; Kliewer and Soleimani, 1972).

The buds on a single vine usually break within a few days in areas with cold winters—provided the winter is not so cold as to damage or kill the buds (see Section 7.4). Budbreak, however, may take 10 times longer in areas with mild winters. It is often erratic in warm temperate and (sub)tropical regions, where most of the world's table grapes are grown. This is why growers in these regions often encourage budbreak by applying hydrogen cyanamide (H_2CN_2), usually in the form of calcium cyanamide ($CaCN_2$), 2–4 weeks before the intended time of budbreak (Dokoozlian et al., 1995; Lavee and May, 1997; Zheng et al., 2018). The higher winter temperatures in such warmer areas lead to higher respiration rates (see Chapter 4), which might be associated with release of hydrogen peroxide (H_2O_2) by the mitochondria. In extreme cases the resulting oxidative stress may lead to bud damage, called bud necrosis (Pérez et al., 2007).

In cooler, temperate climates, mean daily temperatures above 7–10 °C induce budbreak and shoot growth. This temperature threshold, or base temperature, depends on the species and cultivar. For instance, across *Vitis* species the base temperature decreases from *V. berlandieri* to *V. rupestris* to *V. vinifera* to *V. riparia*. Among *V. vinifera* cultivars, Mourvèdre has a high, and Chardonnay a low, base temperature; cultivars with a lower base temperature break buds earlier than cultivars with a higher temperature threshold (Huglin and Schneider, 1998; Pouget, 1972, 1988). Consequently, the time of budbreak at a particular location and in a given year varies by >5 weeks among cultivars of *V. vinifera* (Boursiquot et al., 1995; Ferguson et al., 2014; Pouget, 1988; Zapata et al., 2017). It also varies from year to year by up to 4 weeks for a given cultivar, since the rate of budbreak increases as the temperature rises above the threshold temperature up to approximately 30 °C and then declines again.

However, it is the temperature of the buds themselves, rather than the surrounding air temperature, that determines when and how rapidly those buds break (Keller and Tarara, 2010). During the day, the bud temperature can be several degrees above the ambient temperature, especially when the buds are heated by the sun and when there is little or no wind (Grace, 2006; Peña Quiñones et al., 2019). Yet even at wind speeds >5 m s^{-1} buds may still be 1–2 °C warmer than the air. Conversely, at night the bud temperature is often cooler than the air temperature. Unlike the temperature of the buds, and despite its impact on the rate of bleeding sap flow, the temperature of the roots has only a minor influence on the time and rate of budbreak. The effect of root temperature, moreover, may depend on how the vines were pruned in the winter. Budbreak on spur-pruned vines seems to be rather insensitive to root temperature, whereas budbreak on cane-pruned vines occurs up to several days earlier, and the proportion of buds that do break increases considerably, as the root temperature increases from 10 °C up to a maximum near 30 °C (Field et al., 2009; Kliewer, 1975; May, 2004). Once the buds have grown out, however, subsequent shoot growth is greater at higher root temperature with both pruning approaches. The difference between spur and cane pruning in the response to root temperature of budbreak timing and proportion is probably a consequence of the phenomenon of correlative inhibition discussed below.

Budbreak starts in the distal, or uppermost, buds of a cane or vine, proceeding basipetally toward the base. Yet budbreak of the distal buds often inhibits outgrowth of the basal buds on the same cane (Clingeleffer, 1989; Galet, 2000; Gatti et al., 2016; Keller and Tarara, 2010). Though the preferential outgrowth of distal buds is often attributed to a phenomenon called apical dominance, the more appropriate term to use in this context may be apical control (Cline, 1997). Both apical control and apical dominance are cases of correlative inhibition, but apical control refers to growing shoots inhibiting each other and preventing buds on the same cane from breaking, whereas apical dominance refers to the shoot tip inhibiting the outgrowth of lateral buds on the same shoot (Beveridge, 2006; Ferguson and Beveridge, 2009). Apical control is most conspicuous in early spring when, in cane-pruned vines, the most distal buds of the cane break early and grow vigorously, whereas those toward the base of the cane grow weakly or not at all (Fig. 2.4). This inhibition of basal bud outgrowth is especially prevalent in warm climates, where it can lead to erratic and nonuniform budbreak. It also varies among cultivars; for instance, Carignan (synonym Mazuelo) and Mourvèdre (synonyms Monastrell and Mataro) show particularly strong apical control, whereas Syrah and Cinsaut are at the other end of the spectrum.

FIG. 2.4

Cane-pruned Cabernet Sauvignon vines displaying strong apical control in spring. Note that budbreak is not inhibited on spurs adjacent to the canes.

Photos by M. Keller.

Like the evolution of tendrils, this behavior is an adaptation to the habitat of wild vines, enabling them to access sunlight at the top of tree canopies with a small investment in shoot biomass per unit height gain. This adaptive behavior creates both a viticultural problem and an opportunity for cultivated grapevines. The problem occurs mainly in warm climates and arises from the inconsistent budbreak of cane-pruned vines. Viticultural practices aimed at overcoming apical control and promoting uniform budbreak include spur pruning, bending ("arching") and partial cracking of canes, and application of hydrogen cyanamide (Huglin and Schneider, 1998; Iland et al., 2011; Rosner and Cook, 1983). Spur-pruned vines are typically unaffected by correlative inhibition, and where mixed populations of spurs and canes are retained on the same vine, the buds on the spurs break at the same time as the distal buds of the canes (see Fig. 2.4; Clingeleffer, 1989; Keller and Tarara, 2010). The opportunity arises mainly in cool climates and stems from the delay in budbreak on vines that are spur-pruned only after the canes' distal buds have already broken, which may enable vines grown in frost-prone locations to escape spring frost damage (see Section 7.4). Provided pruning does not occur more than about 3 weeks after normal budbreak, vine development, which now occurs under warmer conditions, can compensate for the late budbreak, so that the berries may still ripen in time for a normal harvest (Keller and Tarara, 2010; Gatti et al., 2016). On the other hand, delaying pruning until >3 weeks after normal budbreak may also delay the onset of fruit ripening, which may be advantageous in regions with hot climates (Frioni et al., 2016; Moran et al., 2017). A potential drawback of such late pruning, however, is that it sometimes leads to abortion of some of the inflorescences, which markedly reduces yield.

2.2.3 Shoot growth

Budbreak is followed by a period of rapid shoot growth, with a new leaf appearing every few days. Grapevine shoots normally grow upward, although there is considerable variation from erect to droopy growth habits among cultivars. For instance, Cabernet Sauvignon shoots are rather erect, whereas Trebbiano or Concord shoots typically grow downward. Upward growth is guided by a negative gravitropic response, whereby the shoot perceives and responds to gravity. Gravitropism includes sedimentation of starch in specialized amyloplasts, called statoliths, inside specialized cells, termed statocytes, in the shoot's endodermis. Starch deposition triggers an increase in auxin and a decrease in extracellular pH on whichever side happens to be lower, causing that side to elongate more than the upper side (Morita and Tasaka, 2004; Petrášek and Friml, 2009). Brassinosteroids, it seems, oppose the action of auxin on gravitropism, perhaps by weakening the cell walls (Vandenbussche et al., 2011). It is possible that the dynamic interplay between auxin and brassinosteroids accounts for the differences in growth habit among cultivars.

In many grape cultivars, the unfolding leaves, and often the emerging inflorescences as well, are reddish before they turn green (Fig. 2.5). Transient accumulation of red anthocyanin pigments in or just below the epidermis probably protects the developing photosynthetic machinery from excess light (Chalker-Scott, 1999; Liakopoulos et al., 2006; Rosenheim, 1920). The anthocyanins usually disappear by the time a leaf has produced roughly half of its maximum amounts of chlorophyll and carotenoid and the palisade cells have expanded to occupy approximately 50% of the mesophyll (Hughes et al., 2007). In addition, a dense covering of "hair," termed pubescence, of the young leaves also fulfills a photoprotective role (Liakopoulos et al., 2006).

Early shoot growth is entirely dependent on nutrient reserves stored in the permanent structure of the vine, until new leaves become photosynthetically competent and begin to produce and export sugar. These reserves comprise mainly carbohydrates, such as starch and sugars, and nitrogen-containing components, such as proteins and amino acids. They are thus depleted and reach a minimum around bloom time or even later, making vines vulnerable to stress around the time of bloom (Holzapfel et al., 2010; Lebon et al., 2008; Weyand and Schultz, 2006b; Williams, 1996; Zapata et al., 2004; Zufferey et al., 2015; see also Section 6.1). Storage reserves are used not only as building supplies to construct new plant material in the absence of photosynthesis but also to generate by respiration the energy required to fuel this construction (see Section 4.4). Such use of reserves, in addition to losses incurred during the earlier bleeding period, means that grapevines literally lose weight during the initial spring growth phase, before the photosynthesizing leaves begin to add new biomass from about the five- to six-leaf stage onward (Eifert et al., 1961; Hale and Weaver, 1962; Koblet, 1969; Pradubsuk and Davenport, 2010). Nitrogen reserves, in addition to carbohydrates, are very important for new growth in spring; early season growth and especially leaf chlorophyll correlate closely with the amount of nitrogen reserves stored in the perennial parts of the vine (Cheng et al., 2004b; Keller and Koblet, 1995a; Treeby and Wheatley, 2006; Weyand and Schultz, 2006b; Zufferey et al., 2015). Other mineral

FIG. 2.5

Emerging Merlot shoot with pink unfolding leaves and inflorescences days after budbreak (left) and Muscat Ottonel shoot with red young leaves (right).

Photos by M. Keller.

nutrients are also remobilized from perennial organs and used to support new spring growth (Pradubsuk and Davenport, 2010, 2011). Higher soil temperatures are associated with more rapid starch remobilization in the roots, higher rates of root growth and branching, and increased water and nutrient uptake, which permits the shoots to grow more rapidly on vines in warm soil than in cool soil (Clarke et al., 2015; Field et al., 2009; Rogiers et al., 2011, 2014; Skene and Kerridge, 1967; Woodham and Alexander, 1966; Zelleke and Kliewer, 1979).

Growth seems to adhere more or less to the principle of compound interest: as new leaves become photosynthetically competent, they permit growth of yet more leaves, so that growth is proportional to present biomass (Turgeon, 2010). Consequently, environments that favor rapid growth early in the spring may maximize seasonal shoot growth and fruit production (Keller and Tarara, 2010; Keller et al., 2010). The initial shoot growth phase is characterized by strong apical dominance, whereby the growing shoot tip suppresses outgrowth of prompt buds by monopolizing sugar supply for its own growth (Mason et al., 2014). The relatively sugar-deficient prompt buds are unable to initiate cell division for bud outgrowth. In addition to this sugar depletion mechanism, slow but active basipetal flow of auxin released by the leaf primordia and young leaves near the shoot apex stimulates internode elongation and contributes to the inhibition of lateral shoot growth (Alleweldt and Istar, 1969; Mason et al., 2014; Müller and Leyser, 2011; Woodward and Bartel, 2005). If auxin flows too rapidly, however, apical dominance is compromised and lateral buds can grow out (Berleth et al., 2007; Dun et al., 2006; Lazar and Goodman, 2006). This might be the case in very vigorously growing shoots, which are characterized by both rapid main shoot elongation and early and strong lateral shoot growth, contrasting with the near absence of lateral shoot growth on weak, slow-growing shoots (Huglin and Schneider, 1998; Keller and Tarara, 2010). The suppression of axillary bud outgrowth by auxin is indirect and may only support or reinforce the low-sugar signal, because the hormone does not enter these buds, which themselves start producing and exporting auxin as soon as they begin to grow (Müller and Leyser, 2011; Ongaro and Leyser, 2008).

Auxin may exert its remote control and determine which buds will grow out by prohibiting auxin efflux from lateral buds and by inhibiting cytokinin production in the roots and in the shoot tissues adjacent to the buds (Beveridge, 2006; Dun et al., 2006; Ferguson and Beveridge, 2009; McSteen and Leyser, 2005; Prusinkiewicz et al., 2009). In addition, strigolactones, which are produced from carotenoids, also inhibit bud outgrowth, probably by decreasing auxin efflux from buds or by blocking cell division, thus preventing bud activation (Crawford et al., 2010; Dun et al., 2012; Gomez-Roldan et al., 2008; Umehara et al., 2008). Like cytokinins, strigolactones are manufactured in the roots and transported in the xylem in addition to their production in vascular tissues throughout the plant and locally in shoot nodes (Ruyter-Spira et al., 2013). But the influence of strigolactone and of cytokinin on bud outgrowth is antagonistic.

Damage to the shoot tip or its removal (e.g., by summer pruning) activates the prompt buds, enabling them to grow out and become lateral shoots. Bud activation occurs rapidly by enhanced sugar availability and, more slowly, by auxin depletion in the shoot, which releases the repression of cytokinin biosynthesis in the shoot and delivery from the roots, and may inhibit strigolactone production (Hayward et al., 2009; Mason et al., 2014; Nordström et al., 2004; Tanaka et al., 2006). Sugar and cytokinins that enter the buds stimulate growth of lateral shoots by promoting cell division, especially near the former shoot tip where auxin is depleted first. In a process termed canalization, active export by means of ATP-fueled transport proteins of newly produced auxin from outgrowing axillary buds occurs down polarized files of cells that then develop into new vascular bundles connecting the lateral shoot to the main shoot's vascular system (Müller and Leyser, 2011; Prusinkiewicz et al., 2009; Ruyter-Spira et al., 2013). After loss of the shoot tip leads to prompt-bud outgrowth, the most distal lateral shoot often gradually assumes the function of the main shoot (Alleweldt and Istar, 1969). The more basal laterals usually stop growing after they have formed just a few leaves that are ordinarily smaller than the leaves of the main shoot (see Fig. 1.8).

Incidentally, removal of shoot tips also temporarily eliminates sugar demand and depletes auxin in the vascular cambium of the shoots, as well as in the trunk and roots, and inhibits their secondary growth and root elongation for a while (Ferguson and Beveridge, 2009; Fu and Harberd, 2003). Growers in the tropics, where strong apical dominance is notorious, employ continued shoot tipping to encourage lateral growth. In temperate climates, the influence of apical dominance decreases once the main shoot has formed about 18–20 leaves, which usually occurs around the time of bloom (Alleweldt and Istar, 1969). Since auxin concentration declines with increasing distance from the shoot tip, the more basal prompt buds are the first to become activated, and lateral shoots begin to grow out from these buds if the main apex is intact (Prusinkiewicz et al., 2009). Such lateral outgrowth in the presence of an actively growing shoot tip is especially pronounced and can occur much earlier than the 18-leaf stage when warm temperatures stimulate rapid growth in spring (Keller and Tarara, 2010). In addition, abundant nitrogen availability also promotes strong and early lateral shoot growth (Keller and Koblet, 1995a). These effects might be related to the stimulation by nitrate of cytokinin production in the roots and by rapid transpiration favoring cytokinin delivery via the xylem from the roots to the shoots (Sakakibara, 2006; Sakakibara et al., 2006).

Gibberellins produced upon stimulation by auxin in or near the shoot tip and brassinosteroids produced in the internode epidermis promote cell elongation in the shoot's internodes, perhaps by activating the wall-loosening protein expansion, in addition to promoting production of the membrane water channel protein aquaporin (Cosgrove, 2000; Frigerio et al., 2006; Maurel et al., 2008; Olszewski et al., 2002; Savaldi-Goldstein et al., 2007; Weiss and Ori, 2007; Yamaguchi, 2008; see also Section 3.1). Meanwhile, the thick, cellulose-reinforced outer cell walls and rigid cuticle of the epidermis act like a corset, forcing the internode cells to elongate rather than simply expanding radially in all directions (Kutschera, 2008a, b; Mirabet et al., 2011; Schopfer, 2006). The turgor pressure of the underlying cells is thus transmitted to the epidermis that acts like a tensile skin and becomes stretched in the process (Javelle et al., 2011).

Gibberellins act by binding to the so-called DELLA proteins in the cell nucleus and inducing the degradation of these growth-suppressing proteins. Although DELLA proteins bind to and thus inactivate transcription factors, which in turn suppresses expression of the transcription factors' target genes, just how these proteins put the brakes on growth remains to be discovered (Achard and Genschik, 2009; Acheampong et al., 2015; Claeys et al., 2014; Hartweck, 2008; Santner et al., 2009). Mutations that render DELLA proteins unable to bind gibberellins and are thus immune to degradation result in dwarf phenotypes that look like bonsai vines (Boss and Thomas, 2002; Boss et al., 2003); such mutations also formed the basis for the short and high-yielding cereal cultivars that gave rise to the agricultural Green Revolution in the second half of the twentieth century (Claeys et al., 2014; Hartweck, 2008).

Shoot growth tends to slow down somewhat during the bloom period but may resume after fruit set if the availability of water and soil nutrients permits (see Sections 7.2 and 7.3). Shortly before or while the grapes begin to ripen, the shoots begin to form a periderm and turn from green to yellowish or reddish brown. This shoot "maturation" proceeds acropetally from the base toward the tip and is accompanied by replenishment of storage reserves in preparation for winter and the subsequent growing season (Eifert et al., 1961; Löhnertz et al., 1989). Abundant supply of water and nutrients delays these processes and, to make matters worse, also encourages continued or renewed shoot growth during grape ripening. This can cause problems in regions with significant summer rainfall, especially on fertile vineyard sites that have high soil water and nutrient storage capacity, where late-season shoot growth may compete with the ripening berries and storage reserve pools for photosynthetic sugar supply (Sartorius, 1973).

In viticultural terminology, the rate of shoot growth or shoot elongation over time is referred to as vigor. In addition to genetic effects of species, cultivar, and rootstock, vigor is strongly influenced by temperature, water and nutrient availability, vine reserve status, pruning level, and even vine age. Some cultivars (e.g., Thompson Seedless, Cabernet franc, or Grenache) or rootstocks (e.g., 140 Ruggeri, 1103 Paulsen, or St. George) are considered to be vigorous, whereas others are thought to be intermediate (e.g., Cabernet Sauvignon, Syrah, Chardonnay, Teleki 5BB, SO-4, or 3309 Couderc) or weak (e.g., Gamay, 101-14 Mgt, or Riparia Gloire). Based on the discovery of an inverse correlation across genotypes between vigor and concentrations of leaf starch, sugars, protein, amino acids, and organic acids, it has been proposed not only that growth drives metabolism rather than the reverse, but also that fast growers may be "spenders," investing their resources in growth, whereas slow growers may be "hoarders" saving their resources and accumulating reserves (Hanson and Smeekens, 2009; Meyer et al., 2007; Sulpice et al., 2009).

Vigor declines with increasing vine size; therefore, shoots of small, young vines often grow more vigorously than those of large, old vines. The size-related decline arises from the increased mechanical cost of maintaining a larger or older vine's permanent structure, from the hydraulics of water transport, and from a higher number of buds that result in more shoots on larger vines (Keller et al., 2015a). Nevertheless, in well-managed vineyards even very old vines (>150 years) can sustain vigorous growth and high yield potential (Grigg et al., 2018). Abundant resource availability, such as high soil moisture and nutrient (especially nitrogen) availability also favor high vigor, as does severe winter pruning, because plants in general try to reestablish a balanced root ÷ shoot ratio (Buttrose and Mullins, 1968; Clingeleffer, 1984; Keller et al., 2004; Poorter and Nagel, 2000). Consequently, reducing the water supply where possible or decreasing the pruning severity by increasing the number of buds per vine retained at pruning are often powerful vineyard management tools to decrease vigor. Grapevines self-regulate their growth to adapt to a wide range of bud numbers, which growers can alter by manual spur or cane pruning or mechanical pruning. For example, as the number of buds increases, vines typically respond with decreases in the proportion of buds that break (percentage budbreak) and in shoot vigor, which leads to a less than proportional increase in total leaf area (Freeman et al., 1979; Keller et al., 2015a; Poni et al., 2016). Nevertheless, the canopy of lightly pruned vines develops earlier and more rapidly than that of heavily pruned vines, because a higher number of buds leads to more shoots growing simultaneously in spring, using a larger pool of reserves accumulated in the previous growing season (Araujo and Williams, 1988; Intrieri et al., 2001; Possingham, 1994; Sommer et al., 1995; Weyand and Schultz, 2006a, b; Winkler, 1929, 1958). On the other hand, the presence of fruit tends to depress vegetative growth, especially of lateral shoots, so that the total biomass production per vine remains approximately constant (Eibach and Alleweldt, 1985; Pallas et al., 2008; Petrie et al., 2000b).

A shoot's growth direction also affects its growth rate (Currle et al., 1983; Kliewer et al., 1989). Upright-growing shoots, such as those in a typical vertical shoot-positioning trellis system, generally grow more vigorously than shoots forced to grow horizontally, which in turn grow more rapidly than shoots pointing downward, such as those in Geneva double-curtain or Scott–Henry systems. The effect on growth of the shoot angle relative to vertical is termed gravimorphism (Wareing and Nasr, 1961). This effect is associated with variations in xylem vessels, whose number and diameter seem to be tailored to optimize water and nutrient supply to the shoot (Lovisolo and Schubert, 2000; Schubert et al., 1999). Accumulation of auxin near the shoot tip may be responsible for the production of more but narrower vessels in downward-pointing shoots (Lovisolo et al., 2002b). Consequently, downward-growing shoots have higher resistance to water flow in the xylem and smaller leaves with lower stomatal conductance and lower photosynthesis compared with upward-growing shoots (Schubert et al., 1995).

Both the rate of leaf unfolding and the extension of internodes are influenced by day length and temperature. Long days and high temperatures favor these processes, so that internode length and leaf number increase as day length and/or temperature increases (Alleweldt, 1957). Growth typically accelerates up to a temperature optimum of 25–30°C but slows as temperature rises still further and ceases above 35°C (Buttrose, 1969c; Currle et al., 1983). This effect partly arises from high temperature stimulating auxin production, which in turn promotes cell elongation. Temperature also influences photosynthesis and respiration: Warm (25–30°C) daytime temperatures promote photosynthetic CO_2 fixation, whereas somewhat cooler (15–20°C) nighttime temperatures limit respiratory CO_2 loss (see Section 5.2). This is especially important because respiration rates are higher at night than during the day (Tcherkez et al., 2017). Consequently, a sufficient but not excessive difference between day and night temperatures, rather than simply a warm daily average temperature, is likely to promote growth, a phenomenon termed thermoperiodic growth (Went, 1944).

Nonetheless, the assumption of a linear increase in growth rate and plant development as the mean temperature rises forms the basis for the calculation of the so-called growing degree days (GDD), thermal time, or heat units. The concept of heat units is based on the recognition that plants measure time as thermal time and pace their development accordingly (Thomas, 2013). Building on what he called a "most natural and most ancient idea," the Swiss botanist Alphonse de Candolle (1855) suggested that a summation of "useful" temperatures, that is the growing-season sum of daily mean temperatures above a base temperature, averaged over many years could be used to compare climatic regions. Although de Candolle had warned that using a 7-month "standard" growing season comprising the months of April through October was unrealistic for grapevines, Amerine and Winkler (1944) proposed that grape-growing regions could be classified by the average heat summation or cumulative GDD for the April–October period over several years. They did not specify the intervals for heat accumulation, but over time the GDD summation has become standardized as follows:

$$\mathrm{GDD} = \sum_{i=1}^{n}(T_i - T_b)$$

where T_i is the mean air temperature of each day for the period 1 April–31 October in the Northern Hemisphere or 1 October–30 April in the Southern Hemisphere ($n=214$ days) and T_b is the base temperature.

Normally, T_i is calculated as the mean of the daily maximum and minimum temperatures. Only daily values of $T_i > T_b$ are added to the cumulative GDD, whereas values of $T_i \leq T_b$ are set to zero. Because very little growth and photosynthesis occur in grapevines below approximately 10°C, this value is usually chosen as T_b, but other thresholds down to 0°C are sometimes used (Amerine and Winkler, 1944; Parker et al., 2011, 2013). In his original comparison of grape-growing regions, de Candolle (1855) used thresholds of both 8 °C and 10 °C, stating that the actual T_b was likely to vary among species, cultivars, and physiological processes. Although the phenological sensitivity to temperature and the T_b for grapevine development may differ among cultivars, T_b is probably within a few degrees of 10 °C (Sadras and Moran, 2013; Zapata et al., 2015, 2017). Some uncertainty arises from the fact that, in the field, plant tissue temperatures are rarely identical to air temperatures (Parent et al., 2019; Peña Quiñones et al., 2019).

Despite some limitations (Dry and Coombe, 2004; Mc Intyre et al., 1987; Parent et al., 2019) and although variations on the GDD theme have been proposed, including a "heliothermal index" (Huglin and Schneider, 1998; Jones et al., 2010), a "latitude-temperature index" (Jackson and Cherry, 1988), and "biologically effective day degrees" (Gladstones, 1992), the "Winkler regions" (Winkler et al., 1974) are still used as a standard today. Evidently, temperature responses of plants are only nearly linear over a relatively narrow range and curvilinear outside that range. But while GDD may not be fully adequate to describe grapevine growth, phenology, and other processes, they serve their stated purpose of capturing differences in overall "heat load" of different vineyard locations or growing seasons. They thus permit direct comparisons among locations or growing seasons. For the most part, different indices seem to be highly correlated, which suggests that they describe essentially similar spatial climate patterns (Badr et al., 2018; Hall and Jones, 2010; Jones et al., 2010).

A lower limit of 850 GDD and an upper limit of 2700 GDD have been proposed for the thermal range that is suitable for wine grape production (Jones et al., 2010). Table grapes and raisins, however, are often produced under considerably warmer conditions. Each 1 °C increment in temperature averaged over the standardized growing season adds 214 GDD, so that vineyards at lower latitude or lower elevation will tend to be classified into higher (warmer) Winkler regions. This applies both across disparate growing regions and within them; whereas latitude is the main driver of differences in GDD between regions (followed by continentality; i.e., distance from an ocean or sea), topography (i.e., changes in elevation) drives most of the GDD differences within a region (Failla et al., 2004; Hall and Jones, 2010; Jones et al., 2010; Keller, 2010). As a rule of thumb, a gain of 100 m in altitude decreases the average annual temperature at the surface by roughly the same extent (0.5–0.6 °C), as does an increase in latitude by 100 km (a 1° difference in latitude is ~111 km). However, the decline in temperature with increasing altitude, the so-called terrestrial lapse rate, is not a universal constant. For example, large mountain masses, such as the Sierra Nevada and the Cascade Range in the western United States, tend to decrease the lapse rate to <0.3 °C per 100 m (Wolfe, 1992).

The ongoing rise in temperature associated with global climate change (Intergovernmental Panel on Climate Change et al., 2013) will likely shift many vineyard sites into the next higher Winkler region within the next 50 years. The predicted increase in average spring and autumn temperatures will also lead to longer actual (as opposed to standard) growing seasons, which are determined by the frost-free period, defined as the number of days between the last spring frost and the first fall frost. Moreover, in addition to its influence on growth rates, a rise in temperature also typically accelerates phenological development, with a consistent trend toward earlier bloom, veraison, and harvest (Alleweldt et al., 1984b; Chuine et al., 2004; Jones and Davis, 2000; Sadras and Moran, 2013; Webb et al., 2007; Wolfe et al., 2005).

The dependence of plant development on temperature is so persistent that the timing of phenological events, such as bloom or veraison, can be predicted with fairly high accuracy using mathematical models based solely on GDD accumulation (Duchêne et al., 2010; Parker et al., 2011, 2013; Zapata et al., 2017). Such models, as well as the observations on which they are based, suggest that temperature accounts for most of the spatial and temporal variation in grapevine phenology and that the temperature-induced variation in phenology is greater for an individual cultivar between years than among many different cultivars within a year. Nevertheless, although the duration of the phase between budbreak and the time the ripening fruit of a given cultivar attains a specific sugar concentration is shorter in a warmer than in a cooler climate, the cumulative GDD during this period is higher in the warmer climate (Mc Intyre et al., 1987). Among cultivars, moreover, the time interval between successive phenological stages is not identical. For example, a cultivar that breaks bud early does not necessarily enter bloom or veraison early (Zapata et al., 2017).

2.2.4 Growth cessation and senescence

Unlike the shoots of many other woody perennials, grapevine shoots do not form terminal buds but continue to grow as long as environmental conditions permit. When growth is no longer possible due to low temperatures or insufficient soil resources, the shoot tip with the apical meristem will die. The sensitivity of the shoot apex to adverse environmental factors may be exploited in areas where the soil dries down sufficiently in early summer so that growth can be controlled by appropriate deficit irrigation strategies (see Section 7.2). Under non-limiting soil moisture, decreasing day length, aided by declining temperatures, eventually bring cell division in the bud apical meristems and the cambium to a halt, which prevents further lateral shoot emergence and induces dormancy (Alleweldt, 1960; Alleweldt and Düring, 1972; Davis and Evert, 1970; Fracheboud et al., 2009; Garris et al., 2009; Rohde and Bhalerao, 2007). The synergistic effect of shorter days and cooler temperatures is evident in warm climates, where shoot growth may continue despite a shortening photoperiod so long as temperatures remain permissive. Growth cessation and periderm formation are brought about by a rapid decline in apex-derived auxin in the cambial region and a concomitant increase in root-derived ABA in the xylem and phloem, where it remains high during the subsequent dormant period (Mwange et al., 2005). In contrast to the buds, the phloem of the shoots and the trunk seems to become dormant only after the first fall frost, and much of the root phloem may not become dormant at all (Esau, 1948). This can lead to cold injury to the phloem when an unseasonable freeze event follows a warm autumn (Keller and Mills, 2007; see also Section 7.4).

In autumn, leaf senescence marks the end phase of differentiation and eventually ends in death and abscission of the leaves, completing the shoot growth cycle. Senescence is accompanied by recycling of foliar nutrients to the canes and permanent parts of the vine; it is followed by leaf shedding and dehydration and cold acclimation of the aboveground woody organs (Conradie, 1986; Dintscheff et al., 1964; Löhnertz et al., 1989; Schaller et al., 1990). Before they are shed, the senescing leaves develop their characteristic and often spectacular yellow, orange, or red autumn color. The coloration occurs because the green chlorophylls are degraded more rapidly than the yellow-orange carotenoids, which are thus "unmasked" in the process. The carotenoids also undergo various chemical modifications that influence their color. In addition, newly produced red anthocyanins may be accumulated once a leaf has degraded roughly half of its original chlorophyll. Nevertheless, whereas white-skinned grape cultivars are unable to produce anthocyanins, even the leaves of most dark-skinned cultivars normally turn yellow rather than red. The red-fleshed cultivar Alicante Bouschet is a vividly red exception, whereas some other *V. vinifera* cultivars, like Tempranillo, Malbec, Carmenère, or Lemberger, develop orange-red to bronze fall colors (Fig. 2.6). Anthocyanin production is triggered by, and is consequently a sign of, sugar accumulation (Gollop et al., 2002; Larronde et al., 1998; Lecourieux et al., 2014;

FIG. 2.6

Brilliant red fall color on Alicante Bouschet leaves, with yellowing Grenache leaves in the background (left) and orange-bronze fall color on Carmenère leaves (right).

Photos by M. Keller.

FIG. 2.7

Red leaves above the point where a tendril has wrapped itself around a Cabernet franc shoot (left), on undercropped shoots growing from a twisted Cabernet Sauvignon cane (center), and pale yellow leaves on a partly broken Muscat blanc shoot (right).

Photos by M. Keller.

Pirie and Mullins, 1976). Foliar sugars tend to accumulate in autumn as phloem export declines, but accumulation can also occur prematurely when phloem efflux is constrained by wind-induced or other injury, shoot girdling, virus infection, or nutrient deficiency (Halldorson and Keller, 2018; Fig. 2.7; see also Section 7.3). After the leaves have been abscised and begin to decompose, the anthocyanins, which are dominated by glucosides of peonidin and cyanidin, and other phenolic components that have been accumulated in the leaves may be released into the soil. Owing to their phytotoxic nature, these chemicals can be regarded as allelopathic (Greek *allelon* = one another, *pathos* = suffering) compounds that are able to suppress the germination of seeds from other plant species the following spring, at least until they are themselves degraded by soil microbes (Bonanomi et al., 2011; Weir et al., 2004). They thus act as pre-emergence bioherbicides that reduce interspecific competition.

It is thought that decreasing day length triggers leaf senescence. Temperature, by contrast, has no effect on the onset of senescence, although lower temperature may accelerate the rate of senescence once initiated (Fracheboud et al., 2009). In other words, for a certain cultivar grown at a certain latitude, leaf senescence should start on approximately the same day each year but progress more slowly in warm than in cool years or on warm than on cool sites. At the other end of the temperature spectrum, heat stress ($\geq 45\,°C$) may also accelerate the senescence rate (Thomas and Stoddart, 1980). Of course, the critical day length necessary to induce senescence and its interaction with temperature likely vary by species and cultivar. During the carefully orchestrated senescence program, which is initiated by ABA and ethylene and normally starts with chloroplast degradation in the mesophyll cells at the leaf margins and then progresses inward and toward the base, the leaves transition from photosynthetic carbon assimilation to the breakdown of chlorophyll, proteins, lipids, and nucleic acids and subsequent export of valuable nutrients to other plant parts (Jones and Dangl, 1996; Lim et al., 2007; Masclaux-Daubresse et al., 2010; Schippers et al., 2015). Before they die, the leaves lose about half of their carbon, much of it due to respiration to generate the energy required for the metabolism of degradation, or catabolism, and export in the face of declining photosynthesis (Keskitalo et al., 2005; Williams and Smith, 1985). The high energy costs of senescence are progressively met by the degradation of cell membranes and use of their lipids for respiration (Buchanan-Wollaston, 1997). Not surprisingly, the nucleus with its complement of genes and the mitochondria with their respiration machinery are the last cell components to be dismantled in the process (Lim et al., 2007; Schippers et al., 2015).

Although the characteristic leaf yellowing during senescence is due to the degradation of chlorophyll, chlorophyll is degraded not because its products are reusable but, rather, because its light-absorbing properties make it a strong phytotoxin in the absence of photosynthesis and to facilitate access to more valuable materials such as chloroplast proteins and lipids. The nitrogen and much of the carbon of the chlorophyll molecules, but not their magnesium, remain in the leaf along with the carbon contained in the cell walls, at least until soil microbes decompose the leaf once it has fallen to the ground (Havé et al., 2017; Hörtensteiner and Feller, 2002). Considering that chlorophyll contains only approximately 2% of the leaf's nitrogen, this loss is a rather modest price the plant pays for recycling of proteins and lipids (Hörtensteiner, 2006). Leaves thus manage to remobilize up to 50–80% of their nitrogen, potassium, and phosphorus, 50% of their sulfur, and 20% of their iron before their cell membranes collapse, marking the end of their life (Keskitalo et al., 2005; Niklas, 2006). Following this resorption process, the phloem is plugged and sealed, and the leaves are shed at predetermined positions of cell separation at the base of the petiole, termed abscission zones, which are differentiated soon after the leaves have fully expanded (Roberts et al., 2002). The formation of abscission zones is induced by the plant hormone ethylene, whereas their position is dictated by auxin, which disappears completely during the subsequent dormancy period. Each year, grapevines invest most of their aboveground carbon and nitrogen in leaf production, which cycles as resorbed material in the perennial plant parts and as dead organic material, the so-called detritus or humus, through the soil.

2.2.5 Bud development and dormancy

The time of rapid shoot growth coincides with secondary growth leading to an increase in diameter of the shoots and the perennial plant organs. This secondary growth ends when the decreasing day length after midsummer leads to cessation of cell division in the cambium, well before the temperature becomes too low to limit cell division (Druart et al., 2007). The shortening day length also decreases gibberellin production in the shoot elongation zone, so that growth rates decline (Cooke et al., 2012). The rapid growth period also coincides with the formation of prompt and compound buds in the leaf axils. However, development of the first compound buds in the basal nodes of a shoot begins as early as 3 or 4 weeks before budbreak or >1 year before these buds will give rise to shoots (Morrison, 1991). Once they have established up to 18 leaf primorida, the compound buds become dormant. During dormancy cell division and expansion cease, whereas metabolic processes such as respiration continue, albeit at a low level. Callose deposits in the plasmodesmata of the apical meristem, provascular tissues, and basal parenchyma block connection and communication of the buds with the shoot (Rinne et al., 2016; Wu et al., 2018). Although the buds are under apical dominance from the growing shoot tip, and thus do not normally break, the prompt buds can break relatively readily to form lateral shoots. Following loss of the shoot tip and removal of lateral shoots or severe defoliation due to drought stress or physical damage caused by hail or pests, even the compound buds are initially able to resume growth (Alleweldt, 1960; Huglin and Schneider, 1998). This phase is variously referred to as paradormancy, predormancy, conditional dormancy, or summer dormancy, defined as the inhibition of growth by distal organs (Galet, 2000; Horvath et al., 2003; Lang et al., 1987; Lavee and May, 1997).

As shoot growth slows down, the buds progressively lose the ability to break even in the absence of apical dominance, beginning with the most basal buds. While the buds' respiration rate declines steadily after their formation, their water content drops quickly from about 80% to <50% as they become hydraulically isolated from the shoot and enter into the phase termed endodormancy, which is also known as true, inherent, organic, deep, or winter dormancy, or rest (Gardea et al., 1994; Huglin and Schneider, 1998; Lavee and May, 1997; Xie et al., 2018). Bud dormancy develops gradually over 2–3 weeks, moving from the base toward the tip of the shoot following behind the wave of brown periderm (Alleweldt, 1960; Cooke et al., 2012; Pouget, 1963). Therefore, the transition from paradormancy to endodormancy is characterized by a color change of the bud scales from green to brown. During this transition the temperature and time required for the buds to break increases greatly and rather suddenly (Camargo Alvarez et al., 2018; Cragin et al., 2017; Pouget, 1963). This shift is induced by decreasing day length and cooler temperatures, and is associated with a strong increase in ABA and decrease in gibberellin in the buds and shoots (Düring and Alleweldt, 1973; Düring and Bachmann, 1975; Koussa et al., 1994; Rohde and Bhalerao, 2007; Zheng et al., 2015, 2018).

The para−/endodormancy transition happens relatively quickly around the time of veraison in regions at higher latitudes but may start after harvest and proceed much more gradually in warm regions at lower latitudes (compare Cragin et al., 2017 and Zheng et al., 2018). It also occurs later, and the endodormancy period is shorter, in cultivars with early budbreak in spring (Camargo Alvarez et al., 2018; Pouget, 1963, 1972). Buds that have become endodormant are now able to acclimate to cool temperatures and acquire cold hardiness (see Section 7.4). The genetic programs for bud dormancy and cold hardiness are superimposed; shorter photoperiods are sufficient to induce dormancy and leaf senescence, whereas low temperatures are required for the dormant organs to acquire cold hardiness (Fennell and Hoover, 1991; see also Section 7.4). Nonetheless, entry into endodormancy is delayed under very warm conditions.

Endodormancy is defined as the inhibition of growth by internal signals. This state of self-arrest prevents budbreak and renewed shoot growth during the favorable growth conditions of late summer and early autumn, which would lead to certain death of the emerging shoots due to freezing temperatures in late autumn and winter (Horvath et al., 2003; Lang et al., 1987). The inability of the buds to grow out arises from carbon starvation that precedes the transition to endodormancy and may lead to inhibition of protein production and cell division (Tarancón et al., 2017). These effects may be linked

FIG. 2.8

Uncoupling of xylem flow between Chardonnay buds and shoots, traced by the red xylem-mobile dye basic fuchsin. Paradormant buds are connected to the shoot (left), endodormant buds are hydraulically isolated (center), and swelling buds reconnect to the shoot before budbreak (right).

Photos by L. Mills.

to the constriction of the plasmodesmata in the buds' apical meristems by callose, whose degradation is induced by gibberellin once the buds have experienced a period of cool temperatures, termed the chilling period, that is followed by growth-conducive warmer temperatures (Burch-Smith and Zambryski, 2012; Rinne et al., 2016; van der Schoot et al., 2014; Wu et al., 2018). Consequently, while gibberellin inhibits budbreak both before and after the release of bud dormancy, it promotes growth once the buds' meristematic activity resumes prior to budbreak in spring (Zheng et al., 2018). In addition, the bud xylem remains uncoupled from the shoot xylem, that is, the dormant buds are hydraulically isolated from the shoot until the resumption of meristem activity (Xie et al., 2018; Fig. 2.8).

A chilling period of approximately 7 consecutive days with mean daily temperatures between about $-3\,°C$ and $+10\,°C$ in combination with short days may irreversibly break endodormancy and stimulate the renewed ability of dormant buds to resume growth when favorable conditions return (Currle et al., 1983; Galet, 2000; Koussa et al., 1994; Pouget, 1963, 1972). The chilling temperature must be perceived directly by the apical meristem inside each bud (van der Schoot et al., 2014). Intermittent warm temperatures cancel the effect of earlier cool episodes lasting less than a week. Consequently, fulfillment of the chilling requirement coincides approximately with leaf fall or soon thereafter in temperate climates but may occur in mid-winter in regions with mild winters (Camargo Alvarez et al., 2018; Cragin et al., 2017; Zheng et al., 2018). It is possible that the chilling requirement itself is longer in warmer regions, because higher rather than lower temperatures induce earlier and deeper dormancy (Cooke et al., 2012).

While chilling releases the buds from endodormancy, it simultaneously enhances their ability to tolerate freezing temperatures (van der Schoot and Rinne, 2011; van der Schoot et al., 2014). Chilling, therefore, has a dual function: to release buds from dormancy and to increase their cold hardiness (see Section 7.4). Like dormancy induction, dormancy release happens gradually. The release might be a consequence of the steady decline with continued chilling of the dormancy hormone ABA in the buds and canes (Cheng et al., 1974; Düring and Alleweldt, 1973; Düring and Bachmann, 1975; Düring and Kismali, 1975; Koussa et al., 1994, 1998; Zheng et al., 2015; see also Section 7.4). At high concentrations, ABA maintains dormancy by suppressing cell division (Vergara et al., 2017). ABA acts by binding to the so-called PYL receptor proteins in the cytosol; the resulting composite molecule in turn binds to and inhibits certain phosphatase enzymes, which allows specific kinase enzymes to phosphorylate, and hence activate or control, ABA's target transcription factors, enzymes, and other proteins that trigger physiological responses (Dejonghe et al., 2018).

Grape species and cultivars differ in their chilling requirement; *V. amurensis*, *V. riparia*, *V. vinifera*, and *V. labrusca* have been classified as low-chill species (<1000 h), whereas *V. rupestris* and *V. aestivalis* have been classified as high-chill species (Londo and Johnson, 2014). Among *V. vinifera* cultivars, the temperature threshold required to break endodormancy is lower for "late" cultivars with a higher base temperature for budbreak and shoot growth in spring and higher for "early" cultivars with lower base temperature. Thus dormancy is less intense and more easily lifted in early cultivars than in late cultivars; in other words, the extent and duration of dormancy are cultivar specific (Cheng et al., 1974; Cragin et al., 2017; Pouget, 1972). Chilling may not be an absolute requirement to break dormancy in all cultivars, and its effect seems to be quantitative rather than qualitative, but no chilling—especially in combination with an extended growing period in late fall—usually leads to delayed, limited, and uneven budbreak (Koussa et al., 1994; Lavee and May, 1997; Mathiason et al., 2009). While 100% of the buds are generally able to grow out after an adequate chilling period, the proportion of buds that do eventually break rarely exceeds 50–70% if the chilling requirement is not met, and the leaves that do emerge from these buds are often malformed and chlorotic (Pouget, 1963).

Inadequate chilling can be a problem in warm subtropical and tropical climates, where growers often induce budbreak by applying H_2CN_2 after pruning. Based on results from studies with such dormancy-breaking agents (Pang et al., 2007; Pérez et al., 2007, 2009), as well as with respiration inhibitors (Pouget, 1963), one might argue that dormancy release may require a temporary decline in bud respiration rates below a cultivar-specific threshold. Indeed, the bud scales form effective barriers to oxygen permeability, and dormant buds are rather hypoxic, that is, oxygen-deprived (Meitha et al., 2015). A requirement for a respiratory minimum might also explain the rather startling finding that a heat shock, that is, a brief (1 h) exposure to very hot temperatures (>40–50 °C), too, is able to break dormancy (Pouget, 1963; Zheng et al., 2018). Longer periods of such high temperatures, however, will kill the dormant buds (Gardea et al., 1994).

Hydrogen cyanamide leads to H_2O_2 accumulation by inhibiting the antioxidant enzyme catalase, but the subsequent movement of calcium ions (Ca^{2+}) from the cell walls to the cytoplasm, similar to the movement induced by chilling, then seems to initiate cellular mechanisms culminating in dormancy release of the buds (Pang et al., 2007). One such mechanism is a reduction in ABA, again like the influence of chilling (Vergara et al., 2017). Another effect may be the activation of glucanase enzymes, which are responsible for the breakdown of winter callose in the phloem and the meristems contained in the buds (Pérez et al., 2009). Like entry into dormancy, dormancy release is a gradual process that occurs over several weeks (Cooke et al., 2012; Pérez et al., 2007).

Once endodormancy has been lifted, low winter temperatures continue to enhance the capacity of buds to resume growth, so long as temperatures are not extreme enough to kill the buds (Cragin et al., 2017). Yet the main factor that prevents budbreak after dormancy release is low temperature (though lack of water can also inhibit budbreak). This last phase is called ecodormancy, also known as enforced, relative, or imposed dormancy, postdormancy, or quiescence. It is defined as the inhibition of growth by temporary, adverse environmental conditions (Galet, 2000; Horvath et al., 2003; Lang et al., 1987). In other words, unfavorable growing conditions simultaneously break endodormancy and impose ecodormancy. The latter is related to cold-induced (or drought-induced) accumulation of ABA and coincides with maximal cold hardiness (see Section 7.4). The return of warmer temperatures in spring, or of unseasonably warm temperatures during winter, eventually enables the buds to break. Budbreak is preceded by a rapid rise in respiration rate, restoration of the buds' vascular connection to the cane (Fig. 2.8), and gain in the buds' moisture content back up to 70–80%, along with a concomitant rapid loss of cold hardiness (Ferguson et al., 2014; Gardea et al., 1994; Lavee and May, 1997; Pouget, 1963; Xie et al., 2018). The increase in respiration rate likely indicates renewed cell division in the buds' apical meristems, and the increase in water content is linked to renewed cell expansion in the compressed internodes. The outward sign of this spring awakening is the swelling of the buds that culminates, a few days later, in the appearance of green tissue that marks budbreak. The date of budbreak of a given cultivar is determined by temperature, or more precisely by thermal time, during the ecodormancy period (Pouget, 1988; Keller and Tarara, 2010).

Like the buds, the cambium of canes, cordons, and trunks also seems to experience similar periods of endodormancy and ecodormancy and is reactivated in spring when the cambial cells resume cell division (Begum et al., 2007). While growing apical and cambial meristems may be regarded as being in "online" (active) mode, endodormant meristems are in "offline" (dormant) mode, and ecodormant ones in "standby" (quiescent) mode (van der Schoot and Rinne, 2011).

2.2.6 Root growth

Dormancy is unknown to roots: low temperatures simply suppress root growth temporarily and reversibly (van der Schoot and Rinne, 2011). The temperature strongly and rapidly modulates root growth, in some cases even more so than it influences shoot growth. A minimum soil temperature of 6°C seems to be necessary for roots to grow, and a 10°C increase in temperature can double or triple the rate of root extension (Richards, 1983; Skene and Kerridge, 1967; Fig. 2.9). The optimum for root growth is close to 30°C, but temperatures above that threshold can kill at least the fine roots within days (Huang et al., 2005b). This can pose problems for grapevines grown in pots, where the soil temperature during the day often exceeds the ambient temperature by several degrees. By contrast, the soil temperature in a vineyard usually fluctuates much less than the aboveground temperature due to the large heat capacity of wet soils, but heavy rainfall or application of cold irrigation water can lead to short-term reductions in soil temperature. Moreover, dry and sandy soils are not as well buffered against temperature fluctuations as are wet and loamy soils.

FIG. 2.9

Riesling roots grown at 15°C (left) or 25°C (center), and Cabernet Sauvignon roots growing down fractures in limestone bedrock (right).

Center: Reproduced with permission from Erlenwein, 1965b; right: photo by M. Keller.

Root growth, especially the formation of lateral roots, is stimulated by auxin that has been delivered from its main place of production in the shoot tips and unfolding leaves to the root tips in the phloem and parenchyma and by gibberellins produced in the root tips (Friml, 2003; Kramer and Bennett, 2006). Changes in auxin flow usually lead to changes in root growth, and interrupting the auxin flow will inhibit root growth completely. Unlike lateral shoots, which are produced by the apical meristem, lateral roots grow from the pericycle following the initiation of renewed cell division by auxin transport from the root tip toward the base in the cortex and epidermis. Following initiation by such basipetal auxin flow, further division of the "founder cells" leads to the formation of a lateral root primordium, which pushes radially out of the parent root via cell expansion (see Fig. 1.5). This cell expansion, too, is stimulated by auxin; it can be repressed by ABA, for example when the soil is too dry for root growth. After emergence, the apical meristem is activated and the new lateral root begins to grow autonomously.

Whereas dormant buds seem to inhibit adventitious root formation on hardwood cuttings, the export of auxin from swelling buds stimulates rooting (Smart et al., 2003). Consequently, disbudded cuttings are often unable or struggle to form roots, unless they are dipped in an auxin solution. Indeed, on single-node cuttings, most new roots form directly beneath the bud (Huglin and Schneider, 1998). Cytokinin and strigolactone contrast with auxin by suppressing adventitious rooting; strigolactone might exert its inhibitory role by reducing auxin flow (Rasmussen et al., 2012). Since cytokinin and strigolactone are typical root signals, their absence in cuttings might induce the cambium to replace the missing roots by initiating new ones.

Similar to shoots, roots also display a kind of apical dominance, whereby the growing tip of a root inhibits the outgrowth of lateral roots from its pericycle. Unlike the shoots, however, the roots use cytokinin rather than auxin to inhibit lateral root initiation too close to the root tip (Aloni et al., 2006b). This enables the root to continue growing down or across the soil profile in search of water and mineral nutrients. If a root fails to detect nutrients such as nitrate or phosphate, its tip ceases to produce cytokinin so that lateral roots can proliferate to explore new territory in the upper soil layers (see Section 7.3). The patchy availability of water and nutrients in the soil therefore forces the roots to move by growing and leads to root systems of varying architecture, length, and density.

Although root tips can extend at rates of several centimeters per day, root growth generally lags behind shoot growth, peaks around the time of bloom and early fruit development, and tapers off during fruit ripening (Reimers et al., 1994). Because early canopy development in spring, such as occurs with lightly or minimally pruned grapevines, promotes early root growth, the growth of roots on lightly pruned vines peaks earlier than that of heavily pruned vines (Comas et al., 2005). The supply of sugars from the leaves is evidently critical to sustain root growth, and loss of leaf area inhibits root growth far more than it does shoot or fruit growth (Buttrose, 1966a). Most new roots are produced while the shoots are growing vigorously, but root growth can continue as long as water and nutrient availability are not overly limiting and the soil temperature remains high enough (Comas et al., 2010; Schreiner, 2005; Williams and Biscay, 1991). In warm climates, or in warm growing seasons in cooler climates, which permit continued sugar production by the leaves, there may be a second flush of new root growth around or just after harvest (Conradie, 1980; Lehnart et al., 2008; McKenry, 1984; Mohr, 1996). Where temperatures are conducive, roots may also grow in the winter after leaf fall; such growth is presumably sustained by internal nutrient reserves (Bauerle et al., 2008b; Schreiner, 2005). Moreover, it appears that drought-susceptible rootstocks (e.g., 101-14 Mgt) tend to produce more new roots in the winter and spring, whereas drought-tolerant rootstocks (e.g., 1103P) grow mostly during the summer by expanding into deeper, moister soil layers (Alsina et al., 2011).

Grapevines, like other plants, are stationary; unlike animals, they cannot move from place to place to find better food sources. Because uptake of water and mineral nutrients from the soil quickly depletes readily available resources in the roots' vicinity, the roots must keep elongating throughout the growing season and throughout the vine's life to maintain the supply of these raw materials. This leads to the formation of a branched root system that is elaborated with root hairs and symbiotic mycorrhiza (Gebbing et al., 1977; Possingham and Groot Obbink, 1971; Schreiner, 2005; Stahl, 1900). Root density increases as a vine grows older but appears to settle and remain more or less constant after the vine reaches 5–10 years of age. It is estimated that vines allocate between 30% and 60% of total net photosynthesis products (see Section 5.1) to the roots under favorable soil conditions, and this portion is even greater in poor soils or during drought (Anderson et al., 2003). The total root biomass increases from growth of permanent roots in length and diameter throughout the growing season in both young and mature grapevines (Araujo and Williams, 1988; Williams and Biscay, 1991). However, most fine roots, which are <1 mm in diameter and are the vine's major water- and nutrient-absorbing structures, are short-lived, so they have to be replaced continuously, probably because the soil around them quickly becomes exhausted (Comas et al., 2000; Comas et al., 2010; Viala and Vermorel, 1901–1909). Only a few of them live longer and, through secondary growth, may eventually become large, woody, structural roots. These structural roots, however, are extremely long lived and may stay alive in the soil for years even after a vineyard has been removed. In wet soils the woody roots can take up water across their suberized bark (Cuneo et al., 2018). Although this is a slow process, it may enable grapevines to absorb soil water in early spring, before new fine roots are produced, or during heavy rainfall or irrigation events following a drought period that may have damaged the fine roots.

New roots are initially white, but they turn brown after approximately 5 weeks and black after an additional 3–6 weeks. The onset of browning generally indicates cessation of root metabolic activity and thus death of the root as a "functional" organ in terms of water and nutrient absorption and as a source for new lateral roots (Comas et al., 2000). Black roots shrivel and eventually disappear altogether. Fine roots produced before bloom have an extremely short life span compared with those produced later in the growing season (Anderson et al., 2003). Fine roots also die sooner in dry soil and at shallower soil depth, and so do roots of heavily pruned vines supporting only few shoots. Such an increase in root mortality can be considered a response to herbivory leading to carbon shortage, whereby plants compensate for the loss of shoot biomass and attempt to restore their root÷shoot ratio to a balance that is more favorable for growth (Bloom et al., 1985). Therefore, mechanical pruning, particularly minimal pruning, tends to result in more and longer-lived fine roots than spur or cane pruning. Even though they generally support a heavier crop, lightly pruned vines balance their larger canopy by producing more fine roots than heavily pruned vines. In addition, infection with mycorrhizal fungi greatly prolongs the life span of fine roots, whereas the supply of nitrogen to the soil shortens it.

Soil is a very porous growth medium: about 40–50% of the volume of a typical soil consists of variably-sized pores containing water and air. To explore the soil, roots ordinarily follow the path of least resistance. They predominantly grow along natural fracture lines of the soil and reinvade macropores whose diameter is much larger than the size of the soil particles. These large pores, which are also called biopores, are created by earthworms or old root debris forming microniches of high organic matter (McKenry, 1984; Passioura, 2002; Smart et al., 2006). Decaying dead roots additionally supply nutrients to the growing new roots, provided these mineralized nutrients are not leached below the rooting zone by excessive rainfall or irrigation. Roots that penetrate smaller soil pores, whose size is mostly determined by the size distribution of the soil particles, must use turgor pressure to exert a growth pressure that is high enough to deform the surrounding soil (Jin et al., 2013). Not surprisingly, therefore, soil compaction, which leads to a decrease in the soil's pore volume, reduces root growth. Pressurizing the soil by 1 bar due to compaction may decrease root elongation by 90% (Chapman et al., 2012). The resistance compaction imposes on root growth increases as the soil dries down and is associated with increased ethylene and ABA production by the root (Bengough et al., 2011; Jin et al., 2013). Although a reduction in root growth normally also slows shoot growth, localized soil compaction, such as that caused by heavy machinery or plow pans, reduces shoot growth only when it decreases the total length of the root system (Montagu et al., 2001). When a single root grows through loose soil into compact subsoil, compensatory growth in the loose soil maintains total root length and thus shoot growth remains unaffected. Such plasticity in root system development in response to heterogeneous soil conditions enables grapevines to acquire soil resources with a conservative investment in root biomass.

The general direction of root growth is downward, guided by a positive gravitropic response. As in the shoot, the "gravity sensors" are starch-filled statoliths that, in this case, settle at the bottom of the root cap's columella cells (Kutschera and Briggs, 2012; Morita and Tasaka, 2004). When rocks or other obstacles force a root is away from vertical, the resettling of the statoliths triggers a redistribution of calcium and movement of auxin to the new lower side of the root tip and back to the elongation zone via the epidermis, so that the cells on that side elongate more than on the upper side, navigating the root around the obstacle (Kramer and Bennett, 2006; Muday et al., 2012; Petrášek and Friml, 2009). The gravity-sensing columella cells are also the ones that produce most of the plant's cell division hormone, cytokinin, which is then distributed throughout the vine via the transpiration stream in the xylem, mainly to rapidly transpiring organs (Aloni et al., 2005). Depending on soil conditions, the roots are usually concentrated in the nutrient- and oxygen-rich surface soil, typically in the top 50 cm, with some roots penetrating to several meters of depth (Celette et al., 2008; Huglin and Schneider, 1998; Morlat and Jacquet, 1993; Schreiner, 2005; Smart et al., 2006). Vineyard floor vegetation, such as cover crops or permanent swards, tends to shift the maximum root density of grapevines downward to 50–100 cm, probably because of competition for water and nutrients (Lehnart et al., 2008; Reimers et al., 1994). Somewhat paradoxically, however, the average rooting depth may be much greater in shallow soils than in deep soils, because the roots in shallow soils can grow down cracks in the underlying bedrock, following the path of water infiltration. Roots can even grow for several days in air, for example, across large cracks, gaps, or macropores. They pull off this feat by relying on water supplied to the growing tip internally by the xylem, the phloem, and/or by cell-to-cell movement (Boyer et al., 2010).

Root growth and distribution are influenced by species, cultivar, and rootstock. Moreover, grafted scion cultivars also alter the root growth of different rootstocks (Erlenwein, 1965a). Because grapevines strive to maintain an equilibrium in their root ÷ shoot ratio, such genetic influences and interactions may be partly responsible for the variation in shoot vigor observed with different scion/rootstock combinations. In fact, partitioning of carbon to the roots is generally proportional to that to the canopy, and total leaf mass roughly scales to the power of ¾ with total root mass (Enquist and Niklas, 2002; Price et al., 2010). Therefore, total root length normally correlates closely with leaf area, and root dry weight correlates with aboveground dry weight in grapevines (Petrie et al., 2000b). However, whereas root density may be dependent on the genotype, the vertical and horizontal distribution of the root system is dictated by soil properties (Mullins et al., 1992; Smart et al., 2006). Root distribution of various rootstocks is mostly a function of soil texture, composition, and water availability, rather than an inherent trait of the rootstock genotype (Smart et al., 2006). Even genetically identical clonal plants can develop vastly different root systems in terms of both size and architecture, depending on microscale variations in soil conditions that the roots encounter during growth. Such differences result from changes in the number, growth rate, distribution, and orientation of lateral roots. For example, whereas mild water deficit has little effect on lateral root growth, more severe deficit suppresses root growth, although not as strongly as it inhibits shoot growth (see Section 7.2). Under water deficit, moreover, lateral roots tend to grow more vertically than with abundant water availability, whereas the opposite effect occurs under phosphorus deficiency (Rellán-Álvarez et al., 2016). In contrast to its effect on shoot growth, decreasing nutrient availability often increases root growth (see Section 7.3).

Roots tend to search out moist soil regions; they grow toward water or down a moisture gradient in a process called hydrotropism, and this ability is so strong that it can override the roots' gravity response (Eapen et al., 2005; Pisciotta et al., 2018). How the roots "sense" the presence of water and how they integrate the resulting signal with the gravity signal remains poorly understood. It seems that, unlike gravity perception, the perception for hydrotropism occurs in the root's elongation zone and involves ABA; the cortex elongates preferentially on the dry side, so that the root tip bends toward the wet side (Dietrich, 2018). Hydrotropism may partly result from the roots' propensity to grow away from regions of high osmotic pressure (Takahashi et al., 2003). Root growth clearly responds to the presence of water tables, the amount and frequency of water supply from rainfall or irrigation, and the type of irrigation system. Grapevines can tolerate considerable drought due to their ability to selectively grow roots in soil patches with available water (Bauerle et al., 2008b). This is why the roots of drip-irrigated vines are often concentrated beneath the drip emitters, especially in sandy soils and dry climates. In rain-fed vineyards or those irrigated with other methods, such as flooding, furrows, or overhead or microjet sprinklers, the root system is typically much more widespread (Pisciotta et al., 2018; Stevens and Douglas, 1994), unless root growth in the midrows is prevented by soil compaction from heavy machinery. Roots of mature vines also change their growth pattern accordingly following conversion from flood or sprinkler irrigation to drip irrigation (Bowen et al., 2012; Soar and Loveys, 2007). Similarly, permanent inter-row cover crops tend to lead to a concentration of vine roots in the bare, herbicide-treated vine rows (Celette et al., 2008; Morlat and Jacquet, 2003). The roots' ability to find water even below the bedrock underlying the "useful" soil horizons can also greatly expand the apparent rootzone or rooting depth as estimated from soil pits (Fig. 2.9). Consequently, grapevines planted on apparently shallow soil are not always less vigorous than those planted on deeper soil, and their growth is not necessarily easier to control using deficit irrigation strategies. Restricting the rooting volume in pots, however, does decrease both root and shoot growth (Poorter et al., 2012; Wang et al., 2001). Roots often become "pot-bound," and as pot size decreases, so does the relative growth rate and overall plant growth (Poorter et al., 2012). These are important yet frequently overlooked considerations in pot experiments and are one reason the results sometimes do not adequately reflect field conditions (e.g., Romero et al., 2017).

2.3 Reproductive cycle
2.3.1 Inflorescence and flower formation

Grapevines grown from seeds, like other woody perennials, have a vegetative juvenile phase of 2–4 years before they reach their reproductive phase and become able to produce fruit. It seems that the degree of expression of the gene that controls the transition to flowering increases gradually during the juvenile phase until it exceeds a threshold that permits the formation of inflorescences (Böhlenius et al., 2006). This mechanism ensures that the vine is large enough and able to produce ample resources to support fruit production. By contrast, vines grown from cuttings can fruit in their first growing season under optimal conditions. Another trait typical of perennial woody species is that they require two growing seasons for flower and fruit production: Buds formed in the first year give rise to shoots carrying fruit in the second year.

Reproductive growth is very similar in the different *Vitis* species and begins with flower formation, which can be divided into three separate processes: inflorescence initiation, flower initiation, and flower differentiation. The first step, which is also called induction, involves the formation of lateral meristems as uncommitted primordia, or initials, by the apical meristem inside the shoot's compound buds from spring through early summer (Alleweldt and Ilter, 1969; Carmona et al., 2008; Gerrath, 1992; Gerrath and Posluszny, 1988b; Meneghetti et al., 2006; Morrison, 1991; Pratt, 1971; Srinivasan and Mullins, 1981). Uncommitted primordia are first formed in the buds at the shoot base, the so-called basal buds, and thereafter they also appear progressively in the buds toward the shoot tip. These primordia are meristems that can differentiate into inflorescence primordia, tendril primordia, or sometimes even shoot primordia (see Fig. 1.16). Initiation of uncommitted primordia occurs from budbreak to bloom, usually after 4–5 leaf primordia have been initiated in a bud, and subsequent inflorescence differentiation begins around bloom of the current season's inflorescences and may continue until the buds enter dormancy (Alleweldt and Ilter, 1969; Morrison, 1991; Pratt, 1979; Snyder, 1933; Swanepoel and Archer, 1988). In other words, inflorescences are formed during the year that precedes the flowering and fruiting year, and the fate of the following year's primordia is determined as early as bloom time of the current year.

The number of inflorescences is lowest in the basal buds and increases with higher bud position before gradually declining again beyond bud positions 10–12 (Sartorius, 1968; Huglin and Schneider, 1998; Eltom et al., 2014). Yet in some cultivars only the basal six to eight buds of a shoot, which had been formed in the previous growing season, are able to form inflorescences; the primordia initiated at younger node positions, which are produced by the actively growing shoot, can only develop tendrils (Morrison, 1991; Sánchez and Dokoozlian, 2005). Thompson Seedless is an exception: Apart from the basal buds, fruitful buds are initiated over the entire length of its shoots (Sánchez and Dokoozlian, 2005). The origin of the shoot is unimportant for the fruitfulness of its buds; shoots arising from 1-year-old spurs or canes produce equally fruitful buds as those growing from older cordons or trunks (Müller-Thurgau, 1883b).

The maximum number of future inflorescences per bud, the so-called bud fruitfulness, appears to be determined by approximately 3 months after budbreak—that is, before veraison (Alleweldt and Ilter, 1969). Depending on species and cultivar, several inflorescence primordia may be formed in the primary buds of *Vitis* species. Most *V. vinifera* cultivars typically initiate two such primordia, but for some, such as Sémillon, Muscadelle, Gamais, or Carignan, three is more common, whereas the fruitful Gouais blanc forms four, *V. labruscana* cultivars form three or four, and *V. riparia* and *V. rupestris* buds form seven or eight inflorescence primordia (Viala and Vermorel, 1901–1909). Bud fruitfulness is linked to shoot vigor. Weak shoots produce buds of lower fruitfulness than do more vigorous shoots (Huglin and Schneider, 1998; Eltom et al., 2014). The degree of bud fruitfulness may be associated with auxin production; more auxin seemingly leads to more inflorescences per shoot (Costantini et al., 2007). Moreover, environmental conditions strongly modulate the number of primordia. Warm, sunny conditions combined with adequate soil water and nutrient availability, along with a sufficiently large, actively photosynthesizing leaf area, are crucial to maximize the number of primordia (see Section 6.1). In addition, the growth direction of the shoot also appears to influence bud fruitfulness, at least in some cultivars: Upright-growing shoots produce buds with more inflorescence primordia than horizontally or downward-growing shoots (Alleweldt and Ilter, 1969).

The formation of inflorescence meristems inside compound buds and the formation of flower meristems are seasonally separated (Carmona et al., 2002, 2008; Considine and Knox, 1979a; Gerrath, 1992; Gerrath and Posluszny, 1988b; May, 2004; Meneghetti et al., 2006; Morrison, 1991; Mullins et al., 1992; Scholefield and Ward, 1975; Snyder, 1933; Srinivasan and Mullins, 1981; Fig. 2.10). Inflorescence primordia grow rapidly during the summer, producing several branch meristems before the buds enter dormancy in late summer. Further inflorescence development is arrested with the onset of dormancy and does not resume until the buds are reactivated in spring. During the period of budswell prior to budbreak, the inflorescence branch meristems produce additional meristems that give rise to clusters of 3–4 flower meristems, called floral primordia. This process is termed flower initiation, and within each of these clusters the floral primordia are initiated "from the top down"; that is, the terminal flower is initiated first and the basal flower last. Flower initiation during the budswell phase happens simultaneously in all parts of the inflorescence primordium and culminates in the formation of a calyx for each flower-to-be. Branching and flower initiation ostensibly cease once a shoot starts to grow out of the bud. The subsequent flower differentiation, or floral organogenesis, which refers to the development of the individual organs of each flower, occurs over approximately 5 weeks during and after budbreak as the inflorescences become visible on the shoot and then separate. The corolla and stamens appear successively under the stimulating influence of gibberellin produced in and released by the ovaries (Binenbaum et al., 2018; Cheng et al., 2004a). The pistil with its reproductive structures does not develop until the individual flowers become visibly separated on the inflorescence. Separate genes or

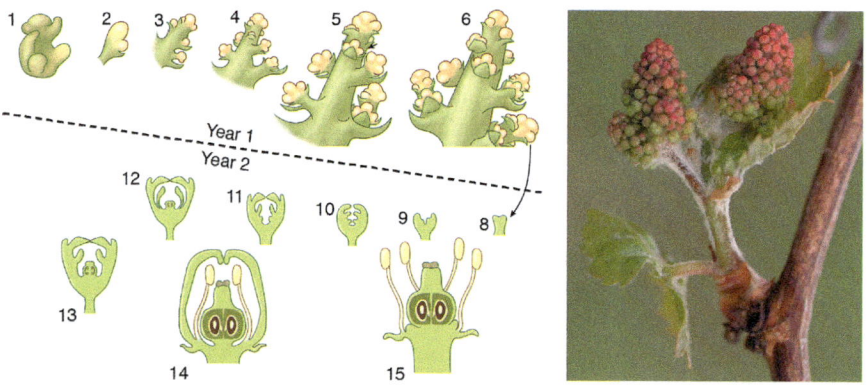

FIG. 2.10

Cluster initiation (1) and differentiation (2–6: numbers indicate increasing orders and positions of branching) and flower development (8: initiation of sepal lobes; 9: initiation of petal lobes; 10: initiation of stamen primordia; 11: initiation of carpel primordia; 12: appearance of locules; 13: ovule initiation; 14: pollen formation; 15: anthesis) of Concord grapevines (left); inflorescences of the rootstock 3309C emerging after budbreak (right).

Left: © Elsevier Inc., illustration after M. Goffinet; right: photo by M. Keller.

groups of genes specify the sequence and timing of flower differentiation and the identity of the individual floral organs, although the mechanism by which they exert their influence is only beginning to be elucidated (Díaz-Riquelme et al., 2009; Fernandez et al., 2010). Intriguingly, the genes that specify the identity of each floral organ are activated only *after* the appearance of the organ primorida (Jenik and Irish, 2000).

A dissenting minority opinion of this developmental process differs mainly in the timing of flower initiation (Agaoglu, 1971; Alleweldt, 1966; Alleweldt and Ilter, 1969; Sartorius, 1968). This view holds that a portion of the calyxes are formed before bud dormancy, some during winter dormancy, and the remainder during the subsequent budswell and budbreak period. It also states that the corolla and stamen appear *before* budbreak, followed by the pistil, whose formation is complete within 10–15 days of the emergence of the inflorescence after budbreak. Evidence supporting both of these views comes from both warm and cool climates, so it is not clear what might cause the disparity in observations. Moreover, inflorescences that arise on lateral shoots, especially on those laterals on the free-growth portion of a shoot that could not possibly have been initiated in the previous growing season, obviously run through their entire developmental program from inflorescence initiation to fruiting within the same growing season.

The extent of branching before and after the dormancy period and the degree to which the individual flowers develop in spring determine the number of flowers that form on each inflorescence. The flower number is highly variable, even on the same shoot: The basal inflorescence of a shoot typically produces the most flowers, and numbers decline with increasing height of insertion on the shoot (Huglin and Balthazard, 1975; Huglin and Schneider, 1998; May, 2004). Branching of the primordia is induced and controlled by two groups of plant hormones, namely auxins and cytokinins. The production of cytokinins, which occurs predominantly in the root tips, and their transport to the shoots in the xylem are regulated by developmental and environmental signals. Auxin is produced and released by the floral meristems at the tip of each floral organ and controls both formation and differentiation of the flowers; it also induces development of their vascular tissues. The anthers produce and release particularly high amounts of auxin, as well as gibberellin, and thereby synchronize the development of the other flower parts so that they become functional just before bloom (Aloni et al., 2006a; Olszewski et al., 2002; Yamaguchi, 2008). Gibberellins help maintain the inflorescence meristems and favor inflorescence elongation (Alleweldt, 1959). Table grape growers sometimes apply gibberellin sprays before bloom to lengthen the rachis and reduce cluster compactness (Mullins et al., 1992). Gibberellin application *during* bloom may lead to lower fruit set, which further promotes loose clusters. However, gibberellins act antagonistically to cytokinins and favor tendril formation by inhibiting the branching process so that bud fruitfulness may decrease (Weaver, 1960; Weyand and Schultz, 2006c). This inhibitory role appears to be somewhat unique to grapes; gibberellins are involved in establishing floral primordia in many other plants.

2.3.2 Pollination and fertilization

Approximately 2 weeks after the ovules have been formed and 5–10 weeks after budbreak, anthesis marks the beginning of bloom, exposing the male and female floral organs (Figs. 2.11 and 2.12). By this time the shoots of *V. vinifera* cultivars have grown to variable lengths depending on vigor but usually have 15–20 unfolded leaves, whereas *V. labruscana* cultivars have fewer than 15 leaves (Pratt and Coombe, 1978). Anthesis theoretically refers to the release of pollen but is commonly

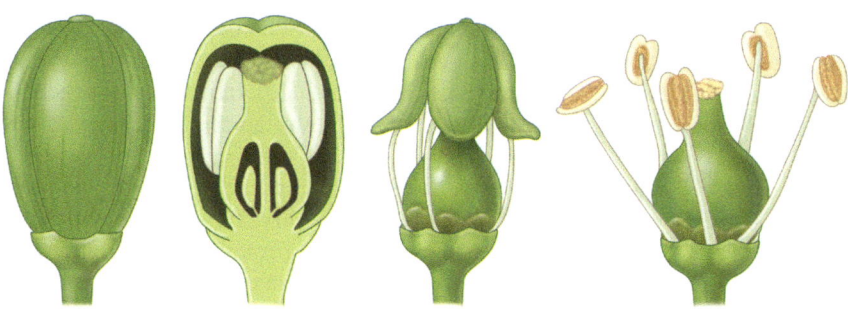

FIG. 2.11

Grape flowers immediately before anthesis (left, with cross section center left), during anthesis (center right), and immediately after anthesis (right).

FIG. 2.12

Chardonnay inflorescence at the beginning of bloom (left) and in full bloom (center), and close-up of a group of flowers (right).

Photos by M. Keller.

regarded as the opening of a flower, which occurs by shedding of the calyptra, or capfall. Like budbreak, anthesis generally begins in the shoots growing from distal buds and progresses basipetally toward the trunk. Within an individual inflorescence anthesis often starts close to the base, and all flowers normally open within 5–7 days. However, capfall can be delayed during cold and rainy conditions. Few flowers open below 15 °C, and the optimum temperature is in the range of 20–25 °C; the rate of capfall slows greatly at 35 °C (Galet, 2000; May, 2004). Each day, anthesis of previously unopened flowers begins around dawn and ceases close to noon (Staudt, 1999). Different inflorescences on the same vine, or on different vines in the same vineyard, follow the same rhythm but may start it on different days so that the entire bloom period in a vineyard may last 2 or 3 weeks. Capfall occurs when the petals detach from the receptacle and are lifted off by the anthers. It is thought that it is the rapid elongation of the anther filaments that pushes the calyptra away from its base once the basal cells have become detached, which is a rare mechanism among angiosperms. Occasionally, the petals instead open at the top and are spread open and pushed downward by the anthers. Such flowers are termed star flowers and can appear in a range of cultivars (Longbottom et al., 2008; Pratt, 1971).

When the anthers burst open, they release their pollen into the air. The production of pollen is stimulated by cytokinin, and pollen release may be facilitated by deliberate weakening of the cell walls.

The latter is achieved by removal of calcium from the cell walls and its sequestration in calcium oxalate crystals (He et al., 2014). The transfer of pollen grains, which are also called male gametophytes or microgametophytes, from the anthers to the stigma is termed pollination and is followed by abscission of the stamens. During anthesis, the stigma releases a sticky sap that retains and rehydrates the pollen for pollination (Considine and Knox, 1979a; Huglin and Schneider, 1998; Meneghetti et al., 2006). This sap is supplied via the newly formed stylar xylem that has previously been induced by auxin released by the stigma's papillae. Because the tiny, dry pollen grains are highly exposed during this process, the pollen wall contains high concentrations of phenolic compounds called flavonols that act as "sunscreen" (Downey et al., 2003b; see also Section 5.2). The flavonols protect the genetic information encoded in the DNA contained in the chromosomes from destruction by UV radiation (Caldwell et al., 1998). It is conceivable, moreover, that the calcium oxalate crystals that are transferred to the stigma along with the pollen grains help scatter excess light. Nevertheless, excessive UVB radiation, such as that resulting from sudden exposure to sunlight of previously shaded inflorescences, especially in conjunction with heat (>33°C), can sometimes decrease pollen germination or pollen tube growth (Torabinejad et al., 1998). In other words, high-light stress reduces pollen viability and leads to male sterility.

Unlike their wild relatives and the few cultivars with female flowers, cultivated grapevines are typically self-pollinated, whereby pollen originates from the flower's own anthers (Mullins et al., 1992; Sartorius, 1926). Wind appears to be responsible for only occasional cross-pollination, where pollen originates from a flower on a different plant than the pollinated one (Scherz, 1939). In fact, in a process termed cleistogamy the anthers often burst 1–24h *before* the calyptra is shed and release their pollen on the flower's own stigma (Heazlewood and Wilson, 2004; May, 2004; Staudt, 1999). Despite this self-pollination strategy, the pollen releases sesquiterpene volatiles that were previously accumulated inside the pollen grains (Martin et al., 2009). In addition, the osmophors on the ovary may serve to attract insects by emitting monoterpenes in the same way that citrus and lavender flowers do, although grape flowers do not offer the insects any nectar (Bönisch et al., 2014a; Brantjes, 1978; Gerrath et al., 2015). It is possible that the release of volatiles might be a defense strategy to deter herbivorous insects or fungi that would feed on the highly vulnerable flowers, or to attract enemies of these herbivores (Gang, 2005). Insect pollination may be much more common and is perhaps required for commercially acceptable fruit set in *Muscadinia* than in *Vitis* (Olien, 1990; Sampson et al., 2001). Even in the latter, however, berries originating from cross-pollinated flowers apparently grow larger than those from self-pollinated flowers, which suggests that cross-pollination improves seed formation.

Regardless of the mode of pollination, the pollen grains absorb water from the moist stigma to rehydrate over a period of approximately 30min, germinate on the stigma whose surface cell layer subsequently suberizes, and form a pollen tube (Considine and Knox, 1979a). The optimum temperature for pollen germination is 25–30°C, whereas temperatures below 10°C and above 35°C inhibit germination. The pollen tube carries two small, fused sperm cells, or male gametes (Greek *gamete* = spouse), surrounded by a large vegetative cell. It penetrates the stigma and grows down through the intercellular spaces of the style like a hydraulic drill, delivering its package of chromosomes into the egg inside the ovule (Zonia and Munnik, 2007). The male and female gametes meet when a pollen tube enters the embryo sac and releases the sperm. The fusion of these sex cells in the ovule is termed fertilization and results in a zygote, which develops into the embryo and leads to seed set and fruit set. Seed set refers to the conversion of an ovule into a seed, and fruit set refers to the conversion of an ovary into a fruit.

Both pollen germination and pollen tube growth are induced by auxin, with which the pollen was "loaded" in the anthers, and stimulated by gibberellin, which is produced in the anthers, perhaps by the pollen itself (Giacomelli et al., 2013; Gillaspy et al., 1993). In a typical grape flower, the pollen tubes cover a distance of approximately 2 mm, which corresponds to a >1000-fold elongation of the pollen cell and is very rapid; rates of $0.3\,\mathrm{mm\,h^{-1}}$ are not uncommon (Staudt, 1982). Not surprisingly, this process consumes a great deal of energy. Growing pollen tubes therefore have high respiration rates fueled by the breakdown of starch reserves that were accumulated in the pollen grains from imported sucrose prior to anthesis. The growth of a pollen tube, moreover, is controlled by passive influx at its tip of calcium ions (Ca^{2+}), which is coupled to efflux of anions, such as chloride (Cl^-), nitrate (NO_3^-), and malate (Gutermuth et al., 2018). Rapid Ca^{2+} influx and anion efflux at the tip enable rapid growth. Ion exchange in the opposite direction, that is, active Ca^{2+} efflux and anion influx, occurs behind the tip and maintains the ion gradients toward the tip.

The speed at which the pollen tubes grow toward the ovules is critical for seed and fruit set because the ovules are receptive for only a limited period after anthesis, after which they can no longer be fertilized. Although only one pollen tube from the myriads of pollen grains landing on the stigma needs to reach one of up to four receptive ovules to make fertilization likely, this outcome nevertheless seems to depend on the pollen density present on the stigma. High pollen numbers may promote fertilization, because the tubes from different pollen grains grow at different rates, and higher numbers mean that there may be more fast-growing pollen tubes than when the pollen density is low. In addition, higher pollen numbers are associated with higher rates of pollen germination. Consequently, fertilization, seed number, and fruit set decline if too few pollen grains land on a stigma (Zhang et al., 2010a). The rate of pollen tube growth also depends on the temperature; at 25–30°C, fertilization can occur within 12 h, at 20°C after 24 h, and at 15°C after 48 h, whereas even lower temperatures do not permit fertilization because the pollen tubes stop growing before they reach the ovules or simply reach the ovules too late (Staudt, 1982). Eggs that are not fertilized within 3 or 4 days after capfall will degenerate (Callis, 1995; Kassemeyer and Staudt, 1981). Therefore, poor fruit set often results from slow pollen tube growth, which can be due to low pollen "fuel status" as a consequence of environmental stress interfering with starch accumulation prior to pollination (Koblet, 1966).

How the pollen tubes find the ovules is not well understood, but it seems that they are guided down the pistil by a combination of physical and chemical mechanisms. Among other cues, a glycoprotein, which is a protein attached to a sugar molecule, establishes an extracellular sugar gradient, which the pollen tubes follow by breaking away the sugar from the protein (Cheung et al., 2010). The tubes' tips are then attracted by the synergid cells (which die after they have served their purpose) flanking the egg cell and follow gradients of γ-amino butyrate (GABA) and other attractants (Berger et al., 2008; Chevalier et al., 2011; Palanivelu et al., 2003). GABA is chemically an amino acid, but it is not referred to as such because it is not incorporated into proteins. In plant cells, it is synthesized from glutamate, and in the human brain GABA released from nerve cells is the predominant inhibitory neurotransmitter. Once a pollen tube reaches the ovule, it stops growing, and its volume and turgor pressure increase rapidly. This leads to an explosive rupture of the tip, which ejects the sperm cells into the ovule (Staudt, 1982). Only a single pollen tube usually penetrates the ovule; those that lose the race seem to be repelled and may continue growing in search of another virgin ovule (Cheung et al., 2010; Chevalier et al., 2011). The four ovules in a flower do not all reach the same stage of development simultaneously: Some embryo sacs may fail to develop or may not mature in time to be fertilized, others are not "found" by a pollen tube, and yet others are fertilized but are slow to develop

(Kassemeyer and Staudt, 1982; Nitsch et al., 1960; Pratt, 1971). This is why grape berries rarely contain four seeds.

Fertilization in angiosperms is more complex than in other plants or in animals; angiosperms require double fertilization. During the process of double fertilization, one haploid ($1n$, i.e., having one set of chromosomes) male gamete fuses with the haploid female gamete, the egg cell or simply egg, to form the diploid ($2n$, i.e., having two sets of chromosomes—one from the mother via her egg and one from the father via his pollen) zygote. The other male gamete, which is delivered by the same pollen tube, simultaneously fuses with the other female gamete, the diploid central cell of the embryo sac, to give rise to the triplod ($3n$, i.e., having three sets of chromosomes—two from the mother and one from the father) endosperm (Berger et al., 2008; May 2004; Raghavan, 2003). Similar to the placenta in mammals, the endosperm protects the embryo developing from the zygote, controls nutrient transfer from the growing berry, and serves as a nutrient store for the developing embryo and the plant organs that will emerge during seed germination. It seems that double fertilization is an all-or-nothing process: Either both the egg cell and the central cell are fertilized, or neither is fertilized (Kassemeyer and Staudt, 1981). The resulting embryo and endosperm together form the filial (daughter) generation, whereas all other berry tissues, including the seed coat and the nucellus, are part of the maternal (mother) generation or mother plant. Although the contribution of the central cell's maternally derived cytoplasm to the developing endosperm is quite significant (Berger et al., 2006), the endosperm is a terminal tissue; it decays while feeding the developing embryo and makes no genetic input to the new plant that develops from the embryo.

The process of embryo formation is called embryogenesis. The embryo should more accurately be called the zygotic embryo because somatic plant cells, for instance from floral organs, shoot tips, or leaf pieces, can be forced in vitro through tissue culture to form so-called somatic embryos. Somatic embryos can go on to develop into whole plants that, barring somatic mutations, are clones of their mother plant (Franks et al., 2002; Martinelli and Gribaudo, 2001; Mullins and Srinivasan, 1976). Somatic embryogenesis or a related technique called shoot organogenesis, in which the embryo stage is bypassed altogether, are exploited in vegetative propagation. Whereas somatic embryogenesis is mostly limited to breeding programs, shoot organogenesis is also used to help eliminate viruses and bacteria by the so-called micropropagation or meristem culture method, whereby surface-sterilized shoot tips <0.5 mm in length are excised under aseptic conditions and grown into full plants (Barlass et al., 1982; Bass and Vuittenez, 1977). Although this practice may occasionally lead to the expression of juvenile leaf characters in the resulting plant, this phenomenon is normally confined to the lower portion of the new shoots; cuttings taken from the distal portion will give rise to true-to-type plants (Grenan, 1984). However, such micropropagation methods are believed to encourage the activation of normally silent jumping genes (see Section 1.2), which can induce occasional somatic mutations and thus introduce additional genetic variation in some of the tissue-cultured vines, leading to novel clones (Benjak et al., 2008; Carrier et al., 2012).

Contrasting with somatic embryogenesis, and although gametes are themselves derived from somatic cells, each egg cell is genetically distinct from all other egg cells in the same flower and on the same vine, and each sperm cell is genetically distinct from all other sperm cells. These differences arise from genetic recombination in addition to occasional mutations. The recombination is basically a random shuffling of paternal and maternal genes, just like the shuffling of a deck of cards and, depending on the cultivar, occurs approximately 1–3 weeks before anthesis (Lebon et al., 2008) during the production of gametes in a special cell division termed meiosis (Greek *meioun* = to make small).

Just before the cell divides, the chromosomes line up alongside each other and randomly exchange pieces of DNA so that the new chromosomes that end up in the gametes constitute a novel and distinctive mix of genes from the mother and the father. This shuffling is the true essence of sex and thus applies only to genes contained in the cell nucleus—that is, on the chromosomes. The other cell organelles with their own genes, the mitochondria and plastids such as chloroplasts, reproduce asexually by simple division, and the male organelles are moreover excluded from or left behind in the pollen (Strefeler et al., 1992). Only very few of them occasionally "slip" through, and some of their genes may escape destruction and be "absorbed" by the nucleus (Birky, 1995; Sheppard et al., 2008; Timmis et al., 2004). Exclusion of the father's organelle genes is the fundamental reason why there are two, and only two, sexes in all sexually reproducing plants and animals. Building and running an organism as complex as a grapevine (or an animal) requires a great deal of cooperation among the genes. The genes of the plastids and mitochondria have never "learned" how to cooperate, and these organelles would often attack and kill each other when two cells fuse—which, in fact, they do in single-celled algae. Assigning these genes to two separate sexes and prohibiting them from mixing prevents this problem while still permitting the chromosomes to take advantage of the costly sexual recombination.

As a result of the chromosome mixing, every single seed, even within the same berry, will give rise to a genetically unique plant, slightly different from all others, even though it inherited all its plastids and mitochondria from the mother plant. This mixing of genes (more accurately, alleles; i.e., slightly different versions of the same gene, one of which may or may not be dominant over the other) produces new genotypes in each generation. This process, in addition to occasional mutations, which generate the different alleles in the first place, provides the endless genetic variety that under natural conditions allows natural selection to act by "weeding out" those individuals (phenotypes) that are less well adapted to the prevailing environmental conditions (Mayr, 2001). As Charles Darwin (1859) stated, "Natural selection acts solely through the preservation of variations in some way advantageous." Because "environment" also comprises other species, including pathogens such as fungi, bacteria, and viruses, the major benefit of recombination is the decreased risk for a plant's offspring to be wiped out by pathogen attack (see Section 8.1). This is the meaning of sex: By producing genetic variation, a species avoids "putting all its eggs in the same basket" and greatly increases the chance that at least some individuals will resist and thus survive pathogen attack or other environmental stresses. Of course, such genetic variation is possible only in wild grapevines and not in cultivated populations of vegetatively propagated, clonal plants. Nevertheless, even the various cultivars originally arose this way, whether by deliberate crossing or by natural pollination (see Section 1.2). Recombination accounts for the vast differences among cultivars in terms of growth, yield formation, fruit composition, pest and disease susceptibility, and reaction to environmental influences. Somatic mutations, however, can arise in vegetatively propagated grapevines and may give rise to clonal variation that can be subject to selection by humans, that is, by artificial as opposed to natural selection.

2.3.3 Fruit set

Within days after fertilization, the central cell, which will become the endosperm, begins to divide under the influence of locally produced auxin, whereas the zygote, which will become the embryo, does not start dividing until 2–3 weeks after fertilization (Nitsch et al., 1960; Pratt, 1971; Zhao, 2018). By this time, the cells of the nucellus have reached their maximum size, and the endosperm, whose nuclei have by now become polyploid, begins to accumulate storage reserves such as starch, proteins, and lipids (Cadot et al., 2006; Kassemeyer and Staudt, 1983). The first few divisions of both the central cell and the zygote seem to be mainly under the control of maternal genes, which may give the mother plant some control over which embryos to develop (Raghavan, 2003). The seeds later take charge of their own maturation by switching on the embryo's genes; in other words, there is a change from maternal to filial control of seed development (Gutierrez et al., 2007). These genes initiate and coordinate the import of nutrients and the accumulation of storage reserves in the endosperm and the subsequent desiccation of the seeds.

The fertilized ovule develops into the seed, whose embryo, endosperm, and seed coat start producing and releasing auxin to the developing pericarp tissues, where it stimulates gibberellin synthesis (Giacomelli et al., 2013; O'Neill, 1997; Pandolfini et al., 2007; Pattison et al., 2014; Serrani et al., 2008). In addition, gibberellin may also be produced in and released by the seeds upon stimulation by auxin (Dorcey et al., 2009; Giacomelli et al., 2013). Auxin synthesis might in turn be stimulated by pollen-derived gibberellin; it is necessary for embryo formation and fruit tissue production and continues until the embryo is mature. Therefore, the interaction, also called cross talk, between auxin and gibberellin plays a key role in the ovary's commitment to initiate fruit development and growth (Gillaspy et al., 1993; Ozga and Reinecke, 2003; Ruan et al., 2012). In other words, fruit set is dependent on the interplay between these two hormones. Auxin reactivates cell division, while gibberellin induces cell expansion (Serrani et al., 2007). The two hormones thereby induce the pistil to develop into the fruit and to differentiate an exocarp (skin) and mesocarp (flesh or pulp).

Although pollination, rather than fertilization, triggers the initial stages of ovary development, perhaps because growing pollen tubes produce gibberellin, subsequent fruit formation depends on a supply of seed-derived auxin to the developing fruit tissues (Kassemeyer and Staudt, 1983; O'Neill, 1997; Pattison et al., 2014; Zhang et al., 2010a). Additionally, the movement of auxin from the berry toward the shoot via the pedicel may prevent premature fruit abscission (Pattison et al., 2014). Therefore, the development of grape berries is usually dependent on pollination, fertilization, and development of at least one seed (Pratt, 1971). Ovaries that do not contain at least one fertilized ovule or whose seed development is aborted early on are abscised from the fruit cluster (Nitsch et al., 1960). Arrested ovule development prior to anthesis prohibits fertilization and also results in flower abortion if all four ovules fail to develop (Kassemeyer and Staudt, 1982). Later degeneration of the embryo and endosperm can result in apparently normal-looking but hollow, nonviable seeds termed "floaters" because they float in water, unlike viable seeds, which are thus named "sinkers" (Ebadi et al., 1996).

Nonetheless, some cultivars can produce apparently normal seedless berries. This phenomenon is called stenospermocarpy or parthenocarpy (Greek *parthenos* = virgin), depending on whether or not fertilization takes place. In parthenocarpic fruit development the berries develop without fertilization—though typically not without pollination—and lack seeds completely. In stenospermocarpic fruit development the berries contain at least one fertilized but subsequently aborted seed so that in many cases all that remains is a small, soft, white rudimentary seed called a seed trace (Barritt, 1970; Ledbetter and Ramming, 1989; Olmo, 1936; Pearson, 1932; Roytchev, 1998, 2000; Staudt and Kassemeyer, 1984). Stenospermocarpy is a quantitative trait: Both size and hardness of the seed traces vary over a wide range (Cabezas et al., 2006; Ledbetter and Shonnard, 1991). Grape cultivars with parthenocarpic fruit are often used to produce seedless raisin grapes, whereas most of the so-called seedless table grapes are stenospermocarpic. Significant examples of the former include Black Corinth or White Corinth, whereas the most important example of the latter is Thompson Seedless.

Parthenocarpy may be caused by a mutation that renders its carriers unable to complete meiosis, which results in degeneration of the embryo sac and in pollen sterility (Royo et al., 2016). By contrast, the mutation that causes stenospermocarpy (see Section 1.2) renders the seed coat unable to differentiate, which leads to an autoimmune response mediated by salicylic acid that halts embryo development and induces disintegration of the endosperm and inhibition of seed coat lignification (Pratt, 1971; Royo et al., 2018). Although the endosperm of stenospermocarpic grapes aborts after fertilization, their embryo survives for up to two months and can be rescued and grown into full plants using in vitro techniques, which permits crossing of seedless table and raisin grapes to breed new cultivars (Cain et al., 1983; Emershad and Ramming, 1984; Emershad et al., 1989; Tang et al., 2009).

Because the lack of growth hormone production in seedless fruit limits cell expansion, both parthenocarpy and stenospermocarpy result in small berries. However, the berry size of stenospermocarpic grape cultivars is less limited than that of parthenocarpic cultivars. Table grape and raisin growers routinely apply gibberellin sprays after fruit set to increase berry size with the added benefit of decreased cluster compactness due to elongation of the rachis and some berry drop (Coombe, 1960; Nitsch et al., 1960; Weaver, 1958, 1960; Weaver and Pool, 1971). However, applying gibberellin *before* bloom induces embryo degeneration and seed abortion after fertilization, and may lead to lower fruit set and stenospermocarpy even in otherwise seeded cultivars (Cheng et al., 2013).

The proportion of flowers that develop into berries following anthesis is typically in the range of 20–50% and is inversely related to the number of flowers present on an inflorescence. Fruit set is highly variable among cultivars and is strongly modulated by environmental conditions and rootstocks (Alleweldt and Hofäcker, 1975; Huglin and Balthazard, 1975; Keller et al., 2001a; Schneider and Staudt, 1978). Cultivars that are susceptible to environmental conditions leading to poor fruit set include Merlot, Grenache, and Gewürztraminer, whereas the Pinots, Chardonnay, Riesling, and Sylvaner seem to be much less susceptible. The process of meiosis is particularly sensitive to stress conditions that curtail the supply of carbohydrates to the inflorescences. Optimum conditions for flowering and fruit set are thought to be similar to those governing inflorescence initiation, namely high light intensity, warm temperature, and adequate soil moisture and nutrient availability. This is the likely reason why years in which the shoots carry more clusters than usual often follow years with above-average fruit set. Adverse environmental conditions, insufficient or inefficient leaf area, shade created by dense canopies, or excessively vigorous shoot growth competing with inflorescences for sugar often result in poor fruit set and loose clusters (see Section 6.1). Because, as a perennial species, the grapevine does not depend on producing seeds each year, such "self-thinning" enables it to adjust its reproductive output to available resources without jeopardizing survival of the plant. By culling a portion of its reproductive structures, the vine may enable the remaining seeds to attain adequate size and maturity.

Poor fruit set due to excessive abortion of flowers and ovaries is also termed shatter, shedding, or "coulure" (French *couler* = to leak, to fall off). Loss of such ovaries can occur for up to 4 weeks after anthesis (Kassemeyer and Staudt, 1983; Staudt and Kassrawi, 1973). The abscised ovaries typically do not contain any fertilized ovules, although the vast majority of them have been pollinated (Kassemeyer and Staudt, 1982; Staudt and Kassemeyer, 1984; Pratt, 1973). Normal fruit set despite inadequate seed development in a portion of the fertilized berries, on the other hand, is called "millerandage" and results in clusters having the appearance of "hens and chicks" (Fougère-Rifot et al., 1995; Galet, 2000; May, 2004). The term hens refers to large, normal berries with at least one viable seed, and chicks describes small berries with tiny, degenerated seeds that often lack an endosperm (Ebadi et al., 1996; Staudt and Kassemeyer, 1984). Although the expansion of chicks is arrested within approximately 3 weeks after anthesis due to seed abortion, these berries apparently ripen normally. Coulure results in very loose clusters, and millerandage leads to highly variable seed numbers and berry sizes on the same cluster. By contrast, the term "shot berries" refers to ovaries that fail to develop into fruit; they remain small, green, and hard; and are high in tannins (Fougère-Rifot et al., 1995). The term is unfortunate: Because these ovaries fail to differentiate a discernible mesocarp and exocarp, shot berries are technically not berries at all (Dry et al., 2010b). These ovaries usually contain at least one sterile ovule, resulting from incomplete or failed meiosis, in addition to degenerated fertile but unfertilized ovules. Shot berries might be prevented from abscising by the initial, albeit very limited, growth of the nucellus of the sterile ovule (Kassemeyer and Staudt, 1982).

Variations in the degree of fruit set together with the extent of inflorescence branching and variations in the rate and/or extent of rachis cell expansion during cluster development before and after bloom lead to wide variation in cluster size and architecture, both between species and cultivars and within them. Such differences in morphological features have implications not only for yield formation (see Section 6.1) but also for the clusters' susceptibility to fungal attack and consequently for disease severity at harvest (Molitor et al., 2012). For instance, cultivars with tight clusters, such as Chardonnay, Riesling, and Pinot, have small rachis cells and a short, compact branching pattern. They are much more susceptible to infection by *Botrytis cinerea* than cultivars with loose clusters, such as Cabernet Sauvignon, Syrah, and Thompson Seedless. This is probably because tight clusters have their berries packed into a small volume and hence retain more rain or irrigation water and dry down more slowly than loose clusters (see Section 8.2). In addition, the pressure of berries against each other may lead to berry splitting, and berries touching and rubbing on each other may lose epicuticular wax and hinder uniform fungicide spray coverage. Yet even cultivars with naturally tight clusters can be manipulated using viticultural practices that tend to decrease cluster compactness, although such practices usually come at the cost of decreased yield. Potential strategies include light or no pruning, resulting in large numbers of clusters and, therefore, relatively poor fruit set (see Section 6.1); removing a portion of the leaves before or soon after bloom, which also diminishes fruit set (Candolfi-Vasconcelos and Koblet, 1990; Palliotti et al., 2011; Poni et al., 2006, 2009; Tardaguila et al., 2010); or cutting or pinching off the distal cluster portion during or after bloom, which often leads to compensatory elongation of the remaining rachis (Winkler, 1958; Molitor et al., 2012). By contrast, when the common yield regulation practice of cluster thinning is applied before rather than after bloom, it tends to enhance fruit set and berry size, and thereby increases cluster compactness and *B. cinerea* infection (Smithyman et al., 1998; Winkler, 1958).

2.3.4 Berry and seed development

In seeded grapes, fertilization is followed by a second period of cell division (the first period occurs before bloom and determines the cell number of the ovary), which is probably stimulated by a brief increase in cytokinin in addition to auxin produced in the developing seeds. In addition to the hormonal influence, cell division in both the seeds and the surrounding pericarp is also promoted by sugar delivered from the leaves (Ruan, 2014). The berry cytokinin concentration, it seems, then declines throughout berry development (Chacko et al., 1976). During the first 3 weeks after anthesis, the polar nucleus of the seeds' endosperm divides several times before the endosperm becomes cellular so that endosperm nuclei become polyploid (Kassemeyer and Staudt, 1983). By contrast, the zygote cells giving rise to the embryo only begin to divide approximately 2–4 weeks after anthesis. The rate and duration of cell division in the berry pericarp are controlled by the embryos so that berries containing more seeds will become larger than berries with fewer seeds (Coombe, 1960; Gillaspy et al., 1993). Whereas the cell number successively doubles approximately 17 times before anthesis, only one or two more doublings occur after anthesis (Coombe, 1976). The innermost cells of the mesocarp stop dividing within 2 weeks after anthesis, and the remaining flesh cells stop dividing 1–2 weeks later, whereas the skin cells may continue to divide for up to 5–6 weeks after anthesis (Considine and Knox, 1981; Harris et al., 1968; Nakagawa and Nanjo, 1965; Pratt, 1971; Staudt et al., 1986). Cell division is completed before the berry enters a lag phase of slow or no fresh weight gain (Ojeda et al., 1999, 2001).

Although it is frequently stated that berry growth is divided into a phase of cell division and one of cell expansion, this is not accurate. Growth is always by cell expansion because organisms cannot add fully grown cells like bricks to a wall. Although dividing cells usually remain small and have small vacuoles, cell division is accompanied and followed by cell expansion (Gillaspy et al., 1993; Pratt, 1971). As the cells expand, their cell walls become thinner and the vacuoles come to dominate the cell volume. Therefore, the vast majority of the volume gain of growing grape berries is due to mesocarp cell expansion as a result of the increase in vacuole volume (see Section 3.1), which is stimulated by seed-produced auxin.

The lag phase is followed by another period of cell expansion so that the volume of mesocarp cells may increase >300-fold from anthesis to maturity or 15-fold between fruit set and maturity (Coombe, 1976). At the same time, the berry surface area may increase approximately 400-fold from anthesis to veraison and almost doubles again from veraison to maturity (Considine and Knox, 1981). The weight of mature berries varies from <0.5 g to >10 g among cultivars of *V. vinifera*, and the berry weight of different clones of the same cultivar can diverge by a factor of two or more (Boursiquot et al., 1995; Dai et al., 2011; Fernandez et al., 2006). But even within the same fruit cluster on a cultivated vine, the weight of individual berries may vary more than twofold (Dai et al., 2011).

Cell division seems to be mostly under genetic control, whereas cell expansion is predominantly driven by environmental factors. This means that berries of different cultivars should have different cell numbers, whereas berries of the same cultivar grown in different environments should differ mostly with respect to cell size. Consequently, fruit size can vary due to variations in both cell number and cell size. Although both processes are dependent on sugar delivery from source organs, cell division is comparatively less sensitive to environmental influences than is cell expansion. Viticultural attempts to manipulate berry size therefore will have to aim mostly at cell expansion, whereas breeding efforts should address cell division.

The maximum extent of cell expansion, and hence berry size, is limited by the elastic properties of the skin, which in turn may be related to the thickness and/or stiffness of the exocarp cell walls or the cuticle (Boudaoud, 2010; Matthews et al., 1987; see Section 6.2). In ripe grape berries, cell wall thickness ranges from >1 µm in the exocarp to <0.1 µm in the mesocarp and, if grapes behave like tomatoes, the resistance of the epidermis—especially that of the cuticle—against deformation increases strongly during fruit ripening (Bargel et al., 2006; see Section 6.2). The growth, or increase in fresh weight, of seeded grape berries traces a double-sigmoid pattern (Winkler and Williams, 1935); that is, the growth curve is bent in two directions twice, like two letters S stacked end-to-end, and can be divided into three more or less distinct phases or stages (Fig. 2.13). These growth stages and their hormonal control are described below, whereas the changes in chemical composition during berry development are discussed in Section. 6.2.

FIG. 2.13

Cabernet Sauvignon berry growth over 3 years (left; note 50% anthesis occurred on day 160 in year 1 and on day 150 in years 2 and 3; M. Keller, unpublished data), and beginning of ripening (veraison) in Concord clusters with berries displaying various stages of color change (right).

Right: photo by M. Keller.

Stage I: This phase is characterized by the initial rapid increase in size of the seeds and pericarp. The first part of seed development is termed morphogenesis and consists of embryo formation and endosperm growth, but there is little growth of the embryo (Ebadi et al., 1996; Nitsch et al., 1960). The seeds are green, and the berry is green and hard and accumulates organic acids but little sugar. The growth hormones auxin, cytokinin, and gibberellin produced by the embryos and released into the pericarp reach high concentrations early on and then decrease (Böttcher et al., 2013; Iwahori et al., 1968; Pattison and Catalá, 2012; Pérez et al., 2000; Sakakibara, 2006; Scienza et al., 1978). Auxin and gibberellin are also exported in the phloem to the pedicel and peduncle, where gibberellin induces cell elongation and auxin induces the production and differentiation of vascular bundles to ensure that the transport capacity of the vascular system does not limit growth of the developing berry. In addition, import from the "mother" vine of the germination-inhibiting hormone ABA prevents seed abortion and promotes normal embryo development (Nambara and Marion-Poll, 2005). Consequently, the ABA concentration in the berry is high during early development but declines as the berry expands (Davies and Böttcher, 2009; Owen et al., 2009; Sun et al., 2010). Stage I lasts 6–9 weeks and ends when the epidermis cells stop dividing (Staudt et al., 1986). By this time, the berry has attained approximately half its maximal size and fresh weight, and the size of the berry at the end of Stage I largely determines its final size.

Stage II: This stage is called the lag phase, because pericarp growth slows down to very low rates. The seeds enter their maturation stage just after the embryo genes take over control from the mother vine. The embryo grows rapidly, and by 10–15 days before veraison the seeds reach their final size, maximum fresh weight, and maximum tannin content (Adams, 2006; Niimi and Torikata, 1979; Ojeda et al., 1999; Pratt, 1971; Ristic and Iland, 2005; Viala and Vermorel, 1901–1909). The cell walls of the seeds' middle integument thicken and become lignified, and the outermost cell walls of the outer integument thicken to form a cuticle; the hard seed coat now restricts further seed expansion (Cadot et al., 2006). The concentration of auxin peaks briefly during this phase and then declines sharply (Alleweldt and Hifny, 1972; Alleweldt et al., 1975; Nitsch et al., 1960). The auxin peak is especially pronounced in the seeds, where it may regulate embryo maturation (Gillaspy et al., 1993; Pattison and Catalá, 2012). Because auxin also acts as a ripening inhibitor, it may serve to prevent premature fruit ripening before the seeds are fully developed (Davies and Böttcher, 2009; Davies et al., 1997; Gouthu and Deluc, 2015; Pattison et al., 2014). The later auxin decrease in the pericarp may be caused mostly by its binding to glucose, which renders the hormone inactive (Fortes et al., 2015). At the same time, there is a fleeting increase in ethylene (though far less pronounced than in the so-called climacteric fruits) that is accompanied by heightened sensitivity to ethylene and a striking and persistent rise in brassinosteroids, a class of growth-promoting hormones probably produced in the epidermis (Chervin et al., 2004, 2008; Pilati et al., 2007; Symons et al., 2006). The influx of ABA and its production in the berry itself, perhaps triggered by ethylene, increase toward the end of this phase, suppressing further embryo growth by blocking gibberellin production in the seeds (Davies and Böttcher, 2009; Fortes et al., 2015; Gutierrez et al., 2007; Nambara and Marion-Poll, 2005; Pérez et al., 2000). The turgor pressure of the mesocarp cells declines approximately 10-fold toward the end of this period (Thomas et al., 2006). The lag phase can last between 1 and 6 weeks, depending primarily on the cultivar. Its duration is important in determining the time of fruit maturity; late-ripening cultivars seem to have a long lag phase (Alleweldt and Hifny, 1972; Currle et al., 1983).

Stage III: The third phase of berry development is the ripening period, which lasts 5–10 weeks. Its onset, which typically occurs over a period of 7–10 days within a grape cluster, is termed veraison and signals a fundamental shift from partly photosynthetic to wholly heterotrophic (Greek *trophé* = nutrition) metabolism. Veraison is marked by berry softening, renewed berry growth, and an increase in sugar content, followed by a rapid change in skin color from green to red, purple, or blue in dark-skinned cultivars (Fig. 2.13), and to more or less yellow in many white cultivars. Softening, which is measured as a decrease in berry firmness or increase in deformability, mainly arises from the disassembly of the mesocarp cell walls (Brummell, 2006; Huang and Huang, 2001). The changes in the expression of many genes that bring about fruit ripening are induced by ABA (Gambetta et al., 2010; Koyama et al., 2009). Together ABA, ethylene, and brassinosteroids may also trigger the changes in cell wall chemistry that lead to berry softening and expansion. Mesocarp cell turgor pressure is low ($P < 0.5$ bar) and varies by cultivar but remains positive and relatively constant throughout ripening (Bernstein and Lustig, 1985; Lang and Düring, 1990; Thomas et al., 2006). Unlike in most other species, whose fruit reach their final size before ripening begins (Gillaspy et al., 1993), grape ripening is characterized by a further increase in berry size. The volume increase is initially very rapid but slows progressively toward fruit maturity; berry size may plateau or decrease due to evaporative water loss during later stages of ripening. The extensibility of the skin temporarily increases at veraison to accommodate berry expansion, but the skin later restricts further expansion (Matthews et al., 1987). Epidermal cells are stretched until the berry reaches its maximum size (Staudt et al., 1986). Sugars accumulate in the pericarp while malic acid and chlorophyll are degraded, and red anthocyanin pigments accumulate in the exocarp and in a few cultivars in the mesocarp as well. Anthocyanin production and other changes, such as sugar transporter and invertase activities, may be triggered by the interplay of sugars and ABA (Davies and Böttcher, 2009; Gambetta et al., 2010; Jia et al., 2011; Koyama et al., 2009; Pirie and Mullins, 1976). ABA acts in a positive feedback loop with sugars, whereby sugars stimulate ABA production and ABA promotes sugar accumulation (Jia et al., 2011). The concentration of ABA increases rapidly after veraison, peaks in the seed and berry flesh during seed maturation, then decreases due to degradation during seed desiccation, and is relatively low in mature seeds and berries (Davies and Böttcher, 2009; Inaba et al., 1976; Lund et al., 2008; Owen et al., 2009; Scienza et al., 1978;

Sun et al., 2010). ABA, which now accumulates mainly as a result of glucose-induced production by the embryo, helps coordinate seed maturation and is necessary to induce seed dormancy (Gutierrez et al., 2007; Nambara and Marion-Poll, 2005; Wobus and Weber, 1999). Intriguingly, the content of phloem-mobile cytokinins, but not that of xylem-mobile cytokinins, also increases in the berry flesh during this stage (Böttcher et al., 2013, 2015). It remains to be seen whether this serves to delay premature senescence of the ripening berry tissues. Unlike in climacteric fruits, the "fruit-ripening" hormone ethylene plays only a minor role in grape ripening, and berry respiration declines during this period (Coombe and Hale, 1973; Fortes et al., 2015; Geisler and Radler, 1963; Harris et al., 1971). Embryo growth, but not seed growth, may continue until berry expansion ceases (Nitsch et al., 1960; Pratt, 1971; Staudt et al., 1986). The seeds turn from green to yellow and finally brown due to oxidation of tannins in the parenchyma cells of the outer integument, and the seeds become hard and desiccated (Adams, 2006; Cadot et al., 2006; Huglin and Schneider, 1998; Ristic and Iland, 2005). Storage reserves, such as starch, proteins, lipids, and mineral nutrients accumulate in the endosperm until the embryo becomes dormant (Cadot et al., 2006; Rogiers et al., 2006). Thus, the seeds' fresh weight declines while their dry weight reaches a plateau or continues to increase until the berry attains its maximum size and weight (Pastor del Rio and Kennedy, 2006; Staudt et al., 1986). Seeds that are hard and brown and have attained their maximal dry weight are called "mature" and enter dormancy during which metabolism ceases until the seeds are rehydrated for germination.

The length of each stage of fruit growth and the final berry size depend on the cultivar but are strongly modified by environmental conditions. Optimum conditions for rapid berry development are similar to those governing the other phases of reproductive development: high light intensity, warm temperature, and adequate soil moisture and nutrient availability (see Section 6.1). The rate of development is also modified by the number of seeds in a berry. Increasing seed number may delay the transition from one growth phase to another (Staudt et al., 1986). As an extension of this trend, seedless grape berries often do not show clearly discernible phases of growth, and their volume increases rather gradually (Iwahori et al., 1968; Nitsch et al., 1960; Winkler and Williams, 1935). They also generally remain smaller than the berries of seeded cultivars (Boursiquot et al., 1995; Huglin and Schneider, 1998).

Harvest of grapes by birds or mammals, including humans, normally completes the reproductive cycle. If grapes remain on the vine for an extended period, the continued cell wall disassembly, leading to diminished cell wall strength and loss of intercellular adhesion, eventually results in tissue failure, cell separation, and senescence, which sets the seeds "free"—often aided by infection by opportunistic fungi such as *B. cinerea* and others (Brummell, 2006, Considine and Knox, 1979a).

Seeds are ready to germinate at veraison; in fact, fruit ripening is not initiated until seed maturation has been completed (Cawthon and Morris, 1982; Gillaspy et al., 1993; Pratt, 1971; Winkler and Williams, 1935). Whereas seeds may germinate immediately if postveraison berries drop to the ground in a moist environment, they lose this ability within a day upon removal from a berry, when their water content declines from approximately 50% to 20% (Currle et al., 1983). Even though the endosperm may lose cellular integrity and die during seed desiccation, grapevine seeds can tolerate water loss to <6% water content without losing their ability to germinate (Orrù et al., 2012; Zeeman et al., 2010). Indeed, drying may enhance a seed's subsequent germination capacity under favorable conditions (Wang et al., 2011). However, it is thought that the change in seed shape from almost spherical with an inconspicuous beak to a pear shape with a pronounced beak that accompanied grape domestication may have reduced the ability of seeds to germinate (Bouby et al., 2013; Terral et al., 2010).

Once dehydrated, the seeds require a stratification period of 4–12 weeks at chilling temperatures of 0–10 °C and at least 40–50% relative humidity to overcome embryo dormancy. The term stratification (Latin *stratum* = blanket) itself is derived from the old practice of spreading seeds between layers of moist, cold soil to break dormancy. Following dormancy release, the base temperature for seed germination, at least in *V. sylvestris*, is around 10 °C, and the germination rate increases up to an optimum of at least 25 °C (Orrù et al., 2012). Light—especially red light—induces water imbibition and subsequent germination by stimulating the embryo cells to produce gibberellin and expand rapidly, which overcomes seed dormancy by accelerating ABA degradation, facilitates breakage of the seed coat, and induces remobilization of nutrients from the endosperm (Nambara and Marion-Poll, 2005; Olszewski et al., 2002). Seeds do not germinate when there is insufficient water for imbibition, which ensures that the young seedlings do not succumb to drought after germination. The capacity for seed germination may also be influenced by the nutritional status of the mother plant. For instance, high nitrate content during seed formation in the fruit increases the germination capacity, apparently because nitrate lessens the maintenance of seed dormancy by accelerating ABA breakdown (Alboresi et al., 2005).

The evolution of a fleshy, sweet, and dark-skinned berry during the late Cretaceous period and the subsequent Paleocene epoch about 100 to 56 million years ago was probably an adaptation to the conversion, driven by climate change and the associated demise of herbivorous dinosaurs, from open to closed, wooded habitats and the simultaneous evolution of birds and mammals (Hardie, 2000; Rodríguez et al., 2013; Seymour et al., 2013). Even though the concept of fleshy fruits per se had already been "invented" earlier by flowerless gymnosperms, such as the maidenhair tree *Ginkgo biloba* L., angiosperms may have exploited similar genetic principles to transform their ovaries into fruits (Lovisetto et al., 2012). Birds are the predominant seed dispersers of European wine grapes, whereas bears and other mammals are major dispersers of American juice grapes, whose distinctive aroma birds find unattractive. The only role grape berries play in the life of a vine is to spread its genes (DNA). The ability to advertise seed maturity by visual and olfactory cues and to reward seed dispersers with energy-rich sugar greatly increases the chances for gene dispersal. As Charles Darwin (1859) remarked, a fruit's "beauty serves merely as a guide to birds and beasts in order that the fruit may be devoured and the manured seeds disseminated." Berries accumulate pigments and aroma volatiles to advertise to potential seed dispersers their content of nutritionally valuable components, such as sugars, amino acids, fatty acids, vitamins, and antioxidants (Goff and Klee, 2006; Rodríguez et al., 2013). These compounds, in turn, constitute the "ticket price" to cover the costs of transportation by means of the wings and legs of seed dispersers.

Low concentrations of volatile ethanol produced by yeasts, which have been living on and in fruits since the Cretaceous period, may also form part of the aroma bouquet of ripe berries (Dudley, 2004). Although fruit color is certainly an important attractant during the day, volatiles may be more important at night, when colors are difficult to see. As a rule of thumb, birds rely more on visual cues to detect fruits, whereas mammals tend to rely more on their sense of smell (Rodríguez et al., 2013). At the same time, tannins and other secondary compounds that accumulate in the grape skin and the seed coat may deter microbes that also would like to consume the fruit but do not disperse the seeds (Levey, 2004). Likewise, the high acidity and the unpleasant aroma caused by methoxypyrazines and other volatile compounds of unripe berries (see Section 6.2) prevents the berries from being eaten before the seeds are mature. Birds learn to shun methoxypyrazine (Siddall and Marples, 2011). In vineyards planted to several different grape cultivars, therefore, those cultivars that accumulate the highest amounts of acids and methoxypyrazines in their immature berries are often the last ones to suffer bird damage.

Apparently, birds prefer Chardonnay (low acid) over Riesling (high acid) and grapes from Burgundy (low methoxypyrazine) over those from Bordeaux (high methoxypyrazine).

The seeds themselves are made unpalatable by the accumulation of bitter and astringent tannins (see Section 6.2), which prevents animals from destroying the seeds by chewing them. Moreover, the seeds survive the digestive systems of their dispersers. The clever strategy of packaging unpalatable seeds in an edible berry enables grapevines to explore new habitats beyond the small area around an existing immobile mother plant that simply drops its seeds to the ground. This strategy ensures that the parent plant does not compete for limited resources with its own offspring. Furthermore, fruit ripening differs from the senescence process in leaves. Unlike ripening, senescence involves disassembly of proteins, which is called proteolysis, and recycling of nutrients to other plant parts (Gillaspy et al., 1993).

CHAPTER 3

Water relations and nutrient uptake

Chapter outline

3.1 Osmosis, water potential, and cell expansion 105
3.2 Transpiration and stomatal action 110
3.3 Water and nutrient uptake and transport 114
    3.3.1 Driving forces and resistances 114
    3.3.2 Root water uptake 119
    3.3.3 Symplast and apoplast 120
    3.3.4 Channels and transporters 122
    3.3.5 Transpiration stream and growth 124

3.1 Osmosis, water potential, and cell expansion

Between 70% and 95% of the fresh mass of a grapevine's organs consists of water (H_2O). Even the woody tissues of trunk and roots have a water content of approximately 60%. Most of this water serves as a solvent for ions and organic molecules in the interior of the vine's cells. Water can diffuse freely, albeit slowly, across the phospholipid bilayer of cell membranes, which are approximately 5–10 nm thick, and diffusion accelerates with rising temperature. More importantly, however, with a diameter of only 0.3 nm the H_2O molecules are small enough to be able to just barely (i.e., in single file), but extremely rapidly (up to 1 billion H_2O molecules per second), penetrate membrane pores formed by water channel proteins called aquaporins (Katsuhara et al., 2008; Maurel et al., 2008; Tyerman et al., 1999, 2009). In contrast to the diffusion pathway, the rate—but not the direction—of water movement across aquaporins can be fine-tuned by active regulation because these proteins are able to open or close their pores in a process called gating (Törnroth-Horsefield et al., 2006). Additional regulation is provided by varying the number of aquaporin proteins in the membranes. Aquaporins are relatively temperature insensitive and can speed up the passage of water across membranes >50-fold; in other words, they strongly raise the membranes' water permeability, or hydraulic conductivity. At the same time, most aquaporins are impermeable to mineral nutrient ions and even to subatomic protons. In fact, cells use protons, in addition to calcium ions (Ca^{2+}) and dephosphorylation, or reversible removal of phosphate groups from the protein, to close aquaporins (Johansson et al., 2000; Luu and Maurel, 2005; Maurel et al., 2008; Tournaire-Roux et al., 2003). In addition to size exclusion, dissolved ions are also repelled by charged residues at the entrance to the pores: Positively charged cations cannot pass through a "gate" guarded by a positive charge, and negatively charged anions cannot pass through a gate guarded by a negative charge. In other words, cell membranes are not only capable of general filtration, but they are selectively permeable or semipermeable (Latin *semi* = half). Specialized gates consisting of transport proteins and ion channels are required for molecules other than H_2O to pass through membranes (see Section 3.3).

In the aqueous solutions of plants and soils water acts as a solvent for solutes, which include dissolved ions from dissociated inorganic and organic salts, and small uncharged organic molecules. The concentration of solutes inside a cell is almost always higher than the solute concentration of the exterior comprising the cell walls, intercellular space, and the space outside the plant body. Because the intracellular solutes are repelled by the pore openings in the cell membrane, their momentum drags water molecules away from the membrane along with the solutes, causing water to move passively through aquaporins and across the membrane into the cell (Kramer and Myers, 2013). This water movement is termed osmosis (Greek *osmos* = impulse, thrust). In other words, the presence of solutes in a cell exerts a "pull" or tension on the water molecules surrounding the cell. The greater the concentration of solutes, the more water will move into the cell to restore the hydraulic equilibrium. Osmotic pressure (π, expressed in MPa) of an aqueous solution is thus caused by dissolved solutes and increases linearly as the solute concentration (c, expressed in mol L^{-1}; i.e., the number of dissolved molecules per unit volume) increases, and as the temperature (T) increases. This relationship is captured by the following equation, for which its discoverer, the Dutch chemist Jacobus van't Hoff (1887), received the first Nobel Prize in chemistry:

$$\pi = RTc$$

where R is the universal gas constant ($= 8.31\,J\,K^{-1}\,mol^{-1} = 0.00831\,MPa\,L\,K^{-1}\,mol^{-1}$) and T is the thermodynamic temperature, or absolute temperature, in Kelvin (K = °C + 273).

Sugars such as sucrose, organic acids such as malate, and inorganic ions such as potassium (K^+) and chloride (Cl^-) are the major osmotic solutes, termed osmolytes, of plant cells. Therefore, solutes have an osmotic function in addition to their metabolic functions. When solutes attract water into a cell, the cell swells and causes the cell membrane to exert a force on the cell wall. This leads to a buildup of pressure, defined as force per unit area, due to the incompressibility of water and the rigidity of plant cell walls resisting an increase in volume. This balancing wall pressure raises the energy of the water inside the cell until it equals that of water outside. At this point, the cell's internal hydrostatic pressure, or more accurately the pressure difference between inside and outside of the cell, termed turgor pressure (P) but often abbreviated as turgor (Latin *turgere* = to be swollen), is equal to the difference in osmotic pressure between the cell and its surroundings ($P = \pi_{inside} - \pi_{outside}$), and net water influx stops. Osmotic pressure, therefore, is defined as the hydrostatic pressure difference required to stop the net flow of water across a membrane separating solutions of differing concentration. In other words, osmosis can only work as a driving force for water flow (see Section 3.3) when two compartments are separated by a semipermeable membrane (Fig. 3.1). Turgor pressure, in turn, is the osmotically maintained pressure inside living plant cells. The turgor pressure of growing cells typically reaches approximately 0.3–1 MPa, or 3–10 times the atmospheric pressure ($P_a \approx 0.1\,MPa$) (Cosgrove, 1997). Opening stomatal guard cells can generate $P > 4\,MPa$, and yet the deformation of these cells and their cell walls is completely reversible (Franks and Farquhar, 2007; McQueen-Mason, 2005; Roelfsema and Hedrich, 2005).

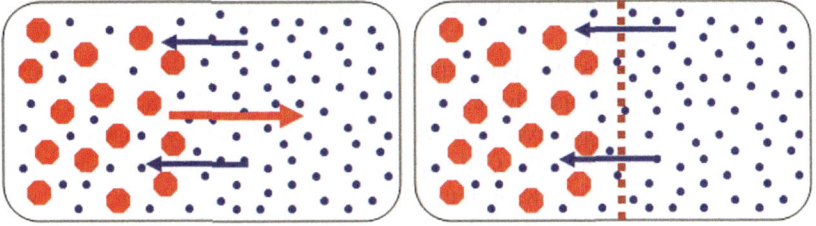

FIG. 3.1
During diffusion, solutes (red) and water (blue) move in opposite directions, and the pressure remains constant (left). During osmosis, water moves toward the solutes across a semipermeable membrane (brown), increasing the pressure in the solute compartment (right).

The solute and pressure forces in plants and soils are conveniently described as free energy per unit volume, which is equivalent to force per unit area, or pressure, and is generated by the random movement and collision of molecules. The free energy of an aqueous solution per unit molar volume of liquid water (~ 18 mL mol$^-$) is termed water potential (Ψ, expressed in J m^{-3} or as its pressure equivalent Pa $= 10^{-5}$ bar $=$ N m^{-2}) and is the sum of the component potentials arising from the effects of turgor pressure (i.e., turgor potential, $\Psi_P = P$) and solutes (i.e., solute potential or osmotic potential, $\Psi_\pi = -\pi$) in addition to interactions with matrices of solids (e.g., cell walls) and macromolecules (i.e., matrix potential, Ψ_M) and the effect of gravity (i.e., gravitational potential, Ψ_G), as described by the following equation (Boyer, 1969):

$$\Psi = P - \pi + \Psi_M + \Psi_G = \Psi_P + \Psi_\pi + \Psi_M + \Psi_G$$

Although Ψ_M is very important in soils due to the adhesion of H$_2$O molecules to soil particles, and in cell walls, it is very close to zero inside the cells of well-watered plant tissues and is therefore negligible unless loss of $>50\%$ of the tissue water renders a tissue severely dehydrated (Hsiao, 1973). The gravitational potential increases with height at a rate of 0.01 MPa m^{-1}. Although this may be of little significance in the relatively short-statured cultivated grapevines that rarely exceed a height of approximately 2 m, the downward pull of Ψ_G is a factor in their taller wild relatives, as well as in the trees serving as their natural trellis system (see Fig. 1.2). Yet even in their natural habitats, grapevines never approach the upper limit of about 130 m that the gain in gravitational potential energy with increasing height imposes on tree height (Koch et al., 2004).

Pure water has the highest possible water potential, which by convention equals zero ($\Psi = 0$) at atmospheric pressure and 25 °C. Therefore, the water potential of unpressurized aqueous solutions is always negative ($\Psi < 0$). In other words, Ψ is a measure of the water concentration, while π is a measure of how much the dissolved molecules decrease Ψ. The Ψ of a cell describes the potential tension, or negative pressure, or suction, that the cell solution exerts on surrounding pure water. In the absence of other forces, the movement of water during osmosis is always from a region of lower solute concentration (i.e., greater Ψ) to one of higher solute concentration (i.e., lesser Ψ). In other words, osmosis is driven by a water potential gradient ($\Delta\Psi$), or a difference in Ψ per unit distance, which comprises both pressure and osmotic components: Water movement across membranes and into cells is determined by $\Psi_P + \Psi_\pi$. In contrast to the movement of water across membranes, its movement outside of cells, through the apoplast (see Section 3.3), is driven solely by pressure differences, because water does not cross membranes in the apoplast.

Plant cells use osmotic solutes, especially sucrose and potassium, to lower their Ψ in order to attract water into the cells. Therefore, solutes have a major role in building P, which is necessary for cell expansion and hence plant and organ growth. In fact, cell growth is mainly a function of turgor and the mechanical properties of the cell walls. Cell expansion is made possible by accumulation of solutes inside the cell vacuole, which draws water into the vacuole to create turgor pressure, and by cell wall loosening. The Ψ_π of the vacuole essentially matches the Ψ_π of the cytoplasm, because membranes cannot sustain differences in P without the aid of cell walls. The P of growing cells irreversibly stretches and pushes the thin primary cell walls outward (Cosgrove, 1997). In engineering terminology, such irreversible extension in response to an applied force is called a plastic deformation, contrasting with elastic deformation, which is fully reversible when the force is removed (Cosgrove, 2018). While P causes tension, however, it does not induce extension of the cell walls. Irreversible, and hence permanent, extension requires loosening of the wall components, which occurs by enzymatic digestion of load-bearing junctions between adjoining cellulose microfibrils, enabling the microfibrils to move and thus relax wall stress (Boyer, 2009; Braidwood et al., 2014; Cosgrove, 2016, 2018; Hamant and Traas, 2010; Schopfer, 2006). Although P must be high enough to expand the cell, it must also be low enough to maintain a gradient of Ψ in order to sustain water influx by osmosis. Continued water influx is necessary in expanding cells because the incompressibility of water means that P would otherwise drop to zero as soon as the cell walls become stretched. Indeed, a decrease in P by as little as 0.02 MPa can stop cell expansion but not reverse it (Taiz, 1984).

While P provides the driving force for cell expansion, the cell walls control the rate and direction of expansion (Winship et al., 2010). Loosening of the cell walls is facilitated by proteins known as expansins that are activated at low pH (Cosgrove, 1997, 2000, 2005, 2018; Li et al., 2003). According to the "acid growth theory," the cell wall is acidified by protons (H^+) pumped from the cell interior to the cell wall in exchange for potassium ions (K^+), which maintain the electrical charge balance inside the cell (Rayle and Cleland, 1992; Stiles and Van Volkenburgh, 2004; van Volkenburgh, 1999). This ion exchange can temporarily decrease the cell wall pH from the normal 5.5–6 to as low as 4.5 (compared with a cytosol pH of about 7.5; Kurkdjian and Guern, 1989), which equates to a >10-fold increase in H^+. The ATP-powered H^+ pumps in turn are activated by auxin, which induces phosphorylation of the H^+-ATPase protein, Mg^{2+} ions, and light, via phytochromes and other photoreceptors (Bassil et al., 2019; Takahashi et al., 2012; see also Section 5.2). Expansins may temporarily disrupt the chemical associations made up of hydrogen bonds between the parallel and inelastic cellulose microfibrils and the cross-linking xyloglucans (hemicelluloses) in the cell walls, which makes the walls flexible enough to allow expansion and, later, insertion of new cell wall material (Braidwood et al., 2014; Cosgrove, 2016, 2018). Additionally, auxin also seems to stimulate the production of reactive oxygen species (see Section 7.1) that are then converted by the multipurpose oxidizing enzyme peroxidase into hydroxyl radicals (OH), which in turn may attack the bonds between cell wall polysaccharides (Møller et al., 2007). Slipping of the parallel microfibrils and production and insertion of new wall material is necessary for the expanding cell to enclose its increasing volume; cell walls usually maintain their thickness during expansion (Boyer, 2009; Cosgrove, 2005).

Because cells must increase in volume before they can divide, cell expansion, in turn, is also a prerequisite for cell division. Since vacuoles occupy >90% of the volume of mature cells, as much as 90% of the entire volume gain during plant growth is due to the expansion of cell vacuoles and the concomitant stretching and addition of cell wall material. Most of the increase in volume during cell expansion is due to water uptake by the vacuole, whereas the amount of cytoplasm increases only insignificantly. The requirement of ATP for cell wall loosening means that cell expansion, and thus growth, are highly dependent on energy provided by respiration and consequently are also sensitive to environmental impacts other than water supply. For example, when low temperatures inhibit respiration, cell expansion ceases very rapidly (Cosgrove, 2016).

The term water potential is convenient because it can be applied not only to cell solutions but also to soil solution and air—that is, to any medium with variable water content. Water potential indicates the availability of water in any aqueous system. The water potential of air can be estimated from the air's relative humidity (RH; expressed in percentage). The RH is the amount of water vapor in air at a particular temperature in relation to the total amount the air could hold at that temperature; the latter is referred to as the saturation vapor density or saturation vapor pressure (VP_{sat}, expressed in kPa). The water-holding capacity of air, and therefore its VP_{sat}, increases exponentially with increasing air temperature (T_{air}, in °C), as described by the following equation (Katul et al., 2012):

$$VP_{sat} = 0.611 e^{\frac{17.5 T_{air}}{T_{air} + 249.93}}$$

For example, in the 10–35 °C range the VP_{sat} doubles with a 12 °C increase in air temperature, which means that air at 22 °C can hold twice as much water vapor as air at 10 °C, and air at 34 °C can hold four times as much. In other words, if water-saturated air of 100% RH is heated by 12 °C, its RH drops to 50%, thus the air develops a saturation deficit or vapor pressure deficit (VPD). The relationship between Ψ_{air} (in MPa) and RH (in %) can be approximated by the following equation (Nobel, 2009; Tyree and Zimmermann, 2002):

$$\Psi_{air} \approx 0.46(T_{air} - 273.15) \ln(RH/100)$$

where T_{air} is the air temperature (in °C) and ln is the natural logarithm.

Because $\ln 1 = 0$, the Ψ of saturated air is always zero, irrespective of temperature, whereas the Ψ of drying air drops dramatically (Table 3.1). Even relatively humid air exerts an enormous tension on the water inside a plant or in the soil. This tension is responsible for transpiration (see Section 3.2). Like water movement into plant cells, therefore, transpiration is driven by a water potential gradient as well. Incidentally, the relationship between temperature and Ψ_{air} is also responsible for the so-called rain shadow effect of the Cascades mountain range in eastern Washington and Oregon and the Andes range in Mendoza (Argentina). As moist air from the Pacific Ocean rises up the western slopes of the

Table 3.1 Typical values of relative humidity (RH) and Ψ for leaves and atmosphere at 25 °C.

	RH (%)	Ψ (MPa)
Air spaces in leaf	99	−1.4
Boundary layer	95	−7.0
Atmosphere	100	0
	90	−14.5
	50	−95.1
	10	−315.9

mountains, it cools and drops excessive moisture as rainfall or snow so that the VPD, or evaporative demand, of the now drier air increases while the air warms as it descends down the eastern slopes of the mountains.

3.2 Transpiration and stomatal action

The evaporation of water from plants is called transpiration (Latin *trans* = on the other side, *spirare* = to breathe). Evaporation constitutes a physical phase change of water from the liquid to the gas form. The term evapotranspiration that is often found in the literature represents the total amount of water evaporated from a land surface, which includes the water transpired by plants (in a vineyard these are grapevines, cover crops, and weeds) and the water evaporated from the soil and surface water bodies (Katul et al., 2012). In a hot climate like central California, such evapotranspiration from well-watered raisin or table grape vineyards can exceed 6 mm d^{-1} (1 mm = 1 L m^{-2}) in the middle of summer and >800 mm over the course of a growing season (Williams et al., 2003). Transpiration from a leaf depends on two major factors: the difference in absolute water vapor concentration, or water potential, between the leaf air spaces (Ψ_{leaf}) and the external air (Ψ_{air}) and the resistance (r) of this pathway to diffusion.

As water evaporates from the epidermal and mesophyll cell walls and is discharged as water vapor into the atmosphere, it meets a series of resistances that oppose its movement (Buckley et al., 2017; see Fig. 1.15). A resistance slows the H_2O molecules down and thus limits their rate of diffusion. The principal resistances are the one at the stomatal pore and the one due to the thin film of still and moist air on the leaf surface. The former is called the stomatal resistance (r_s), and the latter is called the boundary layer resistance (r_b). The boundary layer, which is also known as the unstirred layer, acts like an additional, extracuticular membrane in series with the cuticle. The total resistance to water evaporation is usually dominated by r_s, but wind increases turbulence in the boundary layer, making it thinner and decreasing r_b, which increases transpiration dramatically even as the stomata close to increase r_s (Freeman et al., 1982; Schultz and Stoll, 2010). On the other hand, r_b can be very high in dense canopies so that transpiration of such canopies may be determined mainly by r_b and is not easily influenced by atmospheric conditions.

Since the leaf-to-air vapor concentration difference provides the driving force for transpiration, the transpiration rate per unit leaf area (E, expressed in mmol H_2O m^{-2} s^{-1}) increases as $\Psi_{leaf}-\Psi_{air}$ increases, whereas it decreases as the total resistance increases. This relationship can be described by the following equation, which is analogous to Ohm's law ($I = Vr^{-1}$, where I is the current and V is the electrical potential difference or voltage) used in electricity and thermodynamics:

$$E = \Delta\Psi(r_s + r_b)^{-1}$$

where $\Delta\Psi = \Psi_{leaf} - \Psi_{air}$.

Water vapor concentration is proportional to water vapor pressure, and the difference between the amount of moisture in saturated air and the actual amount of moisture in the air at a given air temperature is called the vapor pressure deficit (VPD). The VPD can be calculated from the air's water-holding capacity or saturation vapor pressure (VP_{sat}, in kPa) and the relative humidity (RH, in %) as follows:

$$VPD = VP_{sat}(1 - RH/100)$$

As discussed in Section 3.1, an increase in air temperature strongly increases VP_{sat} and thus results in a greater VPD. The steeper water concentration gradient equates to a higher $\Delta\Psi$ and stimulates more water evaporation from the leaf. Therefore, temperature is a powerful determinant of the transpiration rate, and transpiration increases with increasing temperature (see Fig. 5.8).

In a typical grapevine leaf, only 5–10% of the total transpiration occurs across the cuticle, although the cuticular portion can reach up to 30% in old leaves that have low overall transpiration rates compared with younger leaves (Boyer et al., 1997). The bulk of the water vapor escapes through the stomatal pores, even though they cover <5% of the leaf surface. Stomata are thus very important for regulating diffusional water loss from a leaf. They function like a pressure regulator, limiting changes in Ψ_{leaf} by controlling the transpiration rate, and they respond very sensitively to environmental variables (Sperry et al., 2002). The most conspicuous of these responses is the light-induced opening of stomata at sunrise and their closing at sunset in response to photosynthetically driven changes in the substomatal CO_2 concentration (Raschke, 1975; Roelfsema and Hedrich, 2005). When gaseous CO_2 diffuses into cell walls and guard cells, it dissolves in water and is partly converted to bicarbonate (HCO_3^-), which binds to and activates proteins that facilitate malate movement across membranes—the so-called malate channels (Engineer et al., 2016; Maurel et al., 2016). The resulting efflux of malate prompts the stomata to close as described below. This mechanism minimizes water loss by transpiration at night, when there is insufficient light energy to fuel photosynthesis (see Section 4.1) and hence no need for CO_2 diffusion into the leaf (see Section 4.2).

Nonetheless, the need to dispose of respiratory CO_2 may prevent the stomata from closing completely at night, which may account for the observation that nighttime transpiration is higher than what would be expected from water diffusion across the cuticle alone (Fricke, 2019). Consequently, at night, when the stomatal conductance ($g_s = r_s^{-1}$) is <50 mmol H_2O m^{-2} s^{-1}, the rate of water evaporation from the leaves is roughly 10% of that during the day (Escalona et al., 2013; Rogiers et al., 2009; Rogiers and Clarke, 2013). When the increasing light in the morning promotes photosynthesis, the stomata open, and stomatal opening then normally follows the daily change in light intensity, which peaks around midday (Düring and Loveys, 1982; Rogiers et al., 2009; Tarara et al., 2011). Similarly, the stomatal aperture is wider in bright light than in dim light such as prevails under cloudy conditions or in the interior of dense canopies (Shimazaki et al., 2007). Conversely, the stomata close when a leaf runs out of water; low Ψ_{leaf}, which signals water stress (see Section 7.2), causes the stomata to close to protect the xylem conduits from failure due to gas embolisms (see Section 3.3). This effect of Ψ_{leaf} is mostly a response to low leaf turgor (Rodriguez-Dominguez et al., 2016). It is strong enough to override the influence of high light intensity on stomatal opening so that even relatively mild soil water deficit in a vineyard often leads to a midday depression of stomatal conductance (Düring and Loveys, 1982; Matthews et al., 2017). This temporary decline in g_s (i.e., increase in r_s) is usually not observed in vines with ample water supply, except for vines growing in hot climates (Downton et al., 1987; Rogiers et al., 2012; Williams et al., 1994). The hot-climate exception occurs because the aperture of the stomatal pores also decreases as the VPD increases (Zhang et al., 2012). The VPD response may actually be a response to transpiration rather than to humidity per se. As would be expected from the preceding "transpiration equation," stomatal closure can compensate for variations in humidity, holding transpiration and Ψ_{leaf} relatively constant. Nevertheless, a decrease in relative humidity from 85% to 45% at constant temperature can decrease the transpiration rate by >80% via partial stomatal closure.

A rise in temperature initially increases stomatal aperture due to its stimulating effect on photosynthesis, but very high temperature leads to stomatal closure, possibly due to its stimulating effect on respiration (see Section 5.2). An increase in leaf nitrogen also leads to a wider stomatal aperture because nitrogen also stimulates photosynthesis (see Section 5.3), but this effect levels off at high nitrogen content.

Stomata are hydraulically driven gas valves because they open when the guard cells' vacuoles take up water and swell, causing their turgor pressure to rise (Franks and Farquhar, 2007). This clever mechanism is possible because the guard cells lack plasmodesmata connecting them to neighboring epidermis and mesophyll cells (Roberts and Oparka, 2003; Turnbull and Lopez-Cobollo, 2013; Wille and Lucas, 1984). In addition, the guard cells of closed stomata contain many small vacuoles that merge into a few large vacuoles when the stomata open (Martinoia et al., 2012). This allows the stomata to open very rapidly, because the fragmentation in the closed state minimizes the vacuoles' volume while preserving their membrane surface. The osmotic uptake of water is driven by a $\Delta\Psi$ (see Section 3.1) caused by movement of potassium ions (K^+) from surrounding cells into the guard cells across K^+ channels in exchange for hydronium ions (H_3O^+) or protons (H^+), which are driven out by a light-activated pump called H^+-ATPase embedded in the plasma membrane (Roelfsema and Hedrich, 2005; Shimazaki et al., 2007; Engineer et al., 2016). The reverse exchange—K^+ out/H^+ in—causes water to move out of the guard cells so that the decrease in turgor causes the stomata to close. The H^+ may come from malate, which is produced from starch stored in the guard cell chloroplasts and, courtesy of its double-negative charge, also serves as the counterion to K^+ and thus escorts K^+ on its movement into and out of the guard cells. The starch, in turn, is produced by guard cell photosynthesis or from sucrose imported across the cell walls from the underlying mesophyll cells (Lawson, 2008). Sucrose derived from starch breakdown also supplements K^+ as an osmoticum, especially in the afternoon (Roelfsema and Hedrich, 2005; Shimazaki et al., 2007; Talbott and Zeiger, 1996). Thus, the guard cells actively and rapidly regulate stomatal opening and closing via a dynamic balance of starch/malate and H^+/K^+.

The hydraulic effect of Ψ_{leaf} on stomatal opening or closing is amplified by abscisic acid (ABA), which is produced from carotenoids in dehydrated cells, but especially in the leaf's vascular tissues, from where it is transported to the guard cells, and in the guard cells themselves (Kuromori et al., 2018; Nambara and Marion-Poll, 2005; Wilkinson and Davies, 2002). As Ψ_{leaf} declines the drying leaves rapidly produce ABA in response to decreasing cell volume, since the leaf cells shrink while they lose water to evaporation (Sack et al., 2018). Such stimulation of ABA biosynthesis may also contribute to the decrease in g_s under high temperature and high VPD (McAdam et al., 2016). In addition to the ABA produced in drying leaves, the amount of ABA arriving from the roots can increase considerably in response to decreasing soil moisture (Bray, 1997; Davies and Zhang, 1991; Davies et al., 2002; Rogiers et al., 2012). Hence, ABA also acts as a messenger from the roots, indicating water stress when the soil dries out (see Section 7.2). As Ψ_{soil} declines, the roots produce increasing amounts of ABA, which is transported in the xylem sap to the leaves' guard cells (Li et al., 2011; Loveys, 1984; Speirs et al., 2013). ABA triggers stomatal closure by inducing an increase in H_2O_2 and other reactive oxygen species that, in turn, prompt a brief rise of calcium (Ca^{2+}) in the cytosol, which blocks influx into and enhances efflux from the guard cells of K^+ (Allen et al., 2001; McAinsh and Pittman, 2009). In other words, ABA modifies the permeability of the guard cell membrane by way of a Ca^{2+} signal. The gaseous signaling molecule nitric oxide (NO) also rapidly accumulates in guard cells and, more slowly, in leaf mesophyll cells when the soil dries down (Patakas et al., 2010). However, it remains to be demonstrated whether NO is necessary to induce stomatal closure in an ABA–H_2O_2–NO–Ca^{2+} signaling cascade or whether it serves to detoxify H_2O_2.

The reaction of stomata to ABA is very rapid. It occurs within minutes, irrespective of Ψ_{leaf} and VPD, and it does not involve starch degradation (Roelfsema and Hedrich, 2005). However, stomatal closure is associated with an increase in the pH of the xylem sap from the "normal" pH 5–6 to pH 6.5–7, which enhances ABA delivery to the guard cells by reducing uptake of ABA by intervening tissues (Bacon et al., 1998; Hartung and Slovik, 1991; Wilkinson and Davies, 2002). At a higher pH in the xylem less ABA becomes "trapped" inside the usually more alkaline parenchyma and phloem cells lining the xylem vessels. Apparently, a decrease in the flow rate of xylem sap leads to an increase in the sap's pH, perhaps because there is more time for the adjacent cells to remove protons from the sap (Jia and Davies, 2007). Furthermore, ABA seems to induce closure of aquaporins in the membranes of the bundle-sheath cells lining the xylem in the leaves, thereby reducing water flow to the mesophyll cells, a hydraulic effect that enhances the "standard" ABA effect on stomatal closure (Pantin et al., 2013; Shatil-Cohen et al., 2011).

Even when only a portion of the roots experience dry soil, the ABA produced by that portion suffices to trigger stomatal closure despite the remainder of the roots taking up enough water to maintain high Ψ_{leaf} (Comstock, 2002; Li et al., 2011; Lovisolo et al., 2002a; Stoll et al., 2000). When rainfall or irrigation replenishes soil moisture, the roots stop producing ABA, whose concentration in the xylem therefore decreases rapidly. Nonetheless, the transpiration-driven rise in water flow up the xylem after such water supply can "flush" residual ABA to the leaves. The residual ABA may keep the stomata partly closed for up to several days after Ψ_{leaf} has recovered, perhaps to enable any gas embolisms that may have formed in the xylem to be repaired (Lovisolo et al., 2008; Romero et al., 2017; see Section 3.3). As the ABA that is present in the leaves is degraded to phaseic acid and other breakdown products, the stomata then reopen (Nambara and Marion-Poll, 2005; Speirs et al., 2013).

The turnover of water in a rapidly transpiring leaf occurs within 10–20 min; that is, the leaf loses the equivalent of its entire water content every 10–20 min (Boyer, 1985; Canny, 1993). Although transpiration is sometimes described as a necessary evil due to a plant's dependence on gas exchange, it has some useful side effects. For instance, the negative pressure or tension in the xylem created by transpiration aids in extracting water from the soil (see Section 3.3) that can be used for plant growth. Water uptake, in turn, attracts soil nutrients, which can then be absorbed by the roots as well. However, although the increased flow of water during transpiration leads to a corresponding increase in the rate at which dissolved nutrients move upward in the xylem conduits, transpiration per se is not necessary for this nutrient movement (Tanner and Beevers, 2001). In fact, plants can absorb and transport nutrients just as well at night when transpiration is minimal as they do during the day. More importantly, the high latent heat of vaporization of water makes evaporation a powerful cooling process, which is why the principle of evaporative cooling is also used in air-conditioning systems. Evaporative heat loss prevents leaves exposed to sunlight from overheating. Though the temperature of actively transpiring leaves is ordinarily no more than 2–3 °C higher than the ambient temperature, it can rapidly fluctuate by $>10\,°C$ in direct sunlight (Peña Quiñones et al., 2019; Sharkey et al., 2008). Water evaporation cannot prevent "heat spikes" completely.

Grapevines use several cooling tactics to dissipate excessive heat gained from absorbed solar radiation. The three main pathways are as follows (Mullins et al., 1992; Taiz and Zeiger, 2006):

Radiation: Direct transfer of heat as long-wavelength radiation to surrounding objects or the sky. Leaves absorb solar radiation in the ultraviolet, visible, and infrared ranges of the spectrum and reradiate infrared energy, which is a cooling process. Such reradiation happens mostly during the night, when the temperature of the surrounding air is cooler than the leaf temperature. Nevertheless, even in bright sunlight, leaves can radiate 50–80% of the energy they absorb. The proportion of incoming radiation reflected by a surface is termed albedo.

Convection: Heating or cooling of ambient air via sensible heat—that is, heat that can be felt and measured with a thermometer. The energy of molecules on the leaf surface is exchanged with that of air molecules with which they are in direct contact. The warmed air becomes lighter and rises, which leads to cooling. Convection is driven by the temperature difference between the air and the leaf, and the boundary layer is the main resistance to this process. Air circulation around the leaf removes heat from the leaf surface so long as the leaf is warmer than the air. Wind decreases the thickness of the boundary layer, which increases convective heat transfer and thus convective cooling.

Transpiration: Release of heat during the evaporation of water into the air, also referred to as transpirational cooling. Water evaporation consumes energy called the latent heat of vaporization ($2.45\,kJ\,g^{-1}$ at $20\,°C$) because water vapor contains more energy than liquid water. Heat removal by evaporation cannot be felt or measured with a thermometer. Transpiration is driven by the water potential difference between the inside of a leaf and the outside. This water potential gradient depends on the VPD, which is a function of temperature, which in turn depends on solar radiation, and relative humidity.

3.3 Water and nutrient uptake and transport

3.3.1 Driving forces and resistances

Although the main force driving water movement through plants is water pressure, water uptake by the roots and transport to the leaves is largely brought on by passive forces because the hydraulic system of plants has no moving parts (Tyree and Zimmermann, 2002). The direction of movement is always from high to low pressure or, if the movement is across membranes, from a region with low solute concentration to one with high solute concentration. As discussed in Section 2.2, water uptake and movement in early spring are driven by positive root pressure induced by remobilization of stored nutrient reserves and release of osmotically active ions and organic molecules into the xylem sap. Under these conditions, water moves from the soil into the root by an osmotically generated gradient of water potential ($\Delta\Psi$) and is then pushed up the vine through the xylem. Although an osmotic gradient ($\Delta\pi$)—that is, a gradient in solute concentration—has virtually no effect on the flow rate of water inside the xylem because vessels have no membranes, it drives water inflow *into* the xylem from the soil across the root cell membranes (Wegner and Zimmermann, 2009). This $\Delta\pi$ usually collapses after leaf expansion (although it can remain important at night or during rainfall or overhead irrigation), and water loss to the atmosphere by transpiration from the unfolding leaves begins to hydraulically produce a negative pressure in the xylem (P_x), which maintains the upward flow of water in the plant (Dixon and Joly, 1895). At night or under light conditions low enough to lead to stomatal closure, such as during overcast periods, and with plentiful soil water supply, P_x can remain positive, albeit below atmospheric pressure ($P_a \approx 0.1\,MPa$ at sea level). However, P_x drops below zero as soon as the stomata open even slightly to let water vapor escape and declines further as light intensity and transpiration rise (Wegner and Zimmermann, 2009).

Approximately 95–98% of all water absorbed by the roots is lost to the atmosphere in the process of transpiration; very little water is required to build the plant body through cell expansion. The tension generated by transpiration is the main driving force for water uptake and movement in the xylem up the vine to the leaves. This is why water flow through the trunk of grapevines oscillates with a minimum at night and a maximum at approximately midday and why vines with a large canopy sustain higher flow rates than do small vines (Pearsall et al., 2014; Tarara and Ferguson, 2006). The transpirational pull is possible because water forms a continuous system from the air near the evaporating surfaces inside the leaves, shoots, tendrils, and flower and fruit clusters to the absorbing surfaces in the roots and out into the soil. This system is termed the soil–plant–air continuum. Water has unique chemical and physical properties, including high surface tension, which it owes to its cohesion, or the strong tendency of H_2O molecules to "stick" together by forming hydrogen bonds between the negatively charged oxygen and the two positively charged hydrogen atoms; and strong adhesion of H_2O molecules to the surfaces of xylem conduits. These properties keep it in this continuum so that water escaping the leaves by transpiration is continually replaced by water being pulled up the xylem from the roots (Dixon and Joly, 1895; Steudle, 2001; Tyree and Zimmermann, 2002).

The transpiration–adhesion–cohesion–tension theory of xylem sap flow states that the evaporation of water from cell walls inside the leaf creates tension, while adhesion of water to the walls of the xylem helps counteract gravity, and cohesion ensures that the tension extends down the xylem, into the roots, and out into the soil. Therefore, plants directly utilize solar energy, which causes water to evaporate from the leaves, to drive water uptake and distribution. Again, the water moves along a gradient of water potential. The magnitude of this gradient is variable within the root xylem (0.3–2.5 MPa m^{-1}, with lower values in well-watered and higher values in water-stressed vines), moderate within a shoot (0.1–0.5 MPa m^{-1}), and quite steep within the petiole (1–5 MPa m^{-1}) of a transpiring leaf (Lovisolo et al., 2008). The flow of water from the soil toward the roots is maintained by suction at the root surface, which is caused osmotically by root pressure or hydrostatically by transpiration. As transpiration removes water, the soil adjacent to the roots, called the rhizosphere, becomes drier than the more distant bulk soil.

For xylem sap to sustain the tension required in a grapevine to pull water from the soil, the hydraulic system has to be airtight. The water's surface tension acts like a seal, keeping water inside the xylem and air out. However, the xylem is a vulnerable pipe, and water is saturated with air at atmospheric pressure (Steudle, 2001; Tyree and Zimmermann, 2002). Sometimes the water column inside the xylem is put under too much tension, for example by excessive transpiration due to drying wind or heat combined with dry soil; the tension can be suddenly relieved, for example by shoot, leaf, or cluster removal; or the column receives some other physical shock, for example due to a bump by machinery. When one of these things happens, the water column can snap back like a piece of elastic in a rigid tube. This phenomenon is termed cavitation, in which the water column breaks and the conduit lumen suddenly fills with H_2O vapor or air bubbles (Fig. 3.2). The air can be sucked to adjacent vessels through the holes of the interconnecting perforation plates and through pits (Brodersen et al., 2013). Such gas blockages are called embolisms and render the vessels nonfunctional; in severe cases, this can lead to leaf wilting and even canopy collapse (Schultz and Matthews, 1988a; Tyree and Ewers, 1991; Tyree and Zimmermann, 2002). For the gases to dissolve and the conduits to refill with liquid water, P_x must rise to >0 MPa, that is, to within about 0.1 MPa of atmospheric pressure or above (De Boer and Volkov, 2003; Sperry et al., 2002, 2003). More accurately, gas emboli dissolve at $P_x > -2 t r^{-1}$, where t is the surface tension of water (72.8 mN m^{-1} = 72.8 mJ m^{-2} at 20 °C) and r is the radius of curvature of the air–water interface. More details of cavitation and its repair are discussed in Section 7.2.

FIG. 3.2

Cavitation of xylem vessel by water vapor bubble (left), bleeding of xylem sap during budbreak (top right), and guttation at night during the growing season (bottom right).

Left: © Elsevier Inc., illustration after Taiz, L., Zeiger, E., 2006. Plant Physiology, fourth ed. Sinauer, Sunderland, MA; right: photos by M. Keller.

In analogy to Ohm's law and to the transpiration equation discussed previously, the flow rate (F, volume per unit time) of water from the soil to the leaves can be described by the following equation:

$$F = \Delta\Psi\, r_h^{-1} = \Delta\Psi\, l_h \left(\text{because } l_h = r_h^{-1}\right)$$

where $\Delta\Psi = \Psi_{soil} - \Psi_{leaf}$, r_h is the hydraulic resistance, and l_h is the hydraulic conductance.

Inside the xylem conduits, the osmotic component of Ψ is insignificant because there are no membranes in the pathway; hence, water flow follows a pressure gradient: $\Delta\Psi \approx \Delta P_x$. Sometimes a negative sign is added on the right side of the equation to indicate that water flows in the opposite direction to that of $\Delta\Psi$—that is, from high to low Ψ (Tyree and Zimmermann, 2002). Although the flow equation is surely an oversimplification, it is quite convenient to explain most phenomena involved in water flow through plants.

The hydraulic conductance is a measure of transport capacity or water permeability. At the whole-plant level, it determines Ψ_{leaf} at any specific transpiration rate. The term "conductance" (l_h) should not be confused with "conductivity" (L_h or K), which refers to the water flow normalized to ΔP over a specific length (x) of the flow path ($l_h = L_h\, x^{-1}$). In addition, "specific conductivity" (L_s) applies to the cross-sectional area (A) of a unit flow path ($L_s = L_h\, A^{-1}$) and is roughly proportional to the number of xylem conduits passing through that cross section and the fourth power of their radius (r^4); the rate of water flow per unit cross-sectional area is termed flux: $J = F\, A^{-1}$ (Tyree and Ewers, 1991; Tyree and Zimmermann, 2002). In other words, a doubling of vessel diameter results in a 16-fold increase in L_s. Therefore, L_h and L_s are independent of the total length of the pathway, whereas l_h decreases with increasing path length, which essentially means with increasing plant height.

The inverse of l_h, the hydraulic resistance (r_h), is a measure of how much a section of the flow path hinders the free flow of water, thereby slowing water flow. It reflects the resistance to water flow due to friction between water and conduit walls and between the H_2O molecules themselves. There are both short-distance radial and long-distance axial, or longitudinal, components of r_h. The axial resistance to water flow from the roots to the leaves is 2–3 orders of magnitude lower than the radial resistance from the root surface to the xylem conduits. In other words, the axial r_h behind the elongation zone of grapevine roots accounts for only 0.1–1% of the total resistance (Gambetta et al., 2013).

The radial r_h is the resistance water encounters when it flows from the root surface across the root and into the xylem conduits, as well as that during flow out of the xylem conduits and into the leaf and fruit cells. Much of the radial r_h is imposed by cell membranes whose permeability is regulated by aquaporins; cell walls are comparatively much more permeable (Maurel et al., 2008; Perrone et al., 2012; Tyerman et al., 2009; Vandeleur et al., 2009). Most of the axial r_h occurs within the xylem conduits and is imposed by the vine's "plumbing layout," termed hydraulic architecture, which is mainly determined by the number, shape, size, and arrangement of xylem conduits and their interconnections, as well as the total length of the flow pathway and the number and shape of bends in the pathway—that is, plant size and shape (Tyree and Zimmermann, 2002). Due to their small size and fine pore meshwork, the pits connecting individual xylem conduits usually account for more than half of the xylem system's total r_h (Choat et al., 2008; Hacke et al., 2006; Nardini et al., 2011). The importance of the pits for overall axial r_h means that longer vessels decrease r_h because the flowing water has to cross fewer high-resistance vessel end walls (Lens et al., 2011). The increase in r_h with increasing plant height partly accounts for the tendency of photosynthesis to decrease as plants grow taller, because the leaves of increasingly taller plants must close their stomata at ever milder water deficit to maintain Ψ_{leaf} above a species- and cultivar-specific minimum (Hubbard et al., 2001; Koch et al., 2004; Ryan and Yoder, 1997). In other words, photosynthesis in tall grapevines, such as those climbing in big trees, may be at least partly hydraulically limited. Note, however, that growth is more sensitive to water deficit than is photosynthesis, so it may be more correct to say that growth, rather than photosynthesis, is hydraulically challenged (Ryan et al., 2006; see also Section 5.2). Moreover, the larger vines' more extensive root system and greater trunk diameter tend to counter this hydraulic limitation trend.

As in electricity, resistances encountered in series are additive (i.e., $r_h = r_{h1} + r_{h2} + \cdots + r_{hn}$), whereas for those acting in parallel, it is the individual conductances that are additive (i.e., $r_h^{-1} = r_{h1}^{-1} + r_{h2}^{-1} + \cdots + r_{hn}^{-1}$). Examples of the former are the resistances as water flows from the roots to the trunk to the cordon to the shoots, and then to the leaves; examples of the latter are the resistances to water flow of multiple shoots on the same cane or cordon. Nevertheless, whole-plant r_h is not constant for a given vine but varies for several reasons. In the long term, r_h usually increases as a vine grows taller and older, but it decreases with increasing shoot number on otherwise similar vines (Keller et al., 2015a). In the short term, a rise in temperature decreases r_h by 2–2.5% per °C due to the decrease in water viscosity, which enables more rapid water delivery to the leaves for transpirational cooling during heat episodes (see Section 3.2). Moreover, transport of nutrient ions, especially of cations such as K^+, decreases r_h by increasing the pit permeability. Cations may induce shrinking of the gel-like polymer matrix, dubbed "hydrogel," in the pit membranes (Nardini et al., 2011; Pérez-Donoso et al., 2010; van Ieperen et al., 2000; Zwieniecki et al., 2001). Because they are attracted to the negative charges of the wall polymers, cations may moreover diminish the extent of the diffuse electrical double layer that lines the pit membrane pores and slows water flow akin to the boundary layer on leaf surfaces (van Doorn et al., 2011). Nitrate (NO_3^-), but not other anions such as phosphate ($H_2PO_4^-$) or sulfate (SO_4^{2-}), also decreases r_h, perhaps by altering the status of aquaporins either directly or via a rise in pH brought about by NO_3^- assimilation (Gloser et al., 2007; Gorska et al., 2008; see also Section 5.3). In addition, higher NO_3^- content is also associated with higher K^+ content in the xylem sap (Keller et al., 2001b). This means that increasing the concentration of some nutrients in the xylem sap increases the sap flow rate, thereby increasing water uptake by the roots; conversely, nutrient deficiency strongly reduces sap flow even under well-watered conditions. It also means that vines may be able to actively, rapidly, and reversibly modify r_h by altering the concentration of ions in the xylem sap, for instance by resorbing nutrients from the surrounding parenchyma cells. Thus the xylem may enhance the delivery of nutrient elements to the leaves by taking advantage of these nutrients' influence on r_h, thereby favoring more rapid water flow.

Grapevines sustain very rapid water flow up their trunks that matches the canopy transpiration rate plus the water "bound" by the growing tissues. The flow rate in field-grown vines has been estimated to fluctuate daily from $<0.1 \, L \, h^{-1}$ at night to $1-10 \, L \, h^{-1}$ during sunny days with high VPD, decreasing to 10–20% of clear-sky values during rainy or overcast days with low VPD (King et al., 2014; Pearsall et al., 2014; Tarara and Ferguson, 2006; Williams et al., 2003). Aquaporin activity in the root cell membranes, and hence radial r_h, approximately matches the diurnal oscillations in flow rate to enable appropriate water uptake by and flow across the roots (Luu and Maurel, 2005; Maurel et al., 2008; Tyerman et al., 2009; Vandeleur et al., 2009). Emplyoing a hydraulic feedback loop that also involves r_s and Ψ_{leaf}, the leaves similarly adjust their r_h to match water supply with VPD-driven transpiration rates (Simonin et al., 2015). The rate of sap flow also varies with canopy size and water availability: For instance, maximum values at midday may be $1-2 \, L \, h^{-1}$ in deficit-irrigated Cabernet Sauvignon wine grapes, declining to $<0.5 \, L \, h^{-1}$ in drying soil, whereas they can reach $3-10 \, L \, h^{-1}$ in large, well-watered Thompson Seedless table grapes or Concord juice grapes (Dragoni et al., 2006; King et al., 2014; Pearsall et al., 2014; Tarara and Ferguson, 2006; Williams et al., 2003; Tarara and Perez Peña, 2015). To overcome the influence of gravity, a pressure gradient $>0.01 \, MPa \, m^{-1}$ is required to maintain such flow rates in the trunk.

Rapid transpiration favors hydrostatic water flow through the root apoplast and thus reduces r_h (Steudle, 2000, 2001; Steudle and Peterson, 1998). Conversely, when there is little or no transpiration, such as at night or with dry soil, r_h increases substantially because water flow is driven by osmotic gradients across cell membranes. Of course, cavitation in the xylem also greatly increases r_h—to infinity in the affected vessel. Cavitation may be a primary reason for the increase in r_h in leaves under drought stress (Johnson et al., 2012). If cavitation is to be avoided, the flow of water from the soil to the leaves must balance the water lost through the stomata. By combining the transpiration equation with the flow equation, this can be written as follows:

$$(\Psi_{leaf} - \Psi_{air})(r_s + r_b)^{-1} = (\Psi_{soil} - \Psi_{leaf})r_h^{-1}$$

As long as transpiration rate and soil-to-leaf flow rate remain constant, Ψ_{leaf} will remain constant. For a given microclimate, which determines VPD or Ψ_{air}, and a given soil water status, which determines Ψ_{soil}, r_b and r_s regulate the transpiration rate, whereas whole-vine r_h determines Ψ_{leaf} at that transpiration rate (Tyree and Zimmermann, 2002). In other words, the difference between Ψ_{soil} and Ψ_{leaf} is determined by the transpiration rate and r_h, which in turn correlates with r_s. Therefore, r_h defines how wide the stomatal pores can be open without desiccating the leaves (Brodribb and Holbrook, 2003; Jones, 1998; Zufferey et al., 2011). A larger r_h will result in a greater decrease in Ψ_{leaf} when transpiration increases.

3.3.2 Root water uptake

Water will flow from the soil to the roots as long as Ψ_{xylem} is lower than Ψ_{soil}, and it will flow from the xylem to the leaf cells as long as Ψ_{leaf} is lower than Ψ_{xylem}. Even in well-watered grapevines, Ψ_{leaf} can fluctuate by as much as 1.5 MPa during the day, depending on the cultivar (see Section 7.2), following both opening and closing of the stomata at dawn and dusk and depending on the evaporative demand (i.e., VPD) of the air that drives transpiration (Rogiers et al., 2012; Williams and Baeza, 2007). Because a change in Ψ_{leaf} implies a change in Ψ of the leaf cell walls, such changes must be balanced by changes in turgor pressure for the leaf cells to maintain their volume and solute concentration, which is aided by vacuoles and rigid but elastic cell walls. Consequently, P_{leaf} declines from a nighttime maximum to a minimum around midday and recovers in the afternoon (Rüger et al., 2010). Because transpiration-driven water flow is minimal at night, Ψ_{leaf} reaches a maximum as well (Keller et al., 2015a; Schultz and Matthews, 1988a). Therefore, Ψ_{soil} determines the baseline Ψ_{leaf} in the near absence of transpiration (Tardieu and Simonneau, 1998).

As a soil dries, the extraction of water by the roots and its transport to the shoots become increasingly more difficult for grapevines. To maintain the water potential gradient that drives water flow, the Ψ in the shoot must decrease. Therefore, the predawn Ψ_{leaf} can be used as a robust indicator of the Ψ_{soil} to which the roots are exposed. Transpiration during the day decreases Ψ_{leaf} and P_{leaf} below the predawn values so that Ψ_{leaf} and P_{leaf} at a given time of day are the result of both soil water status and transpiration driven by evaporative demand (Rüger et al., 2010; Smart, 1974; Romero et al., 2017; Tardieu and Simonneau, 1998; Zufferey et al., 2011). While the absolute values of Ψ_{leaf} and P_{leaf} decrease in drying soil, their daily amplitudes tend to increase, because the decrease is often greater during the day than at night. At the same time, the influence of VPD on Ψ_{leaf} diminishes in drying soil, because the stomata close, so that Ψ_{soil} becomes increasingly dominant in altering Ψ_{leaf} (Williams and Baeza, 2007).

The water potential equation introduced in Section 3.1 also applies to the soil solution, although unlike plant cells, the $\Psi\pi$ of the soil water is generally negligible (approximately -0.01 MPa), except in saline soils (see Section 7.3). Consequently, Ψ_{soil} is determined mainly by Ψ_M, which is close to zero in wet soils but can decrease to -3 MPa in dry soils due to the surface tension resulting from expanding air spaces. In drying soil, a point will eventually be reached at which the resistance to water flow is so great that the vine can no longer maintain a sufficient $\Delta\Psi$ to sustain transpiration and becomes drought stressed (see Section 7.2). Drought stress can also be induced by low soil temperature, which increases the roots' radial r_h due to closure of aquaporins (Aroca et al., 2012; Maurel et al., 2008). Due to the coupling of r_h and r_s, cool soils are associated with higher leaf r_s, leading to lower photosynthesis, compared with warm soils (Rogiers and Clarke, 2013). However, the temperature effect on r_h can cause problems when cold irrigation water is applied to rapidly transpiring vines, especially at high atmospheric VPD (see Section 3.2). The resulting imbalance between water uptake by the roots and water loss from the leaves can induce cavitation in the xylem, which may lead to leaf wilting and injury (Scheenen et al., 2007). Applying cold irrigation water to pot-grown vines can inhibit growth, especially in dark pots whose soil temperature can be several degrees above air temperature when they are heated by the sun (Passioura, 2006). But even field-grown grapevines can experience significant fluctuations in root temperatures, especially in sandy soils and dry climates, where the root zone of irrigated vines is quite shallow (Stevens and Douglas, 1994).

Water influx is proportional to the surface area of the root system (Steudle, 2001). Grapevines often have dense root systems in the topsoil and can extract water effectively from the surface soil layers. Because extracting water from lower soil layers is less effective, the surface soil dries more quickly than the subsoil. Under nonirrigated conditions, roots continue to grow into deeper, wetter soil layers, whereas the roots of irrigated plants proliferate mostly in the topsoil; in both cases, the Ψ of the advancing root tips remains high as long as the roots can find water (Hsiao and Xu, 2000). Consequently, root water uptake shifts to deeper soil layers as the soil dries and back to the upper layers after rainfall or irrigation.

Because plant-available nutrient ions are dissolved in the soil solution, nutrient uptake also depends on water flow through the soil–root–shoot pathway. Although nutrients are often concentrated in the biologically active surface soil, water and nutrient availability vary greatly in both space and time. To complicate matters further, different nutrients are often not readily available in the same location. For example, NO_3^- moves through the soil approximately 10 times faster than K^+ and 500 times faster than $H_2PO_4^-$, so percolating water drains nitrate down to the subsoil much more rapidly than other nutrients. Consequently, shallow roots often take up soil-immobile nutrients, such as potassium and phosphorus, from the topsoil, whereas deeper roots tap water and soil-mobile nutrients, such as nitrate, that leach deeper into the soil profile. The deep roots can also transport water to the surface roots in a process called hydraulic lift or hydraulic redistribution, which is thought to occur mostly at night when transpiration is minimal (Bauerle et al., 2008a; Bleby et al., 2010). Deep roots need not even be grapevine roots; deep-rooting cover crops or weeds might also be able to lift water to the topsoil, where it may be available to the vines, although this does not apply to vineyard floor covers dominated by grass species (Celette et al., 2008; Patrick King and Berry, 2005). Conversely, water that is lost from vine roots by diffusion can also be reabsorbed by competing weeds or cover crops.

Hydraulic redistribution keeps the surface roots alive, at least so long as they do not lose contact with the drying soil, so that they are ready to take advantage of water arriving from rainfall or irrigation (Bleby et al., 2010). By locally increasing soil moisture due to water loss from the roots when $\Psi_{soil} < \Psi_{root}$, hydraulic redistribution even enables the fine roots to sustain nutrient uptake in drying soil. In addition, roots can move water horizontally and downward, for example when rainfall or irrigation make the surface soil wetter than the subsoil, according to local gradients of Ψ_{soil} (Bauerle et al., 2008a; Smart et al., 2005; Stoll et al., 2000). Water redistribution by roots tends to even out local differences in soil moisture arising from variations in soil texture or organic matter, delays soil drying, and may assist in maintaining symbiotic mycorrhiza. Thus hydraulic redistribution can delay the onset of soil water stress by several weeks (Katul et al., 2012). The redistribution process even continues in dormant plants through the winter (Bleby et al., 2010).

3.3.3 Symplast and apoplast

As shown in Fig. 3.3, water that enters the root initially moves radially through the epidermis and cortex tissues along both symplastic and apoplastic routes (Steudle, 2000, 2001; Steudle and Peterson, 1998). The symplast consists of the interconnected cytoplasm of the cells and therefore includes everything within the plant that is bound inside a cell membrane. The symplastic pathway is an intracellular pathway; water and solutes move from cell to cell via small connecting pores called plasmodesmata (Lough and Lucas, 2006; Roberts and Oparka, 2003). Short-distance cell-to-cell movement of solutes within the symplast occurs by diffusion but is also influenced by osmotic gradients across membranes, while

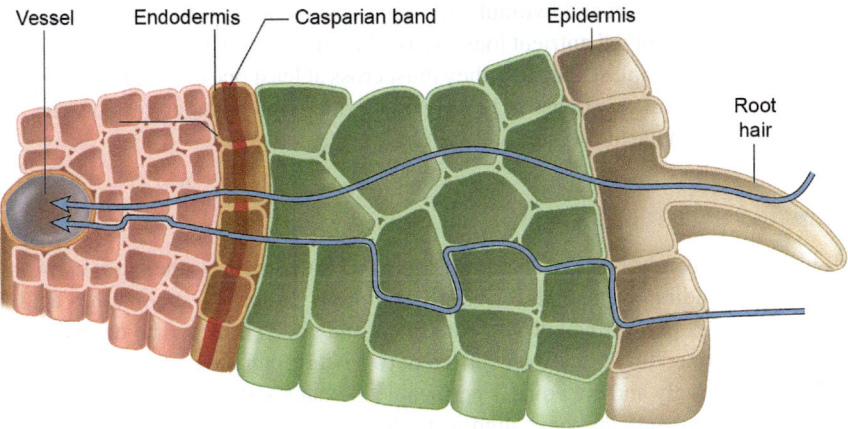

FIG. 3.3
Water and nutrient flow from the soil through a root into the xylem via the symplastic (top blue line) and apoplastic (bottom blue line) pathways.

long-distance movement between symplastic domains occurs by mass flow through the interconnecting phloem (see Section 5.1). The apoplast, on the other hand, comprises everything that is outside the cell membranes and includes cell walls, intercellular spaces, and dead xylem conduits. Cell walls consist of a porous framework of the hydrophilic (Greek *hydros* = water, *philos* = loving) polymers cellulose and hemicellulose embedded in a polymer matrix of pectin, proteins, and in some cells lignin, which is imbued with water like a sponge so that the wettable cell walls resemble dense aqueous gels (Carpita and Gibeaut, 1993; Cosgrove, 1997; Doblin et al., 2010; Steudle, 2001). But although 65–75% of a cell wall is water, the walls have very limited water storage capacity because they comprise only approximately 5% of a tissue's total volume. The cell walls' dense network of microfibrils also acts like a sieve with pore sizes of 4–14 nm, permitting passage only to small molecules such as nutrient ions, amino acids, and sucrose.

The apoplastic pathway is an extracellular pathway; water on an apoplastic route flows around the cell membranes rather than crossing them. Because there are no membranes along the apoplast, hydraulic, nonselective water flow dominates along this pathway (Steudle, 2000, 2001). Rather than moving exclusively in one compartment or the other, a third possibility for water flow is straight across cell walls, membranes, and through cells; in other words, water moves alternately in the apoplast and the symplast. This route is termed the transcellular pathway and is probably the preferred pathway for water on its way to the xylem. However, the endodermis cells that form behind the growing root tip and separate the cortex from the stele have thickened, hydrophobic radial and transverse cell walls called Casparian strips or Casparian bands, which act like gaskets due to their impregnation with lignin (Clarkson, 1993; Geldner, 2013; Perrone et al., 2012; Schreiber, 2010; Steudle, 2000; Fig. 3.3). In addition, access of water and nutrients to channel and transport proteins (see below) in the endodermis cell membranes is blocked by the deposition of suberin in the secondary cell walls, so that nutrients can only move into and out of these cells through plasmodesmata (Geldner, 2013). This waterproofing in the endodermis blocks diffusion and hydraulically separates the apoplast in the cortex from that in the stele, forcing water and dissolved nutrient ions to pass directly through the cell membranes into the cell interior—that is, into the symplast. Because they must cross at least one endodermis cell on their way to the xylem, water and solutes must pass at least two cell membranes, regardless of whether the initial route is symplastic or apoplastic (Tester and Leigh, 2001).

3.3.4 Channels and transporters

Unlike cell walls, cell membranes are semipermeable (see Section 3.1); they are, in principle, impermeable to solutes, including even small ions such as protons. In reality, membranes act like selective sieves, at different times permitting entry to some ions and not to others. The degree of selectivity or semipermeability is termed the reflection coefficient (σ), which can vary from 0 to 1. Membranes with $\sigma=0$ behave nonselectively and are equally permeable to water and solutes, whereas membranes with $\sigma=1$ retain solutes completely and have high water but no solute permeability (Steudle and Peterson, 1998). A high σ in the endodermis ensures, for instance, that nutrients taken up by the roots do not leak back out into the soil solution, which also permits the buildup of root pressure under conditions of slow transpiration. At $\sigma=0$ water moves down a pressure gradient (ΔP), whereas at higher σ osmotic gradients ($\Delta \pi$) increasingly contribute to water movement. Pores created by special channel and transport proteins embedded in the membranes regulate the passage of water and nutrient ions, and each cell type or tissue is equipped with a unique set of these proteins which is, moreover, variable over time. Many channel proteins, especially in the parenchyma tissue around vascular bundles, are specific for water (Maurel et al., 2008; Steudle, 2000; see Section 3.1). These aquaporins control the *rate* of water flow by opening or closing, whereas the *direction* of the flow simply and passively follows the $\Delta \Psi$ (Tyerman et al., 1999). For instance, aquaporins close at night but open during the day; opening decreases the radial r_h, which enables the roots to meet the water demand for transpiration.

Unlike the bidirectional aquaporins, many channels are said to be either inward rectifying or outward rectifying. Inward-rectifying channels facilitate solute movement *into* cells, or solute import, whereas outward-rectifying channels facilitate solute movement *out of* cells, or solute export (Hedrich, 2012; Tyerman et al., 1999). Such controlled one-way movement allows channels to function like valves, which results in concentration gradients across membranes and prevents unnecessary nutrient loss. Because of their ability to control the passage of ions, membranes also act as electrical isolators, which results in gradients of electrical potential across membranes, with negative charges typically dominating inside the cell. The sum of the gradients of concentration and electric charge of an ion is called the electrochemical potential gradient.

Channels are designed to be selective to varying degrees. Some aquaporins permit passage not only to H_2O but also to a few neutral solutes, such as urea, boric acid, or silicic acid, and even to gases, such as CO_2 or NH_3 (Maurel et al., 2008). Other channels, termed ion channels, are specific for cations in general; some ion channels are specific for particular cations or anions; some anion channels transport not only nutrient ions but also organic acids; and solute carriers are responsible for the transport of neutral solutes. A channel is called specific if it is selective, or discriminates, for a particular molecule or group of molecules such that other molecules cannot pass through the channel, regardless of whether their movement is passive or active. Ions moving passively simply diffuse across channels down the electrochemical potential gradient. As with water flow, the direction of movement is from high to low potential, typically from high to low concentration. Ions moving actively must be pumped across membranes by transport proteins called transporters or carriers against their electrochemical potential, ordinarily from low to high concentration (Grossman and Takahashi, 2001). Active movement typically draws on a proton (H^+) gradient across the membrane and requires an input of energy in the form of ATP, which is hydrolyzed to ADP and phosphate (Gilroy and Jones, 2000; Lalonde et al., 2004). Rather than burning ATP directly for nutrient transport, the ATP is used to power pumps, termed H^+-ATPases, which export H^+ from the cell interior to the apoplast at a rate of approximately $1H^+$ per "consumed" ATP (Britto and Kronzucker, 2006; Sondergaard et al., 2004). The resulting decrease in apoplast pH (to pH 4 or 5) generates an electrochemical gradient of -100 to $-150\,mV$ that, in turn, fuels nutrient uptake via the membrane's channels and carriers (Hedrich, 2012). Thus, H^+-ATPases act as energizers that convert the chemical energy released by ATP hydrolysis into chemiosmotic energy; the resulting ADP is then recharged by respiration (see Section 4.4). Whereas in the so-called antiport, one proton is pumped out of the cell for each charge of an incoming cation ($1H^+$ for K^+; $2H^+$ for Ca^{2+}), two protons are needed per charge of an incoming anion that then enters together with the backflowing protons in the so-called proton cotransport or symport (Amtmann and Blatt, 2009).

Active transport enables grapevines to concentrate nutrient ions inside the roots well above their concentration in the surrounding soil solution, although some concentration is also possible simply due to the electrical potential gradients (Amtmann and Blatt, 2009; Keller et al., 1995, 2001b). Most ions occur in the soil water at much lower concentration than would be required for plant cellular functions, and without a concentration mechanism nutrient uptake would not be possible. At the same time, concentrating nutrient ions inside the root cells also generates an osmotic driving force for water uptake by the roots.

Many anions (e.g., NO_3^-, SO_4^{2-}, or Cl^-) are always taken up actively against their electrochemical potential gradient, regardless of their abundance in the soil solution. Some cations (e.g., K^+, NH_4^+, or Na^+), however, are taken up passively across relatively generic, nonselective ion channels when their availability in the soil solution is high and actively and very selectively across carriers when their availability is low (Tester and Leigh, 2001; Véry and Sentenac, 2003). For most macronutrients, the carriers become saturated when the nutrient concentration reaches approximately 1 mM in the soil solution, and the channels are switched on when the nutrient concentration rises above this threshold (Britto and Kronzucker, 2006). Such high nutrient concentrations may be reached after nitrogen fertilizer applications and are relatively common for potassium; moreover, the passive uptake of NH_4^+ and Na^+ can sometimes result in ion toxicity (see Section 7.3). Some transporters, especially those for NO_3^- and K^+, can function in both active and passive mode, depending on the external concentration (Britto and Kronzucker, 2006; Dechorgnat et al., 2011).

Although controlled ion uptake is also important to maintain a neutral electrical charge balance in the plant, uptake rates of cations and anions are rarely precisely equal. Roots restore neutrality by releasing H^+, HCO_3^-, or OH^- into the rhizosphere (Clarkson and Hanson, 1980). In addition, the high uptake capacity of the passive channels is usually accompanied by nutrient efflux from the roots. Whereas anions can leak back passively in exchange for protons, cations must be pumped back out actively (Britto and Kronzucker, 2006). Therefore, because either uptake or efflux is active, even passive nutrient uptake could be called active in terms of energy expenditure for the related active efflux. For most nutrient ions save phosphate, efflux from the roots increases with rising external concentration and becomes almost equivalent to ion uptake at high soil nutrient concentration, resulting in considerable—and apparently futile—nutrient cycling across the membranes (Britto and Kronzucker, 2006).

3.3.5 Transpiration stream and growth

Once inside the stele, most water and dissolved nutrients ultimately return to the apoplast when the water is released into the xylem conduits across the pit membranes separating the vessels and tracheids from the surrounding parenchyma cells (see Section 1.3), which themselves have acquired the solutes via their plasmodesmata from other parenchyma cells or from the cell wall apoplast (Sattelmacher, 2001; Sondergaard et al., 2004; Tegeder and Masclaux-Daubresse, 2018). Because the cell walls of the xylem conduits are waterproof around the pits, release into the xylem is another bottleneck of high radial r_h in the vine's hydraulic system after the passage through the endodermis. In addition, the presence of aquaporins and selective channel or transport proteins in the parenchyma cell membranes provides a chance to actively control access to the xylem (De Boer and Volkov, 2003; Dechorgnat et al., 2011; Tester and Leigh, 2001). Such controlled release and reversible exchange between parenchyma cells and xylem conduits also enables the vine to modify the xylem sap composition and hence to fine-tune nutrient supply with the changing demand of the shoots. Inside the xylem conduits, the negative P_x induced by transpiration and the low axial r_h facilitate rapid transport of water (up to $2\,m\,h^{-1}$ during a sunny day) and its dissolved nutrients to the shoots, where they are delivered initially to the apoplastic compartment of transpiring organs.

Even with the concentration mechanism discussed previously, xylem sap is very dilute compared with phloem sap (see Table 5.1) and the contents of shoot and leaf cells. Nevertheless, due to the transpiration stream, it transports large amounts of nutrient ions by mass flow, although the nutrient concentration can fluctuate severalfold between day and night due to the diurnal change in transpiration rates. Lower concentrations are associated with higher flow rates, which in turn are associated with greater water uptake by the roots; despite the lower concentration during the day, however, the total nutrient flow to the shoots is greater than at night because transpiration-driven water influx is strongly coupled with nutrient uptake (Wegner and Zimmermann, 2009). In addition to nutrient ions, the xylem can also transport sugars, especially during springtime remobilization of storage reserves as discussed previously, as well as other organic molecules such as amino acids as nitrogen carriers (see Section 5.3), organic acids as carriers of metal ions (see Section 7.3), and hormones such as ABA and cytokinins. Solutes can also diffuse in the nonxylem apoplast, but this diffusion is at least one order of magnitude slower than the movement in the xylem (Kramer et al., 2007). This is because cations interact with, and hence are slowed by, the negatively charged carboxyl groups of the cell wall pectins, which function like cation exchangers, and also because the pores in the apoplast hinder the passage of solutes and increase their path length (Sattelmacher, 2001). In addition, xylem-sap water and solutes can readily diffuse to and from adjacent parenchyma tissue en route to the leaves, which helps to buffer fluctuations in xylem sap concentrations and provides nutrients for local metabolism and storage (Metzner et al., 2010a, b; Tegeder and Masclaux-Daubresse, 2018).

The long-distance transport of water and solutes in the xylem, the so-called transpiration stream, occurs by bulk flow (a.k.a. mass flow), in which large numbers of H_2O molecules move together and drag the dissolved ions along, and is driven by a hydrostatic pressure gradient (ΔP_x) caused by transpiration, which "sucks" water up the vine against gravity and friction. While the flow rate increases to the power of 4 with increasing diameter of the xylem conduits, the velocity of the flow increases to the power of 2 (Tyree and Ewers, 1991). Due to the decreasing r_h with increasing diameter, large vessels can transport water much more rapidly than narrower ones.

Older vines with more annual rings and a larger trunk diameter, and hence more large vessels, are able to transport much more water to support a larger leaf area than can young vines, and their generally deeper and more extensive root systems are able to access soil moisture at deeper layers, which is especially important later in the growing season in non-irrigated vineyards as evapotranspiration dries the soil down from the top. The roots, trunks, and cordons also act as important water storage compartments; that is, their parenchyma cells behave as hydraulic capacitors. These reservoirs, which are also called hydraulic capacitance, are depleted in the morning, when the stomata open and transpiration rises, and replenished in the afternoon (Schultz and Matthews, 1988a; Steppe and Lemeur, 2004). Depletion and replenishment of the internal water reserves leads to daily cycles of shrinkage and expansion of the trunk diameter. Older vines can store more water due to their higher number of parenchyma cells. This buffering capacity could make them less vulnerable to xylem cavitation so that they can cope with drought stress better than young vines (Bou Nader et al., 2019). Moreover, the xylem parenchyma cells may also be able to absorb nutrient ions such as nitrate and potassium during periods of abundant supply and release them back into the xylem in times of starvation. In other words, these cells may serve as short-term storage reservoirs of both water and nutrients.

The xylem constitutes >99% of the total length of the transpiration stream; this proportion increases with increasing plant size (Sperry et al., 2003). Nevertheless, a few dozen micrometers in the roots and in the leaves exert the vast majority of whole-vine r_h, even though the entire transpiration stream can be several meters long in tall plants (Sperry et al., 2002). As discussed earlier, one of two major resistances to water flow through a grapevine is not in the trunk or shoots but in the roots, where the radial (i.e., nonxylem) rather than axial (i.e., xylem lumen) resistances dominate (Steudle and Peterson, 1998). The other main resistance is in the leaves, at the terminal section of the transpiration stream, where the water flows through orders of veins in series and in parallel, exits the xylem network at the veins across and around the bundle sheath parenchyma encircling the veins, and moves into and around the mesophyll cells before it evaporates into the intercellular air spaces, changing from the liquid to the vapor phase, and diffuses out of the stomata (Buckley, 2015; Buckley et al., 2017; Comstock, 2002; Sack and Holbrook, 2006; Sack et al., 2003; see also Section 1.3).

Similar to the endodermis in the roots, suberin deposition in the cell walls of the bundle sheath forces water to cross membranes, so that the rate of water flow can be regulated by aquaporins (Shatil-Cohen et al., 2011; Tyerman et al., 2009). In a transpiring grapevine, r_h of the leaves accounts for approximately one-fourth of the whole-plant resistance to water flow, and within the leaf the postvein (i.e., nonxylem) resistance predominates. This means that the distance between the veins and the evaporative surfaces directly correlates with, and controls, leaf r_h, which in turn limits photosynthesis (Brodribb et al., 2007). Rather than remaining constant, however, leaf r_h adjusts to the transpirational water demand, so that more water can be delivered to rapidly transpiring leaves, at least under well-watered conditions (Simonin et al., 2015). Water and solutes move back from the apoplast into the symplast as they cross the pits between the xylem vessels and the neighboring leaf parenchyma cells. Then they move to the mesophyll cells through plasmodesmata or, across the cell membranes, via aquaporins and energy-dependent carrier proteins that are stimulated by auxin. Therefore, the mesophyll is somewhat hydraulically isolated from the remainder of the transpiration stream (Shatil-Cohen et al., 2011).

During the day, grapevines lose most of the absorbed water in transpiration, but a small fraction (~1%) is used for cell expansion, cell metabolism, and phloem transport. At night, when the stomata are essentially closed and transpiration is drastically reduced, those small amounts may become the dominant component of water flow. In fact, water is circulated between the phloem and xylem, and this circulation maintains water flow in the xylem in the absence of any significant transpiration, even at night (Köckenberger et al., 1997; Windt et al., 2006). The water taken up to balance the "growth water" and phloem counterflow amounts to a transpiration-independent water flow that is sufficient to transport nutrients absorbed by the roots; hence, nutrient uptake and long-distance transport in the xylem are not dependent on transpiration (Tanner and Beevers, 2001). Of course, the increase in water flow due to transpiration results in a corresponding increase in the speed at which dissolved molecules move up the xylem (Peuke et al., 2001; Wegner and Zimmermann, 2009). Rapid transpiration thus increases nutrient uptake by the roots, especially when nutrient availability in the soil is high (Alleweldt et al., 1984a; Hsiao, 1973; Keller et al., 1995). Over the course of a day, however, root water uptake must exceed transpiration sufficiently to also supply water for growth because plant growth is mainly caused by cell expansion, which results almost entirely from an increase in the cells' water content (Boyer, 1985; Hsiao and Xu, 2000; Schopfer, 2006). This is why growth is not the same thing as a gain in dry mass. Water import into cells is driven by cell wall loosening and accumulation of solutes inside the cell and thus depends on a water potential gradient between the cell and the supplying xylem ($\Delta\Psi = \Psi_{xylem} - \Psi_{cell}$; see Section 3.1). Therefore, expanding tissues normally have lower Ψ_{cell} than mature, nongrowing tissues. Because growth competes with transpiration for xylem water, cells must also maintain lower Ψ_{cell} during the day than at night if they are to sustain expansion during the day (Boyer and Silk, 2004). Indeed, grape berries, which must remove water from the transpiration stream surging up the shoot and past the fruit clusters, grow mostly at night, especially before veraison and when the water demand of the transpiring canopy is high (Greenspan et al., 1994, 1996; Matthews and Shackel, 2005).

Growth rates of all tissues change rapidly as the extracellular water status, and hence Ψ_{xylem}, fluctuates, and growth is extremely sensitive to water deficit (Boyer, 1985; Boyer and Silk, 2004; Hsiao and Xu, 2000). Shoot growth of grapevines declines with decreasing Ψ_{xylem} and stops completely when Ψ_{xylem} around midday reaches approximately -1.0 to -1.1 MPa (Schultz and Matthews, 1988b). Because water stress decreases Ψ_{xylem}, a major cause for growth inhibition under water deficit may simply be the smaller $\Delta\Psi$, which reduces water uptake by the expanding cells (Nonami and Boyer, 1987; Nonami et al., 1997). Moreover, due to the small, immature xylem conduits near the shoot tip and in young petioles and expanding leaves, the r_h in the "growth zone" is much greater than in the mature portion of the shoot so that a steeper $\Delta\Psi$ is required to sustain growth (Schultz and Matthews, 1993b). If water ceases to move out of the xylem, shoot and leaf growth stop immediately because cells must take up water to expand, whereas root growth is less affected (Boyer and Silk, 2004; Hsiao and Xu, 2000; Wu and Cosgrove, 2000). Lower Ψ_{xylem} is also responsible for the smaller berry size of water-stressed vines. Moreover, it explains why preveraison water deficit has a greater effect on berry size than postveraison water deficit: Water influx into the berry changes at veraison from predominantly via the xylem to predominantly via the phloem (see Sections 1.3 and 6.2).

Photosynthesis of mature leaves is less affected by decreasing Ψ than is expansion of young leaves, and phloem transport can continue at decreasing Ψ (Hsiao and Xu, 2000; Keller et al., 2015b). Consequently, the import of solutes may exceed their use so that solutes often accumulate in sink organs, such as roots or berries, of vines that experience mild water deficit. Therefore, berries grown on mildly water-stressed vines are not only smaller but also have higher sugar concentration. Yet the decrease in photosynthesis under more severe water stress leads to a reduction in berry sugar (see Section 7.2). As in the berries, some of the water for root growth comes from the phloem in addition to direct influx from the soil (Boyer and Silk, 2004). Moreover, because xylem cells are differentiated behind the growing root tip, water uptake for transpiration also occurs behind the expanding cells. Due to this hydraulic isolation, the root tip usually experiences slightly higher soil moisture than the mature root. All of this makes root growth somewhat less vulnerable to fluctuations in Ψ_{xylem} and ensures that it is favored over shoot growth when the soil dries (Boyer and Silk, 2004; Hsiao and Xu, 2000).

CHAPTER 4

Photosynthesis and respiration

Chapter outline

- 4.1 Light absorption and energy capture 129
- 4.2 Carbon uptake and assimilation 134
  - 4.2.1 Gas exchange 134
  - 4.2.2 Carbon assimilation 135
  - 4.2.3 Carbohydrate production 137
- 4.3 Photorespiration 140
- 4.4 Respiration 142
- 4.5 From cells to plants 145

4.1 Light absorption and energy capture

Photosynthesis is the most important way of obtaining energy for all plants including grapevines. As the name suggests, photosynthesis (Greek *photos* = light, *synthesis* = building a whole) is a process by which sunlight is converted into chemical energy that is used to synthesize organic compounds within the plant from inorganic compounds acquired from outside the plant. Sunlight is a form of electromagnetic radiation that has properties of both waves and particles. The wave nature of light can be visualized with a prism separating a ray of visible or white light into a continuous array of different colors according to their wavelengths (Fig. 4.1). Light's particle nature enables us to describe light energy as a stream of energy-carrying particles called quanta (singular: quantum) or photons.

The window of the spectrum called visible light, because our eyes are sensitive to it, also happens to be the range of wavelengths that is important for photosynthesis. This spectral region from about 400 to 700 nm is only a small portion of the entire spectrum of electromagnetic radiation, and other organisms can see light of other wavelengths. Insects and birds, for instance, often distinguish among objects such as flowers or eggs according to their color patterns in the ultraviolet (UV) range, and plants can detect both UV and infrared radiation in addition to visible light (see Section 5.2). All forms of electromagnetic radiation travel at the speed of light ($c \approx 300{,}000 \text{ km s}^{-1}$), but the energy content (E) of photons is higher the shorter their wavelength (λ), or the higher their frequency (υ), within the spectrum. This can be written as follows:

$$E = h\upsilon = hc\lambda^{-1}$$

where h is the Planck constant (6.63×10^{-34} Js).

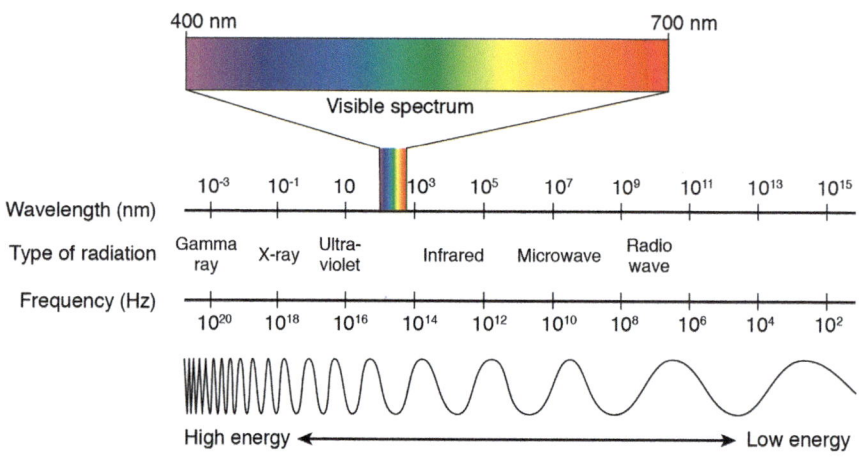

FIG. 4.1

Portion of the electromagnetic spectrum.

The energy contained in the photons is harnessed and exploited by plants during photosynthesis, which proceeds in two stages. The first is the photochemical capture of solar energy, transported in the form of photons, and its temporary storage in the high-energy chemical bonds of adenosine 5′-triphosphate (ATP) and nicotinamide adenine dinucleotide phosphate (NADPH). While ATP is the universal biological energy currency, NADPH is the almost equally universal reducing agent for biochemical reactions. The second stage uses this energy to enzymatically convert carbon dioxide (CO_2) and water (H_2O) to sugar (carbohydrate), which in turn is used to produce all other organic compounds throughout the plant. Light is crucial in this process; without light there is no energy source for photosynthesis, and there is no carbohydrate production.

The two stages of photosynthesis occur in different regions of specialized subcellular organelles termed chloroplasts that contain the photosynthetic machinery. The cells of all green plant organs and tissues have chloroplasts, but the leaves are their main residence, and they are well designed to intercept the maximum amount of light (see Section 1.3). The conversion of light energy into chemical energy happens in the membranes of small, disk-shaped, stacked sacs called thylakoids (Greek *thylakos* = sac) inside the chloroplast, whereas the conversion of CO_2 into sugar (which is discussed in Section 4.2) occurs in the fluid chloroplast matrix referred to as stroma.

This chapter provides a brief summary of the most important aspects of the photosynthetic processes. Because of their importance for life on Earth, most these processes are virtually identical in all plants and are covered in great detail in general texts on plant physiology, from which most of the following information is derived (Atwell et al., 1999; Salisbury and Ross, 1992; Taiz and Zeiger, 2006).

Before the light energy can be used, the light particles must first be absorbed and translated into a flow of electrons that are derived from the splitting of water. The green chlorophyll is the main light-absorbing pigment that makes light energy capture possible. It is the world's most abundant pigment. Chlorophyll is green because it absorbs predominantly blue (430 nm) and red (680 nm) light and

4.1 Light absorption and energy capture

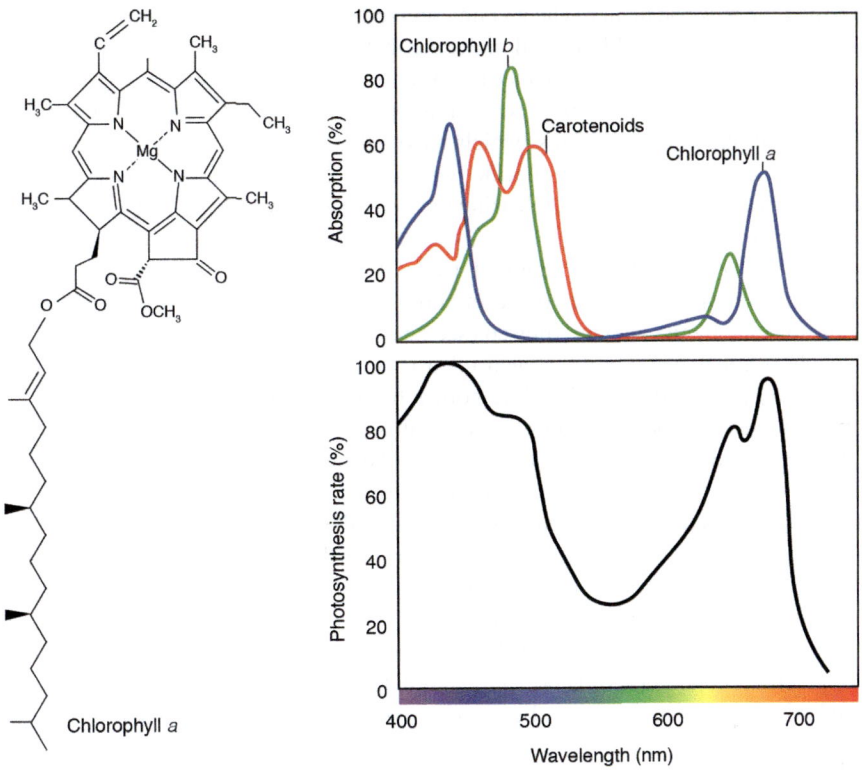

FIG. 4.2

Structure of the chlorophyll *a* molecule (left) and absorption spectra for different light-absorbing pigments (right).

reflects most of the green light with wavelengths intermediate between blue and red (Fig. 4.2). In plants, chlorophyll comes in two slightly different versions called chlorophyll *a* and chlorophyll *b*, both of which, along with various other pigments called accessory pigments, such as carotenes and xanthophylls, are located in the thylakoid membranes. When a chlorophyll molecule absorbs a photon, it is converted from its low-energy state, called ground state, to an excited state by abruptly shifting one of its electrons from a lower energy atomic orbital, or shell, close to the atomic nucleus to a more distant, higher energy orbital. This instantaneous electron "jump" from one energy level to another is termed a quantum leap. The excited chlorophyll is extremely unstable and returns to its ground state within nanoseconds (10^{-9} s) by releasing its available energy as either heat or fluorescent radiation in which the photon is transferred to xanthophylls and reemitted, or by transferring either the energy or an electron to another pigment molecule.

Approximately 250 chlorophyll molecules are grouped together in the so-called photosystem II (PSII), draped on the framework of a multiprotein complex in the thylakoid membrane. Most of these chlorophylls, called antenna pigments, are responsible for "harvesting" light: They absorb photons and transfer their energy from one pigment to another until it arrives at a single specialized chlorophyll molecule called the reaction center. The antenna pigments transfer only their excitation energy so that the reaction center is the only chlorophyll molecule that actually releases an electron in a mechanism called photochemistry. It wins out in the race of possible energy-releasing mechanisms, and thus makes photosynthesis possible, because it occurs at a rate approximately 1000 times faster (i.e., in picoseconds, 10^{-12} s) than the other processes, which makes this energy transfer one of the fastest known chemical reactions. The chlorophyll molecule replaces the lost electron (e^-) by capturing an electron from water, which is thereby oxidized, releasing oxygen (O_2) as a waste product and a proton (H^+). This water-splitting reaction ($2H_2O \rightarrow 4H^+ + 4e^- + O_2$) is catalyzed by a protein complex that contains manganese, calcium, and chloride ions at a ratio of $4 \div 1 \div 1$ (Nelson and Yocum, 2006). Thus, water is the source of all electrons and protons involved in photosynthesis. Both e^- and H^+ are later incorporated into a sugar molecule by the enzyme rubisco (see Section 4.2). For each electron that is released, one water molecule is consumed and its oxygen released to the air. Unlike the hydrogen, the oxygen in photosynthetically produced sugar therefore comes from CO_2, not from H_2O.

The electron released from the reaction center chlorophyll is transferred to an acceptor molecule in a process termed photochemical quenching because it "quenches" the excitation energy of the chlorophyll (Baker, 2008). The acceptor in turn passes the electron on to a secondary acceptor and so on down a cascade of steps called the electron transport chain until the electron arrives at the reaction center of a second photosystem, confusingly named photosystem I (PSI), operating in series with PSII. PSI is an extremely efficient nano-photoelectric machine that is also composed of a multiprotein complex and contains approximately 175 chlorophyll molecules, which also absorb light and transfer its energy (Nelson and Yocum, 2006). The electron having been handed over by PSII is transferred from the excited PSI reaction center to yet more acceptors until it is accepted by an iron sulfur-containing protein called ferredoxin (Latin *ferrum* = iron). Ferredoxin passes the electron on to $NADP^+$ which is thereby reduced to NADPH.

The protons produced during the oxidation of water are released into the thylakoid's interior, the so-called thylakoid lumen (Baker, 2008). In addition, the electron transport chain between PSII and PSI also pumps protons from the stroma into the thylakoids using some of the energy released by the electrons flowing down the chain. This results in a relative shortage of protons in the stroma; the pH difference between the thylakoid (pH 4.5) and the stroma (pH 8) equates to a 3000-fold difference in protons. Use of the term "protons" is a convenient simplification, because every H^+ dissolved in water is attached to a H_2O molecule, which makes it a hydronium ion (H_3O^+). The proton gradient across the thylakoid membrane creates an electrochemical potential gradient which is called the proton motive force and constitutes a source of energy that powers another protein complex in the thylakoid membrane called ATP synthase or ATPase. In a process called photophosphorylation, ATPase produces ATP by attaching inorganic phosphate (P_i) to ADP: $ADP + P_i \rightarrow ATP$. Working like a turbine in a hydroelectric power plant, each rotation of the enzyme ATPase releases three molecules of ATP while transferring 14 protons back out to the stroma. The "waterfall" or flow of protons drives the "turbine" while recycling all protons back to the chloroplast stroma.

In essence, energy transfer and photochemistry convert light energy into chemical energy in the form of NADPH and ATP, which provide the energy for carbon assimilation, as well as for the assimilation of nitrogen (see Section 5.3), phosphorus, and sulfur, and other metabolic processes in the chloroplast (Baker, 2008). The proportion of absorbed photons that are used for photosynthesis is termed quantum yield or quantum efficiency, and it decreases as incident light intensity increases (Baker, 2008). Although inevitable, the release of oxygen so close to the PSII reaction center is problematic because excited chlorophyll can cause this oxygen to be transformed into the very reactive singlet oxygen ($^1O_2$), which can damage membranes and cause mutations or cell death by oxidizing lipids, proteins, and DNA (Apel and Hirt, 2004; Halliwell, 2006; Møller et al., 2007). Indeed, many herbicides (e.g., diuron, diquat, paraquat, simazine) used to kill unwanted plants in vineyards act by blocking the photosynthetic electron transport chain, leading to "self-poisoning" of plants by producing excessive amounts of $^1O_2$ in PSII reaction centers. Spray drift of such herbicides on grapevine leaves thus causes necrotic lesions wherever a spray droplet has landed. As an aside, whereas vegetatively propagated grapevines would have great difficulty evolving resistance to such herbicides, sexually propagated weeds can do so much more readily. This is because even minute mutations that decrease the transport of herbicide molecules or their binding to the thylakoid membrane can markedly reduce herbicide efficacy (Powles and Yu, 2010). Hence, the repeated use of similar herbicides may encourage the evolution of herbicide resistance by increasing the selection pressure on weed populations.

Chlorophyll, then, is a "Jekyll and Hyde" molecule, enabling the crucial light absorption for photosynthesis but acting as a phototoxin when it becomes overly excited (Hörtensteiner, 2009). Excessive oxygen is toxic due to the formation of free radicals termed active (or reactive) oxygen species (see Section 7.1). Fortunately, under normal conditions, the carotenoids associated with the reaction center and antenna complex quickly "scavenge" singlet oxygen in addition to dissipating excess light energy absorbed by the antenna pigments and converting it into heat. This dissipation is termed nonphotochemical, energy-dependent quenching or thermal dissipation, and it employs the so-called xanthophyll cycle, which converts violaxanthin via antheraxanthin to zeaxanthin when there is too much light but runs in the reverse direction when light intensity declines (Baker, 2008; Demmig-Adams and Adams, 1996; Fanciullino et al., 2014; Noctor and Foyer, 1998). Consequently, the amount of the yellow xanthophyll pigments antheraxanthin and zeaxanthin in grapevine leaves often rises in the morning and declines in the evening, whereas violaxanthin follows the opposite trend (Düring, 1999). In addition, carotenoids also assist with light harvesting by absorbing photons at wavelengths not covered by chlorophyll (Bartley and Scolnik, 1995; Fig. 4.2).

When a leaf absorbs more light energy than can be converted to chemical energy and used in CO_2 assimilation, the "energy overload" will damage the photosynthetic apparatus by "knocking out" manganese ions (Mn^{2+}) from the water-splitting protein complex of PSII and inactivating the reaction center. Such damage is an unavoidable ancillary cost of the business of doing photosynthesis and increases in proportion to the light intensity; high doses of high-energy UV light are especially harmful (Takahashi and Badger, 2011; Takahashi and Murata, 2008). It reduces the efficiency of photosynthesis and is called photoinhibition, which also is a form of nonphotochemical quenching (Baker, 2008; Gamon and Pearcy, 1990b). The damaged proteins are normally disassembled in a process termed proteolysis and recycled to build new PSII proteins; thus, photoinhibition results from the balance between protein damage and repair. Excess light, however, suppresses the repair cycle (Takahashi and Badger, 2011).

When the photosynthetic fixation of CO_2 is limited, the demand for ATP and NADPH declines. Oxygen can seize the resulting surplus electrons from PSI, especially in strong light, and the ensuing superoxide ($O_2^{\bullet -}$) is converted to hydrogen peroxide (H_2O_2). These reactive oxygen species strongly inhibit the repair process by interfering with the assembly of proteins, which can cause problems under conditions of environmental stress that curtail CO_2 fixation (Takahashi and Murata, 2008; see Section 7.1).

4.2 Carbon uptake and assimilation
4.2.1 Gas exchange

Just as it does for other plants, uptake of CO_2 by leaves presents a dilemma for grapevines. Because CO_2 is available in the atmosphere around plants and photosynthesis takes place inside the leaf cells, CO_2 must move from the atmosphere to the leaf interior. This happens by way of diffusion through the stomata (see Fig. 1.15), then through the intercellular air spaces, cell walls, and finally across membranes into the cells and chloroplasts. As discussed in Section 3.3, the rate of diffusion is dependent on a driving force, which is provided by a potential gradient or concentration gradient, and the resistance to diffusion. The diffusion path for CO_2 into a leaf is almost entirely through the stomata, whereas the path for water vapor out of a leaf involves both the stomata and the cuticle (Boyer, 2015; Boyer et al., 1997). CO_2 diffusion through the stomata follows the same path as H_2O during transpiration but in the reverse direction (see Section 3.2). While water evaporates from cell walls, CO_2 dissolves in the water of cell walls and is partly converted to HCO_3^- (bicarbonate). However, the subsequent path for CO_2/HCO_3^- is lengthy, because it must cross additional cell walls and membranes, namely the plasma membrane and the chloroplast membranes, before it can be assimilated in the chloroplasts. Thus, although CO_2 may be able to use certain aquaporins (which are normally reserved for water transport and hence have been renamed cooperins if they are permeable to CO_2) to cross membranes, it meets more points of resistance than does water (Evans et al., 2009; Katsuhara et al., 2008; Maurel et al., 2008; Nobel, 2009). About half of the total resistance in the mesophyll is imposed by the cell walls alone (Terashima et al., 2011).

Moreover, the concentration gradient of CO_2 from the outside air to the leaf's interior is very small. The current atmospheric CO_2 concentration (C_a) is approximately 0.04% ($=400\,\mu mol\,mol^{-1} = 400\,\mu bar = 400\,ppm$), whereas the CO_2 concentration inside the leaf (C_i) cannot be <0%. Grapevines literally have to extract their main source of food and energy out of thin air. By contrast, air at 20°C and 50% relative humidity contains approximately 1.25% H_2O, whereas the air inside the leaf is always at almost 100% relative humidity and contains approximately 2.5% H_2O (Buckley et al., 2017). This means that the gradient for H_2O diffusion out of the leaf is >40 times steeper than that for CO_2 diffusion into the leaf. To make matters worse, H_2O molecules are smaller than CO_2 molecules, so H_2O diffuses approximately 1.6 times more easily through air than does CO_2. Consequently, the gas "exchange rate" of mature leaves is only approximately 1 CO_2 molecule fixed for every 200–600 H_2O molecules lost, depending on the cultivar and environmental conditions (Fig. 4.3). This exchange rate is estimated as the ratio of the rate of photosynthesis (A) to the rate of transpiration (E) and is also termed water use efficiency or instantaneous water use efficiency (WUE $= A\,E^{-1}$). Leaves develop fewer stomata when they unfold under elevated atmospheric CO_2 concentrations (Woodward, 1987); the unfavorable exchange rate between CO_2 and H_2O might be the cause of this adaptive response.

The net uptake of CO_2 by the leaf equals the net rate of photosynthesis per unit leaf area (A, usually expressed in $\mu mol\,CO_2\,m^{-2}\,s^{-1}$) and can be approximated by the following equation:

$$A \approx g_l(C_a - C_i)$$

where g_l is the leaf conductance.

The term g_l reflects the combined gas conductances of the boundary layer (g_b) and the stomata (g_s) and increases with increasing stomatal aperture (see Section 3.2). As with water flow, conductance is

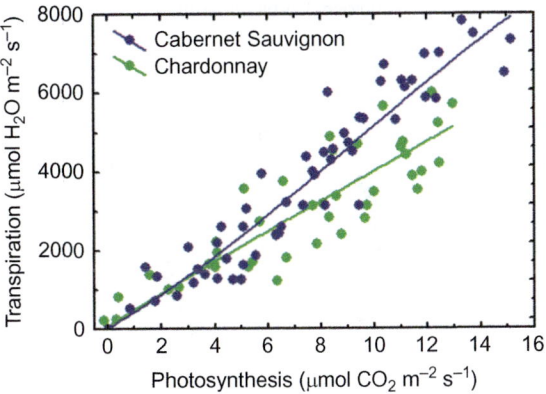

FIG. 4.3
Relationship between photosynthetic CO_2 uptake and transpirational H_2O loss in mature leaves of two grapevine cultivars.

M. Keller, unpublished data.

the reciprocal of resistance ($g = r^{-1}$). The stomata are open in the light, and CO_2 diffuses across their pores into the leaf and then into the mesophyll cells to be assimilated in the process of photosynthesis. In the dark or under drought conditions, the stomata are almost closed, and photosynthesis ceases. As discussed previously, however, opening of the stomata not only allows CO_2 to diffuse into the leaf but also allows water vapor to escape from the leaf. The dilemma for grapevines is that they cannot simultaneously maximize CO_2 uptake and minimize H_2O loss. Consequently, g_s continually adjusts in a manner that optimizes CO_2 acquisition relative to H_2O loss; in economics terminology, the stomata's "job" is to optimize the marginal water cost of carbon gain (Franks et al., 2013).

4.2.2 Carbon assimilation

Once CO_2 has entered the chloroplast, the chemical energy generated by the chloroplast pigments is used to convert CO_2 and water into carbohydrate such as the hexose sugar glucose ($C_6H_{12}O_6$). In the overall process of photosynthesis, the leaf absorbs CO_2 molecules and releases O_2 molecules as a waste product of the process (see Section 4.1), as shown in the following chemical equation:

$$6CO_2 + 12H_2O \rightarrow C_6H_{12}O_6 + 6O_2 + 6H_2O$$

$$\Delta G = +2840 \, kJ \, mol^{-1} \, (1 \, mol \, glucose = 180 \, g)$$

where ΔG is the change in standard (Gibbs) free energy during the reaction, and the "+" sign indicates that the reaction is endothermic, which means it requires addition of energy to proceed.

The necessary energy is provided by ATP, and the ATP-consuming incorporation of CO_2 into carbohydrate is termed carbon assimilation. It is achieved by the action of 13 enzyme proteins in the Calvin cycle located in the chloroplast stroma. The Calvin cycle is named after one of its discoverers, the American biochemist Melvin Calvin, but is also called the photosynthetic carbon reduction cycle or the reductive pentose phosphate pathway. The cycle proceeds in three successive stages

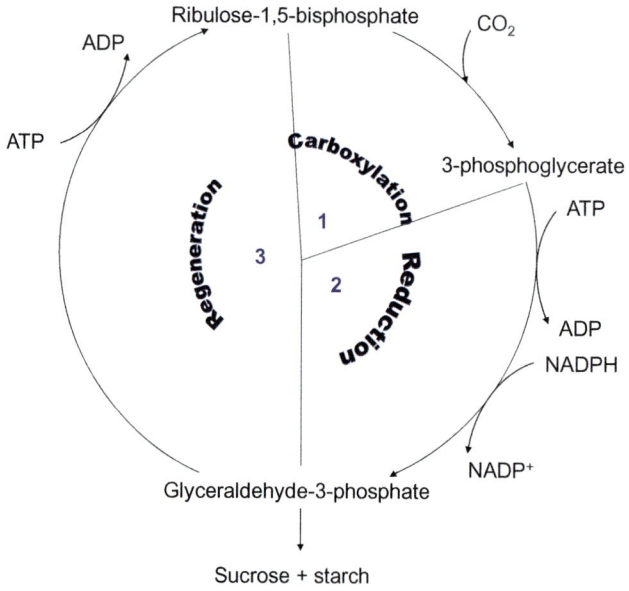

FIG. 4.4

The three stages of the Calvin cycle that converts CO_2 into carbohydrates.

Illustration by M. Keller.

(Fig. 4.4). The first stage is termed carboxylation and combines CO_2 and protons derived from water with ribulose-1,5-bisphosphate, a molecule that specializes as a carbon acceptor, to generate two intermediate molecules of 3-phosphoglycerate, thereby converting the inorganic CO_2 into an organic molecule. As its name suggests, this initial organic compound, like all other intermediates of the Calvin cycle, is a sugar phosphate. Despite the simplified and convenient equation introduced above, glucose is not a direct product of the Calvin cycle. Because 3-phosphoglycerate is the first stable product of CO_2 fixation and contains three carbon atoms, grapevines are grouped with most other crop plants as the so-called C_3 plants.

The carboxylation step is catalyzed by the enzyme ribulose bisphosphate carboxylase/oxygenase (or rubisco for short). Owing to its importance in carbon assimilation, rubisco occurs in chloroplasts at very high concentration—often in substantial excess. It represents approximately one-third of a leaf's total protein complement, appropriating up to half of the leaf's nitrogen, which arguably makes it the most abundant protein nature ever invented (Spreitzer and Salvucci, 2002). Rubisco commandeers an extremely prominent position in grapevine physiology because almost all the carbon that is assimilated is initially "captured" by this protein. As its full name suggests, the enzyme "works" in two opposing directions so that CO_2 competes with oxygen for the same molecular acceptor. Although under normal conditions the forward carboxylation reaction is much faster than the reverse oxygenation reaction, the fact that O_2 is present in the atmosphere at a much higher concentration (21%) than CO_2 (0.04%) means that the competition results in a loss of some of the CO_2 that has entered the cells (Spreitzer and Salvucci, 2002; see also Section 4.4).

The second stage, called reduction, uses ATP and NADPH from photochemistry to form a triose phosphate carbohydrate named glyceraldehyde-3-phosphate. The two reactions involved in this reduction oxidize ATP to ADP and NADPH to $NADP^+$; the ADP and $NADP^+$ then have to be returned to the thylakoid membranes to be "recharged."

During the third and final stage of the cycle, termed regeneration, the initial CO_2 acceptor molecule is restored in a further series of enzymatic reactions involving 10 of the 13 Calvin cycle enzymes and various intermediate sugar phosphates. The last of these steps also uses ATP. Thus, the complete Calvin cycle consumes two NADPH molecules and three ATP molecules for every CO_2 molecule that is fixed into carbohydrate. The regeneration stage allows continued uptake of CO_2, but it also means that only one out of every six triose phosphates produced can be either used for starch production within the chloroplast or exported to the cell's cytosol for the synthesis of sucrose (Paul and Foyer, 2001). Therefore, three turns of the Calvin cycle are required to form one 3-phosphoglycerate molecule, and six turns are required to produce one glucose molecule. However, the cycle operates very rapidly; free sugars such as sucrose appear within $<30s$ of CO_2 entering a leaf. Note that animal cells carry out all reactions of the Calvin cycle except for the first and the last step. Because the two enzymes necessary to carry out those steps are missing, animals cannot convert CO_2 into sugar and rely on plants for their nutrition.

4.2.3 Carbohydrate production

The organic products of photosynthesis are commonly called assimilates or photosynthates; they are often noted as $(CH_2O)n$, which is the basic component of carbohydrates. The simple sugars, such as glucose and fructose, are termed hexoses or monosaccharides and are produced by putting six of the basic units together (i.e., $n=6$). They can be formed either directly inside the chloroplast or in the cytosol from triose phosphates exported from the chloroplast. Several enzymes are involved in hexose biosynthesis, some of which remove the phosphate groups (P_i), which can then be used to recharge ADP into ATP.

Combining glucose and fructose gives the disaccharide sucrose ($n=12$), which is the most abundant sugar found in nature. Sucrose, a glycoside, is the major end product of photosynthesis and the predominant organic transport compound of grapevines. Sucrose is produced in the cytosol rather than the chloroplast. For each sucrose molecule, a glucose phosphate molecule is energized as UDP-glucose before it is linked with a fructose phosphate molecule. In this case, the required energy is not provided by ATP but instead comes from a related compound, uridine triphosphate (UTP). The P_i released during sucrose biosynthesis is transported back into the chloroplast for continued photophosphorylation (Paul and Foyer, 2001). As shown in Fig. 4.5, a specialized protein embedded in the chloroplast membrane, the phosphate translocator, or triose phosphate transporter, is responsible for exchanging triose phosphate (out) for P_i (in). Sucrose is exported from photosynthesizing cells to the phloem and distributed throughout the plant for use in the production of other organic components, maintenance processes, and growth of the various sink organs (see Section 5.1).

In contrast to sucrose synthesis, the production of starch inside the chloroplast requires ATP to form ADP-glucose. The starch "assembly line" links thousands (5000–500,000) of glucose molecules together, resulting in the long, chemically and osmotically inert chains of glucose that form the carbon-storage polymer called starch. Depending on the nature of the bonding between the glucose molecules, the chains can be linear (α-1,4-glucans), in which case they are termed amylose, or branched

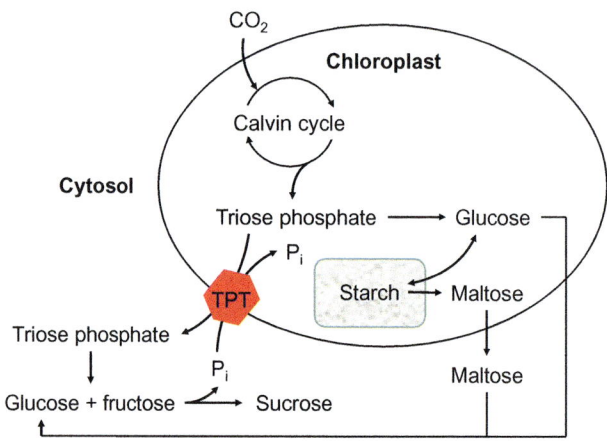

FIG. 4.5

Simplified diagram of the dynamics of sugar production, export, and starch accumulation inside a leaf cell. The triose phosphate transporter (TPT) exports triose phosphate in exchange for inorganic phosphate (Pi) released during sucrose synthesis.

Illustration by M. Keller.

(α-1,4- and α-1,6-glucans), in which case they are called amylopectin. Starch is organized into semi-crystalline grains or granules of alternating amorphous and crystalline regions that can grow from <1 to >100 µm in diameter by adding layers of starch, which results in growth rings (Emes and Neuhaus, 1997; MacNeill et al., 2017; Martin and Smith, 1995; Tetlow et al., 2004; Zeeman et al., 2010).

The Calvin cycle feeds carbon into starch production when the export of sucrose from the cell cannot keep pace with photosynthesis (Paul and Foyer, 2001). This is necessary because osmotic pressure is proportional to the number of dissolved particles (see Section 3.1): n glucose molecules cause n times the osmotic pressure of a polymer composed of n identical glucose units. Accumulation of many osmotically active, small solute particles like sucrose or glucose would therefore cause water to flood the cell and it would burst. To avoid this potentially lethal problem, sucrose that accumulates in the leaf cells slows its own formation in a process called feedback inhibition. The P_i released during sucrose production assumes the function of "regulator"; a high P_i concentration indicates a high rate of sucrose export from the cell and promotes triose phosphate export from the chloroplast (Woodrow and Berry, 1988). Conversely, a low P_i concentration indicates that sucrose export is slow and promotes starch accumulation for temporary storage in the chloroplast (Fig. 4.5). Due to the key role of cellular P_i for this feedback regulation, phosphate deficiency resulting from low P_i uptake by the roots leads to starch accumulation and prevents starch remobilization for sucrose export, which may lead to feedback inhibition of photosynthesis (Hendrickson et al., 2004b; see also Section 7.3). Starch also accumulates in leaves when the phloem of a leaf or shoot is obstructed or injured. This can happen due to physical injury from strong wind or trellis wires, crown gall or virus infection, or tendrils coiling around a shoot. Such shoots can be easily recognized in a vineyard because the leaves turn bright red as the surplus sugars trigger the production of anthocyanin pigments (see Fig. 2.7). Accumulation of light-absorbing anthocyanins may occur in order to protect the leaves' photosynthetic machinery from damage due to excessive light absorption, because sugar accumulation leads to feedback inhibition of photosynthesis to bring the supply of sugar back in balance with the demand for sugar (Ayre, 2011).

The rate of sucrose export is linked to the rate of photosynthesis: The greater the rate of photosynthesis, the greater the rate of concurrent export—and vice versa (Wardlaw, 1990). As discussed previously, grapevines, like other plants, often use starch production as an overflow mechanism when sucrose formation exceeds the leaves' export capacity or the demand for sucrose by the various plant organs (Johnson et al., 1982; Mengin et al., 2017). The starch can be remobilized by removing glucose units simultaneously from the many chain ends in a starch granule and converting them, partly via intermediary maltose, back to sucrose following maltose and glucose export out of the chloroplast (Smith, 2012; Smith et al., 2005; Zeeman et al., 2010; Fig. 4.5). Little or no starch turnover normally occurs in leaves during the day, but degradation is switched on at night when there is no photosynthesis to provide carbon for continued sucrose export and generation of energy and reducing agents in the leaf (Geiger and Servaites, 1991; Geiger et al., 2000; Tetlow et al., 2004).

Therefore, starch acts as a carbon sink in the leaf during the day and switches to a carbon source at night (MacNeill et al., 2017). In the absence of environmental limitations, this transient starch pool reaches a minimum by sunrise and is replenished by sunset, so that sucrose export remains relatively constant over the course of a day–night cycle. An intricate, albeit poorly understood, control mechanism that involves the circadian clock ensures that starch accumulation and turnover are adjusted such that they compensate for changes in day length during the growing season to maintain a steady supply of sucrose over a 24h period (Geiger et al., 2000; Graf and Smith, 2011; Stitt et al., 2007). This means that the *proportion* of assimilates produced by photosynthesis that is allocated to starch increases, and sugar export for growth decreases, as day length decreases to ensure a sufficiently large reserve pool for the longer night (Mengin et al., 2017). Nonetheless, because the total *amount* of starch that accumulates is greater during long days than during short ones, more starch is broken down during the ensuing short nights. In other words, the remobilization of leaf starch reserves is matched exactly to the length of the night (Graf and Smith, 2011). Consequently, the total daily amount of sucrose export from a leaf increases as day length increases so that during much of the growing season more sugar is potentially available for growth, yield formation, and fruit ripening at higher latitudes than at lower latitudes. Daytime starch remobilization, on the other hand, apparently occurs only when dense clouds or canopy shade prevent net carbon assimilation (Smith et al., 2005). Due to their ability to channel assimilated carbon into the fairly large short-term storage pools of starch and sucrose, grapevines can insulate the rate of sucrose supply to other organs from the changes in the rate of photosynthesis that occur during the course of day and night (Woodrow and Berry, 1988).

Glucose molecules can also be linked together as linear β-1,4-glucans forming polymers of 2000–25,000 glucose units called cellulose. Cellulose is arranged in approximately 35 more or less parallel chains that form a network of long and very strong crystalline microfibrils of about 3nm in diameter that are held together by xyloglucan and hydrogen bonds (Doblin et al., 2010; Cosgrove, 2016). These inert, inextensible fibrils, whose tensile strength is greater than that of steel, form the stiff structural elements that make up the basic framework of all plant cell walls, which make up a large portion of a vine's dry weight. In fact, cellulose is the most abundant organic polymer on Earth, accounting for roughly half of the total organic matter. For instance, dry wood consists of 40–50% cellulose in addition to 25% hemicellulose and 25–35% lignin (Plomion et al., 2001), and cotton fibers used in the textile industry are made up of approximately 95% cellulose; cellulose is also the stuff that makes the paper of the print version of this book. The cellulose microfibrils are responsible for the stiffness and mechanical strength of cell walls. Adjoining microfibrils are interconnected with one another and with the wall matrix that fills the space between them (Hamant and Traas, 2010; Cosgrove, 2016, 2018; see also Section 3.1). The viscous matrix is itself mostly comprised of polysaccharides, namely the hydrophilic pectin and the cellulose-binding matrix glycans, also called hemicellulose, which is mainly composed of xyloglucan (Cosgrove, 2005; Doblin et al., 2010; Reiter, 2002; Wolf et al., 2012). Structural proteins and enzyme proteins make up the remainder of the wall matrix.

During periods of rapid growth, cellulose production can be a major drain on a cell's resources because both new, dividing cells and expanding cells require insertion of new primary cell wall material. Indeed, most of the carbon fixed by photosynthesis may be incorporated into cell walls (Reiter, 2002). In secondary cell walls, which fully grown cells insert between their primary cell wall and the cell membrane, the carbohydrate polymers are complemented by the three-dimensional, water-repellent polymer lignin. Lignin consists of simple phenolic building blocks which originate from the aromatic amino acid phenylalanine and are linked together in the process of lignification that often coincides with or even follows cell death. Lignification employs enzymes such as laccase and peroxidase along with oxygen and H_2O_2 as oxidizing agents to activate the lignin monomers that cells have deposited in the secondary cell wall; the activated monomer radicals then self-assemble into the stable lignin polymer (Barros et al., 2015; Boerjan et al., 2003; Schuetz et al., 2013). By displacing water, lignin fills the spaces between the wall polysaccharides, adding extra mechanical support and waterproofing, especially to the xylem cells of the wood. Its importance as a stiffener and strengthener of the cell wall structure makes lignin nature's second most plentiful polymer after cellulose, comprising approximately 30% of all organic carbon (Boerjan et al., 2003). Once lignified, cells become unable to expand further and are glued to their cell neighbors by the pectin-rich middle lamella.

A chloroplast is essentially a solar-powered microscopic factory. In a young, expanding leaf this biosynthetic organelle makes compounds for the leaf's own growth and development. In a mature leaf, on the other hand, a chloroplast assimilates carbon for export to other parts of the plant. Photosynthesis therefore increases with leaf age and in grapevines reaches a maximum approximately 5 or 6 weeks after leaf unfolding, gradually declining thereafter, as senescence sets in (Bertamini and Nedunchezhian, 2003; Williams and Smith, 1985).

4.3 Photorespiration

As discussed in Section 4.2, rubisco, the first enzyme in the Calvin cycle, can catalyze the addition of both CO_2 and O_2 to ribulose-1,5-bisphosphate. The latter reaction is called oxygenation and is the main reaction occurring during a process known as photorespiration (Bauwe et al., 2010; Farquhar et al., 1980; Ogren, 1984; Spreitzer and Salvucci, 2002). Because the Earth's current atmosphere contains vastly more O_2 (21%) than CO_2 (0.04%) and due to the release of O_2 in the water-splitting reaction of photosynthesis (see Section 4.1), the $O_2 \div CO_2$ ratio in the chloroplast fluid is approximately $24 \div 1$ at 25°C. Moreover, though CO_2 is much more water soluble than O_2, increasing temperatures decrease the solubility of CO_2, which increases the $O_2 \div CO_2$ ratio at higher temperatures, thus favoring photorespiration.

Photorespiration is a problem for grapevines because the process works in the opposite direction of photosynthesis and therefore competes with it for ATP and NADPH, produces toxic phosphoglycolate instead of valuable phosphoglycerate, and results in a loss of CO_2 that could otherwise be used for sugar production. A portion of this CO_2 can be recovered in a series of reactions called the photorespiratory carbon oxidation cycle or glycolate pathway. This recovery process is quite complex and involves interconversion and transport of several amino acids and cooperation among three distinct cell organelles—the chloroplast, peroxisome, and mitochondrion (Bauwe et al., 2010; Foyer et al., 2009; Ogren, 1984). It also costs energy in the form of ATP and results in heat loss. In addition, the release of toxic ammonium (NH_4^+) from amino acids requires yet more ATP for detoxification. Although the reassimilation of NH_4^+ is costly, losing the NH_4^+ as gaseous ammonia (NH_3) to the atmosphere and then replacing it with soil-absorbed nitrate (NO_3^-) would be even more expensive. The cellular flux of NH_4^+ due to photorespiration is at least 10-fold higher than that due to root NO_3^- uptake and assimilation (see Section 5.3).

The glycolate pathway, named after the initial product of the oxygenation reaction, recovers approximately 75% of the lost carbon and returns it as phosphoglycerate to the Calvin cycle. Yet approximately one out of every four CO_2 molecules that enter the chloroplast is released back to the atmosphere (Bauwe et al., 2010; Düring, 1988). The upshot is that photorespiration reduces the efficiency of photosynthesis. The biological function, if any, of photorespiration is not understood. One possible function could be the use of electrons when solar energy supply exceeds demand, such as in very bright light, or during drought conditions when the photosynthetic pigments capture more photons than the Calvin cycle can use to fix CO_2 (Apel and Hirt, 2004; Foyer et al., 2009; Lawlor and Cornic, 2002; Niyogi, 2000; Ogren, 1984). The venting of surplus electrons substantially reduces a potential "energy overload" that would otherwise damage the photosynthetic machinery and lead to photoinhibition (Takahashi and Badger, 2011; see also Section 4.1).

Nonetheless, it is likely that this seemingly wasteful pathway may be an evolutionary "hangover." Photosynthesis, and hence rubisco, evolved >3 billion years ago, that is, very early in the existence of life on Earth, when the atmosphere was rich in CO_2 but almost devoid of O_2 (Bauwe et al., 2010; Foyer et al., 2009; Xiong and Bauer, 2002). As a remarkable aside, it was the "invention" of photosynthesis with oxygen as its waste product (in addition to injection of oxygen into the air from abiotic sources) that enabled the evolution of all oxygen-breathing life forms, including humans (Xiong and Bauer, 2002). This innovation was a momentous breakthrough event in the history of life because it enabled the photosynthetically endowed organisms to exploit limitless supplies of water and solar energy. Of course, not all life depends on oxygen for respiration, as viticulturists and winemakers should know. Brewer's or baker's yeast (*Saccharomyces cerevisiae*) is one of many anaerobic microorganisms that cannot, or prefer not to, use oxygen. Although yeast is a facultative anaerobe and thus is not poisoned by oxygen—unlike obligate anaerobes—it relies on energy-inefficient fermentation rather than respiration of sugar for energy generation (Rolland et al., 2006). Yeast's anaerobic nature is why Louis Pasteur described fermentation as "life without oxygen." Stable ethanol production occurs only under anaerobic conditions; in the presence of oxygen, ethanol is oxidized to acetic acid by other, aerobic yeast species or *Acetobacter* bacteria. Without oxygen, grapes are turned into wine, with oxygen, into vinegar.

The rising atmospheric CO_2 concentration ($[CO_2]$) as a consequence of our rapid burning of fossil fuels, forest clearing, and soil cultivation which is responsible for global climate change will, to some extent, turn the clock back, as it were, to benefit the carboxylation function of rubisco and favor photosynthesis over photorespiration (Foyer et al., 2009; Huang et al., 2000; IPCC, 2013; Paul and Foyer, 2001). Whereas throughout most of the last million years, that is, at least five times as long as humans of our species, *Homo sapiens*, have roamed Earth, $[CO_2]$ never exceeded 300 ppm, and was around 200 ppm during the repeated ice ages, it has increased from approximately 270 ppm prior to the beginning of industrialization to >400 ppm today (Franks et al., 2013; Petit et al., 1999). If the rise continues at the current average annual rate of 2–3 ppm, [CO_2] will reach 600 ppm by the end of the 21st century (Franks et al., 2013; IPCC, 2013).

The increase in ambient [CO_2] results in a steeper CO_2 concentration gradient into the leaves, which favors carboxylation by rubisco and enhances photosynthesis while decreasing g_s. Indeed, across all C_3 species studied thus far, an increase in [CO_2] from 370 ppm to 570 ppm was associated with a one-third increase in both light-saturated and daily photosynthesis that was accompanied by a >80% increase in leaf starch content, whereas rubisco content and g_s declined by approximately 20% (Ainsworth and Rogers, 2007; Long et al., 2004). At least some of the decrease in g_s is a consequence of the lower stomatal density of leaves grown at higher [CO_2] (Franks et al., 2013; Woodward, 1987; Engineer et al., 2016). Grapevines are no exception to this general trend (Düring, 2003; Salazar-Parra et al., 2015; Schultz, 2000; Schultz and Stoll, 2010; Tognetti et al., 2005). All other plant processes affected by elevated [CO_2] are thought to be an outcome of the two basic responses of increasing photosynthesis and decreasing g_s (Long et al., 2004). In the long term, however, photosynthesis can only increase if grapevines can utilize the extra sugar produced, for example, by increasing shoot and root growth or yield, all of which increase under higher [CO_2] (Bindi et al., 1996; Johnson et al., 1982). Moreover, a general decrease in g_s decreases the leaves' evaporative cooling ability (see Section 3.2), which leads to an increase in leaf temperature and thus heat stress, especially during water deficit (Engineer et al., 2016).

Some plant species long ago, and often independently, "invented" solutions to the problem posed by high [O_2]: The most prevalent of these are the so-called C_4 plants, which were named for the four-carbon oxaloacetate, rather than the usual three-carbon 3-phosphoglycerate, as the product of CO_2 fixation (Foyer et al., 2009; Lundgren et al., 2014; Ogren, 1984; Sage et al., 2012). Among other adaptations, the C_4 plants use a biochemical pump to concentrate CO_2 near the rubisco enzymes, which almost eliminates photorespiration and speeds up the rates of photosynthesis and growth, especially in high light and at high temperature. Because of their competitive advantage in warm climates, these plants include not only a few important commercial crops, such as maize, sorghum, and sugarcane, but also many highly successful weed species, such as crabgrass, switchgrass, pigweed, and nutgrass. Photosynthesis in C_4 species benefits less from the current rise in atmospheric [CO_2] so that they can be expected to become somewhat less competitive in the future—although it is doubtful that this will make weed control any easier.

4.4 Respiration

In stark contrast with photorespiration, respiration is the series of three processes responsible for releasing the energy stored in carbohydrate and incorporating it into a form that can be readily used to power most energy-requiring metabolic processes in the plant. The only similarity between photorespiration and respiration, and the reason for their similar name, is that both release CO_2. The three sequential ATP-generating processes that make up respiration are glycolysis, citric acid cycle, and electron transport chain. In addition to ATP, respiration produces many intermediate organic components whose "carbon skeletons" can be used for the manufacture of other compounds. During respiration, which occurs in all living tissues, glucose is completely oxidized to CO_2 and its electrons are transferred to oxygen, which is then reduced to water. In the language of chemistry, respiration is the oxidation of carbon reduced by photosynthesis. Therefore, the chemical equation that summarizes this process is basically the reverse of that used to describe photosynthesis:

$$C_6H_{12}O_6 + 6O_2 + 6H_2O \rightarrow 6CO_2 + 12H_2O$$

$$\Delta G = -2840\,kJ\,mol^{-1}$$

where ΔG is the change in standard (Gibbs) free energy during the reaction, and the "−" sign indicates that the reaction is exothermic, which means it releases energy.

Sugars serve as the main carriers of energy that the leaves "extract" from the sun, and the energy is stored in the sugars' chemical bonds. Breaking these bonds consequently releases the chemical energy and makes it available to power other biological processes. Because sucrose, rather than glucose, is the transport sugar in grapevines, and because respiration also occurs in nonphotosynthetic tissues that rely on sucrose import, sucrose is the sugar used as the starting substrate for respiration. However, sucrose must be broken down into its two constituent hexoses, glucose and fructose, both of which can then enter the glycolysis pathway in the cells' cytosol (Fig. 4.6). In addition to hexoses, stored starch can also be used for respiration after it has been hydrolyzed to glucose. When such carbohydrates

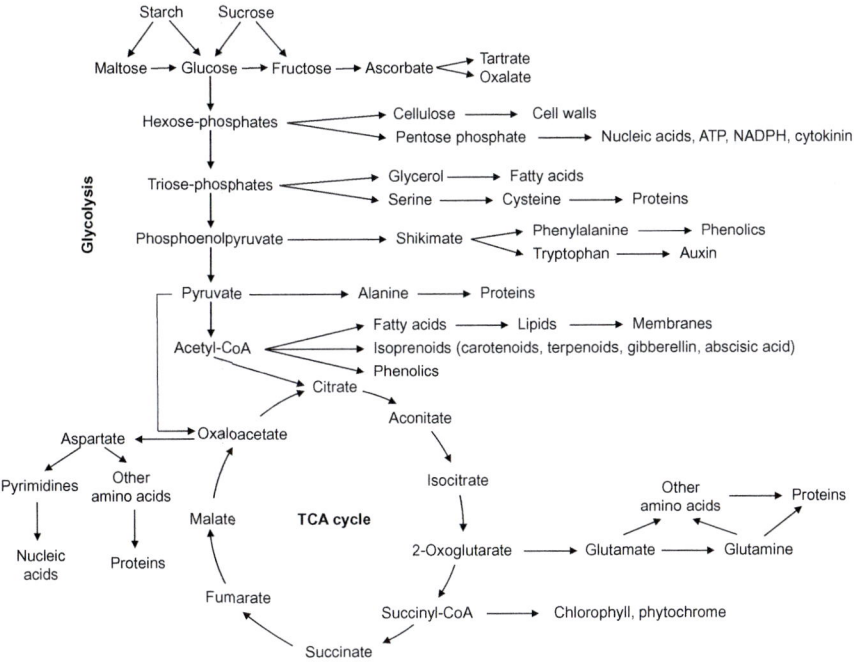

FIG. 4.6

Glycolysis and TCA cycle are of central importance for the manufacture of many organic compounds throughout the plant.

Modified after Salisbury, F.B., Ross, C.W., 1992. Plant Physiology, fourth ed. Wadsworth Publishing, Belmont, CA.

are not readily available, tissues can rely on alternative substrates to fuel respiration (Zell et al., 2010). These include organic acids (e.g., malate), lipids, and even proteins. The nature of these respiratory substrates can be guessed at by measuring the so-called respiratory quotient (RQ), which is the ratio of CO_2 released per O_2 consumed. Only sugars are respired to one molecule of CO_2 for each molecule of O_2 consumed, which results in RQ = 1. For the comparatively more oxidized organic acids, RQ ≈ 2, and for the more reduced lipids and proteins, RQ < 0.5. Although the oxidation of one sucrose molecule with its 12C atoms yields a net energy equivalent of 60–64 ATP molecules, the maximum respiration efficiency is probably slightly <5 molecules of ATP per molecule of CO_2 released (Amthor, 2000).

Glycolysis (Greek *glykos* = sugar, *lysis* = splitting) oxidizes glucose in a series of enzymatic reactions to the three-carbon compound pyruvate (Fig. 4.6). This process initially requires input of energy in the form of ATP but later regenerates twice as many ATP molecules in addition to NADH. Pyruvate still contains >75% of the energy contained in glucose and can be used as the starting material for the subsequent citric acid cycle.

The next steps of respiration are carried out by the citric acid cycle, which is also called the Krebs cycle after its discoverer, the German physician Hans Adolf Krebs; the tricarboxylic acid cycle; or the TCA cycle for short. This cycle then feeds into an electron transport chain known as oxidative phosphorylation. Both reaction sequences occur in specialized organelles termed mitochondria (singular: mitochondrion), which can be regarded as the cell's power plants. The number of mitochondria per cell varies and is directly related to the metabolic activity of the tissue to which the cell belongs; energy generation is driven by demand. However, there are hundreds of mitochondria scattered through most cells. This is possible because mitochondria have their own genome and replicate independently of their "mother cell" by simple division. In other words, they reproduce asexually like bacteria, and they can do so many times while the cell and its nucleus do not divide. Nevertheless, most of the mitochondrial enzymes are now encoded by genes in the nucleus, partly in concert with the few remaining mitochondrial genes (Millar et al., 2011).

To enable ready transfer of their products to the TCA cycle, the enzymes carrying out glycolysis are attached to the outside of the mitochondria (Fernie et al., 2004). The end-product of glycolysis, pyruvate, is transported into the mitochondria, where it is oxidized to acetic acid and linked to a sulfur-containing coenzyme called acetyl coenzyme A, commonly abridged to acetyl-CoA. Acetyl-CoA is then combined with oxaloacetate already present in the cycle or supplied from the glycolysis intermediate phosphoenolpyruvate by the enzyme phosphoenolpyruvate carboxylase (Fig. 4.6). In a series of three enzymatic reactions, the TCA cycle reduces the resulting molecule completely to CO_2 and H_2O, generating a substantial amount of chemical energy or reducing equivalents (10 NADH per glucose) in the process. The remaining three steps of the cycle restore oxaloacetate, which is then ready to accept the next acetyl-CoA from pyruvate, allowing the continued operation of the cycle. Despite its location in the mitochondria, many steps of the TCA cycle can also be performed by enzymes present in the cytosol, and the cycle is often incomplete in that it interacts with the cytosol by means of membrane transporters that permit the bypassing of various mitochondrial steps (Millar et al., 2011).

The subsequent electron transport chain transfers electrons from NADH to oxygen in a stepwise manner involving several carrier proteins bound to the inner membrane of the mitochondrion. This mitochondrial electron transfer chain is structurally very similar to the photosynthetic electron transfer chain in the chloroplast (see Section 4.1) and involves a transfer of protons from the interior of the mitochondria to the cytosol. The electrochemical potential gradient resulting from the movement of protons across the mitochondrial membrane is used to power an enzyme termed ATP synthase that recharges ADP to ATP using P_i (Fernie et al., 2004; Millar et al., 2011). This ATP synthase is life's universal power generator; it produces almost all the ATP of living cells. The production of 15 ATP equivalents per pyruvate molecule by mitochondria, which consume oxygen to burn sugar as fuel, is by far the most efficient means of biological energy generation. However, because respiration also produces a small amount of highly damaging reactive oxygen species, it may slowly but surely lead to oxygen poisoning, which in turn results in aging and, eventually, senescence (Logan, 2006; see also Section 7.1). In addition to the "standard" electron transport chain, the mitochondria employ an alternative electron transport pathway, which uses an enzyme called alternative oxidase to reduce oxygen to water but does not produce ATP, to minimize the production of reactive oxygen species under stress conditions such as low temperature (Millar et al., 2011).

Besides producing ATP, respiration generates various intermediate carbon compounds and thus is central to the production of a wide variety of components of the plant's metabolism (Fernie et al., 2004; Sweetlove et al., 2010; see also Fig. 4.6). These include cellulose, used to build cell walls; nucleotides, used to produce nucleic acids, cytokinins, etc.; amino acids, used to produce proteins, flavonoids, anthocyanins, lignin, etc.; fatty acids, used to produce cell membrane lipids, etc.; isoprenoids, used to produce terpenoids, carotenoids, cytokinins, abscisic acid, gibberrellins, etc.; and porphyrins, used to produce chlorophylls, phytochromes, etc. For example, all 20 standard amino acids formed in plants for the manufacture of proteins are assembled from the carbon skeletons provided by the organic acid products of glycolysis, the TCA cycle, and the Calvin cycle. Therefore, a large portion of the sugar that is metabolized by glycolysis and the TCA cycle is not oxidized to CO_2 but instead is diverted to biosynthetic purposes; sugar metabolites are used to produce new materials. Indeed, it seems that the TCA cycle operates as a full cycle only when a cell has a high demand for ATP. Otherwise it may take various noncyclic "shortcuts" depending on the cell's need for metabolic precursors or intermediates (Sweetlove et al., 2010). Moreover, both ATP and ADP themselves are also used as the original precursor molecules for the assembly of all compounds that comprise the group of cell division hormones called cytokinins (Sakakibara, 2006).

4.5 From cells to plants

Although photosynthesis is a grapevine's ultimate source of carbohydrates and supplies some ATP and NAD(P)H in photosynthetic tissues, respiration is its powerhouse, providing the driving force for biosynthesis, cell maintenance, and active transport into and out of cells. Respiration is generally broken down into two components: growth respiration and maintenance respiration (Amthor, 2000; Penning de Vries, 1975; Penning de Vries et al., 1974). In growth respiration, reduced carbon is processed to fuel the accumulation of new biomass. Growth respiration can thus be viewed as the equivalent of a plant's "construction cost." To enable growth, a plant also incurs a "transport cost," because sugar export from the leaves in the phloem to other growing organs requires energy too. So does the uptake, assimilation, and long-distance transport of mineral nutrients. These processes are sometimes viewed as comprising a third respiratory category that supports growth (Kruse et al., 2011). A substantial portion of growth respiration involves the production of carbon skeletons for nitrogen assimilation (see Section 5.3).

The cost of production for plant tissues varies with the nature of the chemical compounds that are being produced (Table 4.1). For instance, producing proteins or phenolic compounds is much more

Table 4.1 Average construction costs of major chemical plant components.

Component	Cost[a]
Carbohydrates	1.17
Organic acids	0.91
Lipids	3.03
Amino acids, proteins	2.48
Phenolics	2.60
Inorganic compounds (minerals)	0

[a]Expressed as g glucose used to produce 1 g of each compound class.
From Vivin, P., Castelan-Estrada, M., Gaudillére, J.P., 2003. Seasonal changes in chemical composition and construction costs of grapevine tissues. Vitis 42, 5–12.

expensive than producing carbohydrates; thus, plant parts with a high protein content, such as leaves, are relatively expensive to build (Vivin et al., 2003). The rate of growth respiration increases as the rate of growth increases.

The other component, maintenance respiration, is needed to keep existing, mature cells in a viable, functional state and is coupled to energy generation, internal carbon and nutrient transport, protein and lipid turnover, and adjustment or acclimation to changing environmental conditions. Protein turnover, which refers to the production–breakdown–recycling–resynthesis of proteins, and adjustment/acclimation to variable environmental conditions are linked so that plants do not have to maintain a complete set of enzymes and transport proteins at all times just to be ready for potentially changing conditions (Araújo et al., 2011). It is much more economical for them to produce specialized proteins, or groups of proteins, "on demand" and recycle them when not needed so that their amino acid building blocks can be reassigned to other temporary jobs. Scientists do not have a good understanding of energy utilization by maintenance respiration, but estimates suggest that it may represent >50% of the total respiratory CO_2 release.

Growth is not possible without respiration, but respiration consumes some of the carbon that could otherwise be used to "build" the plant body. Therefore, the net amount of carbon accumulated in a grapevine's biomass during a growing season equals the total amount of carbon fixed in photosynthesis minus that "lost in respiration." Respiration can consume a significant portion (30–80%) of the photosynthetically fixed carbon each day over and above the losses due to photorespiration, and it releases large amounts of CO_2 back to the atmosphere (Amthor, 2000; Atkin and Tjoelker, 2003; Atkin et al., 2005). Grapevine leaves respire at a rate of approximately 5–10% of the rate of photosynthesis. However, all plant tissues, whether photosynthetic or not, respire, and they do so 24 h a day; they also grow during both the daily light period and the night (Mengin et al., 2017). At least the green shoots and, to a much lesser degree, even the older woody organs, especially their relatively chlorophyll-rich inner bark, can use photosynthesis to refix some of the lost CO_2 and thereby contribute to their own construction and maintenance costs (Kriedemann and Buttrose, 1971; Pfanz, 2008; Palliotti et al., 2010; Saveyn et al., 2010).

Respiration rates vary with the age of the plant and differ from one organ to another depending on the availability of carbon substrates, stage of development, and temperature. For instance, there is an exponential relationship between respiration and temperature (Zufferey, 2016). The extent to which the respiration rate increases for each 10 °C increment in temperature is termed the Q_{10} factor. A temporary 10 °C rise in temperature, such as may occur during a heat wave, usually leads to a doubling or so of the respiration rate: $Q_{10} \approx 2$.

Young grapevines lose approximately one-third of their daily photosynthate to respiration, and this loss increases in older plants as the amount of nonphotosynthetic tissue in their cordon, trunk, and roots increases. Both growth and whole-plant respiration are approximately proportional to the three-quarter power of whole-plant mass (Price et al., 2010). The total annual respiration of a vineyard soil has been estimated to exceed 5 t of carbon per hectare, with about half of this attributed to grapevine roots (Franck et al., 2011).

As a general rule, the greater the metabolic activity of a given tissue or organ and/or the higher its growth rate, the higher its respiration rate. Meristems have the highest respiration rates so that young, rapidly growing plant parts with high rates of cell division, such as developing buds, shoot and root tips, unfolding leaves, or flowers, can respire up to 10 times more rapidly than older parts that rely mainly on maintenance respiration. Young leaves near the growing shoot tip may respire at twice the rate of mature leaves (Zufferey, 2016). Because grapevine organs contain about 45% carbon (i.e., 0.45 g C g^{-1} dry matter), between 70% and 90% of imported carbon is incorporated in new material in growing organs; the remainder is released as CO_2 (Amthor, 2000). This means that during and after budbreak in spring, vines have a negative carbon balance: They will literally lose weight (dry matter) until the new leaves become photosynthetically competent and begin to replace the lost carbon (Buttrose, 1966b; Holzapfel et al., 2010; Weyand and Schultz, 2006a). The vines' carbon balance may become positive at around the six-leaf stage (Gatti et al., 2016).

It appears that respiration constitutes a compromise between energy generation and carbon gain. A high respiration rate means that more ATP is produced and thus more energy is available for metabolism, but this comes at the expense of carbon lost to the accumulation of biomass. Conversely, a low respiration rate improves the carbon balance but reduces the energy available to utilize this additional carbon. The increase in photosynthesis associated with the rising atmospheric [CO_2] that is driving global climate change (see Section 4.3) may also stimulate respiration by increasing leaf sugar and starch concentrations (Amthor, 2000; Leakey et al., 2009; Salazar-Parra et al., 2015). Although this will diminish the net plant carbon balance, it may nevertheless lead to greater carbon export for growth and yield formation, which is associated with higher growth and maintenance respiration. The rise in temperature that accompanies the higher atmospheric [CO_2] may further stimulate respiration and growth, although probably not as much as do short-term, temporary increases in temperature during a growing season, because plants acclimate to, and hence compensate for, long-term temperature changes (Amthor, 2000).

Although mineral nutrient ions per se do not cost any glucose for their construction since they are inorganic chemicals, grapevines spend much of the energy generated during respiration on nutrient acquisition—that is, on the uptake of nutrient ions from the soil solution and their subsequent assimilation into organic compounds. Acquisition of nitrate and sulfate incurs particularly high respiratory costs. Consequently, high rates of uptake of these nutrients by the roots cause an increase in respiration rates to supply carbon skeletons for the production of amino acids (see Section 5.3).

Vigorous species or cultivars are characterized by spending less respiratory energy for nutrient acquisition and more for growth than do their less vigorous counterparts. In addition, their maintenance costs also seem to be slightly lower, although this difference is small compared with the difference in energy allocated to nutrient acquisition. Thus, rather than having higher photosynthetic rates, vigorous plant species often respire a lower proportion of their acquired carbon than do slow-growing species; that is, the respiration:photosynthesis ratio is smaller in vigorous species (Loveys et al., 2002). However, differences in maintenance respiration become more important when grapevines are grown under unfavorable conditions. For example, low nitrogen availability in the soil decreases the absolute rates of whole-plant respiration but increases the proportion of carbon fixed during the day that is lost during respiration. This happens because vines with a low nitrogen status have reduced photosynthesis and invest more of their photosynthate in the growth of roots, which have higher respiration rates than shoots (Keller and Koblet, 1995a).

CHAPTER 5

Partitioning of assimilates

Chapter outline

- 5.1 **Photosynthate transport and distribution** 149
  - 5.1.1 Sources and sinks 149
  - 5.1.2 Assimilate transport 152
  - 5.1.3 Loading, transport, and unloading 156
  - 5.1.4 Allocation and partitioning 161
- 5.2 **Canopy–environment interactions** 168
  - 5.2.1 Light 170
  - 5.2.2 Temperature 179
  - 5.2.3 Wind 183
  - 5.2.4 Humidity 184
  - 5.2.5 The "ideal" canopy 185
- 5.3 **Nitrogen assimilation and interaction with carbon metabolism** 188
  - 5.3.1 Nitrate uptake and reduction 188
  - 5.3.2 Ammonium assimilation 190
  - 5.3.3 From cells to plants 192

5.1 Photosynthate transport and distribution

5.1.1 Sources and sinks

Grapevines, like all other plants, employ a division of labor between different tissues and organs, which is coupled to their structural framework and relies on an efficient transport system connecting the various parts. Photosynthates or assimilates, which are the organic compounds produced during photosynthesis and mineral nutrient assimilation, and some inorganic nutrients must be transported from their place of manufacture or storage to places of use or storage. The former thus function as organs of supply, called "sources," while the latter function as organs of demand, called "sinks." A source is any plant organ that acquires resources form the environment and exports surplus materials to other organs. The typical source organ for carbon is the mature leaf that produces more photosynthate than it needs for its own growth and metabolism, but all green (i.e., chlorophyll-containing) tissues, including those in a shoot or in a fruit cluster, can contribute to a vine's total assimilate production. The main source organ for mineral nutrients is the growing root. Other, albeit temporary, sources of grapevines include the woody structures of canes, cordons, trunks, and roots, which serve as storage sites for remobilizable assimilates and other nutrients. All sources, however, begin their life as sinks, and all sources for certain classes of compounds are sinks for others. For instance, over their lifetime leaves are net sources of carbon but sinks for mineral nutrients, whereas roots are net sources of mineral nutrients but sinks for carbon. Nevertheless, leaves also become sources of mineral nutrients at the end of their life, and perennial organs are sources of carbon at the beginning of each growing season and at other times to support survival of and recovery from stress conditions when the leaves are unable to meet the sinks' carbon demand. A carbon sink is a nonphotosynthetic plant organ or an organ that produces insufficient photosynthate to supply its own needs. Such sinks can be growing vegetative organs, such as expanding leaves, shoots, and roots; storage organs, such as shoots, canes, trunks, and roots; or reproductive organs, such as flowers, fruits, and seeds.

Sources and sinks have varying strengths and interact by diverse means of communication to regulate resource distribution (Yu et al., 2015). The term source strength refers to a plant organ's net acquisition rate for a particular resource like carbon or nitrogen from the environment and is usually defined as the product of source size and source activity (White et al., 2016). The source size is the dry weight of a source organ, and the source activity is the resource acquisition rate per unit source size. Similarly, the term sink strength refers to a plant organ's net import rate for a particular resource and is defined as the product of sink size and sink activity. The sink size is the dry weight of a sink organ, and the sink activity is the resource import rate per unit sink size. Sink activity comprises the use of imported materials to construct new tissues, to operate, maintain, or modify existing tissues, to accumulate storage reserves, and to lose or dispose of material to the environment. The total source strength of a grapevine is the sum of the source strengths of all source organs and equals the vine's total sink strength, that is, the sum of the sink strengths of all sink organs. Consequently, sources and sinks constantly interact with one another. These source–sink interactions lie at the basis of all aspects of plant growth, and a balance between sources and sinks is critical for efficient plant growth, development, and yield formation (Yu et al., 2015; White et al., 2016). Many viticultural practices are designed to manipulate either the source strength or the sink strength of grapevines.

All leaves are initially sinks because they need to build up their photosynthetic machinery before they can start producing and exporting their own photosynthates. The construction phase entails an investment of carbon and other nutrients in both building material and energy. The time period required for a leaf to make the transition from sink to source is important because it determines how quickly the leaf starts paying dividends on its construction cost. It seems as though leaves unfolding in spring pass through this period more rapidly than leaves unfolding in fall; thus, late-season leaf growth may be too slow to make a significant contribution to fruit ripening or carbohydrate storage. Indeed, late-season shoot growth may compete for resources with the ripening fruit as well as with the replenishment of the reserve pool of the vine (Keller et al., 2010). This is one reason why viticultural practices aim at enhancing shoot growth early but suppressing it late in the growing season.

The sink-to-source transition of a leaf happens gradually, beginning at the leaf tip; a grapevine leaf generally starts exporting assimilates when it has reached approximately one-third its final size but continues to import carbon up to half its final size, which requires bidirectional assimilate transport in the petiole (Koblet, 1969; Taiz and Zeiger, 2006). When a leaf reaches 50% of its final size, it is generally about 10–14 days old, depending on the temperature during leaf expansion, and the oldest, most basal leaf on a shoot typically begins exporting assimilates when the shoot has approximately five or six leaves (Hale and Weaver, 1962; Koblet, 1969; Koblet and Perret, 1972; Yang and Hori, 1979, 1980). Export is initially directed toward the growing shoot tip, but as soon as the next leaf above makes the transition from sink to source, the older leaf also starts to export a portion of its assimilates toward the shoot base and into the permanent structure of the vine. Only a few days later the direction of export from the older leaf becomes exclusively basipetal. This pattern repeats itself as new leaves unfold from the growing shoot tip, although export from leaves below a cluster becomes bidirectional again after fruit set.

Following the sink–source transition, a leaf will no longer import substantial amounts of sugar, even under conditions of heavy shading that prevent photosynthesis and under which a mature leaf acts at most as a very weak sink. A leaf reaches maturity approximately 40 days after it has unfolded from the shoot tip. Thereafter, its chlorophyll content and photosynthetic rate often begin to decline, and the leaf gradually exports less photosynthetically fixed carbon as it ages (Hunter et al., 1994; Keller et al., 2001b; Intrieri et al., 1992, 2001; Poni et al., 1994a,b). Under nonlimiting environmental conditions, however, many leaves, especially those that unfold early in the growing season, can remain fully photosynthetically active for >100 days; indeed, leaf senescence and the associated decline in photosynthesis seemingly set in at about the same time of year in all leaves of a canopy (Schultz et al., 1996). This means that leaves that are formed late in the growing season have a progressively shorter window of opportunity to "pay back" their construction costs. Old leaves, though still photosynthetically active, no longer export assimilates but, rather, retain them for their own metabolic demands (Koblet, 1969). Instead, the leaves become a major source of mineral nutrients, especially nitrogen, phosphorus, and potassium, toward the end of their life as a result of the resorption of proteins and other constituents during senescence (Thomas and Stoddart, 1980; see also Section 2.2).

The woody organs of a grapevine are generally regarded as sinks because their photosynthetic activity is mostly restricted to recycling a portion of the CO_2 lost during respiration, while also increasing the internal O_2 concentration (Pfanz, 2008; Saveyn et al., 2010; Wittmann and Pfanz, 2014). They rely on imported photosynthate and nutrients for growth and metabolism. However, sugar and other nutrients imported and stored as starch, amino acids and proteins in these tissues during the growing season are remobilized in spring to support budbreak and the initial shoot and root growth before the new leaves start to export assimilates (Holzapfel et al., 2010; Wardlaw, 1990; Williams, 1996). Starch remobilization, which entails hydrolyzation of starch to glucose and glucose phosphate and their conversion back to sucrose for export, is stimulated by the plant hormone gibberellin. Although the oldest leaf on a new shoot begins exporting some of its assimilates to the woody parts of the vine as early as 2–3 weeks after budbreak, remobilization and export from the perennial organs to the developing shoots continues through the time of bloom. Consequently, assimilate transport is bidirectional, occurring concurrently in opposing directions up and down a growing shoot and in subtending woody organs. In addition to supporting early spring growth, stored reserves serve as a buffer by providing a temporary backup supply of carbon that can be remobilized during stress periods, such as overcast conditions or loss of leaf area from late spring frost, hail, or pest attack, which curtail photosynthesis (Bloom et al., 1985; Holzapfel et al., 2010).

5.1.2 Assimilate transport

The transportation of assimilates must be sufficiently flexible to supply the growing and metabolizing tissues of roots, shoots, leaves, flowers, and fruits. It therefore must be rapid to keep up with the demand for respiratory fuel and construction material by these growing sink organs, which can be far away from a source. It also must be able to reverse the flow direction depending on the relative needs of different organs at different times. This nimble assimilate transport from sources to sinks occurs in the phloem, or more precisely, in the phloem's sieve elements (see Section 1.3). Because the concentration of assimilates is usually higher in the source phloem than in the sink phloem, transport follows a concentration gradient or chemical potential gradient.

By far the major component besides water that is transported in the phloem of grapevines is sucrose (Koblet, 1969; Swanson and El-Shishiny, 1958). On average, sucrose accounts for 50–70% of the phloem sap osmotic pressure (Table 5.1). However, its concentration declines on its way from a source to a sink, which creates a steep osmotic gradient in the sieve tubes. The resulting mass flow also drags along numerous other solutes, with potassium (K^+), amino acids, glucose, fructose, and malate making up much of the remainder of the osmotic components (Patrick, 1997; Peuke, 2010; Peuke et al., 2001). The concentrations of amino acids, such as glutamine and glutamate, and organic acids, such as malate and ascorbate, vary widely but are normally much lower than the sucrose concentration. Nonetheless,

Table 5.1 Composition of Phloem and Xylem Sap.

Solute	Xylem sap	Phloem sap
Sugars	$\leq 5\,g\,L^{-1}$	$100–300\,g\,L^{-1}$
(mostly sucrose, some glucose and fructose)	$\leq 15\,mM$	$300–900\,mM$
Amino acids	$\leq 2\,g\,L^{-1}$	$5–40\,g\,L^{-1}$
(mostly glutamine, some glutamate, asparagine, and aspartate)	$\leq 15\,mM$	$30–270\,mM$
Proteins		
(mostly glycolytic, structural, and stress-related proteins)	Traces	$0.2–2\,g\,L^{-1}$
Organic acids	$0.1–0.5\,g\,L^{-1}$	$1–3\,g\,L^{-1}$
(mostly malate)	$0.7–4\,mM$	$7–20\,mM$
Inorganic ions	$0.2–4\,g\,L^{-1}$	$1–5\,g\,L^{-1}$
Potassium (K^+)	$1–10\,mM$	$40–100\,mM$
Sodium (Na^+)	$0.1–2\,mM$	$0.1–5\,mM$
Magnesium (Mg^{2+})	$1–2\,mM$	$1–6\,mM$
Calcium (Ca^{2+})	$1–5\,mM$	$0.1–2\,mM$
Phosphate ($H_2PO_4^-/HPO_4^{2-}$)	$0.1–3\,mM$	$7–15\,mM$
Nitrate (NO_3^-)	$0.01–20\,mM$	Traces
Total solute concentration	$10–100\,mM$	$250–1200\,mM$
Osmotic pressure (π)	$0.02–0.2\,MPa$	$0.6–3\,MPa$
pH	$5.0–6.5$	~ 7.5

Modified from Buchanan, B.B., Gruissem, W., Jones, R.L., 2000. Biochemistry and Molecular Biology of Plants. ASPP, Rockville, MD; Peuke, A.D., Rokitta, M., Zimmermann, U., Schreiber, L., Haase, A., 2001. Simultaneous measurement of water flow velocity and solute transport in xylem and phloem of adult plants of Ricinus communis over a daily time course by nuclear magnetic resonance spectrometry. Plant Cell Environ. 24, 491–503, and Keller, M., Kummer, M., Vasconcelos, M.C., 2001b. Soil nitrogen utilisation for growth and gas exchange by grapevines in response to nitrogen supply and rootstock. Aust. J. Grape Wine Res. 7, 2–11.

as the leaves change from carbon sources to mineral nutrient sources during senescence, the phloem sap composition shifts from predominantly sucrose to mainly amino acids (Thomas and Stoddart, 1980).

Besides assimilates, some nutrient ions can also be transported in the phloem. For instance, most of a grape berry's K^+ is imported via the phloem, and phosphate can be retrieved from old leaves and recycled to young, expanding leaves under conditions of phosphorus deficiency. Additionally, the phloem also contains short pieces of RNA, termed microRNA (or miRNA), and several peptides and proteins. An array of phloem-specific proteins are manufactured in the companion cells, and they have diverse functions. For example, sucrose carriers and ATPases are involved in metabolism and transport of sugar, K^+ and Ca^{2+} channels facilitate nutrient transport, and aquaporins speed up water transport across the phloem cell membranes. Other proteins are responsible for the almost instantaneous sealing of the phloem's sieve plate pores upon injury inflicted by summer pruning or removal of leaves or fruit clusters. Long-term plugging is then achieved within minutes by deposition of callose, whose production is induced by a sudden increase in cytosolic Ca^{2+} in the sieve elements (Furch et al., 2007; Kühn et al., 1999; Oparka and Santa Cruz, 2000; van Bel, 2003; see also Section 1.3).

Because the sieve elements lack the necessary machinery to produce proteins, RNA pieces might be used as long-distance signals that promote or, more frequently, inhibit the production of certain proteins in the target sink organs (Kehr and Buhtz, 2008; Lough and Lucas, 2006; Turgeon and Wolf, 2009; Turnbull and Lopez-Cobollo, 2013). Such inhibition is called gene silencing because it essentially "switches off" specific genes by preventing their RNA from being translated into proteins. The RNA signals are astonishingly specific, they even include base sequences that act as a "zip code" specifying the recipient tissue and ensuring proper delivery, even across graft unions.

Like the xylem, moreover, the phloem can also transport almost all plant hormones, including auxin, gibberellins, cytokinins, and abscisic acid. By acting as chemical signals, these hormones exert remote controls on physiological processes in organs and tissues distant from their site of production. Even electrical signals may be transmitted through the phloem, which gives the phloem properties of a neural network (Fromm and Lautner, 2007; Mancuso, 1999). Electrical signals may also be transmitted through the xylem, where they arise locally in response to hydraulic pressure waves or transported compounds due to physical injury. Such signals are generated in response to sudden environmental stimuli, including temperature, light, touch, and wounding, and exert an additional regulation of physiological functions. The long-distance information transmission of electrical signals in the phloem is much more rapid ($>10\,m\,h^{-1}$) than that of chemical signals ($\sim 1\,m\,h^{-1}$).

Unfortunately, some undesired hitchhikers, such as viroids, viruses, and some bacteria, also use the phloem-generated mass flow for their own transport and are able to slip through the intercellular plasmodesmata, enabling them to spread easily throughout the vine (Christensen et al., 2004; Lough and Lucas, 2006; Oparka and Santa Cruz, 2000; Roberts and Oparka, 2003; van Bel, 2003). Finally, the ability of the phloem to transport fungicides, insecticides, and herbicides within the plant is exploited in viticultural practice to combat pathogens and weeds, using so-called systemic pesticides.

The composition of the phloem sap not only depends on the grape species, cultivar, and rootstock but also varies with changing physiological and environmental conditions, location within the plant, and time of the growing season. Although the composition also varies with time of day, the amount of transported solutes remains more or less constant over a 24-h period (Peuke et al., 2001). In addition, solutes can be exchanged between the phloem and the xylem (Lalonde et al., 2004; Metzner et al., 2010b). Xylem-to-phloem transfer occurs especially in the vascular bundles branching away from a shoot into a leaf (i.e., in leaf traces) and in the minor veins of a leaf, but it can also happen in the shoot, especially at nodes. It enhances the delivery of nutrients to rapidly growing, but slowly transpiring, sinks like meristems, fruits, and seeds. Phloem-to-xylem transfer, on the other hand, is important in roots and shoots for redirecting some phloem-delivered surplus solutes, such as amino acids and K^+, back to the shoots via the xylem, and for buffering short-term fluctuations in nutrient concentrations in the xylem.

Phloem sap typically moves at flow velocities of $0.2–1.5\,m\,h^{-1}$, which is much faster than would be expected from simple diffusion (Knoblauch et al., 2016; Koblet, 1969; Windt et al., 2006). Phloem flow can be bidirectional within a single internode but is unidirectional within a given sieve element. In contrast to the xylem, the rate and direction of solute movement in the phloem are under metabolic control and change over time in response to source and sink development. Thus, the direction of phloem flow can be reversed over time and often occurs against the direction of the transpiration stream in the xylem. The pressure flow theory of phloem sap movement developed by Münch (1930) states that sap moves from places of high turgor pressure ($P = \Psi_P$) to places of low pressure within the phloem (Fig. 5.1). The volume flow rate (F_v) varies with the resistance of the flow path (r_h) and can be described by the following equation (Patrick, 1997):

$$F_v = \Delta P\, r_h^{-1}$$

where $\Delta P = P_{source} - P_{sink}$, and r_h is the hydraulic resistance of the phloem path.

The radius of a sieve tube, and hence the cross-sectional area (A), is important insofar as it determines the hydraulic conductance ($l_h = L_h A = r_h^{-1}$, where L_h is the hydraulic conductivity; Patrick et al., 2001). It is important to realize that P_{source} refers to the pressure inside the source phloem and P_{sink} refers to the pressure inside the sink phloem; that is, the hydrostatic pressure gradient (ΔP) occurs within the sieve tubes rather than between a source organ and a sink organ. Therefore, the sieve tubes are pressurized to $P_{source} \approx 0.5–2.5\,MPa$; P_{source} increases with increasing length of the flow path, that is, increasing distance between a source and a sink, which depends on plant size (Knoblauch and Peters, 2010; Knoblauch et al., 2014, 2016). The buildup and dissipation of pressures in the phloem are isolated from the turgor pressures in the source and sink cells to permit continued phloem flow and solute transport. It follows that the flow rate increases as P_{source} increases and/or as P_{sink} decreases. Indeed, sinks compete for assimilates by lowering the pressure in their sieve elements (Wardlaw, 1990). Lower P_{sink} propagates as lower pressure throughout the entire phloem and thus serves as a signal to increase phloem loading in the source (Thompson, 2006; Thompson and Holbrook, 2003). Despite the long-standing and universal acceptance of Münch's pressure flow theory, however, experimental evidence for the necessary ΔP remained elusive until very recently, when it was found to be approximately $0.2\,MPa\,m^{-1}$ from source leaves to root tips of morning glory vines (Knoblauch and Oparka, 2012; Knoblauch et al., 2016).

Osmotically driven water influx progressively dilutes the phloem sap along the transport path (Tyree et al., 1974). Yet inside a sieve tube, the $\Delta \pi$ between source and sink does not oppose the driving force (ΔP) for sap flow. This is because the sieve pores are lined with membranes only in the axial direction. Regardless of pressure flow, the bulk flow rate of solutes (F_s) through the phloem can be written as follows (Lalonde et al., 2003; Patrick, 1997):

$$F_s = F_v c$$

where c is the solute concentration.

Combining the volume flow equation with the bulk flow equation shows that the solute transport rate in the phloem is dependent on solute concentrations and a ΔP. Unlike the xylem (see Section 3.3),

5.1 Photosynthate transport and distribution

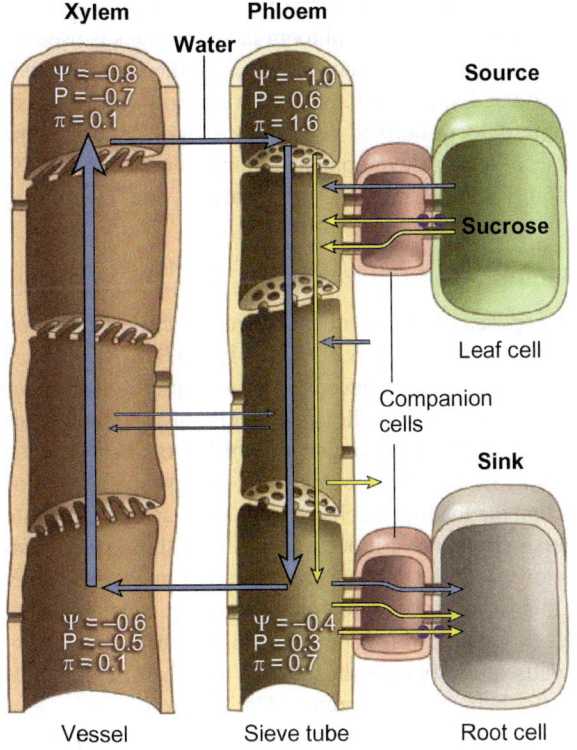

FIG. 5.1

Schematic representation of the movement of phloem sap from regions of high pressure in a source to regions of low pressure in a sink. The purple circles represent solute transporters.

© Elsevier Inc., illustration after Evert, R.F., 2006. Esau's Plant Anatomy. Meristems, Cells, and Tissues of the Plant Body—Their Structure, Function, and Development, third ed. John Wiley & Sons Hoboken, NJ and Nobel, P.S., 2009. Physicochemical and Environmental Plant Physiology, fourth ed. Academic Press, San Diego, CA.

the phloem cannot take advantage of transpiration as the driving force for mass flow but instead relies on a pressure-driven mass flow in the opposite direction (Lalonde et al., 2004; van Bel et al., 2002). Ingeniously, sucrose is not only the major transport form of carbon in the phloem system but also supplies the osmotic driving force for mass flow, thus serving as both cargo and fuel (Hellmann et al., 2000; Münch, 1930; van Bel, 2003). The source cells load sucrose into the sieve elements, which results in an osmotic gradient ($\Delta\Psi_\pi$). The resultant water potential gradient ($\Delta\Psi$) draws water from the nearby xylem into the sieve elements, which increases their turgor pressure (P). In other words, phloem transport depends on the osmotic generation of pressure in the source phloem. The phloem sap moves toward the sink because sucrose is being unloaded there and water passively follows the sugar by osmosis, which leads to a loss of water from the sieve elements so that P is lower at the sink end of the sieve tubes. Water movement into and out of the phloem is driven by $\Delta\Psi$, whereas water movement inside the phloem is driven by ΔP, which in turn is generated by sucrose loading at the source end and unloading at the sink end.

Most of the water entering the phloem in the minor veins of source leaves is absorbed from the xylem, while some of the water leaving the phloem in sink tissues is used for growth of the sink cells, and the remainder enters the xylem to be recycled back to the source (Bradford, 1994; Münch, 1930; Patrick, 1997; van Bel, 2003; see also Fig. 5.1). In transpiring sinks, such as expanding leaves and young grape berries, some phloem-derived water may evaporate to the atmosphere, but this fraction declines over the course of berry development (Blanke and Leyhe, 1987; Düring and Oggionni, 1986; Keller et al., 2015b; Rogiers et al., 2004a; Scharwies and Tyerman, 2017).

5.1.3 Loading, transport, and unloading

The phloem system is composed of three sequential sectors: collection phloem, transport phloem, and release phloem (Lalonde et al., 2003; Patrick et al., 2001; van Bel, 2003). Each of these sectors executes a more or less specific task. Sucrose produced in the mesophyll cells of source leaves diffuses symplastically via plasmodesmata to the bundle sheath cells of the leaf's minor veins. From there it is loaded into the companion cells and sieve elements of the collection phloem (Lalonde et al., 2003, 2004; Ruan, 2014). The way in which grapevines load sucrose into the phloem remains controversial. Some researchers argue that the loading mechanism includes an apoplastic step, as it does in most other crop species, while others maintain that it occurs entirely within the symplast, as it does in many tree species. If it could be shown that the sucrose concentration inside the sieve tubes exceeds that in the surrounding leaf cells, this would support apoplastic loading, since diffusion against a concentration gradient is not possible. For apoplastic phloem loading, sucrose must leak or be pumped out to the solute-permeable cell walls of the bundle sheath cells and then be pumped into the companion cells and sieve elements across their cell membranes. Unlike sucrose, hexoses that may have leaked to the apoplast might be transported back into the cells using hexose transporters to prevent their being loaded into the phloem (Hayes et al., 2007). Whereas sucrose export to the cell walls may passively follow a concentration gradient, active import into the companion cells occurs against a concentration gradient using proton (H^+) symport, a process that requires an input of energy in the form of ATP (Ayre, 2011; Carpaneto et al., 2005; Chen et al., 2012; Geiger, 2011; Lalonde et al., 2003; Patrick et al., 2001). The average energy demand for phloem loading has been estimated at roughly one-third of the entire respiratory energy demand of mature leaves (Amthor, 2000). Despite this high energy demand, active, apoplastic loading may be advantageous, because it can increase P_{source} in the phloem while keeping the sugar concentration in the leaves low, thus increasing carbon export to sinks and reducing feedback inhibition of photosynthesis (Fu et al., 2011; Turgeon, 2010).

In the symplastic loading model, by contrast, sucrose moves entirely passively by diffusion or mass flow along a continuous, membrane-bound pathway from the leaf mesophyll cells to the phloem, which is therefore called "open phloem" (Gamalei, 1989; Rennie and Turgeon, 2009; Turgeon, 2000; Turgeon and Wolf, 2009). In this model, mass flow is generated by a pressure gradient (ΔP) between mesophyll and phloem cells, and movement from cell to cell occurs through plasmodesmata, of which there may be as many as 10,000 per cell. The plasmodesmata are thought to operate as pressure-sensitive valves that close when two neighboring cells experience a $\Delta P > 0.2$ MPa; this enables them to quickly isolate wounded cells from the rest of the symplast (Oparka and Prior, 1992). If it could be shown that the sucrose concentration inside the sieve tubes is lower than that in the surrounding leaf cells, there would be no need for active, energy-dependent transport. Nevertheless, although there is plenty of glucose and fructose in the mesophyll cells, these metabolically unstable hexoses are mostly sequestered in the vacuoles and do not enter the phloem in significant amounts (Koblet, 1969; Turgeon, 2010). One drawback of the open phloem design, however, is that the flow direction of the phloem is reversible, depending on the direction of ΔP. Another drawback is that viral pathogens can also pass through the numerous plasmodesmata and thus become systemic when an insect deposits virus particles near the phloem (see Section 8.2).

It is also possible that both modes of loading operate in tandem or at different times in the leaves (Lalonde et al., 2003; Turgeon, 2000; Turgeon and Wolf, 2009). Regardless of how sucrose is loaded into the phloem, amino acids can be loaded symplastically or apoplastically. Symplastic loading seems to be preferred for amino acids produced in the leaves, whereas those arriving from the roots via the xylem, along with K^+ and other nutrient ions, must be loaded apoplastically (Lalonde et al., 2003). The apoplastic step requires transporters for amino acids and channels for K^+ (see Section 3.3). The energy-dependent transporters are driven by ATPases that move a proton across the membrane along with, or in exchange for, an amino acid, depending on whether a transporter functions in symport or antiport mode (Lalonde et al., 2004). Moreover, loading of K^+ and that of sucrose may be coupled such that K^+ may stimulate sucrose loading into the phloem (Gajdanowicz et al., 2011; Geiger, 2011). Other compounds, such as organic acids and hormones, probably enter the phloem mostly by diffusion, although auxin may also be actively loaded, and are swept along the pathway by mass flow. Finally, the collection phloem in the minor veins also serves as a rerouting system for water; water entering the leaves from the xylem is carried away again in the phloem if it is not used for transpiration or cell expansion (see Section 3.2).

Once the solutes have entered the phloem, they are exported away from the source in the transport phloem, or path phloem, of the main veins, petioles, shoots, trunk, and roots. This long-distance transport occurs by mass flow rather than by diffusion because the pressure inside the phloem is lower in the distant sink organs than in the leaves, with water being provided by the xylem. Because phloem flow is driven by the ΔP between the sieve elements in the source and in the sink, the export rate of sucrose depends on both the photosynthetic sugar production in the source leaf and the capacity of sink tissues to remove sucrose from the phloem. In contrast to the xylem (see Section 3.3), and although it is often symplastically isolated from the surrounding cells, the transport phloem is rather leaky (Ayre, 2011; De Schepper et al., 2013). A considerable portion of the solutes are passively or actively released to the phloem parenchyma cells or their cell walls along the way to supply the needs of axial sinks. For example, sucrose that has leaked to the apoplast along the transport phloem may be transported via interconnecting ray cells into xylem parenchyma cells for respiration, radial growth, long-term storage as starch, and to keep sucrose out of the transpiration stream (Furze et al., 2018; Schmitt et al., 2008).

The leaked solutes can be actively and rapidly retrieved from the parenchyma cells back into the phloem (Furze et al., 2018; Hafke et al., 2005; Lalonde et al., 2003; Martens et al., 2006; Patrick, 1997; van Bel, 2003). Sucrose retrieval may be supported by efflux from the phloem of K^+ ions that are sequestered by the surrounding tissues that invest energy to acquire K^+ from the apoplast. Contrary to the normal K^+/sucrose symport in source leaves and sink organs, in this K^+/sucrose antiport system the energy from the K^+ gradient between the sieve elements and the surrounding apoplast is used to reload sucrose back into the phloem (Gajdanowicz et al., 2011; Dreyer et al., 2017). "Pathway storage" along the transport phloem can buffer diurnal and environmentally induced fluctuations in assimilate supply from the leaves (Wardlaw, 1990). Consequently, the sugar concentration in the phloem entering distant sinks varies much less than that in the collection phloem of the source leaves. Symplastic net release along the transport phloem is favored by high source $\div$ sink ratios, whereas under low source $\div$ sink ratios the plasmodesmata close and solutes can only leave the transport phloem apoplastically and hence slowly (Patrick, 1997; Patrick et al., 2001). High source ÷ sink ratios are characteristic of sink limitation, such as occurs on grapevines with light crop loads, whereas low ratios indicate source limitation, which is typical of overcropped vines.

At any point along the transport phloem, its water potential (Ψ_{phloem}) is coupled with the nearby Ψ_{xylem} because water moves across the separating cells and their membranes (Thompson and Holbrook, 2003; see also Section 3.1). Water enters the transport phloem osmotically from the surrounding apoplast, including the xylem, along the pathway and progressively dilutes the sieve element contents. Consequently, the sucrose concentration in the phloem sap is highest near the source leaves, where it may be around 500 mM, and declines with increasing transport distance. Surprisingly low estimates of 50 mM have been obtained for the pedicel of grape berries (Zhang et al., 2006; Zhang and Keller, 2017). This gradual dilution must be compensated by the declining P_{phloem} to maintain the phloem transport rate. Without such compension, more distant sinks will receive less sugar and will be at a competitive disadvantage compared with more proximal sinks.

Upon arrival in a sink organ, the solutes are unloaded from the release phloem in reverse order of the loading sequence and metabolized or stored in the sink cells (Ayre, 2011; Lalonde et al., 2004; Patrick, 1997; Patrick et al., 2001). In contrast with the dynamic situation along the transport phloem, this release is final and irreversible—that is, retrieval back into the phloem stops in the release phloem. The sucrose concentration in the phloem is generally higher than in the surrounding sink parenchyma cells, so unloading again passively follows a concentration gradient. In sinks such as root and shoot tips, expanding leaves, or young grape berries the unloaded sucrose by and large follows a symplastic pathway through plasmodesmata to the site of usage, although there may also be some additional apoplastic unloading in which sucrose is released to the apoplast (Ayre, 2011; Lalonde et al., 2003; Patrick, 1997; Ruan et al., 2012; van Bel, 2003). The symplastic pathway is one of passive diffusion and bulk flow and is characterized by its large transport capacity and low hydraulic resistance. Nonetheless, most of the resistance that water and solutes encounter between the source and sink cells is imposed by the narrow plasmodesmata that connect the phloem symplast with the surrounding parenchyma symplast. The ability to open or close the plasmodesmata allows sink cells to facilitate diffusion by altering the hydraulic resistance of the unloading pathway, which provides them with a mechanism to control solute import. In young, expanding leaves, assimilate import seems to be restricted to the larger veins, whereas the minor veins later become responsible for assimilate collection for export. Import is maximal at approximately 15–20% of the leaf's final size. The transition from sink to source starts in the leaf margins so that during the transition phase from 30% to 50% of full leaf size (corresponding roughly to leaf positions 4–6 from the shoot tip; Koblet, 1969) there is simultaneous import into the leaf base and export from the leaf tips, with import and export occurring in different vascular bundles (Wardlaw, 1990). The unloading pathway is turned off after a leaf has completed its sink–source transition so that even heavily shaded mature leaves in the interior of dense canopies do not import assimilates, unless there is no competition from sink organs (Lalonde et al., 2003; Quinlan and Weaver, 1969, 1970; Turgeon, 1984).

Grape berries are special in the sense that although unloading in young berries follows mostly the standard symplastic route, the extracellular, apoplastic path of unloading comes to predominate at or just before veraison (Zhang et al., 2006). The pericarp cells seemingly become symplastically isolated from the phloem inside the fruit as callose begins to block the plasmodesmata that connect the sieve element/companion cell complex with the surrounding parenchyma cells inside the vascular bundles and with the berry's flesh cells. Even before veraison, however, there are no plasmodesmatal connections between tissues of different genomes, such as a seed's filial tissues of the embryo and endosperm, and its maternal tissues, which include the seed coat. Therefore, although unloading into the seed coat may be symplastic, unloading into the embryo and endosperm is always apoplastic (Bradford, 1994; Patrick, 1997; Walker et al., 1999). Apoplastic unloading also occurs at the interface between leaf or berry cells and biotrophic pathogens, such as the powdery mildew fungus *Erysiphe necator* (see Section 8.2).

In addition to the facilitated diffusion mentioned previously, apoplastic phloem unloading and post-phloem transport require energy-dependent active transport in which ATPases enable pumping of the solutes across the membranes of the phloem and pericarp cells (Carpaneto et al., 2005; Giribaldi et al., 2007; Martinoia et al., 2012; Sarry et al., 2004; Zhang et al., 2008). These ATPases power sucrose as well as hexose transporter proteins (Ageorges et al., 2000; Davies et al., 1999; Fillion et al., 1999; Hayes et al., 2007; Lecourieux et al., 2014; Vignault et al., 2005). Along with the invertases discussed later, they are activated by sugars and abscisic acid, whose concentration inside the berry increases at or just prior to veraison (Agasse et al., 2009; Castellarin et al., 2016; Pan et al., 2005). Unlike sugars, nutrient ions such as K^+ may be unloaded from the phloem to the apoplast and then taken up by the cells of both preveraison and ripening berries by passive diffusion through channel proteins (Cuéllar et al., 2013; Nieves-Cordones et al., 2019).

As illustrated in Fig. 5.2, the importing berry cells can retrieve sucrose from the apoplast unaltered, aided by sucrose transporters in the cell membranes, or sucrose can be split into glucose and fructose by the enzyme invertase located in the cell walls. Membrane-based hexose transporters can then import these hexoses in a two-step process into the cell cytosol and from there into the vacuole. Alternatively, or additionally, the hexoses may bypass the cytosol via intracellular vesicles—a

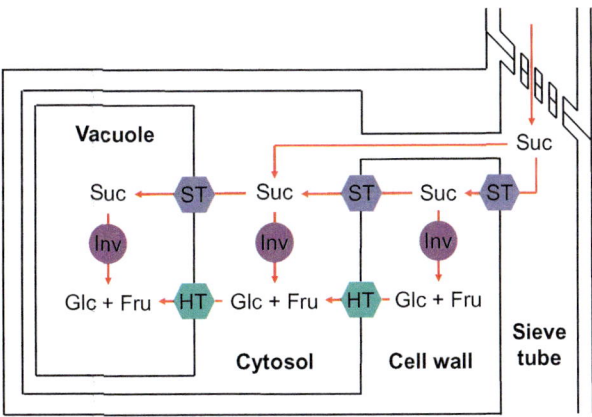

FIG. 5.2

Possible pathways for sugar from the phloem to the vacuole of a grape berry pericarp cell (Glc, glucose; Fru, fructose; HT, hexose transporter; Inv, invertase; ST, sucrose transporter; Suc, sucrose). Specialized forms of invertase work in the cell wall, cytoplasm, and vacuole, respectively.

Illustration by M. Keller.

mechanism termed endocytosis (Etxeberria et al., 2005). Sucrose can also be split inside the cell by a cytoplasmic form of invertase or enter the vacuole unaltered. Inside the vacuole, it can again remain unaltered or be split by a vacuolar invertase (Famiani et al., 2000; Robinson and Davies, 2000; Ruan, 2014). However, most of the sucrose released from the phloem to the apoplast is broken down to hexoses there before being pumped into the cytoplasm, where some of them may be reassembled to sucrose for import into the vacuole, only to be split into glucose and fructose again (Sarry et al., 2004; Terrier et al., 2005). During ripening, surplus hexoses—that is, sugars that are not used in further metabolism—are accumulated for storage inside the vacuoles. If grapes are similar to tomatoes, then cultivars that accumulate different amounts of hexoses in their berries and therefore reach different degrees Brix at maturity should mainly differ in their rate of sugar transport across the mesocarp cell membranes (Patrick et al., 2001). Moreover, due to the dependence of phloem flow on ΔP, berries that are able to rapidly use or accumulate sugars are more competitive than their slower siblings and will tend to develop and ripen more rapidly.

Despite the continued import of sucrose into most sink tissues, its concentration in these tissues remains low because it is "consumed" by respiration for growth and metabolism, used as building blocks in the construction of new cell walls and other cell components, or converted to storage material such as starch. Starch storage is common in shoots, trunks, and roots, but ripening grape berries instead accumulate large amounts of hexose sugars to concentrations that greatly exceed the sucrose concentration in their phloem (Zhang and Keller, 2017). Such accumulation of osmotic solutes to high concentrations without inhibiting phloem flow is made possible by the switch from symplastic to apoplastic phloem unloading before veraison. The sucrose gradient generated by the activity of cell wall invertase may drive the efflux from the phloem of sucrose along with protons against the steep phloem–apoplast proton gradient (Ayre, 2011; Carpaneto et al., 2005; Keller and Shrestha, 2014). In addition, the deposition of hexoses and other solutes in the apoplast lowers the apoplast's osmotic potential (Ψ_π), leading to water efflux from the phloem (Keller and Shrestha, 2014; Wada et al., 2009). This decreases the P_{phloem}, which in turn increases the berry's sink strength and thus the flow rate of phloem sap to the berry (Zhang and Keller, 2017).

Because sugars are actively pumped into the berry vacuoles, they remain there until they are consumed by other organisms. Under conditions that limit sink activity, however, less sucrose is unloaded from the release phloem, which consequently slows the flow rate in the transport phloem. Continued low sink demand leads to sucrose "congestion" in the collection phloem, and sucrose becomes stuck in the leaf apoplast (with apoplastic loading) or the mesophyll cells (with symplastic loading), which ultimately results in feedback inhibition of photosynthesis (Ayre, 2011; see also Section 4.2). Such conditions may include water deficit, nutrient deficiency, low temperatures, or untimely loss of the fruit (see Chapter 7).

Because the pressure flow mechanism requires water to flow out of the phloem along with the unloaded solutes, the bulk flow of phloem sap not only delivers nutrients but also provides a mechanism to meet the water demands of expanding sink cells (Keller et al., 2015b; Lalonde et al., 2003; Patrick, 1997; Patrick et al., 2001). As discussed in Section 3.1, cells, and therefore tissues and organs, grow as water moves across their cell membranes down an osmotically generated water potential gradient. This is especially true of grape berries which expand during ripening (see Section 6.2), but even in growing roots a considerable portion of water is derived from phloem import. This is necessary because transpiration rates of sink organs are typically too slow to drive sufficient water import via the xylem to sustain growth of these organs.

5.1.4 Allocation and partitioning

The amount of photosynthate available for export from a source leaf depends on the leaf's carbon balance, which is determined by its photosynthetic rate and metabolic activity. The process by which the leaf regulates the distribution of fixed carbon to the various metabolic pathways within the leaf is termed allocation. Of course, allocation is also important in the sink organs that import carbon. Allocation can be grouped into three main categories (Taiz and Zeiger, 2006):

1. *Utilization*: Fixed carbon is consumed in respiration to generate energy (ATP) or used as building blocks to produce other components needed by the cell for metabolism and growth; these components comprise mainly polysaccharides for cell walls and amino acids for proteins.
2. *Storage*: Fixed carbon is converted to starch during the day and stored within the chloroplast for remobilization at night or when environmental constraints limit photosynthesis.
3. *Transport*: Fixed carbon is converted to sucrose for export to sink organs or temporary storage in the vacuole as a buffer against short-term fluctuations in sucrose production. Most of the sucrose produced by a source leaf is allocated to transport.

As discussed in Section 4.2, carbon allocation is regulated by the availability of triose phosphate sugars and inorganic phosphate. The coordination of the allocation of fixed carbon to form either sucrose or starch is of particular importance because only sucrose is available for immediate export from a source leaf. This coordination is at least partly driven by the demand for sucrose of all the sinks on a vine. High sink demand removes sucrose from the source, which favors sucrose production over starch production. Consequently, once the source's own needs have been met (allocation to utilization), the proportion of currently produced assimilate allocated to storage in the leaf is determined by the total sink demand for sucrose (allocation to transport). Thus leaf starch accumulates if assimilate production exceeds total sink demand (Johnson et al., 1982). On the other hand, an abundance of fixed carbon in the source can also stimulate sink growth, apparently by inducing the production of additional sucrose transporters. Moreover, both the proportion and the total amount of carbon stored as starch in a leaf also depend on day length—that is, season and latitude (Mengin et al., 2017; Stitt et al., 2007). Although a greater portion of the fixed carbon is allocated to starch during short days, the total amount is smaller than that produced during long days. This is followed by slower rates of starch degradation during the longer nights so that the overall sucrose export over a 24 h period is less in short days. This adjustment of starch turnover to the amount of carbon fixed during the day leads to slower growth rates during short days, which balances the utilization of carbon with its supply (Mengin et al., 2017; Stitt et al., 2007).

The distribution of exported assimilates among the various sink organs is called partitioning (Taiz and Zeiger, 2006). Leaves that have recently become sources initially export their assimilates to the growing shoot tip and unfolding leaves. As new leaves unfold from the shoot tip, the older leaves find themselves at increasing distance from the tip and begin to export assimilates toward the shoot base (Fig. 5.3). Assimilate transport is a dynamic process that can be adapted to the developmental status of the vine and to changing requirements and constraints imposed by the environment. This means that sources do not supply materials equally to all sinks on a plant. Although sources usually supply materials to nearby sinks, and although certain sources may favor certain sinks, the supply pathways are flexible, following variable pressure gradients in the phloem (Wardlaw, 1990).

As the numerous and diverse organs of a vine develop, they compete for space, light, water, and mineral nutrients. Therefore, there is a hierarchy of relative priorities among the sinks, and this

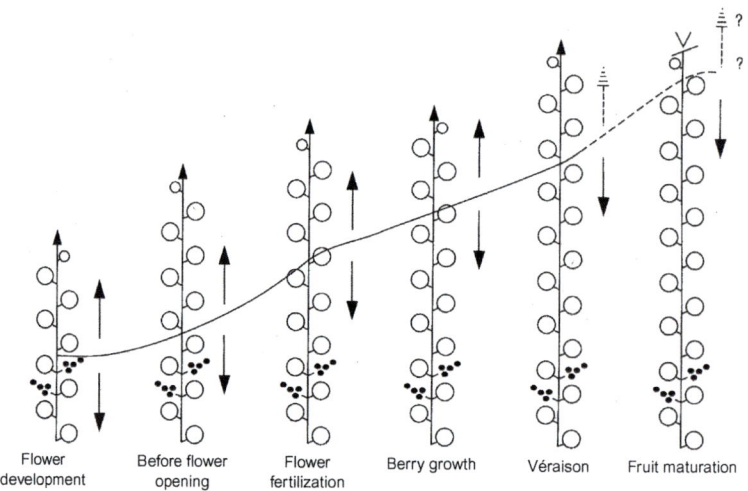

FIG. 5.3

Source–sink relations along a grapevine shoot change over the course of a growing season. Arrows indicate direction of movement.

Reproduced from Koblet, W., 1969. Wanderung von Assimilaten in Rebtrieben und Einfluss der Blattfläche auf Ertrag und Qualität der Trauben. Wein-Wiss. 24, 277–319.

hierarchy is dynamic and sensitive to environmental variables. Flowers are generally poor competitors—especially under source-limiting conditions, whereas after fruit set the berries and seeds dominate the shoots, which in turn often outcompete the roots (Buttrose, 1966a; Hale and Weaver, 1962; Wardlaw, 1990). How much of the available assimilate pool a particular sink organ receives is a matter of supply and demand. The import rate depends on the organ's sink strength (i.e., assimilate demand) relative to the strength of all other sinks on the same vine, as well as on the total amount of assimilate available from the various sources (i.e., assimilate supply). However, low sink strength is not always associated with low sink priority, which refers to the preferential supply of available assimilates between competing sinks (Minchin and Lacointe, 2005). For example, although seeds are too small to be strong sinks, they are generally the top-priority sinks; they are ranked higher than the surrounding berry flesh. Reserve storage pools in the canes, trunks, and roots, on the other hand, often have the lowest sink priority. The relative priority of a particular sink depends on its ability to lower the concentration of sucrose in the phloem—that is, on the rate of phloem unloading—and thereby maintain a favorable ΔP_{phloem} to the source (Patrick, 1997; Wardlaw, 1990).

Sinks usually convert most of the imported sucrose to the hexose sugars glucose and fructose, using invertase and/or sucrose synthase, before they can use the sugar for their own metabolic processes, including the reassembly of starch (Hawker et al., 1991). Modifying either the size or the activity of a sink leads to changes in assimilate transport patterns in a vine. The following patterns are important in partitioning (modified from Taiz and Zeiger, 2006):

1. *Proximity*: The closer a source is to a specific sink, the more likely it is to supply assimilates to that sink (Wardlaw, 1990). The mature leaves closest to the shoot tip generally export their assimilates to the growing tip, whereas the more basal leaves preferentially supply the grape clusters and,

beginning around bloom time, the permanent parts of the vine. The intermediate leaves export in both directions (Hale and Weaver, 1962; Koblet, 1969). Nevertheless, leaves on one shoot can export assimilates to fruit clusters on other shoots, even if they are located several meters away (Koblet and Perret, 1972; Meynhardt and Malan, 1963).

2. *Connection*: A source leaf favors a sink with which it is directly connected via vascular bundles. Any leaf on a shoot is usually connected with the leaves and clusters above and below it on the same side of the shoot. Therefore, a flower or fruit cluster receives assimilates mostly from the leaves located on the same side of the shoot as itself (Koblet, 1969; Motomura, 1990, 1993; Yang and Hori, 1980). This "unilateralism" even applies to the supply of clusters with assimilates derived from lateral shoots (Koblet, 1975; Koblet and Perret, 1971).

3. *Interference*: The "normal" transport pathways can be altered by wounding or pruning, which interrupts the direct connection between a source and a sink. Following such interruption, newly formed vascular interconnections, called anastomoses, can provide alternative connections for cross-transfer of assimilates. This has implications for canopy management. Hedging (shoot tipping or topping) stimulates assimilate cross-transfer to clusters on both sides of the shoot. It also induces young leaves to switch from upward export to downward export that can extend to neighboring nontopped shoots (Koblet and Perret, 1972; Quinlan and Weaver, 1970). This is probably why shoot tipping during bloom time often improves fruit set (Coombe, 1959, 1962; Vasconcelos and Castagnoli, 2000). However, subsequent outgrowth of lateral buds may later reverse assimilate flow again. Defoliation of a shoot triggers compensatory assimilate import into that shoot's clusters from neighboring shoots (Quinlan and Weaver, 1970). Leaf removal or shade in the cluster zone of a canopy induces more distal leaves to export their assimilates to the clusters, whereas loss or shading of younger leaves induces older leaves to reverse the export direction toward the shoot tip at the expense of the perennial parts of the vine. Fruit removal on a shoot induces that shoot to export assimilates to neighboring shoots with clusters (Quinlan and Weaver, 1970). Removing a strip of bark, which contains the phloem, from around a shoot in a practice termed girdling near the shoot base induces the leaves to export assimilates to clusters on both the same and opposite sides of that shoot (Motomura, 1993).

4. *Communication*: Grapevines carefully balance their investment in growth and reproduction. Vegetative growth also must be balanced between shoot growth for photosynthetic productivity and root growth for water and nutrient uptake. As Goethe stated, "in order to spend on one side, nature is forced to economize on the other side" (cited in Darwin, 1859). This requires interaction between centers of supply and demand, which happens by way of phloem pressure gradients, as well as nutrients and hormones. Cytokinins produced by root tips or seeds and transported to the leaves in the xylem may signal sink strength and integrate assimilate supply with sink demand (Ha et al., 2012; Lalonde et al., 2003; Paul and Foyer, 2001). The strongest and most competitive sinks may be those with the highest amount of cytokinins (Thomas, 2013). But the same may be true for auxin produced in shoot tips and seeds and exported in the phloem and parenchyma cells. A developing grape cluster can attract assimilates from nearby shoots, especially when those shoots' assimilate production exceeds their own demand (Currle et al., 1983; Intrigliolo et al., 2009; Koblet and Perret, 1972; Quinlan and Weaver, 1970). Mechanical elimination by shoot hedging of the auxin signal required for root growth may further enhance delivery to the clusters, because lack of auxin flow may temporarily inhibit root growth until outgrowth of lateral buds generates new shoot apices.

5. *Competition*: Competition determines the priority of each sink for assimilate supply relative to all other sinks on the plant. The greater the ability of a sink to store or metabolize imported assimilates, the better is its ability to compete for exported assimilates. A rapidly growing sink is competitive because the consumption of assimilates lowers P_{phloem} in the sink. A competitive sink is said to have high sink strength. Moreover, the greater the number of sinks competing for assimilates, the lower is the availability of assimilates for each individual sink. This has implications for winter pruning and canopy management. Leaving more buds per vine increases the number of shoots and clusters but decreases the vigor of each shoot and the proportion of flowers that set fruit (Keller et al., 2015a). Moreover, increasing the number of berries per vine or per unit leaf area can limit berry growth or slow the rate of ripening (Keller et al., 2008; see also Section 6.3). Conversely, applying gibberellin to table grape clusters increases not only berry size but also the amount of imported assimilates (Weaver et al., 1969).
6. *Development*: The relative importance of different sinks changes during plant development. Shoot tips have high priority after budbreak, when the vine needs to establish its new canopy (Hale and Weaver, 1962). By contrast, clusters are low-priority sinks until the flowers have been fertilized. Consequently, shoot topping or trunk girdling during bloom may improve fruit set, whereas loss of leaf area reduces fruit set (Coombe, 1959, 1962; Poni et al., 2009). After fruit set, however, the clusters become powerful sinks and dominate the sink hierarchy during grape development, especially for nearby leaves (Hale and Weaver, 1962; Williams, 1996). At veraison, the clusters temporarily become extremely strong sinks to meet their demand for sugar, but the vine then gradually shifts its priorities to the woody perennial organs to replenish storage reserves and acquire cold hardiness (Candolfi-Vasconcelos et al., 1994; Keller et al., 2015b; Rossouw et al., 2017).

In summary, for a plant organ to be a strong sink, it pays to be large, active, close to a source, and well connected. However, many other factors also modulate sink strength and assimilate partitioning. Environmental variables play an important role in modifying the pattern of partitioning. For instance, soil water or nutrient deficits decrease overall vine growth but increase the proportion of assimilates partitioned to the roots (see Sections 7.2 and 7.3).

Source and sink organs are parts of a single inseparable system, and an effect on one part is bound to have a consequential and concurrent influence on the others. For instance, removing a sink improves the availability of assimilates to adjacent sinks (Quinlan and Weaver, 1970). But removing too many sinks, or removing them too early, will feed back on the sources and decrease their activity (Keller et al., 2014).

In viticulture, we are mainly interested in maximizing or optimizing the vine's investment in reproduction, and consequently in fruit production and ripening. In mature grapevines, the proportion of total biomass, or dry matter, that is partitioned to the fruit can range from 10% to 70%. Nonirrigated, relatively water-stressed, cluster-thinned wine grapes would be near the low end of this range, while heavily irrigated, minimally pruned juice grapes would be near its high end. The dilemma for grapevines as perennial plants is to maximize annual seed dispersal while at the same time ensuring long-term survival of the plant. In other words, vines must balance their investment in short-term reproductive output and that in long-term reproductive output. They must coordinate the supply of assimilates to seed production and, hence, fruit development such that it does not occur at the expense of other essential processes and structures.

Early shoot growth during and after budbreak is completely dependent on remobilized storage reserves from the permanent parts of the vine, such as canes, spurs, cordons, trunks, and roots, which are therefore depleted beginning at the time of budswell (Eifert et al., 1961; Loescher et al., 1990; Williams, 1996; Zapata et al., 2004). Even unfolding leaves, before becoming sources, must compete with other sinks for assimilates, yet they are stronger sinks than the cambium of the cane, trunk, and roots (Wardlaw, 1990). As more leaves unfold from the shoot tip, they increasingly support the shoot's further growth, which soon becomes independent of the parent plant. The growing shoots also begin to contribute to radial growth via the cambial activity in the permanent parts of the vine. Starting at approximately the time of bloom, assimilate export from the shoots normally begins to replenish the parent plant's storage reserves and also sustains secondary growth of the cordons, trunk, and roots.

The ability to refill the storage pools also depends on the amount of fruit the vine has to support, because the berries generally dominate the hierarchy of sink priorities between fruit set and seed maturity (Hale and Weaver, 1962; Williams, 1996). The replenishment of reserves is often described as an overflow mechanism that directs surplus assimilates to storage only after all other assimilate-requiring processes, such as growth and fruit production, have been satisfied (Lemaire and Millard, 1999). However, this may depend on the vine's developmental stage. In addition to the overflow mechanism, which is also called passive storage, accumulation of reserves may be an active storage process when carbon is diverted to storage at the expense of other processes (Wiley and Helliker, 2012). The woody parts of the vine may become high-priority sinks late in the growing season, when the plant "loses interest" in the fruit and focuses on the replenishment of storage reserves and on cold acclimation (Loescher et al., 1990; see also Section 7.4). This change in sink hierarchy appears to happen after about midripening when berry sugar accumulation begins to slow (Candolfi-Vasconcelos et al., 1994; Rossouw et al., 2017; Wample and Bary, 1992). When adverse environmental conditions limit the availability of resources, this change can clash with a grower's or winemaker's desire to improve fruit quality by delaying harvest.

The net assimilation rate of a grapevine canopy determines the export rate from the canopy, and the export rate increases with increasing source–sink ΔP_{phloem}. Therefore, the best way to increase assimilate export is to increase the rate of photosynthesis in source leaves, because an increase in the amount of photosynthate available for export raises the turgor in the source phloem, which results in a greater ΔP_{phloem} between source and sink. Leaves that unfold under intense competition from other sinks often achieve higher photosynthetic rates than leaves that do not experience competition during development. Even in fully grown leaves, however, photosynthesis can still be influenced by natural or manipulated changes in sink demand (Downton et al., 1987; Flore and Lakso, 1989; Paul and Foyer, 2001). The presence of fruit on a vine tends to increase the leaves' photosynthetic rate, especially in vines with small leaf area that leads to source limitation (Edson et al., 1993, 1995a; Eibach and Alleweldt, 1984; Hofäcker, 1978). The fruit also induces a shift in assimilate supply to reproductive growth at the expense of vegetative growth, especially root growth, keeping whole-vine biomass relatively constant (Edson et al., 1995a,b; Petrie et al., 2000b; Williams, 1996). This shift is particularly pronounced at low soil moisture (Eibach and Alleweldt, 1985).

Destruction or removal of source leaves, for example to enhance fruit exposure to sunlight, also leads to a compensatory increase in photosynthesis of the remaining leaves and a delay of their senescence (Buttrose, 1966a; Candolfi-Vasconcelos and Koblet, 1991; Currle et al., 1983; Hofäcker, 1978; Hunter and Visser, 1988; Iacono et al., 1995). However, this compensation is incomplete, and severe defoliation may retard fruit development and ripening (Petrie et al., 2000a). A late spring frost, by killing leaves and flowers that have been "built" using remobilized storage reserves, can severely deplete the reserve pool and restrict renewed canopy development even in the absence of fruit (Gu et al., 2008). Conversely, if defoliation occurs late in the season, even shortly after harvest, it can result in poor carbohydrate reserve status and in nitrogen deficiency the following growing season (Greven et al., 2016; Loescher et al., 1990; Sartorius, 1973). Nitrogen and other nutrients that could be remobilized from the leaves and transported to the perennial parts of the vine for storage are lost when leaves are removed before normal abscission. This problem also arises when leaves are killed by an early fall frost; nutrients cannot be recycled from dead leaves. Alternatively, inadequate carbohydrate reserves in the roots, by curtailing new root growth and nutrient uptake and assimilation, can also lead to nitrogen deficiency (Loescher et al., 1990). However, even before such nitrogen deficiency may limit plant growth, it is possible that an inadequate reserve status may interfere with the plant's ability to generate adequate root pressure prior to budbreak (Améglio et al., 2001; see also Section 2.2). This would prevent the vine from purging air bubbles from the xylem, potentially leading to poor budbreak or early canopy collapse.

The vine's response to source removal contrasts with its response to sink removal. Excessive removal of fruit, for example by cluster thinning, especially if done early, may result in an assimilate surplus due to a shortage of demand and can decrease photosynthesis because of feedback inhibition from accumulating sugar (Currle et al., 1983; Downton et al., 1987; Iacono et al., 1995; Naor et al., 1997; Paul and Foyer, 2001; Roitsch, 1999). In addition, some of the surplus sugar may be used for more vigorous growth of the shoots, especially lateral shoots, and for root growth (Morinaga et al., 2003; Pallas et al., 2008; Petrie et al., 2000b). Whereas fruit clusters cannot compete with shoot growth before bloom because of their low sink strength, shoots may continue to grow vigorously after fruit set on vines with low amounts of fruit (Keller et al., 2008). This may result in dense, shaded canopies, which can increase disease incidence and compromise fruit quality (see Section 6.2). Vines without fruit also may begin to "shut down" earlier in autumn, which is visible as the leaf chlorophyll content declines in response to decreasing photosynthesis, and the leaves are shed prematurely (Keller et al., 2014; see also Fig. 5.4). Premature leaf senescence may be triggered by the accumulation of sugar in the

FIG. 5.4

Senescing Cabernet Sauvignon leaves in early November (Northern Hemisphere) from plants that had all fruit removed at veraison in mid-August (left) or with the fruit still on the plant (right).

Photo by M. Keller.

leaves, which signals a surplus of uncommitted assimilates (Lim et al., 2007; Rolland et al., 2006; Thomas, 2013; Wingler et al., 2009). It is possible that such sugar accumulation also serves as an internal signal for nutrient recycling during leaf senescence.

Whereas the presence of fruit influences the physiology of the rest of the vine, fruit growth often compensates for changes in sink number, a phenomenon called the yield component compensation principle (see Section 6.1). Early loss of fruit, for example due to poor fruit set or early cluster thinning, is partly compensated by increased growth of the remaining fruit, which tends to lead to an increase in berry size (Dokoozlian and Hirschfelt, 1995; Keller et al., 2008; Kliewer et al., 1983; Winkler, 1958). If the loss in sink number occurs before bloom, then the proportion of the remaining flowers that set fruit often increases in addition to berry growth, which can result in compact clusters with large berries. Such growth compensation is greatly reduced or absent if the sink number decreases only later in the growing season, for example during the lag phase or at veraison, and the remaining fruit may instead accumulate sugar more rapidly (Dokoozlian and Hirschfelt, 1995; Nuzzo and Matthews, 2006). Conversely, grapevines usually self-adjust to a large number of sinks, which is typical of vines with high bud numbers following mechanical or minimal pruning. They do so by lowering the percentage of budbreak and fruit set, and by decreasing berry growth so that there may be many small, loose clusters with small berries (Clingeleffer and Krake, 1992; Intrieri et al., 2001; Keller et al., 2004; Miller and Howell, 1998; Possingham, 1994; Winkler, 1958).

A puzzling phenomenon in grapevines is that different berries on the same cluster develop independently at different rates and accumulate very different amounts of sugar, even though they may be supplied by the same phloem strand through the peduncle (Coombe, 1992; Coombe and Bishop, 1980). The individual berries, rather than the cluster as a whole, are able to control delivery of phloem solutes according to the berries' demand (Keller et al., 2015b; Zhang and Keller, 2017). Berries containing more seeds than others may be better able to attract assimilates, possibly because each seed produces and exports auxin (see Section 2.3). Auxin, in concert with gibberellin, controls cell division and cell expansion in the developing berry (Dorcey et al., 2009; Giacomelli et al., 2013; Ruan et al., 2012). Additionally, auxin also stimulates cambial activity and xylem and phloem development, so that more auxin export from a berry increases the production of vascular tissues in the pedicel, and more water and nutrients can be imported by the fruit (Else et al., 2004). Indeed, berry size and pedicel cross-sectional area are strongly correlated, and so are berry number and peduncle and vascular cross-sectional areas of a cluster (Gourieroux et al., 2016; Theiler and Coombe, 1985). Thus the dry weight of a cluster can be estimated from the diameter of its peduncle (Castelan-Estrada et al., 2002). Because differences in the rate of cell division before bloom can result in differences in ovary size among flowers, berries that begin development with a head start usually remain more competitive throughout their growth and ripening (Coombe, 1976). Berry size is also associated with shoot vigor, perhaps due to common regulatory mechanisms. Berries on vigorously growing shoots tend to be larger than berries on weaker shoots, even though the vigorous shoots may have more berries (Keller et al., 2015a).

Grapevines often form many meristems that can grow under favorable conditions, and they respond strongly to variations in the availability of resources: They are said to be very "plastic" (Lawlor, 2002). Environmental conditions, therefore, also play an important role in the regulation of partitioning. Under low light intensity grapevines favor carbon supply to the shoots, whereas under low nutrient or water availability they favor supply to the roots (Keller and Koblet, 1995a). In the case of water, this does not mean that root growth increases in response to water deficit. On the contrary, root growth decreases, but less so than shoot growth. These relationships are explored in subsequent chapters.

5.2 Canopy–environment interactions

Grapevine productivity ultimately depends on the photosynthetic capacity of the vine's canopy, integrated over the growing season. A canopy in the viticultural sense is defined as the aboveground parts of a vine—that is, its shoots, leaves, fruit, trunk, and cordon. Compared with leaf photosynthesis, shoot and berry photosynthesis is minor and never exceeds the rate of respiration. Therefore, canopy photosynthesis is the sum of the photosynthetic activity of all the leaves minus the respiratory activity of the leaves and all nonphotosynthetic tissues. Although canopy photosynthesis is to a large extent determined by the availability of resources, internal factors also play a role. For example, the length and layout of the water-flow pathway from the roots to the leaves may limit photosynthesis via the effect of hydraulic resistance and gravity on stomatal conductance (see Section 3.3). Because the xylem pressure declines with increasing height, the leaves near the top experience relatively more water stress than those closer to the ground. This influence of a plant's "plumbing" design and length has been put forward as the main reason photosynthesis and vigor tend to decline as plant height and age increase (Koch et al., 2004; Ryan and Yoder, 1997). Alternatively, the decrease in gas exchange in tall plants may be a consequence, rather than the cause, of slower growth because growth is more sensitive to water deficit than is photosynthesis (Ryan et al., 2006). In this case, hydraulics and gravity limit the water potential gradient from the xylem to the growing cells and, hence, the turgor necessary for cell expansion, which decreases sink strength (Bond et al., 2007; see Section 3.1). In support of this sink limitation hypothesis, taller plants tend to accumulate greater amounts of reserve carbohydrates than do shorter plants (Sala and Hoch, 2009). Alternatively again, a lower water potential gradient from the xylem to the phloem might curb phloem export from the leaves (Barnard and Ryan, 2003). The outcome in the latter two scenarios is the same: Sugar accumulates in the leaves and curtails photosynthesis by feedback inhibition. Whatever its direct and indirect consequences, such hydraulic limitation is probably of little importance in cultivated grapevines whose size—in contrast with their tall, tree-climbing wild relatives—is constrained by trellis design and annual pruning practices. A kind of hydraulic limitation does, however, seem to operate in large, lightly pruned vines. Although such vines adjust their capacity for water supply to the large canopy by decreasing their hydraulic resistance, the vines' ability to do so is limited, so that shoot vigor declines as the number of shoots per vine increases (Keller et al., 2015a).

Resource availability is determined by meteorological or climatic (Greek *klima*=surface of the Earth, region) conditions in addition to edaphic (Greek *edaphos*=ground, soil) factors. The impact of variables related to the soil, such as water and nutrient availability, is discussed in Sections 7.2 and 7.3; this section provides an overview of the importance of aboveground climatic variables. Whereas the term "weather" refers to the daily, seasonal, and annual fluctuations of temperature, precipitation, humidity, and wind, "climate" constitutes the summarized and averaged weather situation over a long time. A "long time" is defined by the World Meteorological Organization as 30 years or more. In viticulture, we often distinguish three levels of climate (Jones et al., 2010; Tonietto and Carbonneau, 2004):

1. *Macroclimate*: The climate of a region, which is ordinarily described by data collected at one or several weather stations. It is sometimes viewed as the mean of all the microclimates in a region. It is mainly determined by the geographic location (i.e., latitude, altitude, and distance from large bodies of water) but is independent of local topography, soil type, and vegetation. The size of the

region may extend over hundreds of kilometers. This is the climate that defines a particular grape growing region.
2. *Mesoclimate*: The climate of a site or large vineyard. It is a local variant of the macroclimate modified by topography, and hence also called local climate or topoclimate (Greek *topos* = place). It may differ from the macroclimate because of altitude or elevation from a valley floor. The extent of a particular mesoclimate may be from hundreds of meters to several kilometers. This is the climate that is relevant for vineyard site selection within a region.
3. *Microclimate*: The climate within and immediately surrounding the canopy or within a vineyard. It may differ from the mesoclimate because of aspect, slope, and even soil type. Due largely to the presence of leaves, differences in microclimate may occur over as little as a few centimeters or over hundreds of meters. This is the climate that can be manipulated by vineyard cultural practices.

Compared with the situation near the equator, regions at higher latitudes experience longer periods of summer daylight between the spring and autumn equinoxes. Greater day length enables leaves to be photosynthetically active during a greater portion of each day, but this benefit is partially offset by the lower intensity of solar radiation at higher latitudes. This is because the sun "passes" lower on the horizon—that is, at a greater zenith angle, whereby the zenith is directly overhead, at 0°, and the horizon is at 90°. At the summer solstice, or the "longest" day around 21 June (Northern Hemisphere) or 21 December (Southern Hemisphere), the total daily global irradiance is approximately equal between 30° and 50° latitude, where most of the world's grapes are grown. However, at the two equinoxes around 20 March and 23 September, when the lengths of day and night are equal, the daily irradiance received at 50° latitude is only about 75% of that at 30°.

Altitude (elevation above sea level), aspect (slope direction), and grade (slope steepness) of a vineyard site also determine the radiation received by a vineyard (Failla et al., 2004). The effect of altitude at the same latitude is particularly noticeable as a strong and predictable decrease in temperature by about 0.65 °C for each 100 m gain in elevation, provided the atmosphere is well mixed, which generally occurs courtesy of solar heating and wind. Nonetheless, at night and during the winter the atmospheric mixing is often insufficient to prevent the formation of so-called temperature inversions, which are especially prevalent in sheltered valleys and during clear, calm weather. During such inversions, cold air settles in valley floors and local depressions so that the temperature is coldest at ground level and increases by 2.5–3 °C per 100 m up to the inversion top approximately 200–300 m above the valley floor, above which it begins to decrease at its normal rate (Daly et al., 2008).

A south-facing slope with a grade of 10% at 45°N receives as much radiation, and hence energy, during the April–October growing season as a horizontal plain at 40°N. Grape-growing regions near the 45th parallel, the midpoint between the equator and the poles, include Bordeaux and Rhône Valley in France, Italy's Piemonte, Croatia, Crimea, northern China, Oregon's Willamette Valley, and, in the Southern Hemisphere, New Zealand's Otago region. Examples of regions near 40°N include central Spain and Portugal, southern Italy, northern Greece and Turkey, China's Beijing region, northern California, and near 40°S Hawke's Bay in New Zealand. During the summer, a 50% equator-facing slope receives approximately 25% more solar energy than a north-facing slope, but the difference is much greater in the winter because of the lower elevation of the sun (Holst et al., 2005). Not only does the disparity in radiation translate into warmer, drier soils and earlier budbreak on equator-facing slopes but also, because the energy difference due to vineyard aspect increases during the ripening period, pole-facing slopes can be at a serious disadvantage in marginal climates. Differences are also much larger under clear skies; clouds filter out the direct radiation from the sun so that during overcast days there is only diffuse radiation. Thus, in regions with frequent cloud cover, and therefore few sunshine hours, the incident radiation is similar regardless of aspect. Grapevines perceive differences in solar radiation mainly as differences in temperature, with more radiation leading to higher daytime air and soil temperatures (Peña Quiñones et al., 2019). Light intensities above the canopy, however, are similar regardless of aspect, at least during the growing season (Holst et al., 2004; Mayer et al., 2002).

In addition to the overall climate, annual climate variation and short-term weather fluctuations are also very important in viticulture. Whereas the long-term climatic averages are pertinent to site selection and choice of cultivars and rootstocks when establishing a new vineyard, climate variation among growing seasons and weather variation within seasons determine the weather-associated risks involved in grape production and often influence management decisions in established vineyards (see Sections 7.2 and 7.4). Climatic conditions that are relevant for individual vines are part of the canopy microclimate, which is strongly influenced by the presence of leaves. These conditions include light, temperature, wind, and humidity (Smart, 1985). Leaves alter all these conditions from the exterior of the canopy to the interior and from the top to the bottom.

5.2.1 Light

Light has a more profound effect on plant development than does any other climatic factor or signal. Grapevines, like all plants, use light both as their only source of energy and as a source of information. They can accurately perceive fluctuations in quantity (intensity), quality (spectral composition), directionality, and periodicity (day length) of the incoming light (Fankhauser and Staiger, 2002). Since light has properties of both waves and particles (see Section 4.1), solar radiation is described in both energy terms and photon terms. The former is called irradiance and is measured as the amount of energy received per unit surface area (expressed in $W\ m^{-2}$). The latter is often called light intensity and is measured as photon flux, which is defined as the number of photons received per unit surface area per unit time (expressed in $\mu mol\ m^{-2}\ s^{-1}$). Mathematically, photon flux and irradiance are highly correlated, such that $1\ W\ m^{-2}$ corresponds to $2\ \mu mol\ m^{-2}\ s^{-1}$ (Foyo-Moreno et al., 2017). The amount of incident light on a grapevine canopy varies with latitude, season, time of day, and cloud cover. Leaves effectively absorb sunlight in the visible and ultraviolet (UV) region of the electromagnetic spectrum while reflecting light in the low-energy infrared region (Blanke, 1990a). More precisely, the epidermis absorbs most of the potentially damaging, high-energy UV light but is transparent to visible light, which penetrates the chloroplast-rich mesophyll cells (see Section 1.3). Since light in the visible wavelength range from 400 to 700 nm is utilized in photosynthesis (see Section 4.1), this light is called photosynthetically active radiation (PAR), and the photon flux within this waveband is called photosynthetic photon flux. Therefore, plants exploit the same spectral "window" of light that we see as the colors of a rainbow (see Fig. 4.1). Approximately 85–90% of incident light in the PAR range falling on a grape leaf is absorbed by the leaf; the rest is either reflected at its surface (6%) or transmitted through the leaf (4–9%) (Smart, 1985).

Because absorbing an optimal amount of light is so important for leaves, they have evolved anatomical and physiological strategies that allow them to adapt to a range of light environments. Leaves grown in the shade, for example in the interior of a dense canopy, are often larger but thinner than leaves grown in full sunlight (Keller and Koblet, 1995a; Pallas and Christophe, 2015; Palliotti et al., 2000). The shade leaves owe their greater specific leaf area (ratio of leaf area to leaf dry weight) to their shortened palisade cells and limited investment in structural biomass. As a consequence of their larger size, shade leaves have fewer stomata per unit leaf area than do sun leaves. Shade leaves also have more total chlorophyll per reaction center and per unit nitrogen, but they have less rubisco protein and xanthophyll carotenoids, and up to 50% lower respiration rates than sun leaves (Cartechini and Palliotti, 1995; Evans, 1989; Schultz, 1991; Seemann et al., 1987). These adaptations enhance light absorption and energy transfer in the shade but make these leaves considerably less efficient at higher light intensity (Ortoidze and Düring, 2001; Palliotti et al., 2000; Schultz et al., 1996). Moreover, such leaves are highly light sensitive, and when they are suddenly exposed to the sun, they can suffer from severe oxidative stress and photoinhibition and may even die in extreme cases (Iacono and Sommer, 1996; Triantaphylidès et al., 2008; see also Section 7.1).

The net CO_2 assimilation rate of a leaf is strongly dependent on the intensity of light incident on the leaf. In complete darkness a leaf's net CO_2 assimilation is negative because respiration proceeds in the absence of photosynthesis (see Section 4.4). At night, therefore, the leaf's metabolism and continued sugar export depend on the breakdown of starch accumulated during the day (see Section 4.2). As the photosynthetic photon flux increases, so does the photosynthetic CO_2 uptake. The light intensity at which photosynthetic CO_2 uptake balances respiratory CO_2 release is called the light compensation point; at this point the leaf's net CO_2 assimilation is zero. The light compensation point depends on the species, cultivar, and developmental conditions, but in typical grapevine leaves it is reached at a photon flux of approximately 10–70 $\mu mol\, m^{-2}\, s^{-1}$ (Cartechini and Palliotti, 1995; Düring, 1988; Keller and Koblet, 1994; Rühl et al., 1981). A further increase in photon flux leads to a concomitant, almost linear increase in photosynthesis until it starts to level off and reaches light saturation. Grapevine leaves generally reach light saturation between 700 and 1200 $\mu mol\, m^{-2}\, s^{-1}$ (Fig. 5.5). This is well below the photon flux of full sunlight, which can exceed 2000 $\mu mol\, m^{-2}\, s^{-1}$, but above that under cloud covers, which ranges from about 100 to 1000 $\mu mol\, m^{-2}\, s^{-1}$, depending on cloud type and density. The

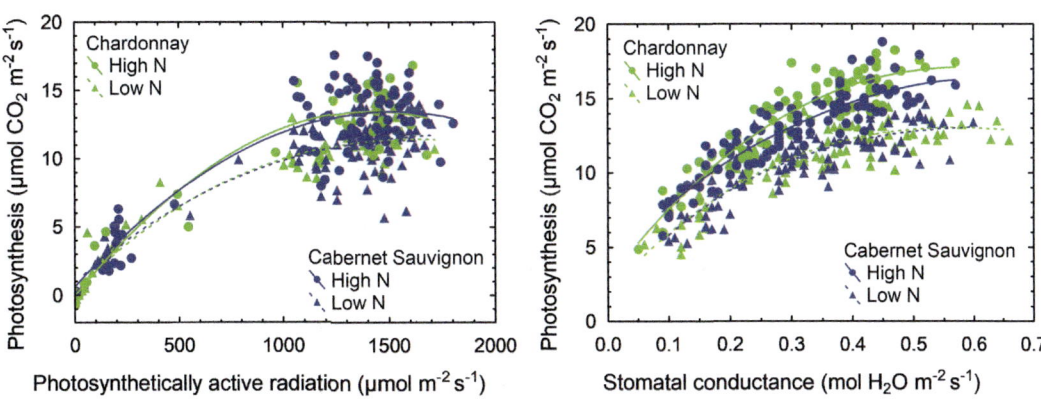

FIG. 5.5

Relationship between light intensity and photosynthesis (left) and between stomatal conductance and light-saturated photosynthesis (right) of mature grapevine leaves. Note that the influence of cultivar is minor compared with that of light and nitrogen status.

M. Keller, unpublished data.

photon flux at which a leaf reach light saturation generally reflects the maximum light intensity to which the leaf was exposed during its growth but varies widely, depending again on species, cultivar, leaf age, developmental stage, temperature, and nutrient status. For example, light saturation may occur near 700 $\mu mol\, m^{-2}\, s^{-1}$ at 20–25 °C, but it almost doubles as the temperature increases by 10 °C (Zufferey et al., 2000). Moreover, at any given stomatal aperture width, light-saturated photosynthesis "runs" faster in vines with high nutrient status than in those with low nutrient status, nitrogen status being especially important (Fig. 5.5).

Photosynthesis below light saturation is referred to as light limited, because there is insufficient light for maximum photochemistry (see Section 4.1). Conversely, photosynthesis above light saturation is referred to as CO_2 limited, because enzymatic reactions cannot keep pace with photochemistry (see Section 4.2). Leaves respond readily to changes in light conditions, such that transient decreases in light due to, say, passing clouds, are associated with equally transient decreases in photosynthesis (Carbonneau and de Loth, 1985). The fact that <10% of the PAR falling on a grape leaf will reach the leaves beneath has implications for canopy photosynthesis. Only the leaves on the exterior of a canopy are exposed to saturating light intensity, so only these exterior leaves will achieve maximum rates of photosynthesis (Schultz et al., 1996; Smart, 1974; Zufferey et al., 2000). The interior leaves receive much less light, often <10 $\mu mol\, m^{-2}\, s^{-1}$ inside a dense canopy; this light is either transmitted through other leaves or passes through gaps in the canopy (Dokoozlian and Kliewer, 1995a,b; Mullins et al., 1992). Most of these canopy gaps are not static because wind moves shoots and leaves. Together with the absorption and reflection of light by leaves and the soil surface this creates an irregular patchwork of highly variable light intensity within the canopy. Light incident on individual leaves, or only on sections of a leaf, can fluctuate within milliseconds to minutes, often resulting in fleeting sunflecks (Kriedemann et al., 1973; Rascher and Nedbal, 2006). The contribution of these sunflecks to the daily photon flux at any given point within a grapevine canopy varies from 20% to 90%. Leaves that experience frequent but brief sunflecks benefit most in terms of photosynthesis, but because shaded leaves can heat up very rapidly upon sudden exposure to sunlight, they are prone to heat damage and wilting from water stress (Pearcy, 1990).

Depending on the trellis and training system, and on vine vigor, light within the fruiting zone of a canopy can range from <1% of ambient in a single-curtain trellis without shoot-positioning to approximately 10% in a vertically upright shoot-positioned system, and to >30% in a downward-positioned double-curtain system (Cartechini and Palliotti, 1995; Dokoozlian and Kliewer, 1995b; Gladstone and Dokoozlian, 2003; Williams et al., 1994). Moreover, the shaded side of a canopy, which faces away from the sun, can receive as little as 3–6%, or 40–100 $\mu mol\, m^{-2}\, s^{-1}$, of the light intercepted by the sunlit side (Smart, 1985). Consequently, whole-canopy photosynthesis is almost never light saturated (Flore and Lakso, 1989; Intrieri et al., 1997; Petrie et al., 2009; Poni et al., 2003, 2008). On the other hand, shade leaves inside the canopy have 30–50% lower respiration rates than sun leaves at the canopy surface, irrespective of temperature and leaf age. This adaptation, which results in a lower light compensation point of photosynthesis, enables many shade leaves to maintain a positive daily net carbon balance, even though this balance is necessarily much lower than that of sun leaves.

The contribution of interior leaves to whole-canopy photosynthesis is greater in loose, open canopies than in dense canopies, because the outermost leaves of the latter absorb almost all the available light (Smart, 1974). A similar effect can be observed when particle films consisting of clay minerals such as kaolin are applied to a canopy in regions with high solar radiation to protect the vines from heat and water stress. Such particle films decrease light absorption by the leaves (Shellie and King, 2013).

Nonetheless, whole-canopy photosynthesis can remain unaffected or even increase following such mineral application, because the particles also reflect light. This reflection may improve the light distribution within the canopy, which tends to compensate for the decrease in photosynthesis of the exterior leaves.

Grapevines also adjust the distribution of proteins, chlorophyll, and photosynthetic capacity to canopy density so that well-exposed leaves have high nitrogen and photosynthetic capacity per unit leaf area (Bowen and Kliewer, 1990; Kriedemann, 1968b; Palliotti et al., 2000; Prieto et al., 2012; Williams, 1987). When photosynthesis in some leaves drops below the light compensation point for some time, the vine sheds these leaves by initiating the process of senescence and abscission in order to prevent a wasteful situation in which the leaves' demand for water and nutrients outweighs their supply of fixed carbon (Flore and Lakso, 1989; Poorter et al., 2006; Taylor and Whitelaw, 2001). In other words, the plant disposes of those leaves that can no longer "pay" for their share of the respiratory costs of the shoots and roots that support them (Reich et al., 2009). Because a decline in photosynthesis is accompanied by a decline in stomatal conductance and consequently in transpiration rate, the delivery to these leaves of root-derived cytokinins via the transpiration stream also decreases, which may serve as a signal to initiate the early senescence program (Boonman et al., 2007, 2009; Buchanan-Wollaston, 1997; Pons et al., 2001).

The ratio of red ÷ far-red light incident on shaded leaves is also lower than at the canopy surface, and this may provide an additional senescence signal (Boonman et al., 2009; Rousseaux et al., 2000). Leaf senescence therefore depends more on a leaf's position than on its age, and abscission due to canopy shading or during stress may be viewed as the elimination of surplus leaves that do not contribute to canopy photosynthesis (Hikosaka, 2005). Senescence is accompanied by remobilization of carbon, proteins, and mineral nutrients from these leaves, although as much as half of a leaf's pool of resources cannot be recycled and is lost (Bertamini and Nedunchezhian, 2001; Hikosaka, 2005). This adaptive strategy enables the plants to survive episodes of limited photosynthate supply by remobilizing buffer reserves from older, senescing leaves and permanent parts of the plant and temporarily supplying resources for maintenance and/or growth processes by recycling the remobilized resources to organs with high sink priority (Geiger and Servaites, 1991; Hunter et al., 1994).

Leaf senescence does not change the direction of phloem transport in the leaf: There is no switch from export to import. There is only a progressive change in the nature of the exported materials from predominantly sucrose to mostly amino acids and other nutrients (Masclaux-Daubresse et al., 2010; Thomas, 2013). In other words, shaded leaves in the interior of dense canopies are not "parasitic" on a vine (Koblet, 1975; Quinlan and Weaver, 1970; Wardlaw, 1990). Such leaves still maintain a positive daily carbon balance and are simply discarded following recycling of their accessible resources: A senescing leaf is still a source, albeit now mostly one of nitrogen instead of carbon (Reich et al., 2009; Schippers et al., 2015; Thomas, 2013). Where the recycled compounds end up depends on the relative strength of the various sink organs. When the shoots are actively growing, remobilized nutrients are redistributed to young, better exposed leaves, but when shoot growth has stopped (e.g., due to water deficit), the fruit clusters or the permanent structures may be the main recipients of recycled nutrients.

Grapevines also adapt to low light by altering leaf and shoot growth, producing longer and thinner internodes, although this occurs at the cost of greater shoot hydraulic resistance (Cartechini and Palliotti, 1995; Schultz and Matthews, 1993b). When entire shoots or vines experience low light, such as during overcast periods, grapevines produce new leaves, especially lateral leaves, rather than maintaining the source capacity of old leaves; however, this response may occur at the expense of fruit production and root growth (Keller and Koblet, 1995a). This adaptive change in carbon partitioning leads to an increase in the leaf area ratio (leaf area per unit whole-plant dry weight), while the leaf weight ratio (leaf dry weight per unit whole-plant dry weight) may remain unaltered. Vines grown under low light also tend to have elevated concentrations of tissue nutrients and low amounts of reserve carbohydrates (Keller et al., 1995).

Although grapevines are rather shade tolerant, wild vines are well adapted for maximum light capture. Their elaborate branching structure, with a large number of short shoots armed with tendrils, enables them to spread their foliage over tree canopies, which results in an enormous increase in surface area for sunlight absorption. This option is not available to cultivated grapevines that are often confined to small trellis systems with foliage concentrated within a more or less defined canopy volume; this is especially true for shoot-positioned vines. In fact, fewer than 20% of the leaves often account for >80% of a vine's total carbon assimilation. Therefore, the canopy surface area of a vineyard, rather than its total leaf area, is important because the more solar energy that is intercepted by foliage, the greater are the biomass production and the yield potential.

Canopy surface area is referred to in terms of exposed surface area or effective surface area. On fully developed canopies, the effective surface area varies from 30% to 85% of the total leaf area depending on trellis design and row spacing. Vines with a small proportion of exposed surface area have many shaded leaves in the canopy interior and may produce less photosynthate for export to be available for fruit ripening, root growth, nutrient uptake and assimilation, and replenishment of storage reserves for cold acclimation and spring growth.

The fact that photosynthetic CO_2 assimilation shows light saturation while the absorption of photons continues to increase with rising irradiance is a potential source of trouble for leaves. When a leaf absorbs more light than it can utilize, some of the excess energy must be dissipated as heat, or else it will cause photoinhibition (see Section 4.1). Photoinhibition in grapevines occurs frequently around midday, when incident light and temperature are at a maximum, leading to a temporary reduction in CO_2 assimilation (Chaves et al., 1987; Correia et al., 1990; Düring, 1999; Iacono and Sommer, 1996). Over the course of a growing season, this temporary depression of photosynthesis, known as dynamic photoinhibition, can result in a roughly 10% loss of potential biomass production. Leaves acclimate to high light by moving the chloroplasts, which are normally aligned at the abaxial surface of mesophyll cells and thus perpendicular to the solar rays to maximize light absorption, to anticlinal positions in the cells and thus parallel to the solar rays. This reversible chloroplast relocation is regulated by a protein photoreceptor termed phototropin that is activated by blue light (450 nm) and is an attempt to avoid photoinhibition (Li et al., 2009; Spalding and Folta, 2005; Takahashi and Badger, 2011). The effect of chloroplast relocation may be reinforced by relocation of calcium oxalate crystals termed druses. Under low light the druses "sit" near the bottom of the leaf's palisade cells, but under excess light they move to the top of these cells where they reflect some of the incoming light (He et al., 2014).

Both low and high temperatures exacerbate the photoinhibitory effect of strong light (Gamon and Pearcy, 1990a,b). This can be problematic in regions with a continental climate, which often experience large diurnal temperature fluctuations so that vines can be exposed to below-optimum temperatures and high light intensities in the morning, especially on the east side of the canopy (Hendrickson et al., 2003). Such conditions are particularly frequent in spring and autumn when clear skies result in radiative heat loss from plant and soil surfaces at night (Peña Quiñones et al., 2019). Untimely cold spells during

FIG. 5.6

Light reflection, absorption, and transmission of a typical leaf (left), and interior of a dense grapevine canopy (right).
© Elsevier Inc., illustration after Taiz, L., Zeiger, E., 2006. Plant Physiology, fourth ed. Sinauer, Sunderland, MA.
Right; photo by M. Keller.

clear-sky conditions can result in irreversible damage to the photosynthetic machinery due to degradation of photosynthetic pigments; this condition is called chronic photoinhibition (see Section 7.4).

Leaf layers alter not only the quantity of light but also its quality. The spectral characteristics of chlorophyll make leaves strong absorbers of photons in the blue (400–500 nm) and red (R: 600–700 nm) wavebands of the solar spectrum (Fig. 5.6; see also Section 4.1). Absorption of green (500–600 nm) and particularly far-red (FR: 700–800 nm) light is weaker, and many photons of these wavelengths are reflected or transmitted and scattered in the form of diffuse radiation (Blanke, 1990a). Thus, light that is reflected from leaves is enriched in the FR region, which lies in the infrared portion of the electromagnetic spectrum, whereas light that is transmitted through leaves is depleted in the R region of the spectrum. The outcome in both situations is identical: The R ÷ FR ratio declines. Leaves detect these changes in the spectral composition of solar radiation, using various photoreceptors, or photosensors, that absorb light and translate its information via signaling networks into physiological responses.

Perhaps the most important photosensor is a pigment system called phytochrome that works like a light-regulated switch (Rockwell et al., 2006; Smith, 2000). Like other plants, grapevines have at least five different phytochrome proteins that are abbreviated phyA–phyE, and each exists as a mixture of two reversible forms—an inactive form (P_r), which absorbs red light of 660–670 nm and has a half-life of approximately 100 h, and an active form (P_{fr}), which absorbs FR light of 725–735 nm and has a half-life of only approximately 1 h. When P_r absorbs red light, it changes into P_{fr}, which induces other proteins to carry P_{fr} from the cytosol to the nucleus. Inside the nucleus, P_{fr} interacts with yet other proteins termed transcription factors to modify gene expression (Bae and Choi, 2008; Fankhauser and Staiger, 2002; Franklin, 2008; Smith, 2000). In other words, red light photoactivates phytochrome which then indirectly switches light-regulated genes on. Conversely, in the dark or when P_{fr} absorbs FR light, it changes back into P_r, switching the light-regulated genes off, as shown in the following diagram (Chory, 1997):

$$P_r \underset{\text{Far-red light}}{\overset{\text{Red light}}{\rightleftarrows}} P_{fr} \longrightarrow \text{Shade-avoidance response}$$

Although light quantity during the day varies as much as 10-fold due to variations in cloud cover and time of year, these conditions have only a minor effect on R ÷ FR, which is remarkably constant at approximately 1.2 (Holmes and Smith, 1977a,b). However, during dawn and dusk R ÷ FR drops temporarily, especially at higher latitudes with longer twilight periods, and the magnitude and duration of this decline inform plants about seasonal changes (Franklin, 2008). The largest variation in R ÷ FR, however, occurs due to sunlight interacting with leaves. As vines grow and their canopy size increases, the R ÷ FR of the light reaching individual leaves decreases—even on the exterior of the canopy. When the plant spacing is wide and/or canopy density low, vines do not shade each other, and the R ÷ FR decreases mainly due to the increase in reflected FR. In grapevines with a loose canopy and vertical shoots, this increase in FR modifies the light environment of the internodes without greatly affecting the spectral balance of the leaves, which is dominated by the contribution of direct sunlight. Nonetheless, the decrease in R ÷ FR is large enough to be sensed by phytochrome molecules located in the shoot tissue and to reduce the proportion of P_{fr}. Above a grapevine canopy or on its sunlit side, R ÷ FR is similar to the ambient value (1.1–1.2), whereas on the shaded side R ÷ FR can drop to 0.3–0.5. Similarly, inside a dense canopy R ÷ FR can be reduced to as little as 0.1 (Dokoozlian and Kliewer, 1995a,b; Smart et al., 1982). In the shade, the decrease in R ÷ FR is due to the selective absorption of R by the leaves. This effect may be more important in vineyards of table grapes, raisin grapes, or juice grapes, which are often grown with considerably larger and denser canopies than are wine grapes.

Together with blue-light and UVA-light receptors named cryptochromes, the phytochrome system enables plants to monitor the quantity and quality of light, as well as its duration and direction. This information allows them to adjust their development in ways that optimize the capture of energy for photosynthesis and synchronize vegetative and reproductive development (Whitelam and Devlin, 1998). Generally, red light induces phytochrome-controlled responses, whereas FR light inhibits these responses. The detection of a low R ÷ FR by phytochrome, translated into less P_{fr}, provides a clear signal of light transmitted through, or reflected from, nearby plants and thus indicates the proximity of potential competitors. Plants can detect the presence of neighbors very early, well before they begin shading each other. To outgrow their competitors and thus avoid being shaded, they respond to low R ÷ FR signals by several morphological changes that involve a reallocation of resources toward shoot elongation at the expense of storage and reproductive growth and are collectively termed the shade avoidance syndrome (Morelli and Ruberti, 2002; Smith and Whitelam, 1997; Whitelam and Devlin, 1998). These changes mainly include stronger apical dominance (less lateral shoot outgrowth), accelerated shoot elongation rates, decreased leaf size and thickness, altered shape and more horizontal orientation of leaf blades, and more vertical shoots. Shoots that grow in the shade tend to be thinner, have longer internodes and thinner leaves with longer petioles, and accumulate less carbohydrate reserves than shoots that grow in the sun (Buttrose, 1969c; Cartechini and Palliotti, 1995; Kliewer et al., 1972; May, 1960; Morgan et al., 1985).

Prolonged low R ÷ FR signals, which indicate that the competitors cannot be outgrown, also lead to developmental responses, such as early flowering or even inhibition of inflorescence initiation, reduced seed and fruit set, truncated fruit development, and often decreased seed viability (Morelli and Ruberti, 2002). These changes are brought about by the interaction of multiple plant hormones. The influence of P_{fr} stimulates auxin production and its distribution in the shoots and roots, and increases ethylene production, which may augment brassinosteroid and gibberellin action and tissue sensitivity to gibberellin, leading to cell wall loosening and enhanced cell elongation in the shade (Franklin, 2008; Gallavotti, 2013; Morelli and Ruberti, 2002; Pierik et al., 2004; Zhao, 2018). The regulation of auxin transport also enables phytochrome to coordinate shoot and root growth, whose reciprocal adjustment allows plants to better utilize the available light. Thus, low R÷FR enhances shoot elongation but reduces root growth, whereas high R÷FR has the opposite effect.

In addition to its involvement in the control of photomorphogenesis and photoperiodism, the phytochrome system also induces changes in the composition of chemicals such as mineral nutrients, chlorophylls, anthocyanins, and other metabolites. Phytochrome seems to play a role in regulating the activities of certain enzymes, such as nitrate reductase, which is important for nitrogen assimilation (Smart et al., 1988; see also Section 5.3), phosphoenolpyruvate carboxylase, which produces oxaloacetate for the TCA cycle (see Section 4.4), and phenylalanine ammonia lyase and chalcone synthase, which are two key enzymes for the biosynthesis of phenolic compounds, including anthocyanins and tannins (see Section 6.2). For example, P_{fr} may inactivate the genes that code for the enzymes that make anthocyanins; thus, a decrease in R÷FR leads to a decrease in anthocyanin production, which impairs fruit color.

The regulation of growth and development by phytochrome in response to changing R÷FR has implications not only for planting density and trellis design in vineyards but also for research studies conducted with artificial light sources. The cool-white fluorescent tubes and high-pressure sodium lamps that are often used in growth chambers and greenhouses have light spectra that differ markedly from the spectra of direct sunlight and even of natural daylight under overcast conditions and are almost devoid of FR (Hogewoning et al., 2010). This and the compounding fact that these lamps generally emit light intensities that are far below photosynthetic light saturation—often as little as 100 μmol m^{-2} s^{-1}—make it doubtful that such studies mimic plant growth under field conditions.

In addition to the effects of visible light, the UV range of the electromagnetic spectrum, which is commonly divided into UVA (320–400 nm), UVB (280–320 nm), and UVC (<280 nm), is also important. Much of the UVB and all the UVC are absorbed by the ozone (O_3) layer in the stratosphere and never reach the surface of the Earth. Just as in the case of visible light, the UVB radiation a vineyard receives depends mainly on the position of the sun, which varies with latitude, altitude, season, and time of day, and on cloud cover. In addition, UVB has been increasing at Earth's surface due to the depletion of stratospheric O_3, although this process is thought to be close to its maximum and is expected to recover by approximately 2050—provided all countries continue to implement the so-called Montreal Protocol that limits the emission of O_3-depleting chemicals, such as chlorofluorocarbons (Madronich et al., 1998; McKenzie et al., 1999). The main problem with UV light is that as the wavelength of electromagnetic radiation declines, its energy content increases (see Section 4.1). Although it comprises only approximately 0.5% of the total solar radiation, the high energy of UVB makes it damaging to membranes, proteins, and DNA. Consequently, plants have evolved mechanisms to screen out UV radiation at or near the surface of their organs (Jansen et al., 1998; Rozema et al., 1997).

Leaf hairs, which are particularly dense on the vulnerable young, expanding leaves of some grape cultivars, scatter and attenuate a large fraction of the UV radiation (Karabourniotis et al., 1999). The epicuticular wax, although itself not a strong UV absorber, also reflects and scatters some of the incident UV light (Kerstiens, 1996; Rozema et al., 1997; Shepherd and Griffiths, 2006). Their higher amount of surface wax and its crystalline structure is why sun-exposed leaves appear more glossy than shaded leaves (Keller et al., 2003a; Fig. 5.7). In addition, the vacuoles of the epidermis cells of leaves and fruits accumulate phenolic compounds known as flavonols and cinnamic acids (see Section 6.2), which act as an optical filter or "sunscreen" that absorbs the damaging UVB radiation and, being strong antioxidants, detoxify the reactive oxygen species generated as a consequence of UV-induced damage

FIG. 5.7

Cabernet Sauvignon leaves grown in full sunlight (right) and with the UV portion of the spectrum filtered out (left) display differences in epicuticular wax and hence in leaf surface glossiness.

Photo by M. Keller.

(Bachmann, 1978; Egger et al., 1976; Hernández et al., 2009; Kolb and Pfündel, 2005; Kolb et al., 2001; Rozema et al., 1997; Yamasaki et al., 1997). Additional phenolics are incorporated into the leaf cell walls (Weber et al., 1995). Together these phenolics effectively screen UV radiation so that virtually none is transmitted through the leaves. Particularly high concentrations of flavonols, in addition to anthocyanins which also absorb UV light, are produced in the unfolding leaves at the shoot tips, giving the shoot tips of many grape cultivars a pink or red appearance (see Fig. 2.5). This protects the UV-susceptible and highly sun-exposed young leaves from photooxidative damage while the photosynthetic apparatus is being assembled in their cells. When the leaves senesce at the end of the growing season, they again accumulate flavonols and, in some cultivars, anthocyanins to protect their cells from sun damage, this time during the disassembly of the photosynthetic apparatus and retrieval of nutrients from the leaves.

Flavonols are also effective inhibitors of auxin transport (Agati and Tattini, 2010; Peer and Murphy, 2007; Winkel-Shirley, 2002). Because basipetal auxin flow from the shoot tip is associated with apical dominance and suppression of lateral shoot growth, UV light may also influence the pattern of shoot growth, and hence canopy architecture. Although the extent of this influence is currently unclear, high UV exposure is indeed associated with a stimulation of lateral shoot growth in grapevines (M. Keller, unpublished data). This effect seems to be especially pronounced if soil moisture does not limit shoot growth, which may explain why overirrigation in climates with high solar radiation effectively promotes dense canopies.

Grape flowers and berries at the exterior of the canopy may be exposed to high intensities of UV radiation, most of which is absorbed by the berry skin (Blanke, 1990a). The production of phenolic compounds, especially flavonols, in the skin is greater in these berries than in berries in the interior of the canopy, which has implications for fruit quality (see Section 6.2). It also has implications for yield formation (see Section 6.1). Similar to the situation in shoots, the need for sun-exposed flowers to produce flavonols for UV protection may interfere with auxin transport. Because auxin flow is necessary for fruit set (see Section 2.3), high UV exposure might lead to poor fruit set, especially in combination with nitrogen deficiency, which further promotes flavonol formation (M. Keller, unpublished data).

5.2.2 Temperature

In general, the temperature of grapevine leaves and other organs more or less tracks the temperature of the surrounding air (Peña Quiñones et al., 2019). Nevertheless, the leaf temperature changes rapidly in response to fluctuations in radiation and air turbulence. Sun-exposed leaves and grape berries are heated by the sun, whereas shaded leaves and berries in the interior of the canopy are generally close to the ambient temperature (Smart and Sinclair, 1976; Spayd et al., 2002; Vogel, 2009). The heat load on a leaf exposed to full sunlight is so great that the leaf would heat up by 1–$2\,°C\,s^{-1}$ to temperatures that would denature proteins and kill the tissues within <1 min if no heat was lost to the environment. This is indeed a problem during sunny days with no wind at all, whereas even slow air movement can prevent lethal leaf heating (Vogel, 2009). Although leaves on the sunlit side of a canopy are often approximately 2–3 °C warmer than leaves on the shaded side, the leaves are generally <5 °C warmer than the surrounding air (Peña Quiñones et al., 2019). Leaves keep from overheating by emission of long-wave radiation, sensible heat loss by convection, and latent heat loss by evaporative cooling due to transpiration (see Section 3.2). Despite these cooling mechanisms, direct sunlight can occasionally heat leaves by as much as 10 °C above air temperature (Sharkey et al., 2008). Conversely, at night the leaves may be several degrees cooler than the surrounding air (Peña Quiñones et al., 2019).

Growing grapevines in a manner that maximizes the canopy surface area for maximum light interception creates a dilemma for the vines, especially in warm climates. Although they can survive brief temperature spikes of up to 60 °C, grape leaves are killed at temperatures above approximately 45 °C due to disintegration of the cell membranes, which leads to membrane leakage and loss of cell contents, and due to thermal denaturation of proteins (see Section 7.4). In a movement termed paraheliotropism, grapevine leaves change their angle to the sun over the course of a day and a growing season, aligning the leaves in parallel with the solar rays during hot periods, to avoid overheating and keep light intensity on the leaf at or slightly below light saturation to maintain photosynthesis while avoiding photoinhibition (Gamon and Pearcy, 1989; Takahashi and Badger, 2011). In addition, transpiration rates increase within a range of <10 °C to >40 °C, although the stomata usually begin to close at approximately 35 °C (Fig. 5.8). The increase in transpiration occurs because warmer air can hold more moisture (by approximately 7% for every 1 °C rise in temperature; IPCC, 2013), thereby increasing the vapor pressure deficit (see Section 3.1), which protects the leaf from overheating so long as it does not run out of water. Soil water deficit decreases transpiration and therefore increases the leaf temperature (Grant et al., 2007).

The stomatal closure at high temperature seems to be a response to heat-induced reduction in photosynthesis rather than the cause of it. Therefore, high temperatures can result in excessive water loss from the canopy while simultaneously reducing CO_2 assimilation. It is possible that the higher transpiration rate of Chardonnay compared to Cabernet Sauvignon at high temperature (see Fig. 5.8) is related to the differences in leaf shape between the two cultivars (compare Figs. 1.14 and 5.7). Local leaf temperature increases approximately with the square root of the distance from the edge, and lobing improves heat transfer (Vogel, 2009). Therefore, Chardonnay's more-or-less entire leaves may need to evaporate more water to keep them from overheating.

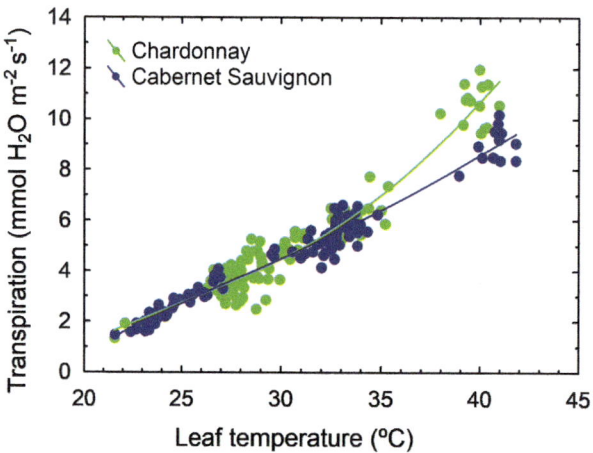

FIG. 5.8

Relationship between leaf temperature and transpiration of mature leaves of two grapevine cultivars.

M. Keller, unpublished data.

As would be expected from a process that involves biochemical reactions, which rely on enzymatic processes, CO_2 assimilation is strongly influenced by temperature (Geiger and Servaites, 1991). But other photosynthetic processes are sensitive to temperature too, especially at high light intensity. For example, electron transport has a pronounced optimum at 30 °C. Below 15 °C, photosynthesis is strongly curtailed by an inhibition of sucrose synthesis, which leads to accumulation of phosphorylated intermediates and prevents the release of phosphate for regeneration of ribulose-1,5-bisphosphate (Hendrickson et al., 2004b; see also Section 4.2). This so-called end-product limitation, or feedback inhibition, occurs because low temperature restricts cell division in sink organs more than photosynthesis; the time it takes for a cell to divide increases exponentially with declining temperature (Körner, 2003). This decrease in sink activity results in surplus sugar accumulating in the leaves and in the perennial organs of the plant. As in the case of very high temperatures, the stomata will partially close in response to the reduced photosynthesis, rather than photosynthesis declining because of lower stomatal conductance.

Similar to light, therefore, rising temperature initially stimulates photosynthesis. However, instead of reaching a saturation point, the temperature response shows a relatively broad optimum, with very high temperatures resulting in a reduction of carbon fixation. In grapevine leaves, there is very little photosynthesis below 10 °C, the optimum ordinarily falls between 25 °C and 30 °C, and photosynthesis declines sharply above 35 °C (Currle et al., 1983; Downton et al., 1987; Gamon and Pearcy, 1990b; Hendrickson et al., 2003; Kriedemann 1968b; Williams et al., 1994). Yet sometimes the photosynthetic rate at 45 °C may still approach half the rate at 30 °C (Greer and Weedon, 2012; Mullins et al., 1992). The optimum temperature range depends on species, cultivar, light intensity, and developmental stage, but it also reflects the maximum temperature experienced during leaf growth. For example, low light intensity leads to a flatter and broader temperature response curve so that shaded leaves have a lower and less pronounced temperature optimum than sun-exposed leaves (Berry and Björkman, 1980;

Gamon and Pearcy, 1990b). Furthermore, in north–south-oriented rows, east-facing leaves typically reach lower daily maximum temperatures than west-facing leaves, which can limit photosynthesis and growth on cool vineyard sites (Hendrickson et al., 2004a). Nevertheless, the west-facing leaves often contribute less photosynthate than their east-facing counterparts because the high vapor pressure deficit, rather than the higher temperature, in the afternoon may lead to partial stomatal closure. On the other hand, high temperatures during leaf development can shift the optimum to as much as 35 °C due to photosynthetic temperature acclimation that makes Calvin cycle enzymes and membranes more heat tolerant. Conversely, low temperatures will shift the optimum downward as the leaves produce more photosynthetic proteins.

In general, the optimal temperature for photosynthesis tends to increase by approximately 1 °C for each 2–3 °C increase in growth temperature—up to the maximum near 35 °C (Berry and Björkman, 1980; Hikosaka et al., 2006; Schultz, 2000). This means that the temperature optimum decreases with increasing vineyard elevation due to the decrease in average temperature by 0.65 °C per 100-m elevation gain. Similarly, within a vineyard site, leaves developing during the hot summer months have a considerably higher temperature optimum for photosynthesis than leaves developing in spring or autumn. But even mature leaves can acclimate to temperature changes within 1 or 2 weeks. The seasonal shift in temperature optimum occurs in both irrigated grapevines and vines growing under water deficit, but water stress can become the dominant factor that determines photosynthetic performance (see Section 7.2), largely overriding the temperature response.

Clouds alter not only the light intensity but also the temperature. Although clouds are usually associated with cooler days than would be the case under clear skies, they also prevent nighttime inversions and thereby dampen the amplitude of the diurnal temperature swings. In other words, clouds keep days cooler and nights warmer. This has implications for the daily carbon balance of grapevines. Because canopy net photosynthesis is light limited and thus lower during cloudy days, whereas whole-plant respiration is higher during cloudy nights, less carbon is available for growth, fruit production, and ripening. Frequent and variable cloud covers, which are indicated by relatively few sunshine hours in climate records, in many cool and maritime growing regions are one reason for the marked interannual fluctuations in yield and fruit quality in such areas.

Both the increasing temperature and rising CO_2 content of the air ($[CO_2]$) associated with global climate change is also influencing photosynthesis. Models as well as empirical observations generally show a relatively steep increase in net CO_2 assimilation rates above approximately 20 °C when $[CO_2]$ rises from 300 to 600 ppm (current ambient values are ~400 ppm) and a shift of the optimum temperature for photosynthesis from 25 to 35 °C to 35–40 °C (Sage and Kubien, 2007). Below 20 °C, there appears to be little effect of rising $[CO_2]$, and above approximately 40 °C photosynthesis drops rapidly. Nevertheless, at 45 °C photosynthesis at 600 ppm CO_2 still runs twice as fast as at 300 ppm and is similar to the rate at 20 °C. Although heat effects will certainly be important in a warmer world with higher $[CO_2]$, climate change will have a greater impact at the lower temperature limits of viticulture, because current and predicted temperature increases will be higher at night, at higher latitudes, and in the winter (IPCC, 2013).

The temperature response of CO_2 assimilation partly reflects a conflict of interest in addition to the dilemma created by the need for evaporative cooling at high temperatures. Enzyme activities are stimulated by increasing temperature, and rubisco is no exception. Unfortunately, rubisco's oxygenation rate increases faster with increasing temperature than does its carboxylation rate (Berry and Björkman, 1980; Foyer et al., 2009; Woodrow and Berry, 1988). The greater rise in CO_2 release than in its fixation is due to temperature-induced increases in both the specificity of rubisco for oxygen and the solubility of oxygen, which results in an increase in photorespiration as the temperature increases (Zufferey et al., 2000). In addition to its effect on photorespiration, temperature also influences mitochondrial respiration—that is, glycolysis and the TCA cycle and the associated electron transport chain (see Sections 4.3 and 4.4). The proportionality factor or temperature coefficient (Q_{10}) of respiration is approximately 2 or 3; in other words, the respiration rate roughly doubles or triples for every 10°C rise in temperature up to about 30°C, above which respiration begins to decline again (Kruse et al., 2008; Mullins et al., 1992; Schultz, 1991; Williams et al., 1994; Zufferey et al., 2000). The Q_{10} value is not constant and must be viewed as an oversimplification of the temperature response of respiration (Kruse et al., 2011). Nonetheless, because of its exponential increase with rising temperature, the respiration rate of grapevine leaves at 10°C is close to zero, whereas at 25°C it can be as high as 2 μmol CO_2 m^{-2} s^{-1}.

The temperature effect on respiration is especially important because photosynthesis "rests" at night, whereas respiration "works" 24h a day in all plant parts, not just the leaves. This temperature effect reduces net CO_2 assimilation because at higher temperatures vines use a greater proportion of their daily fixed carbon for respiration. Therefore, even at modestly high temperatures of 25–30°C there may be less carbon available for vine growth and fruit ripening than at cooler temperatures near 15–20°C. Incidentally, water-stressed plants often have a lower Q_{10} than plants with abundant water supply (Atkin et al., 2005). In other words, water deficit decreases the temperature sensitivity of respiration, although this effect is often complicated by an increase in soil temperature as the soil dries. Because the Q_{10} declines when the temperature increases only briefly (Atkin and Tjoelker, 2003), a hot day may stimulate respiration more in vines growing in cool climates than in vines growing in warm climates. After a few days, however, respiration acclimates somewhat to above- or below-average temperatures to compensate for the change in temperature. Thus, heat waves lasting several days decrease the respiration rate, whereas cold waves increase the respiration rate (Atkin and Tjoelker, 2003). Above approximately 35°C, the capacity of the electron transport system becomes limiting to photosynthesis, and at very hot temperatures above about 45°C membranes become increasingly fluid and leaky so that respiration declines rapidly (Sage and Kubien, 2007). The fact that the Q_{10} is higher when there is abundant sugar for respiration suggests that a low crop load might stimulate respiratory CO_2 release more in a warm climate than in a cool climate.

The temperature that a leaf experiences during its growth also affects leaf growth. Meristem temperature, through its effect on cell division, is an important determinant of the rate of leaf appearance and the rate of leaf expansion, as long as the temperature causes neither chilling stress nor heat shock (see Section 7.4). Leaf area development on a grapevine canopy depends on three separate processes: leaf initiation, leaf expansion, and outgrowth of lateral shoots. Increasing temperature accelerates all three of these processes so that higher temperature leads to more rapid canopy development, longer shoots, and denser canopies (Keller and Tarara, 2010; Lebon et al., 2004). Indeed, a fundamental response of plants to high temperature appears to be a shift in carbon partitioning to favor shoot growth at the expense of fruit growth and ripening and, probably, storage reserve accumulation (Richardson et al., 2004). These changes in plant development, collectively termed thermomorphogenesis, are rather similar to the shade avoidance syndrome that occurs in response to shade and are similarly mediated by the phytochrome system and its effect on auxin production (Zhao, 2018).

5.2.3 Wind

Wind with a speed $>6\,m\,s^{-1}$ can inflict physical damage to plants in addition to a reduction of shoot length, leaf size, and stomatal density. However, even if it is not strong enough to induce visible damage (Fig. 5.9), wind mechanically disturbs shoot growth by repeated bending, often at rates of dozens to hundreds of times per minute, leading to shorter but thicker shoots. This so-called thigmomorphogenetic response intensifies with increasing height above the ground (Braam, 2005; Niklas and Cobb, 2006). Wind during the day seems to inhibit shoot growth more than at night (Hotta et al., 2007). The reduction in shoot growth is also more severe on the windward (into the wind) side of the canopy than on the leeward (away from the wind) side and intensifies with increasing number of wind perturbations rather than with increasing wind speed in each single event (Tarara et al., 2005; Williams et al., 1994). Moreover, the shoots are often displaced away from the wind so that the canopy becomes lopsided, which has consequences for fruit exposure to sunlight. Clusters on the windward side of the canopy may receive more intense light for longer duration than do clusters on the leeward side of the canopy (Tarara et al., 2005). Thigmomorphogenetic responses that resemble responses to wind may also be triggered by other physical influences, such as repeated bending or touching by passing vineyard workers, animals, or machinery. Shoot elongation ceases within minutes of a mechanical stimulus and is accompanied by cessation of phloem flow due to rapid callose deposition in the sieve tubes. This response is followed by a temporary increase in radial growth, callose removal, and recovery of elongation growth over the next few days (Coutand, 2010).

As discussed in Section 3.3, wind decreases a leaf's boundary layer resistance, which increases transpiration and evaporative cooling. This is an advantage under warm conditions because leaves can heat up rapidly when the wind speed drops below approximately $0.5\,m\,s^{-1}$ (Vogel, 2009). Under otherwise similar conditions, the temperature of sun-exposed leaves tends to vary inversely with wind speed, whereas shaded leaves track the air temperature (Vogel, 2009). To avoid excessive water loss and

FIG. 5.9

Physical wind damage on Merlot canopy from shoot being knocked against a foliage wire (left) and leaf being twisted at the petiole junction (right).

Photos by M. Keller.

dehydration in stronger wind, however, grapevines respond to wind speeds >2.5 m s^{-1} by partly closing their stomata (Campbell-Clause, 1998; Freeman et al., 1982; Williams et al., 1994). Although this strategy may conserve water, it also reduces photosynthetic CO_2 assimilation and increases leaf temperature. This may at least partially explain why vines growing in areas with frequent strong winds often produce fewer and smaller clusters and berries with lower soluble solids (Williams et al., 1994). Conversely, reduced wind speed and air mixing in sheltered vineyards could also decrease photosynthesis because leaves may deplete the CO_2 in the air surrounding the foliage. Moreover, wind speeds below 0.5 m s^{-1} result in humid canopies, which favors disease development. For instance, the powdery mildew fungus *E. necator* requires only 40% relative humidity for germination; this threshold is easily exceeded within the leaf boundary layer, where the fungus resides (Keller et al., 2003a; see also Section 8.2).

Wind moving down the rows of a vineyard creates less turbulence and movement of leaves than wind moving across rows (Weiss and Allen, 1976). Consequently, orienting the rows in the direction of the major wind load may decrease evaporative water loss, especially under dry conditions, and reduce the negative effect of stomatal closure on photosynthesis. In addition, because leaves slow wind down, the wind speed in the center of a canopy is often <20% of the speed at the exterior. Although this may not be important in terms of stomatal effects on CO_2 assimilation, which is light-limited in the canopy interior, it has implications for drying of leaves and fruits after rain: interior surfaces dry more slowly than exterior surfaces. Nevertheless, even slow wind speeds can move leaves in a canopy sufficiently to briefly increase the light exposure of otherwise shaded leaves. The resulting and continuously changing sunflecks on these leaves can account for a temporary rise in photosynthesis, improving the leaves' photosynthetic efficiency and the overall carbon balance of the canopy (Intrieri et al., 1995; Kriedemann et al., 1973).

5.2.4 Humidity

Intuitively, one would expect transpiration by leaves and berries to increase the humidity inside the canopy, with subsequent implications for the development of fungal diseases (see Section 8.2). This is a subject that has been little studied. Increases in air humidity in the canopy interior of <10% have been recorded, and the significance of these increases is not well understood. However, the humidity of the air is an important driver of transpirational water loss from a canopy. A decrease in relative humidity from 95% to 50% increases the vapor pressure deficit (VPD) of the atmosphere surrounding a leaf >10-fold (see Section 3.1). Moreover, humidity strongly depends on air temperature because increasing temperature also sharply increases VPD (see Section 3.1). As the VPD increases, the leaves' stomatal conductance declines to control excessive water loss by transpiration, which reduces photosynthesis. In other words, a decrease in relative humidity effectively increases the transpiration rate while decreasing CO_2 assimilation. Due to the modulating effect of leaf layers, changes in humidity affect exterior leaves more than interior leaves of a canopy.

Humidity also affects leaf growth. A high VPD reduces the growth rate of leaves by decreasing the rate of cell division and cell expansion, even when there is no soil water deficit (Lebon et al., 2004). Therefore, leaves growing in low humidity remain smaller than leaves grown in high humidity. Due to the decrease in leaf growth at high VPD, grapevines growing in dry climates tend to have more open canopies than vines growing in more humid climates, even when they are equally well watered.

In addition to the effects of air humidity, even small changes in vine water status can alter the canopy microclimate (Keller et al., 2016a). Water-stressed vines often have higher canopy temperatures than fully irrigated vines because water-stressed leaves close their stomata, which reduces transpirational cooling, and because such vines have a sparser, more open canopy (Grant et al., 2007; see also Section 7.2).

5.2.5 The "ideal" canopy

The upper limit on vineyard productivity is set by the total seasonal amount of PAR intercepted by the vines planted in the vineyard. Canopy structure, and especially the spatial distribution of leaves, has important consequences for canopy light interception and hence productivity (Carbonneau et al., 2007; Prieto et al., 2012). As discussed previously, the acquisition of energy and carbon by a canopy depends on total leaf area, leaf surface distribution, canopy architecture, and photosynthetic capacity of individual leaves. A canopy that has an ideal microclimate in terms of maximum light interception for vine productivity has the following features (modified from Smart, 1985 and Smart et al., 1990):

- Rows should be oriented from north to south to maximize light interception by both sides of the canopy for some part of the day. However, row direction is of minor importance for fruit and wine quality compared to the features described next, and the north–south requirement may well be overridden by other considerations, such as topography, economic row length, or prevailing wind. In a high-irradiance environment, it may be beneficial to deviate from the north–south orientation, because rows shifted somewhat to a northeast–southwest orientation may protect grape berries from overheating on the vulnerable west side of the canopy.
- The ratio between canopy height and row width should be $1 \div 1$. For any one canopy height, as the distance between rows decreases, the percentage of light interception by the canopy increases. However, decreasing row width increases the likelihood of one row shading the lower canopy of the neighboring row. Similarly, for any one row width, as the height of the canopy increases, the percentage of light interception by the canopy increases, but again the potential for cross-row shading also increases. The $1 \div 1$ ratio is a compromise between the canopy's intercepting as much light as possible and avoiding one row shading another.
- Shoots should be trained vertically to avoid shading on one side of the canopy and promote sun exposure of and hence light interception by leaves and fruit clusters. However, this requirement is less important in regions with high temperature and high irradiance during the growing season than in regions with frequent cloudy and cool conditions. In a high-irradiance environment, it may be beneficial to deviate from the I-shape associated with vertical shoot positioning, because somewhat sprawling shoots that form a V-shape may protect grape berries from overheating.
- The canopy surface area should be approximately $21,000 \, m^2 \, ha^{-1}$, and 80–100% of the leaves should be on the outside of the canopy. The larger the surface area, the more light is intercepted, and hence the potential for photosynthesis and yield production is increased. However, if the surface area is too large, then the canopy-height $\div$ row-width ratio of $1 \div 1$ is not adhered to.
- The canopy should be approximately 30–40 cm wide with 1–2 layers of leaves horizontally across the canopy. This leaf layer number is the number of leaves from one side of a canopy to another. Higher values are associated with shading and reduced fruit quality, whereas lower values are

associated with incomplete light interception. More leaf layers can be tolerated in a high-irradiance environment than in a light-limited environment.
- There should be approximately 15 shoots per meter of canopy length. A higher figure means that shoots may be crowded and shading may occur. A lower figure means that light interception is suboptimal. The value varies according to cultivar and vine vigor and can be considerably higher without detriment in warm regions with plenty of sunshine compared with cool regions where light is often limiting. Maintaining ideal shoot spacing is a key underlying theme of canopy management.
- A total of 20–40% of the canopy should be composed of gaps, that is, the neighboring row should be visible across the canopy. Too many gaps reduce the yield potential, whereas too few gaps indicate that shading in the canopy is likely.
- Shoots should stop growing when they are approximately 15 nodes or 1–1.2 m long to provide sufficient leaf area to ripen the fruit. A leaf-area ÷ fruit-weight ratio of approximately 10–15 $cm^2\,g^{-1}$ is adequate for ripening in most cases. If shoots grow beyond this length and are not supported or trimmed, they will fall across each other and create shade. Excessive and repeated trimming, however, wastes the vine's resources and is an indication of too much vigor.
- Lateral shoot growth should be limited to <10 lateral nodes per main shoot. For example, 10 lateral nodes could be made up of 5 lateral shoots of 2 nodes each or 2 lateral shoots of 5 nodes each. Some lateral growth up to veraison is beneficial, as the recently-mature lateral leaves contribute to whole-canopy photosynthesis during fruit ripening. Excessive lateral growth, however, indicates high shoot vigor, and growing shoot tips after veraison compete with the fruit for assimilates.
- The fruit zone should be near the top or the outside of the canopy so that 50–100% of the fruit is exposed to the sun. This promotes anthocyanin and tannin production in the berry skins and also improves disease control. High temperature (>33 °C) and the UV radiation of bright sunlight can inhibit the spread of powdery mildew colonies, and exposed clusters also dry more quickly after rain, which reduces bunch rot infections. Fruit sun-exposure is more beneficial for red wine grapes than for white wine grapes, because tannins and other phenolic components are not desirable in white wines.
- Fruit sun-exposure on the east side of north–south-oriented rows should be near 100%, whereas on the west side it should be closer to 50% because of the increased heat load in the afternoon. Excessive fruit exposure can result in heat damage, sunburn, and impaired anthocyanin accumulation while increasing phenolics other than anthocyanins beyond desirable levels, especially in white grapes (see Section 6.2).
- The renewal zone, which is the part of the shoot that will become the fruit-bearing unit in the following year, should be near the top or the outside of the canopy to promote bud fruitfulness.
- The pruning weight, which is the total fresh weight of all canes removed during winter pruning, should be in the range 0.3–0.6 $kg\,m^{-1}$ of canopy length, and the average cane weight should be 20–40 g. Pruning weight values higher than 1 $kg\,m^{-1}$ indicate overly vigorous vines, whereas values lower than 0.2 $kg\,m^{-1}$ indicate insufficient shoot vigor.
- The ratio of fruit yield to pruning weight per vine should be in the range 5–10. The yield ÷ pruning-weight ratio is a measure of the crop load of a grapevine and of its balance between reproductive growth and vegetative growth. Values higher than 10 may be associated with overcropping and delayed ripening, whereas values lower than 5 are indicative of low yield and high shoot vigor.

Vineyard sites with low to moderate vigor potential tend to produce canopies that are close to this "ideotype." In regions where rainfall is not limiting to vine growth and productivity, such sites are often characterized by relatively shallow rootzones and well-drained soils with somewhat limited water and nutrient storage capacity. In more arid regions that require supplemental irrigation, by contrast, irrigation management rather than soil type and depth is the main factor determining vine vigor and yield formation. Provided vines are not spaced too closely at planting and are not grafted to high-vigor rootstocks, they can be easily trained to a vertical shoot-positioning trellis system or sprawl-trained with little shoot positioning. The former is better suited to cool climates where light is sometimes limiting, whereas the latter is well suited to warm climates which often experience high irradiance during the growing season.

Deep and fertile soils with high water storage capacity are less desirable in humid climates, because such soils may result in overly vigorous vines and dense canopies. Although achieving ideal canopy characteristics is more difficult on such sites, it is not impossible. Vines can be grafted to low-vigor rootstocks (see Section 1.2), planted at greater distances that accommodate lighter pruning levels (see Section 6.1), or grown together with cover crops that compete with the vines for soil resources (see Section 8.1). In contrast, close plant spacing tends to exacerbate the vigorous-canopy problem due to shoot crowding and competition among shoots for access to light rather than competition between roots for water and nutrients (Falcetti and Scienza, 1989). In fact, high planting density may result in a so-called "tragedy of the commons," whereby plants that compete for the same soil resources do not limit their root production to match resource availability (O'Brien et al., 2005). Instead they may increase both root and shoot growth—thereby keeping the root $\div$ shoot ratio constant—at the expense of fruit production.

Vines growing on sites with high vigor potential can also be trained to trellis systems that are vertically divided (e.g., Scott Henry, Smart–Dyson, Ballerina) or horizontally divided (e.g., Geneva Double Curtain, Lyre). Such canopy division increases the total length of the canopy per plant and is advisable when the pruning weight exceeds $1\,\text{kg}\,\text{m}^{-1}$ canopy length. This will allow greater numbers of shoots to be spaced more ideally, which in turn will improve light interception and fruit sun-exposure and accommodate higher yields without sacrificing fruit quality (Reynolds and Vanden Heuvel, 2009; see also Section 6.2).

The impact on vigor and canopy density of the shoot number per unit canopy length may also depend on *when* the shoot density is established. Heavy winter pruning, in which most buds produced in the preceding growing season are removed, tends to result in compensatory budbreak of latent buds and vigorous growth of the few shoots that are allowed to grow (Keller et al., 2004, 2015a). Similar considerations apply to shoot thinning, which refers to the viticultural practice of removing growing shoots. Early shoot removal, especially excessive shoot removal, soon after budbreak often leads to compensatory growth of the remaining shoots such that the total leaf area per vine is similar in shoot-thinned and nonthinned plants (Bernizzoni et al., 2011). For example, shoot thinning approximately 3 weeks before bloom time led to compensatory growth of lateral shoots, which negated the intended beneficial influence of lower shoot density on canopy microclimate, whereas shoot thinning at veraison had no such effects (Reynolds et al., 1994a; Smart, 1988). However, delaying shoot thinning or not thinning dense canopies at all can nevertheless be associated with delayed fruit ripening in situations where the canopy is overly vigorous (Bernizzoni et al., 2011; Reynolds et al., 2005).

5.3 Nitrogen assimilation and interaction with carbon metabolism

Nitrogen (N) assimilation is the process by which inorganic N acquired by a plant from the soil or from foliar fertilizers is incorporated into the organic carbon-based compounds necessary for growth and development. Nitrogen is generally the fourth most abundant chemical element in grapevines after hydrogen (H), carbon (C), and oxygen (O), although calcium (Ca) can exceed N in vines grown on calcerous soils (see Section 7.3). Typical concentrations per unit dry matter, or biomass, are roughly 60 mmol H g^{-1}, 40 mmol C g^{-1}, 30 mmol O g^{-1}, and 1 mmol N g^{-1}, and the production of 1 kg biomass requires between 20 and 50 g N (Xu et al., 2012). Therefore, N is the mineral nutrient for which plants have the highest demand and the nutrient that most often limits growth. This makes N the most important mineral nutrient for grapevines. Moreover, when grapes are harvested, some of the fixed N is permanently removed from the vineyard soil. This loss amounts to 2–3 kg t^{-1} of fruit removed but can be reduced to <1 kg t^{-1} if stalks (grape peduncles after destemming) and pomace (solid remains of grapes after pressing of juice or must) are recycled back to the vineyard. To sustain vineyard productivity, any such loss must be replaced by addition of organic or mineral fertilizer or by biological N fixation using leguminous cover crops.

Although N makes up almost 80% of the Earth's atmosphere, grapevines, unlike legumes whose roots "employ" symbiotic bacteria called rhizobia for the task of fixing atmospheric nitrogen gas (N_2), cannot directly utilize this N_2. Instead, they generally rely on uptake by their roots of nitrogen ions—mostly in the form of nitrate (NO_3^-)—dissolved in the soil water. These nitrate ions are then reduced to ammonium (NH_4^+) and assimilated into amino acids in the roots and leaves for transport as both nitrate and amino acids throughout the vine to be used in growth, metabolism, or storage. Amino acids are N-containing organic acids that are the units or building blocks from which protein molecules are manufactured by cellular ribosomes. Ribosomes, composed of proteins and N-containing RNA, utilize the DNA's genetic code to assemble amino acids into proteins, thereby translating the RNA template that is transcribed from the plant's genes. Each 1 g protein typically contains 0.16 g N, and plants invest approximately 55% of their total N content in proteins (Niklas, 2006).

5.3.1 Nitrate uptake and reduction

Close to 99% of all the nitrogen present in the soil is bound in organic matter, which cannot be taken up directly by the roots. Due to rapid nitrification of NH_4^+ derived from organic sources to NO_3^- by microbes in most aerobic and sufficiently moist soils, NO_3^- is the primary source of N for grapevines (Bair et al., 2008; Keller et al., 1995, 2001b). However, roots are also capable of taking up NH_4^+ and amino acids as well, especially in acidic soils that are rich in organic matter (Fischer et al., 1998; Grossman and Takahashi, 2001; Jiménez et al., 2007; Tegeder and Masclaux-Daubresse, 2018). In fact, where they are available, roots may prefer amino acids over NO_3^- for uptake (Miller et al., 2008). Nitrate is dissolved in the soil water and taken up across the root's epidermis and cortex (see Sections 1.3 and 3.2). The concentration of NO_3^- in the soil water can vary from 10 µM to 100 mM, but it is generally orders of magnitude lower than the concentration inside the xylem stream (Crawford, 1995; Crawford and Glass, 1998; Keller et al., 1995, 2001b; Robinson, 1994).

Although the removal of NO_3^- by nitrogen assimilation and xylem transport should favor passive diffusion into the root, the concentration gradient in the "wrong" direction is far too great for this to occur under most conditions. Therefore, roots absorb NO_3^- actively by means of proton/nitrate (H^+/NO_3^-) cotransport using an ATP pump, a protein called H^+-ATPase embedded in the cell membranes (Barbier-Brygoo et al., 2011; Crawford, 1995; Crawford and Glass, 1998). In this process, energy from ATP consumption is used to pump protons out of the root cells into the soil water, generating a proton gradient across the membrane. The protons diffuse back into the cells, dragging negatively charged nitrate molecules along with them. The micronutrient boron (B) is essential to keep the ATP pump going so that insufficient B availability strongly decreases in the roots' ability to absorb NO_3^- and, consequently, may lead to secondary N deficiency and accumulation of sugars and starch in the leaves (Camacho-Cristóbal and González-Fontes, 1999; see also Section 7.3).

Two distinct groups of transporter proteins are responsible for NO_3^- uptake into the roots. The so-called low-affinity nitrate transporters operate when the external NO_3^- supply is high (in the mM range), whereas under low NO_3^- availability (in the μM range) the high-affinity transporters are switched on (Barbier-Brygoo et al., 2011). Some nitrate transporters can even switch back and forth between low-affinity mode and high-affinity mode, using a phosphorylation switch that removes or adds a phosphate group from or to the transport protein (Tsay et al., 2011; Wang et al., 2012). Without the phosphate group the transporter operates in low-affinity mode, with it, in high-affinity mode.

Once inside the root cells, the NO_3^- ions can be moved to the vacuoles for temporary storage or loaded into the xylem and transported to the shoots and leaves as shown in Fig. 5.10. Much of the NO_3^-, however, is assimilated into organic N compounds—that is, amino acids (Loulakakis et al., 2009; Roubelakis-Angelakis and Kliewer, 1992; Tester and Leigh, 2001; Xu et al., 2012). Nitrogen assimilation in grapevines occurs in both the roots and the leaves (Llorens et al., 2002; Perez and Kliewer, 1982; Stoev et al., 1966; Zerihun and Treeby, 2002). The following overview of nitrogen assimilation mainly summarizes aspects of the detailed reviews provided by Loulakakis and Roubelakis-Angelakis (2001) and Loulakakis et al. (2009). The first step of this process is the reduction of NO_3^- to nitrite (NO_2^-) by the enzyme nitrate reductase (NR) using two electrons provided by NADH or NADPH:

$$NO_3^- + 2H^+ + 2e^- \rightarrow NO_2^- + H_2O$$

There are two slightly different forms of NR. One is located in the cytosol of root epidermis and cortex cells and of leaf mesophyll cells. The other, which appears to be restricted to roots, is bound to the outer surface of the plasma membrane—and thus located in the apoplast. The cytosolic NR is active only during the day, whereas the apoplastic NR "operates" day and night but peaks during the night. The apoplastic version of NR also normally prefers electrons from succinate over those from NADH. Nitrate reductase is the main molybdenum (Mo)-containing protein in plants (Schwarz and Mendel, 2006), and one symptom of Mo deficiency is the accumulation of nitrate due to diminished NR activity (see Section 7.3). The presence of NO_3^- stimulates NR, but its full induction in the roots requires sugars, especially sucrose, whereas in the leaves the enzyme is induced by light via the action of phytochrome (see Section 5.2). The hormone cytokinin promotes NR activity, while the amino acid glutamine suppresses it.

In addition to assimilating NO_3^-, grapevines can load large amounts of NO_3^- into the xylem for distribution or store it for later use in the cell vacuoles without deleterious effects (Roubelakis-Angelakis and Kliewer, 1992). This vacuolar storage pool buffers the NO_3^- concentration in the cytosol and may even participate in maintaining cell turgor (Dechorgnat et al., 2011). Unlike NO_3^-, NO_2^- is a toxic ion

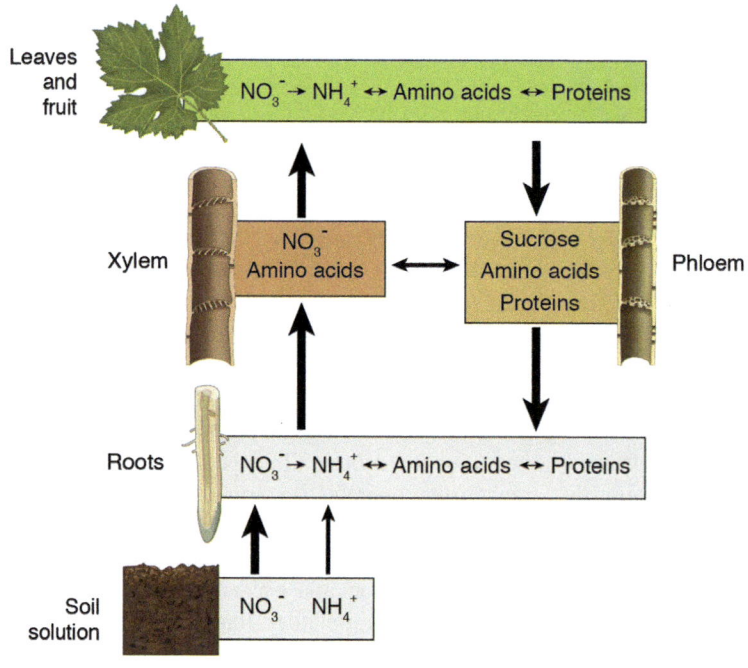

FIG. 5.10

Simplified diagram of nitrogen uptake, assimilation, and circulation in grapevines.

that can induce mutations in plant tissues. Upon its production, NO_2^- is therefore immediately transported from the cytosol into the root plastids or leaf chloroplasts, where it is reduced to NH_4^+ by the enzyme nitrite reductase (NiR). This reaction uses six electrons provided by reduced ferredoxin (Fd_{red}), which is thus oxidized to Fd_{ox}, in the following process:

$$NO_2^- + 8H^+ + 6e^- \rightarrow NH_4^+ + 2H_2O$$

The sulfur (S) and iron (Fe) containing protein ferredoxin is produced in the photosynthetic electron transport chain in the chloroplasts (see Section 4.1) or by NADPH generated by the oxidative pentose phosphate pathway in the nonphotosynthetic plastids (Crawford, 1995). Because NO_2^- is so toxic, plants maintain an excess of NiR whenever NR is present. In the roots, NO_3^- is sufficient to activate NiR, whereas in the leaves light is required in addition to NO_3^-.

5.3.2 Ammonium assimilation

Like NO_2^-, NH_4^+ is toxic to plants and is either rapidly incorporated into amino acids or, during surplus supply, temporarily stored in the cell vacuoles. The incorporation of NH_4^+ into amino acids is called ammonium assimilation and is normally catalyzed by the two enzymes glutamine synthetase (GS)

5.3 Nitrogen assimilation and interaction with carbon metabolism

and glutamate synthase, also known as glutamine-2-oxoglutarate aminotransferase (GOGAT), in a cycle consisting of two sequential reactions (Fig. 5.11):

$$\text{Glutamate} + NH_4^+ \rightarrow \text{Glutamine}$$

$$\text{Glutamine} + 2\text{-oxoglutarate} \rightarrow 2\,\text{Glutamates}$$

The first step of the GS/GOGAT cycle requires energy provided by ATP and involves a divalent cation (Mg^{2+}, Mn^{2+}, Ca^{2+}, or Co^{2+}) as a so-called cofactor of GS; the obligatory cofactor assists in the proper function of the enzyme (Roubelakis-Angelakis and Kliewer, 1983). There are two main forms of GS: GS1 is located in the cytosol of all plant organs and in the phloem companion cells, and GS2 is located in the chloroplasts of photosynthetic tissues and in the plastids of roots (Grossman and Takahashi, 2001). The cytosolic GS1 is central to NH_4^+ assimilation in the roots, and its activity increases with increasing sugar content. In contrast, GS2 predominates in NH_4^+ assimilation in the leaf mesophyll.

The glutamine produced by GS in the first reaction stimulates the activity of the protein glutamate synthase which contains Fe and S as cofactors. This enzyme produces two molecules of glutamate in a reaction that consumes two more electrons. Plants contain two types of glutamate synthase: One is called NADH-GOGAT because it accepts electrons from NADH, and the other is termed Fd-GOGAT because it accepts electrons from Fd_{red} (Temple et al., 1998). The NADH type is located only in plastids of nonphotosynthetic tissues such as roots or vascular bundles of developing leaves, whereas the ferredoxin type is located in both chloroplasts and nonphotosynthetic plastids but dominates in leaves. The Fd-GOGAT is very important in the recapture of NH_4^+ released during photorespiration (see Section 4.3) in addition to its role in assimilating NH_4^+ derived from NO_3^- reduction—the so-called primary NH_4^+ assimilation. The former role is very important because up to 90% of the NH_4^+ that flows through the GS/GOGAT cycle is generated during photorespiration (Stitt, 1999; Stitt et al., 2002). Both GS and GOGAT are stimulated by light and sucrose and inhibited by amino acids (Grossman and Takahashi, 2001). The carbon "backbone" 2-oxoglutarate, also known as α-ketoglutarate, is an organic acid provided by the TCA cycle (see Section 4.4) and from other, stored organic acids, such as malate and citrate (Gauthier et al., 2010; Tcherkez et al., 2017).

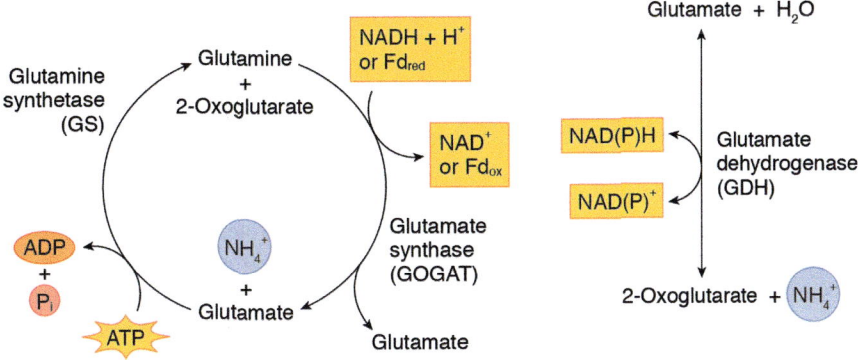

FIG. 5.11

Nature and reactions of compounds involved in ammonium metabolism.

One of the two glutamate molecules produced by GOGAT is used to regenerate the cycle, and the other is used to supply amino acids for general metabolism. In the roots, glutamate can also be transported back to the cytoplasm, where it is converted back to glutamine by a slightly different form of GS for export in the xylem to the shoots. Glutamine thus provides N groups, either directly or via glutamate, for the production via enzymes termed amino transferases of virtually all organic nitrogenous compounds in grapevines. Some such N groups might additionally be derived from asparagine, which can apparently be produced by the addition of NH_4^+ to aspartate by the ATP-powered asparagine synthetase (Masclaux-Daubresse et al., 2010).

In an alternative pathway for NH_4^+ assimilation, the enzyme glutamate dehydrogenase (GDH) catalyzes a reversible reaction that can either form or degrade glutamate (Fig. 5.11):

$$2-\text{Oxoglutarate} + NH_4^+ \leftrightarrow \text{Glutamate} + H_2O$$

The forward reaction is called amination, whereas the reverse reaction is called deamination. The GDH also comes in two main forms, one localized in the mitochondria of all organs, but especially in phloem companion cells, and the other in leaf chloroplasts. The mitochondrion version uses electrons supplied by NADH, whereas the chloroplast version uses electrons donated by NADPH. As a stress-related enzyme GDH, activated by Ca^{2+}, may participate in NH_4^+ assimilation in tissues with excessive NH_4^+ concentration and in senescing leaves, where it is thought to recycle and thereby detoxify the NH_4^+ that is released during protein remobilization (Loulakakis and Roubelakis-Angelakis, 1990; Loulakakis et al., 2002; Masclaux et al., 2000; Skopelitis et al., 2006). Thus, GDH may complement the GS/GOGAT cycle, whose activity declines during senescence in parallel with the decrease in rubisco, photosynthesis, and chlorophyll. But the main role of GDH seems to be very different. In its reverse mode, the enzyme can oxidize glutamate when fixed carbon is depleted, for example, as a result of restricted photosynthesis. Under these conditions, GDH participates in the remobilization and breakdown of proteins in a process termed proteolysis and in the subsequent degradation of amino acids to supply carbon skeletons back to the TCA cycle (see Section 4.4) for continued ATP regeneration (Aubert et al., 2001; Miyashita and Good, 2008; Robinson et al., 1991). Some of the "leftover" NH_4^+ that forms by the action of GDH during senescence or that escapes the action of GS during photorespiration may be accumulated or lost as volatile ammonia (NH_3) to the atmosphere (Farquhar et al., 1979; Xu et al., 2012).

5.3.3 From cells to plants

Nitrogen is a structural component of nucleic acids, which form the organic bases in DNA, RNA, and ATP, as well as of chlorophyll, hormones, and amino acids. Many amino acids are ultimately assembled into proteins such as enzymes, which catalyze or increase the rates of biochemical reactions, and are essential to plant metabolism and energy generation. Other proteins become channels and transporters that carry molecules across membranes, and others store nitrogen for future use. Nitrogen uptake and assimilation cost large amounts of energy to convert stable, low-energy inorganic compounds present at low concentration in the soil water into high-energy organic compounds with high concentrations inside the plant. Nitrate assimilation uses 2.5 times the NADPH required for CO_2 assimilation. The reduction of one molecule of NO_3^- to NO_2^- and then to NH_4^+ requires the transfer of 10 electrons and often accounts for 10–25% of the total energy expenditures in both roots and shoots. This means that a grapevine may use as much as one-fourth of its energy to assimilate N, a constituent that comprises <2% of the total plant dry weight.

The distribution of available N in the soil is extremely heterogeneous. Nitrate concentrations in the soil solution can vary by several orders of magnitude, both temporally and spatially, even over short distances (Crawford, 1995; Keller et al., 1995, 2001b; Robinson, 1994). Because N is so important to them, plants have evolved mechanisms that remove NO_3^- from the soil solution as quickly as possible. The root system can react to a patch of NO_3^- by rapidly activating NO_3^- uptake and, more slowly, initiating lateral root proliferation within the nitrate-rich zone (Forde, 2002; Gastal and Lemaire, 2002; Nibau et al., 2008; Scheible et al., 1997). These responses are especially pronounced when vine N status is low (see Section 7.3). To support such localized NO_3^- uptake, those roots in contact with the rich N source also augment water influx, whereas in a compensatory response the roots that happen to grow in N-poor soil patches simultaneously take up less water (Gloser et al., 2007; Gorska et al., 2008). It seems obvious that this mechanism would improve NO_3^- delivery to those roots that are best positioned for immediate uptake by accelerating mass flow to the root surface. Accordingly, the rates of NO_3^- uptake and respiration decline rapidly as newly formed roots age and deplete the soil N in their vicinity (Volder et al., 2005).

When photosynthetic electron transport generates more energy than is needed by the Calvin cycle (see Section 4.2), some of this energy becomes available for N assimilation. Because the production of amino acids also requires a supply of carbon through the TCA cycle, the incorporation of inorganic N into amino acids is a dynamic process that is regulated by both internal and external factors. The former include the availability of carbon and N metabolites, whereas the latter include the availability of light and inorganic N. When a grapevine takes up NO_3^- through the roots, this NO_3^- functions as a signal that induces the plant to switch on the pathway of N assimilation and divert carbon away from starch production toward the manufacture of amino acids and organic acids, such as malate and citrate, that substitute for NO_3^- to maintain charge neutrality following its reduction (Foyer et al., 2003; Huppe and Turpin, 1994; Jiménez et al., 2007; Stitt, 1999; Stitt et al., 2002). Consequently, export from the leaves of malate in the phloem increases as NO_3^- supply in the xylem increases. At the same time, K^+ flow in both the xylem and phloem also increases, because K^+ acts as a counterion that helps to neutralize the negative electrical charge of both NO_3^- and malate (Keller et al., 2001b; Peuke, 2010).

Nitrogen uptake and assimilation are linked to sugar availability. Assimilation proceeds rapidly in plants with high carbohydrate status and slows as the photosynthetic sugar supply declines (Perez and Kliewer, 1982; Wang and Ruan, 2016). A decrease in carbon status can occur due to inclement weather, drought stress, loss of leaf area due to hail or insects, or infection by pathogens and can result in NO_3^- accumulation in the vine's tissues (Keller, 2005). Energy-dependent NO_3^- uptake also declines in vines with low carbon status, but uptake is less limited than is N assimilation (Keller et al., 1995; Oaks and Hirel, 1985; Rufty et al., 1989). Similarly, low temperatures reduce protein production and enzyme activity while NO_3^- uptake continues, so that amino acids and NO_3^- accumulate in the leaves (Lawlor, 2002). Because cool conditions limit organ growth more than N uptake and assimilation, the N ÷ C ratio in a grapevine's organs is higher than that seen in warm conditions. In contrast, cell division is particularly sensitive to N status, and inadequate N supply generally suppresses growth more than photosynthesis so that starch accumulates in the leaves and the N ÷ C ratio is low (Gastal and Lemaire, 2002). Although foliar carbohydrate accumulation in response to N deficiency ultimately leads to suppression of photosynthesis through feedback inhibition, accumulation is often greater in the roots than in the leaves.

Both roots and leaves of grapevines can assimilate NO_3^- to amino acids. Because it depends on photosynthetic energy, the production of glutamine and glutamate in the leaves occurs mainly during the day, and all the requisite enzymes are induced by the active form of phytochrome (see Section 5.2). In the roots, on the other hand, N assimilation proceeds both day and night. The relative extent to which N assimilation takes place in the roots or the leaves depends on several factors, including species, cultivar and rootstock, weather conditions, carbohydrate availability, and the amount of NO_3^- the roots take up (Alleweldt and Merkt, 1992; Keller et al., 1995, 2001b; Llorens et al., 2002; Zerihun and Treeby, 2002; see also Fig. 5.12). Species native to temperate regions usually rely more heavily on root N assimilation than do species of tropical or subtropical origin. Despite the variation among grape cultivars in their capacity to assimilate NO_3^-, neither total N nor NO_3^- concentrations in the leaf blades vary much among cultivars, although NO_3^- does fluctuate in the petioles (Christensen, 1984). Petioles, it seems, serve as temporary NO_3^- storage sites, from where the leaves retrieve the nutrient on demand (Wang et al., 2012).

When environmental factors permit rapid photosynthesis resulting in high vine carbon status, and when NO_3^- availability in the soil water is relatively low, root-absorbed NO_3^- is rapidly assimilated in the roots and transported in the xylem as metabolically active glutamine (Fig. 5.12). As the supply of N increases, so does root uptake, and increasingly more NO_3^- is transported to the shoots in addition to glutamine (Alleweldt and Merkt, 1992; Keller et al., 1995, 2001b; Pate, 1980; Stoev et al., 1966). This makes sense for the vine because reduction and assimilation of one NO_3^- molecule in the roots cost the equivalent of 12 ATP molecules, while it is much cheaper in the leaves. The reason for this discrepancy is that N assimilation in the roots depends entirely on phloem-supplied sucrose for both carbon skeletons and energy (Huppe and Turpin, 1994; Oaks and Hirel, 1985; Wang and Ruan, 2016). The roots'

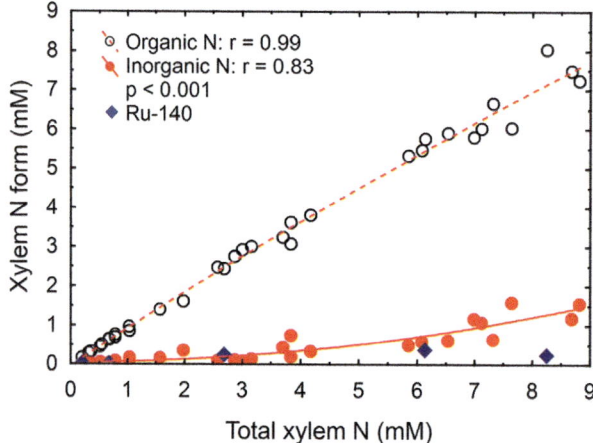

FIG. 5.12

Major nitrogen (N) forms in the xylem sap of *V. vinifera* scion grafted to six different rootstocks. The sap composition is dominated by amino acids, especially glutamine, but the contribution of inorganic N, especially nitrate, increases as the total N increases—except in scions grafted to 140 Ruggeri.

Modified from Keller, M., Kummer, M., Vasconcelos, M.C., 2001b. Soil nitrogen utilisation for growth and gas exchange by grapevines in response to nitrogen supply and rootstock. Aust. J. Grape Wine Res. 7, 2–11.

capacity to assimilate N is directly related to their carbohydrate status, and the carbon skeletons for glutamine production are derived from recently imported sucrose or from stored starch (Emes and Neuhaus, 1997). Therefore, the attendant increase in N content under high carbon availability may be greater in the roots than in other plant parts (Zufferey et al., 2015). However, high soil N availability, which is common after fertilizer application or tilling of cover crops, can lead to lower starch reserves in the vine's permanent organs, especially under conditions that limit photosynthesis (Bair et al., 2008; Cheng et al., 2004b; Guerra and Steenwerth, 2012; Keller et al., 1995; Müller, 1985).

Even when photosynthesis is not limiting, root absorption can exceed the capacity of a vine for N assimilation at very high soil NO_3^- availability. This is especially true under conditions that favor rapid transpiration, such as warm and sunny days, and therefore enhance nutrient uptake. In such situations glutamine and even NO_3^- can be transported back from the leaves to the roots in the phloem as signals to limit NO_3^- uptake (Dechorgnat et al., 2011; Gessler et al., 1998; Lemaire and Millard, 1999; Wang et al., 2012). This feedback regulation normally coordinates NO_3^- uptake with the vine's demand for N. However, sometimes grapevines have access to large amounts of soil NO_3^- during periods of low solar radiation. Because NO_3^- reduction and subsequent assimilation are more sensitive to sugar supply than are NO_3^- uptake or long-distance transport, low carbohydrate status leads to export of NO_3^- to the shoots (Keller et al., 1995). When NO_3^- assimilation cannot keep pace with uptake, NO_3^- accumulates in the leaves and petioles, as well as other organs once available starch reserves have been exhausted (Keller et al., 1995, 2001b; Perez and Kliewer, 1982; Scheible et al., 1997; Smart et al., 1988; Zerihun and Treeby, 2002).

High leaf NO_3^- reduces the amount of sucrose available for export, depleting root starch reserves and inhibiting root growth, which decreases the root $\div$ shoot ratio (Keller and Koblet, 1995a; Keller et al., 1995; Scheible et al., 1997). This has special implications during the establishment phase of a vineyard. Heavily fertilizing and irrigating young vines in order to harvest a crop in the year after planting may curb root development, which might be detrimental to vineyard productivity in the long term. The increase in leaf NO_3^- under cool or overcast conditions also has implications for the interpretation of the results from analysis of leaf samples collected in the vineyard to diagnose plant nutrient status: Sampling during cool or cloudy periods may result in the appearance of incorrectly high vine N status.

Leaves gradually acquire an increased capacity for N assimilation as they unfold and mature along with an increase in photosynthesis. Accordingly, the amounts and activities of GS and GOGAT initially increase with leaf age but decrease in senescing leaves. This is especially true for GS2, whereas GDH follows the opposite trend. Unlike roots, leaves can use excess photosynthetic energy for N assimilation (Huppe and Turpin, 1994). Indeed, higher leaf N status due to increasing N supply via the xylem stimulates light-saturated photosynthesis in leaves (Keller, 2005; Keller et al., 2001b; Schreiner et al., 2018; see also Fig. 5.5). However, high N status also increases the respiration rate due to the energy demands of N assimilation. The qualifier "light-saturated" is important because light intensities below saturation limit photosynthesis directly, and N has no effect (Keller and Koblet, 1994; see also Sections 4.1 and 5.2). The leaves' photosynthetic capacity also seems to increase with increasing leaf N content, mainly because rubisco, the enzyme responsible for CO_2 fixation, comprises approximately one-third of a leaf's protein and up to 25% of its N (Evans, 1989; Seemann et al., 1987; see also Section 4.2). However, the total amount of carbon fixed per day saturates at high N levels so that there is an optimal leaf N content that maximizes carbon gain. The optimal N concentration is higher in leaves grown at high light intensity than in those grown under cloudy conditions or in the shade (Hikosaka et al., 2006). Plants therefore tend to distribute leaf N so that it closely follows the light distribution within a canopy in order to maximize whole-canopy photosynthesis (Gastal and Lemaire, 2002; Prieto et al., 2012). This non-uniform distribution of N has to be taken into account when leaves are sampled for nutrient analysis to determine vineyard fertilizer requirements. On the other hand, when N becomes limiting, rubisco activity declines more than electron transport capacity (see Section 4.1). To avoid excessive light absorption that would damage the photosynthetic apparatus via oxidative stress, the leaves decrease their chlorophyll content and activate their energy-dissipation and antioxidant systems (Chen and Cheng, 2003a; Keller, 2005). In addition, they decrease the angle between the leaf blade and the petiole, which further reduces the number of light photons captured by a leaf (see Section 7.3).

Much of the N that continuously arrives, mainly in the form of glutamine and NO_3^-, via the xylem in the leaves is redistributed as glutamine via the phloem to the shoot tips, fruit clusters, and woody organs for further use in growth and metabolism or storage. Since the phloem can transport only small amounts of NO_3^-, the leaves must assimilate most of the xylem-delivered NO_3^- to glutamine or store it in the vacuoles to buffer temporary shortages in N supply (Peuke, 2010; Tegeder and Masclaux-Daubresse, 2018). Nonetheless, vacuolar NO_3^- may be remobilized and exported to sink organs from older and senescing leaves via the phloem (Fan et al., 2009; Havé et al., 2017; Wang et al., 2012). In addition, nitrogenous compounds, as well as other mineral nutrient ions, are remobilized from proteins and nucleic acids in senescing leaves at the end of the growing season and exported to the perennial parts of the vine for storage (Conradie, 1986; Dintscheff et al., 1964; Kliewer, 1967a; Loulakakis et al., 2002). This process is also called resorption and is associated with a decline in photosynthesis and a transition from nutrient assimilation to nutrient remobilization (Hoch et al., 2003; Lim et al., 2007). As much as 90% of leaf N may be recycled during senescence, and the GS1 form of glutamine synthetase produces glutamine for export from the leaves (Löhnertz et al., 1989; Tegeder and Masclaux-Daubresse, 2018). Some of the remobilized protein is not broken down completely and may be exported by the phloem in the form of small peptides two or three amino acid units in length.

With the help of GDH, remobilization also occurs from shaded leaves for recycling of the recovered N to light-exposed leaves and to the shoot tips for production of new leaves that are better exposed to sunlight (Hikosaka, 2005; Keller and Koblet, 1995a). Moreover, the ability of GDH to oxidize glutamate during periods of carbon starvation may have implications for yield formation in grapevines. When adverse environmental conditions, such as overcast skies or unseasonably cool weather, restrict photosynthesis (see Section 7.4), vines can experience severe carbon shortage. Meanwhile, N assimilation is reduced when photosynthesis declines, which leads to an inhibition of amino acid production. Consequently, plants may become starved for both C and N (Stitt et al., 2002). This may be especially critical during bloom, when storage reserves in the permanent organs are at their lowest. In addition, local carbon starvation can also occur inside dense canopies or due to stunted shoot growth limiting leaf area. In such situations, GDH participates in the resorption of proteins from relatively unimportant sinks that may subsequently be discarded. The amino acids derived from this proteolysis can serve as alternative electron donors for respiration in the absence of sufficient carbohydrates (Araújo et al., 2011).

Protein resorption may be the main reason that high amounts of NH_4^+ are sometimes found in organs that suffer from carbon starvation due to insufficient light or other stresses. For example, the syndrome of inflorescence necrosis (a.k.a. early bunchstem necrosis), which is an extreme form of poor fruit set that culminates in abscission of portions of or whole inflorescences, is probably induced by carbon starvation (Jackson, 1991; Jackson and Coombe, 1988; Keller and Koblet, 1994, 1995b; Keller et al., 2010; Rogiers et al., 2011; see also Section 6.1). The attendant NH_4^+ accumulation in such inflorescences is thus a by-product of the organ's stress-induced senescence and retrieval of its valuable resources, rather than being the toxic culprit for senescence (Keller and Koblet, 1994, 1995b). Organ abscission, preceded by nutrient recycling, can be viewed as a sacrifice by the vine of comparatively less important organs that will permit more important sinks to survive episodes of stress-induced carbon starvation. Consequently, if the incident light is insufficient to sustain growth of all vine organs, leaf growth, especially on lateral shoots, may be enhanced to capture more light at the expense of inflorescence survival (Keller and Koblet, 1994, 1995a; Keller et al., 2001a).

The primary amino acids glutamine and glutamate can be converted to many other amino acids by the action of transaminases. These glutamate-utilizing enzymes recycle 2-oxoglutarate so that it can be used for continued NH_4^+ assimilation. Under favorable conditions that are associated with high rates of C and N assimilation, grapevines convert surplus glutamine, which contains about 19% N, to arginine. At 32% N, arginine has the highest $N \div C$ ratio of all amino acids and is the major and most efficient N storage compound in grapevines, both as soluble amino acid and incorporated into proteins (Kliewer, 1967a; Kliewer and Cook, 1971; Schaller et al., 1989; Xia and Cheng, 2004). Accordingly, the amount of stored arginine in the perennial parts of a vine increases as external N supply increases (Perez and Kliewer, 1982). Yet proteins form the largest pool of N in all plant organs, contributing significantly to N storage (Tegeder and Masclaux-Daubresse, 2018). During drought or salt stress, and possibly under N deficiency too, a larger proportion of glutamate may be converted to the amino acid proline. Much of this proline synthesis results from glutamine produced by the GS1 form of glutamine synthetase in the phloem's companion cells. Proline is highly soluble and can accumulate in cells to high levels without disrupting their metabolism. This allows plants to lower their tissue water potential while maintaining turgor pressure during periods of drought (see Section 7.2) or salinity (see Section 7.3).

Storage reserves are available as a buffer during periods of low N supply and to support new growth in spring. In fact, reserve N may be at least as important as reserve carbohydrates in supporting spring growth of grapevines (Cheng et al., 2004b; Keller and Koblet, 1995a; Schaefer, 1981). Up to half of the developing canopy's N demand is supplied from such reserves. Because of the rapid shoot growth in spring, the nutrient demand of the canopy is greatest between budbreak and bloom, even though most of the N uptake from the soil may occur after bloom, provided there is sufficient soil moisture (Conradie, 1986; Keller, 2005; Löhnertz et al., 1989; Peacock et al., 1989; Pradubsuk and Davenport, 2010). Limited availability of N reserves due to inadequate refilling during the previous growing season can restrict early shoot growth and canopy development, and may lead to poor fruit set (Celette et al., 2009; Keller and Koblet, 1995a; Keller et al., 2001a). The reserve pool is depleted during and after budbreak, reaching a minimum by bloom time or, sometimes, as late as veraison, and it is replenished later in the growing season (Eifert et al., 1961; Kliewer, 1967a; Löhnertz et al., 1989; Pradubsuk and Davenport, 2010; Schaller et al., 1989; Schreiner et al., 2006; Zapata et al., 2004; Zufferey et al., 2015). Insufficient N availability at this time may not allow the vine to fully replenish its N reserves, a situation that may occur because of late-season growth of a competing cover crop (Celette et al., 2009).

Accumulation of storage reserves occurs mostly when other plant requirements, such as growth and fruit production, have been satisfied—that is, when the supply of resources exceeds demand (Lemaire and Millard, 1999; see also Section 5.1). This may be one reason that the N concentration in the roots tends to decrease as the yield per vine increases (Zufferey et al., 2015). In warm climates, where leaves remain photosynthetically active for several weeks or even months after harvest, much of the N taken up by the roots after harvest may be directly incorporated into the reserve pool (Conradie, 1986; Peacock et al., 1989). Such postharvest nutrient uptake may not occur in cool climates, where the leaves are abscised soon after harvest or succumb to the first fall frost (Pradubsuk and Davenport, 2010). Late-season foliar application of N fertilizer in cool climates may nevertheless increase the reserve N pool, albeit at the expense of carbohydrate reserves (Xia and Cheng, 2004). Clearly, then, late-season fertilizer application or cover crop tillage can be used to increase the amount of reserve N available the following spring if necessary. Increasing the size of the reserve pool will increase the N content of the emerging leaves and enhance early season growth, which in turn boosts photosynthesis and may lead to greater seasonal demand for N due to the larger canopy (Cheng et al., 2004b; Keller and Koblet, 1995a; Treeby and Wheatley, 2006).

CHAPTER

Developmental physiology

6

Chapter outline

6.1 Yield formation 199
    6.1.1 Yield components and compensation 199
    6.1.2 Yield potential and its realization 203
6.2 Grape composition and fruit quality 211
    6.2.1 Cell walls and membranes 211
    6.2.2 Water 215
    6.2.3 Sugars 219
    6.2.4 Acids 223
    6.2.5 Nitrogenous compounds and mineral nutrients 226
    6.2.6 Phenolics 230
    6.2.7 Lipids and volatiles 245
6.3 Sources of variation in fruit composition 251
    6.3.1 Fruit maturity 254
    6.3.2 Light 255
    6.3.3 Temperature 259
    6.3.4 Water status 264
    6.3.5 Nutrient status 268
    6.3.6 Crop load 272
    6.3.7 Rootstock 275

6.1 Yield formation
6.1.1 Yield components and compensation

The amount of fruit production in a given year and over the lifetime of grapevines determines both their reproductive success as a species and their agronomic trait of yield potential. Viticultural yields are determined by the amount of sugar partitioned to the fruit rather than to other organs. Yield formation is often referred to as cropping, with the crop being the amount of fruit borne on a vine or produced by a vineyard. The "crop level" is the amount of fruit (sometimes the number of clusters) per shoot or per unit of canopy length, the "crop size" is the yield per vine or per unit of land area. The terms crop level and crop size are often used synonymously. By contrast, the term "crop load" refers to the crop size relative to vine size. Vine size is usually estimated as pruning weight or leaf area; thus, crop load is a measure of the canopy sink ÷ source ratio and is often called "vine balance" by viticulturists. "Overcropping" refers to the production of more crop than a vine can bring to acceptable maturity by normal harvest time (see Section 6.2). The concept of crop load is similar to the concept of harvest index used in cereals and other crops; the harvest index is defined as the amount of harvested product relative to the total above-ground biomass.

Grapevine yield is made up of a number of different components. Yield components are those factors in grapevine reproduction that, multiplied together, total the yield obtained from a single vine or an entire vineyard. For a single vine, this can be written as follows (Coombe and Dry, 2001):

$$\text{Yield} = \frac{\text{buds}}{\text{vine}} \times \frac{\text{shoots}}{\text{bud}} \times \frac{\text{clusters}}{\text{shoot}} \times \frac{\text{berries}}{\text{cluster}} \times \text{berry weight}$$

The vineyard yield is the sum of the yield of all individual vines and depends on the planted land surface, the number of bearing vines per unit land area, and the size of each vine. Vine size in turn depends on the planting density, as well as on the trellis and training system and the pruning method (e.g., spur, cane, mechanical, or minimal). The upper limit on the number of fruitful shoots per vine is largely determined by the number of buds left after winter pruning—that is, on the pruning severity. However, the maximum number of fruit clusters available for harvest is limited by the number of inflorescences initiated in those buds during the previous growing season. The weight of each cluster depends on the number of berries on the cluster and their final weight. The berry number, in turn, is determined by the number of flowers that set fruit.

The discussion of the reproductive cycle in Section 2.3 is mainly concerned with anatomical and temporal aspects involved in the initiation and differentiation of grape clusters and their subsequent development to fruit maturity. But yield formation also depends on several internal and external factors and the interactions among them. They include the following: the genetic makeup, or genotype, of the plant, which is determined by species, cultivar, clone, and rootstock; the vineyard site, which determines soil type, water and nutrient availability, and climate; seasonal weather patterns, which comprise variations in light, temperature, rainfall, and humidity; the trellis and training system, which results in single or divided canopies; cultural practices, such as pruning, canopy management, irrigation, nutrition, and pest control; as well as legal and economic aspects, such as desired wine style, winery demand, or yield regulation.

Although interactions between genotype and environment lead to tremendous spatial and temporal variation in grape yield, both within and between vineyards, not all components of yield respond equally to environmental conditions. The planting density may decrease over the life of a vineyard because a grower may choose not to replace some of the vines that die for one reason or another. This change constitutes the difference between number of vines per unit of land area and number of bearing vines per unit of area. It is often said that the potential yield of a vineyard increases as the planting density increases, but this is not necessarily true (Winkler, 1959). The total number of shoots growing on a vine is determined primarily by planting density, trellis and training system, and the number of buds retained at winter pruning. For a given planting density and pruning level, the berry weight is relatively highly conserved, and the number of clusters per shoot ordinarily varies far less than the number of berries per cluster (Currle et al., 1983). This is true for both the variation between cultivars and the variation between growing seasons for the same cultivar.

At the genotype level, a few wine grape cultivars have been grouped according to their "reproductive performance" (Dry et al., 2010b). In this classification, Cabernet Sauvignon and Merlot were characterized by high flower numbers per inflorescence (>300) and low fruit set (~30%); Syrah, Tempranillo, Pinot noir, Chardonnay, and Sauvignon blanc were characterized by low flower numbers (~200), high fruit set (~45%), and low berry numbers per cluster (~100); whereas Sangiovese, Nebbiolo, and Zinfandel were characterized by low cluster numbers, high flower numbers (>300), high fruit set (~45%), and high berry number (>200). Despite this variation, average berry weight (~1 g) and yield (3–4 kg m^{-1} of cordon) were similar among these cultivars. Vine growth and survival on the one hand, and reproduction on the other hand, both require resources that the environment provides in limited supply. Therefore, there is an optimal balance between a vine's vegetative and reproductive growth that is the outcome of a trade-off between survival and reproduction. The interaction between these two goals is related to the interdependence of vine capacity and vigor (Huglin and Schneider, 1998; Winkler, 1958; Winkler et al., 1974).

Capacity is defined by the total annual growth of a grapevine, including its production of fruit, leaves, shoots, and roots in one growing season. Capacity, which is termed annual net primary production in the general plant ecology literature, is thus a measure of the net resource gain from the environment, analogous to profit in economic theory (Bloom et al., 1985). Accurate estimates of vine capacity require measurements of total plant dry mass at the beginning and at the end of the growing season—that is, before budbreak and again before leaf fall. Nevertheless, capacity can be approximated from a vine's total annual weight of fruit and canes. For manually pruned vines, therefore, capacity roughly equals yield plus pruning weight. In practice, pruning weight is often incorrectly used as the sole indicator of vine capacity. Vine capacity increases with increasing size of the root system, number of shoots, and leaf area. Consequently, within a specific vineyard the total annual growth is roughly proportional to the vines' ability to intercept sunlight, which depends on their total and sun-exposed leaf area (see also Section 5.2). Capacity is also thought to be approximately proportional to the three-fourths power of total plant dry mass ($C \sim m^{3/4}$), which is a case of the law of diminishing returns: Large plants grow more than small plants, but the extra gain diminishes as plant size increases further (Niklas, 2006; Niklas and Cobb, 2006).

Vigor, on the other hand, is defined as the rate of shoot growth and can be measured by the change in shoot length over time (see also Section 2.2). Alternatively, vigor is sometimes defined as a vine's annual production of vegetative biomass. Ignoring root growth and the secondary growth of the canopy's perennial structure, the latter definition integrates the growth of all shoots over an entire growing season and can be estimated from a vine's pruning weight. However, although grapevines with more shoots tend to produce more total biomass, the vigor of individual shoots declines as the shoot number per vine increases (Keller et al., 2015a). Pruning weight, moreover, is not an appropriate measure of vine vigor for mechanically or minimally pruned grapevines, because insufficient amounts of cane material are pruned off to permit meaningful estimates.

Vigor and capacity are interrelated and are ultimately determined by vine size, pruning level, the amount of stored reserves present at the beginning of the growing season, seasonal weather patterns, and availability of water and nutrients in the rootzone. Thus, a young vine may grow vigorously but have a low capacity, whereas a mature vine with many buds can give rise to shoots of low vigor but can have a high capacity. The main factors involved in the regulation and interdependence of vine capacity, vigor, and crop load are as follows (Coombe and Dry, 2001; Downton and Grant, 1992; Huglin and Schneider, 1998; Winkler et al., 1974):

- Grapevines have a finite capacity because of the limitations imposed by the amount of external resources available. Thus, vines can support a limited number of shoots and ripen a limited amount of fruit in any one growing season. Considering the large size of wild grapevines, however, these limits are seldom reached in production viticulture.
- Vine capacity is proportional to total potential growth. Thus, older or larger vines can support more shoots and more fruit than younger or smaller vines.
- Pruning tends to depress vine capacity because it reduces the number of shoots and leaves per vine, delays the development of maximum canopy leaf area, and reduces the total amount of fruit produced. Vines typically respond to more severe pruning with increased vigor and production of more total 1-year-old wood.
- The production of fruit tends to depress vine capacity by reducing vigor and limiting the amount of stored carbohydrates, which can lead to lower capacity during the following year. This relationship is not clear-cut, however, since the presence of fruit may also stimulate higher leaf photosynthesis, and fruit removal may result in earlier leaf senescence.
- Shoot vigor is inversely proportional to the number of shoots per vine and to the vine's crop load. Thus, an increase in shoot number and an increase in crop load both lead to a decrease in vigor.
- The direction of shoot growth influences vigor. Shoots growing upward are most vigorous, and shoots growing downward are least vigorous.
- Fruitfulness increases with shoot vigor to a maximum and then levels off, and it may even decline as vigor increases further due to deterioration in canopy microclimate.
- Vines self-regulate by adjusting the proportion of buds that break to their bud number. Severe winter pruning stimulates budbreak of noncount buds, which are buds not deliberately retained at pruning, and thus results in many double shoots, water shoots, and suckers. >100% budbreak, which results in more than one shoot per retained bud, is an indication that the vine has the capacity to support more growth.

Siquidem luxuriosa vitis nisi fructu compescitur, male deflorescit, et in materiam frondemque effunditur; infirma rursus, cum onerata est., affligitur.

(Indeed, a vigorous vine, unless restrained by cropping, aborts its flowers and teems with vegetative growth; a weak vine, on the other hand, declines when burdened with fruit.)

—Columella: De Re Rustica.

These relationships were obviously known to the Romans, as the quotation from Columella (AD 4–ca. 70) shows. They have important implications for the choice of trellis design and training system to be used, as well as pruning and other cultural practices. For example, the balance-pruning concept is based on the assumption that larger vines can support more shoots with a heavier crop. Indeed, larger vines within a vineyard generally maintain higher productivity across years than do smaller vines of the same age (Keller et al., 2004, 2005, 2012). The idea of balance pruning eventually led to the recommendation to retain approximately 30 buds for each kilogram of canes removed during winter pruning in order to make optimal use of the capacity of most wine grape cultivars. In addition, irrigation and nutrition can also modulate the relationship among vine balance, vigor, and capacity because vegetative growth often reacts more strongly to water or nutrient supply than does reproductive growth, at least after fruit set (see Sections 7.2 and 7.3).

Grapevines have an inherent propensity to self-regulate the balance among the growth of roots, shoots, and fruit. For example, effects of high planting density are partly offset by production of fewer buds per vine and, sometimes, fewer shoots per bud. Similarly, leaving twice as many buds per vine during winter pruning will normally not double the leaf area nor the yield because shoot vigor and the number of berries and their average size will decrease. As the number of buds increases, grapevines typically respond with decreases in percentage budbreak, shoot vigor, bud fruitfulness, percentage fruit set, and berry growth, which leads to a less than proportional increase in total leaf area and yield comprised of more but smaller and less compact clusters (Keller et al., 2004, 2015a; Poni et al., 2016). Generally, as the number of clusters per vine increases, the number of berries per cluster and the average weight of these berries tend to decrease. Consequently, the yield component compensation principle states that changing the level of one yield component will induce compensatory changes of other yield components, resulting in a subproportional change in yield.

6.1.2 Yield potential and its realization

The yield potential of a crop plant can be roughly defined as the yield of a cultivar when grown in environments to which it is adapted, with abundant external resources, and with pests, diseases, weeds, and other stresses effectively controlled (Evans and Fischer, 1999). In viticulture, the yield potential is typically regarded as the maximum amount of fruit a grapevine or vineyard could potentially produce once the numbers of clusters and flowers are set at the beginning of each growing season. Taking all external factors into account, the largest yield potential is often achieved on vines of medium vigor because excessive vigor tends to impair bud fruitfulness, whereas very low vigor is generally a sign of limitation by some stress. A grapevine bud is said to be fruitful if it contains at least one inflorescence primordium that develops into a cluster. Using appropriate magnification, such clusters "in the making" can be seen in longitudinal cuts of winter buds, and this enables growers to conduct an initial assessment of a vineyard's yield potential prior to winter pruning. Based on the observed average number of inflorescence primordia per shoot, the pruning strategy can be adapted to compensate for annual changes in bud fruitfulness and bud injury. Bud fruitfulness can also be estimated retroactively by counting the number of clusters per shoot during the growing season or at harvest, but this ignores clusters that have been abscised after budbreak due to environmental stress. Fruitfulness is an inherited characteristic that is modulated by environmental factors at the time of inflorescence initiation, which occurs before bloom time, and subsequent differentiation, beginning around bloom time (see Section 2.3). Yield per node is different from bud fruitfulness because it is a function of the number of shoots per node, the number of clusters per shoot, the number of berries per cluster, and the size of these berries.

Once the maximum number of potentially fruitful shoots has been determined by the number of buds retained at winter pruning, bud fruitfulness sets the vine's yield potential for the following growing season. It must be kept in mind, however, that leaving more buds does not lead to proportionally more shoots, because fewer of the extra buds will break (Jackson et al., 1984; Keller et al., 2004). The extent to which this initial yield potential is realized by the vine depends on how many flowers develop on each inflorescence and how many of these flowers set fruit and develop into berries (Alleweldt, 1958).

Both the number of inflorescences and the extent of branching of these inflorescences are often lower in the buds produced at the shoot base than at higher bud positions (Vasconcelos et al., 2009). Consequently, when their bud and shoot numbers are similar, cane-pruned vines tend to have a greater yield potential than spur-pruned vines. However, the presence of markedly more basal, noncount buds often results in more shoots growing from spur-pruned vines than from cane-pruned vines when the same number of count buds are retained. This can lead to higher numbers of clusters of more variable size on spur-pruned vines, and it can also result in higher canopy density (Jones et al., 2018). Moreover, inflorescences on shoots arising from spurs sometimes produce more flowers than inflorescences from the same bud positions on canes (Huglin and Schneider, 1998).

Irrespective of the pruning method, environmental conditions required for the formation of the maximum number of inflorescence primordia are similar to those required to maximize photosynthesis rates of leaves, namely high light intensity, warm temperature, and adequate water and nutrient supply (Currle et al., 1983; Meneghetti et al., 2006; Mullins et al., 1992; Sartorius, 1968; Sommer et al., 2000; Vasconcelos et al., 2009). In addition, long days ostensibly favor the induction of inflorescence primordia in American *Vitis* species and their cultivars and in some *V. vinifera* cultivars, such as Riesling and Muscat of Alexandria (Mullins et al., 1992). However, the influence of day length on bud fruitfulness is far less important than the impact of light intensity (Srinivasan and Mullins, 1981). The effect of light intensity is independent of the effect of temperature, although the two are usually linked: cloudy or shady conditions are often associated with cool temperatures. Overcast conditions during the bloom–fruit set period decrease bud fruitfulness (Keller and Koblet, 1995a; May and Antcliff, 1963). A marked reduction in fruitfulness occurs when daily total photosynthetic photon flux is reduced to one-third or less of that reached under clear skies.

Furthermore, buds that develop inside dense canopies are less fruitful than the buds near the exterior of the canopy (Buttrose, 1974a; Dry, 2000; May et al., 1976; Sánchez and Dokoozlian, 2005; Shaulis et al., 1966; Sommer et al., 2000). This effect may be part of the shade avoidance syndrome that is mediated by phytochromes (Morelli and Ruberti, 2002; see also Section 5.2). In extreme cases, such shading may even cause bud necrosis, which looks like—and thus is sometimes misdiagnosed as—winter freeze damage (Perez and Kliewer, 1990; Vasudevan et al., 1998). However, bud fruitfulness appears to be independent of the light exposure of the buds themselves; instead, fruitfulness depends strongly on the supply of assimilates to the buds (Keller and Koblet, 1995a; May and Antcliff, 1963; Sánchez and Dokoozlian, 2005). These assimilates may come from stored reserves or from leaf photosynthesis, but they are mostly supplied to a bud by the leaves on the same side of the shoot (Hale and Weaver, 1962). Consequently, low light on the leaves above and below a bud reduces bud fruitfulness. This has implications for the design of trellis and training systems in viticulture (Reynolds and Vanden Heuvel, 2009). Systems that maximize the potential crop have a renewal zone, which is the region of the canopy that contains the buds that produce next year's shoots and crop, with sun-exposed leaves, ideally situated close to the top of the canopy, and a canopy that is not too dense.

In summary, the main reason for the pronounced light effect on bud fruitfulness is the influence of light on leaf photosynthesis; low light limits the amount of assimilates produced by the leaves and thus the amount supplied to the developing buds. In addition to high light intensity, relatively high temperatures are also required for maximum inflorescence initiation and development (Buttrose, 1969a, 1970b, 1974a; Sommer et al., 2000; Srinivasan and Mullins, 1981). Warm temperatures promote the induction of inflorescences, whereas temperatures below 20°C promote tendril formation (Buttrose, 1969a, 1970b). Some cultivars, such as Thompson Seedless and Muscat of Alexandria, appear to be unfruitful below 20°C. In general, however, bud fruitfulness increases as the temperature during the inflorescence initiation phase increases from 10°C to almost 35°C and then drops abruptly to zero at even higher temperatures (Vasconcelos et al., 2009).

Severe loss of leaf area resulting from pest or disease infection or hail can reduce inflorescence initiation and differentiation in the buds (Bennett et al., 2005; Candolfi-Vasconcelos and Koblet, 1990; Lebon et al., 2008; May et al., 1969). Because of the importance of functional leaf area for inflorescence formation, the number and size of the future inflorescences decrease as the severity of leaf destruction increases and the earlier the loss occurs. But a negative effect on bud fruitfulness still persists if loss of leaf area occurs as late as 2 months after bloom time.

Water deficit normally reduces bud fruitfulness (Buttrose, 1974b; Srinivasan and Mullins, 1981; Junquera et al., 2012). Nevertheless, a mild water deficit reducing canopy density can actually increase fruitfulness due to the improved canopy microclimate and thus better light exposure (Keller, 2005; see also Section 7.2). Mineral nutrition also influences bud fruitfulness, but little is known about what these effects are. It is likely that they partly involve changes of the canopy microclimate in addition to more direct effects, for example via the influence of nutrients such as nitrogen (N) on cytokinin. Soil N availability during bloom time is important because the response of cluster initiation to vine N status shows a clear optimum, with both N deficiency and N excess resulting in a reduced number of inflorescences per shoot (Alleweldt and Ilter, 1969; Currle et al., 1983; Keller and Koblet, 1995a; Srinivasan and Mullins, 1981). Phosphorus and potassium deficiency also decrease bud fruitfulness (Vasconcelos et al., 2009).

Although the number of flowers per inflorescence of a particular cultivar can vary from <10 to >1000 in the same vineyard, the environmental causes of this variation, beyond those involved in bud fruitfulness, are mostly unknown. In addition to the variation introduced during inflorescence differentiation in the buds, the temperature during budswell and budbreak may alter the number of flowers that form on an inflorescence. Higher temperatures during this brief period are associated with fewer flowers per inflorescence compared with lower temperatures (Eltom et al., 2017; Keller et al., 2010; Petrie and Clingeleffer, 2005; Pouget, 1981). One possible explanation for this effect is that the faster growth at higher temperatures might lead to more intense competition between inflorescences and shoot tips for remobilized storage reserves (Bowen and Kliewer, 1990). On the other hand, high air temperatures and warm, moist soil conditions accelerate the rate of reserve mobilization in spring and favor both shoot growth and flower development (Keller et al., 2010; Rogiers et al., 2011). Flower differentiation and early development during and after budbreak are entirely dependent on the availability of stored nutrient reserves and the rate of their remobilization because the unfolding leaves act as sinks during this period.

The need to supply carbon and other nutrients to both shoots and flowers may be one reason that fruit-bearing vines tend to grow less vigorously than do vines without fruit (Greer and Weston, 2010a). The more shoots that are growing on a vine, the more their demand for stored nutrients may compete with the formation of flowers. The amount of reserves stored during the previous growing season is therefore critical for maximum flower formation. Overcropping, excessively cool or hot growing seasons, untimely loss of active leaf area, and poor water and nutrient management all contribute to limiting—and sometimes depleting—the reserve pool in the vine's permanent organs and thus interfere with flower formation. Consequently, the number of flowers formed per inflorescence and per vine increases as the reserve status of a vine increases (Bennett et al., 2005; Holzapfel et al., 2010). Adequate reserve status may be particularly important to sustain productivity of high-yielding vineyards, where functional leaves seem to be required after fruit harvest to replenish the reserve pool (Holzapfel et al., 2006, 2010). Compared with lightly cropped vines and vines whose shoot growth has ceased by veraison, therefore, heavily cropped and vigorously growing vines are more dependent on favorable weather conditions and adequate leaf area late in the growing season to refill their nutrient reserves before winter (Holzapfel et al., 2006; Zufferey et al., 2012).

As discussed in Section 2.3, bloom occurs within 5–10 weeks after budbreak. The actual timing depends on the cultivar and weather conditions. Anthesis typically does not occur until the mean daily temperature exceeds approximately 18 °C. Under favorable conditions of high light intensity and high temperature, an inflorescence completes flowering within a few days, and the bloom period of an individual vine lasts approximately 7–10 days (Currle et al., 1983). Adverse weather conditions that are associated with low temperatures or rainfall can extend the bloom period to >1 month, and the abscised calyptra may remain trapped on the flowers. This does not prevent pollination, but it can interfere with fertilization by reducing pollen viability or germination rate, leading to excessive abscission of flowers and poor fruit set, thereby decreasing the number of berries per cluster (Koblet, 1966).

The proportion of flowers that set fruit is often between 20% and 50% (Currle et al., 1983; Mullins et al., 1992). Limited assimilate supply is the main cause of flower abortion and poor fruit set (Ruan et al., 2012). Because early shoot growth in spring depletes storage reserves, grapevines with a low carbohydrate status can experience poor fruit set if photosynthesis becomes limiting during bloom (Keller and Koblet, 1994; Rogiers et al., 2011). Thus the final fruit yield is strongly dependent on the photosynthetically active leaf area early in the growing season. Loss of leaf area during this period generally leads to poor fruit set due to low pollen viability resulting from carbon starvation of the flowers, which is a consequence of their low sink strength before fruit set (Candolfi-Vasconcelos and Koblet, 1990; Caspari et al., 1998; Coombe, 1959; Winkler, 1929; see also Section 5.1). A leaf area of less than about 5 cm^2 per flower can severely limit fruit set (Keller et al., 2010). Although the total requirement for assimilates by flowers and small berries is quite low, the developing flowers cannot compete with the growing shoot tips if resources become limiting. This makes the flowers vulnerable to environmental stress. Ovaries use glucose derived from the breakdown of phloem-imported sucrose as a signal: High amounts of glucose promote cell division and thus fruit set and growth, whereas low amounts of glucose activate the process of programmed cell death, culminating in flower or fruitlet abortion (Ruan et al., 2012; Ruan, 2014).

Ideal conditions during bloom, leading to maximum fruit set, are practically identical to those required for maximum inflorescence initiation. The optimum temperature range for bloom and fruit set in grapevines appears to be 20–30 °C, and both high (>35 °C) and low (<15 °C) temperatures reduce fruit set (Currle et al., 1983; Kliewer, 1977a). Heat stress reduces pollen viability, and at >40 °C, fruit set can be completely inhibited, at least in vulnerable cultivars like Sémillon, which may lead to abortion of inflorescences (Greer and Weston, 2010b; Pereira et al., 2014). Overcast, cool, or hot weather and water or nutrient deficits limit the amount of photosynthate available to the flower clusters and can lead to embryo abortion and hence poor fruit set (Keller, 2005; Lebon et al., 2008; Müller-Thurgau, 1883a; Roubelakis and Kliewer, 1976). Water deficit and other stresses that reduce sugar supply to and thus starch accumulation by the developing pollen before anthesis interfere with meiosis and result in male sterility, which also curtails fruit set (De Storme and Geelen, 2014; Dorion et al., 1996; Goetz et al., 2001; Saini, 1997; Zinn et al., 2010). Although ovaries are thought to be less vulnerable than anthers, they too can suffer from sugar starvation, which leads to flower abortion (Ruan, 2014). Even moderate water deficit at the time leading up to anthesis can significantly reduce fruit set and final yield (see Section 7.2). Fruit set can also suffer in the interior of dense canopies, which might be associated with the shade avoidance syndrome that is mediated by phytochromes (Morelli and Ruberti, 2002; see also Section 5.2). The vulnerability to stress of grape flowers, and especially of the anthers, could be an adaptation to domestication and vegetative reproduction (McKey et al., 2010). Because sexual reproduction is a metabolically expensive process, clonally propagated plants are thought to reduce over time their investment in male function. It is possible that such a bias against male function was aided unconsciously by early viticulturists who selected hermaphroditic variants from their dioecious progenitors during *V. vinifera* domestication.

Unfavorable bloom conditions and carbon shortage may also result in a high number of so-called "shot berries," which are in fact ovaries that fail to develop properly. In extreme cases, such stress culminates in a syndrome termed inflorescence necrosis or early bunch stem necrosis, whereby portions of or entire inflorescences die and are abscised (Caspari et al., 1998; Jackson, 1991; Jackson and Coombe, 1988; Keller and Koblet, 1994; Keller et al., 2001a, 2010; Winkler, 1929; Fig. 6.1). The term "necrosis" (Greek *nekrós* = dead) may be unfortunate because it implies an unprogrammed, chaotic cell death without nutrient recycling (Müntz, 2007). But death and abscission of reproductive organs are likely to be organized and genetically programmed and to involve remobilization and recycling of nutrients to other sinks. Deliberate shedding of inflorescences allows grapevines to adjust their reproductive output according to the prevailing external conditions. Such self-thinning avoids a situation whereby resources are overinvested in reproduction at the expense of the long-term survival of the vine. Moreover, early abortion of some fruits may enable the survival of others; that is, by limiting its investment in the number of reproductive organs, the vine has a greater chance to support the remaining fruits to seed maturity. Consequently, the number of berries per cluster is most susceptible to environmental stress at or just before anthesis. Removing the shoot tips, especially those of vigorously growing shoots, or girdling the canes or trunks during bloom can sometimes ameliorate the impact of adverse conditions by temporarily eliminating strong competitors for assimilates, whereas removing leaves before bloom accentuates fruit loss (Caspari et al., 1998; Coombe, 1959, 1962; Koblet, 1969; Müller-Thurgau, 1883a; Vasconcelos and Castagnoli, 2000). Indeed, where conditions favor high fruit set, especially of cultivars with compact clusters, growers increasingly use judicious prebloom leaf removal in the fruit zone to decrease cluster compactness by deliberately reducing fruit set (Acimovic et al., 2016; Palliotti et al., 2011; Poni et al., 2006).

In addition to its effect on fruit set, bloom-time temperature might also somehow influence bunch stem necrosis, a physiological disorder that is associated with necrotic lesions on the rachis (see Section 7.3). The lesions develop only after the berries begin to ripen, often beginning around stomata or dead bracts at branching points or beneath the epidermis, and interrupt assimilate supply to the berries distal to a lesion, which interferes with ripening and may lead to abortion of the afflicted

FIG. 6.1

Inflorescence necrosis on Müller–Thurgau following a cool and overcast bloom period (left) and on nitrogen-deficient Cabernet Sauvignon following water deficit during bloom, with normal fruit set on nitrogen-sufficient vines (center and right).

Photos by M. Keller.

cluster portion (Bondada and Keller, 2012; Currle et al., 1983; Hifny and Alleweldt, 1972; Theiler, 1970). Compared with nonaffected clusters, even on the same vine, berries on symptomatic clusters are typically lighter, often due to shriveling, and have lower amounts of sugars, potassium, and anthocyanins but higher amounts of organic acids, especially tartrate, and calcium, and lower pH (Keller et al., 2016b; Morrison and Iodi, 1990; Ureta et al., 1981). A negative correlation has been found in some circumstances between the average daily maximum temperature during the bloom period and the incidence of bunch stem necrosis (Theiler and Müller, 1986, 1987). However, a satisfactory explanation for this phenomenon has yet to be found.

Plants, like animals, regulate organ size and shape at the organ level rather than at the level of individual cells (Anastasiou and Lenhard, 2007). Yet the final size of grape berries that set after bloom is determined by the number of cell divisions before and after bloom, the extent of expansion of these cells, and the degree of shrinkage due to water loss prior to harvest. A mature berry is thought to contain approximately 600,000 cells compared with 200,000 cells in the ovary (Harris et al., 1968). This implies that there are many more cycles of cell division before bloom than after bloom. Consequently, there is ample opportunity for early season environmental factors to influence berry size (Coombe, 1976; Keller et al., 2010, 2012). Because cell division is dependent on carbon supply, stress factors that diminish assimilate supply will also limit the cell number and hence berry size (Caspari et al., 1998; Keller et al., 2010; van Volkenburgh, 1999).

The initial postbloom rates of cell division and cell expansion in a berry are controlled by the number of seeds that have set following fertilization, perhaps because more seeds make more auxin and stimulate more gibberellin production, making the berry a stronger sink (Cawthon and Morris, 1982; Scienza et al., 1978). Thus, as the number of seeds per berry increases, so too does the final berry size (Gillaspy et al., 1993; Huglin and Schneider, 1998; Olmo, 1946; Winkler and Williams, 1935). It is possible that a loss of ovule viability at temperatures above approximately 35 °C might be associated with fewer seeds per berry and thereby constrain berry size (Williams et al., 1994). The effect of seed number on berry size is most pronounced at the lower end so that an increase in seed number from three to four is associated with only a minor increase in volume.

The absence of seeds in many table grape cultivars limits berry growth, which conflicts with the general consumer demand for large table grapes. Growers therefore often stimulate berry expansion by applying gibberellin sprays at bloom and fruit set and/or by removing a strip of bark around the canes or trunk at fruit set (Currle et al., 1983; Williams and Matthews, 1990). Whereas gibberellin may stimulate the berries' sink strength, bark removal, called girdling, eliminates phloem flow from the leaves to the perennial parts of the vine so that more sugar is available to the developing fruit clusters (Roper and Williams, 1989; Weaver et al., 1969). In addition, girdling increases leaf water status due to a decrease in stomatal conductance that may be associated with feedback inhibition of photosynthesis (Williams et al., 2000). Girdling typically constitutes only a temporary interruption of phloem flow because callus, ostensibly formed by nearby phloem parenchyma cells, heals the wound, leading to reestablishment of the phloem connection (Sidlowski et al., 1971). Application of gibberellin to seeded cultivars often decreases fruit set and can also lead to lower bud fruitfulness and even bud necrosis (Huglin and Schneider, 1998). Although this carryover effect is typically regarded as undesirable, it may be exploited as a means to limit the crop size of large, minimally pruned grapevines (Weyand and Schultz, 2006c).

Maximum cell division rates occur at approximately 20–25 °C, and temperatures below 15 °C or above 35 °C lead to a marked reduction in the rate of cell division, resulting in smaller berries (Kliewer, 1977a). Even a few days with temperatures exceeding 40 °C after fruit set can limit final berry size (Matsui et al., 1986). The decrease in cell division due to both cold and heat stress may be caused by the thermosensitivity of cytokinin production and transport. Cytokinins may be inactivated at low temperatures by conversion of the active form to a storage form (Mok and Mok, 2001). During a heat episode, on the other hand, the seeds may produce less cytokinin and the xylem transports less cytokinin from the roots (Banowetz et al., 1999). It is possible that the need for leaf cooling, which causes rapid transpiration rates in leaves during heat stress (see Section 5.2), reduces the availability of cytokinin for fruit development.

Cell division is generally less sensitive to temperature than is cell expansion. When low temperatures inhibit respiration, cell expansion ceases very quickly (Cosgrove, 2016). Conversely, heat stress also decreases cell expansion. Whereas high temperatures before the berry enters the lag phase of growth irreversibly limit berry size, the temperature experienced after veraison often has no noticeable effect on berry size (Hale and Buttrose, 1974; Kliewer, 1973). Nevertheless, temperatures above 40°C during ripening may restrict berry expansion in sensitive cultivars (Greer and Weston, 2010b).

Because heat stress has a negative effect on both fruit set and berry growth, poor set and small berries are typical of years that experience a "heat wave" during the bloom–fruit set period (Kliewer, 1977a). The temperature sensitivity of berry cell division and cell expansion also has implications for vineyard management: Removing leaves in the cluster zone early during berry development may result in smaller berries due to solar heating of the sun-exposed fruit. Warm temperatures in the range of 25–32°C between bloom and the lag phase of berry growth are furthermore thought to promote the capacity of seeds to germinate (Currle et al., 1983). In addition to their effect on berry size, both cool and hot temperatures also delay fruit development, especially by prolonging the lag phase of berry growth (Hale and Buttrose, 1974). In a cool climate, however, the duration of phase I of berry development also has been found to be inversely correlated with the average temperature during this period (Alleweldt et al., 1984b).

Although the berries and their seeds are predominantly sink organs, it should be remembered that they are green—at least before veraison—and their skin absorbs much of the incident light (Blanke, 1990a). The pericarp cells, and even the seed cells, of preveraison berries contain chloroplasts with functional and shade-adapted photosynthetic machinery that participates in berry and seed metabolism (Blanke and Leyhe, 1989b; Famiani et al., 2000; Goes da Silva et al., 2005; Kriedemann, 1968a; Martínez-Esteso et al., 2011). Pericarp and seed photosynthesis may help to maintain adequate internal O_2 concentrations, refix some of the CO_2 that would otherwise be lost in respiration, produce some of the NADPH and ATP required for energy-intensive metabolism, and contribute to the accumulation of malate as a by-product of CO_2 refixation (Koch and Alleweldt, 1978; Palliotti and Cartechini, 2001; Palliotti et al., 2010). Moreover, the epidermis of young developing grape berries has a few stomata for gas exchange, even though the stomatal density on the berry surface is only $1-2\,mm^{-2}$, compared with $100-400\,mm^{-2}$ in leaves. As the berry expands, the stomatal density decreases because no new stomata are produced after anthesis, and the stomata are progressively occluded by epicuticular wax, suberin, and other components so that they become nonfunctional by veraison (Bessis, 1972b; Currle et al., 1983; Rogiers et al., 2004a; Swift et al., 1973).

Only early in its development does the berry fix an appreciable amount of carbon, which it can use for growth and metabolism; photosynthetic enzymes and chlorophyll are dismantled as the berry begins to ripen (Famiani et al., 2000; Geisler and Radler, 1963; Martínez-Esteso et al., 2011). Consequently, although immature berries may recycle approximately 20–30% of the daily respiratory carbon loss, and up to 65% in the light, via photosynthesis, the photosynthetic rate declines to near 10% of the respiration rate during ripening (Koch and Alleweldt, 1978; Ollat et al., 2002). The respiration rate, too, decreases throughout berry development and ripening (Geisler and Radler, 1963; Palliotti et al., 2010). Therefore, grape berries rely heavily on continued import of sugar for proper development and metabolism.

Any limitation in assimilate supply during bloom and early berry growth can lead to abortion of embryos. Thus, early stress also leads to fewer seeds per berry and strongly reduces the size of mesocarp cells but does not impair fruit ripening (Ollat et al., 2002). The amount of photosynthates that are partitioned to the fruit, especially early during fruit development, is critical for final berry size (Candolfi-Vasconcelos and Koblet, 1990; May et al., 1969; Ollat and Gaudillere, 1998). Insufficient photosynthate supply due to inadequate leaf area or low rates of photosynthesis limits both cell division and cell expansion, with very little compensation later in the season so that final berry size is irreversibly reduced. Therefore, berry size at harvest also tends to be inversely correlated with the number of berries per vine (Keller et al., 2008; Kliewer et al., 1983). Conversely, limiting assimilate supply after veraison has little to no effect on berry size but restricts ripening and may curtail the realization of the following year's yield potential by limiting the replenishment of storage reserves (Holzapfel et al., 2010; Zufferey et al., 2012).

Water deficit or nutrient deficiency typically reduce yield, particularly if a deficit occurs early in the growing season, when the inflorescences are not competitive in comparison with the shoot tips (Williams and Matthews, 1990; see also Sections 7.2 and 7.3). Inadequate water or nutrient supply before bloom can lead to abortion of entire inflorescences, which limits cluster number; stress during bloom results in poor fruit set, which limits berry number; and stress during the early stages of fruit development restricts berry growth, which limits berry size. Although water deficit may reduce the rates of both cell division and cell expansion, cell division is much less sensitive to changes in water status than is cell expansion (Ojeda et al., 2001). If the stress occurs later, then it will limit only cell expansion, which still may result in somewhat smaller berries. However, in contrast to the weak sink strength of the flowers prior to fruit set, the developing seeds and berries are stronger sinks than the shoots and, especially, the roots. Therefore, although stress does reduce seed and berry size, the decrease is less than would be expected from the reduction in photosynthate availability. This is probably due to remobilization of stored reserves in both the leaves and the permanent parts of the vine. Indeed, as grape berries begin to ripen, they are such strong sinks that they can accumulate sugar and resume expansion even in the face of severe water stress (Keller et al., 2015b).

Grape berries grow mostly at night, and they often lose some weight during the day, especially during periods of water deficit (Greenspan et al., 1994, 1996; Keller et al., 2006). Whereas such daytime berry shrinkage is common before veraison, it seems to occur only under severe water deficit after veraison, at least until the mesocarp cell membranes of some cultivars begin to fail during the late stages of ripening (Tilbrook and Tyerman, 2008). Once that happens, shrinkage may proceed rather rapidly, especially under warm, dry conditions (Bonada et al., 2013; Fuentes et al., 2010). The same extent of water deficit occurring before the lag phase of berry growth normally curtails berry size more than if it occurs during or after that phase. This means that the maximum berry size is mostly determined before veraison and that increasing the supply of irrigation water later in the growing season cannot compensate for the limitation imposed on berry size by an early deficit (Fig. 6.2). Although late-season irrigation may lessen berry shrinkage due to drought-induced dehydration, irrigation and even rainfall shortly before harvest alter berry size and composition far less than is often feared (Coombe and Monk, 1979; Huglin and Schneider, 1998; see also Section 6.2).

At every step during the yield formation process, there are management tools available to maximize or optimize final yield according to the desired specifications. For example, the timing and rate of water or nutrient application may be varied according to the intended end use of the grapes produced in a particular vineyard block. The same principle applies to canopy management practices such as leaf

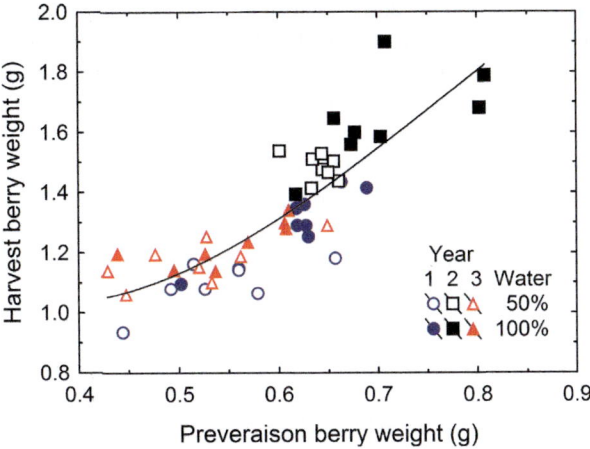

FIG. 6.2

Weight of Sauvignon blanc berries at the end of the preveraison lag phase and at harvest in 3 years. The closed symbols represent vines that were irrigated throughout berry development to replace 100% of the evaporative water loss from the vineyard, and the open symbols represent vines irrigated at 50% of that amount between fruit set and veraison and 100% during ripening (Keller, unpublished data).

or fruit removal. Seasonal weather patterns, as well as economic and legal considerations, often determine whether and when these cultural practices are applied. Growers also always must keep in mind the yield component compensation principle when deciding on strategies aimed at optimizing yield and fruit quality. In addition, because berry growth is most sensitive to stress factors during the early stages of development and because final berry size is largely predetermined by veraison, yield predictions are generally much more accurate after this stage than before it.

Cultivars, clones, and rootstocks differ in their sensitivity to changes in environmental conditions (e.g., Molitor and Keller, 2016). Such genetic differences also influence yield formation and consequently yield variation among vineyard sites and between years. This has implications for the selection and layout of planting material during vineyard establishment. Because genetically heterogenous plants typically experience lower yield variation than homogenous plants, it may be prudent to block vineyards into smaller units of different clones, cultivars, and rootstocks, rather than maximizing block size for each clone–rootstock combination.

6.2 Grape composition and fruit quality

6.2.1 Cell walls and membranes

The physical and chemical composition of grape berries at harvest is responsible for their quality characteristics and, consequently, the quality attributes of the wine or grape juice produced from the fruit. Much of what we call "quality" in grapes, juice, and wine has to do with our perception of color, taste, and flavor, which is the outcome of the processing in our brain of stimuli received by light receptors in our eyes, taste receptors on our tongue, and olfactory receptors in our nose. Compounds ranging from sugars to acids and phenolics to volatile chemicals all contribute interactively to the overall taste and flavor perception—that is, to the sensory quality. As grape berries ripen, they undergo a multitude of both physical and chemical changes, and yet many changes and processes important to fruit quality occur long before ripening begins (Brummell, 2006; Conde et al., 2007; Coombe, 1992; Deytieux et al., 2007; Hrazdina et al., 1984; Kanellis and Roubelakis-Angelakis, 1993; Sarry et al., 2004). These developmental changes are in turn coordinated by the interplay and cooperation of thousands of genes (Boss et al., 1996; Castellarin and Di Gaspero, 2007; Davies and Robinson, 2000; Deluc et al., 2007; Goes da Silva et al., 2005; Pilati et al., 2007; Robinson and Davies, 2000; Terrier et al., 2005). The expression of many of these genes is initiated or increases at the beginning of ripening, whereas that of many other genes decreases, and it is thought that abscisic acid (ABA), probably in concert with sugars, plays a central role in bringing about these changes (Gambetta et al., 2010; Jia et al., 2011; Koyama et al., 2009; Lijavetzky et al., 2012; see also Section 2.3). However, many more genes are downregulated (i.e., their expression decreases) than are upregulated (i.e., their expression increases) during the transition to ripening (Massonnet et al., 2017). At the level of gene expression, the ripening program of a berry starts in internal tissues around the seeds toward the end of the lag phase of berry development and then radiates outward, consistent with a role for the seeds in controlling ripening initiation (Shinozaki et al., 2018; see also Section 2.3).

A general overview of important physiological changes during berry development is shown in Fig. 6.3. The beginning of grape ripening is easily recognized by the sudden softening and the change of skin color of the berry (see Section 2.3). The color change occurs due to the degradation of the green chlorophylls, which unmasks the previously invisible yellow to orange-red carotenoids, and due to the simultaneous accumulation of red, purple, or blue anthocyanin pigments in dark-skinned cultivars (Hardie et al., 1996a; Uhlig and Clingeleffer, 1998; Young et al., 2012). White grapes, of course, do not make anthocyanins (see Section 1.2); they owe their skin color to a combination of chlorophylls and carotenoids, as well as the pale yellow, but barely visible, flavonols. A minor portion of the chlorophylls may be converted in the skin and, to a lesser extent, in the pulp into colorless tetrapyrroles, which are identical to those produced in senescing leaves and are strong antioxidants (Müller et al., 2007). Additional changes include expansion of the berry volume; structural changes in the skin and pulp; a switch in metabolic pathways; rapid accumulation within cell vacuoles of sugar and other solutes leading to dry mass gain; and decrease in acidity and increase in pH.

Berry softening, which can be measured as deformability, coincides with the beginning of sugar accumulation but precedes by several days the change in skin pigmentation and resumption of berry growth (Coombe, 1992; Coombe and Bishop, 1980; Creasy et al., 1993; Terrier et al., 2005; Wada et al., 2009). Softening occurs due to the gradual disassembly of the mesocarp cell walls and the decline in mesocarp cell turgor toward the end of the lag phase of berry growth (Huang and Huang, 2001; Thomas et al., 2008). Although the cell walls remain functionally intact, they become more open, more hydrated, less flexible, and weaker so that they can no longer hold the cells together very well (Brummell, 2006). Cell wall loosening proteins such as expansin, whose activity peaks at the end of the lag phase, and an assortment of enzymes secreted to the cell walls modify the cell wall polysaccharide components by "unlocking" the hydrogen bonds between the stiff cellulose microfibrils and the hemicellulose matrix just before and during ripening to enable the berry to soften (Deluc et al., 2007; Ishimaru and Kobayashi, 2002; Nunan et al., 2001; Pilati et al., 2007; Rose and Bennett, 1999; Rose et al., 1997, 1998; Schlosser et al., 2008). Other enzymes attack and solubilize the branched pectin

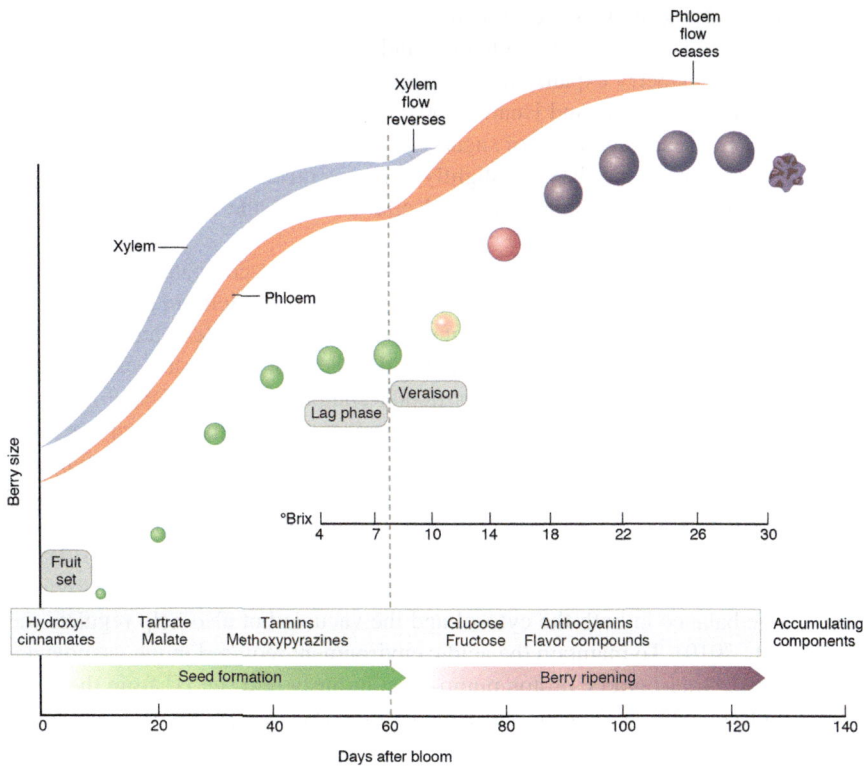

FIG. 6.3

Diagrammatic representation of the relative size and color of grape berries at 10-day intervals after bloom and of the major changes occurring during berry development.

© Elsevier Inc., illustration after Coombe, B.G., 2001. Ripening berries—a critical issue. Aust. Vitic. 5, 28–34.

polysaccharides that are integrated within the cellulose network and help to hold the microfibrils together; such solubilization is especially prevalent in cell wall junctions and the middle lamella (Wang et al., 2018). In addition to this enzyme action, the hydroxyl radical (OH), produced from other reactive oxygen species with the help of apoplastic peroxidase enzymes, may sever the bonds between cell wall polysaccharides nonenzymatically, which may enhance the cell wall loosening and degradation process required for berry expansion and softening (Møller et al., 2007; Rose et al., 2004). Peroxidase appears to be activated in the skin and, to a lesser extent, in the mesocarp at veraison (Calderón et al., 1994; Ros Barceló et al., 2003). Hydrogen peroxide (H_2O_2) serves as the oxidant for peroxidase, and H_2O_2 thus transiently accumulates in the skin during berry softening; due to its potential toxicity, H_2O_2 is soon degraded to water and oxygen by catalase: $2H_2O_2 \rightarrow 2H_2O + O_2$ (Pilati et al., 2014).

Loosening of the skin cell walls and incorporation of proteins, such as hydroxyproline-rich extensins for added tensile strength, enable the skin cells to become tangentially stretched as the flesh cells resume expansion after the lag phase (Chervin et al., 2008; Huang et al., 2005a; Nunan et al., 1998; Ortega-Regules et al., 2008; Vicens et al., 2009). Wall loosening is aided by the release of calcium and modifications in polysaccharides such as cellulose, hemicellulose, and pectin. Thus, changes in the flesh cell walls are responsible for berry softening, and soon afterward cell wall loosening in both the flesh and the skin enables berry expansion, which occurs despite the >10-fold decline in mesocarp turgor just before veraison (Huang and Huang, 2001; Huang et al., 2005a; Thomas et al., 2006). This turgor loss results from the accumulation of sugars and other solutes not only in the mesocarp cell vacuoles but also in the apoplast, which starts slightly before veraison as phloem unloading in the berry switches from a symplastic to an apoplastic route (Keller and Shrestha, 2014; Wada et al., 2008, 2009; Zhang et al., 2006; see also Section 5.1).

Solute accumulation begins in the stylar end of the berry and is accompanied by a decrease in xylem influx that also progresses from the stylar end to the pedicel end (Castellarin et al., 2011; Keller et al., 2006; Zhang and Keller, 2017). Transport of solutes into the berry vacuoles occurs via an assortment of active transporters and passive channels embedded in the vacuolar membrane, called tonoplast. Many of these transport proteins are specific for a particular compound or for closely related compounds (Fontes et al., 2011b; Martinoia et al., 2012). For instance, some transport proteins are responsible for hexose movement, others transport malate, others carry cations such as potassium (K^+), sodium (Na^+), or calcium (Ca^{2+}), or anions such as phosphate (HPO_4^{2-}), and still others move anthocyanins. Some of these transporters move their cargo into the vacuole along with protons (H^+), whereas many others move H^+ out of the vacuole in exchange for the incoming solute (Hedrich, 2012). For example, the so-called cation/H^+ antiporters exchange H^+ (out) for K^+ or Na^+ (in); the metal cations not only help maintain the charge balance in both the cytosol and the vacuole but also help regulate the vacuole's acidity (Bassil et al., 2019). To maintain the acidic environment required in the vacuole and to permit H^+ export to power the transporters, proton pumps continuously relocate H^+ from the cytosol into the vacuole. These proton pumps are named inorganic pyrophosphatase, or PPase for short, and ATPase (Fontes et al., 2011b; Terrier et al., 1998).

The volume and weight of grape berries increase during ripening, but they cannot do so indefinitely because the skin sets an upper limit to the expansion of the mesocarp cells (Matthews et al., 1987). Following a temporary increase at veraison, the extensibility of the skin later decreases in parallel with the stiffening of the epidermis and, especially, the cuticle. In other words, the skin becomes stiffer or more resistant to deformation. Stiffness is measured as the elastic modulus or modulus of elasticity, which is defined as the so-called stress ÷ strain ratio, meaning the ratio of force per unit area (i.e., pressure) to relative deformation (Cosgrove, 2016). The cuticle stiffness of mature fruit is similar to that of the common plastic polyethylene yet lower than the stiffness of leaf cuticles (Wiedemann and Neinhuis, 1998). Cuticle stiffening occurs partly due to a change in the composition of long-chain fatty acids, namely a decrease in the unsaturated and elastic C_{18} cutin monomers and a concomitant increase in the saturated and rigid C_{16} monomers (Bargel et al., 2006; Marga et al., 2001). In addition, the intracuticular wax embedded in the cutin matrix acts to constrain the cuticle's ability to expand (Khanal et al., 2013).

Stiffening of the epidermis occurs as a result of the activity of pectin methylesterase, which enhances binding of Ca^{2+} ions to the pectin's newly exposed carboxyl groups; the resulting calcium-pectate gel then stiffens the epidermal cell walls (Goldberg et al., 1996; Hyodo et al., 2013; Mirabet et al., 2011). The incorporation of Ca into the cell walls can deplete the soluble apoplastic Ca during fruit expansion—at a time when Ca influx via the xylem declines rapidly (De Freitas et al., 2012; Rogiers et al., 2006). Furthermore, additional polysaccharides and phenolics such as hydroxycinnamic acids and flavonoids, including tannins, may be deposited in the skin cell walls

(Amrani Joutei et al., 1994; Domínguez et al., 2009, 2011; Lecas and Brillouet, 1994; Schlosser et al., 2008). The phenolics are bound to cell wall polysaccharides and proteins by peroxidase, which further stiffens the cell walls, especially in the epidermis. This peroxidase activity limits cell expansion and thus counteracts the initial wall-loosening properties of peroxidase (Bargel et al., 2006; Potters et al., 2009; Thompson et al., 1998). Because many environmental stresses increase the amount of phenolics in the apoplast, one way by which such stresses often limit berry size might be due to the influence of these phenolics on cell wall stiffness. By the same logic, the bonding of phenolics to cell wall polymers should diminish their extractability during fermentation.

At the same time, protein incorporation and polysaccharide modifications also occur in the mesocarp cell walls (Ortega-Regules et al., 2008). Peroxidase-mediated inclusion of soluble proteins, especially extensin glycoproteins, reinforces the cell walls (Nunan et al., 1998; Ros Barceló et al., 2003). Despite these massive changes, the thickness of the epidermis, hypodermis, and mesocarp cell walls remains essentially constant during ripening (Hardie et al., 1996b; Nunan et al., 1998). Moreover, the ultrastructure, cell walls, and membranes of the berry cells remain intact (Diakou and Carde, 2001; Fontes et al., 2011a; Fougère-Rifot et al., 1995; Hardie et al., 1996b). Cell compartmentation is retained, mitochondria remain functional, and plasmodesmata maintain their cell-to-cell connections—except the ones in the phloem membranes, which are plugged to facilitate apoplastic phloem unloading (see Section 5.1). Only the locule area around the seeds appears to be subject to some degradation, especially toward the end of the ripening phase (Krasnow et al., 2008). In the remainder of the pericarp, senescence and loss of cell compartmentation only seem to set in once berries have attained their maximum weight, at least in some cultivars; this loss of membrane integrity is sometimes associated with subsequent berry shrinkage (Fuentes et al., 2010; Hardie et al., 1996a; Tilbrook and Tyerman, 2008). Loss of membrane integrity culminates in cell death and may be associated with oxidative stress as a result of local oxygen deprivation in the central mesocarp region between the peripheral and central vascular bundles (Xiao et al., 2018). Oxygen deprivation, called hypoxia, also leads to the production of minute amounts of ethanol as respiration shifts to fermentation.

The shrinkage of berries that occurs near the end of their lifecycle is a result of water loss from the berries. Large berries, which have a large surface ÷ volume ratio, evaporate more water and thus shrink more rapidly than do small berries (Rogiers et al., 2004b; Zhang and Keller, 2015). Nonetheless, the low gas permeability of the berry skin ensures that shrinkage due to dehydration occurs only slowly, which is why the production of raisin grapes is essentially restricted to hot climates where a high vapor pressure deficit (VPD) provides the necessary driving force for water evaporation.

6.2.2 Water

The final size of the berries of a particular grape variety is mainly influenced by cell enlargement, because cell division determining the number of cells is comparatively less affected by environmental conditions. The rapid initial growth of the berry during ripening is mostly due to water import and retention by the mesocarp cell vacuoles. Because these vacuoles come to occupy 99% of the mesocarp cell volume, grape juice essentially consists of vacuolar sap of mesocarp cells (Diakou and Carde, 2001; Storey, 1987). Thus, water is by far the dominant chemical component of a ripening berry (Matthews and Anderson, 1988). Due to the accumulation of sugars and other solutes during ripening the water content is highest before veraison (~90%), declining to approximately 75–80% at maturity and sometimes to <70% in overripe grapes with >30°Brix. Consequently, 1 kg of grapes yields approximately 0.6–0.8 L of juice or wine, depending on species, cultivar, berry size, maturity, and extent of pressing (Viala and Vermorel, 1901–1909). However, the water content can decrease to <15% when grapes are dried to produce raisins of about 68°Brix; thus, 1 kg of grapes yields <250 g of raisins (Bongers et al., 1991; Parpinello et al., 2012).

Xylem sap is the main source of water for the berry before veraison; it is thought to account for approximately 75% of the total water influx (Ollat et al., 2002). At veraison, water import via the phloem increases rapidly to satisfy the berry's high demand for sugar (water being the solvent for the incoming sugar), and water import through the xylem declines while the berry expands and changes color (Zhang and Keller, 2017). The rate of carbon import into a berry increases three- to fourfold at veraison, and virtually all of this carbon is imported via the phloem (Ollat et al., 2002). Therefore, phloem sap becomes the primary or only source of berry water during ripening because excess water is delivered to the berry in the phloem along with assimilates such as sucrose and amino acids (Greenspan et al., 1994; Keller et al., 2006; Keller et al., 2015b). The xylem now serves mostly to recycle surplus phloem water back out of the berry (Keller et al., 2006, 2015b; Patrick, 1997; Zhang and Keller, 2017).

Berry transpiration accounts for the bulk of water loss from the berry, and the evaporation rate from the berry surface increases with increasing berry size. However, transpiration rates of grape berries are 10–50 times lower than those of leaves and decline during berry development (Becker and Knoche, 2011; Blanke and Leyhe, 1987; Düring and Oggionni, 1986; Rogiers et al., 2004a; Scharwies and Tyerman, 2017). A brief rise in berry transpiration at veraison, which is mostly related to the concomitant increase in berry size, is followed by a further decline later on during ripening (Zhang and Keller, 2015). Indeed, grape berries are designed to minimize evaporative water loss. First, they are typically round or nearly so, a geometry that translates into the least amount of surface area relative to volume; this in turn translates into a large water storage capacity compared with the evaporative surface. Second, per unit surface area the berries have about 100 times fewer stomata than do leaves, and these stomata become nonfunctional by veraison (Blanke et al., 1999; Swift et al., 1973). Finally, the berry cuticle is impregnated with approximately 10 times more wax than the leaf cuticle (Possingham et al., 1967; Radler, 1965a, 1965b, 1970). The berry cuticle, with its intra- and epicuticular waxes, is a formidable barrier to the diffusion of water vapor and other gases (Becker and Knoche, 2011; Leide et al., 2007; Zhang and Keller, 2015). Yet the decrease in transpiration during fruit development is not caused by an increase in the amount of cuticular waxes (Grncarevic and Radler, 1971; Rogiers et al., 2004a). Instead, changes in the composition of these waxes, such as the incorporation of long-chain ($>C_{28}$) n-alkanes, seem to be responsible for the developmental increase in cuticular resistance to water vapor (Leide et al., 2007).

In addition to evaporating from the berry surface, water can also flow back through the xylem from the berry to the shoot (Keller et al., 2006, 2015b; Lang and Thorpe, 1989; Zhang and Keller, 2017). Due to the low transpiration rate of the berry, some of the water received by a preveraison berry via the xylem and phloem may be recycled back to the vine, especially in the afternoon, when Ψ_{leaf} may be lower than Ψ_{berry}. The decreasing Ψ_{berry} after veraison due to sugar import changes the water potential gradient in favor of the berry so that water that has already entered the berry cell vacuoles becomes increasingly protected from being "pulled back" by the leaves. Instead, surplus phloem water, which moves to the berry cell walls osmotically during apoplastic phloem unloading (see Section 5.1), may be recycled out of the berry via the xylem without entering the berry cells (Keller et al., 2006, 2015b; Zhang and Keller, 2017). It appears that cultivars differ in the extent of such xylem backflow, perhaps due to variation in the hydraulic resistance of the pedicel or berry xylem (Scharwies and Tyerman, 2017; Tilbrook and Tyerman, 2009; Tyerman et al., 2004).

Although it used to be thought that grape berries become hydraulically isolated from the vine at veraison, the berry xylem remains functional during ripening (Bondada et al., 2005; Chatelet et al., 2008b; Keller et al., 2006). Nevertheless, the change from xylem to phloem water import at veraison affords the berries some independence from the fluctuations in Ψ experienced by the leaves due to transpiration. Consequently, the sensitivity of berry water status to soil and plant water status declines greatly at veraison, and the daily cycle of berry shrinkage during the day and expansion at night becomes much less pronounced (Greenspan et al., 1994, 1996; Matthews and Shackel, 2005). Indeed, the berries become such strong sinks for sugar at veraison that they are able to attract phloem water and expand even in the face of severe water deficit that leads to leaf wilting (Keller et al., 2015b). The sugar in this case is likely supplied from starch reserves in the perennial organs of the vine (Rossouw et al., 2017). Furthermore, although berry transpiration is strongly driven by, and hence fluctuates with, atmospheric VPD, the water supplied by the phloem normally compensates for any evaporative water loss (Keller et al., 2015b; Zhang and Keller, 2015, 2017). Accordingly, whereas preveraison berries shrink and expand readily with fluctuating vine water status, postveraison berries are much less subject to such fluctuations (Keller et al., 2006).

If the results obtained with algae are applicable to grape berries, then the increase in solute concentration, leading to "osmotic dehydration," should lead to closure of aquaporins, which would decrease membrane water permeability (Ye et al., 2004). This not only could contribute to the high resistance to dehydration of ripening berries but also may support the changes in skin stiffness in limiting berry growth. Although the berries will still shrink during prolonged episodes of water stress, shrinkage becomes limited because phloem water supply is much less prone to changes in vine water status than is xylem water supply (Keller et al., 2006, 2015b). It is only when the mesocarp membranes begin to fail in some cultivars—Syrah is a classic example—during late ripening that the leaves are able to extract water rather rapidly from the berries, and berry shrinkage sets in or accelerates (Bonada et al., 2013; Clarke et al., 2010; Fuentes et al., 2010). Because high temperatures increase both the rate of water evaporation from the berries and the rate of membrane failure, hot and dry conditions hasten berry shrinkage (Bonada et al., 2013; Fuentes et al., 2010; Zhang and Keller, 2015).

Berry expansion, too, becomes limited because the increasing stiffness of the thick epidermal cell walls and the cuticle decrease the skin's elasticity as the fruit matures (Bargel and Neinhuis, 2005; Bargel et al., 2006; Matthews et al., 1987). A rigid, or unelastic, skin may be necessary for grapes to limit expansion of mesocarp cells while they accumulate sugars to high concentrations, which raises their osmotic pressure (π) to as much as 4 MPa (Keller et al., 2015b). Simultaneous accumulation of solutes in the cell walls and recycling of surplus phloem water in the xylem keep the mesocarp turgor pressure (P) low (Keller and Shrestha, 2014; Thomas et al., 2006). But the increase in surface area of the expanding berry causes mechanical stress due to tension in the skin cells that also bear the P that is transmitted to them from the interior mesocarp cells (Considine and Brown, 1981; Considine and Knox, 1981). Because the skin cell walls are >10-fold thicker than the mesocarp cell walls, the mechanical structure of the berry is often approximated as an inflated elastic shell, in which the tensional stress (s) is needed to balance the outward P (Boudaoud, 2010; Lustig and Bernstein, 1985). The balance of forces in a berry can be stated as follows:

$$s = P d h^{-1}$$

where d = berry diameter, h = skin thickness.

In other words, given the constant P during ripening, the tension in the skin cells increases as the berry diameter increases and as the skin thickness decreases. As the berry expands, the skin would have to become thicker to support the same P and keep its cells from being pulled apart (Boudaoud, 2010). Because stiff materials are easily fractured, the stiffening of the cuticle and epidermis and the decrease in skin extensibility and thickness seem to come at the cost of increased vulnerability to cracking. Berries are generally thought to crack when the mesocarp P exceeds the cell wall resistance of the skin cells, a relationship that is cultivar specific (Bernstein and Lustig, 1985; Considine, 1981; Lang and Düring, 1990; Lustig and Bernstein, 1985).

Cracking can occur during periods of high humidity or rainfall when water may be absorbed through the skin (Considine and Kriedemann, 1972; Lang and Thorpe, 1989; Meynhardt, 1964). The permeability of the berry cuticle for water uptake appears to be severalfold higher than that for transpirational water loss, and the driving force for osmotic water uptake through the skin increases as fruit sugar concentration increases (Becker and Knoche, 2011; Beyer and Knoche, 2002). In addition, water uptake can also occur through the berry pedicel and receptacle, especially in the corky scar area left from the abscised flower cap (Becker and Knoche, 2011; Becker et al., 2012).

Cracking can ruin the quality of table and juice grapes and increase the incidence of bunch rot (see Section 8.2). It is also a common problem in tomato, cherry, and blueberry production (Ehret et al., 1993; Marshall et al., 2007; Sekse, 1995). Warm and moist conditions favor cracking because the water permeability of the cuticle increases with rising temperature, while the cuticle's strength decreases (Becker and Knoche, 2011; Beyer and Knoche, 2002; Domínguez et al., 2011; Lopez-Casado et al., 2010; Schönherr et al., 1979). Cuticular permeability also increases with increasing relative humidity of the atmosphere, probably because absorbed water molecules lead to swelling of the cuticle (Schönherr, 2006; Schreiber, 2005). In addition, moisture increases the cuticle's vulnerability to cracking, perhaps because water acts as a plasticizer that softens the cuticle, increasing its flexibility but decreasing its strength (Domínguez et al., 2011; Lopez-Casado et al., 2010; Wiedemann and Neinhuis, 1998). Nonetheless, although the cuticle constrains berry expansion, the mechanical properties of the skin, including its vulnerability to cracking, may be determined by the cell walls of the epidermis and hypodermis, not the cuticle (Brüggenwirth et al., 2014). When berries absorb surface water, their epidermal cell walls swell and eventually fail, resulting in cracking (Brüggenwirth and Knoche, 2017).

The uptake of liquid surface water or water vapor by grape berries in a humid environment, is one reason why removal of leaves around the fruit clusters is a key canopy management practice, especially in regions that typically experience rainfall during the ripening phase. The increased air flow and, once sunshine returns, higher berry surface temperature are conducive to rapid drying of the clusters after rainfall, which limits the amount of water that can be absorbed through the skin and receptacle, as well as the extent of dilution of berry solutes. However, cracking may also ensue under conditions of root pressure that may prohibit recycling of phloem-derived water out of the berry via the xylem (Keller et al., 2006, 2015b; Zhang and Keller, 2017). At advanced stages of maturity, the vulnerability to cracking may decline in cultivars that are prone to shrinking and especially once the mesocarp cell membranes begin to break down, and hence lose osmotic competence (Clarke et al., 2010).

The slow transpiration rate of grape berries strongly limits transpirational cooling. This results in marked day–night fluctuations of berry temperature and permits solar radiation to raise the skin temperature of sun-exposed berries by 10–15 °C, and occasionally even more, above ambient temperature (Cola et al., 2009; Keller et al., 2016a; Millar, 1972; Smart and Sinclair, 1976; Spayd et al., 2002).

The postveraison decrease in berry transpiration also slows the drying rate of raisin grapes, especially if the grapes are harvested late in the season, when the temperature and VPD decline. To facilitate drying and mechanical harvesting, raisin growers often cut off the fruiting canes with clusters still attached to the shoots in late summer, a practice known as "drying on the vine" or "harvest pruning." Yet such drying takes much longer than drying manually harvested clusters spread on the ground because the temperature in the canopy is lower than that on the ground, so late-ripening and slow-drying cultivars such as Thompson Seedless are not well suited to this practice (Ramming, 2009). Because severing canes eliminates photosynthesizing leaf area at a time when inflorescence primordia may still be differentiating, this practice can sometimes come at the cost of some reduction in yield potential for the subsequent year (May, 2004; Scholefield et al., 1977a,b, 1978).

6.2.3 Sugars

Sugars, especially glucose and fructose, are among the principal grape ingredients determining fruit and wine quality because they are responsible for the sweet taste of the fruit, and thus of grape juice and wine; they mask the perception of sourness, bitterness, and astringency (which is the main reason many people add sugar to tea and coffee); and they enhance the "mouthfeel," "texture", "body," or "balance" of wines (Hufnagel and Hofmann, 2008b). Most important for wine, anaerobic fermentation converts each molecule of hexose ($C_6H_{12}O_6$) sugar into two molecules each of ethanol (CH_3CH_2OH), which is colloquially called alcohol, and CO_2:

$$C_6H_{12}O_6 \rightarrow 2CH_3CH_2OH + 2CO_2$$

This process is accomplished by various species of yeast, but mostly by *Saccharomyces cerevisiae*, which live on the surface of grape berries (Barata et al., 2012b; Belin, 1972). The sugar concentration in grape juice can be estimated from the concentration of total soluble solids, which is measured in °Brix (g/100 g of liquid, or % w/w soluble solids) by refractometry at a standard temperature of 20°C. Yeast converts sugar into ethanol at a ratio of approximately 0.6% (v/v) ethanol per 1°Brix (Jones and Ough, 1985; Ough and Amerine, 1963). Ethanol is important for wine quality partly because most aroma volatiles are more soluble in ethanol than in water, so that their aroma impact declines with increasing ethanol content, and partly because it can suppress the perception of a wine's overall "fruity" character (Escudero et al., 2007). As the grape sugar concentration increases during fruit ripening, there is not only more sugar available to be converted into ethanol, but the influence of sugar (and other berry constituents) on yeast metabolism also alters the content in wine of yeast-derived metabolites, such as higher alcohols, esters, and sulfur-containing volatiles (Bindon et al., 2013). In addition to producing ethanol and CO_2, yeasts convert a portion of the sugars into glycerol and higher alcohols as a by-product of fermentation, which adds to the perception of sweetness, mouthfeel, and body of wines (Hufnagel and Hofmann, 2008b; Ribéreau-Gayon et al., 2006; Noble and Bursick, 1984; Robinson et al., 2014).

The anaerobic conditions that predominate during alcoholic fermentation contrast with the oxygen-rich environment of a vineyard. Under such aerobic conditions, yeasts meet their energy demand by respiration rather than fermentation so that the sugars present in unharvested grapes are turned into CO_2 and water. The little ethanol that is being produced is further oxidized to acetic acid (CH_3COOH) and water by aerobic bacteria of the genus *Acetobacter*:

$$CH_3CH_2OH + O_2 \rightarrow CH_3COOH + H_2O$$

The end-product of acetic acid degradation is CO_2 as well. However, employing the same respiratory process in the presence of oxygen, these acetic acid bacteria can also turn wine into vinegar (French *vin* = wine, *aigre* = sour), especially at high temperature and high pH.

Developing and ripening grape berries import sugar as sucrose via the phloem (see Section 5.1). Although grape flowers and young berries may contain some starch due to their photosynthetic activity, starch does not accumulate to any great extent and becomes limited to the skin by veraison (Fougère-Rifot et al., 1995; Hardie et al., 1996a; Kriedemann, 1968a; Palliotti et al., 2010; Swift et al., 1973). Starch does, however, accumulate between bloom and veraison in the pedicels and rachis of a grape cluster (Amerine and Bailey, 1959; Amerine and Root, 1960; Gourieroux et al., 2016). In addition, the seed endosperm accumulates starch that serves as a source of energy and carbon during seed germination.

While the soluble solids of grape berry juice are dominated by organic acids up to veraison, this changes rapidly as sugar accumulation sets in at veraison (Keller and Shrestha, 2014; Keller et al., 2015b, 2016b). Consequently, sugars account for >90% of the soluble solids in mature grapes, with much of the remainder being organic acids. In the berries of most *Vitis* species and their cultivars, 95–99% of these sugars are present in the form of the hexoses glucose and fructose, with the remainder predominantly made up of sucrose (Dai et al., 2011; Hawker et al., 1976; Hrazdina et al., 1984; Kliewer, 1967b; Liu et al., 2006; Lott and Barrett, 1967). After veraison the two hexoses are responsible for >80% of the osmotic pressure (π) of the mesocarp cells and almost 60% of the π in the exocarp (Keller et al., 2015b; Storey, 1987). In contrast, sucrose may account for up to 30% of the sugar content in mature *Muscadinia* berries (Carroll et al., 1971; Carroll and Marcy, 1982). With regard to the taste of sweetness, fructose > sucrose > glucose; the fact that fructose is almost twice as sweet as sucrose may explain why inexpensive high-fructose corn syrup is so widely used by the food industry. At low temperatures around ~10°C, preveraison *V. vinifera* berries contain approximately equal amounts of glucose and fructose. However, the glucose ÷ fructose ratio rises as the temperature increases, reaching 3–10 at 35°C, whereas the proportion of sucrose increases at low temperatures (Currle et al., 1983; Findlay et al., 1987; Harris et al., 1968; Kliewer, 1964, 1965a). Irrespective of temperature, the glucose ÷ fructose ratio declines to approximately 1 at veraison and remains more or less constant or decreases slightly thereafter (Keller and Shrestha, 2014; Kliewer, 1964, 1965a, 1967b,c; Viala and Vermorel, 1901–1909).

A few days before a berry begins to change color at veraison, hexose sugars begin to accumulate in the cell vacuoles and the apoplast of the mesocarp and, with a slight delay, in the exocarp (Castellarin et al., 2011, 2016; Coombe, 1987; Keller and Shrestha, 2014; Keller et al., 2015b; Wada et al., 2009). Just as it does in tomatoes, hexose accumulation may start near the stylar end of a grape berry and then proceeds toward the pedicl end (Castellarin et al., 2011; Giovannoni et al., 2017; Zhang and Keller, 2017). Sugar accumulation is dependent on import of sucrose from photosynthesizing leaves or woody storage organs via the phloem. Because the phloem supplies surplus water that needs to be discharged from the berry, hexoses accumulate more rapidly in berries that transpire more rapidly—that is, under warm and dry conditions that are associated with a high VPD (Rebucci et al., 1997; Zhang and Keller, 2017). Up until veraison, most of the imported sugar does not enter the vacuoles; instead it is used for energy generation via respiration (see Section 4.4) and to produce other organic compounds. Accumulation in the vacuoles starts suddenly and is concomitant with berry softening. By the time they begin changing color, grape berries generally have reached a soluble solids concentration of 9–10°Brix. During ripening, there is a rapid increase in hexose sugars due to import of sucrose and its subsequent breakdown by invertases (see Section 5.1).

Sugar accumulation proceeds most rapidly during the early stages of ripening but can continue for some time after the stiffening cuticle puts an end to berry expansion. Depending on the species and cultivar, the sugar content of grape berries reaches a maximum at about 22–25°Brix, when the berries contain 0.6–0.7 M each of glucose and fructose. The maximum sugar content may be associated with a gradual loss in mesocarp membrane integrity that leads to an increase in membrane leakiness (Keller and Shrestha, 2014; Caravia et al., 2015). A sugar concentration of 25°Brix results in an ethanol concentration of approximately 15% (v/v) in wine made from such grapes, which also happens to be close to the upper limit of survival for the most ethanol-tolerant yeasts. Berries at 25°Brix have a total solute concentration of approximately 1.6 M, which is equivalent to an osmotic pressure of 5.4 MPa (Keller and Shrestha, 2014; Keller et al., 2015b). Although sugar import ceases around this stage, the sugar concentration can increase further due to berry dehydration and shrinkage (Fig. 6.4). Thus, the final sugar concentration at harvest varies widely; table grapes and grapes destined for juice production may be harvested at 15–18°Brix or even less, whereas overripe, late-harvest wine grapes may reach twice that sugar concentration. Dried raisin grapes or wine grapes that undergo a postharvest drying phase for the production of straw or passito wines and similar products are much more concentrated still. Since the berries shrink during the drying process, their solute concentration tends to increase by roughly 2°Brix for every 10% loss in berry weight (Muganu et al., 2011).

Sugars are carbohydrates, a class of chemicals that also includes the mostly insoluble polysaccharides contained in the cell walls. The contribution of these polysaccharides to a berry's total sugars is small, for cell walls make up only 1.5% of the berry weight at veraison, decreasing to 1% during ripening (Barnavon et al., 2000). The bulk of the cell wall polysaccharides in grape berries is composed of cellulose and pectins with smaller amounts of hemicelluloses, most of the latter being made up of xyloglucans (Hanlin et al., 2010; Lecas and Brillouet, 1994; Nunan et al., 1997; Vidal et al., 2001).

FIG. 6.4

Syrah berries reach a maximum amount of sugar when their sugar concentration approaches 25°Brix, but the concentration may continue to increase as berries shrink due to water loss; notice the large variation in sugar content due to variation in berry size (left; Keller, unpublished data) and mature Cabernet Sauvignon berries shrinking from dehydration

Right; photo by M. Keller.

Hundreds to thousands of glucose units are assembled in the long cellulose chains, whereas pectins and hemicelluloses also contain other sugar building blocks, such as arabinose, xylose, galactose, mannose, and rhamnose in addition to glucose, as well as their oxidized forms, or sugar acids, such as glucuronic acid and galacturonic acid (Brummell, 2006; Doco et al., 2003; Vidal et al., 2001). The latter forms the "backbone" of pectin that is interspersed with rhamnose to which other sugars are attached as side chains. In contrast, the highly branched hemicellulose does not have a backbone proper and consists of a mixture of different sugar units.

Rapid shortening of hemicellulose, referred to as depolymerization, and disassembly of pectin in the mesocarp cell walls enable grape berries to soften abruptly at veraison as they start to accumulate sugar but before they change color. At the onset of ripening, β-galactosidases, aided by other enzymes, pry a portion of the pectin away from its bondage with cellulose in the cell wall and cut its sugar chains into shorter pieces, making them water soluble; this depolymerization process intensifies during ripening due to the activity of pectinases, such as polygalacturonase and pectate lyase (Barnavon et al., 2000; Nunan et al., 2001; Robertson et al., 1980; Rose et al., 1998; Vicens et al., 2009). Cell expansion during the early ripening phase is accompanied by production of additional pectin so that the total amount of pectin per berry increases while its concentration declines (Silacci and Morrison, 1990). At the same time, the xyloglucans are also depolymerized, and their content decreases under the influence of enzymes that degrade and modify the cell walls (Brummell, 2006; Deytieux et al., 2007; Doco et al., 2003; Ishimaru and Kobayashi, 2002; Rose and Bennett, 1999; Schlosser et al., 2008). Because the mesocarp cells are very large and have thin walls, most of a berry's cell wall material is in the skin, where xyloglucans tightly bind the cellulose microfibrils and pectins, forming an extensive and tough polysaccharide network (Doco et al., 2003; Hardie et al., 1996b). In addition to their cross-linking function, xyloglucans may also keep the cellulose fibrils properly spaced and porous (Hamant and Traas, 2010; McCann et al., 1990; Wolf et al., 2012). Although the mesocarp accounts for approximately 75% of the fresh weight of a mature berry, 75% of the berry's total pectin is located in the skin (Vidal et al., 2001).

In contrast to the simple sugars, cell wall polysaccharides do not taste sweet, but they are nonetheless important for fruit and wine quality. The cell wall matrix, especially the shorter pectin fragments that result from cell wall disassembly and the integrated extensin proteins, may bind other berry ingredients, such as tannins and anthocyanins, during fruit ripening and postharvest processing (Cadot et al., 2006; Hanlin et al., 2010; Kennedy et al., 2001). The binding capacity of pectin arises from the multitude of negative charges of pectin and its degradation products, which confer considerable cation binding and exchange capacity on these cell wall materials (Nobel, 2009; Sattelmacher, 2001). When pectic polysaccharides are released by the cells to build their walls they are initially highly methylated, but the methyl groups are quickly removed by pectin methylesterases, which exposes the reactive pectate backbone (Wang et al., 2018). The binding of tannins to pectin fragments might be responsible for the decrease in the perception of astringency in ripening fruit, because the change in tannin solubility that results from such binding may make tannins less available to bind to salivary proteins when berries are chewed (Macheix et al., 1990; Soares et al., 2012). However, the pectin binding capacity probably exerts its major influence after berries are damaged and their cell compartmentation is broken to release the juice from the vacuoles during the juicing and winemaking processes. Alterations in cell wall polysaccharides may account for some of the changes in extractability of quality-relevant compounds observed during ripening that are discussed later. Moreover, yeast enzymes can degrade and extract some of the pectins during fermentation. Because most white grapes are fermented *after* pressing—that is, after the skins and seeds have been removed by destemming—they result in wine with much lower polysaccharide content than red wines, which are typically fermented "on the skins" (O'Neill et al., 2004).

6.2.4 Acids

Acidity plays an important part in the perception of fruit and wine quality because it affects not only the sour taste, or tartness, but also sweetness by masking the taste of sugars and thus gives grapes, juice, and wine their fresh, crisp taste. In addition, acids such as malate also taste astringent and increase the overall perception of astringency (Hufnagel and Hofmann, 2008b). This is why acidity in wine is sometimes difficult to distinguish from astringency due to tannins. Wines with high acidity and low pH are often perceived as "sour," unless the acidity is masked by residual sugar, whereas wines with low acidity and high pH are described as "flat," because our tongue functions essentially as a pH meter, detecting the concentration of hydronium ions (H_3O^+) as a sour taste. Similarly, the acidity of grape juice or wine is estimated by titration against an alkaline solution, usually made with sodium hydroxide (NaOH). Giving rise to the term titratable acidity, this volumetric analysis measures the concentration of the acids' rather weakly bound hydrogen ions or protons (H^+) that are released by the addition of the strong base NaOH.

Organic acids are chemically termed carboxylic acids because they contain at least one carboxyl group (COOH) that can be ionized to COO^- by losing a proton. Their high concentration in grape berries and the active transfer into the fruit vacuoles of protons by H^+ pumps are responsible for the low pH of grape juice. The pH falls as the H^+ concentration rises as defined by the relationship $pH = -\log_{10}[H^+]$, where $[H^+]$ is the H^+ concentration (in mol L^{-1}). Although the pH is generally inversely related to acid concentration (Fig. 6.5), there is no simple relationship between titratable acidity and pH nor between "total acidity" and pH (Iland et al., 2011; Smith and Raven, 1979). In fact, the organic acids tend to buffer and hence stabilize the pH (Shiratake and Martinoia, 2007). Although plants generally regulate their cell's cytoplasmic pH within narrow limits, the pH inside the vacuoles can vary considerably. The vacuolar pH increases during fruit ripening due to both a decrease in

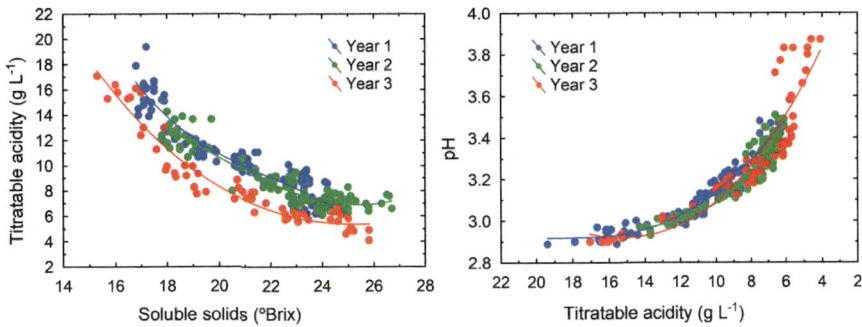

FIG. 6.5

The acidity declines as the concentration of sugars (expressed as soluble solids) increases (left), and the pH rises as the acidity decreases (right) during ripening of Cabernet Sauvignon berries over 3 years; year 2 had the coolest and year 3 the warmest ripening period (Keller, unpublished data).

organic acids and an increase in metal cations such as potassium (K^+) and sodium (Na^+). These cations counter the influence of organic acids by substituting for H^+, thereby raising the pH (Boulton, 1980a,b). This effect becomes dominant toward the end of the ripening phase, because K^+ may continue to move into the berries until phloem influx ceases, whereas the rate of malate respiration decreases as most malate has already been respired and the temperature typically declines in autumn. Consequently, juice expressed from harvested grapes typically has a pH between 3.0 and 3.5, but the pH can sometimes exceed 4.0 in overripe fruit. Values in excess of approximately 3.6 are undesirable because they lead to decreased color intensity and microbial stability, which increases the potential for spoilage, and increased susceptibility to oxidation in wine and other grape products.

Juice from mature grapes typically contains between 5 and $10\,g\,L^{-1}$ of organic acids. A titratable acidity concentration of $6.5–8.5\,g\,L^{-1}$ is considered optimal for the production of well-balanced wines (Conde et al., 2007). Although acidity is commonly expressed as tartaric acid equivalents, the organic acids are present as a mixture of several free acids and their salts; the latter are usually ionized (Etienne et al., 2013; Kliewer et al., 1967; Smith and Raven, 1979). For example, malic acid ($C_4H_6O_5$) also occurs in the interconvertible forms $C_4H_5O_5^-$ (the malate anion) and $C_4H_4O_5^{2-}$ (the malate dianion), while tartaric acid ($C_4H_6O_6$) may also exist as $C_4H_5O_6^-$ (the tartrate anion, a.k.a. bitartrate) or $C_4H_4O_6^{2-}$ (the tartrate dianion). The relative proportions of the various forms are dependent on the pH, since ionization increases with rising pH. Though this book uses the terms malate and tartrate for all forms of these acids, almost all the malate and much of the tartrate are present as free malic acid and tartaric acid, respectively, in the low-pH grape vacuoles (Iland and Coombe, 1988).

In most fruit species, malate and citrate are the dominant organic acid. Grapes of both *Vitis* and *Muscadinia* species are unusual because tartrate is much more abundant than citrate. In fact, tartrate and malate are the major soluble organic components that accumulate in a preveraison grape berry and together comprise 70–90% of the berry's total acids (Dai et al., 2011; Jackson, 2008; Keller and Shrestha, 2014; Kliewer, 1966, 1967b; Ruffner, 1982a). Other organic acids mainly include oxalate, citrate, succinate, and fumarate, as well as phenolic acids, amino acids, and fatty acids. Citrate is an important flavor precursor in grapes because it can be converted during malolactic fermentation of wines to diacetyl, a compound imparting a "buttery" or "butterscotch" flavor. The tart malate, on the other hand, can be produced or degraded by various yeasts—which also independently produce succinate—and is converted to the much milder-tasting lactate and CO_2 by lactic acid bacteria, such as *Oenococcus oeni*, during anaerobic malolactic fermentation. In contrast, tartrate is rather metabolically and microbially stable, which is why it is often added in winemaking to adjust the juice and wine pH. Nonetheless, the bunch-rot fungus *Botrytis cinerea* (see Section 8.2) and some lactic acid bacteria are able to degrade tartrate, especially when the pH exceeds 3.5. Also, in the presence of iron or copper molecules, tartrate can be oxidized to glyoxylate (Monagas et al., 2006).

Most organic acids are accumulated early in berry development and stored in the vacuoles of the mesocarp and skin cells following biosynthesis inside the berry (Hale, 1962; Ruffner, 1982a,b; Sweetman et al., 2009). Additionally, some malate may be imported via both the phloem and, in small amounts, the xylem (Hardy, 1969; Peuke, 2010). The acids serve important roles in regulating osmotic pressure, at least before sugars assume this role at veraison, as charge-balancing anions for K^+, Na^+, Ca^{2+}, and Mg^{2+}, and as complexing agents for essential but toxic heavy metals, such as Fe^{2+}, Zn^{2+}, and Cu^{2+} (Keller et al., 2015b; Meyer et al., 2010).

Tartrate and oxalate are both products of the breakdown of ascorbate (vitamin C), which occurs at least partly in the cell walls (DeBolt et al., 2004, 2006; Green and Fry, 2005; Saito et al., 1997).

Ascorbate can be produced by oxidation of sugars, such as glucose, mannose, and galactose, or of galacturonic acid derived from pectin breakdown in the berries or imported from the leaves via the phloem; some may even be produced within the phloem (DeBolt et al., 2007; Melino et al., 2009). During berry ripening, the tartrate concentration decreases due to dilution as the berry expands, whereas the amount of tartrate per berry generally remains almost constant (Hale, 1977; Keller and Shrestha, 2014; Possner and Kliewer, 1985). Some tartrate turnover or respiration does, however, seem to occur (Drawert and Steffan, 1966a; Keller et al., 2016b; Takimoto et al., 1976). Contrary to widespread opinion, tartrate does not form crystals with K^+ or Ca^{2+} in intact grape berries; thus, it does not become insoluble and precipitate during ripening. Despite the continued influx of K^+ in the phloem, crystals are formed between oxalate and Ca^{2+} rather than tartrate and K^+ (DeBolt et al., 2004; Fougère-Rifot et al., 1995). Of course, once the cell compartmentation breaks down when grapes are crushed, tartrate may indeed crystallize with both K^+ and Ca^{2+}, but this process is very slow. Indeed, the discovery of Ca-tartrate crystals in an ancient jar was used as evidence that Neolithic people made wine at least 7000 years ago (McGovern et al., 1996).

Crystallization is a biomineralization process that takes place in hypodermis cells called idioblasts specializing in Ca-oxalate production (Franceschi and Nakata, 2005; Loewus, 1999). Its insoluble products are deposited as bundles of needle-like Ca-oxalate crystals termed raphides in the idioblast vacuoles. The idioblasts probably serve as a storage compartment for surplus Ca^{2+} and can release Ca^{2+} on demand, which is important after veraison, when Ca^{2+} influx into the berry through the xylem has mostly ceased. Raphides may play an antifeeding role by discouraging insects and mammals from eating unripe grapes. Raphides are also a cause of contact dermatitis among people harvesting agave for tequila production, while Ca-oxalate is a major component of kidney and gall stones, and both oxalate and tartrate are believed to be the dominant pain-inducing toxins of stinging nettles. The granular Ca-oxalate crystals termed druses that accumulate in the berry endocarp, on the other hand, may serve to store and later release oxalate. The druses might also help scatter and thereby distribute sparse light to the interior of the berry, as they do in leaves (He et al., 2014). Whereas the production of tartrate and malate ceases by veraison, oxalate continues to accumulate throughout ripening, at least in some cultivars (Keller and Shrestha, 2014; Keller et al., 2016b). Its concurrent breakdown releases H_2O_2 and CO_2 after veraison. Aided by the enzymes polyphenol oxidase and peroxidase, the H_2O_2 oxidizes the phenolics in the seed coat so that the green seeds gradually turn brown (Braidot et al., 2008; Hanlin et al., 2010).

Malate is produced from phosphoenolpyruvate as a by-product of glycolysis and the photosynthetic refixation of CO_2 released during berry respiration or, in a reversible reaction, from oxaloacetate in the TCA cycle and is initially stored in the berry's vacuoles (Etienne et al., 2013; Martínez-Esteso et al., 2011; Moskowitz and Hrazdina, 1981; Ruffner, 1982b; Sweetman et al., 2009; Terrier et al., 1998). Preveraison malate accumulation has a temperature optimum of 20–25 °C and declines sharply above 38 °C (Kliewer, 1964; Lakso and Kliewer, 1975, 1978). Like tartrate, most malate is formed during the preveraison period (Kliewer, 1965b; Ruffner and Hawker, 1977). There is very little biosynthesis after veraison, although this "production stop" may be somewhat delayed in the skin (Gutiérrez-Granda and Morrison, 1992; Iland and Coombe, 1988). Berry acidity therefore peaks immediately before veraison, a fact that is exploited in the production of verjuice or verjus (French $vert$ = green, unripe; jus = juice). Verjuice has been used as an acidifying cooking ingredient and substitute for vinegar since ancient times and is made by pressing fresh, unripe grapes harvested around veraison for this purpose or collected during cluster thinning for crop load adjustment.

The amount of malate per berry declines after veraison due to inhibition of glycolysis and remetabolism of stored malate for respiration through the TCA cycle and for gluconeogenesis (Drawert and Steffan, 1966b; Martínez-Esteso et al., 2011; Ruffner, 1982b; Ruffner and Hawker, 1977; Ruffner et al., 1975; Steffan and Rapp, 1979). It is thought that the activation of hexose storage in the vacuoles at veraison triggers malate efflux from the vacuoles and its breakdown in the outer mesocarp near the peripheral vascular bundles (Gutiérrez-Granda and Morrison, 1992; Sarry et al., 2004). Since vacuolar sugar storage may deplete sugars in the cytosol, the respiratory carbon flux through glycolysis (see Section 4.4) also declines at veraison, at least in the mesocarp (Giribaldi et al., 2007; Harris et al., 1971). Hexose storage and malate breakdown are activated simultaneously as a berry begins to ripen, and malate compensates for the lack of sugar as the substrate for berry respiration. Consequently, the berry switches its carbon supply for respiration from glucose to malate at veraison so that the juice acidity declines as the sugar content rises (Fig. 6.5). Provided that low temperatures do not slow down malate catabolism, most of the decrease in mesocarp acidity occurs early in ripening, slowing around 16–18°Brix, and often becoming insignificant above 20–21°Brix—even as the pH continues to rise (Keller et al., 2012, 2016b; Keller and Shrestha, 2014). In contrast, the rate of glycolysis in the skin may increase during ripening, and there seems to be little, if any, postveraison malate decrease in the skin (Gutiérrez-Granda and Morrison, 1992; Iland and Coombe, 1988; Negri et al., 2008; Possner and Kliewer, 1985; Storey, 1987).

Gluconeogenesis is the formation of glucose through a reversal of the glycolytic pathway (see Fig. 4.6). Such a reversal can occur if a cell's energy level exceeds demand so that the carbon of pyruvate can be recycled back to glucose instead of entering the respiratory TCA cycle. Gluconeogenesis has a temperature optimum near 20°C and decreases to half the maximum rate at both 10°C and 30°C (Ruffner et al., 1975). While respiratory malate degradation increases with increasing temperature up to 50°C, gluconeogenesis is more important at lower temperatures, when sugar import and energy demands are low. Nonetheless, the contribution of gluconeogenesis to total sugar accumulation is minor and seems unlikely to exceed approximately 5% (Ruffner and Hawker, 1977).

6.2.5 Nitrogenous compounds and mineral nutrients

Nitrogen-containing compounds are important sources of food for yeast during fermentation, and the nitrogen (N) content and composition of grape musts greatly impact fermentation rates and wine aroma profiles. Yeast can utilize most free amino acids in addition to ammonium (NH_4^+); the N from these nitrogen sources is therefore collectively termed "yeast-assimilable nitrogen" (YAN) in contrast to the N forms that yeast cannot metabolize, namely proteins, peptides, and the amino acid proline. An adequate range for YAN in grape musts destined for fermentation is 150–250 mg L^{-1}, but the yeast's N requirement inceases as the amount of sugar in the must increases. Musts with <150 mg YAN L^{-1} are prone to sluggish or stuck fermentations and associated problems with hydrogen sulfide (H_2S) production (Bell and Henschke, 2005). The H_2S is liberated from proteins that are broken down by the starving yeast cells and has a "rotten egg" odor. Consequently, the aroma intensity and quality of wines tend to decrease as the N content of the grapes decreases (Rapp and Versini, 1996). Grape musts are therefore often supplemented with diammonium phosphate (($NH_4)_2HPO_4$) to enhance yeast nutrition.

Although the N content of grape berries is extremely variable, they may contain approximately half the total canopy N at maturity (Pradubsuk and Davenport, 2010; Wermelinger and Koblet, 1990;

Williams and Biscay, 1991). Between 50% and 90% of the mesocarp N is present in the form of free amino acids (Roubelakis-Angelakis and Kliewer, 1992). The remainder is made up mostly of proteins, NH_4^+, and nitrate (NO_3^-). Whereas amino acids predominate in the mesocarp, proteins are much more abundant in the skin and, especially, the seeds (Yokotsuka and Fukui, 2002). The various berry tissues contain hundreds of different proteins, most of which are enzymes that are involved in the construction and transport of the broad range of berry metabolites and in respiration that provides the energy required for these activities (Goes da Silva et al., 2005; Martínez-Esteso et al., 2011; Sarry et al., 2004). The seeds accumulate large amounts of storage proteins, which make up close to 90% of the total seed protein and serve as a source of energy, N, and sulfur during and after seed germination. Moreover, proteins are also embedded in the pectin matrix of mesocarp and skin cell walls (Lecas and Brillouet, 1994; Saulnier and Brillouet, 1989; Vidal et al., 2001). Grape berry proteins can be extracted during fermentation and make up as much as 90% of the total protein content of wine; the remainder of the proteins is derived from yeast (Fukui and Yokotsuka, 2003).

Import of nitrogenous compounds into a grape berry before veraison can occur in both the xylem, predominantly in the form of glutamine and NO_3^-, and the phloem, mostly as glutamine. At veraison import becomes essentially restricted to the phloem. Nevertheless, more than half of a berry's total N is normally imported after veraison, and N accumulation may continue throughout ripening (Conradie, 1986; Kliewer, 1970; Löhnertz et al., 1989; Pradubsuk and Davenport, 2010; Rodriguez-Lovelle and Gaudillére, 2002; Schaller et al., 1990). Although grape berries can assimilate NO_3^- (Schaller et al., 1985), glutamine is the major nitrogen transport compound (see Section 5.3); enzymes called aminotransferases convert glutamine in the berry into other amino acids. Of these, arginine and proline together account for 60–80% of the free amino acids in mature grape berries (Bell and Henschke, 2005; Goes da Silva et al., 2005; Hernández-Orte et al., 1999; Roubelakis-Angelakis and Kliewer, 1992; Stines et al., 2000; van Heeswijck et al., 2001). However, the concentrations of arginine and proline may vary by an order of magnitude among grape cultivars grown under similar conditions (Huang and Ough, 1991; Kliewer, 1970). Furthermore, the release of amino acids associated with protein breakdown during ripening may contribute to the increase in amino acids after veraison (Martínez-Esteso et al., 2011).

Arginine can be accumulated in the skin, flesh, and seeds throughout fruit development, although in some cultivars (e.g., Chardonnay and Cabernet Sauvignon) accumulation appears to cease at veraison, whereas in yet others (e.g., Müller-Thurgau) it may not begin until veraison (Kliewer, 1970; Schaller and Löhnertz, 1991). In contrast, most of the proline is accumulated after veraison, and accumulation seemingly continues throughout ripening (Coombe and Monk, 1979; Kliewer, 1968; Lafon-Lafourcade and Guimberteau, 1962; Miguel et al., 1985; Stines et al., 2000). Proline often serves to protect cells from osmotic stress; thus, proline accumulation during ripening could be related to the osmotic pressure caused by the accumulating hexose sugars (van Heeswijck et al., 2001). Proline accumulation might also contribute to the rise in juice pH because its production from glutamate may involve hydroxide (OH^-) release (Smith and Raven, 1979).

Some arginine can be converted to polyamines, which are defined as compounds with more than one amino group (Moschou et al., 2008). Although the production of polyamines such as putrescine and spermidine occurs throughout berry development, their concentration declines during ripening due to simultaneous catabolism (Fortes et al., 2015). Polyamine breakdown releases H_2O_2, which might assist polyphenol oxidase and peroxidase in oxidizing seed coat phenolics, leading to seed browning.

Arginine and other amino acids and protein residues, but not proline, can react with sugars in the so-called Maillard reaction to produce brown polymers termed melanoidins. This process leads to slow nonoxidative, nonenzymatic browning during grape processing—for instance, during drying for raisin production and subsequent storage (Frank et al., 2004). Melanoidins can have pleasant (malty, bread crust–like, caramel, and coffee) or unpleasant aromas (burnt, onion, solvent, rancid, sweaty, and cabbage) and may taste bitter. Note that beer, coffee, chocolate, toasted bread, and caramel owe much of their color and aroma to melanoidins.

At very high concentrations, amino acids such as proline, threonine, glycine, serine, alanine, and methionine, impart a sweet taste, whereas arginine, lysine, histidine, phenylalanine, and valine, appear to taste bitter, and still others, such as glutamine, glutamate, asparagines, and aspartate, have an umami (Japanese *umami* = savory) taste (Hufnagel and Hofmann, 2008b). However, apart from proline and, perhaps occasionally, glutamate, amino acids are present in wine at concentrations that are far too low to have any sensory impact. Moreover, the amino acid profile of a wine differs considerably from that of its grape constituents and may reflect metabolic activities of yeasts and other microorganisms. For instance, arginine is the major yeast-assimilable amino acid, whereas proline cannot be readily used by yeast (Bell and Henschke, 2005). Some of the "fermentation aroma" of wines—that is, the bouquet caused by flavor and aroma compounds generated during fermentation—stems from the conversion by yeasts of amino acids to the so-called higher alcohols and their esters, which are defined as alcohol molecules with more than the two carbon atoms of ethanol (Hirst and Richter, 2016; Robinson et al., 2014; Swiegers et al., 2005). For example, leucine yields isoamyl alcohol with a "harsh," "burnt," or "whiskey" odor, and valine yields ethyl isobutyrate with an aroma of "strawberry."

Several minor amino acids in grapes have key implications for fruit quality and human health. The key position of phenylalanine as a chemical precursor for phenolic compounds is discussed in Section 6.2.6. Phenylalanine is an aromatic amino acid, in the chemical rather than the sensorial sense, produced by the shikimate pathway. Its aromatic sibling tryptophan not only is one of the key precursors for the plant hormone auxin but also can be degraded to the aroma volatile methyl anthranilate or further metabolized to the neurotransmitter serotonin and the human hormone and antioxidant melatonin (Iriti et al., 2006; Wang and De Luca, 2005). Fermentation yeast can metabolize phenylalanine and tyrosine, another shikimate offshoot, into phenylpropanoid alcohols and acetates with aromas of "honey" or "rose" (Robinson et al., 2014). In addition, lactic acid bacteria may convert tyrosine during malolactic fermentation into the neurotransmitter tyramine, which may lead to increased heart rate and higher blood pressure in humans, although the amounts of tyramine found in wine are rarely high enough to cause problems. Histidine is important because wine spoilage bacteria can convert it to histamine, a biogenic amine that at high concentrations is suspected to trigger headaches, including migraine, and asthma attacks. Unfortunately, ethanol inhibits histamine degradation in the body, which tends to exacerbate the adverse reaction. Nonetheless, except at extremely high concentrations, the scientific evidence for a link between biogenic amines and adverse reactions in humans appears to be inconclusive (Jansen et al., 2003; Skypala et al., 2015).

In addition to proline, certain proteins also accumulate in grape berries after veraison and throughout ripening, probably in response to the osmotic stress arising from increasing sugar concentration (Giribaldi et al., 2007; Monteiro et al., 2007; Murphey et al., 1989; Negri et al., 2008; Salzman et al., 1998). These are the chitinases and thaumatin-like proteins, which are members of the so-called pathogenesis-related (PR) protein families (see Section 8.1). As a berry ripens, PR proteins occupy a growing fraction of the total berry proteins and can make up half or more of the total soluble protein in mature grape berries, with roughly equal amounts in the skin and the pulp. These PR proteins contribute to protein instability in white wines because they are responsible for haze formation or clouding in wine

(Pocock et al., 2000; Tattersall et al., 2001). Due to the rise in protein content, the potential for protein instability increases with advancing grape maturity. This is exacerbated by the concomitant increase in mesocarp pH, which makes PR proteins progressively easier to extract. In addition, protein extraction into the juice from damaged machine-harvested grapes is much greater than that from undamaged or hand-harvested grapes, especially if the grapes are left standing for prolonged periods or transported over great distances before processing (Pocock et al., 1998). Red wines, however, are not normally subject to this protein haze problem because the tannins extracted from the berry skins during fermentation precipitate these and other proteins (Bindon et al., 2016).

Mineral nutrients other than N also accumulate in grape berries. Potassium (K^+) is by far the most important of these and is the major cation in grapes; its concentration in expressed juice often exceeds $1\,g\,L^{-1}$ and can be substantially higher in the skin (Coombe, 1987; Donèche and Chardonnet, 1992; Hrazdina et al., 1984; Mpelasoka et al., 2003; Possner and Kliewer, 1985; Rogiers et al., 2006; Storey, 1987). Cells use K^+ to neutralize the negative charge of inorganic and organic anions and anionic groups of macromolecules, because the total charge of all anions and cations inside a cell must balance (Clarkson and Hanson, 1980). Because it can substitute for protons in grape juice, K^+ impacts the juice pH: A 10% increase in K^+ concentration is associated with a rise in pH by approximately 0.1 units (Boulton, 1980a). This is significant because the K^+ concentration in the berry juice often approximately doubles during ripening (Keller and Shrestha, 2014; Rogiers et al., 2006). Although K^+ can be transported in both the xylem and the phloem, its concentration in the xylem is orders of magnitude lower than that in the phloem, and import into the berry probably occurs predominantly in the phloem (see Section 5.1). Import occurs throughout berry development, but it sometimes increases markedly at veraison (Rogiers et al., 2006). Along with K^+, the amount of sodium (Na^+) also increases in both the pulp and the skin during ripening, but the Na^+ content per berry is at least an order of magnitude lower than the K^+ content (Etchebarne et al., 2010; Iland et al., 2011; Iland and Coombe, 1988). Most of the K^+ is stored in the vacuoles of the inner hypodermis and the mesocarp. It has been proposed that K^+ contributes to berry cell expansion by increasing the cells' osmotic pressure (Davies et al., 2006; Mpelasoka et al., 2003; Rogiers et al., 2006). Yet this seems rather unlikely given that even before sugars assume this role at veraison, the contribution of K^+ to mesocarp cell osmotic pressure is <5% and declines further during ripening (Keller et al., 2015b).

As indicated in Section 6.2.4, free K^+ ions, liberated when the cell membranes are ruptured during grape processing, can form crystals with tartrate that are called potassium bitartrate, potassium hydrogen tartrate, or cream of tartar ($KC_4H_5O_6$). The resulting precipitation of the poorly soluble K-tartrate will decrease juice and wine acidity and may increase or decrease the pH depending on the initial pH and the presence of other acids (Allen, 1982). Extraction of K^+ from the skins during fermentation amplifies this effect in red wines (Mpelasoka et al., 2003). Note that cream of tartar, along with other components such as anthocyanins and other phenolics, is also sometimes extracted industrially from pomace remaining after pressing of juice or must.

Free calcium ions (Ca^{2+}) also combine with tartrate to form insoluble calcium tartrate in juice from crushed grapes and in wine, but this crystallization is very slow and may occur over a period of months or years. In contrast to K^+, Ca^{2+} import into grape berries is largely restricted to the xylem. Consequently, 75–100% of a berry's final Ca^{2+} content is accumulated before veraison (Creasy et al., 1993; Esteban et al., 1999; Rogiers et al., 2006). Magnesium (Mg^{2+}), on the other hand, can be transported in both the xylem and the phloem, and approximately half the Mg^{2+} is imported into a berry after veraison (Bertoldi et al., 2011; Esteban et al., 1999; Pradubsuk and Davenport, 2010; Rogiers et al., 2006). Rapid berry transpiration rates favor the accumulation of K^+, Mg^{2+}, and especially Ca^{2+}, even after veraison (Düring and Oggionni, 1986). Warm, dry, and windy conditions and other factors that accelerate berry transpiration could therefore be associated with higher fruit mineral nutrient content.

Both Ca^{2+} and Mg^{2+} are important components of cell walls, and Ca^{2+} also accumulates to high concentrations in the cell vacuoles in the form of Ca-oxalate crystals. The highest concentrations of these ions are found in the seeds, which is also true for phosphorus (P) and sulfur (S). However, due to its much larger volume, the mesocarp can sometimes be an equally important storage site for Mg^{2+}, P, and S in terms of total amount per berry (Bertoldi et al., 2011; Rogiers et al., 2006). Like K^+, P accumulation in grape berries occurs throughout berry development and ripening (Bertoldi et al., 2011; Pradubsuk and Davenport, 2010). Moreover, S residues on the berries from foliar applications after veraison to control powdery mildew may lead to raised H_2S production by yeast during fermentation (Kwasniewski et al., 2014; Thomas et al., 1993; see also Section 7.3).

Micronutrients, or trace metals, such as iron (Fe), copper (Cu), manganese (Mn), and zinc (Zn), are mostly accumulated in the seeds (Bertoldi et al., 2011; Rogiers et al., 2006). Nevertheless, the mesocarp and, in some instances, even the skin may also contribute substantial amounts on a per berry basis. Boron (B) accumulates in both the seeds and the skin, which probably reflects the importance of B as a cell wall component (see Section 7.3). The amount of most trace metals generally increases throughout berry development and ripening, but most of the Mn and Zn is accumulated before veraison (Pradubsuk and Davenport, 2011; Rogiers et al., 2006). Import in the phloem of Cu and B may be so strong that not only their amount per berry but also their concentration increases during ripening, whereas the concentration of other micronutrients tends to decline. Unlike the pericarp, however, the seeds essentially stop accumulating phloem-mobile nutrient ions other than K^+ at veraison (Rogiers et al., 2006). Trace metals may be important in altering the color hue of grape juice and wine. For instance, a small increase in Fe concentration can lead to an increase in the blue hue and a corresponding decrease in red hue. Moreover, dark pigments may form through interactions of proline and other amino acids with phenolics in the presence of Fe. Moreover, traces of Fe and Cu can lead to oxidation of tartrate to glyoxylate (Monagas et al., 2005). As a component of polyphenol oxidase, Cu is also involved in the enzymatic oxidation of phenolics. In addition, fungicide-derived Cu residues may lead to an increase in H_2S formation by yeast during fermentation (Eschenbruch and Kleynhans, 1974).

Minute quantities of rare earth elements have also been detected in grape berries (Bertoldi et al., 2009, 2011; Yang et al., 2010). Most of these belong to the lanthanoid series and are thought to have biological properties that resemble those of Ca^{2+}, with whose effects they may interfere. In grapes, the rare earth elements are dominated by cerium (Ce), neodymium (Nd), and lanthanum (La). Most of them are present in the skin, with slightly smaller amounts in the mesocarp and only traces in the seeds. Rare earth elements accumulate predominantly before veraison, but accumulation may continue at a slower rate during ripening (Bertoldi et al., 2011).

6.2.6 Phenolics

Compounds other than water, sugars, acids, and minerals account for only a small proportion of the berry weight but contribute significantly to fruit and wine quality. Among these, the class of phenolics is one of the most important contributors. All phenolic compounds are made up of an "aromatic ring" consisting of six carbon atoms with one or more hydroxyl (OH) groups or derivatives of this basic structure (Table 6.1). They are very important for the color and astringency of red wine, contribute to grape

Table 6.1 The main classes of phenolic compounds in grapes.

Class name	Basic chemical structure	Examples
Flavan-3-ols		Catechin (left); Epicatechin (right)
Tannins		Hypothetical proanthocyanidin tetramer with (from top to bottom) epigallocatechin, epicatechin, catechin, and epicatechin gallate (note: the subunits shown are a combination of those found in both seeds and skins)

Continued

Table 6.1 The main classes of phenolic compounds in grapes—cont'd

Class	Structure	Compounds
Anthocyanins		Cyanidin-3-GLU[a] ($R_1 = OH$, $R_2 = H$); Delphinidin-3-GLU ($R_1 = OH$, $R_2 = OH$); Peonidin-3-GLU ($R_1 = OCH_3$, $R_2 = H$); Petunidin-3-GLU ($R_1 = OCH_3$, $R_2 = OH$); Malvidin-3-GLU ($R_1 = OCH_3$, $R_2 = OCH_3$)
Flavonols		Kämpferol-3-GLU ($R_1 = H$, $R_2 = H$); Quercetin-3-GLU ($R_1 = OH$, $R_2 = H$); Myricetin-3-GLU ($R_1 = OH$, $R_2 = OH$); Isorhamnetin-3-GLU ($R_1 = OCH_3$, $R_2 = H$); Laricitrin-3-GLU ($R_1 = OCH_3$, $R_2 = OH$); Syringetin-3-GLU ($R_1 = OCH_3$, $R_2 = OCH_3$)
Hydroxycinnamic acids		Cinnamic acid ($R_1 = H$, $R_2 = H$); Coumaric acid ($R_1 = H$, $R_2 = OH$); Caffeic acid ($R_1 = OH$, $R_2 = OH$); Ferulic acid ($R_1 = OCH_3$, $R_2 = OH$)
Hydroxybenzoic acids		Protocatechuic acid ($R_1 = H$, $R_2 = OH$); Gallic acid ($R_1 = OH$, $R_2 = OH$); Syringic acid ($R_1 = OCH_3$, $R_2 = OCH_3$)
Stilbenes		Resveratrol ($R_1 = OH$, $R_2 = OH$); Pterostilbene ($R_1 = OCH_3$, $R_2 = OCH_3$); Piceid ($R_1 = OH$, $R_2 = GLU$); Viniferins (resveratrol polymers)

[a]*GLU*, glucose.
Chemical structures courtesy of J. Harbertson.

and wine flavor and aroma, and are the main substrates for juice and wine oxidation (Macheix et al., 1991; Singleton, 1992). Their susceptibility to oxidation, which they owe to their hydroxyl groups and unsaturated double bonds, is what makes phenolics such good antioxidants (Danilewicz, 2003; Rice-Evans et al., 1997).

Nonflavonoids and flavonoids

Phenolics are produced inside grape berries by several different routes (Fig. 6.6). They form a diverse group from a metabolic standpoint and are usually classified into nonflavonoids and flavonoids. Nonflavonoids accumulate mainly in the mesocarp, whereas flavonoids accumulate mainly in the skin, seeds, and stem. Like other biochemical processes throughout the plant, the location, timing, and extent of the production of the multifarious phenolic compounds is tightly regulated by transcription factors, which function as genetic "switches" and "throttles" that control the activity of the genes involved in the various biosynthetic pathways (Guo et al., 2014; Hichri et al., 2011). In addition to their production in the berries, some flavonoids might also be imported via the phloem from the leaves (Buer et al., 2007; Zhao et al., 2010). The two main "assembly lines" are the shikimate pathway and the malonate pathway. The shikimate pathway, which is named after one of its intermediates, is responsible for the biosynthesis of most plant phenolics, whereas the malonate pathway is less important in plants but is essential in fungi and bacteria. The shikimate pathway operates within chloroplasts and converts simple carbohydrate precursors, namely phosphoenolpyruvate derived from glycolysis and erythrose-4-phosphate derived from the pentose phosphate pathway, into the aromatic amino acid phenylalanine while releasing phosphate (Weaver and Herrmann, 1997). The fact that the systemic broad-spectrum herbicide glyphosate kills plants by blocking a step in this reaction sequence highlights the central importance of the shikimate pathway for plant survival (Holländer and Amrhein, 1980; Powles and Yu, 2010).

The shikimate pathway feeds phenylalanine into the downstream phenylpropanoid pathway that converts the amino acid into cinnamic acid by the deaminating enzyme phenylalanine ammonia lyase (PAL). Sitting at a branch point between protein and phenolic metabolism, PAL exerts a kind of metabolic traffic control by routing phenylalanine away from the manufacture of amino acids and proteins toward the phenolics assembly line. Roughly 30–40% of a plant's fixed carbon flows down the phenylpropanoid pathway, mainly because this pathway also produces the phenolic building blocks of the cell wall structural polymer lignin (Barros et al., 2015; Humphreys and Chapple, 2002). The ammonium released by PAL is continuously recycled via the nitrogen-assimilating GS/GOGAT cycle (see Section 5.3) to regenerate more phenylalanine, according to tissue-specific demands (Weaver and Herrmann, 1997). This recycling mechanism ensures the sustained production of phenolics from an amino acid even in the face of nitrogen deficiency. Additional phenylalanine might be derived from the breakdown of proteins under such conditions (Margna et al., 1989).

The slightly volatile cinnamic acid has a "floral" and "honey" odor and forms part of the flavor of cinnamon. It is further modified and activated by the addition of a coenzyme A thioester called S-CoA for short, which is assembled from ATP and pantothenate (a.k.a. vitamin B_5), to yield 4-coumaryl-CoA. The first step specific to flavonoid biosynthesis and hence to the flavonoid pathway, catalyzed by the enzyme chalcone synthase (CHS), is the condensation of 4-coumaryl-CoA with three molecules of malonyl-CoA, which is derived from acetyl-CoA produced during respiration (Yu and Jez, 2008). The product of this reaction, chalcone, is rapidly isomerized to the flavanone naringenin, which is

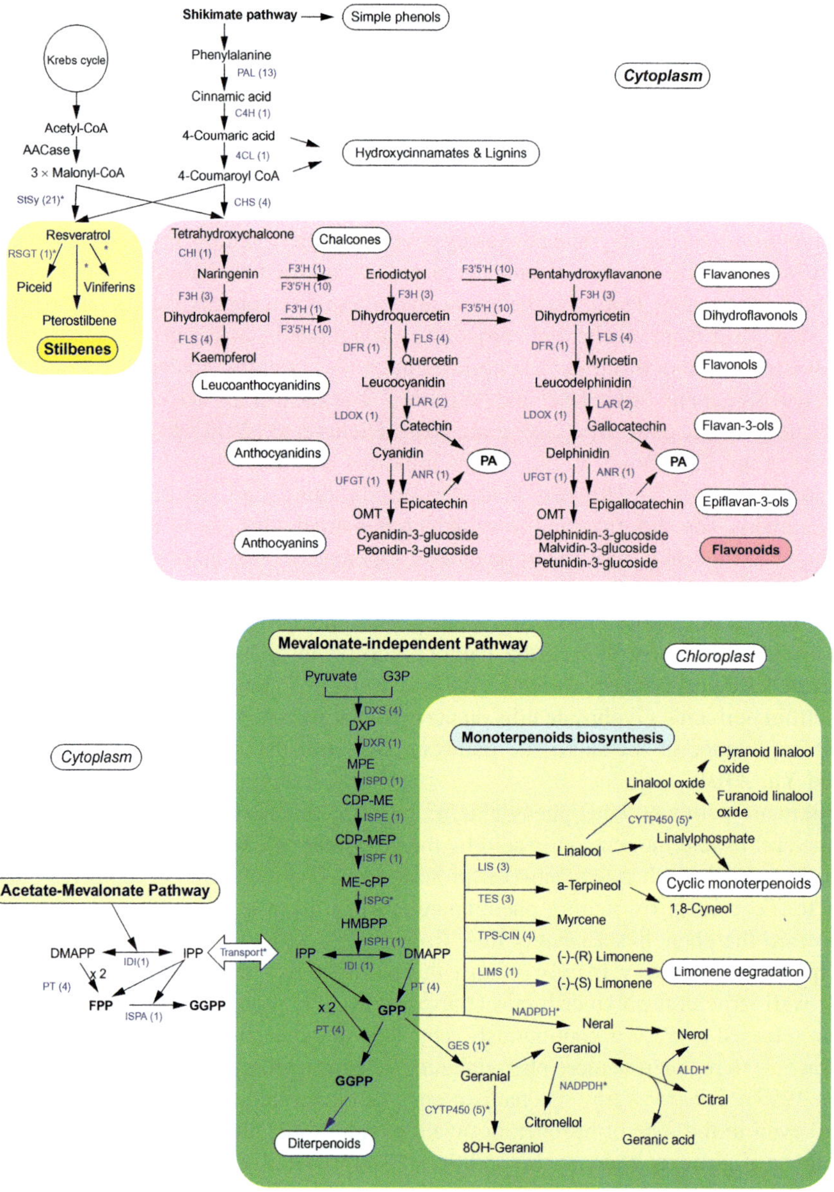

FIG. 6.6

Biosynthetic pathways for the production of phenolics (A) and terpenoids (B) in grapevines.

Reproduced from Velasco, R. Zharkikh, A., Troggio, M., Cartwright, D.A., Cestaro, A., Pruss, D., et al., 2007. A high quality draft consensus sequence of the genome of a heterozygous grapevine variety. PLoS ONE 2, e1326.

the precursor of a large number of flavonoids with a basic chemical structure of C6–C3–C6 (Table 6.1). The flavonoids are the most important and most wide-ranging group of phenolics.

Alternatively, stilbene synthase—using the same substrates with the same stoichiometry as CHS, from which it may have evolved—catalyzes the synthesis of resveratrol (Schröder, 1997; Tropf et al., 1994). Resveratrol can be converted to the various stilbene derivatives, such as the resveratrol-glucoside piceid, pterostilbene, and the polymeric viniferins (Jeandet et al., 2002; Monagas et al., 2006; Romero-Pérez et al., 1999; Schmidlin et al., 2008). These nonflavonoids are produced in the skin in response to injury due to wounding or pathogen infection and are important components of the vine's defense against attacking pathogens (see Section 8.2). Stilbenes have moreover been implicated in numerous beneficial effects of grapes or their products on human health. For instance, besides its antimicrobial properties, at very high doses and by acting as an antioxidant and antimutagen among other activities, resveratrol appears to be capable of reducing heart disease, preventing many types of cancer and memory loss, delaying type II diabetes, and even delaying aging and diseases related to aging (Baur and Sinclair, 2006; German and Walzem, 2000; Guilford and Pezzuto, 2011; Pezzuto, 2008).

In addition to the stilbenes, nonflavonoid phenolics also include the phenolic acids of the two groups hydroxybenzoic acids and hydroxycinnamic acids. High concentrations of tartrate esters of hydroxycinnamic acids, such as caffeic, coumaric, or ferulic acid, are accumulated in the vacuoles of the berry mesocarp and skin during and after bloom, but their concentration often declines during berry development (Monagas et al., 2006; Singleton et al., 1986). They may, in part, be used to manufacture lignin for the berry's vascular bundles and the seed coat (Braidot et al., 2008; Currle et al., 1983; Humphreys and Chapple, 2002). Lignin monomers, such as coumaryl-, coniferyl-, and sinapyl-alcohol, are exuded to the secondary cell walls of the seed coat, where they are activated by the H_2O_2-dependent peroxidase and especially the O_2-dependent laccase. This oxidative activation enables the monomers to polymerize into lignin, which provides mechanical strength and gas- and water-proofing to the seed coat (Barros et al., 2015). Some ferulic acid also becomes linked to the glycerol/fatty acid polyester suberin (Franke and Schreiber, 2007). Suberin, in addition to wax, plugs the berry's stomata before ripening begins—suberin is also the compound that seals wine bottles in the form of cork derived from the bark of the cork oak, *Quercus suber* L., after which the chemical is named.

Hydroxycinnamic acids are the major phenolic constituents of free-run grape juice and white wines; at high concentrations they may impart a slightly astringent mouthfeel (Hufnagel and Hofmann, 2008a,b; Ong and Nagel, 1978; Sáenz-Navajas et al., 2010). Free hydroxycinnamic acids may also contribute to the heart disease-decreasing properties of grapes and grape products, but their esterification with ethanol may render hydroxycinnamates somewhat bitter (Hufnagel and Hofmann, 2008a,b). However, their main enological importance may lie in the fact that *Brettanomyces* and some other yeasts can convert them during fermentation to volatile phenols such as ethyl or vinyl guaiacol and eugenol, which are odor active and at low concentrations smell "smoky," "woody," "leathery," or "peppery" (Chatonnet et al., 1992; Rapp and Versini, 1996). The common esterification with tartrate, however, seems to protect hydroxycinnamates from this unfavorable conversion by *Brettanomyces* (Schopp et al., 2013).

Guaiacol ("smoky"), eugenol ("clove"), and related compounds are also extracted into wine from toasted oak barrels as oxidative breakdown products of lignin, and eugenol and other phenylpropanoids are important aroma components of basil leaves and of wines made from grapes of *Vitis cinerea* and, to a lesser extent, *Vitis riparia* (Chatonnet and Dubourdieu, 1998; Maga, 1989; Sun et al., 2011b). At higher concentrations, the odor of volatile phenols becomes unpleasantly "pharmaceutical," "medicinal," or "burnt" at the expense of varietal fruit aroma. This poses a problem not only following contamination with *Brettanomyces* during the winemaking process but also when grapes have been exposed during ripening to smoke from wildfires. Smoke-derived phenols can be absorbed by grape berries and bound to various sugar molecules, perhaps in the cell walls, which leads to accumulation of glycosylated phenols in the skin and mesocarp. These phenols may subsequently be released, making them volatile, during alcoholic and malolactic fermentation and wine storage (Hayasaka et al., 2010; Kennison et al., 2008). Just like in toasted barrels, the volatile phenols in the wildfire smoke originate from the thermal degradation of lignin (Edye and Richards, 1991; Fine et al., 2001). Grapes appear to be most sensitive to the adverse influence of smoke exposure during the initial stages of fruit ripening (Kennison et al., 2009, 2011).

Rapid oxidation of hydroxycinnamic acids during grape drying or processing results in browning of both raisins and grape juice, including the grapes used for the sweet, fortified Spanish dessert wine Pedro Ximenez (Sapis et al., 1983; Serratosa et al., 2008; Singleton et al., 1985). Such enzymatic oxidation becomes possible when mechanical damage before or after harvest or bunch rot infection physically disrupts berry cell compartmentation, which brings the phenolics into contact with oxygen and two groups of enzymes termed polyphenoloxidases and peroxidases. These enzymes convert phenolics to quinones, which can then polymerize with other phenols to form brown pigments (Hernández et al., 2009; Macheix et al., 1991; Pourcel et al., 2007; Singleton, 1992). This reaction is particularly dominant if the grapes contain plenty of hydroxycinnamic acids and little glutathione (Greek *thion* = sulfur) that would otherwise protect the phenolics, or at least the colorless quinones, from oxidation by binding to them (Kritzinger et al., 2013; Monagas et al., 2006).

Many other phenolics are glycosylated—that is, they are attached to a sugar molecule—to increase their solubility, prevent their free diffusion across membranes, and decrease their reactivity toward other cellular components, which permits accumulation and storage in the cell vacuoles (Bowles et al., 2006; Rice-Evans et al., 1997). Glycosylation also decreases a phenol's antioxidant activity. Glucose, following its activation by uridine diphosphate to the so-called UDP-glucose, is usually the sugar of choice for attachment by a family of enzymes called glycosyltransferases to one of the phenol's oxygen atoms—hence the term *O*-glucoside. These phenolics include flavonoids such as anthocyanins and flavonols (Table 6.1). Like other phenolics, flavonoids have also been implicated in the protective activity of grapes and their products against heart disease among other health benefits. Indeed, they have important pharmacological properties, reducing oxidative stress by acting as antioxidants that are more powerful than ascorbate and α-tocopherol, which are colloquially known as vitamins C and E, respectively (Rice-Evans et al., 1997). This is one of the supposed main reasons why eating grapes and drinking wine and grape juice may be good for our health (Dixon et al., 2005; German and Walzem, 2000; Guilford and Pezzuto, 2011). Nonetheless, human physiologists have noted that our bodies normally balance reactive oxygen species (see Section 7.1) and antioxidants so carefully that consuming antioxidants may not decrease oxidative damage (Halliwell, 2006, 2007).

Tannins

Among enologically important phenolic compounds, flavonoids called tannins are the most abundant class; the lignin formed in the seed coat from cinnamic acids is enologically unimportant because it is not extracted during fermentation. Tannins are divided into two classes: hydrolysable and condensed (nonhydrolysable) tannins. The major building blocks or subunits of the hydrolysable tannins are gallic acid and ellagic acid, with small amounts of hydroxycinnamic acids, bound to glucose. The much longer condensed tannins are composed of flavan-3-ols, or flavanols, such as catechin and epicatechin, both of which taste bitter when they occur as monomers (Dixon et al., 2005). Grape berries contain only condensed tannins; the hydrolysable tannins present in some wines are extracted from wood, for example, during barrel aging (Puech et al., 1999). Condensed tannins are also called proanthocyanidins because they break down to anthocyanidins when heated in acids. Indeed, grape berries manufacture the flavanol tannin precursors via a branch in the anthocyanin-forming pathway by enzymes called leucoanthocyanidin reductase and anthocyanidin reductase. The former synthesizes catechin from the cyanidin precursor leucocyanidin, and the latter produces epicatechin directly from cyanidin and epigallocatechin from delphinidin (Bogs et al., 2005; Dixon et al., 2005; Terrier et al., 2009; Xie et al., 2003). In other words, the same anthocyanidins can be converted either to tannins, by reduction and subsequent polymerization, or to anthocyanins, by glucose addition.

Upon extraction from grapes, tannins can polymerize with anthocyanins and among themselves and thus are important for color stability in grape juice and wine (Jackson, 2008; Zimman and Waterhouse, 2004). Mostly, however, they are responsible for the bitter taste and the astringent tactile sensation of grapes and wines because they bind to and precipitate proteins that are much larger than the tannins (Gawel, 1998; Robichaud and Noble, 1990). Precipitation of salivary proteins increases the friction of the tongue gliding over the inner surfaces of the mouth, resulting in a drying, rough feeling—a property that gave rise to the terms mouthfeel and texture. In fact, the name tannin derives from these molecules' ability to tan leather—that is, to convert animal hide into leather by interacting with the hide's proteins. The perceived quality of red wines tends to increase as their tannin content increases, but tannins are undesirable in most white wines (Mercurio et al., 2010; Sáenz-Navajas et al., 2010).

The colorless tannins are assembled from simple flavanol building blocks in the chloroplasts, wrapped in outgrowths of the chloroplast membrane, and "shuttled" to the vacuoles and cell walls of the seed coat and, to a lesser extent, the skin cells of the berries of both red and white cultivars (Amrani Joutei et al., 1994; Braidot et al., 2008; Brillouet et al., 2013; Gagné et al., 2006; Gény et al., 2003; Joslyn and Dittmar, 1967; Souquet et al., 1996; Thorngate and Singleton, 1994). Additionally, low amounts of tannins may accumulate in the mesocarp (Terrier et al., 2009). Although tannins and their precursors are also found in the rachis tissues, this source normally contributes only a minor portion of the tannins extracted during winemaking because most grapes are destemmed prior to fermentation (Souquet et al., 2000; Sun et al., 1999). Sealing off the forming tannins within a membrane at their production site in the chloroplasts ensures that they do not precipitate—and thereby denature—enzymes on their way to and inside the vacuoles and prevents them from reacting with other vacuolar components in intact berries (Brillouet et al., 2013).

Tannins are the true polyphenols of grapes; by definition they consist of at least two flavonoid subunits and can be composed of long, often branched, chains of almost identical subunits, varying in length over two orders of magnitude. Tannins with two to four subunits are usually classified as oligomers, and tannins with more than four units are classified as polymers. Though catechin is the major flavan-3-ol in grape skins, epicatechin and, to a lesser extent, epigallocatechin provide most of the extension subunits for skin tannins. Epicatechin dominates in the seeds, followed by smaller portions of epicatechin gallate (Bogs et al., 2005; Fournand et al., 2006; Hanlin and Downey, 2009; Hanlin et al., 2011; Prieur et al., 1994; Souquet et al., 1996). In the seeds, but generally not in the skin, catechin and epicatechin are often esterified with gallic acid; this so-called galloylation increases their astringency or "coarseness" (Brossaud et al., 2001; Dixon et al., 2005; Vidal et al., 2003). Epicatechin gallate is usually only a minor tannin component in the skin and, in a twist of confusing terminology, epigallocatechin is normally absent in the seed

(Downey et al., 2003a; Gagné et al., 2006; Hanlin et al., 2010; Mattivi et al., 2009). The same tannin subunits are present in the seeds of diverse *Vitis* species, albeit in somewhat variable proportions (Liang et al., 2012a).

The shortest of these flavonoid polymers taste bitter but are usually confined to the seeds. Skin tannins tend to be longer (4 to >100 subunits) than seed tannins (2–20 subunits) and are thought to provide a better mouthfeel (Hanlin et al., 2011; Kennedy et al., 2001; Monagas et al., 2006; Souquet et al., 1996). Overall, bitterness decreases while astringency increases with increasing degree of polymerization—that is, with increasing molecule size or chain length (Cheynier et al., 2006). The perceived astringency also increases as the tannin concentration rises and varies noticeably depending on the relative composition of the various subunits. Higher proportions of epicatechin extension subunits and of gallocatechin in terminal positions are believed to enhance astringency, whereas higher proportions of epigallocatechin in both extension and terminal positions may decrease astringency (Quijada-Morín et al., 2012). Nevertheless, although the total amount of tannins present in a wine is certainly related to the wine's astringency, small and mostly unknown chemical changes, especially during wine aging, may account for the perception of tannins as "fine," "soft," "green," or "harsh." For instance, the incorporation of galloylated procyanidins extracted from grape seeds seems to increase astringency (Brossaud et al., 2001). The perception of astringency, but not bitterness, also increases with increasing acidity, whereas ethanol has the opposite effect.

Although the amount of tannins contained in the seeds is typically several times higher than that in the berry skin, grape cultivars differ in the amount of both seed and skin tannins they produce. For example, the seeds of Cabernet Sauvignon, Merlot, and Syrah seem to contain much lower tannin concentrations than those of Cabernet franc, Pinot noir, Grenache, or Tempranillo. On the other hand, and perhaps surprisingly, not only Tempranillo and Nebbiolo but also many table grape cultivars (e.g., Red Globe, Flame Seedless, and Ruby Seedless) may have far higher skin tannin concentrations than Cabernet Sauvignon, Merlot, Syrah, Pinot, and even Tannat, which are all relatively similar. Barbera and Malbec skins, along with those of other table grapes (e.g., Muscat Hamburg), are at the low end of the spectrum. Although most of them lack the ability to produce anthocyanins, white grapes accumulate the same amount of tannins as their dark-skinned siblings (De Freitas and Glories, 1999; Rodríguez Montealegre et al., 2006; Viala and Vermorel, 1901–1909).

Because the genes responsible for the biosynthesis of tannin subunits in seeds are switched on soon after fertilization, seed tannins accumulate from fruit set through veraison or soon thereafter, and then their concentration, or at least their extractability, declines (Bogs et al., 2005; Dixon et al., 2005; Downey et al., 2006; Kennedy et al., 2000). Seed tannins also become increasingly polymerized, and thus less bitter, during berry ripening and seed desiccation. Polymerization is accompanied by oxidation, which is responsible for the browning of the seed coat after veraison (Adams, 2006; Braidot et al., 2008; Kennedy et al., 2000; Ristic and Iland, 2005). Oxidized tannins bind strongly to cell walls, providing some "stress-proofing" to the seed coat and limiting the extractability of seed tannins (Cadot et al., 2006; Gény et al., 2003; Pourcel et al., 2007). The suberized cuticle of the seed coat further limits tannin extraction during winemaking. Consequently, the amount of seed tannins per berry and the amount extracted into wine depend mainly on the number of seeds, rather than the amount per seed, and on fruit maturity but are relatively unaffected by environmental factors.

The tannin subunits in the skin are mostly produced and assembled in the developing flowers and young berries up to, or somewhat before, veraison (Adams, 2006; Bogs et al., 2005; Downey et al., 2003a; Hanlin and Downey, 2009; Harbertson et al., 2002; Kennedy et al., 2001). This means that the production of flavanols from which tannins are assembled occurs during the warmest part of the growing season and before anthocyanin accumulation begins, a developmental "switch" that grapes share with strawberries (Fait et al., 2008). Indeed, anthocyanins cannot be produced while the production of tannin precursors is in full swing, because the competing flavanol-forming enzymes suppress the attachment of glucose to anthocyanidins (Fischer et al., 2014). It is this competition among biosynthetic enzymes that makes flavonoid production a biphasic process: flavanols for tannins before veraison, and anthocyanins for red pigmentation beginning at veraison. As a result, the concentration, though not the absolute amount, of skin tannins declines during the postveraison berry expansion phase. At the same time tannin polymerization, which is thought to be an antioxidant mechanism, increases at veraison and may or may not continue through late ripening (Bachmann and Blaich, 1979; Bindon and Kennedy, 2011; Hernández et al., 2009; Kennedy et al., 2001). In addition, a portion of the anthocyanins, or of their nonglycosylated anthocyanidin precursors, produced from veraison onwards might be incorporated into tannins. Such changes may account for most of any apparent tannin accumulation in the skin after veraison and for the decrease in free anthocyanins later in the ripening phase (Fournand et al., 2006).

With structural differences, such as different chain length, charge, and stereochemistry, come differences in functional properties among tannins and hence in their reactivity toward other molecules (Kraus et al., 2003). This may be one reason why a considerable fraction of the skin tannins can be bound to the insoluble matrix, consisting of cell wall pectins and glycans, of grape berries (Amrani Joutei et al., 1994; Kennedy et al., 2001; Lecas and Brillouet, 1994; Pinelo et al., 2006). Binding to cell walls reduces tannin extractability into wine. The cell walls' binding capacity seems to increase strongly during early grape ripening and then to decrease as the fruit approaches maturity and cell walls are dismantled and pectins degraded (Robertson et al., 1980). However, while some studies suggested a general decrease in tannin and anthocyanin extractability during grape ripening, others found no change in extractability, and still others found an increase in extractability (Bindon et al., 2013; Downey et al., 2006; Fournand et al., 2006). Nonetheless, longer-chain—and hence more astringent—tannins, it seems, are not as readily extracted as their shorter-chain counterparts (Fournand et al., 2006). The larger tannins may be increasingly bound to de-esterified cell wall pectins as berry ripening progresses (Bindon et al., 2012). Moreover, when grapes are crushed for winemaking, tannins may also bind to, and precipitate with, proteins extracted from the mesocarp cells (Bindon et al., 2016). It would be surprising if variations in functional properties among the tannins of different cultivars or arising during fruit ripening did not account for at least some of the differences observed in tannin extractability, subsequent reactions during winemaking and, ultimately, astringency.

Depending on grape maturity and binding capacity, between 25% and 75% of the berry tannin may be extracted during winemaking, 50–80% of which is derived from the skins (Cerpa-Calderón and Kennedy, 2008). Such a wide range of extractability means that the tannin content of a wine often cannot be predicted from the tannin content of the grapes. Indeed, the tannin concentration in red wines of the same cultivar varies more than two orders of magnitude, although the variation in skin tannins is within less than one order of magnitude (Harbertson et al., 2002, 2008; Ristic et al., 2010; Seddon and Downey, 2008). Because an increase in temperature weakens the association between tannins and cell walls (Hanlin et al., 2010), it is possible that high fermentation temperatures lead to more complete tannin extraction than do low fermentation temperatures.

Due to this incomplete extraction, which applies to other phenolic compounds, organic acids, and other ingredients as well, grape pomace of both red and white cultivars contains considerable amounts of nutritionally valuable organic and inorganic compounds (Arvanitoyannis et al., 2006; González-Centeno et al., 2013; Kammerer et al., 2004; Lu and Foo, 1999; Mazza and Miniati, 1993; Monagas et al., 2006). These chemicals can be further extracted by distillation to produce brandy, such as marc or grappa, and by pressing to produce grape seed oil. Also, they can be used to produce dietary supplements or phytochemicals, as well as for composting or direct recycling back to the vineyard.

Anthocyanins

Anthocyanins (Greek *anthos* = flower, *kyanos* = blue) are the second class of phenolics of major sensory and enological importance after the tannins. They are responsible for the red, purple, or blue coloration of dark-skinned grapes and of grape juice and wine produced when such skins are extracted. Anthocyanin accumulation in grape berries starts at veraison. It is probably triggered by the increasing sugar concentration in the berry skin and mediated by calcium signals; the threshold for switching on the genes involved in anthocyanin production seems to be approximately 9–10°Brix (Castellarin et al., 2007a,b; Keller and Hrazdina, 1998; Larronde et al., 1998; Lecourieux et al., 2014; Pirie and Mullins, 1977; Vitrac et al., 2000; Fig. 6.7). In other words, accumulation of anthocyanins begins *after* the onset of sugar accumulation and berry softening and after the production of the flavanol tannin subunits in the skin has ceased (Bogs et al., 2005; Castellarin et al., 2011, 2016; Fournand et al., 2006; Pirie and Mullins, 1980; Sadras and Moran, 2012).

Although it is commonly used to describe the beginning of grape ripening, the term veraison refers to the color change of a population of berries in a fruit cluster or vineyard rather than to an individual berry. At veraison, each cluster can have berries ranging from green to pink to red to purple to blue, often spanning a two- to fourfold range in sugar and acid concentration (Keller and Shrestha, 2014; Zhang and Keller, 2017; Fig. 6.7). Moreover, the green berries can be hard to the touch or at varying stages of softness, whereas all other berries are soft. This implies that the different berries of a cluster are at different developmental stages and that the profound changes in the expression of ripening-related genes, whereby many genes are being "switched on" while many others are being "switched off," occur rapidly but asynchronously in these berries.

FIG. 6.7

Change in the pigmentation of Merlot (left and center) and Pinot noir (right) grape skins due to anthocyanin accumulation with increasing sugar concentration at veraison

Biondi and Keller, unpublished data; photos by M. Keller.

Anthocyanins are generally confined to the vacuoles of the outer hypodermis, except in teinturier (French *teinturier* = dyer) cultivars whose mesocarp is red, and in overripe grapes with senescing skin cells that leak vacuolar components into the mesocarp (Fontes et al., 2011b; Moskowitz and Hrazdina, 1981; Viala and Vermorel, 1901–1909). But even in teinturier grapes, such as Alicante Bouschet (synonym Garnacha Tintorera), Gamay Fréaux, Dunkelfelder, or Rubired, the anthocyanin content in the skin is severalfold higher than that in the mesocarp (Falginella et al., 2012). Storage inside the acidic cell vacuoles protects anthocyanins from oxidation and ensures that they function as red pigments; at the neutral pH of the cytosol they would be mostly colorless or blue and unstable. More accurately, however, anthocyanins accumulate mostly within special, membrane-bound structures or vesicles named anthocyanic vacuolar inclusions (formerly called anthocyanoplasts) that trap and stabilize the pigments and may also contain long-chain tannins (Conn et al., 2003, 2010; Falginella et al., 2012; Markham et al., 2000). These vesicles originate from the endoplasmic reticulum, on whose surface the anthocyanins are produced, and may serve as transport vessels that carry the pigments to their final destination in the vacuoles with which the vesicles fuse (Gomez et al., 2011). Both the total amount of anthocyanins and their relative composition vary among *Vitis* species and cultivars.

As a rule, anthocyanidins of *V. vinifera* cultivars are bound to a single glucose molecule turning them into 3-glucosides. Only traces of anthocyanidins are sometimes bound to two glucose molecules, which makes them 3,5-diglucosides (Liang et al., 2011; Xing et al., 2015). Most other *Vitis* species, whether from Asia or North America, as well as the *Muscadinia* species, contain considerable amounts of 3,5-diglucosides in addition to the 3-glucosides (Acevedo De la Cruz et al., 2012; Dai et al., 2011; Hrazdina, 1975; Liang et al., 2012b, 2013; Mazza and Miniati, 1993; Ribéreau-Gayon and Ribéreau-Gayon, 1958). Interspecific hybrid cultivars, such as Kyoho, Catawba, or Concord, contain both 3-glucosides and 3,5-diglucosides as well. The inability of most *V. vinifera* cultivars to attach a second glucose to anthocyanidins arises from two mutations that disrupt the gene prescribing the necessary 5-*O*-glucosyltransferase enzyme (Jánváry et al., 2009; Yang et al., 2014).

The glucose addition stabilizes the metabolically unstable anthocyanidins, enhances their solubility, and turns them into anthocyanins; grapes cannot accumulate anthocyanidins (Bowles et al., 2006; He et al., 2010b). The glucose, in turn, can bind to acetic, coumaric, or caffeic acid to form acylated anthocyanins, but these may be less stable. Thus, the stability in solutions, such as grape juice or wine, is highest for the 3,5-diglucosides and decreases for the 3-glucosides, then the acylated 3,5-diglucosides, and, finally, the acylated 3-glucosides. In addition, anthocyanin stability is also influenced by the type and number of functional groups attached to the phenolic carbon skeleton: More methoxy groups (-OCH$_3$) enhance stability, whereas more hydroxyl groups (-OH) reduce stability. Functional groups, moreover, are important for the color of anthocyanins: Addition of a hydroxyl group at the so-called R_1 position (Table 6.1) or addition of methoxy groups shifts the color toward orange-red, whereas hydroxyl groups at both the R_1 and R_2 positions turn anthocyanins blue (He et al., 2010b; Jackson, 2008; Lillo et al., 2008; Tanaka et al., 2008). The genes encoding the enzymes that are responsible for adding these hydroxyl groups, the flavonoid hydroxylases, therefore determine the color of anthocyanins. Due to the variation in functional groups, glucosylation, and acylation, the dark-skinned *V. vinifera* cultivars typically have between 5 and 20 anthocyanins, while the number may approach 40 in other *Vitis* species that accumulate diglucosides in addition to the monoglucosides (Acevedo De la Cruz et al., 2012; He et al., 2010a,b; Liang et al., 2012b; Mazza and Miniati, 1993). Moreover, like the tannin-forming flavanols, two or three anthocyanin molecules can also bond together to form oligomeric pigments in grape berries (Vidal et al., 2004). However, there is no close relationship between the amounts of anthocyanins and tannins in the berry skin.

Anthocyanins are mostly glucosides of the anthocyanidins malvidin, peonidin, delphinidin, petunidin, and cyanidin, whereas most flavonols are glycosides of quercetin, kämpferol, myricetin, isorhamnetin, syringetin, and laricitrin. Although grapes produce at most traces of pelargonidin-glucoside (He et al., 2010a,b), which is the principal anthocyanin of strawberries, the major flavonols are the same in the two species. The two independently produced "starter" compounds of the anthocyanin-specific metabolic pathway are cyanidin and delphinidin (Fig. 6.6). Once glucosylated, cyanidin-3-glucoside can be methoxylated to become peonidin-3-glucoside, while methoxylation turns delphinidin-3-glucoside into either petunidin-3-glucoside or malvidin-3-glucoside (Boss and Davies, 2009). In dark-skinned grapes, the pigments based on malvidin are normally more abundant than those based on other anthocyanidins, and this dominance tends to increase during fruit ripening (Keller and Hrazdina, 1998; Liang et al., 2011; Pomar et al., 2005; Wenzel et al., 1987). Nonetheless, the relative proportion of the different anthocyanins varies by species and cultivar (Acevedo De la Cruz et al., 2012; Cantos et al., 2002; Carreño et al., 1997; Dai et al., 2011; Mattivi et al., 2006; Mazza and Miniati, 1993; Pomar et al., 2005). For instance, malvidin derivatives dominate in the berry skin of Malbec, Pinot noir, Syrah, Tempranillo, and many red table grapes; peonidin derivatives hold the majority in Nebbiolo; cyanidin and/or peonidin derivatives prevail in Pinotage and some table grapes, as well as in the berry flesh of Alicante Bouschet; delphinidin and cyanidin derivatives are dominant in Concord; whereas the distribution is comparatively balanced in Merlot, Mourvèdre, and Sangiovese. Moreover, Pinot noir and some *Muscadinia* varieties lack the acylated forms of anthocyanins that most other cultivars contain. At least in Pinot noir a mutation in the gene that codes for the required acyltransferase renders this gene nonfunctional (Rinaldo et al., 2015). Even clones of a cultivar may vary somewhat in their relative anthocyanin composition (Muñoz et al., 2014; Revilla et al., 2009).

The berries of some dark-skinned cultivars (e.g., Syrah, Durif, Petit Verdot, and Tempranillo) accumulate anthocyanin pigments to much higher concentrations than those of other cultivars (e.g., Malbec, Pinot noir, Nebbiolo, and many table grapes), with still others (e.g., Cabernet Sauvignon, Merlot, and Barbera) being intermediate (Ferrandino et al., 2012; Mattivi et al., 2006). Anthocyanins are also partly responsible for the somewhat reddish or bluish berry skin of cultivars that are not considered to be dark-skinned. For example, many Muscat cultivars, as well as Pinot gris and Gewürztraminer berries accumulate small amounts of anthocyanins. The anthocyanin profile of Pinot gris is very similar to that of Pinot noir, albeit on a much lower level (Castellarin and Di Gaspero, 2007; He et al., 2010b). Furthermore, it is possible that the bronzelike color of Pinot gris and Gewürztraminer arises from an interaction between anthocyanins and carotenoids (Forkmann, 1991; Liang et al., 2011). Among the Muscats, cyanidin-glucoside often dominates in the more lightly colored cultivars—as it does in Gewürztraminer, whereas malvidin- and peonidin-based pigments tend to be more abundant in the blue/black-skinned cultivars (Cravero et al., 1994).

Anthocyanins are probably involved in photoprotection by complementing the spectral absorption profile of chlorophylls. The skin of red cultivars absorbs much more visible and, especially, ultraviolet (UV) light than the skin of white cultivars (Blanke, 1990a). A dark (red, purple, blue, or black) berry skin is the "default" version for grapes of all *Vitis* species (Cadle-Davidson and Owens, 2008; This et al., 2007; Viala and Vermorel, 1901–1909). Most so-called white grapes, whose skin is green or yellow, have evolved from a dark-skinned relative by two mutations that prevent the activation of the gene prescribing glucosyl transferase, the enzyme that attaches a glucose molecule to the anthocyanidins (Boss and Davies, 2009; Kobayashi et al., 2004; Walker et al., 2007). The same skin color mutation is also responsible for the conversion of Pinot noir to Pinot blanc. The inability to attach glucose probably impedes the entire branch of the phenylpropanoid and anthocyanin-producing pathways through feedback inhibition, which prevents the accumulation of hydrophobic and possibly toxic anthocyanidins (Yin et al., 2012).

Crushing the grapes during or after harvest brings anthocyanins in contact with other juice components, such as tannins and organic acids, with which they react to form polymeric pigments (Monagas et al., 2006; Wollmann and Hofmann, 2013; Zimman and Waterhouse, 2004). Between 50% and 90% of the total anthocyanins present in the skins are extracted during fermentation. The large variation in anthocyanin extraction is partly caused by differences in fruit maturity (extractability may decline as grapes mature) and amount and composition of cell wall material (e.g., cellulose content and degree of pectin methylation), which are themselves influenced by maturity and probably sun exposure (McLeod et al., 2008; Ortega-Regules et al., 2006b). It is also partly due to differences in skin contact time, fermentation temperature, and additions such as macerating enzymes. Despite the incomplete extraction during winemaking, the variation in wine color is preset by the composition of the grapes at harvest, and winemaking practices can do little to change this variation.

All phenolic compounds extracted into wine are subject to a variety of enzymatic and chemical reactions that modify wine color and decrease astringency during aging (Cheynier et al., 2006; Fulcrand et al., 2006; Jackson, 2008; Monagas et al., 2006; Ribéreau-Gayon, 1973; Somers, 1971; Waterhouse and Laurie, 2006). Enzymatic oxidation processes dominate during the early stages of aging, while chemical processes become increasingly dominant during later stages. The type and quantities of the different pigments influence the intensity, hue, and stability of wine color. Moreover, in aqueous solutions anthocyanins come in five main molecular forms; they coexist in a dynamic equilibrium, but most of them are actually colorless at juice or wine pH (Jackson, 2008). Above pH 3, <30% of an anthocyanin exists as the red-colored flavylium cation, and its proportion declines further with increasing pH to almost nothing at pH 4. Even more important, sulfur dioxide (SO_2), used in winemaking to control microbial growth, effectively bleaches anthocyanins. Consequently, the colorless forms account for 75–95% of the anthocyanins in wine.

Flavonols

Flavonols are present in grapes at much lower concentrations than tannins and anthocyanins. They are produced in the epidermis throughout flower and berry development and act as "sunscreen," protecting the berry from harmful UVB radiation (Downey et al., 2003b; Keller and Hrazdina, 1998; Keller et al., 2003b; Kolb et al., 2003; Ribéreau-Gayon, 1964). Consequently, sun-exposed grapes contain several-fold higher amounts of flavonols than shaded grapes (Downey et al., 2004; Liu et al., 2015; Price et al., 1995). Even within the same berry, the side facing the sun can have a much higher skin flavonol concentration than the side facing away from the sun (Lenk et al., 2007). In addition to the skin, the seed coat also contains flavonols (Liang et al., 2012a).

Flavonol glycosides may contribute some astringency to wines (Hufnagel and Hofmann, 2008a,b; Sáenz-Navajas et al., 2010). Grape cultivars with dark-skinned fruit contain flavonols based on quercetin and myricetin, with small amounts of kämpferol, laricitrin, isorhamnetin, and syringetin (Table 6.1). White grapes, however, apparently produce only quercetin, kämpferol, and traces of isorhamnetin, which indicates that an entire branch of the flavonol pathway is switched off in white grapes (Ferrandino et al., 2012; Mattivi et al., 2006; see also Fig. 6.6). Despite the dominance of malvidin-based compounds in the anthocyanin profile of many dark-skinned cultivars, the structurally corresponding flavonol, syringetin, usually occurs only in trace amounts. Instead, and although some cultivars accumulate more myricetin than quercetin, quercetin derivatives typically dominate the flavonol profile, whereas the "matching" cyanidin often contributes little to the anthocyanin pool (Adams, 2006; Cheynier and Rigaud, 1986; Downey and Rochfort, 2008; Liang et al., 2011, 2012b; Mattivi et al., 2006). However, cyanidin is the preferred substrate for anthocyanidin reductase and, thus, for tannin formation (Dixon et al., 2005).

Once extracted from grapes into juice or wine, flavonols can be partially hydrolyzed, whereby their glucose "tail" breaks away (Monagas et al., 2006). The resulting flavonol aglycones, together with cinnamic acids and catechin, can serve as the so-called cofactors, which stack together with anthocyanins in wine. This short-lived association of anthocyanin pigments with colorless phenolics is termed copigmentation. Copigmentation stabilizes color by protecting the relatively unstable anthocyanins from SO_2 bleaching and high pH while shifting the color toward purple or blue and intensifying it (Scheffeldt and Hrazdina, 1978; Tanaka et al., 2008). It is responsible for the temporary purpleness and color stability of young red wines and is said to account for 30–50% of their color (Boulton, 2001). Indeed, the color of a young wine seems to be limited more by cofactors than by anthocyanins. Moreover, anthocyanins can also join together with other anthocyanins in the so-called self-association, which also stabilizes color. The reaction of anthocyanins with acetaldehyde or pyruvate extracted from grapes or produced during fermentation to form orange-red pyranoanthocyanins confers additional stability. Yet red wines owe most of the gradual change in their color from red-purple to tawny during aging to conversion of anthocyanins to polymeric pigments that form by condensation between anthocyanins and tannins and other components (Bindon et al., 2013; Monagas et al., 2006; Ribéreau-Gayon, 1973; Somers, 1968). However, polymerization does not always increase color stability. Unlike young red wine color, therefore, aged red wine color is determined by the grapes' anthocyanin and tannin composition, and both of these phenolic classes can limit wine color. In other words, if grapes are rich in anthocyanins but poor in tannins, wine color will be no better than if the reverse is the case. Tannins also polymerize with other tannins, and there is no upper limit to this polymerization (Jackson, 2008; Wollmann and Hofmann, 2013). Even long polymeric tannins, as long as 70 subunits, remain in solution; tannins precipitate only when they bind to proteins in wine. Moreover, polymerization is a dynamic process; polymeric tannins are susceptible to cleavage so that their degree of polymerization can increase or decrease during wine aging.

The grapes' anthocyanin, flavonol, and tannin profile determines the color potential of the wine made from these grapes, and color is a main driver of perceived red wine quality. Color hue and intensity are obviously important for quality in their own right, and they also influence the subjective perception of quality in other ways. Intriguingly, people associate certain flavors with specific colors, and our perception of taste and/or flavor intensity and complexity tends to increase as the color intensity increases. In other words, if two otherwise identical grape juices or wines differ in color, the more deeply colored one is perceived as being more complex and as having more "body" and more dark-fruit aromas (Delwiche, 2003; Morrot et al., 2001). In the brain, it is clearly all about perception. In contrast, phenolic compounds are far less important for white grape and wine quality, which is determined mainly by the grapes' flavor and aroma potential. In white wines, phenolics may even contribute to undesirable astringency and bitterness—traits that are particularly unattractive in sparkling wines. Because the phenolics are concentrated in the berry skin and the seeds, skin contact is usually minimized in white winemaking to limit their extraction (Singleton and Trousdale, 1983).

6.2.7 Lipids and volatiles

Lipids in grape berries consist primarily of cuticular waxes, fatty acids, membrane lipids, and seed oils. They are present throughout berry development and ripening. Fatty acids produced by epidermal cells supply the biosynthetic precursors for the cuticle's cutin as well as for its waxes (Lara et al., 2015). Isoprene is another important precursor for cuticular wax, namely for the triterpenoid oleanolic acid, which may be the dominant lipid in the cuticular wax of grape berries despite being an ineffective barrier to water permeability (Radler, 1965a,b, 1970). Contrary to widespread belief, plant membranes also contain cholesterol in addition to many other sterols and their derivatives, called steroids or phytosterols (Fujioka and Yokota, 2003). However, the contribution of cholesterol to the total lipids is about two orders of magnitude less than in animals. For example, cholesterol and other sterols are common, albeit minor, constituents of fruit cuticular waxes (Leide et al., 2007; Pensec et al., 2014). The polyunsaturated fatty acid linoleic acid accounts for up to half of the fatty acids in the berry skin and mesocarp, followed by the saturated palmitic acid, the polyunsaturated linolenic acid, and the monounsaturated oleic acid characteristic of olives (Roufet et al., 1987). In most grape cultivars, the concentration of these lipids in the skin is up to three times higher than in the mesocarp and changes little during fruit ripening. Similarly, up to 17% of the fresh weight of grape seeds consists of linoleic acid and other polyunsaturated fatty acids in addition to oleic, palmitic and stearic acids (Currle et al., 1983; Miele et al., 1993; Rubio et al., 2009).

When seeds and skins are damaged during the postharvest destemming and crushing process, the ensuing catabolism of fatty acids involves the formation of so-called C6 compounds with a "fresh green," "grassy," or "herbaceous" aroma. For instance, linoleic acid may be oxidized to hexanal, and linolenic acid to hexenal. These C6 compounds are also called "green leaf" volatiles because they are produced by leaves as a generic wound response and because of their characteristic smell of freshly cut grass (Dudareva et al., 2013; Rodríguez et al., 2013). They are typical aroma components of tomatoes and strawberries and contribute to the aroma of unripe grapes, including table grapes harvested at low sugar concentration (Maoz et al., 2018). Such C6 compounds are also present in the rachis in addition to the berries, but their extraction during fermentation is very limited (Hashizume and Samuta, 1997; Schwab et al., 2008). Alcohol dehydrogenase enzymes, some of which are particularly active after veraison, might convert some of the C6 aldehydes to their more "fruity" alcohol derivatives; for instance, hexanal can be converted to hexanol during grape ripening (Kalua and Boss, 2009; Tesniere et al., 2006). During fermentation, yeast may further convert these C6 compounds to hexyl acetate with an odor of "fruit," "apples," or "herb" (Dunlevy et al., 2009). Yeast also converts lipids to esters and acetates, which impart an overall "sweet-fruity" aroma (Robinson et al., 2014).

However, the most important group of lipids in terms of fruit and wine quality are yellow or orange pigments termed carotenoids. β-Carotene (a.k.a. provitamin A) and the xanthophyll (Greek *xanthos* = yellow) lutein, which is produced from α-carotene, account for approximately 80% of the carotenoids in grape berries (Young et al., 2016). Minor compounds include the β-carotene-derived xanthophylls neoxanthin, violaxanthin, and zeaxanthin. The hydrophobic carotenoids belong to the class of isoprenoids, which are also called terpenoids or terpenes. Most carotenoids are C40 compounds; that is, they contain a "backbone" of 40 carbon atoms and are therefore tetraterpenes. They are assembled from isopentenyl pyrophosphate (a.k.a. isopentenyl diphosphate), which is derived from glyceraldehyde-3-phosphate and pyruvate through the methyl-erythritol 4-phosphate (MEP) pathway (Cazzonelli and Pogson, 2010; Della Penna and Pogson, 2006; Hirschberg, 2001; Tanaka et al., 2008).

In addition to carotenoids, the MEP pathway also produces the chlorophyll "backbone," the hormones gibberellin and ABA, and other compounds.

Similar to their role in leaves (see Section 4.1), carotenoids are accumulated in skin chloroplasts of young grape berries as part of the photosynthetic machinery to protect berry tissues from oxidative stress, especially stress caused by high light (Düring and Davtyan, 2002). Carotenoids have pharmacological properties as antioxidants and provitamins and contribute to the yellow to reddish skin color of some white grape cultivars when the chlorophylls are degraded during and after veraison (Bartley and Scolnik, 1995; Della Penna and Pogson, 2006; Uhlig and Clingeleffer, 1998). At this time the chloroplasts are converted into chromoplasts (Greek *chróma* = color) that specialize in carotenoid accumulation (Fanciullino et al., 2014). During this transformation, carotenoids become confined to oil bodies termed plastoglobuli, which enhances their light stability (Cazzonelli and Pogson, 2010).

However, the concentration of most carotenoids declines during grape ripening, which is in sharp contrast to the situation in tomatoes that owe their skin color at maturity mainly to high concentrations of red lycopene and orange β-carotene (Fanciullino et al., 2014; Fraser et al., 1994; Razungles et al., 1988; Young et al., 2012). This decrease is at least partly due to conversion of carotenoid pigments to glycosylated norisoprenoids. These C13 ketones, or apocarotenoids, are precursors of potent grape and wine aroma and flavor components (Baumes et al., 2002; Razungles et al., 1996). Carotenoid cleavage dioxygenases are the enzymes that oxidatively "slice" the various carotenoids, thus converting them into norisoprenoids whose further conversion products, following hydrolysis of the glucose unit in juice or wine, contribute to floral and fruity aroma attributes (Dunlevy et al., 2009; Mathieu et al., 2005; Robinson et al., 2014; Winterhalter and Schreier, 1994). Like other volatile organic compounds, norisoprenoids have poor water solubility, relatively low molecular weight, and high vapor pressure in the ambient temperature range. These properties allow them to easily cross membranes and evaporate. When such volatiles escape into the air, they can be detected by the olfactory receptors of the nose so that we can smell them as an odor; they sometimes impart taste as well (Schwab et al., 2008). Our perception of a specific aroma volatile's odor, however, may change depending on its concentration, which is why the same volatile is often given varying sensory descriptors.

It is thought that the carotenoid composition determines the resulting profile of aroma volatiles produced from the carotenoids or their breakdown products (Klee, 2010). Because grape cultivars differ in their fruit carotenoid composition, they also differ in their bouquet of the corresponding flavor and aroma compounds. This means that part of the "varietal aroma" may be derived directly from variations in skin color, some of which arise from the amounts and relative proportions of different carotenoids (Lewinsohn et al., 2005; Schwab et al., 2008). Examples of norisoprenoids (apart from the hormone ABA) include β-damascenone, which imparts a "rose," "dried fruit," "exotic fruit," or "tropical flowers" aroma, and β-ionone, which imparts a "violet" or "raspberry" odor contributing to "floral" or "fruity" aromas. These norisoprenoids are found in most, if not all, grape cultivars (Robinson et al., 2014; Sefton et al., 2011). Chardonnay is a prime example of a cultivar whose varietal aroma is dominated by norisoprenoids, although in wine these are often masked by components derived from the oak wood that is used during fermentation and/or wine aging (Gambetta et al., 2014; Sefton et al., 1990, 1993; Simpson and Miller, 1984). Both β-damascenone and β-ionone strongly enhance the overall fruity character and reduce or mask vegetative notes of wines (Escudero et al., 2007; Pineau et al., 2007). Though these two volatile components have little aroma on their own, at low concentrations they may enhance the "berry fruit" aroma and at high concentrations lead to "plum" and "dried fruit" aromas. β-Damascenone is also an important aroma component of brandies, beer, rose petal oil, tobacco, and tea leaves, whereas β-ionone is one of the most important volatiles contributing to tomato fruit flavor and to the fragrance of many flowers (Sefton et al., 2011).

The production of these norisoprenoids appears to be directly dependent on the amount of their respective carotenoid precursor formed during fruit development. Although the carotenoids are largely accumulated before veraison, the production of the resulting apocarotenoids does not peak until late in ripening. Moreover, the bulk of these norisoprenoids are glycosylated, and only their release from the sugar molecule during fermentation and wine storage leads to volatilization and "smellability" (Sefton et al., 1993; Mathieu et al., 2005). Owing to competing chemical processes of norisoprenoid production and destruction, however, the concentration of these fruity aroma compounds may increase or decrease during wine aging (Sefton et al., 2011).

Other volatile isoprenoid compounds include the highly aromatic monoterpenes, such as geraniol, linalool, nerol, and citronellol and its derivative rose oxide. These C10 compounds are reminiscent of "rose," "lilac," "pine," or "citrus," and accumulate in postveraison grapes of Muscat cultivars, Gewürztraminer and, to a lesser extent, Riesling, Viognier, and Chenin blanc (Bayonove et al., 1975b; Gunata et al., 1985; Guth, 1997a,b; Ribéreau-Gayon et al., 1975; Robinson et al., 2014; Wilson et al., 1984). Although monoterpenes are important in Riesling, the typical Riesling aroma may be dominated by the norisoprenoids vitispirane ("floral-fruity"), trimethyl-dihydronaphthalene (TDN, which smells of kerosene), and β-damascenone (Simpson and Miller, 1983; Skinkis et al., 2008; Strauss et al. 1987). Many, if not all, other grape cultivars also produce monoterpenes, but at much lower concentrations than these aromatic cultivars. The "muscat" aroma is sought after in table grapes, and the same and other monoterpenes also contribute to the characteristic fragrances and varietal differences of many flowers (e.g., rose, lilac, and jasmine), mint leaves, and conifer needles, whereas Gewürztraminer, like citrus fruits and basil leaves, owes its aroma predominantly to a combination of mono- and sesquiterpenes.

Although grape leaves are capable of producing monoterpenes, the flowers and berries generally manufacture their own complement of these aroma compounds (Bönisch et al., 2014b; Gholami et al., 1995; Luan and Wüst, 2002). The production of geraniol and nerol is mainly restricted to the berry skin, whereas linalool and rose oxide biosynthesis occurs in both the skin and the mesocarp, especially in the Muscats (Gunata et al., 1985; Luan and Wüst, 2002; Luan et al., 2005; Park et al., 1991; Wilson et al., 1986). The sesquiterpene rotundone, which is the key aroma compound of both black and white pepper, also imparts a "peppery" and "spicy" aroma to grapes and the resulting wines of Syrah, Mourvèdre, Durif and, to a lesser extent, other cultivars as well (Wood et al., 2008). Other sesquiterpenes, such as farnesene, cadinene, and germacrene, are found in the berries of many grape cultivars, notably Gewürztraminer, Riesling, Syrah, and Lemberger (May and Wüst, 2012; Versini et al., 1994). Like the monoterpenes, sesquiterpenes accumulate transiently in the flowers and then mostly in the berry skin, but also in the rachis, beginning at veraison and continuing through advanced stages of fruit maturity (Caputi et al., 2011; May and Wüst, 2012; Zhang et al., 2016). However, while the glycosylated C10 monoterpenes probably accumulate in the vacuoles, the C15 sesquiterpenes, which are also produced in the skin cells themselves, are mostly exported to the epicuticular wax from where they may be partly emitted to the atmosphere (May and Wüst, 2012; Schwab and Wüst, 2015).

All volatile grape terpenoids are monoterpenes, sesquiterpenes, or norisoprenoid terpenes—the diterpenoid hormone gibberellin is not volatile. Two compartmentally separated biochemical assembly lines operate in parallel with some interaction to produce a wide variety of terpenoids (Dudareva et al., 2013; Dunlevy et al., 2009; McGarvey and Croteau, 1995; Schwab et al., 2008): the mevalonate pathway in the cytosol and the deoxy-xylulose 5-phosphate/methyl-erythritol 4-phosphate (DOXP/MEP) pathway in the plastids (Fig. 6.6). Although the mevalonate pathway uses citrate derived from malate breakdown in the TCA cycle as a starting substrate, this pathway is probably limited to the production of sesquiterpenes and a few other compounds (Dudareva et al., 2013; Sweetlove et al., 2010). Grape berries make monoterpenes mainly through the mevalonate-independent DOXP/MEP pathway (Luan and Wüst, 2002; Schwab and Wüst, 2015). The berries of different cultivars manufacture specific bouquets of monoterpenes from the original geraniol, and much of this interconversion occurs only late during the ripening phase (Bönisch et al., 2014b; Luan et al., 2005, 2006; Schwab and Wüst, 2015).

Most isoprenoids (e.g., 80–90% of the terpenes) found in grapes are glycosylated by UDP-glucosyltransferases and may be regarded as potential aroma or flavor precursors that can be released enzymatically or chemically during winemaking and aging (Bönisch et al., 2014a,b; Park et al., 1991). The sugar of choice for the glycosylation is glucose, which in turn is often bound to other, minor sugar molecules to form disaccharide glycosides (Dunlevy et al., 2009; Hjelmeland and Ebeler, 2015). Depending on the cultivar, accumulation of terpenoid glycosides may start in the developing flowers but usually peaks during fruit ripening (Bönisch et al., 2014b). Once extracted, many such glycosylated precursors continue to release aroma volatiles during the winemaking and aging processes until they are exhausted. Although terpene glycosides may add a bitter taste, it is the volatile aglycones that are central to the distinctive fruity aroma and flavor, and many of these are slowly released in aging wines in part due to the activities of glycosidase and other metabolic enzymes of yeast and lactic acid bacteria (Gambetta et al., 2014; Jackson, 2008; Noble et al., 1988).

Among the >20 red-colored Muscat cultivars, the amount of terpenes, especially linalool, tends to vary inversely with the anthocyanin content (Cravero et al., 1994). In other words, the more deeply colored cultivars tend to be less aromatic. Indeed, it appears that the DOXP/MEP pathway and the shikimate/phenylpropanoid pathway are in competition for access to carbon substrates (Dudareva et al., 2013; Xie et al., 2008). This competition occurs even though the carbon itself is abundant due to the accumulation of hexoses during grape ripening. Interestingly, carvone, a limonene-derived monoterpene isolated from the oil of caraway or dill seeds, is used in biochemical experiments as an inhibitor of PAL. It is possible, therefore, that some monoterpenes might directly suppress the production of phenolics. Consequently, cultivars that generate and accumulate high amounts of terpenoids may tend to produce lower amounts of phenolics and vice versa, and viticultural attempts to enhance the production of one class of compounds might come at the cost of lower production of the other. If confirmed, this would be good news for the production of aromatic white wine grapes in which high amounts of phenolics are generally undesirable. There may be trade-offs with aroma potential that concern yield formation as well: Among cultivars with muscat aroma, the more aromatic ones seem to be much more vulnerable to adverse conditions that cause poor fruit set than are the less aromatic cultivars (Huglin and Schneider, 1998).

Both grapes and wines contain hundreds of different volatiles in addition to the isoprenoids. Although these belong to many different classes of chemicals, including alcohols, esters and aldehydes, ketones, acids, lactones, and furanones, the majority of them (the sugar-derived furanones are one exception) are produced by oxidation of fatty acids by enzymes such as lipoxygenase, isomerases, alcohol dehydrogenases, and alcohol acyl transferases (Jackson, 2008; Kalua and Boss, 2009; Robinson et al., 2014; Schwab et al., 2008). For example, fatty acids from cell membrane lipids are degraded during advanced stages of fruit maturity and may subsequently be converted to volatile esters and aldehydes. Alcohol dehydrogenases further convert aldehydes to alcohols, while alcohol acyl transferases convert alcohols to esters. In addition, the catabolism of amino acids, often by yeast during fermentation, contributes smaller portions of the same chemical classes—for instance, the leucine-derived methylbutanal and methylbutanol, which are also important flavor compounds of apples and strawberries. Cultivars differ greatly in the type and amounts of volatile constituents they produce, and although most volatiles have no odor and many esters simply smell fruity, these differences largely account for the characteristic varietal aroma and flavor that can be directly attributed to volatiles present in grapes (Conde et al., 2007; Jackson, 2008). It is thought that the more diverse the mix of volatiles, the less important will be the role of the individual components and the more complex will be the overall aroma and flavor.

Like the phenolics and isoprenoids, these components are normally accumulated in glycosylated form, which prevents the otherwise volatile aglycones from prematurely escaping to the atmosphere. This is why the nonvolatile glycosides are odorless and are generally called aroma or flavor precursors, or potential aroma or flavor compounds (Parker et al., 2018). Consequently, the majority of potentially important wine flavor and aroma components cannot be perceived when we eat grape berries or taste their juice, because the saliva of most people lacks sufficient quantities of the glycosidase enzymes necessary to break away the sugar molecules (Kennison et al., 2008; Stradwick et al., 2017). Moreover, although grape berries do contain glycosidase, their high glucose content and low pH greatly slow the enzyme's activity (Aryan et al., 1987; Robinson et al., 2014).

The main exceptions to this principle are Gewürztraminer and Muscat cultivars whose grape juice really does smell like the wine made from it. The few exceptions aside, when harvest decisions are based on tasting fruit in the vineyard, they often consciously or unconsciously become decisions on the basis of the perception of sweetness and sourness in addition to the disappearance of "veggie" aroma (discussed later). At least for cultivars that do not accumulate such "veggie" components to any great extent, it seems that predicting wine flavor, or even wine style, from a sensory assessment of grapes is very challenging at best (Niimi et al., 2018). Moreover, there is substantial fruit-to-fruit variation within a cultivar due to, among other variables, differences in fruit location, growth temperature and light, nutrition, harvest date, and postharvest handling and storage. For instance, the continued metabolism in harvested grapes, especially machine-harvested grapes that are damaged in the process, can lead to substantial aroma and flavor loss if the fruit is picked, transported, or left standing in the heat for prolonged periods. On the other hand, the continuation of berry metabolism, much of it altered in response to dehydration-induced osmotic stress, together with the concentration effect from water loss can also be exploited during a deliberate and prolonged postharvest drying phase to produce distinct wine styles such as the dry or sweet straw wines (Costantini et al., 2006; Zamboni et al., 2008). Amarone is the classic example of the former, while examples of the latter include Recioto, Vin Santo, and Vins de Paille.

In contrast to the glycosylated components discussed previously, the distinctive "foxy" *Vitis labrusca* aroma components methyl anthranilate, aminoacetophenone, and the furanone furaneol are produced in the free volatile form during grape ripening and are emitted into the air, which leads to the characteristic smell of Concord vineyards in autumn (Shure and Acree, 1994). Methyl anthranilate is also an important aroma component of orange flowers and Chinese jasmine green tea and, along with furaneol, of old strawberry varieties. Birds dislike the aroma and generally avoid Concord and similar grapes; therefore, synthetic forms of methyl anthranilate are used in bird repellents in a variety of crops. Methyl anthranilate seems to be assembled in the berry's outer mesocarp from methanol and a breakdown product of tryptophan (Wang and De Luca, 2005). Methanol may be either directly released from the cell walls or produced from methane (CH_4) released from cell wall pectins when the berries soften during and after veraison (Keppler et al., 2008; Lee et al., 1979). Tryptophan is thought to enable this reaction by acting as a photosensitizer that confers on pectin the ability to emit CH_4 upon UV irradiation via the production of reactive oxygen species (Messenger et al., 2009).

A close relative of methyl anthranilate, 2-aminoacetophenone, is present in Concord and other cultivars with *V. labrusca* parentage in both free and glycosidically bound form and is reminiscent of "acacia flower," "naphthalene," or "floor polish." Like methyl anthranilate, it repels birds (Mason et al., 1991). Although aminoacetophenone is a regular aroma component of beer and corn chips, as well as of the pheromone of honey-bee queens, its formation, along with that of the ascorbate-derived furanone sotolon, during wine storage is thought to contribute to the unpleasant "atypical aging" off-flavor of *V. vinifera* wines (Kritzinger et al., 2013; Robinson et al., 2014).

Because of its volatile nature, the concentration of methyl anthranilate begins to decrease in mature grapes. Other free aroma volatiles begin their decline already at veraison. Such is the case for the "veggie," "herbaceous," "bell pepper," or "asparagus"-like methoxypyrazines typical of several kindred Bordeaux cultivars (e.g., Sauvignon blanc, Sémillon, Carmenère, Cabernet franc, Cabernet Sauvignon, Merlot) and a few other red and white cultivars—as well as of bell pepper, green pea, and asparagus (Bayonove et al., 1975a; Hashizume and Samuta, 1999; Koch et al., 2010; Lacey et al., 1991; Ryona et al., 2010). Incidentally, coffee beans also produce methoxypyrazines that, at high concentrations, can cause an off-flavor in roasted beans called the potato-taste defect (Frato, 2019). Apart from Malbec and Petit Verdot, Bordeaux cultivars produce the highest amounts of methoxypyrazines among *V. vinifera* cultivars. Many other cultivars produce their immediate biosynthetic precursor, hydroxypyrazine, but are unable to convert it to the more potent methoxypyrazine (Dunlevy et al., 2013a). Compared with *V. vinifera*, however, berries of *V. riparia* and *V. cinerea* contain particularly high amounts of methoxypyrazines (Koch et al., 2010; Sun et al., 2011b). It is possible that such differences among species and cultivars arise from the presence of several jumping genes that lead to differences in the expression of genes encoding enzymes involved in methoxypyrazine production (Dunlevy et al., 2013a; Guillaumie et al., 2013).

The cyclic, nitrogen-containing hydroxy- and methoxypyrazines are probably produced from the amino acids leucine, glycine and valine. Although the leaves produce methoxypyrazines as well, the clusters manufacture their own complement, almost exclusively in the berry skin and seeds and in rachis tissues (Koch et al., 2010; Roujou de Boubée et al., 2002). Unlike methyl anthranilate, however, methoxypyrazines are not released into the air. Instead they are accumulated in the berries as free volatiles that are released only when the berry cells are disrupted by frugivores, which they help discourage from eating unripe berries. Hydroxy- and methoxypyrazines accumulate early during berry development to a maximum before veraison and are then degraded during ripening, often to $<10\%$ of their maximum concentration (Hashizume and Samuta, 1999; Helwi et al., 2015; Koch et al., 2010; Lacey et al., 1991; Ryona et al., 2010). Nevertheless, the methoxypyrazine concentration at or just before veraison is thought to be a good predictor of its concentration at harvest (Ryona et al., 2008). The postveraison decrease, it seems, is similar to that of malate: Most of the decline in methoxypyrazines occurs during the initial ripening phase, with only a slow further decrease above approximately 20°Brix (Lacey et al., 1991; Roujou de Boubée et al., 2000). In other words, grapes of the same cultivar that contain high amounts of malate at harvest also tend to be high in methoxypyrazine. Methoxypyrazines are extracted extremely rapidly during juice processing, before fermentation even starts, and are stable in wine (Hashizume and Samuta, 1997; Roujou de Boubée et al., 2002). Consequently, the methoxypyrazine content of wine is closely correlated with that of the grapes used to make the wine (Ryona et al., 2009).

Certain sulfur-containing compounds, namely volatile thiols, which are chemicals with a sulfhydryl (-SH) group, or mercaptans, also contribute to the aroma complex in wine but generally not in grapes. Although they are partially bound by saliva proteins, they usually have a very unpleasant smell of rotten eggs; in fact, skunks use sprays of volatile thiols as a defense method (Starkenmann et al., 2008; Wood et al., 2002). At very low concentrations, however, such thiols can have an enjoyable aroma impact: Mercaptobutanol and mercaptopentanol are reminiscent of "blackcurrant," which is characteristic of wine made from ripe Sauvignon blanc and Cabernet Sauvignon grapes, while mercaptohexanol imparts "passion fruit" and "grapefruit" aromas, especially in Sauvignon blanc and Gewürztraminer but also in Merlot, Muscat, Riesling, Sémillon, and other cultivars (Dunlevy et al., 2009; Guth, 1997a; Tominaga et al., 1998, 2000). These compounds may be produced from breakdown products of fatty acids and are bound to a nonvolatile thiol termed glutathione and its breakdown product, cysteine, in the grape skin and, to a lesser extent, the mesocarp (Kritzinger et al., 2013; Peña-Gallego et al., 2012; Peyrot des Gachons et al., 2002a,b; Robinson et al., 2014). The amount of glutathione itself increases in concert with sugars during grape ripening, and so does the concentration of most of the bound thiols, especially in cultivars such as Sauvignon blanc (Adams and Liyanage, 1993; Cerreti et al., 2015; Okuda and Yokotsuka, 1999). However, the concentration of these aroma precursors in the berries seems to be subject to diurnal fluctuations, with peak concentrations occurring near sunrise (Kobayashi et al., 2012). Although the binding to glutathione and cysteine renders these thiols odorless in grapes, the odor-active thiols are released by enzymes of bacteria in the human mouth and by yeast enzymes during fermentation (Dubourdieu et al., 2006; Starkenmann et al., 2008). The concentration of volatile thiols in wine is directly related to the concentration of their precursors in the grapes, even though <10% of these precursors is actually converted to aroma-active compounds.

Yeast also seems to be able to produce another S-containing compound, termed dimethyl sulfide ((CH_3)$_2$S, often abbreviated as DMS), perhaps from the S-containing amino acids cysteine and methionine or from the cysteine-containing peptides glutathione and cystine (De Mora et al., 1986). The DMS concentration further increases during wine aging and may enhance a wine's fruity character, especially of "blackcurrant" and "strawberry/raspberry," and add "truffle" (of which it is a major aroma component) and "olive" notes (Escudero et al., 2007; Segurel et al., 2004). At higher concentrations, however, DMS has an unpleasant "cabbage" odor, which is considered an off-flavor.

Other aroma components do not enter wine from the grapes at all. Instead, they have their origin in storage or packaging materials (Parker et al., 2018). Examples include oak lactone, which has a "coconut" aroma, as well as vanillin and eugenol, all of which are derived from oak wood. Another example is the infamous 2,4,6-trichloroanisole (TCA), which imparts a musty or moldy odor at extremely low concentrations in the parts-per-trillion range and is responsible for the majority of cork taints in wine. Most of the TCA is produced by microorganisms in corks that have been bleached with chlorinated compounds or in oak wood that has been treated with chlorine-containing insecticides or fungicides.

6.3 Sources of variation in fruit composition

Uniform fruit composition is often described as a critical factor for premium winemaking and is equally desirable for the production of grape juice, table grapes, and raisins. Nevertheless, when put to the test, no differences in fruit uniformity were found among vineyards producing fruit that sold for high, medium, or low prices (Calderon-Orellana et al., 2014b). It could be argued, moreover, that less uniform fruit may sometimes result in more complex wine, and this might be one of the, perhaps often unconscious, motivations for blending wines from different vineyard sources. In many European wine regions, such blending was, and in some cases still is, traditionally done in the field, where a range of clones or cultivars were interplanted, sometimes systematically and sometimes randomly. Of course, the grape genotype is a major source of differences in fruit composition (Dai et al., 2011; Kliewer, 1967b; Kliewer et al., 1967).

Whereas traits related to color, astringency, acidity, and flavor are key quality attributes of wine grapes, table grape quality depends more on size, sugar ÷ acid ratio and firmness as signs of freshness, texture for crunchiness or crispness, juiciness, visual appearance including color, and shelf life, although low astringency and pronounced Muscat flavor are often sought after as well (Mullins et al., 1992; Piva et al., 2006; Rolle et al., 2012; Winkler, 1958). Seedlessness is important in some table grape markets and is usually desirable for raisin grapes. High sugar content, moreover, is much more important for raisin grapes than it is for table grapes, not only because consumers prefer sweeter raisins, but also because a higher sugar concentration (20–22°Brix is considered ideal) at harvest cuts down on drying time (Christensen et al., 1995; Parpinello et al., 2012; Ramming, 2009).

Fruit composition changes over time during berry development and ripening as part of the grapevine's developmental program and is therefore under genetic control. In addition, like phenology and yield formation, the extent of these changes may also be modulated by environmental factors and by the interaction of these factors with the genotype of the vine. However, it is often difficult to separate the influence of one environmental factor from that of another. For example, solar radiation affects both incident light and tissue temperature. A decrease in water or nutrient supply leads to a decrease in shoot growth, which can increase the proportion of exposed leaves and clusters, which in turn alters both light quantity and light quality and may also lead to other changes in canopy microclimate (see also Section 5.2). Furthermore, the harvest sugar concentration is higher and the malate concentration lower in berries that have fewer seeds, because veraison occurs earlier in such berries due to the earlier decrease in auxin and increase in ABA (Currle et al., 1983; Gouthu and Deluc, 2015; Rapp and Klenert, 1974). Thus, even within a grape cluster, there is often a natural variation of sugar concentration in the range of 5–7°Brix that results from asynchronous development of individual berries (Coombe, 1992; Pagay and Cheng, 2010). This range may be twice as wide across berries harvested on the same day within a vineyard (Kasimatis et al., 1975; Singleton et al., 1966). Yet the variation in berry weight and fruit composition among clusters on a single vine can be greater than the variation among vines in the same vineyard (Calderon-Orellana et al., 2014a,b). Due to asynchronous berry development, neither time nor thermal time after bloom adequately reflects the developmental stage and maturity of individual berries in a sample from the entire population of berries on a vine, let alone in a vineyard. However, some synchronization does occur during berry development and ripening. For example, berries that originate from late-blooming flowers may develop somewhat more rapidly than berries that originate from early-blooming flowers, and berries that enter the maturation phase later than their earlier neighbors may progress through ripening more rapidly than the early berries (Gouthu et al., 2014; Keller and Tarara, 2010; Vondras et al., 2016). Since most ripening-related processes slow down after an initial phase of rapid change that occurs as a berry changes color, it is possible that, at least in source-limited vines, the late berries benefit from a decrease in overall sink demand for photosynthates and thus catch up with the early berries. Consequntly, the variation in fruit composition generally decreases as the grapes ripen and more and more of the berries approach their sugar maximum and acid minimum (Calderon-Orellana et al., 2014a; Rankine et al., 1962; Trought and Bramley, 2011).

Any factor that influences vine growth and metabolism either directly or indirectly impacts fruit composition, and this leads to large variation among growing seasons in terms of fruit quality. Wine grape production is especially sensitive to climate variability. This sensitivity is the source of considerable variation in vintage quality, which in turn forms one of the bases for the existence of an entire "industry" of wine judges, consumer magazines, and related services. Most fluctuations in fruit composition within a vineyard are caused by climate variability. Consequently, weather differences among years in addition to vineyard location rather than, say, within-vineyard differences in soil composition, are by far the strongest determinants of fruit composition and wine quality (Downey et al., 2006; Noble, 1979; Reynolds et al., 2007, 2013). Such climate variation often trumps differences in soil moisture in both dry-farmed and irrigated vineyards—except at the far low and high ends of the moisture spectrum (Bowen et al., 2011; Herrera et al., 2017; Keller et al., 2008; Pereira et al., 2006b; van Leeuwen et al., 2004). For instance, annual variation accounts for an almost twofold range in sugar concentration, a two- or threefold variation in acid and anthocyanin content, and at least a 10-fold range in some aroma components (Cacho et al., 1992; Castellarin et al., 2007b; Downey et al., 2004; Viala and Vermorel, 1901–1909; Wood et al., 2008). Even in climates that permit harvest at identical target sugar concentrations across vintages, there is still marked variation in other components, such as organic acids and anthocyanins, between years (Calderon-Orellana et al., 2014a; Keller et al., 2004, 2005).

The anthocyanin profile—that is, the relative proportion of individual pigments for a given cultivar—varies more among years than from veraison to harvest in the same year (Pomar et al., 2005). The concentration of carotenoids, it seems, varies less among diverse grape cultivars than it does among growing seasons for the same cultivar (Oliveira et al., 2004). Even the amount of skin and seed tannins varies as much between years as it does between cultivars (Harbertson and Keller, 2012; Liang et al., 2012a). Vintage differences also occur at the level of molecular biology: Differences in gene expression in grape berries between growing seasons can be greater than those between pre- and postveraison berries within a season (Pilati et al., 2007). On the other hand, comparing the profiles of both metabolites and gene expression in Cabernet Sauvignon and Pinot noir berries from fruit set through maturity across three growing seasons at the same site in a warm climate found developmental stage to be the most important source of variation, followed by cultivar, and finally vintage (Fasoli et al., 2018). In a comparison of the sources of variation in gene expression in the berries of Cabernet Sauvignon and Sangiovese from the pea-size stage to maturity across three distinct vineyard sites and two growing seasons, developmental stage was the most influential, followed by the interactions of cultivar with vintage and/or site, and cultivar, while vintage and, especially, location were comparatively less important (Dal Santo et al., 2018). Cultural practices can at best be used to fine-tune what nature imposes on any given vineyard site or in any given growing season. The same applies to clones of a cultivar and to the rootstock that is used as a grafting partner: Growing season and vineyard site still vastly outweigh the influence of these on fruit composition (Harbertson and Keller, 2012; Keller et al., 2012; Schumann, 1974). Naturally, such variation results in different sensory properties in wines made from the grapes (Forde et al., 2011).

For a specific clone of a cultivar, variation in grape composition therefore occurs within a cluster, within a vine, between vines in a vineyard, between vineyards, and between vintages (Calderon-Orellana et al., 2014a,b; Rankine et al., 1962). This variation arises due to the factors discussed in the following sections, as well as to other factors—some effects of pathogen infection, for example, are discussed in Section 8.2 (see also Jackson, 2008). However, although wines labeled as having been made from so-called "old vines" are often viewed as being of higher quality than wines from young vines, there is currently no scientific evidence for the belief that vine age impacts fruit composition (Bou Nader et al., 2019).

6.3.1 Fruit maturity

Contrary to widespread misunderstanding, grape berries are physiologically mature when their seeds attain the ability to germinate, which is at veraison (see Section 2.3), not at some arbitrarily defined point thereafter. Subsequent alterations in fruit composition may nevertheless be beneficial for the seeds and hence for the species because they help attract seed dispersers. As argued in Section 2.3, the accumulation of red pigments and aroma volatiles serves as an "advertisement," signaling to potential seed dispersers the availability of food material of high nutritional and health value. Indeed, most pigments and volatiles are manufactured from nutritionally valuable components such as sugars, amino acids, fatty acids, and carotenoids, and thus serve as positive nutritional signals (Goff and Klee, 2006). Many of the associated changes during grape ripening were discussed in Section 6.2 for individual compounds or specific groups of compounds. The changes in the concentrations of sugars, organic acids, amino acids, and phenolic compounds (especially anthocyanins in red grapes), and in pH, are rather well known. Sugars and anthocyanins accumulate and malate declines rapidly during the early ripening phase, and the sugar concentration may continue to increase due to berry shrinkage from water loss well beyond the maximum of 23–25°Brix at which phloem inflow gradually ceases. The same concentration effect occurring late in the growing season may apply to other berry flesh components, such as organic acids or K^+, as well (Bondada et al., 2017). Moreover, skin tannins may continue to polymerize, perhaps along with the incorporation of anthocyanins to form polymeric pigments. The amount of skin anthocyanins, however, sometimes declines at these late stages of ripening (Holt et al., 2010). In addition to these major compounds, flavor and aroma components also undergo changes as grapes ripen. For instance, methoxypyrazine, whose production can vary by an order of magnitude among vineyard sites, declines rapidly during and after veraison, but its degradation slows considerably later in the ripening phase (Lacey et al., 1991; Marais et al., 1999; Roujou de Boubée et al., 2000; Ryona et al., 2008). Whereas esters may dominate the volatile profile of preveraison grape berries, grassy-herbaceous C6 aldehydes peak during the early ripening phase but may be converted to more fruity alcohols during later stages of fruit maturation (Kalua and Boss, 2009; Maoz et al., 2018). Norisoprenoids and most other terpenes, or their glycosylated aroma precursors, appear to continue to increase even at very advanced stages of maturity above 30°Brix (Fang and Qian, 2006; Hardie et al., 1996a; Park et al., 1991).

Such changes during extended grape ripening form the basis for delaying harvest, a practice dubbed "hang time" that aims to minimize vegetal aromas and maximize terpenoid-based and other fruity aroma components (Kalua and Boss, 2009; Ryona et al., 2008; Wilson et al., 1984). Whereas the concentration of some volatile phenols, such as guaiacol, may increase during maturation, perhaps due to conversion of hydroxycinnamic acids, others, such as eugenol, may remain constant or even decrease (Fang and Qian, 2006). The precursors of some aroma-active fruity esters also appear to decrease over time, but most of them show no consistent trend. While some of these changes late in the growing season may be due to the continued production or modification of quality-relevant components in the grape berries, others are simply brought about by the concentration effect that results from the slow dehydration of the berries—weather permitting. Moreover, physical and chemical changes in fruit composition also lead to changes in the number and composition of microorganisms that call the fruit home. As the sugar concentration and the pH increase, both the number and the diversity of this microflora increase. Yet the consequences of these population dynamics for grape juice and wine composition are not well understood and little appreciated.

Clearly, some of the alterations in fruit composition that occur with increasing grape maturity are desirable—for example, more sugar, less acid and bitter tannin, and more "good" and fewer "bad" flavors. Other changes, however, are undesirable—for example, even more sugar, even less acid, high pH, and less "good" and more "bad" flavors. Moreover, the desired level of fruit maturity depends on the intended use of the grapes. Taking soluble solids as a simple measure of maturity, table and juice grapes may be harvested when they have accumulated just 16–18°Brix, which is immediately after they have fully turned color. Grapes destined for sparkling wine production are considered optimally ripe at approximately 18–20°Brix (Iland et al., 2011). And while the typical harvest "window" for many white wine grapes ranges from 20 to 24°Brix, red wine grapes and white wine grapes destined for sweet "dessert" wines are often left on the vines well beyond the physiological sugar maximum of 23–25°Brix when and where possible. The factors described in the following sections exert much of their influence on fruit composition by accelerating or retarding grape ripening and hence maturity, some directly through their influence on the biosynthesis or degradation of particular compounds, and some indirectly through their influence on berry size or canopy structure, and hence microclimate, or other secondary effects. Disentangling direct effects from indirect effects is not straightforward, because changes in canopy microclimate in turn also alter fruit biochemistry.

6.3.2 Light

As the source of energy for photosynthesis, sunlight is the most important climatic factor affecting grape composition. Under clear skies the photosynthetic photon flux can exceed $2000\,\mu mol\,m^{-2}\,s^{-1}$, but clouds can decrease the light intensity on the canopy to <10% of this value, and overcast conditions may result in light intensities below the light compensation point of photosynthesis. The effect of light is often expressed in terms of the effects of shade, especially as it refers to the within-canopy microclimate (Jackson and Lombard, 1993; Smart, 1985; see also Section 5.2). Apart from weather conditions, this microclimate, and hence the degree of sun exposure of the fruit clusters, is a function of planting distance, trellis design and training system, canopy density, shoot length, and shoot architecture and can be modified by canopy management practices such as shoot thinning and positioning, hedging, or leaf removal (Reynolds and Vanden Heuvel, 2009). It is important to remember, however, that shade decreases both light intensity and temperature during the day, and it is very difficult to disentangle the separate effects of each of these variables (e.g., Spayd et al., 2002). By reducing the vapor pressure deficit, the cooling effect of shading also reduces fruit transpiration and consequently water loss from the fruit. Because berry transpiration is an important avenue for ripening grapes to discharge excess water derived from phloem import, reduced transpiration may slow sugar and anthocyanin accumulation (Keller et al., 2015b; Rebucci et al., 1997; Zhang and Keller, 2017). Consequently, severe cluster shading, for example due to high canopy density and shoot vigor, generally tends to delay fruit ripening. Berry growth can also be slower in shaded clusters than in sun-exposed clusters, especially if shade occurs early during berry development (Dokoozlian and Kliewer, 1996). Exposing shaded clusters to sunlight at veraison or later by removing leaves in the fruit zone does not lead to compensatory berry growth but may increase the incidence of sunburn on the berries. On the other hand, where fruit exposure is associated with high temperatures, sun-exposed berries may remain smaller than shaded berries (Bergqvist et al., 2001; Crippen and Morrison, 1986; Reynolds et al., 1986).

Shade on the leaves depresses the rate of photosynthesis, thus limiting the rate of sugar export to grape berries, which may curtail berry growth and sugar accumulation (Cartechini and Palliotti, 1995; Rojas-Lara and Morrison, 1989; Smart et al., 1988). When shaded leaves die, their nutrients are recycled to important sinks. If the fruit happens to be important while leaves are senescing, especially during the early ripening phase, they may receive a dose of nitrogen, K^+, and other nutrients, which may raise the juice pH (Morrison and Noble, 1990; Smart et al., 1985a,b). In situations that favor high vigor and dense canopies, hedging or removal of leaves and/or lateral shoots in the fruit zone soon after fruit set may open up the canopy and reduce K^+ accumulation and the associated pH increase in the berries (Coniberti et al., 2012). Increases in K^+ and pH do not occur in shaded fruit as long as the leaves are exposed to sunlight and remain photosynthetically active (Crippen and Morrison, 1986).

Among organic acids, tartrate seems to be relatively insensitive to low light intensity. If anything, its production decreases in low light, perhaps because the biosynthesis of its precursor, ascorbate, is a light-sensitive process (DeBolt et al., 2007; Kliewer and Schultz, 1964). In contrast, shaded berries often contain higher amounts of malate, which is partly due to preveraison malate production being favored over tartrate production in low light and partly due to postveraison malate catabolism being slowed by the lower temperatures that accompany shade (Friedel et al., 2015; Kliewer and Schultz, 1964; Pereira et al., 2006a; Viala and Vermorel, 1901–1909). The impact of canopy shade on total nitrogen and amino acids remains unresolved. Both lower and higher nitrogen contents have been found in shaded berries (Friedel et al., 2015; Pereira et al., 2006a). Similarly, although shade may generally increase the concentration of amino acids, especially arginine and glutamine, shaded berries occasionally contain less proline than sun-exposed berries.

Because light, partly via phytochrome, stimulates CHS and many other enzymes of the phenolics assembly line, shading reduces the production of phenolic components. However, different classes of phenolics respond differently to specific portions of the light spectrum. Anthocyanins and hydroxycinnamic acids are stimulated primarily by visible light, whereas flavonols accumulate mainly in response to UVB (Berli et al., 2011; Gregan et al., 2012; Liu et al., 2015). Anthocyanin accumulation is not only responsive to light intensity but also to day length: Longer days, it seems, are generally associated with more intense anthocyanin production than are shorter days (Jaakola and Hohtola, 2010). In addition to the influence of cooler temperatures, this photoperiod effect—or the cumulative effect of more light each day—may be one reason that anthocyanins tend to accumulate more rapidly and to higher concentrations at higher than lower latitudes, at least before the autumnal equinox. But although light is necessary for anthocyanin production, direct sun exposure of the fruit is not nearly as important for anthocyanin accumulation as for flavonol production; indeed, flavonols may be useful markers of light exposure (Downey et al., 2004; Friedel et al., 2015; Haselgrove et al., 2000; Keller and Hrazdina, 1998; Macheix et al., 1990; Pereira et al., 2006a; Price et al., 1995; Spayd et al., 2002). However, high amounts of flavonols may lead to noticeable astringency or even bitterness in grapes and wines (Gawel, 1998).

Because the synthesis of all phenolics is dependent on imported sucrose, episodes of cloudy or overcast conditions slow anthocyanin accumulation and may result in fruit with lower anthocyanin content (Cartechini and Palliotti, 1995; Herrera et al., 2017; Keller and Hrazdina, 1998; Smart et al., 1988). Cultivars differ in their fruit's susceptibility to low light. Compared with sunlit fruit, the anthocyanin content is often considerably lower in shaded Pinot noir, Cabernet Sauvignon, or

Malbec grapes than in Merlot, whereas coloration in varieties such as Syrah, Petit Verdot, or Nebbiolo appears to be almost unaffected by shade. Clouds and shade may also induce a shift in the relative proportion of individual anthocyanins. In grapes grown under overcast conditions, malvidin-based anthocyanins may come to dominate the anthocyanin profile because their production seems to be less susceptible to low solar radiation than that of other anthocyanins (Keller and Hrazdina, 1998). Moreover, acylated anthocyanins may be less sensitive to canopy shade than are their nonacylated counterparts; thus the relative proportion of acylated anthocyanins is higher in heavily shaded grapes than in partially or fully sun-exposed grapes (Ristic et al., 2010). Shading of clusters by leaves only interferes with pigmentation if it is severe, as is the case inside overly dense canopies, apparently by suppressing the genes that code for anthocyanin biosynthesis (Jeong et al., 2004; Koyama and Goto-Yamamoto, 2008; Ristic et al., 2010). It appears that anthocyanin formation in the berry skin reaches a plateau at a photon flux of approximately $100 \, \mu mol \, m^{-2} \, s^{-1}$ in the fruit zone; above this light intensity, temperature becomes the dominant factor in berry coloration (Bergqvist et al., 2001; Downey et al., 2006; Spayd et al., 2002; Tarara et al., 2008). Too much radiation can even inhibit anthocyanin production or induce degradation, perhaps due to UVB-induced formation of hydrogen peroxide (H_2O_2) as a result of oxidative stress. The H_2O_2 in combination with peroxidase might degrade anthocyanins and other phenolics in the vacuoles. Nonetheless, all else being equal, grapes produced at higher altitudes tend to have higher concentrations of total phenolics and anthocyanins than their lower elevation counterparts, although it is not clear if this is due to the increase in UVB with increasing altitude (Berli et al., 2008). Sun exposure may also improve the extractability of anthocyanins during fermentation (Cortell and Kennedy, 2006). Taken together, it appears that sun exposure of grapes may enhance the color of young red wines mainly by increasing the amount of flavonol cofactors and less so that of anthocyanins. In aged wines, however, anthocyanin-tannin interactions become increasingly dominant.

Visible light may enhance tannin accumulation in the skin, but usually not in the seeds, of sun-exposed berries up to veraison and increases the average length of the tannin polymers (Cortell and Kennedy, 2006; Downey et al., 2006; Koyama and Goto-Yamamoto, 2008). This light effect seems to be more prominent in cultivars that produce low amounts of tannins than in high-tannin cultivars. As in the case of anthocyanins, however, only severe shade may reduce tannin production in the skin (Ristic et al., 2010). But contrary to its influence on anthocyanin extractability, sun exposure appears to decrease the extractability of tannins during ripening, which tends to cancel out at least part of the prior gain in tannin content (Downey et al., 2006). The nature of this change in tannin extractability remains unknown, but considering that high temperature and UV radiation enhance methane release from cell wall pectins, possibly via the generation of reactive oxygen species (Keppler et al., 2008; McLeod et al., 2008), it is conceivable that the resulting pectate may bind some of the tannin. Nevertheless, compared with vigorous vines with limited fruit sun exposure, grapes grown on less vigorous vines tend to have greater amounts of skin tannins and anthocyanins, which may result in higher concentrations of stable polymeric wine pigments (Cortell et al., 2005, 2007a,b). In vines with very low vigor, however, anthocyanin accumulation can be impaired due to insufficient leaf area and excessive fruit sun exposure.

Carotenoids act as photoprotectants whose synthesis is regulated by phytochrome (Cazzonelli and Pogson, 2010). Therefore, their preveraison production is induced by light exposure, whereas episodes of low light due to overcast conditions are associated with a temporary decline of xanthophyll carotenoids (Düring and Davtyan, 2002; Joubert et al., 2016; Young et al., 2016). Given that xanthophyll production apparently does not respond to temperature, one might therefore conclude that grape berries are most vulnerable to sunburn if bright, hot days suddenly follow a period of dense cloud cover or if shaded berries are suddenly exposed to the sun. Sudden sun exposure occurs, for example, following leaf removal in the cluster zone too late in the growing season in an attempt to improve the microclimate in the fruit zone. If grapes behave like apples, then only cultivars that are resistant to sunburn should respond to high light exposure with elevated carotenoid production, whereas carotenoids should decrease in exposed fruit of susceptible cultivars (Merzlyak et al., 2002). Cabernet Sauvignon, for example, accumulates much higher amounts of lutein but not β-carotene in preveraison berries under high UV light than under low UV light (Steel and Keller, 2000). Another reason for the vulnerability to sudden sun exposure is that shaded berries do not produce enough UV-protective flavonols in their skin (Kolb et al., 2003).

The DOXP/MEP pathway that produces isoprenoids in grape berries does not operate in the dark (Mongélard et al., 2011). Therefore, it is conceivable that long days favor the production of carotenoids and terpenoids. The productivity of the pathway also increases with increasing light intensity. Sun exposure not only enhances the preveraison production of carotenoids but also accelerates their postveraison degradation, which is associated with increased conversion to aroma-active norisoprenoids (Baumes et al., 2002; Razungles et al., 1998; Schultz, 2000; Young et al., 2016). Although this may result in high norisoprenoid concentrations in exposed fruit, removing leaves to increase light exposure can sometimes—for unknown reasons—diminish the amount of norisoprenoids (Lee et al., 2007). In some cases, grape norisoprenoids were found to be insensitive to canopy shade and light exposure (Ristic et al., 2010). Light might also promote methyl anthranilate formation in sunlit Concord berries because higher UVB radiation favors the release of its precursor, methane, from cell wall pectins, possibly as a result of oxidative stress (McLeod et al., 2008; Messenger et al., 2009). Furthermore, UVB light may enhance the production and emission of terpenes (Joubert et al., 2016; Loreto and Schnitzler, 2010). Although the net effect on aroma precursors in harvested grapes is poorly understood, the concentration of glycosylated monoterpenoids is reduced in shaded fruit (Reynolds and Wardle, 1989b; Young et al., 2016). In contrast, shading below about 50% of full sunlight decreases terpenoid production while substantially increasing methoxypyrazine accumulation in young berries, and differences induced early during berry development often persist through most or all of the ripening period (Joubert et al., 2016; Koch et al., 2012; Marais et al., 1999; Ryona et al., 2008). Consequently, growers often remove leaves in the fruit zone soon after fruit set to increase sun exposure of the developing clusters and thereby enhance berry monoterpenoids and reduce methoxypyrazine (Gregan et al., 2012; Reynolds and Wardle, 1989a; Reynolds et al., 1996; Scheiner et al., 2010). It is not clear whether this is due to a direct effect of light on aroma biosynthesis or degradation or an indirect effect of higher berry temperature, or a combination of light and temperature effects.

Postveraison fruit sun exposure cannot compensate for inadequate exposure before veraison (Koch et al., 2012). Conversely, even relatively brief periods of low light during ripening may be detrimental to the flavor quality of the fruit. The sensory perception in wine of "fruitiness" is positively correlated and that of "vegetal/asparagus/green pepper" is negatively correlated with the average light intensity during the ripening period (Marais et al., 2001; Ristic et al., 2010). Indeed, it is quite common for "fruity" and "vegetal" characters to be inversely correlated in wines (Escudero et al., 2007; Heymann and Noble, 1987; Noble and Ebeler, 2002).

In addition to the direct effects discussed previously, light exposure of grape clusters could also have indirect effects on wine quality. It appears that yeasts are among the fungi that are most susceptible to UVB radiation (Newsham et al., 1997). Therefore, light exposure might alter the yeast microflora on the surface of grape berries. Indeed, the distribution of yeast species and strains, as well as that of other fungi and bacteria, on grape berries is influenced by environmental effects (Bokulich et al., 2014; Longo et al., 1991; Stefanini et al., 2012; Vezinhet et al., 1992). This may have implications for wine composition, because even in musts inoculated with single-strain yeast cultures the native yeasts may be important during the initial stages of fermentation before the inoculated strain takes over.

6.3.3 Temperature

Sunlight heats plant tissues. The dependence of cellular integrity and enzymatic reactions on conditions that are neither too cold nor too hot makes temperature a very important factor influencing grape composition. In general, higher temperatures accelerate plant development and are associated with a more rapid succession of phenological stages, including earlier veraison (see Section 2.2). In cool regions, low temperatures often limit photosynthesis and sugar production in the leaves. Conversely, in hot regions, temperatures often exceed the photosynthetic optimum during a large portion of the day; thus, for a considerable fraction of the growing season the temperature can be too high for maximal photosynthesis. In addition, high nighttime temperatures in hot regions increase the proportion of assimilated carbon that is lost through respiration, which reduces the total amount of sugar available to the clusters. Global climate change is associated with higher temperature increases at night than during the day (IPCC, 2013). Consequently, nighttime respiratory carbon loss is increasing as well. The long-distance transport of assimilates in the phloem is relatively insensitive to temperature but is inhibited by prolonged periods above 40 °C, probably due to temporary blockage of sieve plate pores by callose.

Temperature, in addition to seasonal rainfall, is thought to be the main factor driving fluctuations in grape quality between years; sugars may accumulate most rapidly in the temperature range 20–30 °C, provided soil moisture and other factors are not limiting (Alleweldt et al., 1984b; Hofäcker et al., 1976; Kliewer, 1973). On a global scale, the average growing season temperature for individual wine regions correlates positively with average wine vintage ratings, with a clear temperature optimum that varies by region and wine style (Jones et al., 2005). Heat summation, usually expressed as growing degree days (see Section 2.2), has been found to be a good predictor of grape maturity, especially if a lower temperature threshold of 18 °C is applied rather than the 10 °C threshold generally used in grapevine growth models. Nevertheless, the link between temperature and fruit quality is not straightforward, in part because the temperature of individual berries is also important. Because photosynthesis is less restricted by low temperatures than are most other plant processes, lack of sugar supply is probably not the primary reason for the slow ripening during unseasonably cool or hot periods or the slower sugar accumulation in cool compared with warm regions; sink activity and its direct control by temperature seem to be much more important. Although both low and high temperatures do indeed limit photosynthesis (see Section 5.2), this limitation may be due to feedback inhibition from sugars accumulating in the leaves, because fruit growth and sink activity are more sensitive to temperature than is photosynthesis (Klenert et al., 1978; Körner, 2003; Wardlaw, 1990). Berry expansion and sugar accumulation may stall at temperatures below 10 °C and above 40 °C (Greer and Weston, 2010b; Matsui et al., 1986; Sepúlveda and Kliewer, 1986; Hall and Keller, unpublished data). Perhaps the ripening berries are lower in the sink hierarchy of vines that experience cold or heat stress. Such stress seems to have a much less marked effect on shoot growth, which implies that assimilates are preferentially supplied to the shoot tips under such conditions (Greer and Weston, 2010b; Sepúlveda et al., 1986).

Berries in different positions of a grapevine canopy experience different temperatures. The surface of green berries exposed to full sunlight can be 12 °C warmer than the surrounding air, and that of dark-colored berries can be as much as 17 °C warmer, whereas shaded berries are usually close to the ambient temperature (Bergqvist et al., 2001; Smart and Sinclair, 1976; Spayd et al., 2002; Tarara et al., 2008). Berries facing the sun on a sun-exposed cluster are also heated much more than the nonexposed berries on the same cluster (Keller et al., 2016a; Kliewer and Lider, 1968). At night, however, the exposed berries can be several degrees cooler than their shaded counterparts; shade dampens the berries' diurnal temperature range. In addition to shade, the cooling effect of wind during the day also reduces the difference between berry skin and ambient temperature, because higher wind speed reduces the boundary layer resistance and increases convective heat loss from the berries (Cola et al., 2009). Moreover, berries on loose clusters are heated less by solar radiation than those on tight or compact clusters, because berries that touch one another conduct more heat and lose less heat to convection (Smart and Sinclair, 1976). A viticultural strategy employed to diminish cluster compactness is the practice of cutting through inflorescences at bloom, which in some cultivars (e.g., Riesling) results in compensatory stretching of the remaining portion of the cluster but also diminishes yield. Another strategy, used mainly in some table grape production, is the application of gibberellin sprays at bloom, which leads to decreased fruit set and elongation of the rachis and increased berry size (Currle et al., 1983; Weaver and Pool, 1971; Williams, 1996).

Vineyard row direction has a major influence on heating of exposed fruit (Fig. 6.8). In east–west-oriented rows, the berries on the south side of the canopy in the Northern Hemisphere and on the north side in the Southern Hemisphere can be at higher than optimal temperature during much of the day (Bergqvist et al., 2001). In north–south-oriented rows, berries on the west side of the canopy can be considerably warmer than those on the east side, because ambient temperatures are generally highest after solar noon (Reynolds et al., 1986; Spayd et al., 2002; Zarrouk et al., 2016). Strong solar heating of berries may be advantageous in cool climates but may hinder fruit ripening in warm climates, so the common viticultural practice of removing leaves in the fruit zone must be applied judiciously depending on vineyard location. Whereas fruit sugar content is relatively insensitive to temperature, the berry temperature affects the metabolic processes that convert sugars to the various chemical components that are important for fruit quality. Sun-exposed berries are vulnerable to sunburn or sunscald; symptoms include discoloration and dead, brown patches on the skin, followed by berry shriveling (see Section 7.4). Such damage may be due to overheating at skin temperatures above 42 °C and excess UV and/or visible light leading to photooxidative damage to the epidermis and hypodermis. Excessive temperatures adversely impact the quality of table grapes due to discoloration and fruit shriveling (Mullins et al., 1992).

Both low (<10 °C) and high (>40 °C) temperatures can limit berry size and delay fruit ripening. In warm, sunny climates, berries on the west side of the canopy in north–south-oriented rows can remain considerably smaller than berries on the east side, likely because the high afternoon temperature of west-exposed berries limits their growth, especially if seasonal water deficit leads to high sun exposure (Shellie, 2011). Very high fruit temperature limits sugar accumulation in sun-exposed berries but may have little effect on shaded berries, which also seem to have less variation in sugar concentration than their exposed counterparts (Kliewer and Lider, 1968). Conversely, although acidity in mature grape berries is lower if the grapes ripened at higher temperature because high temperatures hasten malate respiration, this effect is much more pronounced in shaded berries than in sun-exposed berries

FIG. 6.8

Sun-exposed grape berries may heat up considerably above the ambient temperature.

Photo by M. Keller.

(Kliewer, 1973). Nevertheless, due to the pronounced temperature difference between shaded and sun-exposed berries, the latter generally have lower amounts of malate, though not of tartrate, and higher pH at maturity (Kliewer and Lider, 1968; Reynolds et al., 1986).

The temperature optimum for preveraison malate accumulation is 20–25 °C, and there is a sharp drop in accumulation above approximately 40 °C (Kliewer, 1964; Lakso and Kliewer, 1975, 1978). Therefore, grapes grown in warm climates or warm growing seasons may have higher acidity at the beginning of ripening compared with grapes grown in cooler climates or cool seasons, unless excessive sunlight exposure heats the berries above the optimum in the warm climate (Klenert et al., 1978). But because higher temperatures accelerate malate respiration during ripening, the acidity declines rapidly after veraison in warm-climate regions or warm growing seasons, whereas the decrease is more gradual in cool-climate regions or cool growing seasons (Sweetman et al., 2014). Malate degradation not only proceeds more rapidly but also seems to start earlier under warm conditions (Buttrose et al., 1971; Ruffner, 1982b). Therefore, the malate concentration at harvest is usually higher and the pH lower during cool growing seasons than during warm seasons (Currle et al., 1983; Hofäcker et al., 1976; Jackson and Lombard, 1993; Klenert et al., 1978).

The French saying, *"C'est l'août qui fait le goût"* ("It's August that makes the taste"), along with a neat linguistic detail (French *août* = August, *aoûté* = mature), points to the importance of temperature conditions during veraison and early ripening for harvest fruit quality. In hot regions, veraison occurs early and the temperature is often too high during ripening for optimal fruit quality. Grape growers in such regions sometimes employ the strategy of double-pruning, whereby green shoots that grew after the normal winter pruning are hedged severely and leaves, laterals, and clusters are removed soon after fruit set to break paradormancy (see Section 2.2). Such severe interventions induce renewed budbreak that will result in a crop that ripens during the cooler fall season (Gu et al., 2012; Iland et al., 2011). Conversely, in climates close to the latitudinal or altitudinal limits of grape production, early fall frosts may occur before the grapes are ready for harvest. Some growers, however, deliberately delay harvest until the grapes are frozen to produce ice wine. The freezing process has little effect on subsequent juice or wine composition, except, of course, when grapes are pressed while frozen or are left on the vine for prolonged periods during which they may be invaded by pathogens. The sugar concentration is rather immune to freezing and thawing, while the titratable acidity decreases somewhat, and K^+ and pH increase (Spayd et al., 1987). The concomitant rise in K^+ and pH might be mostly a result of K^+ extraction from the skin during thawing.

Temperature is one of the main climatic variables affected by changes in the amount of atmospheric CO_2 (Arrhenius, 1896). The forward shift of phenological development resulting from current and future climate change (see Section 2.3) may be especially important, because earlier veraison implies that the ripening period shifts toward the hotter part of the growing season (Duchêne et al., 2010; Keller, 2010). That this has already occurred during the second half of the twentieth century has been documented in France, where the period between budbreak and harvest has become shorter and ripening is occurring under increasingly warm conditions (Duchêne and Schneider, 2005; Jones and Davis, 2000). Although this trend has been correlated with greater fruit sugar and lower acid concentrations, and generally better wine quality, even warmer does not necessarily mean even better. For one thing, if veraison occurs earlier, there is less time for the preveraison production of organic acids and of flavanols that can be assembled into tannins. Moreover, with early and rapid sugar accumulation that is not necessarily paralleled by accumulation of anthocyanin pigments and flavor components, growers—or their winemaking customers—may be tempted to wait for flavors to "catch up" so that the grapes may be harvested at very high sugar concentrations that result in wines with high alcohol contents. This tendency may be exacerbated by an adherence to the adage of "less is more" with respect to a perceived yield—quality dichotomy.

Accumulation of some amino acids, especially proline and γ-aminobutyric acid, in ripening berries accelerates with increasing temperature, but arginine is less responsive to temperature (Buttrose et al., 1971; Kliewer, 1973; Sweetman et al., 2014). The increase in amino acid production under high temperature might be a consequence of the greater respiratory flux of malate through the TCA cycle (Sweetman et al., 2014). Thus, warm growing seasons may be associated not only with lower acidity but also with higher amounts of amino acid in grape berries (Pereira et al., 2006b). Similarly, the K^+ content also rises with temperature, which in combination with reduced malate raises the juice pH.

In *V. vinifera* cultivars, anthocyanin production increases up to an optimum berry temperature near 30 °C but may be inhibited above 35 °C, whereas the optimum seems to be somewhat lower in *V. labrusca* and its hybrids with *V. vinifera* (Kliewer, 1977b; Kliewer and Torres, 1972; Spayd et al., 2002; Yamane et al., 2006). Excessively high temperatures, like excessive light, may also induce oxidative stress, which can lead to peroxidase-mediated anthocyanin degradation in addition to the inhibition of anthocyanin production (Gouot et al., 2019; Mori et al., 2007; Pastore et al., 2017; Zarrouk et al., 2016). Consequently, if the air temperature is 25 °C and berries are at 40 °C, red pigmentation will be impaired. This sensitivity to daytime temperature—night temperatures seem to be much less important—at least partially explains the poor skin color often observed in warm climates or in warm growing seasons, especially when the grapes are directly exposed to the sun (Gouot et al., 2019; Haselgrove et al., 2000; Ortega-Regules et al., 2006a; Shellie, 2011). It also explains why, at the same sugar concentration, red coloration is lower under hot compared with cool conditions (Sadras and Moran, 2012). Among *V. vinifera* cultivars, Pinot noir, Sangiovese and Tempranillo seem to be among the most susceptible, whereas Cabernet Sauvignon, Malbec, and Syrah may be comparatively less sensitive to heat-induced color impairment (Gouot et al., 2019). Owing to their lower temperature optimum for anthocyanin accumulation, cultivars with *V. labrusca* parentage are especially sensitive to heat; red pigmentation stalls at high temperatures, which is why Concord production for grape juice is mostly confined to relatively cool-climate areas.

Since anthocyanin production proceeds most rapidly during the early ripening period, this is also the most sensitive time window for temperature to impact final skin color. Maximum skin coloration often occurs in sun-exposed grapes at cool ambient temperature, and minimum coloration in shaded fruit at high or very low temperatures, but daytime temperature, rather than light, appears to be the main driver of anthocyanin accumulation (Spayd et al., 2002; Tarara et al., 2008). Moreover, high temperature leads to shifts in the relative amounts of individual anthocyanins. Because the formation of malvidin-based anthocyanins seems to be far less responsive to temperature than that of other anthocyanins, the former become dominant in grapes grown under hot conditions (Gouot et al., 2019; Mori et al., 2005; Ortega-Regules et al., 2006a; Tarara et al., 2008). Incidentally, cool autumn temperatures also lead to more intense coloration of the leaves, especially in combination with high light that results in energy overload in the leaves. This effect is particularly intense in grapevines infected with leafroll viruses (Gutha et al., 2010). Unfortunately, the virus impairs grape ripening and leads to poor fruit color even under optimum temperatures (see Section 8.2).

The formation of tannins or their flavanol precursors appears to increase with increasing temperature in preveraison grape berries, whereas the temperature has little effect during ripening, when flavanols are no longer produced (Gouot et al., 2019; Pastore et al., 2017). Flavonol production may be less sensitive to temperature than is anthocyanin production but instead is more responsive to light, especially UV light. Nonetheless, temperatures below 20 °C may stimulate flavonol biosynthesis, an effect that is apparently enhanced by low nitrogen status, whereas temperatures above 35 °C strongly reduce flavonol production (Olsen et al., 2009; Pastore et al., 2017). Consequently, cool growing seasons, especially those with lower temperatures before veraison, are associated with lower amounts of tannins and higher amounts of flavonols in the berry skin than are warm growing seasons (Ferrandino et al., 2012; Pastor del Rio and Kennedy, 2006).

The impact of temperature on aroma and flavor compounds is not well understood and probably varies depending on the chemical nature of the diverse compounds and their precursors. For instance, higher preveraison temperatures are associated with lower methoxypyrazine production, and higher postveraison temperatures are associated with more rapid methoxypyrazine degradation. Methoxypyrazine production prior to veraison may be more responsive to temperature than is the degradation of these compounds after veraison. Therefore, methoxypyrazine concentrations are highest when berries develop under very cool conditions and, thus, in more northerly latitudes and/or higher altitudes, especially in cool growing seasons. Even if grapes from a warm and a cool region, a warm and a cool vineyard site within a region, or a warm and a cool growing season, were harvested at the same sugar concentration, the "cool" grapes would still typically contain severalfold more methoxypyrazine than their "warm" counterparts (Helwi et al., 2015; Lacey et al., 1991).

The activity of the DOXP/MEP pathway increases strongly with increasing temperature, even above 35 °C (Mongélard et al., 2011). Nevertheless, terpenoid accumulation in grape berries has a relatively broad temperature optimum from approximately 10 °C to 20 °C. Fruit monoterpene concentrations may be inversely correlated with the average daily maximum temperature over the ripening period (Marais et al., 2001). It is possible that this is because terpenes are increasingly volatilized as the temperature increases: terpene emission has a $Q_{10} = 2$–4 in the range 20–40 °C (Loreto and Schnitzler, 2010). Thus, although the production of terpenes increases as the temperature rises, so does their loss through volatilization. Above about 45 °C, however, terpene biosynthesis becomes rapidly inhibited. Norisoprenoids such as β-damascenone or β-ionone appear to be rather insensitive to temperature, but under cool conditions they can be masked by elevated amounts of methoxypyrazine. The "peppery" rotundone, on the other hand, may accumulate faster and to higher concentrations when grapes ripen under cool conditions (Caputi et al., 2011).

6.3.4 Water status

Water supply from both the soil and the atmosphere is very important for fruit composition due to its influence on vine growth, yield formation, and fruit ripening. Most of the differences in grape composition and perceived wine quality caused by soil-related differences among vineyard sites may be attributable to differences in soil moisture rather than, for example, parent rock or soil type and composition (Huglin and Schneider, 1998; van Leeuwen et al., 2004). Nonetheless, deep, fine-textured soils not only hold more water than do shallow, coarse-textured soils, but they also tend to be more fertile, because clay has a higher capacity for nutrient storage and exchange than does sand. At a given vineyard site, the absolute amounts and seasonal changes of soil moisture depend on the annual amount and temporal distribution of rainfall and on temperature-mediated water evaporation. Site-specific and local variation in soil moisture due to differences in effective rootzone, water-holding capacity, and drainage can modulate the response of individual grapevines to climatic factors and has a pronounced impact on vine-to-vine variation within a vineyard. In fact, the variation in vigor and yield among vines in the same vineyard is often related to the variation in plant-available water, in both hot and cooler climates (Cortell et al., 2005; Hall et al., 2002; Lamb et al., 2004). However, the spatial variation in fruit quality is not necessarily the same as the spatial variation in yield.

Water availability influences shoot vigor and thus canopy microclimate (Keller et al., 2016a). Severe water deficit may cause senescence and subsequent abscission of older leaves, which will increase sun exposure of the fruit (Romero et al., 2010; Zufferey et al., 2017). In contrast, high soil moisture stimulates vigor, which can lead to a dense canopy and shaded fruit. A larger total leaf area also has a greater water demand due to transpiration, which in turn increases the vine's vulnerability to drought stress (Keller et al., 2015a; Mirás-Avalos et al., 2017). While abundant water supply favors shoot and berry growth, it delays veraison and slows the rate of fruit ripening (Alleweldt et al., 1984b; Hepner et al., 1985). Although this is generally true for both water supplied by rainfall and that coming from irrigation, it is important to remember that rainfall, unlike irrigation, is also normally associated with both lower temperature and lower light intensity due to cloud cover. Moreover, rainwater adversely affects the canopy microclimate by leading to surface wetness on leaves and fruit, and by increasing the humidity around them. Surface wetness and high humidity are associated with increased disease pressure (see Section 8.2), while water uptake across the berry skin can lead to fruit splitting (see Section 6.2). Moisture also hinders drying of raisin grapes. Thus, at comparable amounts of excess water supply, rainfall should be expected to be more detrimental to fruit composition than is irrigation. There are differences among irrigation methods, too. Supplying irrigation water via overhead sprinklers is similar to rainfall in terms of canopy wetness but not in terms of cloud cover. Flood, furrow, or drip irrigation, on the other hand, do not add water to the canopy surface. Unlike drip irrigation, however, flood and furrow irrigation may still be associated with higher humidity in the canopy due to surface water evaporation. Even less water evaporates from the soil surface when irrigation water is applied by subsurface drip irrigation than by aboveground drip irrigation (Pisciotta et al., 2018).

Large, dense canopies that result from abundant water and nutrient availability are associated with reduced fruit sugar, high acidity, and poor color (Dry and Loveys, 1998; Jackson and Lombard, 1993; Salón et al., 2005). Water deficit, by contrast, typically reduces yield (see Section 7.2) and can increase or decrease berry sugar content, acidity, pH, and color, depending on the extent and timing of the deficit. Some degree of water deficit is regarded as beneficial for fruit composition and wine quality. Grapes, especially wine grapes, have been grown successfully for many centuries on rather marginal sites with infertile, shallow soils of low water-holding capacity, and controlled application of water stress is the premise of modern deficit irrigation strategies (Dry and Loveys, 1998; Dry et al., 2001; Keller, 2005; Kriedemann and Goodwin, 2003; see also Section 7.2).

In rainfed (unirrigated) vineyards located in regions where evapotranspiration exceeds summer rainfall, soil moisture often starts out near field capacity and then dries down more or less gradually, so that the soil water deficit becomes more pronounced as the growing season progresses (Intrigliolo et al., 2012; Zufferey et al., 2017). However, water deficit generally seems to be most effective if applied soon after bloom time by limiting shoot and berry growth and canopy density (Dry et al., 2001; Huglin and Schneider, 1998; Iland et al., 2011; Salón et al., 2005). Considerable improvements in fruit composition under mild to moderate water deficit may be attributable to indirect effects of a smaller, more open canopy and, occasionally, reduced crop load, but water deficit may also have more direct effects on berry size and composition. When water deficit is associated with lower canopy density, the improved light penetration into the fruit zone also raises daytime berry temperature (Keller et al., 2016a; Santos et al., 2005).

Whereas any plant water deficit almost always limits berry size (see Section 7.2), its influence on sugar accumulation is less pronounced (Hepner et al., 1985; Hofäcker et al., 1976; Matthews and Anderson, 1988; Santesteban and Royo, 2006; Stevens et al., 1995; Williams and Matthews, 1990). This apparent difference might arise from a decrease in the ability of berries growing under limited sugar supply to accumulate water so that the sugar concentration, albeit not the amount per berry, remains almost constant. The same principle may apply to the accumulation of mineral nutrients such as K^+ and Ca^{2+} in the berries. Although the amount of these nutrients may be reduced under water deficit, the concentration in the juice may remain unaffected due to the concomitant decrease in berry size (Esteban et al., 1999; Etchebarne et al., 2009; Freeman, 1983; Keller et al., 2008). Indeed, the juice K^+ concentration may even increase with decreasing plant water status, leading to an increase in pH (Dundon and Smart, 1984; Keller et al., 2012). In addition, the decrease in berry size may be mostly due to a smaller mesocarp, whereas the weight of the skin and seeds seems to be less affected by water deficit (Roby and Matthews, 2004). The resulting increase in the skin ÷ juice and seed ÷ juice ratios increases the relative contribution of skin- and seed-derived compounds during winemaking (Casassa et al., 2013).

The decrease in photosynthesis and sugar export from the leaves under more severe stress can curtail berry sugar accumulation, especially if the deficit occurs during ripening (Currle et al., 1983; Dry et al., 2001; Hardie and Considine, 1976; Matthews and Anderson, 1988; Peyrot des Gachons et al., 2005; Quick et al., 1992; Rogiers et al., 2004b; Santesteban and Royo, 2006). Thus, although mild or moderate water deficit may limit berry size and improve fruit composition by restricting shoot growth or by reducing canopy density (Kennedy et al., 2002; Ojeda et al., 2002; Shellie, 2014; van Leeuwen et al., 2004), severe stress, contrary to winemaking lore and European regulations, can delay berry development and ripening because of a reduction in photosynthesis or even leaf abscission (Intrigliolo et al., 2012; Romero et al., 2010, 2013; Williams and Matthews, 1990; Williams et al., 1994). The growth of ripening berries is much less responsive to soil water status than that of preveraison berries, but postveraison berries can shrink, even if slowly, due to dehydration under severe water deficit (Keller et al., 2006; Ojeda et al., 2001; Salón et al., 2005). Such weight loss leads to a concentration effect on berry solutes, so that there is an apparent gain of about 2°Brix for every 10% loss in berry weight (Muganu et al., 2011).

Small berry size is less important for most white wine grapes than for red wine grapes. Skin and seed components are not usually extracted during white winemaking, and the yellow-green color of the so-called white wines is rarely viewed as a limiting trait that requires improvement in the vineyard. This is why in the grape and wine quality literature the term "color" is typically, though inaccurately, used as an acronym for red color. Constituents of the skin, moreover, are not nearly as critical for the quality of nonwine grapes, whose surface $\div$ volume ratio is therefore of little concern. On the contrary, large rather than small berries are typically desired for juice, table, and raisin grapes. Consequently, insufficient water supply, especially before veraison, to vines producing raisin grapes can lead to undesirable loss of berry size and sugar content (Christensen, 1975).

Soil moisture has little effect on tartrate content per berry, although water deficit early during berry development has been found to limit tartrate accumulation (Eibach and Alleweldt, 1985; Esteban et al., 1999). Preveraison malate accumulation, too, tends to decline with a decrease in soil moisture so that the juice titratable acidity is often lower at harvest (Esteban et al., 1999; Keller et al., 2008; Salón et al., 2005; Shellie and Bowen, 2014; Stevens et al., 1995; van Leeuwen et al., 2009). Though postveraison malate degradation may also be accelerated under water deficit, the reduction in berry malate content is more pronounced when the deficit occurs before veraison than after veraison (Matthews and Anderson, 1988; Williams and Matthews, 1990; Zufferey et al., 2017). The amount of ABA in grape berries correlates positively with the concentration of ABA in the xylem sap at all stages of berry development, and ABA may inhibit enzymes involved in malate respiration and gluconeogenesis (Antolín et al., 2003; Palejwala et al., 1985). However, the ABA effect on malate seems to be in the "wrong" direction: More ABA would be expected to preserve malate rather than reduce it. At least part of the decrease in malate with increasing water deficit may therefore be associated with higher berry temperatures due to a decrease in leaf area and canopy density (Intrigliolo and Castel, 2010; Romero et al., 2010).

The concentration of arginine, and hence of yeast-assimilable nitrogen, is lower in berries of water-stressed vines, whereas the proline concentration may increase or decrease, perhaps depending on the extent or timing of the stress (Coombe and Monk, 1979; Freeman and Kliewer, 1983; Matthews and Anderson, 1988; Savoi et al., 2017; Zufferey et al., 2017). Despite these changes in amino acid composition, the amount of PR proteins in grape berries seems to be rather insensitive to water deficit (Pocock et al., 2000). It is thought that the elevated protein concentration that is sometimes found in juice or wine made from water-stressed grapes is due to the smaller size of those berries rather than an increase in protein accumulation.

The improved red pigmentation often observed with mild to moderate water deficit, especially preveraison water deficit, is to some extent simply due to smaller berry size, which increases the skin÷pulp ratio (Freeman, 1983; Intrigliolo et al., 2012; Kennedy et al., 2002; Romero et al., 2010; Salón et al., 2005; Shellie, 2014). In addition, there is also a more direct influence of water deficit enhancing the production of anthocyanins (Dry et al., 2001; Savoi et al., 2017). This effect may arise as a combination of greater fruit sun-exposure due to lower canopy density and of the stimulating effect of root-derived ABA on the genes stipulating flavonoid biosynthesis and on the activity of the corresponding enzymes (Castellarin et al., 2007a,b; Jeong et al., 2004; Keller et al., 2016a; Romero et al., 2010; Fig. 6.9). Exploiting this effect of ABA, table grape growers sometimes apply the hormone as a spray at or just before veraison to boost red coloration of the fruit (Kataoka et al., 1982; Peppi et al., 2008). Even wine grapes, grown in either hot or cool climates that limit anthocyanin production in the skin, may benefit from ABA sprays applied at veraison, which vines may interpret as a perceived water deficit (Balint and Reynolds, 2013; Gu et al., 2011). Application of the synthetic growth regulator ethephon also improves red color formation, perhaps because its breakdown product, ethylene, stimulates ABA production (Szyjewicz et al., 1984).

In addition to enhancing anthocyanin accumulation, water deficit may also shift the relative proportion of individual anthocyanins, favoring the production of pigments with higher degrees of hydroxylation and methoxylation, notably malvidin- and petunidin-based anthocyanins (Castellarin et al., 2007a,b). However, irrigation using the technique of partial rootzone drying (see Section 7.2) may leave malvidin-based anthocyanins unchanged while enhancing accumulation of the glucosides of the other four anthocyanins (Bindon et al., 2008). Contrary to the effects of mild to moderate water deficit, severe water stress, especially before veraison, can lead to uneven ripening with some berries remaining green or coloring poorly.

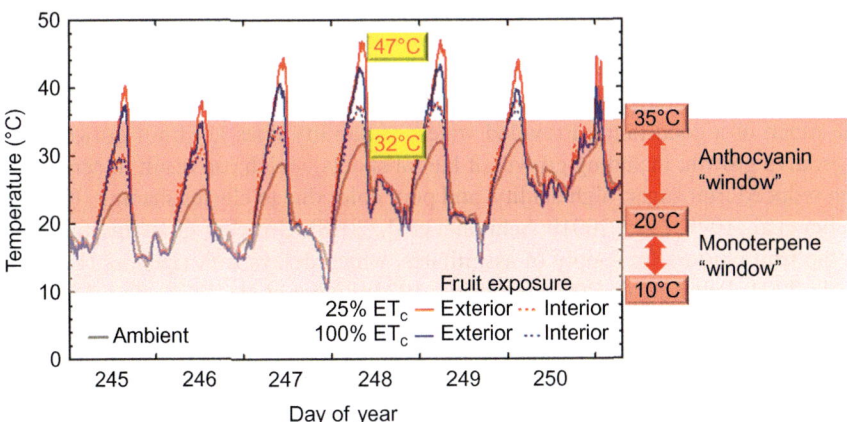

FIG. 6.9

Water deficit decreases canopy size and density, enhancing sun-exposure of grape clusters which, in turn, increases berry temperature during the day. Diurnal ambient and cluster temperature profiles during the first week of September for Cabernet Sauvignon vines irrigated replacing either 100% of crop evapotranspiration (ET_c) or 25% of ET_c.

Modified from Keller, M., Romero, P., Gohil, H., Smithyman, R.P., Riley, W.R., Casassa, L.F., Harbertson, J.F., 2016. Deficit irrigation alters grapevine growth, physiology, and fruit microclimate. Am. J. Enol. Vitic. 67, 426–435.

Unlike anthocyanin accumulation, the accumulation of tannins, flavonols, and cinnamic acids appears to be rather insensitive to changes in plant water status, both before and after veraison, unless the water deficit enhances fruit exposure to sunlight (Bucchetti et al., 2011; Downey et al., 2006; Gény et al., 2003; Martínez-Lüscher et al., 2014). However, low plant water status and low vigor tend to be associated with low methoxypyrazine concentrations in the fruit (Scheiner et al., 2012). Water deficit may accelerate or advance by a few days the postveraison breakdown of methoxypyrazine (Sala et al., 2005). Furthermore, the carotenoid concentration is often lower in grapes grown at low soil moisture; much of this effect can probably also be attributed to higher sun exposure due to the reduced canopy density, which accelerates carotenoid degradation (Savoi et al., 2016). This degradation may be associated with enhanced conversion of carotenoids to norisoprenoids: Water deficit often increases the concentration of β-damascenone in wine (Ou et al., 2010). Mild water deficit is thought to enhance the grapes' aroma potential, for instance by increasing the amounts of monoterpenes and volatile thiol precursors, irrespective of the influence of water status on berry size (Peyrot des Gachons et al., 2005; Savoi et al., 2016). More severe water stress, however, may curtail the berries' aroma potential. This outcome may be particularly relevant for white wine grapes, for which a high aroma potential is much more desirable than a high amount of potentially bitter or astringent phenolics. Indeed, the quality of grapes destined for white wines, and especially sparkling wines, may be impaired by even moderate water deficit (Bellvert et al., 2016). Clearly, grape quality benefits from carefully controlled irrigation in dry climates or dry growing seasons. Perhaps surprisingly, however, despite reducing berry sugar concentration, heavy rainfall during the ripening period has been found not to alter the profile of aroma-active components in grapes (Yuan et al., 2018b).

6.3.5 Nutrient status

Of all mineral nutrients, nitrogen (N) is the most potent in its ability to influence grapevine growth, canopy structure, yield formation, and fruit composition (Bell and Henschke, 2005; Jackson and Lombard, 1993). Increasing N availability enhances photosynthesis, which means that more sugar is available for growth and fruit production and development. Just as with water, however, there can be too much of a good thing. Provided water availability does not limit growth, high N supply favors vegetative growth, including growth of lateral shoots, which can result in dense canopies associated with reduced fruit sugar, high acidity, and poor color due to cluster shading (Keller and Koblet, 1995a; Keller et al., 1998, 1999, 2001b; Schreiner et al., 2018). Growing shoot tips may moreover compete with the fruit clusters for supply of assimilates, which delays fruit ripening (Keller et al., 2010; Lebon et al., 2004; Pallas et al., 2008; Sartorius, 1973; Spayd et al., 1994). In addition, high vine N status increases yield by promoting fruit set, and yet sometimes it also increases berry size, especially if N is applied soon after fruit set (Choné et al., 2006; Keller et al., 2001a; Schreiner et al., 2018).

In addition to these indirect effects on fruit composition, the effects of N can also be more direct because nitrate (NO_3^-) uptake leads to a reprogramming of the expression of many genes involved in metabolism (Scheible et al., 2004). High N status stimulates the production but not the respiration of organic acids including malate, which substitutes for NO_3^- to prevent tissue alkalinization due to NO_3^- reduction (Scheible et al., 2004; Stitt et al., 2002). Although this is sometimes associated with higher malate concentrations in grape berries, juice from berries grown on a high N "diet" nevertheless tends to have a higher pH, possibly because rapid NO_3^- uptake by the roots is often associated with more K^+ uptake (Keller et al., 1999, 2001b; Ruhl et al., 1992; Schreiner et al., 2018; Spayd et al., 1994). Since K^+

helps to neutralize the negative electrical charges of both NO_3^- and malate, uptake of K^+, its transport to the leaves in the xylem, and reexport in the phloem along with malate are all correlated with NO_3^- uptake (Peuke, 2010). Moreover, carbon skeletons are necessary to assimilate NO_3^- to amino acids (see Section 5.3). Increased formation of organic acids and amino acids diverts carbon away from sugar production and impedes the formation of phenolic components, such as anthocyanins, tannins, flavonols, and hydroxycinnamic acids. Even more directly, whereas N deficiency induces enzymes of the shikimate pathway, NO_3^- supply suppresses the expression of genes involved in phenolics production (Fritz et al., 2006; Olsen et al., 2009; Scheible et al., 2004; Soubeyrand et al., 2014; Weaver and Herrmann, 1997). In other words, NO_3^- acts like a dimmer light switch: Abundant NO_3^- keeps the phenol-metabolism genes partly "switched off" or, if they were "on" before NO_3^- supply, it partly switches them off, which hinders the genes' ability to design and build the necessary enzymes. In contrast, carotenoid accumulation and terpenoid release are increased by high N status, but it is not known whether this enhances or suppresses isoprenoid aroma production and accumulation. It is, however, clear that high vine N status stimulates the production of cysteine- and glutathione-bound thiols in grape berries (Choné et al., 2006). Glutathione concentrations themselves are also much higher in high- N grapes, and this may stabilize both aroma volatiles and phenolics in wine by limiting their oxidation (Choné et al., 2006; Monagas et al., 2006). Moderate N availability is thought to maximize the flavor and aroma potential of white grapes, whereas both N deficiency and N excess ostensibly diminish the aroma potential (Peyrot des Gachons et al., 2005). Although each methoxypyrazine molecule comprises two N atoms, however, N supply apparently does not influence the amount of methoxypyrazine in grape berries, at least so long as high N supply is not associated with high vigor (Helwi et al., 2015).

The shade problem created by excessive N supply cannot be overcome by the popular canopy management technique of leaf removal in the cluster zone in an attempt to improve fruit exposure to sunlight, because the berries' high N and low flavonol and carotenoid contents make them more susceptible to sunburn (Düring and Davtyan, 2002; Kolb et al., 2003). Another common "Band-Aid" action is to tip or hedge the excess shoot tips, which can temporarily improve assimilate supply to the fruit clusters and eliminate shade from overhanging shoots. However, repeated hedging on fertile sites or in areas that receive sufficient summer rainfall to permit continued shoot growth throughout the growing season may only make matters worse because it wastes the vine's resources and eliminates young, photosynthetically active leaves while leaving old, less efficient leaves behind (Coombe, 1959; Keller et al., 1999). Moreover, the generation of multiple new shoot apical meristems by repeated stimulation of lateral shoot growth due to the elimination of apical dominance may draw resources away from the clusters and the reserve storage pool in the permanent organs of the vine. In dry climates with seasonal water deficits, however, a somewhat taller canopy that transpires more water may result in greater plant water deficit, which may be beneficial for fruit quality, provided irrigation water can be applied to prevent excessive water stress (Mirás-Avalos et al., 2017).

High soil N also inhibits root growth, which could make vines more vulnerable to water stress in subsequent growing seasons—a consideration that is especially important during the establishment phase of young vines. Nitrogen deficiency, on the other hand, generally reduces fruit set and bud fruitfulness, leading to loss of yield potential. Unless the yield is drastically reduced, however, fruit sugar concentration may be inadequate because there is insufficient N available for efficient photosynthesis (Keller et al., 1998; Schreiner et al., 2013). Whereas soil N status apparently has little effect on berry tartrate, malate concentration tends to be lower at low N availability, but this response may be modified by other environmental variables (Choné et al., 2006; Keller et al., 1998, 1999; Spayd et al., 1994).

A typical response of plants to low N status is the accumulation of phenolic compounds. The stimulation of phenolics production is a response to low NO_3^- content rather than low amino acid status. Therefore, anthocyanin accumulation in dark-skinned grapes is maximized at low to moderate N availability, and it is minimized when high vine N status due to high soil fertility or heavy fertilizer applications coincides with cloudy skies during fruit ripening (Hilbert et al., 2003; Keller and Hrazdina, 1998; Kliewer and Torres, 1972; Soubeyrand et al., 2014). This interaction of N with light affects not only total skin color but also the relative distribution of individual anthocyanins. Conditions that favor overall color accumulation also lead to the most balanced distribution of pigments (Keller and Hrazdina, 1998; King et al., 2014). Because the formation of malvidin-based anthocyanins is less susceptible to stress conditions than that of other anthocyanins, the former can become dominant in grapes grown with high N status, particularly in combination with poor light or excessive heat. Under non-limiting light conditions, and in contrast to the situation with water deficit, low N status favors the production of pigments with lesser degrees of hydroxylation and methoxylation, such as cyanidin- and peonidin-based anthocyanins. Vine N status therefore has a direct influence on the production of individual pigments in the berry skin in addition to the indirect effect brought about by its influence on vigor and fruit set. Moreover, low N status also favors the accumulation of tannins, flavonols, and cinnamic acid derivatives in preveraison berries (Keller and Hrazdina, 1998; Schreiner et al., 2014).

The optimum N supply in a vineyard also depends on the intended use of the grapes. The stimulating action of low N status on phenolics accumulation is certainly desirable for red wine and juice grapes but can be detrimental in white wine grapes and table grapes. Because of the potential of phenolic compounds to increase the perception of astringency and bitterness, accumulation of phenolics is much less desirable in table and raisin grapes and in grapes destined for white wine production than in those destined for red wine production. Indeed, the restriction on phenolics accumulation placed by higher N availability may be advantageous for white wine quality (Choné et al., 2006). In analogy to the irrigation practice of regulated deficit irrigation (see Section 7.2), regulated deficit nutrition rather than simply starving grapevines of N would seem to be a promising vineyard nutrition strategy that can be adapted according to the final use of the grapes (Keller, 2005).

Nitrogen supply influences the transport of NO_3^- and amino acids to grape berries and amino acid production and accumulation inside the berries (Bell and Henschke, 2005). It has been estimated that approximately two-thirds of the berries' total N is derived from the leaves after veraison, and that the crop's demand for N is mainly a function of the number of berries per vine (Treeby and Wheatley, 2006). Moreover, fruit N at harvest correlates positively with leaf N at veraison (Holzapfel and Treeby, 2007). Because NO_3^- transport in the phloem is very limited, almost all the leaf-derived N would have to be in the form of amino acids (see Section 5.3). Arginine accumulation in grapevines is restricted at low soil N availability but responds strongly to N supply (King et al., 2014; Kliewer and Cook, 1971; Schreiner et al., 2014). Arginine increasingly accumulates in the berries as N supply rises, whereas the production of proline and other amino acids may be less responsive to N status (Bell and Henschke, 2005; Bell et al., 1979; Kliewer and Torres, 1972; Spayd et al., 1994). When grapes from vines with high N status are dried during raisin production, they tend to turn dark brown due to the involvement of arginine in Maillard reactions, which produce pigmented melanoidins from sugars.

Yeast-assimilable N increases strongly as soil N status increases, and N supply during ripening sometimes raises fruit N more effectively than N supply earlier in the growing season (Choné et al., 2006; Rodriguez-Lovelle and Gaudillére, 2002; Schreiner et al., 2013, 2018). Consequently, application of soil or foliar fertilizers at veraison is a possible strategy to enhance the N content of the berries if necessary (Keller, 2005). When dry conditions limit root nutrient uptake, supplying N via foliar applications during ripening may have a greater effect on fruit N than supplying it as fertilizer to the soil (Hannam et al., 2016). The response of berry N content to fertilizer application has implications for winemaking: Because yeast cells cannot metabolize proline, N-deficient grapes often lead to sluggish or stuck fermentations (Bell and Henschke, 2005; Spayd et al., 1995). Such fermentations can be associated with malate production by yeast and often result in the formation of H_2S from sulfur (S), SO_2, and the S-containing amino acids cysteine and methionine. H_2S is perceived as a "reduced" smell at low concentration but worsens to a smell of cabbage or even rotten eggs at high concentration. Its production increases in proportion to the amount of S present in the grapes as well (Schütz and Kunkee, 1977). In addition, during fermentation of must of either low or very high N concentration, yeast may also form acetaldehyde and acetic acid. Moreover, low plant N status, especially in combination with water stress, is suspected to be one reason for premature aging of white wines, although increasing N fertilizer application has sometimes exacerbated the problem (Linsenmeier et al., 2007). This phenomenon is termed atypical aging and is associated with the development of an off-odor reminiscent of "acacia flower," "naphthalene," or "floor polish" at the expense of varietal aroma.

A threshold of approximately 150 mg YAN L^{-1} seems to be required for grape must to readily complete fermentation (Bell and Henschke, 2005). Higher concentrations of amino acids in the berries often result in higher concentrations of fruity and floral esters, such as ethyl acetate, ethyl butyrate, and other similar compounds, as well as thiols in wine. In contrast, high amino acid status leads to lower concentrations of higher alcohols known as fusel alcohols, such as propanol, butanol, hexanol, and related compounds, which typically enhances the wine's sensory properties. The β-damascenone concentration in grape berries and the resulting wine also increases as vine N status increases, but so does the amount of "green" C6 components (Yuan et al., 2018a,b). On the other hand, high concentrations of juice N often result in the yeast producing more SO_2 during fermentation. Because SO_2 inhibits bacterial growth, lactic acid bacteria (*O. oeni*) may struggle to carry out malolactic fermentation in such wines. Grape berry proteins (e.g., PR proteins) also increase at high plant N status, which can increase the requirement for bentonite fining in the resulting wines (Spayd et al., 1994).

Indirect effects of high plant N status include heightened vulnerability to fungal diseases, such as *Botrytis* bunch rot and powdery mildew, which may be detrimental to grape and wine quality (Bell and Henschke, 2005; Christensen et al., 1994; Valdés-Gómez et al., 2008; see also Section 8.2). The optimum availability of soil N to supply the vine's demands and satisfy fruit quality requirements depends on the cultivar, intended use of the grapes, and climatic conditions. In a warm, dry, sunny climate or growing season, the optimum N supply may be higher than that under cool, humid, and cloudy conditions. Nitrogen availability also depends on soil moisture because the roots can take up only N ions dissolved in the soil water (see Section 5.3). Whether the N that vine roots take up is derived from synthetic fertilizers or from organic amendments such as manure or compost makes little difference to grapevines. High rates of organic amendments have the same deleterious consequences for fruit composition as do high rates of N fertilizer: low sugars, tannins, and anthocyanins, and high K^+, pH, and vegetal aroma components (Morlat and Symoneaux, 2008).

Mineral nutrients other than N usually have a less pronounced effect on grape composition. Because grape juice consists mostly of the sap from mesocarp vacuoles, the juice pH is somewhat influenced by changes in soil conditions, vine nutrition, and nutrient transport (Conradie and Saayman, 1989). Abundant K^+ availability in the soil may be associated with greater K^+ uptake and transport to the berries, which tends to increase the berry and juice pH, and consequently the wine pH (Jackson and Lombard,

1993; Mpelasoka et al., 2003; Walker and Blackmore, 2012). But even the incorporation into the soil of excessive rates of K fertilizer (up to 900 kg K ha^{-1} for 5 years) does not always lead to a higher juice pH (Dundon and Smart, 1984; Huglin and Schneider, 1998; Morris et al., 1980, 1983, 1987). One reason for this inconsistency might be that higher K status is often associated with higher berry malate concentrations (Hale, 1977; Hepner and Bravdo, 1985; Ruhl et al., 1992). In any case, the presence of high amounts of K$^+$ may lead to precipitation of K-tartrate during juice or wine storage (Morris et al., 1980, 1983). Unlike N supply, K supply may have little impact on the amount of amino acids in grape berries (Schreiner et al., 2014).

High rates of K$^+$ uptake by the roots may decrease Mg^{2+} uptake (Morris et al., 1983; see also Section 7.3). In contrast, high soil Mg or Ca may decrease K$^+$ uptake by the roots and sometimes lead to lower juice and wine pH. The effect of high Ca and Mg availability on juice pH appears to be enhanced by a concomitant increase in the amounts of tartrate and malate in the berries, although the reasons for this effect are unclear. Therefore, contrary to intuition, high soil pH, which is associated with high soil Ca status, may be coupled with high rather than low juice and wine titratable acidity, which may or may not be associated with a lower pH (Noble, 1979). Juice from mature grapes harvested from vines grown on a soil with pH 8.5 usually has a pH of approximately 3.5, much like that from vines grown on a soil with pH 5.5, although the H$^+$ concentration in the two soils differs by a factor of 1000.

We know very little about the influence of soil nutrients other than N on the accumulation of phenolic components and flavor volatiles in grape berries. The stimulation of anthocyanin production in leaves of plants deficient in K, P, or Mg (see Section 7.3) suggests that low nutrient status in general might also enhance the production of phenolics in the fruit. Low sulfur (S) supply also leads to higher flavonoid production (Takahashi et al., 2011). Nevertheless, anthocyanins, tannins, and cinnamic acid derivatives in grape berries are rather insensitive to K or P supply (Schreiner et al., 2014). Similarly, perhaps with the exception of β-damascenone, whose concentration may sometimes increase with higher K supply, other aroma-active compounds in the berries and the resulting wine also appear to be unresponsive to changes in vine K or P status (Yuan et al., 2018a,b). High amounts of Mg or Mn apparently protect anthocyanins from degradation in cell vacuoles, perhaps by stabilizing the pigment molecules via formation of blue anthocyanin–Mg complexes (Sinilal et al., 2011). High vine S status, for example as a result of foliar S applications for powdery mildew control, may increase the concentration of volatile thiols in wine (Lacroux et al., 2008). Although S inhibits microbial growth, S residues on grape berries decrease rapidly following treatment. Consequently, S applications up to about veraison are not usually a concern for fermentation (Thomas et al., 1993). Foliar S applications within approximately 50 days of harvest, however, may lead to elevated S residues on the berries. When such residues exceed 1 μg g^{-1} in the must, they may be associated with elevated H$_2$S production by yeast during fermentation, especially when fermentation is conducted on the skins (Kwasniewski et al., 2014).

6.3.6 Crop load

The classical view of the relationship between grape yield and quality is that of a linear decrease in quality with increasing yield per vine (Currle et al., 1983). However, while lower yield often tends to accelerate sugar accumulation by grape berries, this does not imply greater quality (Matthews, 2015). The "less-is-more" mantra is an oversimplification. There are many instances in which the quantity and the quality of the crop are not related or vary in the same instead of opposite directions (Bowen et al., 2005; Bravdo et al., 1985a,b; Chapman et al., 2004b; Hofäcker et al., 1976; Keller et al., 2005, 2008). In a study conducted over 4 years with 80 individual Riesling vines, the between-vine variation in yield was approximately fivefold higher than that in either juice soluble solids or titratable acidity (Geisler and Staab, 1958). A meta-analysis of Riesling clonal trials conducted at 16 locations over 37 years found that yield as well as soluble solids increased over time, whereas titratable acidity decreased (Laidig et al., 2009). Much of the variation in yield was attributed to vineyard site effects, whereas the change in fruit composition was clearly linked to the rise in average temperature during the same period. Although yields and mean temperatures continued to vary considerably from year to year, the variation in fruit composition declined over time, indicating that composition was not greatly affected by yield and that adavnces in vineyard management also contributed to the change in fruit composition. Similarly, seasonal weather, vineyard site, or other factors were found to be much more important than yield in determining fruit composition or wine quality of Chardonnay, Chenin blanc, Cabernet Sauvignon, or Merlot (Bowen et al., 2005, 2011; Keller et al., 2005, 2008; Ough and Nagaoka, 1984).

It is not so much the crop size or yield per se that is important but, rather, the crop load, which is a reflection of a vine's sink÷source ratio (Bravdo et al., 1984, 1985a; Jackson and Lombard, 1993; Santesteban and Royo, 2006; Winkler, 1958; see also Section 6.1). For instance, an increase in planting distance is typically associated with a higher yield per vine, but vine size, and thus the leaf area per vine, also increases, so fruit composition and wine quality may be completely unaffected (Winkler, 1969). Conversely, when a high planting density, which is usually associated with low yields per vine, leads to competition among shoots for sunlight, the fruit may have higher titratable acidity while at the same time the juice pH is also higher (Falcetti and Scienza, 1989). Simultaneous increases in both organic acids and pH may be caused by recirculation of K^+ from shaded, aging leaves to the clusters. Vines planted at higher density also tend to produce fruit that is higher in methoxypyrazine content (Sala et al., 2005). Furthermore, water availability may be far more important than crop load in determining fruit and wine quality. Excess water supply can compromise quality even in vines whose yield is severely reduced by cluster thinning after bloom, whereas quality may be high in high-yielding vines that experience seasonal water deficit (Bravdo et al., 1985b).

As a general rule, a leaf area of $1-1.5\,m^2$ is required to fully ripen 1 kg of fruit, and this normally results in a yield÷pruning-weight ratio in the range 5–10 (Kliewer and Dokoozlian, 2005; Kliewer and Weaver, 1971). If the crop load is lower than this, grapevines are said to be undercropped or sink limited. Sink-limited vines will invest comparatively more resources in vegetative growth, which can, in extreme cases, delay ripening and reduce fruit quality via the follow-on effects of a dense canopy (Bravdo et al., 1985b). Moreover, berry size may increase to compensate for the low number of berries relative to leaf area (Keller et al., 2008; Kliewer et al., 1983; Santesteban and Royo, 2006). This increase may be a result of elevated rates of sugar import by the berries due to the low sink÷source ratio. Sink limitation may also lead to a strong increase in the import of amino acids from the leaves via the phloem (Do et al., 2010). Consequently, the concentration of arginine and, especially, proline is high in grapes of lightly cropped vines (Kliewer and Ough, 1970; Kliewer and Weaver, 1971). Conversely, if the crop load is so high that there is insufficient leaf area to ripen the fruit, grapevines are said to be overcropped or source limited. Source-limited vines will have slow ripening rates. This is why vines of intermediate vigor often produce both higher yields and better-quality fruit than vines at either end of the vigor spectrum. Such vines are said to be balanced—that is, their crop size matches their vegetative growth and leaf area development. Furthermore, the leaf area required to ripen 1 kg of fruit also depends on the trellis and training system; it appears to be higher in vines with a typical vertically shoot-positioned canopy than in vines that are sprawl trained or trained to divided canopies (Kliewer and Dokoozlian, 2005). In the latter, $<0.8\,\text{m}^2\,\text{kg}^{-1}$ may be a sufficient ratio for adequate fruit ripening.

Overcropping ordinarily delays fruit maturation and therefore decreases grape sugar, pH, amino acids, and anthocyanins, and often increases titratable acidity if harvest cannot be delayed (Bravdo et al., 1984; Huglin and Schneider, 1998; Intrieri et al., 2001; Kliewer and Ough, 1970; Miller and Howell, 1998; Weaver and Pool, 1968; Weaver et al., 1957; Williams, 1996; Winkler, 1954, 1958). Berries on heavily cropped vines begin to ripen later (i.e., veraison is delayed) or they ripen more slowly (i.e., the rate of ripening is low) compared with berries on lightly cropped vines (Nuzzo and Matthews, 2006). Delayed ripening can sometimes be beneficial, especially for white wine grapes in warm climates or warm growing seasons, which may benefit from enhanced acid retention, lower pH, and improved production of aroma precursors in the cooler autumn weather. However, a late harvest also increases the risk of damaging rainfall, the amount and frequency of which typically increase in autumn. It should be noted, moreover, that if equal numbers of berry or cluster samples are collected from both lighly cropped and heavily cropped vines in a nonuniform vineyard to assess fruit maturity, then the sample will overestimate the true maturity status of the vineyard (Rankine et al., 1962). This is because the fruit of lightly cropped vines tends to be more mature at a specific time point than the fruit of heavily cropped vines. To avoid this bias, the sampling strategy should take into account the differences in crop load among vines such that proportionally more berries are sampled from high-yielding than low-yielding vines.

The effect of crop load on berry composition depends on how a difference in crop load is achieved. A pest or disease outbreak or a hail storm can reduce photosynthetically active leaf area after the yield potential has been established, which results in reduced sugar accumulation. If an increase in yield is accompanied by deterioration of the canopy microclimate, then fruit and wine composition will suffer (Reynolds et al., 1994b). In other words, the so-called overcropping effects are often actually shade effects caused by poor pruning practices or other vineyard management errors. For example, if pruning is too light (i.e., too many buds retained), then there may be too many shoots, which leads to dense canopies. If pruning is too severe (i.e., too few buds retained), however, then the few remaining shoots may grow too vigorously and produce too many laterals, which leads to shade in the fruiting zone. In contrast, when an increase in yield is accompanied by an improvement in the canopy microclimate or by a decrease in berry size, fruit composition may be improved as well. For example, the concentration of undesirable methoxypyrazine in grapes grown on high-yielding minimally pruned vines can be strikingly reduced compared with lower yielding, aggressively spur-pruned vines. Indeed, when yields are increased by leaving more buds at pruning, the resulting wines may have more intense red color and are often fruitier and less vegetal, perhaps due to the reduced shoot vigor and smaller berry size (Chapman et al., 2004a,b; Freeman, 1983; see also Section 6.1). If, on the other hand, yields are stimulated by abundant water supply, the result is usually the opposite. Like heavy pruning, excessive early cluster thinning may also increase both berry size and methoxypyrazine concentration (Dunlevy et al., 2013b). Carotenoids are another class of compounds whose production may be higher at higher rather than lower crop loads (Fanciullino et al., 2014). It is unknown, however, whether this results in differences in norisoprenoid accumulation during fruit ripening.

Balancing shoot growth and fruit production is an important viticultural goal. For an individual grapevine, the relationship between crop load and fruit quality generally follows an optimum curve:

The berries' desired attributes increase as the crop load is increased from a very low level, then reach an optimum or plateau, and finally decline when the crop load is further increased (Carbonneau et al., 2007). Under changing external conditions, which include cultural practices, this curve can be shifted upward or downward. Rather than setting a specific, inflexible target yield, economically minded vineyard owners and managers aim for the highest possible crop load that does not compromise quality. Moreover, where weather conditions permit, delayed ripening may be compensated by delayed harvest (Keller et al., 2005; Nuzzo and Matthews, 2006; Matthews, 2015).

On overcropped vines, cluster thinning is typically employed to reduce the crop load and enhance ripening. This can be especially beneficial in cultivars that are prone to overcropping due to their large clusters, such as Syrah, Mourvèdre, Grenache, or Zinfandel. In addition to such wine grape cultivars, heavily cropped table grapes may also benefit from cluster thinning, especially if done soon after fruit set (Dokoozlian and Hirschfelt, 1995). One form of thinning consists of cutting through the clusters during or shortly after bloom to eliminate the distal one-third or half of each cluster. This strategy not only decreases yield but also often leads to less compact clusters due to compensatory stretching of the rachis (Winkler, 1958). The advanced maturity observed following such early thinning might be partly attributed to the tendency for berries of the proximal portion of a cluster to ripen more rapidly than those of the distal portion (Weaver and Ibrahim, 1968). In table grapes, where large berries are desirable, thinning of flower or fruit clusters is sometimes supplemented by berry thinning. A different, more indirect approach developed for wine grapes, exploits the relationship between leaf area and fruit set as discussed in Section 6.1, while avoiding the compensatory berry growth that often results from early cluster thinning. Leaf removal from the fruit zone before or during bloom may reduce fruit set and result in less compact clusters, smaller berries, faster accumulation of berry sugars and anthocyanins, and higher titratable acidity (Gatti et al., 2012; Palliotti et al., 2011; Poni et al., 2006, 2009; Tardaguila et al., 2010). Removing green clusters at veraison, while increasing the uniformity of fruit composition at the time of thinning, does not necessarily improve composition or its variability at harvest, at least not in vines that are sink-limited rather than source-limited (Calderon-Orellana et al., 2014a).

Berry size is often more important in determining fruit composition than is the crop level or even the crop load. Berry development and size are influenced by the number of seeds per berry. While higher seed numbers usually result in larger berries, they also tend to slow down berry development and delay the beginning of ripening (Gouthu and Deluc, 2015; Staudt et al., 1986). The concentration of K^+ in the berries increases as the crop load decreases and berry size increases (Hepner and Bravdo, 1985). However, because tartrate biosynthesis in the berries ceases at veraison, an increase in berry size after veraison can lead to a substantial decrease in tartrate concentration due to a dilution effect. More K^+ and less tartrate in the berries will result in a corresponding increase in juice and wine pH. In addition, larger berries have a relatively smaller skin $\div$ pulp ratio, which has implications for red wine composition and quality due to the importance of the extraction of skin-derived compounds, such as anthocyanins, tannins, and flavonols, during fermentation (Casassa et al., 2016). In contrast to grapes used for red winemaking, large size and crispness are important quality traits of table grapes.

6.3.7 Rootstock

Grafting to a rootstock does not directly affect the quality-relevant traits of the grapes produced by a scion cultivar, because the chemicals responsible for these traits are produced inside the berries and are therefore determined by the genotype of the scion (Gholami et al., 1995; Koch et al., 2010). For example, the aroma profile of Riesling grapes remains the same, regardless of whether Riesling is grown on its own roots or grafted to rootstocks derived from different *Vitis* species. This is true even though a rootstock and a scion may be able to exchange DNA pieces at the graft union (Stegemann and Bock, 2009). Because long-distance transfer of such genetic material beyond the immediate contact zone has never been observed, the only way in which such hybrid cells could become part of a new shoot apical meristem, and thus heritable, is through the emergence of shoots from the graft union itself (Bock, 2010; Stegemann and Bock, 2009). However, because the rootstock to which a scion cultivar is grafted may alter water and nutrient uptake and distribution, plant growth, and yield formation, it seems logical to expect the rootstock to influence fruit composition as well. Consequently, an indirect effect on fruit composition, especially on acidity, is possible due to the potential influence of the rootstock on scion vigor, canopy architecture, and yield components (Keller et al., 2001a; Olarte Mantilla et al., 2018; Ruhl et al., 1988; Schumann, 1974). Rootstock effects on yield and its components, however, are often minor and variable from year to year (Di Filippo and Vila, 2011; Nuzzo and Matthews, 2006; Stevens et al., 2008, 2010; Williams, 2010). Furthermore, rootstock effects on different yield components sometimes cancel each other out (Keller et al., 2012).

It is instructive to contemplate that the adoption of the practice of grafting European wine grapes onto American rootstocks near the end of the nineteenth century (see Section 8.1) did not perceptibly alter the cultivar profile in any European wine region, as it should have if rootstocks notably altered grape and wine quality. Moreover, despite considerable variation in yield, the effects of climate variation, vineyard location, and soil properties on fruit composition still trump the variation introduced by rootstocks (Harbertson and Keller, 2012; Keller et al., 2012; Kidman et al., 2014; Schumann, 1974). For example, in a soil with very high nematode population, Chardonnay vines grafted to 15 different rootstocks were found to produce up to 7 times more fruit than own-rooted vines, with considerable variation from year to year, and yet there was no difference in fruit soluble solids at harvest (McCarthy and Cirami, 1990).

Rootstocks vary in their capacity to take up mineral nutrients from the soil solution (Gautier et al., 2018; Keller et al., 2001b). Because of the association between juice K^+ concentration and pH, it is possible that rootstocks with a lower capacity for K^+ uptake could be used to control wine pH. The main difference, however, may be between own-rooted vines and vines grafted to rootstocks in general. The pH and K^+ are often higher in grapes and wines from own-rooted than from grafted vines (Harbertson and Keller, 2012; Keller et al., 2012; Ruhl et al., 1988; Walker et al., 1998, 2000). In contrast, the ranking of rootstocks among different studies, sometimes even for the same scion cultivar in the same environment, has been anything but consistent (Walker and Blackmore, 2012). Nevertheless, rootstocks may alter amino acids, especially arginine, in the berries of their grafting partners. For example, 140 Ruggeri and 101–14 Mgt may sometimes lead to considerably lower berry amino acid concentrations than some other rootstocks and, particularly, own-rooted vines (Treeby et al., 1998).

One of the more important uses of rootstocks to alter fruit composition is in areas affected by salinity (see Section 7.3). Because rootstocks such as 1103 Paulsen, 110 Richter, 140 Ruggeri, or 101–14 Mgt can exclude much of the salt dissolved in the soil solution from root uptake and xylem transport, the scions grafted to them accumulate less Na^+ and Cl^- in the fruit (Walker et al., 2000, 2010). The differences in fruit ion concentrations also translate into the wines made from such fruit, and they are sometimes associated with differences in anthocyanin concentration as well; lower Na^+ and Cl^- tends to correlate with higher anthocyanins (Stevens et al., 2016; Olarte Mantilla et al., 2018). However, since *V. vinifera* roots are equally strong, if not stronger, excluders of salt ions, or at least Na^+, compared with these rootstocks, grafting to rootstocks may not provide any benefit over own-rooted vines in terms of salt accumulation (Henderson et al., 2018).

In hot and dry climates, where excessive sunlight exposure of the clusters may be associated with poor berry color due to impaired anthocyanin accumulation, relatively vigorous rootstocks may be desirable. Unlike in cool climates, high-vigor rootstocks that are associated with earlier canopy development may enhance anthocyanin accumulation in hot conditions (Nelson et al., 2016). Effects of rootstocks on other aspects of fruit and/or wine composition and sensory quality are as inconsistent as are their effects on yield components, often even more so. Here, too, the climate variation from year to year within a vineyard typically outweighs the influence of any given rootstock (Harbertson and Keller, 2012; Olarte Mantilla et al., 2018).

CHAPTER

Environmental constraints and stress physiology

7

Chapter outline

7.1 Responses to abiotic stress	279
7.2 Water: Too much or too little	284
7.2.1 Water surplus	285
7.2.2 Water deficit	287
7.3 Mineral nutrients: Deficiency and excess	301
7.3.1 Macronutrients	307
7.3.2 Transition metals and micronutrients	322
7.3.3 Salinity	330
7.4 Temperature: Too hot or too cold	335
7.4.1 Heat acclimation and damage	335
7.4.2 Chilling stress	340
7.4.3 Cold acclimation and damage	342

7.1 Responses to abiotic stress

Grapevines, like other plants, obtain their energy from sunlight and require just three categories of resources, or raw materials, to grow and produce fruit: carbon dioxide (CO_2), water (H_2O), and mineral nutrients (Bloom et al., 1985). Nonetheless, they are often exposed to suboptimal growing conditions that cause environmental stress. Such stress is said to be abiotic because it arises from nonbiological constraints, such as an overcast or too-bright sky, heat or cold, water surplus or deficit, and nutrient deficiency or toxicity. These environmental conditions limit either the availability of one or several resources to the plant or the plant's ability to put these resources to use. Although there is a common perception that "stressing" grapevines in the vineyard will improve fruit quality, stress adversely affects plant growth, development, and productivity.

However, the environment per se does not constitute abiotic stress; external conditions are neutral. It is the ability of plant mechanisms to function in a particular set of environmental conditions that determines whether these conditions are "stressful." Depending on the severity and duration of a stressful event and on when it starts and ends, stress can trigger acclimation processes in the entire vine. Shoot and root apical and lateral meristems give vines the flexibility to integrate developmental decisions in response to continuous changes in the environment. Their high degree of developmental and morphogenic plasticity enables grapevines to maximize their photosynthetic efficiency, long-term survival, and reproductive potential. Such plasticity is characteristic of plants that evolved in resource-rich environments (Bloom et al., 1985).

The optimum resource allocation hypothesis holds that plants respond to insufficient resource availability by investing biomass in those organs and processes that enhance the acquisition of the resource that most strongly limits growth, often at the expense of investment in plant parts that have a high demand for that resource (Bloom et al., 1985; Poorter and Nagel, 2000). In general, plants exposed to carbon limitation (e.g., due to cloudy weather) often increase partitioning to the shoots to grow more leaves, whereas plants exposed to nutrient deficiency typically increase root growth or at least limit shoot growth more than root growth (Chapin, 1991; Keller and Koblet, 1995a). Consequently, if a nutrient deficiency is relieved by fertilizer application, grapevines may respond by shifting their carbon investment to favor shoot growth at the expense of the roots. This allows them to enlarge their leaf area in order to enhance acquisition of the now more limiting carbon. The same response would be expected when irrigation water is applied to relieve drought stress—provided mineral nutrients are not limiting. Moreover, the number and size of sinks competing for carbon during stress periods, and their developmental stage or relative priority, are important because they determine which sink is preferentially supplied with the remaining resources and which ones are abandoned in order to guarantee the survival of the plant (Geiger and Servaites, 1991; see also Section 5.1). Thus, many plant reactions to stress involve morphogenic responses that are characterized less by an overall cessation of growth and more by a redirection of growth that entails inhibition of cell expansion, local stimulation of cell division, and changes in cell differentiation (Potters et al., 2007, 2009). Plant hormones are important mediators of these responses, and the various organs of a grapevine respond differently to different hormones.

The auxin ÷ cytokinin ratio, rather than their absolute concentrations, may determine which organs are to be favored during a particular environmental constraint, and tissue auxin and cytokinin concentrations are usually inversely correlated. For example, a low auxin ÷ cytokinin ratio stimulates shoot growth over root growth, whereas a high auxin ÷ cytokinin ratio favors root growth and results in shorter shoot internodes and smaller leaves—but probably stronger apical dominance. In addition, auxin also appears to suppress cytokinin biosynthesis and to promote gibberellin and ethylene biosynthesis, while ethylene interferes with auxin transport toward the roots and instead promotes transport toward the shoots (Negi et al., 2008; Woodward and Bartel, 2005). Whereas auxin produced by the expanding leaves near the shoot tip and transported basipetally from cell to cell and in the phloem inhibits lateral shoot growth, a local increase in shoot-derived auxin in the roots prompts pericycle cells to resume cell division and produce lateral and adventitious roots while at the same time inhibiting root elongation. Conversely, cytokinins produced in the root tips and transported acropetally in the xylem inhibit lateral root initiation but activate paradormant lateral buds and stimulate lateral shoot growth, thus counteracting apical dominance (Ha et al., 2012). Cytokinins apparently can also be produced in very young, expanding leaves, probably by the dividing cells (Nordström et al., 2004). They generally increase sink strength by promoting cell division and differentiation (except in roots, where they inhibit cell division) and delay senescence. The ability of auxin to promote root formation and that of cytokinin to induce shoot formation are used to regenerate plants from undifferentiated callus in tissue culture. In addition, the stress hormone abscisic acid (ABA) normally inhibits growth by inhibiting production of the cell elongation hormone gibberellin and by stiffening the cell walls, although at low concentration

ABA may instead activate growth (Del Pozo et al., 2005; Hartweck, 2008). An organ's sensitivity to ABA may be modulated by cytokinin: Low cytokinin contents increase ABA sensitivity and high cytokinin contents reduce it (Ha et al., 2012). As leaves expand and mature, their sensitivity to ABA increases markedly (Chater et al., 2014); perhaps this is because their cytokinin content decreases.

Various abiotic stress factors, including drought, nutrient excess, salinity, and chilling, cause cell dehydration and hence osmotic stress. One universal mechanism by which plants cope with this challenge is the accumulation of so-called compatible solutes inside their cells (Bohnert et al., 1995; Bray, 1997; Zhu, 2002). Compatible solutes are relatively small, highly water-soluble, stable organic compounds that cannot be easily metabolized and do not disrupt cell functions. These compounds include sugars (mainly sucrose and fructose), sugar alcohols (e.g., mannitol or glycerol), and amino acids (e.g., proline). They probably serve to lower the cells' osmotic potential (Ψ_π). This osmotic adjustment favors continued water uptake or prevents excessive water loss and helps plant tissues to maintain a higher turgor even as the water potential (Ψ) decreases (Tardieu et al., 2018). Compatible solutes also act as osmoprotectants, which stabilize enzymes and membranes and thereby protect them from osmotic stress. In addition, compatible solutes may play a role in protecting the tissues against oxidative stress.

When grapevines experience environmental constraints, they generally suffer from oxygen toxicity, termed oxidative stress (Apel and Hirt, 2004). Oxidative stress normally is a secondary stress that develops as a consequence of the effects of primary stresses, namely osmotic and ionic stresses. Oxidative stress can result from an inability to use photosynthetic energy, because light capture proceeds while carbon fixation declines, which is why stress factors that curtail photosynthesis typically increase the leaves' susceptibility to bright light. The excess energy then interacts with oxygen to form so-called reactive oxygen species. The molecular oxygen (O_2) that is abundant in the atmosphere is rather inactive—if this were not so, we would all burst into flames. But most of the aptly named reactive oxygen species are indeed highly reactive. They are formed as intermediates and by-products of the reduction of O_2 to water (H_2O), in which electrons (e^-) are added in a stepwise manner. This can be written in simplified form as follows [modified from Lane (2002) and Apel and Hirt (2004)]:

$$O_2 + e^- \rightarrow O_2^{\bullet -} + e^- + 2H^+ \rightarrow H_2O_2 + e^- \rightarrow {}^\bullet OH + e^- + H^+ \rightarrow H_2O$$

The reactive agents of decay include free radicals, such as superoxide ($O_2^{\bullet -}$), hydroxyl radical ($^\bullet OH$), and singlet oxygen ($^1O_2$), as well as somewhat less unstable nonradicals, such as hydrogen peroxide (H_2O_2). The production in the photosynthetic chloroplasts of O_2 and its associated dangers are discussed in Section 4.1. In the mitochondria, the reduction of O_2 to H_2O is carried out by the enzyme cytochrome oxidase as part of ATP generation (see Section 4.4) without releasing any reactive oxygen species, but other respiratory reactions are not so "electron-tight." Thus, reactive oxygen species are normal by-products of photosynthesis and, to a small extent, respiration (Halliwell, 2006; Møller et al., 2007; Noctor and Foyer, 1998).

Plants use H_2O_2 and $O_2^{\bullet -}$ as key components of the cascade of signals involved in the adaptation to changing environments (Foyer and Noctor, 2005; Yang and Poovaiah, 2002). However, the formation of reactive oxygen species increases under stress, which can lead to oxygen toxicity (Apel and Hirt, 2004; Mittler, 2002). Unlike the comparatively "lazy" O_2, these highly reactive molecules are very electrophilic: They crave electrons because their oxygen atom is just one electron short of filling its outer shell. To satisfy their demand, they "steal" electrons from other cellular components, which are thereby oxidized. Oxidation of cell membrane lipids, proteins, cell wall polysaccharides, and DNA damages or even kills cells (Lane, 2002; Møller et al., 2007). Oxidative damage to DNA is a major cause of mutations, whereas damaged membranes become leaky, and damaged enzyme and transporter proteins are inactivated. In leaves, it appears that regardless of the initial source and type of reactive oxygen species, they all ultimately result in overproduction of $^1O_2$ by excess light energy, which oxidizes membrane lipids and triggers the processes culminating in cell death (Triantaphylidès et al., 2008). Note that the contact herbicide paraquat acts by inducing $O_2^{\cdot-}$ and H_2O_2 production by stealing electrons from photosystem I (Halliwell, 2006). Failure to quickly scavenge and dispose of the reactive culprits leads to inhibition of photosynthesis and development of chlorotic and necrotic leaves. In addition, overproduction of reactive oxygen species also leads to callose deposition in plasmodesmata, which curtails cell-to-cell movement and phloem transport (Wu et al., 2018).

To cope with oxygen toxicity, plants have evolved a suite of defense systems designed to prevent oxidative damage by detoxifying reactive oxygen species (Apel and Hirt, 2004; Halliwell, 2006; Lane, 2002; Mittler, 2002; Noctor and Foyer, 1998). These defense systems employ antioxidants, which are defined as chemicals that "consume" or quench reactive oxygen species without themselves being converted to damaging radicals in the process. Several of these systems use the amino acid cysteine, which is a component of proteins and peptides and is easily and reversibly oxidized (Møller et al., 2007). One system employs the enzymes superoxide dismutase and catalase to convert superoxide into water. Another, the so-called ascorbate–glutathione cycle, turns H_2O_2 into H_2O by allowing it to oxidize the antioxidant ascorbic acid and then recovering the vitamin C by reducing its oxidized form, partly with the help of the peptide glutathione. Indeed, plants detoxify many electrophilic chemicals by attaching glutathione to them. Because some plant species have "learned" to use this mechanism to detoxify triazine herbicides, they have become herbicide-resistant weeds (Powles and Yu, 2010).

Chemically, oxidation typically means loss of an electron from a compound, whereas reduction means gain of an electron so that the compound's charge is reduced; reactions involving reduction and oxidation are termed redox reactions. The so-called redox state of a cell or tissue is often estimated by the proportion of reduced ascorbate and/or glutathione relative to their total amounts (Potters et al., 2009). Furthermore, the production of carotenoids and flavonoids in addition to compatible solutes and stress proteins also contributes to the detoxification process by scavenging reactive oxygen species or preventing them from damaging the cells' structures. Flavonoids may be oxidized and hence degraded by peroxidase, thereby consuming H_2O_2 (Pérez et al., 2002; Ros Barceló et al., 2003). Oxidative stress, therefore, results from an imbalance between production and degradation of reactive oxygen species.

Although many stressful events reduce the leaves' photosynthetic rate or the total photosynthetic leaf area, the causes of these decreases and the impact of a stress factor vary with the type of stress. For instance, whereas low nutrient availability decreases photosynthetic rates in the leaves, this decrease is not as great as and starts later than the reduction in growth (Chapin, 1991). This is because nutrients are needed to produce proteins and carbohydrate building blocks for the addition of new biomass by the vine's sinks, whereas the photosynthetic apparatus in the source leaves is already established. Therefore, although grapevines grown under conditions of nutrient deficiency show diminished growth, they experience a relative excess of photosynthate rather than a shortage, and they accumulate sugar and starch in their leaves (see Section 7.3). Low water availability also decreases photosynthesis, but this does not result in starch accumulation in the leaves because the decrease in growth due to water deficit often approximately matches the decrease in photosynthesis (Poorter and Nagel, 2000; see Section 7.2).

Eventually, stress may induce a depletion of assimilates available for export from the leaves. Carbon depletion in turn stimulates processes involved in photosynthesis and reserve remobilization, whereas abundant availability of carbon favors carbon utilization, export, and storage (Hellmann et al., 2000). The former is termed a famine response and the latter a feast response. When they are starved for sugar, plant cells initially consume their available starch reserves and then sacrifice their cell membrane phospholipids, especially linoleic and linolenic acids, to recycle fatty acids and other metabolites (Aubert et al., 1996). Under prolonged and severe stress that occurs as a consequence of a deficit or surplus of water or nutrients, or extremes of temperature, senescence of cells and, eventually, whole organs sets in. Senescence is defined as the organized degradation of cell constituents that ends in death and is generally followed by shedding of the dead organs by breakdown of the cell walls in the preformed abscission zones. When a grapevine sacrifices an organ or part of its structure in such an ordered and genetically programmed way, this constitutes an adaptive strategy to survive a stress period.

Several plant hormones interact to bring about and coordinate the premature abscission of leaves, flowers, or fruit. ABA stimulates the senescence process, and ethylene operates as an accelerator of abscission, whereas auxin acts as a brake (Roberts et al., 2002). Therefore, ABA and ethylene can trigger senescence and abscission only in organs whose auxin concentration is low. For instance, although a variety of soil-related stress factors result in ABA production in the roots and ABA transport to the shoots, developing leaves and flowers, as well as grape berries during the period of seed development, are usually protected from shedding by their high auxin levels. This ensure that only fully developed leaves or flowers, or berries with mature seeds, are typically abscised. However, the original trigger and partial executioner of the senescence program again appears to be oxidative stress, namely in the form of $^1O_2$ accumulation (Triantaphylidès et al., 2008).

Different tissues within the same organ do not die at the same time. In senescing leaves, the mesophyll cells are the first to be disassembled, followed by the epidermis and, finally, the vascular bundles. The moment of death of a cell occurs when the vacuolar membrane ruptures (van Doorn and Woltering, 2004). But the earliest and most dramatic change in the cellular structures during leaf senescence is the breakdown of the chloroplasts, which contain the cell's photosynthetic machinery and most of the leaf protein and are the site of major biosynthetic processes. This means that photosynthesis declines due to rubisco and chlorophyll degradation. Rubisco thus serves a dual role: it is both the main photosynthetic enzyme and the main storage protein for nitrogen remobilization (Thomas, 2013). Meanwhile, the chloroplasts in the guard cells remain green and functional, probably to keep the stomata closed to prevent desiccation before nutrient recycling is complete (Tallman, 2004). The carbon, nitrogen, and other nutrients stored in the leaf's proteins, nucleic acids, polysaccharides, and lipids are rapidly remobilized during senescence and can be used to sustain the growth and metabolism of important sink organs such as young leaves, clusters, or roots. Of course, the distribution of nutrients recovered from senescing leaves depends on the relative strength or importance of the various sinks of the vine. Moreover, strong sink demand may increase this nutrient recycling, whereas weak demand will decrease it. For example, strong sinks that cannot satisfy their demand for nitrogen by import from the roots will induce nitrogen remobilization from the reserve pools in older tissues (Thomas, 2013). Nevertheless, overall lack of sink demand can also induce early leaf senescence, which can be triggered by a surplus of uncommitted assimilates (Keller et al., 2014; Thomas, 2013; see also Fig. 5.4).

7.2 Water: Too much or too little

Water availability influences canopy development and microclimate, yield formation, and fruit ripening, and lack of water constitutes the overwhelming limitation to plant growth and yield formation (Kramer and Boyer, 1995). Under natural conditions, water is supplied by snow and rainfall, and the portion of water that is not drained by gravity is temporarily stored in the soil for extraction by plant roots or surface evaporation. Following each recharge, the stored water reserves decrease progressively as water evaporates from the soil surface or is transpired by plants; this evapotranspiration (ET) increases as the temperature rises (see Section 3.2). Where water is not limiting, the total water use of a vineyard over the course of a growing season in a cool climate or short season varies from 300 to 600 mm (3–6 ML ha^{-1}), depending on the cultivar, planting density, canopy size and configuration, and seasonal temperature patterns (Evans et al., 1993; Keller et al., 2016a; Williams, 2014). Vineyard water use in a warm climate with longer growing seasons ranges approximately from 400 to 800 mm, or 4–8 ML ha^{-1} (Williams and Baeza, 2007).

The amount of rainfall varies greatly from region to region and from year to year, which may impair vine performance and the economics of viticulture in some regions and some years. Water availability depends not only on how much rainfall a vineyard receives but also on when the rain falls and how rapidly the water evaporates. In addition, soil water-holding capacity and hence the amount of plant-available water vary with soil depth, texture (i.e., the relative proportions of clay, silt, and sand particles), and organic matter content. For example, a loamy soil can hold up to 25% water (0.25 g H$_2$O per gram soil), a clay soil even 40%, whereas a sandy soil may contain only 12% water at field capacity. The field capacity is also called drained upper limit, whereby the soil pores with a diameter of <30 μm retain water against gravity after all the water in the larger pores has drained away and the soil $\Psi_M \approx -0.01$ MPa. The loam's higher water-holding capacity compared with sand is counteracted to some extent by the fact that a larger fraction of the water is held too tightly by the loam to be accessible to the roots. Although roots can extract water more easily from water-saturated sand than from loam, a coarse soil dries much more rapidly so that water extraction becomes much more difficult in the sand as the soil water potential (Ψ_{soil}) decreases (Chapman et al., 2012). This is because the hydraulic resistance to water movement in the soil (r_{soil}) is approximately 10 times lower in a water-saturated sand than in a loam but increases in a drying soil by the same mechanism that causes cavitation in the xylem (see Section 3.3): The pores of a drying soil fill with air, which results in failure of capillary forces. Just as larger xylem vessels are more vulnerable to cavitation, so too are larger soil pores more susceptible to capillary failure.

The more sand a soil contains, the larger are its pores, and the coarser the soil type, the more limiting the soil is to plant water use, whereas in a fine-textured soil with high clay content the xylem's vulnerability to cavitation is more limiting (Sperry et al., 1998). Grapevines growing in sandy soils therefore often invest more resources in root growth than vines growing in loamy soils. This increases the root $\div$ shoot ratio, which compensates for the lower soil water extraction potential. Similarly, a coarse soil favors greater rooting depth than a fine-textured soil because less water is available for extraction at low Ψ_{soil} in the surface layers of a coarse soil. But even for the same soil type there can be large variation in pore size and hence physical properties depending on soil management practices (Chapman et al., 2012). Soil compaction, for example, decreases pore size and increases r_{soil}, which curtails the soil's ability to transport water and nutrients and constrains root growth. Spatial and temporal variation in soil moisture due to differences in water-holding capacity and effective rootzone strongly influences vine performance both between and within vineyards (Hall et al., 2002; Lamb et al., 2004).

7.2.1 Water surplus

When a soil is saturated with water, its moisture content is well above field capacity and the soil is said to be waterlogged. Grapevines can be grown hydroponically, which shows that they do not suffer from an excess of water during waterlogging. However, at high soil moisture, such as during flooding due to heavy rainfall or excessive irrigation, especially on poorly drained soils and soils with high ground water levels, water drives gases out of the soil pores so that there is insufficient oxygen (O_2) for proper root function and growth (Kreuzwieser and Rennenberg, 2014; Mancuso and Boselli, 2002). The condition of O_2 deficiency is termed hypoxia or, in extreme cases of O_2 being absent altogether, anoxia. This is because gases diffuse four orders of magnitude more slowly in water than in air. Due to the dependence of respiration on O_2 supply, waterlogged roots can become so O_2 deficient that they cannot respire. To continue producing ATP via glycolysis the roots switch to anaerobic fermentation of pyruvate to recycle NADH to NAD, which is far less efficient than respiration and uses up stored carbohydrates while producing lactate and, later, ethanol (Bailey-Serres and Voesenek, 2008; Kreuzwieser and Rennenberg, 2014; Ruperti et al., 2019; Zabalza et al., 2009). Much of the ethanol may be transported via the xylem to the leaves where most of it is converted to acetate for recycling back into primary metabolism, and the remainder may evaporate from the leaves (Kreuzwieser and Rennenberg, 2014). The depletion of storage reserves under hypoxia is aggravated by the susceptibility of phloem transport to low ATP availability, which impairs phloem unloading (van Dongen et al., 2003).

Inhibition of respiration is accompanied by an increase in reactive oxygen species and a decline in the pH of the root cells' cytosol, and the excess protons in turn block the aquaporins in the cell membranes so that the roots become less water permeable (Kreuzwieser and Rennenberg, 2014; Tournaire-Roux et al., 2003). Consequently, but somewhat paradoxically, waterlogging leads to a rapid increase in root hydraulic resistance, which slows water uptake and is associated with a decline in stomatal conductance, transpiration rate, and Ψ_{leaf} (Aroca et al., 2012; Flore and Lakso, 1989; Keller and Koblet, 1994; Striegler et al., 1993). Thus, similar to water deficit, an excess of water can lead to partial closure of the stomata, which over time diminishes photosynthesis and may damage the photosynthetic apparatus (Else et al., 2009; Kreuzwieser and Rennenberg, 2014; Stevens and Prior, 1994). In addition, energy-intensive processes such as cell division come to a halt. Cell division also ceases when uncontrolled acidification of the cytosol, due to leakage of acids from the vacuoles and slow escape of CO_2 produced during fermentation, leads to death of the root meristem cells (Roberts et al., 1984).

Excessive soil water, similar to the effect of soil compaction, also seems to increase ethylene production by the roots and prevents the diffusion of this gas away from the roots (Feldman, 1984; Kreuzwieser and Rennenberg, 2014; Sharp and LeNoble, 2002). Small doses of ethylene produced by the roots from the amino acid methionine normally promote root growth, perhaps by increasing tissue responsiveness to auxin and gibberellin (Kende and Zeevaart, 1997; Potters et al., 2009). Together with auxin, ethylene also induces the formation of lateral roots and root hairs, so long as the ethylene gas can easily diffuse away from the roots. If diffusion is inhibited, however, ethylene interferes with cell division in the root apical meristem and delays the inactivation by gibberellin of the growth-suppressing DELLA proteins so that root growth slows (Dugardeyn and Van Der Straeten, 2008; Feldman, 1984; Mancuso and Boselli, 2002).

Moreover, anoxia increases the xylem sap pH, which is a response similar to that in vines experiencing water deficit, and leads to reduced shoot growth. The leaves of waterlogged vines may turn red and roll downward as a result of the accumulation of surplus sugar in the leaves (Kreuzwieser and Rennenberg, 2014; Fig. 7.1). In some cases, the leaf symptoms due to waterlogging are indistinguishable from those due to leafroll virus infection (see Fig. 8.3) or other factors that lead to foliar sugar

FIG. 7.1

Merlot leaves in late summer, showing symptoms of drought stress (left) or waterlogging (right), with an unstressed leaf in the center.

Photo by M. Keller.

accumulation (see Fig. 2.7). Inflorescence differentiation as well as pollen germination and pollen tube growth decrease as soil hypoxia worsens, so flooding can be detrimental to bud fruitfulness and fruit set; the latter may be reduced to zero if the soil O_2 concentration decreases below 5% (Kobayashi et al., 1964; Stevens et al., 1999).

If only a portion of the root system experiences waterlogging, then grapevines may compensate by increasing root growth in the nonaffected portion, which is typically in the surface soil (Kawai et al., 1996; Stevens et al., 1999; Striegler et al., 1993). *Vitis* species vary in their susceptibility to waterlogging (Mancuso and Boselli, 2002; Mancuso and Marras, 2006). For instance, *Vitis riparia* can tolerate root flooding for several days by rapidly suspending root metabolism, especially nutrient uptake and assimilation, in order to reduce ATP demand and thereby conserve energy. On the other hand, susceptible species such as *V. rupestris* are unable to adapt their root metabolism and literally run out of energy. In addition, the inability of susceptible species to suppress nutrient uptake may lead to K^+ poisoning, membrane leakage, and death.

Flooding during the dormant winter season has little effect on grapevines, but it may slow the rate of budbreak and initial shoot growth if it occurs around the time of budbreak (Ruperti et al., 2019; Williams et al., 1994). Many wild grapes that have evolved alongside or near riverbanks frequently experience flooding during or soon after budbreak, when the melting snow in the mountains upstream leads to an annual peak in river runoff. Beyond periodic spring floods, overly abundant water supply during the growing season is a predicament that mainly concerns coastal regions and areas on the windward side of mountain ranges, as well as some subtropical and tropical regions that experience frequent summer rainfall. Many Chinese grape growing areas, for example, suffer from the fact that the rainy season coincides with the growing season. Because rainfall requires a cloud cover, such regions also often experience low light intensities that curtail photosynthesis (see Section 5.2). In Bordeaux, France, soils with a propensity to waterlogging are associated with poor wine quality compared with free-draining soils, and interannual variation in water availability under such conditions may be more important for wine quality than is variation in growing-season temperature (Renouf et al., 2010; van Leeuwen et al., 2009). Furthermore, grapevines grown in pots are quite often at risk from hypoxia.

The soil water content after drainage of irrigation water from pots is usually much higher ($\Psi_{soil} \approx -0.001$ to -0.003 MPa near the surface) than that in a vineyard at field capacity and depends on the height of the pot, which is why frequently irrigated potted plants are more likely to suffer from waterlogging than from drought (Passioura, 2006).

7.2.2 Water deficit

Many of the world's grapes are grown in more or less Mediterranean climates, which experience cool, moist winters and warm, dry summers with high evaporative demand from the atmosphere. Thus, across regions the natural supply of water and the supply of carbon for growth and fruit production are often inversely related (Bloom et al., 1985). In Mediterranean regions and regions with continental climates on the leeward side of mountain ranges, winter precipitation often, but by no means always, saturate the soil profile to field capacity (Celette and Gary 2013; Davenport et al., 2008b). Dormant-season refill of soil water stores can be enhanced by cover crops that reduce runoff and improve infiltration, even though the cover crop may later compete with grapevines for water during the growing season (Battany and Grismer, 2000; Celette et al., 2008; Klik et al., 1998). The soil then progressively dries down as temperatures rise and water evaporates, and the growing season is often characterized by declining soil moisture and high vapor pressure deficit (VPD) due to warm, dry air and limited summer rainfall. The resulting increase in air spaces in the soil and water retention by successively smaller pores leads to an increase in surface tension, which in turn is responsible for the development of increasingly negative pressure in the soil. In arid soils, Ψ_{soil} can decline to -3 MPa, compared with -0.01 to -0.03 MPa in a soil at field capacity. Water in soil pores $<0.2\,\mu m$ wide cannot be extracted by the roots and thus constitutes the lower limit of plant-available water (Watt et al., 2006). A suction of about 1.5 MPa would be required to overcome the soil Ψ_M in these pores and extract water from them; this lower threshold of soil water availability is termed the permanent wilting point.

Vitis vinifera is rather drought-tolerant and can be grown without irrigation in areas that receive as little as 300 mm annual precipitation, although growth and crop yield are very limited under such conditions (Huglin and Schneider, 1998). The water balance of grapevines and their cells is determined by the amount of water lost by evaporation to the atmosphere (see Section 3.2) and the amount of water absorbed from the soil. A vine can become water-stressed as a result of both decreased Ψ_{soil}, which generally occurs progressively over time, and fluctuating transpiration rate (E), which occurs with daily and seasonal changes in VPD due to fluctuations in radiation and humidity (Williams et al., 1994). The relationship between leaf water potential (Ψ_{leaf}) and these variables can be expressed as follows:

$$\Psi_{leaf} = \Psi_{soil} - E r_h$$

where r_h is the total resistance to water flow in the vine (see Section 3.3). In the plant physiology literature, the predawn Ψ_{leaf} is often used as a proxy for Ψ_{soil}, assuming the plant water status equilibrates with Ψ_{soil} before transpiration intensifies at sunrise (see Section 3.3).

When transpiration exceeds water absorption by a vine's root system, cell turgor (P), relative water content (RWC; percentage of tissue water relative to the water content of the same tissue at full turgor) and cell volume decrease, whereas the concentration of solutes in the cell increases, reducing the cell's osmotic potential (Ψ_π) and Ψ. Consequently, the response of the cells to water deficit involves a response to osmotic stress. Mild plant water stress has been defined as a decrease of Ψ_{leaf} by several bars or of RWC by 8–10% below corresponding values in well-watered plants under mild evaporative demand (Hsiao, 1973). Moderate water stress corresponds to a decrease of Ψ_{leaf} by more than several bars but less than 1.2–1.5 MPa or of RWC by 10–20%, and severe water stress corresponds to a lowering of Ψ_{leaf} by more than 1.5 MPa or of RWC by more than 20%. Because declining water availability reduces the leaves' stomatal conductance (g_s) for water vapor, the severity of water stress has also been classified according to the decrease in g_s (Flexas et al., 2002; Lovisolo et al., 2010): g_s of well-watered grapevines is usually in the range 200–500 mmol H_2O m^{-2} s^{-1}; under mild water stress, g_s declines to near 150 mmol H_2O m^{-2} s^{-1}; under moderate water stress, to 50–150 mmol H_2O m^{-2} s^{-1}; and under severe stress, to <50 mmol H_2O m^{-2} s^{-1}. In nonstressed grape leaves, g_s typically falls below 50 mmol H_2O m^{-2} s^{-1} only at night (Rogiers et al., 2009; Rogiers and Clarke, 2013).

Stomatal limitation and metabolic limitation

Stomatal closure diminishes water loss by transpiration but also limits photosynthesis (Fig. 7.2), because CO_2 diffusion into grape leaves is much more dependent on open stomata than is H_2O vapor diffusion out of the leaves (Boyer et al., 1997; Escalona et al., 1999; Flexas et al., 1998, 2002; see also Section 4.2). This is called stomatal regulation or stomatal limitation of photosynthesis and is accompanied by a decrease in ATP and NADPH consumption because less energy is needed for CO_2 fixation. Not surprisingly, the effect on photosynthesis and assimilate export of short episodes of high VPD is similar to the effect of longer periods of soil water deficit (Shirke and Pathre, 2004). However, even a stress that is mild enough not to diminish photosynthesis can reduce shoot growth, and hence canopy development, because a reduction in cell expansion usually occurs before the stomata begin to close (Hsiao, 1973; McDowell, 2011; Muller et al., 2011). The combination of smaller total leaf area and decreased photosynthesis will result in reduced daily assimilate production by the grapevine canopy (Perez Peña and Tarara, 2004; Poni et al., 2008; Tarara et al., 2011). A reduction in canopy photosynthesis will reduce the amount of photosynthate available for export to the vine's sink organs. Nonetheless, as long as the water deficit is mild enough to restrict growth more than photosynthesis, there is relatively more sugar available for other sinks (Lawlor and Cornic, 2002; McDowell, 2011; Muller et al., 2011). This may benefit fruit ripening and carbohydrate storage in the vine's perennial organs. Whereas leaf sugar concentrations may remain high for some time, starch becomes depleted and

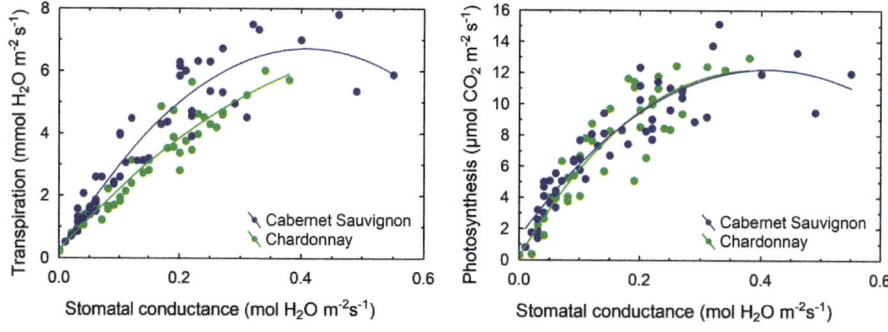

FIG. 7.2

Relationship between stomatal conductance and transpiration (left) and photosynthesis (right) of mature leaves of two grapevine cultivars during soil drying (Keller, unpublished data). Note the steeper decline of photosynthesis compared to transpiration at low stomatal conductance.

assimilate export decreases progressively with increasing severity of water deficit (Dayer et al., 2016; Düring, 1984; Quick et al., 1992). Consequently, more severe water stress will reduce sugar supply to the berries and prevent replenishment of storage reserves, especially in the roots (Holzapfel et al., 2010; Keller et al., 2016a; see also Section 6.2). At veraison, when the sink strength of grape berries reaches its peak, the berries may attract sugar remobilized from storage reserves to satisfy their heavy demand even under severe water deficit (Keller et al., 2015b). Although the sugar concentration in the phloem sap increases as water availability declines, severe water stress may eventually lead to phloem transport failure, because the phloem may no longer be able to maintain turgor (Sala et al., 2010).

Under severe water stress, and once g_s has fallen below about 50 mmol H_2O m^{-2} s^{-1}, the photosynthetic metabolism is progressively impaired, initially by a reduction of electron transport and ATP synthesis that inhibits the Calvin cycle and, at very low RWC, by inhibition of light harvesting and photosystem function (Escalona et al., 1999; Hsiao, 1973; Tezara et al., 1999; see also Section 4.1). The amount of rubisco protein in the leaves also declines under severe drought stress (Bota et al., 2004). Yet this might be a consequence rather than a cause of reduced photosynthesis. This depression of photosynthesis is called metabolic regulation or metabolic limitation, and in contrast to stomatal limitation, it is irreversible (Escalona et al., 1999; Lawlor and Cornic, 2002). The decrease in photosynthesis is accompanied by only a slight decrease in respiration and, possibly, an increase in the proportion of photorespiration relative to photosynthesis, which may serve to counter an energy overload by "consuming" excess electrons (Cramer et al., 2007; Düring, 1988; Lawlor and Cornic, 2002; Lawlor and Tezara, 2009; see also Section 4.3). Therefore, as water stress intensifies, an increasing proportion of fixed CO_2 is lost from the leaf. The more severe the stress, the more of this CO_2 is derived from stored carbohydrates (Lawlor and Cornic, 2002; McDowell, 2011). As internal reserves are depleted, amino acids, particularly proline and glutamate, accumulate because their production (supply) exceeds consumption (demand). This in turn may lead to nitrate accumulation in the leaves, possibly due to feedback inhibition of nitrate reductase (Patakas et al., 2002; see also Section 5.3). Unlike under nonstress conditions, the carbon skeletons for the production of amino acids in water-stressed leaves are provided by glycolysis and organic acids from the citric acid cycle, rather than by assimilates directly from the chloroplast, so that the amount of organic acids decreases (Lawlor and Cornic, 2002).

When the leaves lose turgor due to severe water stress, they will wilt; the leaf surface area decreases, and the leaves hang down and become parallel to the solar rays (Lawlor and Cornic, 2002; Romero et al., 2017; Smart, 1974). By limiting light absorption, wilting is an effective mechanism to reduce a potential "energy overload" that would damage the photosynthetic system (Flexas et al., 1999b; Lawlor and Cornic, 2002). Even before they wilt, leaves move downward via bending of the petiole due to reversible K^+ redistribution from the abaxial to the adaxial side, which leads to osmotic water redistribution and, consequently, cell expansion on the petiole's upper side and shrinkage on its lower side (Nieves-Cordones et al., 2019). Nevertheless, water stress can be especially damaging at high light intensity that is associated with a relative energy surplus in the chloroplasts (Lawlor and Tezara, 2009).

Moreover, because the cuticle allows small amounts of water vapor to pass through by diffusion, leaves cannot avoid water loss completely, no matter how tightly the stomata are closed (Boyer, 2015; Boyer et al., 1997; see Fig. 7.1). In grapevines, the cuticular conductance or water permeability is of the order of 5 mmol H_2O m^{-2} s^{-1} (Escalona et al., 2013.) The cuticular transpiration increases exponentially with rising temperature because both the water permeability and the atmospheric VPD that provides the driving force for water loss increase as the temperature rises (Riederer and Schreiber, 2001). The problem is particularly acute when drying wind reduces r_b and stimulates transpirational water loss. As a leaf dries out, however, its declining turgor seems to be associated with a tightening of the cuticle, and hence increasing cuticular resistance to water vapor (Boyer, 2015). Although young leaves may be more sun-exposed and thus experience higher evaporative demand than older leaves, the old leaves wilt at a higher Ψ_{leaf} than the young leaves because the latter are better able to osmoregulate (Hochberg et al., 2017; Patakas et al., 1997). The wilting older leaves produce plenty of ABA, which is exported to the young leaves to reduce water loss.

In addition, ABA can result in oxidative stress via increased production of reactive oxygen species that, when produced in excess, damage and eventually kill the plant cells due to uncontrolled oxidation (see Section 7.1). Accumulation of reactive oxygen species stimulates the vine's antioxidant defense system to prevent oxidative damage, which results in higher foliar concentrations of antioxidants such as carotenoids (especially xanthophylls), glutathione, ascorbate, and α-tocopherol (Cramer et al., 2007). If the drought stress is not relieved by rainfall or irrigation, the increased ABA and reduced cytokinin contents accelerate leaf aging and lead to senescence of older leaves (Romero et al., 2010; Yang et al., 2002; Zufferey et al., 2017). Senescence, which is initiated when leaf cytokinins fall below a threshold level, involves the disassembly of the chloroplasts with their photosynthetic machinery and is accompanied by a decline in chlorophyll and remobilization of carbon, proteins, and mineral nutrients from these leaves. It culminates in leaf abscission, which can also be induced by excessive xylem cavitation in the leaves and petioles, beginning near the shoot base, whereas the meristems inside the buds remain alive (Hochberg et al., 2017; Romero et al., 2017). This drought-induced deciduous behavior protects the shoot and trunk system from hydraulic failure, which enables vines to survive severe drought by drastically reducing water evaporation and conserving resources while also promoting early inception of bud dormancy.

Once initiated, the rate of leaf senescence accelerates with increasing temperature so that a heat wave following drought stress can lead to severe defoliation of grapevines. The cytosolic glutamine synthetase (GS1; see Section 5.3) in the phloem may be involved in the production of glutamine for nitrogen export and of proline for nitrogen storage during water stress or senescence. Some of the sugars, amino acids, and phloem-mobile mineral nutrients (see Section 7.3) can be recycled to the fruit, which may partially sustain ripening. Such nutrient recycling may lead to an undesirable increase in fruit K^+ content that may subsequently increase juice and wine pH (see Section 6.2). This effect is probably most pronounced if severe water stress occurs during a period in which the clusters' sink strength is at its maximum—that is, during and immediately after veraison. Even under drought conditions leading to complete defoliation, the perennial parts of grapevines rarely die. Restoring water supply to defoliated plants usually results in budbreak and regrowth, presumably aided by the generation of root pressure (see Section 3.3). Nevertheless, partial or complete dieback, beginning in the youngest portion of the shoots and roots, can occur if complete hydraulic failure persists for prolonged periods.

Xylem cavitation and repair

The primary function of stomata is to avoid damaging water deficits that would cause xylem cavitation (Brodribb and Holbrook, 2003; see also Section 3.3). Nonetheless, cavitation is thought to be quite common in grapevines and might be important in inhibiting shoot growth at moderate water deficits (Lovisolo et al., 2008; Lovisolo and Schubert, 1998; Schultz and Matthews, 1988a). Most of the embolisms that result from cavitation occur in the leaves and petioles rather than in the shoots or trunks. Within the leaf, the large major veins cavitate before the small minor veins do (Brodribb et al.,

2016; Hochberg et al., 2017). In the relatively vulnerable major veins and petioles, a significant rise in cavitation events, which is associated with a rapid increase in r_h, typically occurs only *after* the stomata have closed completely in response to water deficit (Hochberg et al., 2016a,b, 2017; Romero et al., 2017). In addition, the strong increase in a leaf's r_h outside the xylem, perhaps because aquaporins in the cell membranes close as the cells dehydrate and lose turgor, tends to protect the leaf xylem itself from cavitation during mild to moderate water deficit (Bartlett et al., 2016; Scoffoni et al., 2017).

The shoots and trunk, at least after they have produced some lignified secondary xylem, are much less prone to cavitation than the leaves (Choat et al., 2005; Hochberg et al., 2016a; Lovisolo et al., 2008; Nardini et al., 2001; Zufferey et al., 2011). The proportion of cavitated vessels in a shoot apparently does not exceed 50% until the xylem water potential (Ψ_x) declines below -1.7 to -2 MPa, whereas in the petioles this proportion may surpass 50% at $\Psi_x \approx -1.5$ MPa, depending on cultivar and extent of secondary growth, and approaches 100% at $\Psi_x = -2$ MPa (Charrier et al., 2016; Choat et al., 2010; Hochberg et al., 2016b; Zufferey et al., 2011). The primary xylem near the pith of a shoot is the most vulnerable to cavitation, and cavitation spreads radially outward as drought stress intensifies (Brodersen et al., 2013).

Although the leaf veins may suffer from embolisms during days with high VPD, the high concentration of inorganic ions and organic molecules in the leaves may generate the positive pressure needed to refill the veins relatively easily (De Boer and Volkov, 2003; Trifilò et al., 2003; see also Section 3.3). Moreover, the vessels and tracheids in leaves and petioles are much shorter than those in the rest of the plant, and the pits in the adjoining end walls of these tracheary elements normally do not allow air to pass into the next vessel—that is, pits function as hydraulic safety valves (Choat et al., 2008). This prevents spreading of embolisms to the shoot xylem network following leaf damage and abscission, a sacrificial protective mechanism that acts as a circuit breaker and has been dubbed the "lizard's tail" strategy (Hochberg et al., 2017; Romero et al., 2017; Tyree and Zimmermann, 2002; Zufferey et al., 2011).

The high number of parallel tracheary elements makes it unlikely that the entire transport pathway is interrupted at any one time. Indeed, the parenchyma cells in the rays act like a physical barrier that tends to contain cavitated areas within sectors that look like spokes of a wheel (Brodersen et al., 2013). The fact that roots, trunks, and shoots are not simply hollow pipes ensures that some residual water flow persists under all but the most severe drought conditions (see Section 7.2), and this flow can rapidly rehydrate the leaves upon restoration of soil moisture (Lovisolo et al., 2008; Romero et al., 2017). Consequently, Ψ_{leaf} begins to recover within hours after a rainfall or irrigation event, regardless of the severity of drought stress that existed prior to rewatering. Moreover, during rainfall, overhead irrigation, or even fog that leads to a surface water film, leaves can take up water directly across their surface. Although such water uptake can rehydrate the leaves, this high-resistance route is roughly 10 times slower than the process of water supply via the xylem (Guzmán-Delgado et al., 2018).

If cavitation occurs in the shoots or trunk of a grapevine, repair may occur by different processes. *Vitis* species can generate root pressure in excess of 0.1 MPa, which suffices to push water to a height of over 10 m above ground level (Tibbetts and Ewers, 2000). In addition, the parenchyma cells surrounding a cavitated xylem vessel may remobilize starch and actively release sugar and K^+ into any small water droplets that remain along the embolized vessel's walls, which lowers their Ψ_π (Charrier et al., 2016; Secchi and Zwieniecki, 2012). Water from the parenchyma cells then follows the solutes osmotically across aquaporins in the cell membranes (Knipfer et al., 2016; Lovisolo and Schubert, 2006; Secchi and Zwieniecki, 2010). The droplets in the embolized vessel thus grow and eventually coalesce, forcing the air back into solution or pushing it out into the pores of the vessel walls (Brodersen et al., 2010; Charrier et al., 2016; Knipfer et al., 2016). The small parenchyma cells in turn are supplied with water by the phloem. Whereas moderately water-stressed vines may be able to restore water flow quickly after water supply and while the leaves are transpiring, complete embolism repair in more severely stressed plants may require both a rise in Ψ_{leaf} and a cessation of xylem flow so that complete repair occurs only at night or during rainfall (Holbrook et al., 2001; Secchi and Zwieniecki, 2010).

The root xylem, or at least the fine-root xylem, of grapevines is much more vulnerable to cavitation than their shoot xylem (Lovisolo and Schubert, 2006; Lovisolo et al., 2008). This can be a problem with shallow root systems, because roots in the surface soil are the first to experience dry soil, and their failure can trigger leaf wilting and canopy collapse. If deeper roots are present, cavitation in the surface roots will simply shift water uptake down to wetter soil layers so that shoot damage can be avoided (Sperry et al., 2002). Furthermore, the expansive mycelium network created by symbiotic mycorrhizal fungi facilitates water uptake into the roots and helps avoid cavitation by osmotic water lifting on account of the fungi's high carbohydrate content supplied from the plants (Stahl, 1900). Plants, it seems, can reciprocate the favor at night by transferring water from the roots to the symbionts (Querejeta et al., 2003). Fungal colonization of the roots tends to increase as soil moisture, and hence root growth, decrease and vice versa (Schreiner et al., 2007). The fungi can even transfer water, as well as sugar and other nutrients, between neighboring vines and between vines and weeds or cover crops, either directly through the mycelium connecting the roots of two plants, or indirectly through uptake of water that was released by the roots of another plant (Egerton-Warburton et al., 2007).

Grapevines growing under water deficit develop narrower xylem vessels than vines growing under abundant water supply (Lovisolo and Schubert, 1998; Mapfumo et al., 1994b). Narrower vessels might be a consequence of curtailed supply of assimilates to the cambium resulting from lower photosynthesis due to higher r_s. Although a smaller vessel diameter increases r_h, which may reduce water loss and increase cavitation resistance, this effect may be limited because the few large vessels typically present in the xylem dominate its overall r_h (Tibbetts and Ewers, 2000). The vulnerability to cavitation also varies among species and cultivars (Schultz, 2003). Less vulnerable cultivars, such as Syrah, may have denser and stronger wood and less permeable pits, making the xylem more resistant, than do more vulnerable cultivars, such as Grenache—which nevertheless thrives in hot, arid, and windy conditions.

From Isohydric to anisohydric behavior

Grapevines are often regarded as so-called isohydric (Greek *isos*=equal, *hydros*=water) species whose sensitive stomata rapidly decrease g_s to limit transpiration rates in response to low Ψ_{soil}, which enables them to maintain relatively constant Ψ_{leaf} throughout the day regardless of Ψ_{soil} (Düring, 1987; Galmés et al., 2007). This effect can override the influence of high light intensity on stomatal opening so that soil water deficit in a vineyard often leads to a midday decrease in g_s (Correia et al., 1990, 1995; Loveys and Düring, 1984; Lovisolo et al., 2010). Nonetheless, pronounced diurnal changes in Ψ_{leaf} in addition to g_s have been reported; the difference between predawn and midday Ψ_{leaf} sometimes exceeds 1 MPa in vines growing in dry soil (Keller et al., 2015a; Romero et al., 2017; Williams and Matthews, 1990). Because *Vitis* species and cultivars differ in their vulnerability to cavitation, they also vary in their stomatal sensitivity to drought (Chaves et al., 2010; Currle et al., 1983; Escalona et al., 1999; Liu et al., 1978; Lovisolo et al., 2010; Schultz, 2003; Soar et al., 2006b; Winkel and Rambal, 1993).

Isohydric species or cultivars have more sensitive stomata than anisohydric species or cultivars. The latter maintain higher g_s and higher transpiration rates, and they markedly decrease Ψ_{leaf} during the day and in response to soil or atmospheric water deficit. Such differences among cultivars are also reflected in their roots' radial r_h and aquaporin activity, which are regulated to adjust water flow to transpiration rate (Perrone et al., 2012; Tyerman et al., 2009; Vandeleur et al., 2009). However, the isohydric/anisohydric distinction is mostly a matter of degree and not nearly as clear-cut as the terminology implies. Because Ψ_{leaf} never remains absolutely constant, it may be more useful to think of grapevine species and cultivars as varying on a spectrum from near-isohydric or pseudo-isohydric to near-anisohydric or pseudo-anisohydric. *Vitis berlandieri*, *V. rupestris*, and some *V. vinifera* cultivars are closer to the isohydric end of this spectrum, whereas *V. californica* and other *V. vinifera* cultivars are closer to the anisohydric end (Table 7.1). Some cultivars, such as Cabernet Sauvignon, Pinot noir, Tempranillo, and Concord have been variously described as isohydric or anisohydric, depending on environmental conditions and/or developmental status (Chaves et al., 2010; Lovisolo et al., 2010). Others, such as Merlot, have been reported to change from anisohydric in wet soil to isohydric as the soil dries down (Zhang et al., 2012). Moreover, even cultivars like Syrah that close their stomata readily under water deficit cannot prevent a strong decline in Ψ_{leaf} and eventual leaf wilting and abscission when drought stress becomes too severe (Romero et al., 2017). Only a few species and cultivars have been carefully studied, and it is likely that the category boundaries will become increasingly blurred as more information becomes available.

Isohydric species may have evolved in wet habitats; they may be regarded as "pessimists" that modify their behavior to conserve resources. Although this conservative strategy keeps the canopy's water demand below levels that would jeopardize the xylem's supply capacity, the decrease in g_s is associated with a decline in photosynthesis (McDowell, 2011; Palliotti et al., 2014; Sperry, 2004; Tarara et al., 2011). Consequently, the stomatal restriction of water loss during drought comes at the cost of a loss in carbon gain. Anisohydric plants, on the other hand, may originate from more arid regions; they may be viewed as "optimists" that use available resources in expectation of more arriving. To them, the soil water "glass" remains half full when their isohydric counterparts already perceive it as half empty.

Table 7.1 Some near-isohydric and near-anisohydric grapevine cultivars.

Near-isohydric	Near-anisohydric
Grenache	Cabernet franc
Tempranillo	Cabernet Sauvignon
Trincadeira	Carignan
	Chardonnay
	Merlot
	Nebbiolo
	Riesling
	Sangiovese
	Sémillon
	Syrah
	Thompson Seedless
	Touriga Nacional

From Chaves, M.M., Zarrouk, O., Francisco, R., Costa, J.M., Santos, T., Regalado, A.P., 2010. Grapevine under deficit irrigation: hints from physiological and molecular data. Ann. Bot. 105, 661–676 and Lovisolo, C., Perrone, I., Carra, A., Ferrandino, A., Flexas, J., Medrano, H., 2010. Drought-induced changes in development and function of grapevine (Vitis spp.) organs and in their hydraulic and non-hydraulic interactions at the whole-plant level: a physiological and molecular update. Funct. Plant Biol. 37, 98–116.

Due to the narrow safety margin from hydraulic failure, where xylem cavitation becomes irreversible, the anisohydric approach is riskier, and such vines may pay by shedding leaves when a drought becomes too severe (Palliotti et al., 2014; Romero et al., 2017). However, it may enable a plant to more efficiently match water supply with demand so that it can maximize CO_2 uptake by pushing the xylem to its full capacity, at least so long as the plant can retain its leaves (Sack and Holbrook, 2006; Sperry, 2004). Not surprisingly, therefore, the responsiveness of shoot and leaf growth to water deficit varies among cultivars (Tardieu et al., 2018). For example, the anisohydric Syrah maintains vigorous shoot growth under water-deficit conditions that would completely inhibit shoot growth in the isohydric Cabernet Sauvignon. Moreover, if left to its own devices, rather than being constrained by a trellis system and foliage wires, the droopy growth habit of Syrah shoots also differs from that of Cabernet Sauvignon or Grenache, whose shoots are much more erect (Louarn et al., 2007). Anisohydric plants also would be expected to have a larger root $\div$ shoot ratio than isohydric plants, which may be a drought-avoidance strategy. These differences among cultivars complicate irrigation management in a vineyard, because the same, regulated decrease in water supply will not result in the same outcome across all cultivars.

To complicate matters further, rootstocks derived from American *Vitis* species, too, differ in their susceptibility and response to drought, potentially altering Ψ_{leaf}, g_s, and photosynthesis of the scion leaves via the production and export in the xylem of ABA and via differences in root growth (Alsina et al., 2011; Bauerle et al., 2008b; Knipfer et al., 2015; Lavoie-Lamoureux et al., 2017; Soar et al., 2006a). Nevertheless, such differences among rootstocks do not necessarily result in differences in cavitation and gas exchange of *V. vinifera* scions grafted to them, at least not in very young vines (Barrios-Masias et al., 2019). It furthermore appears that own-rooted *V. vinifera* vines tend to be somewhat better able to tolerate water deficit than are *V. vinifera* scions grafted to American rootstocks, especially drought-intolerant *V. rupestris* rootstocks such as St. George (Ezzahouani and Williams, 1995; Keller et al., 2012; Kidman et al., 2014). Differences in scion behavior imparted by the rootstock, however, are rather small, and it may be the genotype of the scion, rather than that of the rootstock, that drives a plant's overall response to drought stress (Kuromori et al., 2018). Keeping in mind that scion–rootstock studies into hydraulic behavior are almost always done with very small, young vines, this leaves open the possibility that plant responses might change as vines age. Moreover, results obtained with young, pot-grown, ungrafted rootstock genotypes alone cannot be readily translated into behavior of grafted scion–rootstock combinations in a vineyard.

Drought signals and growth

In addition to its obvious consequences for Ψ, dehydration of plant cells also stimulates the biosynthesis of ABA from the carotenoid zeaxanthin in the parenchyma cells, especially near the cambium, of the root, shoot, and leaf vascular system (Endo et al., 2008; Nambara and Marion-Poll, 2005). The main function of ABA is to regulate the plant water balance and osmotic stress tolerance. The role of ABA in water balance is mainly through the regulation of g_s (see Section 3.2), whereas its role in osmotic stress tolerance is through the production of dehydration-tolerance proteins in most cells. In addition, ABA may induce closure of aquaporins in the leaf's bundle-sheath cells, which increases r_h and decreases water flow to the mesophyll cells (Shatil-Cohen et al., 2011). This hydraulic reinforcement of the ABA effect on g_s may provide an explanation for the isohydric behavior discussed above (Pantin et al., 2013). Indeed, even though their stomata are equally sensitive to ABA, leaf r_h of the near-isohydric Grenache increases in response to ABA, whereas leaf r_h of the near-anisohydric Syrah is not only relatively insensitive to ABA but is also lower than that of Grenache (Coupel-Ledru et al., 2017). Perhaps plants that perceive ABA as a signal to both their stomata and their leaf aquaporins are more ABA-sensitive and respond to water deficit by closing their stomata sooner and more tightly than plants that perceive ABA only as a stomatal signal.

Another function of ABA is to oppose the stimulatory role of auxin in cell-wall loosening for cell expansion during growth (see Section 3.1). Moreover, by interfering with the cell cycle machinery, high amounts of ABA inhibit cell division (Del Pozo et al., 2005). In addition to the ABA produced in drying leaves, the amount of root-sourced ABA in the xylem sap may increase substantially under soil water deficit (Loveys, 1984; Lovisolo et al., 2010). In contrast to ABA, the production of cytokinins in the roots and their transport in the xylem sap decrease in response to soil drying (Davies et al., 2002; Ha et al., 2012; Yang et al., 2002). Both ABA and cytokinin can act as stress signals that mediate conditions in the rootzone via the xylem to the shoots, leaves, and clusters. The production of ABA in the roots rises rapidly in response to water stress, whereas cytokinin production decreases much more gradually. However, some root-produced ABA may be leached out into the soil solution, especially in alkaline soils (Wilkinson and Davies, 2002). Consequently, a high soil pH might weaken the root-sourced ABA signal.

When soil water stress occurs suddenly, production and transmission of the root ABA signal may be too slow to enable the stomata to close in time to avoid leaf dehydration. Under these conditions, the shoots and leaves ramp up ABA production in response to a hydraulic signal from the roots (Christmann et al., 2007; Soar et al., 2004). Changes in root xylem pressure are transmitted virtually instantaneously to the shoot, where they alter the turgor pressure of the leaf cells. Nonetheless, a decrease in cell volume, which accompanies the turgor decline, may suffice to trigger ABA biosynthesis in the leaves; thus, leaf cell volume is a key signal of water stress (Sack et al., 2018).

Moreover, the pH of the xylem sap and other apoplastic sap generally increases in response to soil drying (Hartung and Slovik, 1991; Li et al., 2011; Wilkinson and Davies, 2002). The xylem sap pH of well-watered plants is approximately 6.0 (see Section 5.1), whereas that of water-stressed plants increases to about 7.0. This increase in xylem sap pH constitutes yet another root-sourced signal to the leaf, inducing stomatal closure and reducing growth by slowing the rate of cell expansion. Cell expansion involves cell wall loosening, which requires acidification by protons (H^+) pumped from the cell interior to the cell wall in exchange for K^+ maintaining the cell's electrical charge balance (Bassil et al., 2019; Stiles and Van Volkenburgh, 2004). Therefore, the increase in apoplast pH due to water deficit may directly interferes with cell expansion (Bacon et al., 1998).

Roots must absorb more water than what is lost in transpiration to enable plant growth because growth is caused mainly by cell expansion due to water import. Cell water uptake is driven by an osmotically generated $\Delta\Psi$ from the exterior to the interior of the cell (see Section 3.1). Young, expanding leaves and growing root tips that experience water deficit commonly accumulate solutes, especially inorganic ions such as K^+ and Ca^{2+}, but also sugars such glucose and fructose and, to a lesser extent, amino acids such as proline, and organic acids such as malate (Cramer et al., 2007; Patakas et al., 2002; Sharp et al., 2004). While this osmotic adjustment, or osmoregulation, maintains cell P as Ψ decreases, this can be insufficient to permit continued growth (Düring, 1984; Morgan, 1984; Schultz and Matthews, 1993a; Tardieu et al., 2018). Fully grown leaves, moreover, gradually lose their ability to osmoregulate (Patakas et al., 1997). However, leaves that develop later in the season can achieve a greater capacity for osmotic adjustment than leaves that are formed earlier. Because water stress reduces Ψ_{xylem}, a major cause for growth inhibition under water deficit may be the smaller $\Delta\Psi$, which reduces cell water uptake and consequently hampers the generation of the necessary P (Nonami and Boyer, 1987; Nonami et al., 1997; Hsiao and Xu, 2000). Shoot elongation and leaf expansion decline linearly with decreasing Ψ and stop completely at $\Psi = -1.0$ to -1.2 MPa, even though leaf expansion is relatively independent of P (Schultz and Matthews, 1988b; Shackel et al., 1987; Sweet et al., 1990; Williams et al., 2010). The growth of lateral shoots appears to be even more constrained by water deficit than is the growth of main shoots (Lebon et al., 2006). Therefore, shoot elongation and leaf expansion change rapidly with fluctuating Ψ_{xylem} and pH_{xylem}, and a decline in shoot growth and leaf appearance is the first visible sign of water deficit; shoot growth is more sensitive to water stress than is photosynthesis (Hsiao and Xu, 2000; Stevens et al., 1995; Sweet et al., 1990; Williams et al., 1994). Tendrils are another sensitive indicator of vine water status: On nonstressed, vigorously growing shoots, the uppermost tendrils extend beyond the shoot tip. As water stress sets in and growth begins to slow, new tendrils remain small so that the shoot tip catches up with them. With more severe stress, the shoot tip stops producing new leaves and tendrils, and the youngest leaf extends beyond the shoot tip. Young, green tendrils are highly sensitive to water stress and start wilting before the leaves do.

Root growth and nutrient uptake

Root growth of water-stressed vines declines due to a combination of plant water deficit and increased penetration resistance of the drying soil (Bengough et al., 2011). But the decrease in root growth is less marked than that in shoot growth, because roots can grow at lower Ψ than shoots, partly because higher activity of expansin proteins may increase cell wall extensibility in the root tips (Erlenwein, 1965a; Sharp et al., 2004; Wu and Cosgrove, 2000). In addition, whereas ABA is involved in inhibiting shoot growth of water-stressed plants, it appears to have the opposite function in roots. Only at high concentrations due to more severe water stress does ABA inhibit root growth (Zhang et al., 2010b). The effect of water stress is more pronounced on a root's diameter than on its length: A stressed root becomes thinner (Mapfumo et al., 1994a). These adaptations increase the root $\div$ shoot ratio and maintain some water and nutrient supply to the shoots (Hsiao and Xu, 2000). By inducing a blockage of ion channels in the stele (see Section 3.3), ABA reduces the release of K^+ and possibly other nutrient ions, such as $H_2PO_4^-$, from the root cortex into the xylem conduits (De Boer and Volkov, 2003; Roberts and Snowman, 2000). Because ABA does not affect root K^+ uptake, this traps K^+ taken up from the soil and that delivered from the shoots via the phloem, which leads to K^+ accumulation in the roots. The osmotic activity of K^+ lowers Ψ_{root}, which may enhance phloem water import (see Sections 3.2 and 5.1). This helps grapevines to maintain root growth while avoiding excess K^+ transport to the leaves, where it would only worsen the water deficit by lowering Ψ_{leaf}. In addition, the growing root tips osmotically adjust, or osmoregulate, by accumulating sugars and amino acids in order to lower Ψ_{root} to favor water uptake and continued growth, and possibly to reduce the risk of xylem cavitation (Düring and Dry, 1995; Schultz and Matthews, 1988a). Nevertheless, the young fine roots, though not the older and suberized woody roots, may experience partial collapse of their cortical cells, which leads to air spaces, called aerenchyma, that increase radial r_h and thus limit both water transport to the xylem and water loss to the soil (Rellán-Álvarez et al., 2016; Cuneo et al., 2016). When the fine roots fail, the woody roots may assume a greater importance for overall root water uptake to partly compensate for the loss in the fine roots' uptake capacity (Cuneo et al., 2018). During more severe water stress, however, the vessels cavitate, which interrupts water flow in the affected xylem conduits altogether.

Soil drying also reduces the microbial activity (which is responsible for transforming mineral nutrients locked up in organic matter into plant-available forms) in the soil and the rate of nutrient transport to the root surface. This decreases overall nutrient availability and uptake, especially if the water deficit slows root growth (Freeman and Kliewer, 1983; Hsiao, 1973; McDowell, 2011; see also Section 7.3). Thus, soil drying makes nutrient uptake increasingly difficult, and a greater proportion of the nutrient demand of the developing canopy and especially the ripening fruit may have to be supplied from stored reserves in the woody parts of the vine. Under prolonged and relatively severe drought stress, however, plant growth may decrease more than nutrient uptake, so that the overall plant nutrient concentration may actually increase (He and Dijkstra, 2014). A decline in soil moisture is also associated with a decrease in root respiration and, under severe drought conditions, leads to a loss of membrane integrity so that roots begin to die back (Hernández-Montes et al., 2017; Huang et al., 2005b).

Under dry conditions, moderate nitrogen (N) supply via application of organic or synthetic fertilizers or other soil amendments may conserve soil water in the short term. The increase in nitrate concentration in the xylem sap that accompanies abundant N availability (see Section 5.3) is associated with an increase in xylem sap pH, which in turn raises the stomata's sensitivity to ABA (Jia and Davies, 2007; Wilkinson and Davies, 2002). Therefore, the stomata may close sooner and more rapidly when a high-N soil dries compared with a relatively N-poor soil. In the long term, however, a decrease in the root ÷ shoot ratio due to N addition can increase a vine's susceptibility to drought stress, particularly in irrigated vineyards, if irrigation water is suddenly unavailable (Erlenwein, 1965a; Sperry et al., 2002). Vigorous shoot growth under high N availability is associated with wider xylem vessels and lower r_h, which favors rapid water delivery to the canopy but could render vines vulnerable to xylem cavitation and, consequently, canopy desiccation (Plavcová et al., 2012; Sperry et al., 2002). A small root system supplying water to a large canopy results in a steep water potential gradient ($\Delta\Psi = \Psi_{soil} - \Psi_{leaf}$; see Section 3.3). To balance water loss by transpiration with water supply by root uptake, such vines must either close their stomata or use osmotic adjustment to lower their Ψ_{leaf}, especially around midday when evaporative demand peaks. Fertilizer application to nonirrigated plants is thought to enhance the capacity for osmotic adjustment and the resistance to xylem cavitation, which tends to counter the lower root ÷ shoot ratio (Bucci et al., 2006; Sperry et al., 2002). Excessive N supply, however, can reduce water uptake due to the buildup of solutes in the soil water, similar to the effect of salinity discussed in Section 7.3 (Keller and Koblet, 1994).

Yield formation

It is often stated that grapevine vegetative growth is more sensitive to water deficit than is reproductive growth (Williams et al., 1994). However, the sensitivity of reproductive growth, and therefore fruit production and yield formation, depend on the developmental stage of the vine. Water deficit typically reduces yield, particularly if the deficit occurs early in the growing season (Junquera et al., 2012; Williams and Matthews, 1990). Soon after budbreak, the developing flower clusters begin competing successfully with the growing shoots for limited water and assimilates (Hale and Weaver, 1962). Once the inflorescences contribute assimilates via their own gas exchange (Leyhe and Blanke, 1989), they may survive even relatively severe drought episodes sufficient to stunt shoot growth. Closer to and during bloom, the flowers' metabolic requirements far exceed their own contribution—after all, most stomata are located in the cap that is shed at anthesis (Blanke, 1990b; Blanke and Leyhe, 1988, 1989a; Vivin et al., 2003). Nevertheless, the flower clusters continue to be weak sinks until fruit set.

The female floral organs are thought to be relatively insensitive to water stress, but smaller ovaries are more susceptible to stress than are larger ovaries within the same fruit cluster. In contrast, water deficit during meiosis in the anthers may cause pollen sterility (Dorion et al., 1996; Saini, 1997).

Even brief and moderate water stress during male meiosis, which occurs 1–2 weeks before anthesis, can hamper pollen development because water stress may inhibit sugar transport to and starch accumulation in the pollen grains (De Storme and Geelen, 2014; Dorion et al., 1996; Saini, 1997). Starch is required to fuel respiration for pollen development and germination and for pollen tube growth, so their low "fuel" status prevents the pollen tubes from reaching the ovary in time for fertilization, if they can germinate at all (see Section 2.3). Reduced pollination and fertilization result in poor fruit set or, in more severe cases, abscission of inflorescences (Callis, 1995; Smart et al., 1983). Consequently, reproductive development is most sensitive to water stress from meiosis to fruit set; water stress during this time can markedly reduce fruit set in grapevines (Hardie and Considine, 1976; Wenter et al., 2018). Unlike the inhibition of shoot growth and leaf expansion, fruit abscission is irreversible.

The effect of water deficit on the flowers is likely mediated by a combination of low Ψ and changes in xylem sap pH, ABA, and cytokinins in addition to the supply of photosynthate via the phloem. For instance, elevated ABA may reduce fruit set in grapevines (Padmalatha et al., 2017). Although ABA enhances vacuolar invertase activity in the leaves in order to maintain P to sustain photosynthesis and assimilate export, it suppresses the same enzyme in flowers and young fruits, which are subsequently aborted (Roitsch and González, 2004). An involvement of ABA and cytokinin is also consistent with the observation that PRD, which maintains high Ψ_{leaf}, may be associated with poor fruit set (Rogiers et al., 2004b). It is further possible that when embolisms develop in the inflorescence xylem, they may not be as easy to repair as they are in leaves.

In addition to reducing berry number, drought stress during bloom could also have implications for a grapevine's yield potential for the following year. Water deficit normally decreases cluster initiation, which can reduce bud fruitfulness and thus result in low numbers of clusters per shoot (Alleweldt and Hofäcker, 1975; Buttrose, 1974b; Junquera et al., 2012). However, if the water deficit is just mild enough to decrease canopy density, it may instead enhance cluster initiation and differentiation due to the improved canopy microclimate (Smart et al., 1983; Williams and Matthews, 1990; Williams et al., 1994).

Once fruit set has passed and grapevines have made a significant investment in seed production, vines experiencing water deficit generally maintain fruit development at the expense of shoot and root growth and replenishment of storage reserves (Eibach and Alleweldt, 1985; Mullins et al., 1992; Williams et al., 1994). Favoring the current year's crop under water stress can have long-term implications because grapevines are heavily dependent on stored reserves for budbreak the following spring. Although fruit growth may be less sensitive to drought stress compared with shoot growth, limited water supply during the period of berry cell division and cell expansion restricts berry enlargement, which limits berry size (Roby and Matthews, 2004; Romero and Martinez-Cutillas, 2012; Shellie, 2014; Williams and Matthews, 1990). Whereas cell division in young grape berries is relatively insensitive to water deficit, cell expansion responds readily to changes in water supply (Ojeda et al., 2001). In keeping with the effects on shoot growth, reduced Ψ_{xylem} and xylem sap pH could also be partly responsible for the smaller berry size of water-stressed vines, particularly if the deficit occurs before veraison. Since the limitation on berry size imposed by early-season stress is irreversible, it seems likely that it involves changes in the composition of the cell walls (Ojeda et al., 2001). Water deficit curtails cellulose production and cell wall extensibility in expanding leaves (Schultz and Matthews, 1993a; Sweet et al., 1990), and the same changes probably occur in berries too.

Yield reductions due to drought stress can still be severe after fruit set, whereas postveraison berries rapidly become rather insensitive to water deficit. Thus, the same extent of water deficit occurring during the preveraison phase of berry growth normally constrains final berry size much more than if it occurs after veraison (Currle et al., 1983; Hardie and Considine, 1976; Hofäcker et al., 1976; Matthews and Anderson, 1989; McCarthy, 1997; Williams and Matthews, 1990). Applying more water later in the growing season cannot compensate for the decrease in berry size due to early-season water deficit (Fig. 6.2). Increased phloem flow into the berry, combined with the stiffening skin, after veraison strongly decrease the sensitivity of berry water status and berry size to soil and plant water status (Creasy and Lombard, 1993; Düring et al., 1987; Findlay et al., 1987; Greenspan et al., 1994, 1996; Ollat et al., 2002; Rogiers et al., 2001). Despite the fears of many winemakers, drip or flood irrigation during fruit ripening does not increase berry size or "dilute" fruit composition (Coombe and Monk, 1979; Keller et al., 2006, 2015b). The same cannot be said for rainfall or overhead irrigation, however, because uptake of water through the berry skin may induce berry splitting (see Section 6.2). Nevertheless, due to the reversal of berry xylem flow at veraison, postveraison water deficit, especially prolonged deficit coupled with high VPD, can lead to berry shrinkage by dehydration when xylem efflux plus berry transpiration exceed phloem influx (Keller et al., 2006; Zhang and Keller, 2015, 2017). Thus, the extent of preharvest shrinkage is inversely related to the water status of the vine at that time.

The response of grape berry growth to water deficit also depends on the number of berries per vine and on the crop load. Berry size may be restricted more on plants bearing a heavy crop than on vines with a light crop load, because less sugar import by each berry also limits berry water import (Fishman and Génard, 1998; Santesteban and Royo, 2006). If a drought stress event occurs early enough to reduce fruit set, the size of the remaining berries may increase and partially compensate for the untimely loss in yield potential, which would offset prospective benefits for wine quality of smaller berries (see Section 6.2). Conversely, water deficit after fruit set may prevent compensatory berry growth on vines that start the growing season with a small crop load (Keller et al., 2008). Although symptoms of plant water stress often become apparent when >50% of the plant-available water has been extracted from the soil, both pre- and postveraison grape berries do not begin to shrink until 80% of the available soil water has been transpired by the vine (Keller et al., 2006).

Deficit irrigation

Deliberate application of water deficit is an important vineyard management strategy in premium wine grape production in regions with low summer rainfall. The time between fruit set and veraison is the most effective period to impose water deficit to control shoot growth and berry development. This principle is exploited by regulated deficit irrigation (RDI), whereby water deficit is imposed as soon as possible after fruit set (Dry et al., 2001; Keller, 2005; Kriedemann and Goodwin, 2003). Deficit irrigation means that less water is applied than is necessary to balance vineyard ET, so that soil moisture and plant water status decline. To help growers estimate required irrigation water amounts, weather station networks often provide measures of daily ET, but these are valid only for so-called reference crops such as grass or alfalfa. This baseline or reference ET is referred to as ET_0, and a crop factor (K_c) is applied to estimate crop or vineyard water use (ET_c) as follows:

$$ET_c = K_c \, ET_0$$

The K_c usually increases from <0.1 to about 0.8 in wine grapes, and up to 1.3 in large table grapes, depending on seasonal shoot growth and canopy size (Evans et al., 1993; Williams and Ayars, 2005;

Williams, 2014). By applying amounts of water that are lower than ET_c, the soil is allowed to dry down during the RDI period until shoot growth ceases. Obviously, this is possible only in areas with sufficiently low seasonal rainfall or high evaporative demand and on soils with limited water storage capacity. In addition to limiting berry size, and at least as importantly, RDI also reduces the canopy density and increases the sun exposure of the clusters (Keller et al., 2016a; Romero et al., 2010). Once shoot growth stops and especially after veraison, the vines are stressed only sufficiently to discourage renewed shoot growth. In regions with cold and/or dry winters, growers irrigate to refill the rootzone to near field capacity at the end of the growing season in order to avoid cold injury to roots (see Section 7.4).

Rootzone conditions are generally heterogeneous, and soil moisture is no exception to this rule. When roots of the same plant experience areas of both wet and dry soil, the influence on shoot growth depends on whether the difference in soil moisture is perceived by the same or different roots. If the surface soil is dry while the subsoil remains wet, shoot growth continues so long as the roots have access to the subsoil water (Phillips and Riha, 1994). Shoot growth slows or stops, however, if separate roots of the same vine experience dry and wet soil columns (Dry and Loveys, 1999; Dry et al., 2001; Lovisolo et al., 2002a). This can be exploited in another irrigation strategy termed partial rootzone drying (PRD), in which water is supplied alternately to only one side of a vine while the other side is allowed to dry down (Dry and Loveys, 1998). This strategy attempts to separate the physiological responses to water stress (e.g., ABA production and decreased g_s) from the physical effects of water stress, such as low Ψ_{leaf} (Davies et al., 2002; Dry et al., 2001; Kriedemann and Goodwin, 2003). The ABA produced by the drying roots induces partial stomatal closure and reduced shoot growth while the fully hydrated roots maintain a favorable plant water status. Water delivery to the shoots by the wet roots may increase during PRD, and the wet roots also sustain the drying roots by supplying water to them (Kang et al., 2003; Stoll et al., 2000). Eventually, however, water flow from the dry roots becomes so slow that ABA delivery from these roots ceases and the stomata begin to reopen (Dodd et al., 2008; Dry et al., 2000a). To prevent such adaptation by the vines, irrigation is alternated every week or two between the two sides of the root system. While the drying roots increase ABA production, they decrease cytokinin production, which reduces canopy size by inhibiting lateral shoot growth (dos Santos et al., 2003; Dry et al., 2001; Stoll et al., 2000). Provided Ψ_{leaf} is maintained under PRD, berry size and yield are also maintained, whereas the lower canopy density often results in improved fruit composition.

The fundamental difference between RDI and PRD is that RDI imposes a soil water deficit over time, whereas PRD imposes a deficit over space (Keller, 2005). If rainfall during the growing season is low enough, RDI always results in a plant water deficit, whereas PRD does not if managed properly (De Souza et al., 2003; Dry et al., 2001; Kriedemann and Goodwin, 2003). Nonetheless, whether the effects of PRD differ from those of RDI may also depend on the total amount of available soil water (Romero and Martinez-Cutillas, 2012). The distinct PRD effect may be lost if the switching interval is too long, the wet soil volume is too small for a given vine size, two discrete root systems cannot be established, vines have access to a high water table, or in poorly structured clay soils. Moreover, whereas under standard drip irrigation the roots are often concentrated in the surface soil, grapevines under PRD shift root growth to deeper soil layers (Dry et al., 2000b, 2001). Although this adaptive behavior makes vines more drought resistant, it also interferes with the application of the PRD principle on soils with a deep rootzone and ready access of the roots to subsurface water.

7.3 Mineral nutrients: Deficiency and excess

In a growing grapevine, each new cell has a relatively specific requirement for a well-defined set of inorganic mineral nutrients in addition to water and carbon. Some of these nutrients, especially during budbreak and early-season growth, are supplied from reserves stored in the permanent structure of the vine. But most nutrients throughout the growing season must be acquired by the roots from the soil. Contrary to surprisingly widespread belief, roots cannot absorb soil particles, such as clay minerals or sand grains, directly. Wine writers and consumers sometimes talk about wine "minerality" as a taste or flavor originating from bits of soil or underlying bedrock. These particles, however, are mostly composed of aluminum-silicates—and aluminum (Al) is toxic to plants. Minerals are chemical compounds, normally crystalline, that have been formed by geological processes. To make it across the root cell walls and membranes, minerals must be broken down into their chemical elements and/or converted into charged ions. Even silicon dioxide, or silica (SiO_2), the most abundant component of rocks, must be converted to water-soluble silicic acid ($Si(OH)_4$) before it can be taken up by roots. Most other mineral nutrients available for root uptake are present in the form of ions dissolved in the soil water—which is therefore called the soil solution. As discussed in Section 3.3, uptake is quite selective and occurs by channel and transport proteins embedded in the root's cell membranes.

Roots can either grow toward nutrient-rich areas or patches in the soil in a process called root interception or wait for the nutrients to arrive at their surface with the soil water. In the latter case, the nutrients must move through the soil either by diffusion or by mass flow. Diffusion describes the movement of dissolved nutrients toward the root down a concentration gradient generated by nutrient uptake at the root surface (see Section 3.3). Mass flow, on the other hand, is the process of convective nutrient transport toward the root in the "stream" of water down a water potential gradient generated by the transpiration stream, which results in water uptake by the root (Chapman et al., 2012). Diffusion is often the dominant process, but mass flow is more important when plants transpire rapidly and thus remove copious amounts of soil water. In fact, one adaptive advantage of transpiration may be its capacity to enhance mass flow of nutrients to the root surface, which may be important in nutrient-poor soils—provided water is not limiting (Cramer et al., 2008). Accordingly, grapevines seem to transpire more rapidly when the nutrient concentration in the soil solution is low than when nutrients are abundant, but the leaves of the nutrient-rich vines wilt more readily when they run out of water (Keller and Koblet, 1994; Scienza and Düring, 1980).

Soils vary not only in their capacity to store water but also in the amount and composition of mineral nutrients they contain and in the extent to which these nutrients are available for uptake by the roots. Nutrient storage capacity and accessibility are influenced by soil texture, rooting depth, and organic matter content, but nutrient availability is modified by soil moisture and pH (Fig. 7.3). For instance, just as a loamy soil can store considerably more water than a sandy soil (see Section 7.2), it also adsorbs, and thus retains, more nutrient ions than sandy soils in which the nutrients are easily leached away by percolating water. Soil organic matter, or humus, on the other hand, increases water and nutrient storage capacity in both loamy and sandy soils due to its beneficial effect on soil structure via the formation of stable aggregates. Although the nutrient concentration in a dry soil is usually higher than that in a wet soil, these nutrients are less available in the dry soil because the lack of water slows diffusion and mass flow; the diffusion rate decreases roughly with the square of the soil water content (Marschner, 1995). Moreover, mineralization of organic matter and nitrification ($NH_4^+ \rightarrow NO_3^-$) also

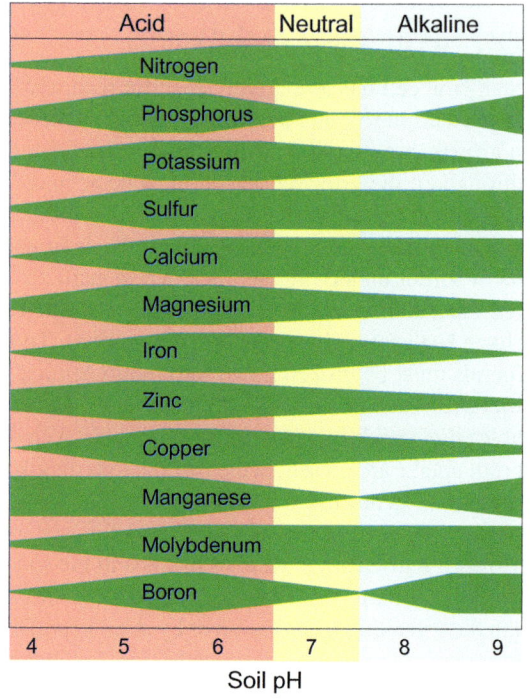

FIG. 7.3

Nutrient availability (indicated by width of green bands) in the soil is dependent on soil pH.

Illustration by M. Keller, after Lucas, R.E., Davis, J.F., 1961. Relationships between pH values of organic soils and availabilities of 12 plant nutrients. Soil Sci. 92, 177–182.

slow down in drying soil, and the concentration of exchangeable cations (e.g., calcium) rises at the expense of anions (e.g., phosphate). Roots that shrink as a soil dries may lose contact with the soil particles so that nutrient availability to the roots declines steeply in a drying soil (Bloom et al., 1985; Jackson et al., 2008). Although this diminishes nutrient uptake and delivery of nutrients to the shoots, the leaf nutrient concentration may nevertheless remain constant or even increase, because water stress curtails shoot growth and carbon assimilation more than nutrient uptake (Bloom et al., 1985; Davies et al., 2002; Mpelasoka et al., 2003; Zufferey et al., 2017; see also Section 7.2). Nonetheless, both the concentration and the total amount of nitrogen in the shoots, leaves, and berries tend to be lower in grapevines whose growth is slowed by competition with cover crops for access to both water and mineral nutrients (Celette et al., 2009; Celette and Gary, 2013; Guerra and Steenwerth, 2012; Reeve et al., 2016; Tesic et al., 2007).

Nitrogen (N) availability fluctuates widely throughout the growing season, whereas the availability of the relatively mobile calcium (Ca) and magnesium (Mg, and of the nearly immobile phosphorus (P) and potassium (K), is much more constant (Nord and Lynch, 2009). Potassium and phosphate ($H_2PO_4^-$) can be taken up by mature root sections, whereas absorption of Ca^{2+} and Mg^{2+} seems to be restricted to the young fine roots. Iron (Fe) uptake is confined to the growing root tips, whereas sulfate (SO_4^{2-})

uptake is concentrated in the elongation zone immediately behind the meristematic region. The fibrous fine-root system is most effective in acquiring mineral nutrients, but the older sections of grapevine roots usually form mutually beneficial units with mycorrhizal fungi (Gebbing et al., 1977; Menge et al., 1983; Possingham and Groot Obbink, 1971; Schreiner, 2005; Stahl, 1900). The roots exude carotenoid-derived strigolactones to attract these symbiotic microbes and induce branching of the fungal hyphae (Akiyama et al., 2005; Gomez-Roldan et al., 2008; Ruyter-Spira et al., 2013). The fungi in turn stimulate lateral root formation in their future plant hosts by secreting auxin and other signaling molecules such as polyamines (Bellini et al., 2014). They then infect the roots, generally behind the root-hair zone, and provide an additional link between the vine and the surrounding soil, living partly within the soil and partly within the root cortex, whereby the fungal and plant cytoplasms remain separated by the cell membranes and cell walls of the two partners (Bucher, 2007; Smith and Smith, 2011). The fungi can absorb water and nutrients relatively far away from the root and thereby greatly extend the soil volume exploited by the roots by forming an extensive and interconnected hyphal network that can even link different plants together. This increases the roots' absorbing capacity well beyond the depletion zones caused by the growing root tips. The fungi transfer most of the absorbed nutrients to the plant roots in exchange for sugar, especially glucose, which provides the energy and building material for mycelium growth, spore production, and active nutrient uptake.

Mycorrhizae are mainly known for their ability to enhance $H_2PO_4^-$ and zinc (Zn) uptake, but they also transfer sulfur (S) and copper (Cu) to the roots (Marschner, 1995; Menge et al., 1983; Smith et al., 2001). In addition, they may facilitate acquisition of N, K, Ca, Mg, and Fe under certain conditions. The symbiotic microorganisms grow readily toward nutrient-rich organic patches in the soil, where their mycelium proliferates for the benefit of their host plant. Some mycorrhizae can apparently tap directly into otherwise plant-unavailable organic material, such as decaying plant or animal remains, where N predominantly occurs in the form of proteins. The fungi promote decomposition and mineralization of such material and then capture and transfer to the roots the resulting inorganic N in addition to P (Hodge, 2006; Hodge et al., 2001; Leigh et al., 2009).

However, their carbon requirement makes the symbiotic mycorrhizae sinks for photosynthetic assimilates—there is no free lunch after all (Hall and Williams, 2000; Smith et al., 2001). The supply of sugar to the fungi represents the price paid by the vine for enhanced nutrient uptake. Once established inside a root, the fungi ostensibly suppress lateral root growth, making the plant host more dependent on nutrients supplied by the fungal guests (Osmont et al., 2007). Because of the carbon cost to the plant of establishing and maintaining the symbiosis, young plants initially tend to grow more slowly in mycorrhiza-contaminated soil than in fumigated soil, especially if the soil is adequately supplied with mineral nutrients (Clarkson, 1985). In nutrient-rich soils, root colonization by mycorrhizae is usually much slower than in nutrient-poor soils, because in an environment with plentiful resources there is no need for a plant to "waste" carbon on the symbionts.

Nutrient ion uptake is mainly controlled by the demand of a grapevine and varies according to growth requirements, even as soil nutrient availability varies over several orders of magnitude (Keller et al., 1995). Thus, growth is the "pacemaker" for nutrient uptake, and the rates of uptake and root growth reflect the demand created by plant growth (Clarkson, 1985; Clarkson and Hanson, 1980; Gastal and Lemaire, 2002; Tester and Leigh, 2001). Such demand also includes the growing fruit, so that seasonal nutrient uptake by different grapevines varies according to their crop load (Pradubsuk and Davenport, 2010, 2011). Positive correlations between grapevine growth and nutrient supply often indicate that nutrient availability limits growth. Insufficient supply of a nutrient ion slows shoot growth to a rate consistent with supply (Clarkson, 1985; Gastal and Lemaire, 2002). Leaf expansion, for example, is particularly sensitive to fluctuations in N supply (Forde, 2002). The decrease in growth due to water deficit also decreases the nutrient requirements of grapevines (see Section 7.2).

Both nutrient uptake and shoot growth of grapevines show saturation-type responses to increasing soil N and P (Bell and Robson, 1999; Keller and Koblet, 1995a; Keller et al., 1995; Linsenmeier et al., 2008; Schreiner and Osborne, 2018; Spayd et al., 1993; Williams and Matthews, 1990). Recycling in the phloem of xylem-delivered inorganic nutrients or of their assimilated organic versions from the leaves to the roots acts as a feedback signaling system to regulate nutrient uptake. For example, nitrate transporters in the roots can be "shut" in response to high concentrations of amino acids or sugars, and sulfate transporters are repressed by cysteine and glutathione (Amtmann and Blatt, 2009; Forde, 2002; Grossman and Takahashi, 2001; Tsay et al., 2011). Such feedback regulation also fine-tunes K uptake according to shoot demands (Tester and Leigh, 2001; Véry and Sentenac, 2003). Furthermore, a plant's carbon status impacts nutrient uptake, at least partly because of the need for energy to run ATP pumps. Under conditions of carbon depletion—for instance, when photosynthesis is limited by environmental constraints such as low light—vines reduce nutrient uptake and transport in the xylem (Keller et al., 1995). A glucose phosphate is thought to be the sugar signal that synchronizes nutrient uptake to photosynthesis (Lejay et al., 2008). In addition, the phloem can recycle nutrients from aging leaves back to the shoot xylem, which helps buffer short-term changes in xylem nutrient concentrations (Metzner et al., 2010b).

Nevertheless, nutrient uptake can exceed growth requirements when nutrients are abundantly available in the soil solution—for example, after fertilizer application or incorporation and decomposition of cover crops. Although growth and crop yield may not increase further, the extra supply can lead to nutrient accumulation inside the vine (Delas and Pouget, 1984; Keller et al., 1995; Pouget, 1984). Because grapevines cannot uproot themselves and relocate to search for new resources, this inappropriately named "luxury consumption" can be useful for temporary storage of nutrients, such as nitrate or phosphate, that are subject to wide temporal and spatial fluctuations in the soil (Clarkson, 1985; see also Section 5.3). Storage within plant cells occurs in the vacuoles, whereas the cytoplasm is well buffered against changes in nutrient concentration (Forde, 2002; Tester and Leigh, 2001). As a result, the concentration of nutrient ions may be three to four orders of magnitude lower in the cytoplasm than in either the vacuole or the apoplast. The storage pools serve as an insurance against temporary shortages in supply; they can be accessed to sustain growth when nutrient availability is low or to permit regrowth in spring or following catastrophic events such as defoliation by pests, frost, drought, or fire (Bloom et al., 1985; Pradubsuk and Davenport, 2010, 2011). However, even if their availability increases in unison, all nutrients are not necessarily taken up and accumulated equally; for example, N, P, and K may accumulate throughout the vine, including the fruit, at the expense of Ca and Mg (Delas and Pouget, 1984).

When vacuolar nutrients fall to a critical minimum concentration, the cytoplasmic concentration cannot be sustained, and metabolism is disturbed. Nutrient deficiency triggers a range of responses in the vine, some of which are general stress responses and some of which are specific for the nutrient in question (Grossman and Takahashi, 2001). General responses include reduced growth via cessation of cell division and cell expansion, changes in vine morphology (e.g., increased root ÷ shoot ratio), accumulation of carbohydrates, decrease in photosynthesis, and modification of metabolism to adapt to the limited nutrient supply. The decline in growth is due to a decline in sink activity when nutrients are deficient. Consequently, reduced growth is responsible for the accumulation of foliar sugar and feedback inhibition of photosynthesis, rather than the decrease in photosynthesis limiting growth (Peuke, 2010). Indeed, it appears that elevated sugar concentration in the phloem is indicative of nutrient deficiency, whereas high concentrations of other nutrients in the phloem indicate abundant nutrient supply; thus phloem sap composition might be used as a marker of plant nutrient status (Peuke, 2010). Accumulation of reactive oxygen species, moreover, is a general response that is common to at least N, P, K, and S deficiency (Schachtman and Shin, 2007). Another frequent reaction to mineral nutrient deficiency is the accumulation of amino acids such as arginine, glutamine, and asparagine, as well as the polyamine putrescine (Rabe, 1990).

Specific responses to limited supply of a particular soil nutrient include induction of transport systems to enhance nutrient uptake and remobilization of stored reserves of that nutrient. In addition, because plants can make optimum use of a particular nutrient only if no other nutrient is limiting, deficiency in one nutrient often also impacts the uptake and transport of other nutrients (Amtmann and Blatt, 2009; Keller et al., 2001b). When a root has depleted a nutrient in a soil region, plants often enhance root growth and uptake mechanisms in another soil region where the availability of that nutrient is still adequate. Such local increases in nutrient acquisition can compensate for differences in nutrient supply arising from highly heterogeneous soil conditions (Nibau et al., 2008; Scheible et al., 1997). Nevertheless, a decline in shoot growth similar to that in response to PRD (see Section 7.2) often occurs when nutrients such as N or P are available to only a portion of the roots (Baker and Milburn, 1965; Robinson, 1994). Nutrient deficiency, particularly N and P deficiency, also trigger other plant responses that are very similar to the effects of soil water deficit, namely increased ABA and reduced cytokinin production, increased xylem sap pH raising guard cell sensitivity to ABA, increased r_h and r_s, decreased transpiration, restricted leaf expansion and senescence of older leaves, and increased root $\div$ shoot ratio (Clarkson et al., 2000; Wilkinson and Davies, 2002). These responses lead to a decrease in water uptake and transport, which is probably mediated by closure of aquaporins in the roots (Maurel et al., 2008). If different roots of the same vine experience differences in nutrient availability, it makes sense for the plant to reduce water uptake from the depleted soil and compensate by increasing water uptake from the nutrient-rich areas.

Meristematic cells are not directly connected to the vine's vascular tissues, and the very small vacuoles of these small, dividing cells permit little nutrient storage. Therefore, meristematic cells are more sensitive to fluctuations in nutrient delivery via the transpiration stream than are mature cells. Nutrient limitation interferes with cell division by preventing the normal progression of the cell cycle, which prolongs the time it takes for a cell to divide (Gastal and Lemaire, 2002; Grossman and Takahashi, 2001; Lemaire and Millard, 1999). In developing leaves, both cell division and cell expansion decline rapidly in response to N deficiency. Given that nitrate and phosphate stimulate cytokinin production, this reaction is partly caused by a decrease in the production of cytokinin from isoprenoids in the root tips and its transport in the xylem to the shoots, in addition to reduced cytokinin production in the young, expanding leaves (Coruzzi and Zhou, 2001; Forde, 2002; Kakimoto, 2003; Takei et al., 2002). Lack of cytokinin also seems to inhibit lateral bud outgrowth so that low-N vines produce fewer and shorter lateral shoots (Ferguson and Beveridge, 2009; Keller and Koblet, 1995a).

Unlike shoot growth, root growth may increase under nutrient deficiency, notably in response to limiting supplies of N and P but probably also of K and S. In contrast to their action in shoot organs, cytokinins inhibit lateral root growth, perhaps by stimulating the production of the growth-inhibiting gas ethylene (Dugardeyn and Van Der Straeten, 2008; Ha et al., 2012; Kakimoto, 2003; Werner et al., 2003). As the amount of cytokinins declines, root growth increases. This means that when nutrients limit overall plant growth, roots become relatively stronger sinks than shoots for assimilates in order to alleviate the deficiency by improving nutrient uptake from previously untapped soil regions where higher nutrient concentrations may be found (Clarkson, 1985). Additionally, cytokinins seem to hinder the uptake of nutrient ions by repressing the ions' transport proteins in the root epidermis and cortex (Kiba et al., 2011). Local production and movement of other hormones, such as auxin and ABA, also change in response to alterations in nutrient availability and uptake. Transport of these diverse hormones from the roots to the shoots in the xylem or from the shoots to the roots in the phloem mediate and integrate information on nutrient status in plants that is then translated in each organ into a response specific to each nutrient ion (Kiba et al., 2011; Krouk et al., 2011).

Slow growth and gradual appearance of chlorosis, necrosis, or senescence of leaves are typical symptoms of increasingly severe mineral nutrient deficiency in grapevines. Chlorosis refers to leaf yellowing associated with the breakdown of chlorophyll, whereas necrosis refers to uncontrolled and often patchy or marginal cell death associated with the failure of cell membranes. In contrast to necrotic death, the shedding of older leaves following their orderly senescence is typically associated with remobilization of nutrients in those leaves and redistribution to developing sink organs such as young leaves and grape berries. Remobilization is particularly true of the phloem-mobile ions of N, P, K, S, Mg, Na, and chlorine (Cl), whereas phloem-immobile nutrients, such as Ca and boron (B), cannot be readily redistributed. Consequently, deficiency symptoms of phloem-mobile mineral nutrients are first apparent on older leaves, whereas deficiency symptoms of phloem-immobile nutrients are usually confined to young, growing organs.

The amount of minerals accumulated by different *Vitis* genotypes can vary severalfold (Christensen, 1984; Gautier et al., 2018). Moreover, rootstocks may influence the concentration of nutrient ions in the xylem sap and leaves of the scion grafted to them (Delas and Pouget, 1984; Grant and Matthews, 1996a; Keller et al., 2001b; Trieb and Becker, 1969). This has implications for fertilizer recommendations based on tissue nutrient tests. Species, cultivars, and rootstocks that are more efficient than others in nutrient uptake can produce more growth from a given amount of absorbed nutrient ion (Clarkson and Hanson, 1980; Grant and Matthews, 1996a; Scienza et al., 1986). This is especially important when nutrient availability is limited. Since nutrient use efficiency is the reciprocal of the nutrient cost of growth, efficient species or cultivars often have lower average tissue concentrations of mineral nutrients than do inefficient species or cultivars (Bloom et al., 1985). However, the former often produce larger root systems and hence a larger root $\div$ shoot ratio (Clarkson and Hanson, 1980). When grafted grapevines are planted, rootstocks with efficient nutrient uptake could be chosen to minimize fertilizer input. On the other hand, inefficient rootstocks might reduce scion vigor on fertile sites.

Among mineral nutrients required for plant growth, N, P, K, Ca, Mg, and S are usually classified as macronutrients, and the remaining essential nutrients Cl, B, Fe, manganese (Mn), Zn, Cu, molybdenum (Mo), and nickel (Ni) are classified as micronutrients. This division is arbitrary and merely reflects the fact that different nutrients are required in vastly different amounts to satisfy vine growth and metabolism. The tissue concentrations of both macro- and micronutrients vary severalfold across different grape cultivars (Christensen, 1984). Depending on the cultivar, rootstock, and seasonal weather conditions, the amounts of macronutrients removed from a vineyard per ton of harvested grapes are in the range 1–3 kg N, 0.2–0.4 kg P, 1.5–4 kg K, 0.2–1 kg Ca, and 0.05–0.2 kg Mg (Bettner, 1988; Conradie, 1981a; Currle et al., 1983; Mullins et al., 1992; Schreiner et al., 2006). The lower end of these ranges applies only where stalks and pomace are recycled to the vineyard. Additional amounts of each nutrient are incorporated into the growing permanent structure of the vines.

Deficiency of micronutrients, especially B, Fe, Mn, and Zn, is relatively common in vineyards throughout the world (Mullins et al., 1992). Although grapevines normally acquire mineral nutrients via root uptake from the soil solution, it is possible to alleviate nutrient deficiencies in the short term with foliar sprays. Nutrient ions that are dissolved in water can diffuse through plant cuticles, probably via aqueous polar pores, and can then be taken up by leaf cells and distributed via the apoplast (Schönherr, 2000, 2006). Diffusion increases in proportion to the amount of nutrient ions applied and is independent of temperature but increases with increasing air humidity (Schönherr, 2000; Schreiber et al., 2001). In addition, nutrients may also be taken up through the stomata (Eichert and Goldbach, 2008).

7.3.1 Macronutrients

Nitrogen

The role of N and its uptake, assimilation, and transport in grapevines are discussed in Section 5.3. Grapevines normally acquire N in the form of inorganic nitrate (NO_3^-), although most of the N in a typical vineyard soil is bound in organic matter, which is unavailable for direct uptake by the roots. Soil microbes decompose and mineralize organic material, thereby converting the bound N into plant-available NO_3^-. In addition, rain and lightning may deposit some N that is present in the atmosphere due to emissions from volcanic activity and fossil fuel. However, the distribution of both organic matter and plant-available inorganic N in the soil is extremely heterogeneous over both time and space. Nitrate concentrations in the soil water can vary over several orders of magnitude, from a few micromoles to approximately 100 mM, even over short distances and time spans (Keller et al., 1995, 2001b).

Because of the rapid shoot growth in spring, a vine's N demand is greatest between budbreak and bloom, even though most of the N uptake from the soil occurs after bloom, provided there is sufficient soil moisture (Currle et al., 1983; Peacock et al., 1989). During this period, the vine is heavily dependent on the N reserves stored in its permanent structure, because the roots absorb very little N from the soil before five or six leaves have unfolded on the shoots (Löhnertz et al., 1989; Pradubsuk and Davenport, 2010; Treeby and Wheatley, 2006). In fact, nutrient application has little effect on bleeding sap composition, and the bleeding sap exuded during budbreak of grapevines fertilized with a whopping 672 kg N ha^{-1} contained similar amounts of NO_3^- to that of vines that did not receive any N fertilizer (Andersen and Brodbeck, 1991; Roubelakis-Angelakis and Kliewer, 1979). Even newly planted rooted cuttings initially grow almost exclusively from stored reserves (Groot Obbink et al., 1973). Remobilization from the reserve pool seems to be independent of soil N availability, and poor N reserve status due to inadequate refilling in the previous growing season can restrict early shoot growth and canopy development and probably flower development as well (Keller and Koblet, 1995a).

Uptake from the soil increases progressively through bloom, fruit set, and the first phase of berry growth and may increase further after veraison. In some cases, uptake was found to pause once shoot growth ceased, but in other cases the uptake rate was at its peak during this preveraison period and was sometimes—especially in warm climates—followed by another rise after harvest (Conradie, 1980, 1986; Löhnertz et al., 1989; Peacock et al., 1989). It seems likely that such discrepancies reflect differences in soil moisture during summer, because both shoot growth and nutrient uptake slow in drying soil (Keller, 2005). When water is not limiting, maximum N uptake may occur during the warmest portion of the growing season. In any case, uptake until the end of bloom constitutes <30% of the seasonal demand, although the leaves may accumulate up to 60% of their seasonal N requirement. Thus, storage reserves reach a minimum around bloom time or even later, which makes vines vulnerable to deficiency if insufficient N is available in the soil (Schaller et al., 1989; Weyand and Schultz, 2006b). However, the strong dependence on reserve N after budbreak also means that fertilizer application is ineffective at the start of the growing season to alleviate an existing deficiency that arises from low reserve status, whereas the ability of vines to absorb N after harvest in sufficiently warm climates permits enhanced refilling of storage reserves through late-season fertilizer supply (Conradie, 1980, 1986; Holzapfel and Treeby, 2007; Peacock et al., 1989; Treeby and Wheatley, 2006).

High soil N status inhibits lateral root growth and leads to a reduction in the root ÷ shoot ratio (Erlenwein, 1965a; Keller et al., 1995; Zerihun and Treeby, 2002). The inhibition of lateral root growth is mediated by ethylene, whose production increases when certain root NO_3^- transporters sense high external NO_3^- concentrations (Wang et al., 2012). A depression of root growth also occurs when cuttings are rooted that are derived from vines with a very high N status; this effect is exacerbated by high water status of the mother vines (Alleweldt, 1964). Nevertheless, localized N supply, such as occurs through fertigation via drip irrigation, to a low N soil stimulates lateral root growth in the new high N zone to enhance N uptake there (López-Bucio et al., 2003; Nibau et al., 2008; Scheible et al., 1997). This so-called systemic N acquisition response enables the root system to compensate for N shortage in one part of the soil by increasing uptake in an area that contains more N (Tegeder and Masclaux-Daubresse, 2018). Whereas new lateral roots proliferate in N-rich soil regions, the extension of existing lateral roots is stimulated under N-limiting conditions, which allows the roots to search for N further away from the plant (Bellini et al., 2014; Gifford et al., 2008). This response is mediated by the assimilated form of N, glutamine: A high glutamine status leads to lateral root proliferation, whereas low glutamine leads to lateral root elongation. In addition, an increase in auxin in the root tip inhibits its growth but stimulates outgrowth of lateral roots behind the tip (Tsay et al., 2011; Wang et al., 2012). Despite the increase in root growth under low N, water flow through the roots generally decreases due to higher hydraulic resistance (Chapin, 1991). The higher r_h results from a combination of decreased cell expansion, which leads to narrower xylem vessels, and delayed lignification and death of xylem cells (Plavcová et al., 2012).

Like other plants, grapevines with an optimum supply of N and water make longer use of the growing season by extending the effective life of their leaves (Alleweldt et al., 1984a; Lawlor, 2002). Nitrate uptake leads to increased production and export from the roots of the antisenescence hormone cytokinin (Hwang and Sakakibara, 2006; Sakakibara et al., 2006; Takei et al., 2002). Therefore, when vines are grown on N-rich soil, the chlorophyll degradation that accompanies autumn senescence of the leaves is delayed (Keller et al., 2001b). Conversely, N starvation results in reduced plant growth and yield, gradual chlorosis of older leaves, and early abscission of those leaves (Scienza and Düring, 1980; Schreiner et al., 2013; see also Fig. 7.4). It seems likely that such premature leaf senescence is triggered by a combination of sugar accumulation and low N and cytokinin status in the leaves (Wingler et al., 2009). Old leaves become an alternative source of N for the roots and other sink organs when N uptake by the roots fails to meet sink demand (Hikosaka, 2005; Schippers et al., 2015).

Nitrogen starvation increases ABA production in the roots and decreases the amount of cytokinins, whereas N resupply increases cytokinins (Chapin, 1991; Wilkinson and Davies, 2002). Although cytokinins are predominantly produced in the roots, they can also be produced in the shoots and leaves, especially in the phloem (Kiba et al., 2011). Additionally, N deficit leads to an increase in auxin flow from the shoots to the roots, perhaps because high sugar availability stimulates auxin production

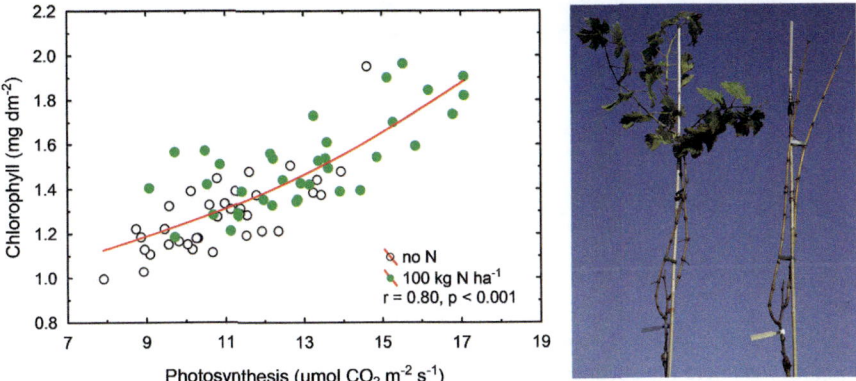

FIG. 7.4

Association between leaf photosynthesis and chlorophyll content in Müller-Thurgau grapevines grown under two long-term N fertilizer regimes (left), and increased lateral shoot growth and delayed leaf senescence on a vine with high N status (left plant on right) compared with a low-N vine (right plant on right; photo by M. Keller).

Left: from Keller, M., 2005. Deficit irrigation and vine mineral nutrition. Am. J. Enol. Vitic. 56, 267–283; right: photo by M. Keller.

(Krouk et al., 2011; Wang and Ruan, 2016). Enhanced cytokinin delivery to the shoots may reduce xylem formation and lignification (Groover and Robischon, 2006). Because cytokinins stimulate cell division, the growth-depressing effect of N deficiency may be partly related to the effect of N on cytokinin biosynthesis: All cytokinin molecules contain five N atoms within the structure formed by the attachment of an isoprene functional group to ATP or ADP. Through the suppression of cell division, N deficiency can inhibit shoot growth at any time during the growing season (Gastal and Lemaire, 2002). While cytokinins stimulate cell division in the shoot apical meristem and promote lateral bud outgrowth, they inhibit cell division in the root apical meristem (Ferguson and Beveridge, 2009). A decrease in cytokinin production thus decreases main and lateral shoot growth and enhances root growth, and an increase in auxin supply stimulates the outgrowth of lateral roots. The sensitivity of leaf initiation and expansion to fluctuations in N supply results in marked changes in the leaf area ÷ plant weight ratio. In addition, cytokinins may also function to inhibit NO_3^- uptake by the roots, whereas auxin seems to have the opposite effect (Kiba et al., 2011; Wang and Ruan, 2016).

Approximately half of the leaves' N is invested in photosynthetic hardware, especially rubisco and other enzymes, and the photosynthetic capacity is strongly correlated with N supply (see Section 5.2 and Fig. 5.5). By limiting the amount and activity of these proteins, N deficiency restricts photosynthesis (Chen and Cheng, 2003b). This enzymatic inhibition in turn increases r_s and also results in a potential energy overload in the leaves. To avoid excessive light absorption that would damage the photosynthetic apparatus, leaves disassemble a portion of their chlorophylls and activate their thermal energy-dissipation and antioxidant systems (Chen and Cheng, 2003a; Keller, 2005). In other words, the reduced chlorophyll content of N-starved vines (Keller et al., 2001b; Spayd et al., 1993) is a consequence, not a cause, of the decrease in photosynthesis (Fig. 7.4). The N contained in the leaf proteins, as well as any NO_3^- remaining in the vacuoles, are recycled to young leaves so that older leaves are the first to become chlorotic (see Section 5.3).

FIG. 7.5

The angle between leaf blade and petiole in a nitrogen-sufficient (left) and a nitrogen-deficient Müller-Thurgau grapevine (right). Also note the absence of lateral shoot growth and the red pigmentation of the petiole on the right.

Photos by M. Keller.

Because inadequate N supply decreases the demand for carbon skeletons for N metabolism and generally suppresses growth more than photosynthesis, sugar export declines and starch accumulates in the leaves (Chapin, 1991; Chen and Cheng, 2003b; Gastal and Lemaire, 2002; Lawlor, 2002; Schreiner et al., 2018). At least part of the decrease in photosynthesis of N-deficient leaves may be caused by feedback inhibition as a consequence of carbohydrate accumulation (Hermans et al., 2006; Schreiner et al., 2013). When this is accompanied by oxidative stress arising from too much light absorption, the leaves also produce photoprotective anthocyanins in the petioles, leaf veins, and sometimes in the leaf blades and cluster peduncles, which therefore turn red—even in some white cultivars (Currle et al., 1983; Gärtel, 1993; Lillo et al., 2008; Matus et al., 2017; Fig. 7.5). Conversely, the yellowing of chlorotic leaves is caused by unmasking and partial retention of carotenoids (see Section 4.1) that accompanies chlorophyll breakdown rather than by new production of yellow pigments.

Surplus sugar in the leaves may become available for export to the roots (Lemaire and Millard, 1999). In contrast to the response to water deficit, the leaf sugar is not needed for osmotic adjustment. Similar to their reaction to water stress, however, leaves decrease the angle between leaf blade and petiole (Fig. 7.5), which further curbs light absorption to avoid an energy surplus. In addition, leaves of N-deficient grapevines tend to have a thicker epidermis (Rühl and Imgraben, 1985). Another similarity with water deficit is that under prolonged N deficit grapevines start shedding older leaves following the remobilization of nutrients in those leaves and redistribution to sink organs (Currle et al., 1983; Gärtel, 1993; Lawlor, 2002). The similarity of vine responses to water and N deficit suggests that the combined effects of these two stress factors are predominantly additive rather than interactive. Thus, grapevines growing under low soil moisture, present during a drought or deficit irrigation, may require a somewhat higher N supply to support photosynthetic rates as high as those of fully irrigated vines (Alleweldt et al., 1984a; Rogiers et al., 2004b). Insufficient N also appears to sensitize plants to subsequent water stress, possibly because the increased ABA production induces the stomata to close at higher Ψ_{leaf}, and accelerates nutrient remobilization followed by leaf senescence and abscission during water stress. On the other hand, abundant water and/or N may make vines more susceptible to stress when one of the two resources suddenly is in short supply.

Insufficient N availability during bloom interferes with flowering and reduces fruit set and cluster initiation in the buds, but too much N decreases cluster initiation too (Ewart and Kliewer, 1977; Keller and Koblet, 1995a; Keller et al., 2001a; Spayd et al., 1993). In addition, low plant N status heightens the vulnerability of fruit set to water deficit (Keller et al., 1998). Low N status may sometimes, though not always, lead to bigger berries (Hilbert et al., 2003; Keller et al., 1998, 2001a; Rogiers et al., 2004b; Schreiner et al., 2018). Greater berry growth is probably a consequence of diminished sink number due to poor fruit set. An early restriction of the crop load tends to result in compensatory growth of the remaining berries, unless such compensation is prohibited by water deficit (Keller et al., 2008). Compensatory berry growth might also be possible because of limited competition from shoot meristems under N deficiency. This may leave more photosynthates available for export to the fruit clusters, which in turn would also increase water uptake by the berries. In any case, N deficiency leads to grape berries with a low N content, which often results in musts that are prone to sluggish or stuck fermentations (Bell and Henschke, 2005; Spayd et al., 1995; see also Section 6.2).

Excessive soil N availability can pose problems too, especially in young vines whose small root systems may temporarily be exposed to very high concentrations of N following fertilizer application. Very high NO_3^- concentrations may decrease the osmotic potential (Ψ_π) of the soil solution, which can inhibit shoot growth and decrease stomatal conductance due to insufficient water uptake (Keller and Koblet, 1994). This effect of high soil N is similar to the impact of salinity, discussed later. High N also leads to larger xylem vessel and fiber cells with thinner cell walls whose lignification is delayed and reduced (Plavcová et al., 2012). Though this effect decreases r_h, it increases the xylem's vulnerability to cavitation and may weaken the structural support that the fibers lend the wood. In bearing vines, the vigorous growth and often dense canopies that accompany high soil N status can have undesirable consequences for fruit composition (see Section 6.2) and heighten the vulnerability to pathogens (see Section 8.2). In addition, vigorous vines with high N status may be more vulnerable to physiological ripening disorders such as bunch stem necrosis (Christensen and Boggero, 1985; Currle et al., 1983; Keller et al., 2001a). In some cases, however, N deficiency may be associated with a higher incidence of bunch stem necrosis too (Capps and Wolf, 2000).

Soil N status may also have implications for the availability of other mineral nutrients. High amounts of available N can increase the concentrations of cations such as K^+, Ca^{2+}, and Mg^{2+} in the soil solution and reduce that of anions such as $H_2PO_4^-$ (Keller et al., 1995). Higher soil cation availability may be coupled with enhanced uptake by grapevines, but it may also increase the potential for rainfall-driven leaching of these nutrients (Keller et al., 2001b; Perret and Roth, 1996). Over time, excessive N application to vineyard soils might therefore increase the likelihood of nutrient deficiencies developing. This may explain the observation that leaf Ca, Mg, Mn, and B are sometimes increased in N-deficient vines (Schreiner et al., 2013). In contrast, abundant N supply sometimes, but not always, decreases P uptake (Hilbert et al., 2003; Keller et al., 1995, 2001b; Schreiner et al., 2018; Spayd et al., 1993). Perhaps a reduction in root carbohydrate status of high-N vines occasionally limits carbon availability for mycorrhizal fungi. Furthermore, because NO_3^- uptake by the roots is coupled to influx of protons (H^+) and release of OH^-, the pH in the rhizosphere rises as N uptake increases. Depending on the soil pH before the N-induced alkalinization, this may enhance or hinder the availability of other mineral nutrients (Fig. 7.3).

Phosphorus

The anion phosphate ($H_2PO_4^-$) is often abbreviated P_i for inorganic phosphate. It plays an important role in photosynthesis, respiration, and the regulation of many enzymes and signal receptors. Phosphate is a major substrate in energy metabolism, contributing three P_i groups to ATP (see Sections 4.1 and 4.4), and in the biosynthesis of nucleic acids, contributing P_i groups to both DNA and RNA. In addition, P_i is a component of cell membranes as a constituent of phospholipids and supplies starter compounds for the production of carotenoids and gibberellins. Reversible phosphorylation occurs through addition or removal of a P_i group by kinases or phosphatases, respectively, and activates ($+P_i$) or deactivates ($-P_i$) proteins. In other words, P_i addition switches proteins on, while P_i removal switches them off. Other proteins are labeled by phosphorylation for later disassembly. One of the most ubiquitous groups of phosporylated proteins are the H^+-ATPases that are essential for cellexpansion (see Section 3.1) and nutrient transport (see Section 3.3) via the reversible reaction $ATP + H_2O + H_{in}^+ \leftrightarrows ADP + P_i + H_{out}^+$. Furthermore, phosphorylation of aquaporins regulates water transport across membranes: Phosphorylated aquaporins are open (Maurel et al., 2008). In addition, inorganic pyrophosphate (PP_i; $P_2O_7^{4-}$), a by-product of starch production, is essential for phloem function (Koch, 2004).

Nevertheless, about 90% of a cell's P is located in the vacuole (Shen et al., 2011). Because of its importance, the leaf P content increases roughly as the four-thirds power of the leaf N content: $P \sim N^{4/3}$ (Niklas, 2006; Niklas and Cobb, 2006). This relationship also holds approximately true across different grape cultivars (calculated from data in Christensen, 1984). Consequently, the seasonal pattern of P uptake by grapevines from the soil more or less traces the N uptake pattern (Conradie, 1980, 1981b; Pradubsuk and Davenport, 2010; Schreiner, 2016).

Although most soils contain abundant P, it is one of the least plant-available of all essential nutrient elements. This "phosphorus paradox" arises because most P is extremely insoluble, and hence immobile, due to its affinity to cations such as Ca^{2+}, Mg^{2+}, Al^{3+}, and Fe^{2+}, with which it forms insoluble complexes; and due to its conversion by soil microorganisms into organic forms that plant roots cannot take up (Marschner, 1995; Raghothama, 1999; Rellán-Álvarez et al., 2016; Shen et al., 2011). In addition, the negative charge of the water-soluble $H_2PO_4^-$ means that it is readily leached out from the rooting zone. The P_i concentration in the soil solution is generally $<10\,\mu M$, which is below that of many micronutrients, whereas the concentration in the cell cytosol is 5–20 mM. Many of the heavily leached and often coarse-textured Australian soils are especially poor in P. In weakly alkaline soils, P occurs mostly as hydrogen phosphate (HPO_4^{2-}), whereas in weakly acidic soils the predominant form is dihydrogen phosphate ($H_2PO_4^-$). In acid soils (pH ≤ 5.5), which are common in Bordeaux, France, and in the northeastern United States, P is the major limiting nutrient (Kochian et al., 2004). This is one reason why such soils are often ameliorated before vineyard establishment by the application of lime ($CaCO_3$) to increase the soil pH. Moreover, P availability is normally highest in the topsoil and decreases sharply with soil depth, which is why high-phosphate fertilizers are often applied to improve P availability. Unfortunately, however, a considerable portion of this supplementary P can leach from vineyards into nearby rivers and lakes, where it may cause eutrophication and proliferation of algae (so-called algal blooms), oxygen depletion, and death of fish.

Roots take up P mainly as $H_2PO_4^-$, using high-affinity transporters (Gojon et al., 2009; Grossman and Takahashi, 2001; Shen et al., 2011). Growing roots initially absorb P in the root tips and root hairs, but this uptake rapidly depletes available P in the rhizosphere (Bucher, 2007; Gilroy and Jones, 2000). Decreasing P availability leads to a rapid decrease in the growth of main roots in the depleted soil area and an increase in the growth of lateral roots and root hairs (Hermans et al., 2006; López-Bucio et al., 2003; Nibau et al., 2008; Osmont et al., 2007; Raghothama, 1999). This shift in root growth pattern may be mediated by sucrose and microRNA pieces imported from the leaves via the phloem (Hammond and White, 2011; Lei et al., 2011). Upon arrival in the roots these molecules may induce accumulation of auxin and ethylene and depletion of cytokinin and gibberellin, which ultimately results in a denser but shallower root system and an increase in the root $\div$ shoot ratio (Chiou and Lin, 2011;

Grant and Matthews, 1996a; Malamy, 2005; Potters et al., 2009). Although this adaptive strategy enables grapevines to better explore the rather more P-rich surface soil, it also makes them more vulnerable to drought stress when the soil dries out. This may be why the change in root growth is limited to situations in which the entire root system, rather than just portions of it, experiences P deficiency (Schachtman and Shin, 2007).

The roots of some P-deficient species also exude organic acids such as citrate and malate that acidify the rhizosphere and act as anion exchangers that chelate metal cations and solubilize P from rock phosphate (Dakora and Phillips, 2002; Hammond and White, 2008; Marschner, 1995). Such organic acid extrusion, however, comes at a high cost: it can drain >10% of a plant's dry matter (Plaxton and Tran, 2011). In addition, P deficiency results in proton extrusion by the roots to acidify the soil. The lower soil pH greatly enhances P availability while also increasing the availability of several other mineral nutrients as well (Fig. 7.3). In addition, grapevines rely heavily on mycorrhizal fungi for P uptake, and P-starved plants ostensibly encourage mycorrhizal colonization of their roots by exudation of strigolactone sesquiterpenes that help the fungi to establish root contact (Akiyama et al., 2005; Bais et al., 2006). By modulating auxin flux, the strigolactones also stimulate lateral root growth in P-deficient plants, while they suppress lateral root growth under P-sufficient conditions (Ruyter-Spira et al., 2011, 2013). At the same time, strigolactones are transported via the xylem to the shoots, where they inhibit lateral shoot growth (Gomez-Roldan et al., 2008; Kohlen et al., 2011; Umehara et al., 2008). Indeed, a restriction of leaf number and leaf size is one of the earliest and most reliable symptoms of P deficiency. Thus, the increase in carbon partitioning to the roots at the expense of the shoots results in a change in plant architecture that may be coordinated by strigolactones and favors P uptake while limiting its use for canopy construction (Grant and Matthews, 1996a; Ruyter-Spira et al., 2013).

Low P supply slows cell division, which restricts leaf initiation in the shoot apical meristem and expansion of newly developed leaves, thus limiting plant leaf area (Chiera et al., 2002; Schreiner and Osborne, 2018). Reduced cell division may additionally be responsible for the inhibition of flower cluster initiation or differentiation in P-deficient vines (Gärtel, 1993; Grant and Matthews, 1996b; Skinner et al., 1988). Phosphate deficiency can also increase r_h and r_s so that the reduction in water supply to growing organs restricts cell expansion, which in turn strongly limits leaf expansion (Clarkson et al., 2000). Therefore, although as a phloem-mobile nutrient P can be remobilized and recycled from old leaves and woody structures to support growing organs, P deficiency primarily restricts the sink activity of shoot meristems (Grant and Matthews, 1996a; Hermans et al., 2006; Kochian et al., 2004). The resulting lack of demand for assimilates is associated with sugar and starch accumulation in the source leaves. Excess leaf sugar eventually leads to feedback inhibition of photosynthesis, which seems to be especially pronounced at temperatures below approximately 15 °C (Hendrickson et al., 2004b). An additional decrease in photosynthesis may come about because there is insufficient P_i for ATP synthesis and because the leaves do not fully activate their rubisco (Paul and Foyer, 2001; Woodrow and Berry, 1988). Nonetheless, sugar export via the phloem may continue to stimulate and sustain lateral root branching (Hammond and White, 2008, 2011; Lei et al., 2011).

Phosphorus deficiency leads to salvage of P_i from nucleic acids and phospholipids throughout the plant, and the latter are replaced by galacto- and sulfolipids (Plaxton and Tran, 2011; Schachtman and Shin, 2007). Moreover, P-deficient leaves remain small and dark grayish-green rather than turning chlorotic as in other cases of nutrient starvation (Gärtel, 1993; Hammond and White, 2008). Severely P-deficient leaves may turn red due to anthocyanin production from surplus sugars that are not exported from the leaves (Currle et al., 1983; Grossman and Takahashi, 2001; Lillo et al., 2008). The red pigmentation is probably a photoprotective reaction to limit damage from oxidative stress. Such stress ensues because the decline in photosynthesis is associated with an energy surplus, since light absorption and electron transfer proceed even as CO_2 fixation declines (Feild et al., 2001; Hoch et al., 2003).

Phosphorus availability has consequences for the uptake and use of other mineral nutrients as well. Insufficient P supply appears to restrict Mg transport in the xylem, which can lead to symptoms of Mg deficiency (Skinner and Matthews, 1990). Due to the propensity of P and Fe to precipitate together in the soil, P deficiency may be associated with increased Fe uptake and subsequent accumulation in the plant. Excess P, by contrast, may lead to Zn and Fe deficiency through complexation (Gärtel, 1993; Lei et al., 2011; Rellán-Álvarez et al., 2016).

Potassium

As K^+, potassium (K) is the most abundant cation in plant cells. Its concentration in grape leaves is approximately half that of N (Gärtel, 1993). In contrast with all other macronutrients except calcium, K is not incorporated into organic compounds but remains in its ionic form (Robinson, 1994; Tränkner et al., 2018). As such, it is one of the cells' major osmotic solutes and plays a key role in cell expansion (see Section 3.1) and stomatal movement (see Section 3.2). In addition, K is required for the production of proteins (Tränkner et al., 2018). Therefore, rapid plant growth and development require large K^+ fluxes to provide this ion to the growing tissues. This means that growth in general is sensitive to a vine's K status. Cells also use K^+ to neutralize the negative charges of anions, which helps maintain membrane potential and counterbalance the movement of other cations such as protons (H^+). The protons are needed for the activity of enzymes such as ATP synthase, as well as H^+-ATPases and other transport proteins (Amtmann and Blatt, 2009; Bassil et al., 2019; Clarkson and Hanson, 1980; Terrier et al., 1998). By maintaining the electrical neutrality that is necessary to generate pH gradients by H^+-ATPases, K^+ also participates in sucrose loading into the phloem for export from source organs (Gajdanowicz et al., 2011; Lalonde et al., 2004; Marten et al., 1999; see also Section 5.1).

The K^+ concentration in the soil solution varies from approximately 0.1–6 mM, whereas its concentration in the cell cytosol is approximately 100 mM (Ashley et al., 2006). The initial growth of grapevines after budbreak is dependent on K reserves remobilized from the vine's permanent structure. But uptake from the soil increases well before bloom time and, provided soil moisture is not limiting, proceeds more or less linearly throughout the growing season until a few weeks before leaf fall (Conradie, 1981b; Pradubsuk and Davenport, 2010; Schreiner, 2016; Williams and Biscay, 1991). Developing and ripening grape berries are particularly strong sinks for K. Between bloom and harvest, the berries accumulate far more K than any other nutrient. Indeed, mature grapes contain almost twice as much K as N and approximately 10 times as much as P and Ca, which are the next most abundant mineral nutrients in grape berries. Depending on the crop level, the harvested fruit contains 50–75% of a vine's total amount of K (Conradie, 1981b; Rogiers et al., 2006).

Rootstocks differ in their ability to take up K^+ and load it into the xylem; for instance, rootstocks derived from *V. berlandieri* may not take up K^+ as readily as those derived from *V. champinii*. Transport of K^+ in the xylem affects r_h and, thus, sap flow rate (see Section 3.3). Inadequate K supply strongly reduces xylem sap flow even under well-watered conditions, which limits shoot and fruit growth and greatly increases the risk of drought stress (Currle et al., 1983). Potassium deficiency also suppresses sugar loading into the phloem and may result in sucrose accumulation in the leaves to substitute for the missing K^+ as osmoticum (Cakmak et al., 1994; Hermans et al., 2006; Tränkner et al., 2018; Tsay et al., 2011). This explains not only why root growth cannot increase—and often stops—in

K-deficient plants but also why deficiency leads to feedback inhibition of photosynthesis. With sucrose being trapped in the leaves and photosynthesis declining, the consequences for fruit production and ripening can be severe. At the same time, its mobility in the phloem enables K^+ to be recycled from older organs to active sinks and to move preferentially toward growing tissues or organs, including the fruit and the permanent vine structure (Conradie, 1981b; Currle et al., 1983; Mpelasoka et al., 2003). While root growth is restricted, ethylene production induced by K deficiency stimulates the outgrowth of root hairs (Tsay et al., 2011). If the deficiency is not severe enough to stop root growth, the roots often respond to inadequate K by growing sideways: They deviate from their normal gravitropic growth, which may help them to explore and exploit previously untapped soil patches that contain more K (Ashley et al., 2006).

Visual deficiency symptoms include glossy leaves, often followed by bronzelike discoloration from dying epidermis cells and pale-yellow leaf margins (Fig. 7.6). In more severe cases, the leaf margins roll upward in a process termed "cupping" and eventually become necrotic before the leaves are shed (Currle et al., 1983; Gärtel, 1993). These symptoms develop more quickly under intense sunlight because the excited surplus electrons lead to oxidative stress due to accumulation of H_2O_2 and O_2^- (Marschner and Cakmak, 1989; Tränkner et al., 2018). Leaf symptoms are also more severe on heavily cropped vines due to the high demand for K by the growing berries (Currle et al., 1983; Gärtel, 1993). Under continued K deficiency, shoot vigor declines gradually over several years. Deficient vines also become more susceptible to powdery mildew and to winter injury (Currle et al., 1983; Gärtel, 1993). Moreover, due to the importance of K^+ as an osmoticum for pollen hydration and germination and for pollen tube growth after pollination, K deficiency interferes with fertilization, which results in poor fruit set.

Foliar deficiency symptoms are usually associated with low protein content and accumulation of amino acids and polyamines (especially putrescine) in the shoots and leaves (Gény et al., 1997b; Tränkner et al., 2018). It has been proposed that it is this putrescine buildup that induces the visible leaf symptoms. Nonetheless, putrescine application can improve fruit set, perhaps because a putrescine metabolite named hypusine-eukaryotic translation initiation factor 5A is required to transfer RNA from the cell nucleus to the cytoplasm during embryo development (Gény et al., 1997a, 1998; Takahashi and Kakehi, 2010). Polyamines are ultimately breakdown products of the amino acid arginine, and as K

FIG. 7.6

Leaf symptoms of potassium deficiency in Syrah.

Photo by M. Keller.

deficiency grows more severe, putrescine may become the main organic soluble N compound (Kusano et al., 2008; Takahashi and Kakehi, 2010). The cause and/or function of polyamine accumulation during K deficiency is unknown. However, given that polyamines can block cation channels and boost H^+-ATPases in plant cells (Alcázar et al., 2010; Kusano et al., 2008; Takahashi and Kakehi, 2010), it is possible that they serve to prevent compensatory but detrimental accumulation of Na^+ in the absence of K^+. Despite the decrease in arginine that accompanies polyamine accumulation, must derived from grapes harvested from K-deficient vines is not prone to sluggish or stuck fermentation because yeast can readily metabolize polyamines.

Putrescine also accumulates in leaves, especially those of vigorously growing young vines, during the stop-and-go growth caused by widely fluctuating spring temperatures (Adams et al., 1990). This can lead to a syndrome referred to as "spring fever," whose symptoms are virtually identical to those of K deficiency, even though the K content of these leaves is almost normal—while their N content is elevated. It is possible that the accumulation of putrescine during spring fever is part of a cold acclimation response that enhances the leaves' ability to survive freezing temperatures (Cuevas et al., 2008). The link between putrescine and low temperature may also be the reason why K deficiency symptoms appear to be more severe under cool conditions. In contrast to K deficiency, however, spring fever symptoms are confined to the oldest leaves, and the shoots usually outgrow the condition within a few weeks once warmer temperatures prevail. Nevertheless, the berries often remain small, and ripening is delayed.

Potassium deficiency can result from competition for root uptake by other cations present at high concentration in the soil solution (Wang and Wu, 2013). This occurs in saline soils, where sodium (Na^+) is the predominant dissolved cation, and in acid soils, which are defined as soils with a pH ≤ 5.5, where NH_4^+ is the predominant nitrogen form (Kochian et al., 2004). In turn, NH_4^+-based fertilizers contribute to soil acidification. In contrast, excess soil K^+ competes with Ca^{2+} and Mg^{2+} for root uptake, which can decrease the concentrations of these divalent cations in a vine's organs and induce symptoms of Mg deficiency (Morris et al., 1980, 1983). This effect is particularly pronounced in young, nonfruiting vines, and especially on sandy soils with a low pH (Gärtel, 1993). At tissue concentrations exceeding about 100 mM, K^+ can even inhibit enzymes, thus becoming toxic.

Sulfur

Sulfur (S) is an essential constituent of the amino acids cysteine and methionine, lipids, intermediary metabolites in energy generation and electron transport, and molecules involved in the protection of tissues against oxidative stress (Kopriva, 2006; Takahashi et al., 2011). The S-containing acetyl-CoA provides starter components for the citric acid cycle and the production of fatty acids, phenolics, and terpenoids. Both cysteine and methionine can be incorporated in proteins, and cysteine also is a building block of the peptide glutathione. Cysteine and glutathione are used by cells to counteract reactive oxygen species whose production is brought on by such environmental stressors as drought, cold, heat, high light, and fungal attack (Noctor and Foyer, 1998). Via the so-called ferredoxin/thioredoxin system S-containing proteins in concert with Fe-containing proteins regulate starch production and remobilization in response to light and sugar supply. In addition, attachment of a sulfate (SO_4^{2-}) group deactivates the hormones brassinosteroid and jasmonate (Takahashi et al., 2011).

The seasonal pattern of S uptake by grapevines is rather similar to that of N and P (Schreiner, 2016). Plant growth depends on sustained S uptake, usually as SO_4^{2-}, from the soil and transport in the xylem to the leaves, where it is assimilated or partially reexported in the phloem as SO_4^{2-} or glutathione

(Gojon et al., 2009; Grossman and Takahashi, 2001; Leustek and Saito, 1999). Following its uptake into the roots by SO_4^{2-}/H^+ cotransport, SO_4^{2-} can be transported in both the xylem and phloem and temporarily stored in the cell vacuoles before it is reduced in the plastids to sulfite (SO_3^{2-}) and further to sulfide (S^{2-}). The toxic S^{2-} is assimilated mostly in the cytosol to cysteine, which can be used for protein biosynthesis or converted to other organic S-containing compounds such as methionine and glutathione (Amâncio et al., 2009; Rennenberg and Herschbach, 2014; Takahashi et al., 2011). Both uptake and reduction of SO_4^{2-} require energy in the form of ATP; reduction additionally requires iron in the form of ferredoxin as an electron donor. Glutathione, which is produced by combination of the three amino acids cysteine, glutamate, and glycine, is not only the general scavenger of oxygen radicals but also the major transport and storage form of reduced S; storage occurs mostly in the bark (Cooper and Williams, 2004; Kopriva, 2006). In addition, long-term S storage also occurs in the form methionine-containing storage proteins (Rennenberg and Herschbach, 2014).

Under high S supply, feedback inhibition of SO_4^{2-} uptake by the roots ensures that uptake is in step with S demand (Takahashi et al., 2011). The roots decrease the activity of their high-affinity transporters when S availability in the soil is high and increase such activity when soil S becomes low (Amâncio et al., 2009). Insufficient S supply, however, leads to a depletion of cysteine and glutathione pools, which is associated with an increase in reactive oxygen species (Schachtman and Shin, 2007). The lowered capacity to assimilate S results in a decrease in photosynthesis and an increase in photorespiration, flavonoid production, and lipid breakdown (Takahashi et al., 2011). Deficiency symptoms include chlorosis of young leaves and stunted shoot growth. Growth might be especially compromised when oxygen radicals accumulating due to other stresses place a high demand on glutathione production, which diverts cysteine away from growth-promoting protein production (Speiser et al., 2018). While S-starved plants decrease shoot growth, they normally accelerate root elongation and form prolific lateral roots close to the root tip to enhance S uptake so that the root $\div$ shoot ratio increases, perhaps in response to increased auxin production (López-Bucio et al., 2003; Schachtman and Shin, 2007).

The production or release via glutathione or cysteine degradation of elemental S or hydrogen sulfide (H_2S) may be involved in the vine's defense strategy against certain fungal and bacterial pathogens (Cooper and Williams, 2004; see also Section 8.1). Sulfur deficiency could thus result in increased susceptibility to pathogen attack, because only plants growing at high SO_4^{2-} availability are able to release enough S to counter invasion. Consequently, application of S-based fungicides may have the added benefits of enhancing the grapevine's pathogen defenses and at the same time countering S deficiency (Cooper and Williams, 2004; see also Section 8.2). However, foliar S application at temperatures above 32 °C can induce burn symptoms on leaves, shoots, and berries. Moreover, S residues on grape berries following late foliar applications may have implications for fermentation and wine composition (see Section 6.3). Over the long term, frequent application of S-based fungicides, which comprise predominantly elemental S, can increase soil acidity because soil microorganisms convert S to sulfuric acid (H_2SO_4), which alters the plant availability of many mineral nutrients (Fig. 7.3).

Calcium

Calcium (Ca) is unique among macronutrients in that a high proportion of a plant's Ca^{2+} is located in the cell walls, where it serves as a reinforcing agent. Cell walls probably owe their rigidity not just to the strong and stiff cellulose microfibrils but also to the cross-linking of pectins by Ca^{2+}, which also holds adjacent cells together and prevents them from sliding (Ferguson, 1984; Hamant and Traas, 2010; Jarvis, 1984). Indeed, the Ca concentration in the apoplast is 300–1000 times higher than that in the cytoplasm, and most of this Ca^{2+} is tightly bound as Ca-pectate in the cell walls (Boyer, 2009). Although, like K, Ca is not actually incorporated into organic molecules, the ionic bonding of Ca^{2+} contrasts with the mostly soluble apoplastic K^+ (Amtmann and Blatt, 2009). Another important site of Ca accumulation and storage, in the form of Ca-oxalate, is the cell vacuole. Crystals of Ca-oxalate, especially the needle-shaped raphides, may serve a defense role by deterring herbivores.

Because of its ability to connect lipids and proteins at membrane surfaces, Ca^{2+} is important in maintaining membrane integrity; it stabilizes membranes by bridging lipid and protein phosphate and carboxylate groups and prevents membrane damage and leakiness to solutes (Clarkson and Hanson, 1980; Hirschi, 2004). Since membrane leakage is synonymous with the loss of semipermeability and results in cell death, Ca delays senescence and organ abscission (Poovaiah and Leopold, 1973). Moreover, Ca^{2+} ions inhibit the Mg^{2+}-dependent reactions of respiratory and intermediary metabolism and alter the water permeability of membranes by modulating aquaporin activity (Clarkson and Hanson, 1980; Johansson et al., 2000).

Plants also require Ca to produce electrons, protons, and oxygen during photosynthesis (see Section 4.1). In addition, Ca is involved in intracellular signaling in response to osmotic stress that arises from many environmental impacts (Plieth, 2005; Zhu, 2002). Calcium serves as a second messenger, whereby an external stimulus leads to an increase in reactive oxygen species that results in temporary oscillations of the cytosolic Ca^{2+} concentration. The period, frequency, and amplitude of these oscillations amount to characteristic "Ca^{2+} signatures" that encode information about the nature and strength of the stimulus (Hetherington and Brownlee, 2004; McAinsh and Pittman, 2009). For instance, ABA promotes release of Ca^{2+} from the guard cell vacuoles and influx from the apoplast to induce stomatal closure in response to water deficit (Allen et al., 2001; Amtmann and Blatt, 2009). Auxin and gibberellin coordinate cell expansion by inducing a rise in cytosolic Ca^{2+} released from intra- and extracellular stores. Calcium binds to a small protein termed calmodulin to activate other proteins by inducing their phosphorylation (Scrase-Field and Knight, 2003; Snedden and Fromm, 2001; Veluthambi and Poovaiah, 1984). Moreover, Ca^{2+} influx and a cytosolic Ca^{2+} gradient with increasing concentration toward the tip are necessary for the directional growth of root hairs and pollen tubes (Dutta and Robinson, 2004; Holdaway-Clarke and Hepler, 2003; Véry and Davies, 2000).

At high concentration, Ca^{2+} can disrupt metabolism by precipitating phosphate as $CaHPO_4$; thus, high amounts of Ca are extremely toxic to cells (He et al., 2014; Plieth, 2005). Plants utilize ATP-driven pumps, which in turn are often regulated by calmodulin, to keep the concentration of cytosolic Ca below $0.1\,\mu M$, allowing only small Ca oscillations potentially generated as signals by these pumps (Hetherington and Brownlee, 2004; Hirschi, 2004; McAinsh and Pittman, 2009).

Uptake of Ca from the soil occurs mostly near the root tips across nonselective cation channels, increases with Ca availability in the soil, and varies along with plant transpiration (Gilliham et al., 2011; Karley and White, 2009; Pradubsuk and Davenport, 2010). As plants have little control over Ca uptake, they sequester Ca as Ca-oxalate crystals inside specialized cells called idioblasts to prevent toxicity (Franceschi and Nakata, 2005; He et al., 2014; Storey et al., 2003b). The idioblasts are located in roots, leaves (especially along veins), petioles, and fruit. This biomineralization takes the soluble Ca out of circulation, and the crystals may serve as a storage form that can be rapidly remobilized during Ca starvation and during early spring growth. Accumulation, in the form of raphide crystals, in the root cortex occurs especially near the root tip, from where Ca may be supplied to the xylem (Storey et al., 2003b).

The Ca concentration and amount in grape leaves and other organs is similar to or higher than that of N, but in contrast to N, the Ca concentration increases as leaves age (Gärtel, 1993; Pradubsuk and Davenport, 2010; Schreiner, 2016). Because Ca^{2+} is poorly phloem-mobile, it is preferentially supplied to rapidly transpiring organs such as mature leaves and mostly becomes immobile once deposited in an organ; in addition to the leaves, the bark seems to be a major sink for Ca in grapevines (Conradie, 1981b; Gilliham et al., 2011; Karley and White, 2009; Peuke, 2010).

Similar to the impact of other stresses, a shortage in Ca supply leads to downregulation of aquaporins, which in turn increases the roots' radial r_h and interferes with water flow to the xylem (Luu and Maurel, 2005). In addition, Ca deficiency makes plants more susceptible to damage by salinity or low soil pH (Hirschi, 2004; Plieth, 2005). Plants on a low-Ca "diet" grow smaller and fewer leaves than plants with abundant Ca supply (Suárez, 2010). Symptoms of Ca deficiency appear mainly in young, expanding leaves and may include marginal leaf necrosis (Gärtel, 1993). The symptoms develop when Ca supply to cells cannot keep pace with Ca movement to the vacuoles. Deficiency causes cell walls to disintegrate and membranes to become leaky, which leads to cell death and can culminate in the collapse of affected tissues (Hirschi, 2004). A low Ca status can also compromise plant survival after freezing (see Section 7.2).

Due to its importance for pollen tube growth, Ca deficiency is also detrimental to fertilization and fruit set. After fruit set, xylem-supplied Ca tends to accumulate in the stomatal regions of grape berries (Blanke et al., 1999). But local deficiency in the fruit may be promoted by excessive canopy transpiration and rapid shoot growth diverting xylem flow, and thus Ca supply, away from the fruit. This effect could be especially pronounced during periods of rapid berry cell expansion requiring Ca^{2+} for incorporation in new cell walls and membranes and as an intracellular signal. Therefore, Ca deficiency can lead to berry shrinkage beginning at the tip of fruit clusters, which resembles symptoms of bunch stem necrosis (Gärtel, 1993). Indeed, the appearance of bunch stem necrosis has been associated with insufficient Ca (and often low Mg too) relative to K in rachis tissues, especially in the epidermal and subepidermal layers (Currle et al., 1983; Feucht et al., 1975). Excessive Ca supply, on the other hand, may retard the rate of fruit ripening, possibly by preventing cell wall disassembly (Ferguson, 1984).

Natural Ca deficiency is rare and occurs primarily on soils with very low pH because the highly toxic aluminum (Al^{3+}) becomes very soluble at pH < 5 and blocks Ca^{2+} uptake. Excessive soil K, Na, and/or Mg may also induce Ca deficiency (Delas and Pouget, 1984). However, the availability of Ca at the root surface often greatly exceeds the demand of grapevines, especially in high-pH soils (Storey et al., 2003b). Because Ca uptake is linked to the transpiration rate of the canopy, vines growing on calcareous soils generally have considerably higher tissue Ca contents (and often higher Mg but lower K contents) than vines growing on other soils. On high-Ca soils, the amount of Ca in a vine's permanent structure can even exceed its content of N, which is normally the most abundant mineral element in grapevines. Moreover, in soils whose pH exceeds 7.5, the soils' calcium carbonate ($CaCO_3$) can lead to precipitation of P, Fe, Zn, Cu, and Mn, making these essential nutrients less plant-available. Depending on their origin, different *Vitis* species are adapted to a wide range of soil Ca. Nevertheless, due to our propensity to classify things into simple categories, they are often grouped as being either calcicole or calcifuge. Calcicole species are lime-tolerant that are adapted to calcareous or alkaline soils, whereas calcifuge species are lime-sensitive and adapted to neutral to acidic soils. The calcicole category includes species like *V. vinifera* and *V. berlandieri*, whereas the calcifuge category includes species like *V. labrusca* and *V. riparia*.

Magnesium

Although magnesium (Mg) is a structural component of chlorophyll, <20% of a grapevine's total Mg content is bound in chlorophyll. As much as 75% of the Mg in leaves may be involved in the manufacture of proteins (Karley and White, 2009; Tränkner et al., 2018). As a divalent cation, Mg^{2+} is an important cofactor required to activate many enzymes and transport proteins (Cakmak and Kirkby, 2008; Clarkson and Hanson, 1980; Tränkner et al., 2018). The former include such key players as rubisco and glutamine synthetase (see Sections 4.2 and 5.3), and the latter include the ATPase proton pumps responsible for phloem loading (see Section 5.1) among many others. In addition, Mg seems to be involved in protecting anthocyanins from degradation in the cell vacuoles (Sinilal et al., 2011).

In contrast with most other nutrient ions, little is known about how roots take up Mg^{2+} and how they regulate this uptake and subsequent transport. Uptake and transport to the xylem do seem to be mostly active, via transport proteins, while Mg delivery to the leaves varies with transpiration rates (Karley and White, 2009). The seasonal pattern of Mg uptake by grapevine roots is rather similar to that of K, starting soon after budbreak and continuing until shortly before leaf fall (Conradie, 1981b; Pradubsuk and Davenport, 2010; Schreiner, 2016). Grape leaves contain approximately 10-fold less Mg than N (Gärtel, 1993; Pradubsuk and Davenport, 2010).

Magnesium deficiency is a common predicament of vines growing in sandy, very acidic (pH < 4.5) soils, where high concentrations of Al^{3+}, NH_4^+, and H_3O^+ tend to inhibit Mg^{2+} uptake (Gärtel, 1993). However, high soil Ca^{2+} common in soils with high pH, high K^+, and/or high Na^+ characteristic of saline soils can also curb Mg^{2+} uptake and induce Mg deficiency due to competition among these cations for root uptake (Delas and Pouget, 1984; Shaul, 2002). Grapevines grafted on rootstocks derived from American *Vitis* species may be more prone than own-rooted *V. vinifera* cultivars to interference of high soil K with Mg uptake (Mullins et al., 1992). In addition, although application of N fertilizers can enhance the short-term availability of Mg in the soil solution and its uptake by the vine, it may also increase the long-term risk for Mg leaching and depletion of the surface soil (Keller et al., 1995, 2001b).

Because Mg is much more phloem-mobile than Ca, symptoms of Mg deficiency first appear as chlorotic discoloration of the interveinal areas of old leaves as chlorophylls are being dismantled (Currle et al., 1983; Karley and White, 2009; Fig. 7.7). However, even before any symptoms become visible,

FIG. 7.7

Leaf symptoms of magnesium deficiency in the white cultivar Müller-Thurgau (left) and the red cultivar Pinot noir (center), and cluster abscising due to severe bunch stem necrosis (right).

Photos by M. Keller.

the inhibitory effect of Mg starvation on phloem loading leads to accumulation of sucrose and starch in the leaves, while the export of sucrose and amino acids declines (Cakmak and Kirkby, 2008; Cakmak et al., 1994; Hermans et al., 2006; Tränkner et al., 2018). This explains why the roots of Mg-starved plants stop growing, similar to the situation with K and Zn deficiency but in stark contrast with the response to N and P deficiency. The link between Mg status and sucrose export may also be the reason for the observed correlation between petiole Mg content and grape berry sugar (van Leeuwen et al., 2004). The buildup of sugar in the leaves, in turn, leads to feedback inhibition of photosynthesis and results in chlorophyll degradation, probably to mitigate excess light absorption leading to oxidative stress from accumulating reactive oxygen species such as H_2O_2 and O_2^- (Tränkner et al., 2018; see also Section 7.1). Indeed, Mg-deficient plants are extremely light-sensitive; high light intensity accelerates the appearance of the characteristic interveinal chlorosis of Mg-deficient leaves, and the deficiency symptoms are more apparent on sun-exposed than on shaded leaves (Marschner and Cakmak, 1989). The surplus sugar is used to produce anthocyanins as a photoprotectant in the interveinal areas of the leaves of some red but not white cultivars (Currle et al., 1983; Fig. 7.7). In addition, even before the destruction of chlorophyll, the leaves activate their antioxidant system so that ascorbate and glutathione concentrations increase (Cakmak and Kirkby, 2008). It therefore appears that vines grown in areas that typically experience high irradiance may require more Mg to avoid the risk of enduring oxidative stress (Cakmak and Kirkby, 2008). Some *V. vinifera* cultivars (e.g., Barbera, Zinfandel) seem to be more vulnerable to Mg deficiency than are other cultivars, but such differences may often be masked by rootstocks.

In addition to the detrimental consequences for yield formation and fruit ripening due to the decrease in phloem export from the leaves, insufficient Mg availability—sometimes in conjunction with low Ca and high K supply—has also been implicated in the development of bunch stem necrosis (BSN), a physiological disorder that may develop during ripening (Cocucci et al., 1988; Keller and Koblet, 1995b). The rachis of affected fruit clusters develops reddish-brown to black necrotic lesions, beginning in some stomatal guard cells or a few subepidermal cells following collapse of their cell walls (Hall et al., 2011; Hifny and Alleweldt, 1972; Jähnl, 1967, 1971; Theiler, 1970). The lesions probably reflect the oxidation of phenolic compounds and rapidly spread into, along, and/or around the stem and in the process may girdle the phloem, which effectively interrupts import of assimilates into the berries distal to the necrosis, while initially leaving the xylem intact. The girdling effect stops sugar accumulation in the berries, whereas the degradation of malate continues, which can lead to poor fruit quality and berry shrinkage (Bondada and Keller, 2012; Keller et al., 2016b). The rachis eventually dries up, and in severe cases the distal portion of affected clusters may fall off. Clusters appear to be susceptible to this syndrome before and during bloom, at which time the disorder is also termed inflorescence necrosis (see Section 6.1), and after the beginning of sugar accumulation at veraison (Jackson and Coombe, 1988; Keller et al., 2001a). Cabernet Sauvignon, Riesling, and Gewürztraminer are particularly vulnerable to BSN, whereas the Pinots and Chardonnay are relatively insensitive (Currle et al., 1983). Affected rachis tissues may be invaded by saprophytic pathogenic fungi such as *Botrytis cinerea*.

It is conceivable that the role of Mg in sugar export from the leaves may be to blame for this disorder, especially because symptom development is often related to adverse environmental conditions that limit photosynthesis and assimilate supply, and the incidence tends to be higher on cool compared with warm vineyard sites (Jackson, 1991; Jackson and Coombe, 1988; Keller and Koblet, 1994, 1995b; Pérez Harvey and Gaete, 1986). In addition, loss of leaf area during the ripening period also seems to favor the appearance of BSN (Redl, 1984). Finally, drying winds and high VPD following even brief episodes of rainfall, as well as high light intensity, may favor the sudden appearance of necrotic lesions on the rachis. Whether irrigation can substitute for rainfall in dry climates is unknown. In Europe, Mg sprays, typically in the form of dissolved Mg-sulfate or Epsom salt ($MgSO_4 \cdot 7H_2O$), are sometimes applied into the fruit zone at veraison to increase rachis Mg contents and partially alleviate the incidence of BSN.

Excessive soil Mg, defined as >40% of the soil's cation exchange capacity, is rare but occurs in serpentine soils that are poor in silicates but rich in ferromagnesium minerals. Such soils occur mainly in the North American Pacific coastal ranges and California's Sierra Nevada foothills. They are often associated with high pH; very low K, Ca, and P; and high Fe, B, cobalt (Co), and Ni. Although no symptoms of Mg toxicity are known, it is thought that very high amounts of Mg in leaves may impair photosynthesis (Shaul, 2002). Moreover, excessive Mg can lead to poor soil structure and may induce K and, sometimes, P deficiency, especially toward the end of the growing season. Conversely, in Na-rich soils, high Mg contents may ameliorate some of the adverse effects of Na on soil structure, provided there is enough Ca in the soil.

7.3.2 Transition metals and micronutrients

Iron

Iron (Fe) is the most abundant metal on Earth but is extremely insoluble in oxygen-rich environments, which includes most soils (Schmidt, 2003). It is generally the most abundant micronutrient in grapevines (Gärtel, 1993). Due to its ability to occur in two different ionic states, namely Fe^{2+} and Fe^{3+}, it is a cofactor or ingredient of proteins such as ferredoxin that are involved in electron transfer and of many enzymes that catalyze reduction/oxidation (redox) reactions (Clarkson and Hanson, 1980; Curie and Briat, 2003; Curie et al., 2009). As such, it is involved in chlorophyll production, photosynthesis and respiration and, via enzyme activation, in C, N, and S assimilation, lipid and hormone biosynthesis and degradation, DNA biosynthesis and repair, as well as in detoxification of reactive oxygen species. In addition to its multifaceted roles in leaves, Fe is also abundant in seeds and pollen and may be necessary for pollen production (Bertoldi et al., 2011; Curie et al., 2009). Its profusion in soil particles is the source of many erroneously high apparent tissue Fe concentrations, and consequently inaccurate fertilizer recommendations, resulting from analysis of leaf samples that are contaminated with dust.

Since oxygen is usually abundant in the upper soil layers, Fe generally occurs in the oxidized form ferric iron (Fe^{3+}) in the rootzone, although the reduced form ferrous iron (Fe^{2+}) is the soluble cation and is the form required by plants. However, Fe^{3+} forms soluble complexes called chelates with many organic and inorganic molecules, including humic and fulvic acids, which are important components of soil organic matter, but also with tannins and phosphate. Like other plants termed "strategy I" species, *Vitis* species absorb iron molecules as Fe^{3+}–chelates, which the enzyme ferric chelate reductase (FCR) reduces to Fe^{2+} on the root plasma membranes (Bavaresco et al., 1991; Briat and Lobréaux, 1997; Kobayashi and Nishizawa, 2012; Schmidt, 2003; Varanini and Maggioni, 1982). The poorly soluble yet highly reactive Fe^{2+} ions are moved across the membranes by Fe^{2+} transporters and are "picked up" in the apoplast by citrate. The soluble and metabolically inactive Fe^{3+}–citrate complexes are then transported to the leaves in the xylem sap (Curie and Briat, 2003; Kobayashi and Nishizawa, 2012). Upon arrival in the leaves, the iron is reactivated to Fe^{2+} by FCR before it can be taken up into the leaf mesophyll. The FCR protein is highly pH-sensitive, and enzyme activity declines as the apoplast pH increases (Nikolic et al., 2000). Inside the cells, any unused Fe is again chelated, this time by a nonprotein amino acid called nicotianamine that serves to keep Fe in a soluble form and ensures its correct distribution among the various cell organelles and in the phloem (Curie and Briat, 2003; Curie et al., 2009; Schmidt, 2003).

Although Fe^{3+} can be stored in a protein termed ferritin, approximately 80–90% of the Fe in a leaf is located in the chloroplast as part of the photosynthetic machinery (Marschner, 1995). Therefore, Fe deficiency impairs electron transfer and results in oxidative stress from an excess of absorbed light. This in turn restricts photosynthesis and sugar production for export to the vine's sink organs and strongly limits grapevine growth and yield (Bavaresco et al., 2003; Bertamini and Nedunchezhian, 2005; Briat et al., 2015; Gruber and Kosegarten, 2002). When a vine's crop load is high, developing grape berries may satisfy some of their Fe demand by importing Fe remobilized from leaves and even from the vine's permanent structure (Pradubsuk and Davenport, 2011). It is possible that oxidative stress is the reason for the decline in leaf chlorophyll that is characteristic of Fe deficiency (Fig. 7.8). Thus, deficiency symptoms include leaf chlorosis from gradual loss of chlorophyll or lack of its production beginning at the margins and progressing to the interveinal areas of young leaves, followed by marginal necrosis and, finally, leaf abscission. Even though Fe is phloem-mobile and can be remobilized from older and senescing leaves, developing lateral shoots are stunted, often show pink internodes, and have small, chlorotic leaves that may fail to unfold (Curie and Briat, 2003). Under severe deficiency, the tendrils and inflorescences also become chlorotic, and fruit set is poor, probably because Fe is required for pollen development (Gärtel, 1993). Furthermore, Fe-deficient plants also tend to accumulate Zn, Mn, Co, and cadmium (Cd) due to the general increase in the activity of divalent metal transporters under Fe deficiency (Curie and Briat, 2003; Kobayashi and Nishizawa, 2012).

Like Fe deficiency, excess Fe can also cause oxidative stress, reduce leaf chlorophyll, and result in chlorotic or necrotic spots on the leaves (Curie and Briat, 2003). Because Fe^{2+} reacts rapidly with H_2O_2 to form reactive oxygen species in the Fenton reaction, plants tightly control the availability of both Fe^{2+} and H_2O_2 (Curie et al., 2009; Halliwell, 2006; Lane, 2002). Excess Fe can be stored in vacuoles, where it is chelated by nicotianamine or citrate, or in plastids, where it is bound up by ferritin and from where it can also be retrieved on demand to keep cytosolic Fe concentrations stable. Consequently, Fe toxicity is very rare under natural conditions and is mainly limited to acid soils (Kochian et al., 2004).

FIG. 7.8

Leaf symptoms of lime-induced chlorosis in Concord.

Photos by M. Keller.

Toxicity may, however, occur as a result of incorrect applications of foliar nutrient sprays, often due to errors in dose calculations.

Leaf chlorosis due to Fe deficiency, on the other hand, frequently develops when grapes are grown on calcareous soils. Such lime-rich soils are prevalent, for example, in Burgundy and Champagne in France, much of Hungary, and eastern Washington. Diagnosis of Fe chlorosis by tissue analysis is difficult because the total Fe concentration of chlorotic leaves is often similar to or even higher than that of "healthy," green leaves, partly because the chlorotic leaves are usually smaller (Sattelmacher, 2001). Moreover, the Fe content of a leaf is not closely related to its physiologically active Fe content. Although Fe may be abundant in calcareous soils, it is often precipitated as insoluble Fe^{3+} oxides and hydroxides, making it unavailable for the roots. A one-unit increase in soil pH can decrease Fe solubility 1000-fold (see also Fig. 7.3).

The ability of grapevines to cope with such Fe unavailability depends on the species and cultivar. Species that have evolved in regions with calcareous soils include *V. vinifera* and, to a lesser extent, *V. berlandieri* and *V. champinii*. Such species can pump out protons (H^+) and organic acids such as malate and citrate from the roots, which acidifies the soil solution and improves Fe solubilization and uptake (Brancadoro et al., 1995; Jiménez et al., 2007; Mengel and Malissiovas, 1982). Iron-inefficient species, on the other hand, have evolved in regions with more acidic soils and include *V. labrusca*, *V. riparia*, and *Muscadinia rotundifolia*. These species are unable or less able to release H^+, but they, too, can release organic acids (Ollat et al., 2003).

The H^+-release strategy works only for species that are not sensitive to high soil pH. The culprit is the high amount of bicarbonate (HCO_3^-) that forms from $CaCO_3$ in calcareous soils. Although some H^+ release may enhance Fe uptake and transport to the leaves in sensitive species as well, and although the apoplast pH seems to be unaffected by soil pH, uptake of HCO_3^- leads to changes in the apoplast that inhibit conversion of the inactive Fe^{3+} to the active Fe^{2+}. Consequently, Fe becomes locked up in the apoplast (Mengel and Bübl, 1983; Mengel et al., 1984). This Fe cannot enter the mesophyll cells and is unavailable for metabolism, which leads to yellowing between the veins of young leaves and eventually whitening of the whole leaf. Nevertheless, because the affected leaves ostensibly signal the roots to increase Fe uptake, the total leaf Fe concentration may be as high as or higher than that in nonchlorotic plants; in addition, Fe may also accumulate in the roots (Currle et al., 1983; Gruber and Kosegarten, 2002).

This "iron chlorosis paradox" is frequent in own-rooted Concord grapes whose soil pH optimum is about 5.5, and in *V. vinifera* cultivars grafted to *V. riparia* rootstocks, grown on high-pH soils. Hence, the condition is also termed lime-induced chlorosis, although it should probably more appropriately be called bicarbonate-induced chlorosis (Currle et al., 1983; Gärtel, 1993; Mengel, 1994; Ollat et al., 2003). Due to their diverse origins, *Vitis* species and thus rootstocks differ in their tolerance of calcareous soils (see Section 1.2). Therefore, where own-rooted *V. vinifera* is not an option, the chlorosis problem may be alleviated by grafting susceptible cultivars to tolerant rootstocks, such as those derived from *V. berlandieri* and, to a lesser extent, *V. rupestris* (Bavaresco et al., 2003; Schumann and Frieß, 1976). Since grass roots exude Fe^{3+} chelators termed phytosiderophores, cover crops containing grass species may also enhance the availability of Fe to grapevines (Briat et al., 2015).

Excessive soil moisture, especially waterlogging, due to abundant rainfall or overirrigation also reduces Fe availability—along with the availability of N, K, and Mn—in addition to restricting root growth. In addition, the poor aeration of waterlogged soils and of compacted soils also increases the soil HCO_3^- content due to reduced diffusion of CO_2 from the soil water: $CO_2 + H_2O \leftrightarrow H_2CO_3 \leftrightarrow HCO_3^- + H^+$.

These effects may combine to induce or exacerbate chlorosis in calcareous soils and in soils with a high P content that leads to precipitation of insoluble Fe-phosphates (Currle et al., 1983; Williams et al., 1994). Chlorosis is further exacerbated if excessive soil moisture coincides with low soil temperatures (Davenport and Stevens, 2006).

Zinc

Zinc (Zn) is the only metal ion that is present in all six classes of enzymes, and much of the Zn in plants is strongly bound in proteins (Broadley et al., 2007). Most Zn-binding proteins are involved in the regulation of gene transcription via, among others, effects on DNA and RNA binding and RNA metabolism. Zinc, along with copper, is a key constituent of one form of the antioxidant enzyme superoxide dismutase, which helps protect plant tissues from excess reactive oxygen species that accumulate in response to many environmental stress factors (Apel and Hirt, 2004; see also Section 7.1).

The concentration of Zn in grape leaves is approximately one-third to one-half that of Fe, but the concentrations of these two micronutrients may be similar in grape berries, where the concentration, though not the absolute amount, of the phloem-immobile Zn declines during ripening (Bertoldi et al., 2011; Gärtel, 1993; Rogiers et al., 2006). Seasonal Zn uptake by grapevines more or less follows the same pattern as that of K and Mg (Schreiner, 2016). More than 90% of the Zn in a soil is insoluble and hence not plant-available. Up to half of the soluble fraction is in the main plant-available form Zn^{2+}, although some Zn may be taken up by the roots as organic complexes (Broadley et al., 2007). The solubility of Zn is strongly dependent on the pH, and like Fe availability, Zn availability is low in calcareous soils with a pH > 7 (see Fig. 7.3) and high bicarbonate content. Prolonged flooding, high soil organic matter content, and high Mg ÷ Ca ratios also seem to decrease Zn availability (Broadley et al., 2007). Moreover, excessive P supply can immobilize Zn due to the formation of insoluble Zn phosphate ($Zn_3(PO_4)_2$) (Currle et al., 1983; Gärtel, 1993). Low Zn availability can lead to P accumulation in the plant to the point of toxicity if P availability is high (Cakmak and Marschner, 1986). Although insufficient Zn availability is the most widespread cause of micronutrient deficiency, especially in sandy soils with high pH, plants respond readily to foliar Zn application (Broadley et al., 2007).

Zinc deficiency results in a decline in protein and starch production, accumulation of sugar in the leaves, and stunted shoots with short internodes. The leaves remain small and are malformed; they are asymmetrical with a wide petiolar sinus and sharply toothed margins. In addition, they develop a mosaic-like chlorotic pattern between the veins called "mottle leaf"; the veins may become clear with green borders (Currle et al., 1983; Gärtel, 1993). High light intensity seems to accelerate symptom development, likely as a result of oxidative stress (Marschner and Cakmak, 1989). More severe deficiency stimulates lateral shoot growth and delays periderm formation in the shoots, while the leaves' interveinal areas become reddish-brown or bronze and then necrotic, leading to rolling of the leaf blades (Broadley et al., 2007; Currle et al., 1983). Furthermore, insufficient Zn supply impairs pollen formation and, therefore, pollination, which leads to poor fruit set and a "hens and chicks" appearance of the clusters, mimicking what occurs during B deficiency. Severe Zn deficiency eventually culminates in necrotic root tips, which is lethal.

Unlike Zn deficiency, Zn toxicity is rare and occurs predominantly in low-pH soils treated with sewage sludge or contaminated with other man-made Zn-containing waste products, such as corroding galvanized objects (Broadley et al., 2007). It decreases uptake of P, Mg, and Mn, leads to stunted growth and reduced yield, and can result in chlorosis caused by Fe deficiency.

Copper

Copper (Cu) is an integral component of several proteins, including plastocyanin, cytochrome oxidase, ascorbate oxidase, superoxide dismutase, polyphenol oxidase, and laccase. Therefore, Cu participates in photosynthetic and respiratory electron transfer, ethylene perception, cell wall metabolism, lignification, oxidative stress protection, and molybdenum cofactor biosynthesis (Burkhead et al., 2009). Its ability to switch between Cu^{2+} and Cu^+ not only makes this metal cation an essential cofactor in many oxidase enzymes that catalyze redox reactions but also contributes to its inherent toxicity (Clarkson and Hanson, 1980; Yruela, 2009).

Grape leaves contain Cu at concentrations that are approximately 15 times lower than those of Fe, but in grape berries they are in the same range, and the content of both metals continues to increase as the berries ripen (Bertoldi et al., 2011; Gärtel, 1993; Rogiers et al., 2006). Developing berries are strong sinks for Cu, and by harvest up to one-third of a vine's Cu may be in the fruit clusters (Pradubsuk and Davenport, 2011). The pattern of Cu acquisition by grapevine roots during the growing season resembles that of K, Mg, Zn, and Mn (Schreiner, 2016). Roots absorb Cu^{2+} using high-affinity transporters and probably transport it in the xylem in the form of Cu-nicotianamine chelate (Curie et al., 2009; Martins et al., 2014a). High plant N status may increase the demand for Cu (Yruela, 2009). The phloem-mobile Cu can be remobilized from old leaves for redistribution to sinks in need of extra Cu, although such recycling may not be very efficient so that young leaves are more impacted by Cu deficiency than old leaves (Burkhead et al., 2009; Marschner, 1995).

Deficiency symptoms include inhibition of root growth; stunted shoots; small, misshapen, pale-green or chlorotic leaves, often with curled leaf margins; and decreased fruit set due to poor pollen and embryo viability (Marschner, 1995). However, Cu deficiency is rare and is mainly restricted to soils that are very high in organic matter which strongly binds Cu. Moreover, Fe may substitute for Cu, because the tasks of many Cu proteins can also be carried out by equivalent Fe proteins (Yruela, 2009).

When present in excess, Cu can be toxic and, despite its linkage with antioxidant enzymes, may cause oxidative damage due to the production of reactive oxygen species. Due to the damage it causes to the photosynthetic machinery, Cu can exacerbate the impact of high light intensity in leaves. Consequently, excess Cu may cause leaf chlorosis or even necrosis, and it may also inhibit root growth (Nogales et al., 2019; Yruela, 2009).

The availability of Cu for grapevines increases at low soil pH (Fig. 7.3). Moreover, long-term applications of Cu-based fungicides can lead to Cu accumulation in the surface soil (Gärtel, 1993). For instance, downy mildew has been controlled since the late nineteenth century by the copper sulfate–lime blend ($CuSO_4 \cdot 3Cu(OH)_2 \cdot 3CaSO_4$) called Bordeaux mixture and, more recently, by copper oxychloride ($CuCl_2 \cdot 3Cu(OH)_2$) and copper hydroxide ($Cu(OH)_2$), which has led to >10-fold increases in Cu in many vineyard soils (see also Section 8.2). Such Cu accumulation may diminish uptake of P, Fe, and, in sandy soils, Mg and Ca as well. It can also lead to Cu toxicity, especially in vineyards planted on sandy and acid soils (Toselli et al., 2009). Whereas Cu toxicity is extremely rare in established vineyards because of the older vines' deeper root system, it can result in stunted growth of newly planted vines, whose roots are concentrated in the surface soil. In extreme cases, young vines may die after replanting of old vineyard sites or in nurseries. Moreover, the fungicidal activity of Cu may have other undesirable side effects, since application of Cu-based fungicides may lead to excessive Cu accumulation in the berries (Martins et al., 2014b). High Cu content of harvested grapes can inhibit yeast growth, leading to sluggish or stuck fermentations and, sometimes, poor wine quality (Tromp and De Klerk, 1988).

Manganese

Like other transition metals, such as Fe and Cu, manganese (Mn) exists in various oxidation states, but only Mn^{2+} can be taken up by roots and transported throughout the plant in both the xylem and the phloem (Pittman, 2005). Leaf Mn concentrations are similar to those of Zn, and so is the seasonal pattern of Mn uptake by the roots (Gärtel, 1993; Schreiner, 2016). In ripening grape berries, the Mn concentration, but not its content, may decline during ripening, like that of the phloem-immobile elements Ca and Zn (Bertoldi et al., 2011; Rogiers et al., 2006). Like Zn, Mn has antioxidative functions in plant tissues. It can serve as an antioxidant by being oxidized from Mn^{2+} to Mn^{3+} and is also a structural component of antioxidant enzymes, such as superoxide dismutase and catalase, as well as of the water-splitting enzyme in photosynthesis (Pittman, 2005). As a redox shuttle, Mn supports laccase and peroxidase in oxidizing, and thus activating, lignin monomers for polymerization in the cell walls, especially those of the xylem vessels, the root endodermis, and the seed coat (Barros et al., 2015). In addition, Mn is required for the function of such enzymes as glucosyltransferases, which attach a glucose molecule to phenolics, terpenoids, and other compounds (Marschner, 1995).

Insufficient Mn supply may increase tissue sensitivity to oxidative stress resulting from a variety of environmental stress factors (see Section 7.1). Deficiency symptoms are consequently more severe on sun-exposed leaves that may suffer from an energy overload. Chlorotic leaves in response to insufficient Mn availability occur first in the basal portion of the shoots, soon after budbreak. Chlorosis appears mosaic-like, and the leaves later acquire a reddish or bronze color, whereas lateral leaves often remain green, and fruit ripening is delayed (Currle et al., 1983; Gärtel, 1993).

The plant availability of Mn is highly pH dependent and is minimal around or slightly above pH 7 (Fig. 7.3). Deficiency can therefore be a problem in neutral soils and especially in soils with high amounts of sand and organic matter. However, Mn deficiency is often masked in such soils by the simultaneous development of lime-induced Fe chlorosis (Gärtel, 1993). On the other hand, plants can take up much more Mn than they require, and excess Mn can be extremely toxic (Pittman, 2005). Despite its antioxidative properties at normal concentrations, excess Mn induces oxidative stress and results in symptoms such as stunted growth, leaf chlorosis, and necrotic lesions (Kochian et al., 2004). Plants therefore sequester and store Mn^{2+} ions and Mn-chelates in the vacuoles in addition to the Mn required in the chloroplasts. However, Mn toxicity is rare and occurs predominantly on acid soils (pH < 5.5) with high Mn availability (Kochian et al., 2004). Uptake of Mn may also be increased under conditions of Fe and Zn deficiency, whereas excess Fe inhibits Mn uptake (Pittman, 2005).

Molybdenum

Molybdenum (Mo) is available to grapevine roots mostly as molybdate oxyanion (MoO_4^{2-}). Although it is an essential element, it is required only in minute amounts by grapevines, mainly as a cofactor that forms the active site of a few proteins called molybdoenzymes (Schwarz and Mendel, 2006). Nitrate reductase is the main molybdoenzyme in plants, and NO_3^- induces Mo uptake (Clarkson and Hanson, 1980; Kaiser et al., 2005; Reid, 2001). Therefore, a major effect of Mo deficiency is the accumulation of NO_3^- due to reduced nitrate reductase activity and lack of amino acid biosynthesis (Currle et al., 1983; see also Section 5.3). Consequently, the visual deficiency symptoms resemble those of N deficiency: reduced growth and yield. Another molybdoenzyme is aldehyde oxidase, which is involved in the production of both auxin and ABA (Schwarz and Mendel, 2006).

Shortage in Mo supply may result in reduced fruit set and clusters displaying "hens and chicks," especially in susceptible cultivars such as Merlot (Kaiser et al., 2005; Longbottom et al., 2010; Williams et al., 2004). Other deficiency symptoms include short internodes, zigzag growth, and pale-green leaves with necrotic margins. Mo-deficient grapevines may furthermore show flaccid and cupped leaves that wilt easily due to excessive transpiration (Kaiser et al., 2005). This is because the hampered ABA production renders Mo-starved vines vulnerable to losing stomatal control under water deficit (see Sections 3.2 and 7.2). Merlot appears to be less able to take up Mo from the soil than many other *V. vinifera* cultivars, but grafting on rootstocks can overcome this shortcoming.

Although Mo deficiency is relatively rare, it can occur on acid soils because Mo availability strongly declines below a soil pH of about 5.5 (Fig. 7.3). Deficiency may also occur on soils with high amounts of Fe, especially in cool climates. During a cool, wet spring, release of MoO_4^{2-} in the soil may be restricted, as is root growth. In addition, high sulfate availability can inhibit molybdate uptake; the two anions compete for root uptake due to their similar size. However, because MoO_4^{2-} is highly mobile in plants, foliar applications, mostly in the form of sodium molybdate ($Na_2MoO_4 \cdot 2H_2O$), can be very effective in distributing Mo throughout the plant, alleviating deficiency symptoms, and improving fruit set (Kaiser et al., 2005; Longbottom et al., 2010; Williams et al., 2004). Toxicity due to excess Mo is even rarer than deficiency; symptoms can include purple leaves due to anthocyanin accumulation from surplus foliar sugar.

Boron

Although boron (B) is an essential element for plants, its functions are poorly understood. As borate ($B(OH)_4^-$), it binds strongly to pectic polysaccharides and hence supports the function of Ca^{2+} in facilitating cross-linking of the pectin network in the primary cell walls (O'Neill et al., 2004). Therefore, B is essential for cell expansion, and almost all the B in plants is located in the apoplast. To sustain cell wall structure and growth, B must be continually transported to the plant's growing regions. The B requirement of pollen tubes is especially high because they are rich in pectin, particularly in the tip. In addition, B is required to sustain the activity of NO_3^- uptake transporters in the roots (Camacho-Cristóbal and González-Fontes, 1999; see also Section 5.3).

Boron is available in the soil solution as boric acid ($B(OH)_3$). Thus, it differs from other micronutrients in that it exists as a neutral molecule at physiological pH and appears to be taken up into plants mostly by simple diffusion, especially at high external concentration (Clarkson and Hanson, 1980; Reid, 2001). Consequently, the roots may be unable to exclude excess B, which increases the risk of B toxicity that arises from the ability of B to bind to ATP and NAD(P)H. In addition to passive diffusion, active transport into the roots and the xylem probably occurs when the B concentration in the soil water is low (Reid, 2001; Takano et al., 2008).

Boron continues to build up in the leaf blades and petioles as the growing season progresses (Christensen, 1984). Therefore, in contrast to other mineral nutrients, the tissue B concentration is often highest at the end of the season and in the oldest leaves. Accumulation also continues in grape berries throughout their development and ripening (Bertoldi et al., 2011; Rogiers et al., 2006). Indeed, the developing berries appear to be strong sinks for B, so that by harvest more than one-third of a vine's B may be in the grape clusters (Pradubsuk and Davenport, 2011).

The tissue concentration range between B deficiency and B toxicity is very narrow (Currle et al., 1983; Takano et al., 2008). Moreover, B toxicity symptoms resemble deficiency symptoms (Fig. 7.9). Boron toxicity can be induced by undue B fertilization and is a particular peril of arid regions.

FIG. 7.9

Leaf symptoms of boron deficiency in Chardonnay (left) and boron toxicity in Cabernet Sauvignon (right).

Photos by M. Keller.

Insufficient B availability, on the other hand, is frequent on very acid soils (pH < 4.5), especially under dry conditions, and rapidly results in cessation of cell division (Clarkson and Hanson, 1980; Currle et al., 1983). The inhibition of meristem activity leads to stunted shoot and root growth, with shoots developing "swollen" internodes and often displaying a zigzag appearance and petioles remaining short and thickened (Currle et al., 1983; Gärtel, 1993). Symptoms are usually confined to young tissues; although B is phloem mobile, it apparently cannot be recycled from old leaves.

Bushy, branched shoot growth, resembling symptoms of fanleaf virus infection, and poor bud fruitfulness may be carryover effects of insufficient B available for proper primordium formation during the previous growing season (Gärtel, 1993). In addition, reduced N uptake in B-deficient grapevines leads to low leaf N status and sugar and starch accumulation in the leaves (Camacho-Cristóbal and González-Fontes, 1999; see also Section 5.3). Root growth also ceases, and roots may swell and crack. Because B is important for pollen germination and pollen tube growth, B deficiency interferes with fertilization (Currle et al., 1983; May 2004; O'Neill et al., 2004). This results in poor fruit set and clusters with a "hens and chicks" appearance (Gärtel, 1993). In addition, seed development is impaired, and the seeds remain small (Hardie and Aggenbach, 1996). Inadequate B supply probably also induces oxidative damage to cells, and high light intensity consequently exacerbates the symptoms of B deficiency (Marschner and Cakmak, 1989).

Nickel

Compared with the minute amount required by plants, nickel (Ni) is relatively abundant in almost all soils. Although many proteins contain Ni, very little is known about the element's function in plants. Nickel activates several enzymes, including urease, which is involved with N metabolism required to process urea, for example, during remobilization of the storage amino acid arginine (Witte et al., 2005). Another such enzyme is superoxide dismutase, which plays an important role in the defense against oxidative stress (see Section 7.1). Following uptake of Ni^{2+} by the roots, it is probably transported in the xylem in the form of Ni-nicotianamine chelate (Curie et al., 2009). Due to its partial mobility in the phloem, Ni can also be remobilized from old leaves and redistributed to high-priority sinks. Nickel deficiency results in reduced shoot vigor following budbreak; dwarfed, thick leaves with cupped tips—perhaps caused by oxalate accumulation; loss of apical dominance; and brittle wood due to poor lignification.

Silicon

Silicon (Si) is the second most abundant element after oxygen in soils, making up approximately 28% of the Earth's surface. As a major component of minerals, it is present in clay, silt, and sand particles and in rocks in the form of silicon dioxide (SiO_2), which accounts for 50–70% of the soil mass and also serves as the building block of quartz and glass (Ma and Yamaji, 2006). Roots, but apparently not root hairs, absorb Si passively or via transporters as soluble silicic acid ($Si(OH)_4$), which is then released to and transported in the xylem sap (Liang et al., 2005; Ma and Yamaji, 2006, 2015). Thus, Si resembles B in that they are both taken up and distributed with the transpiration stream as undissociated and neutral molecules. Because $Si(OH)_4$ polymerizes at concentrations >2 mM, it is ultimately precipitated throughout the plant as amorphous, hydrated silica bodies (SiO_2-nH_2O), which are also variously called silica gel, plant opal, or phytoliths (Ma and Yamaji, 2006). Due to its immobility in the phloem, xylem-supplied Si accumulates in older tissues, including the stomatal regions of grape berries (Blanke et al., 1999). Its existence as an undissociated molecule combined with its ability to polymerize makes Si the only mineral element for which excessive uptake and accumulation does not result in detrimental effects to plants.

Plants use polymeric silicates to impregnate and strengthen the cell walls of epidermis and vascular tissues, thereby reducing water loss and hindering invasion by fungal pathogens (Clarkson and Hanson, 1980; Ma and Yamaji, 2006, 2015). For example, penetration by the powdery mildew fungus *Erysiphe necator* leads to localized accumulation of Si in the cell walls (Blaich and Wind, 1989). Accordingly, foliar Si applications can reduce powdery mildew infections. In addition, Si also stimulates the antioxidant systems, which may avert stress-induced oxidative damage (see Section 7.1). Therefore, although deemed a nonessential nutrient (i.e., one that is not required to complete the plant's life cycle), Si provides many benefits, such as improved resistance to pests and diseases; tolerance of drought, salinity, heavy metals, and high temperatures; and even higher yield and, by enhancing flavonoid production, improved fruit quality (Currie and Perry, 2007; Epstein, 1999; Ma and Yamaji, 2006). For this reason, Si is integrated in many fertilizers. Most of its stress-alleviating effects result from the strengthening of cell walls by Si and its ability to enhance cation binding to the cell walls (Saqib et al., 2008). Such immobilization of cations such as Na^+ or Mn^{2+} may prevent their buildup to toxic concentrations inside the cells. In addition, plants that are rich in Si may be able to reduce Na^+ uptake from saline soils and its transport to the leaves.

7.3.3 Salinity

The term salinity describes the occurrence of high concentrations of ionic forms of soluble salts in water and soils. The gradual development of salinity is termed salinization and occurs in regions where water evaporation from the soil exceeds precipitation so that salts dissolved in the soil solution tend to become concentrated at the soil surface. Whereas this process is characteristic of arid environments, the opposite process, called acidification, occurs in regions where rainfall consistently exceeds evaporation—especially in the tropics and subtropics. Abundant rainfall leaches cations such as K^+, Ca^{2+}, and Mg^{2+} so that the soil pH decreases and the highly toxic Al^{3+}, along with Mn^{2+} and Fe^{2+}, becomes soluble in the soil solution. Rainwater also contains some salt ($\leq$50 mg NaCl L^{-1}), especially in coastal areas (Munns and Tester, 2008). Moreover, wind may blow seawater inland and lead to salt deposits on the soil surface.

Irrigated vineyards are at much greater risk from salinization than nonirrigated vineyards, because irrigation water is enriched in dissolved salts compared with rainwater, and because irrigation tends to raise water tables. Therefore, irrigation in arid and semiarid regions over prolonged periods can lead to a buildup of salt near the soil surface. Most table and raisin grapes are grown in rather dry and warm climatic regions, such as southwestern Asia, California, Chile, or Australia, and are thus especially threatened by salinity.

The dominant soil salts are cations, such as Na^+, K^+, Ca^{2+}, and Mg^{2+}, and their associated anions, such as chloride (Cl^-), sulfate (SO_4^{2-}), carbonate (CO_3^{2-}), and bicarbonate (HCO_3^-). Small amounts of other ions are also present in the soil solution. The relative amounts of different ions vary between water sources and soil types, but the ions most often associated with the effects of salinity in grapevines are Na^+ and Cl^-. Dissolved ions increase the electrical conductivity of water, and thus the degree of salinity of irrigation water or soil water extracts is expressed in electrical conductivity units measured in decisiemens per meter (dS m^{-1}). The threshold above which salinity begins to affect *V. vinifera* growth and yield formation is approximately 2 dS m^{-1}, and above 16 dS m^{-1} vines cannot survive (Zhang et al., 2002). Because dissolved ions decrease the osmotic potential (Ψ_π) of water, electrical conductivity is also a measure of Ψ_π: 2 dS m^{-1} corresponds to approximately 20 mM NaCl generating a $\Psi_\pi \approx -0.1$ MPa.

Sodicity is related to salinity and refers to the presence of sodium relative to calcium and magnesium in the soil. As a measure of sodium hazard, sodicity is expressed as the sodium adsorption ratio (SAR) because most cations in the soil are attracted to the negative charges of clays. The sodicity of irrigation water or soil water extracts is estimated as follows (Lesch and Suarez, 2009):

$$SAR = \frac{[Na^+]}{\sqrt{([Ca^{2+}] + [Mg^{2+}])}}$$

where [...] denotes the concentration of an ion in millimoles per liter (mM).

In addition to their elevated Na^+ content, sodic soils are afflicted with a deterioration of structure due to clay dispersion and a rise in hydraulic resistance. Saline and sodic soils are usually classed together as salt-affected soils. Such soils contain a sufficiently high concentration of soluble salts or exchangeable Na^+ to interfere with plant growth. But while salinity impacts plants directly through the effect of ions on plant physiology, the influence of sodicity is mostly indirect due to its deleterious effects on the soil's physical properties.

The most common cause of salt stress is a high concentration of Na^+ and Cl^- in the soil solution. Both ions can serve as plant nutrients but become toxic at much lower concentrations than other mineral nutrients. Plant damage due to salt-affected soils is the outcome of a combination of hyperosmotic stress and hyperionic stress (Greek *hupér* = over) due to a disruption of homeostasis (Greek *homois* = similar, *stasis* = stand still, steady) in water status and ion distribution (Hasegawa et al., 2000; Zhu, 2001). Initially, a buildup of salt ions in the soil decreases Ψ_π of the soil solution; the Ψ_π in "normal" soils is generally approximately -0.01 MPa but can drop to less than -0.2 MPa in saline soils. The resulting decrease in Ψ_{soil} impedes water uptake by the roots, increases root r_h due to closure of aquaporins, and results in plant water deficit and a decline in Ψ_{leaf} (Cramer et al., 2007; Downton and Loveys, 1981; Luu and Maurel, 2005; Shani et al., 1993; Walker et al., 1981). Consequently, the initial effects of a rise in soil salinity are identical to the effects of drought stress (see Section 7.2). The ensuing collapse of the water potential gradient necessary for cell expansion curtails shoot growth and leaf expansion. Therefore, the first visible sign of salt stress is an inhibition of shoot and leaf growth (Munns and Tester, 2008; Walker et al., 1981).

The decrease in Ψ_{leaf} and the associated increase in root-derived and locally produced ABA also induce closure of the stomata, which in turn decreases transpiration and photosynthesis and consequently the production of sugar for export to other plant parts (Downton et al., 1990; Shani and Ben-Gal, 2005). Rising Cl^- concentration in the leaves further decreases g_s, although Na^+ tends to counter this effect by replacing K^+ in the guard cells and thereby keeping the stomata partially open (Walker et al., 1981). However, elevated foliar Cl^- reduces the rate of photosynthesis well before any visible symptoms of salt damage become apparent (Downton, 1977b). Thus the salinity-induced growth reduction is in part due to the decrease in photosynthesis and in part due to inhibition of cell division and cell expansion (Zhu, 2001). Growth may also slow through deactivation of gibberellins under salt stress (Yamaguchi, 2008).

The roots are the first and most important organs to experience salinity, which decreases their ability to explore the soil for water and nutrients. Yet root growth is usually less sensitive to salt stress than is shoot growth—although the opposite has also been reported for grapevines (Hawker and Walker, 1978; Munns and Tester, 2008). One reason for a decrease in root growth may be that the decline in soil Ψ_π limits the turgor pressure of the expanding cells in the root tip, so that the roots can no longer push their way through the soil (Bengough et al., 2011). Where possible, however, root growth also changes its direction away from high salt concentrations in the soil. This propensity, termed halotropism, is strong enough to override the normal gravity response of roots and comes about by active, salt-induced redistribution of auxin in the root tip (Galvan-Ampudia et al., 2013).

Impaired root growth, respiration, and water uptake further restrict vine growth, yield, and fruit quality (Shani and Ben-Gal, 2005; Shani et al., 1993). In some cases, salinity is associated with changes in the fruit that are typical of mild water deficit, such as earlier veraison, higher fruit sugar, proline, K^+ and Cl^-, and lower acidity (Downton and Loveys, 1978; Walker et al., 2000). As the salt stress becomes more severe, however, fruit set, berry size, and sugar and anthocyanin accumulation are increasingly curtailed (Hawker and Walker, 1978). Moreover, Cl^- and Na^+ ions also accumulate in the berries, especially in the skin, and from there are readily extracted into wine during fermentation (Downton, 1977a; Gong et al., 2010; Walker et al., 2010).

The challenge for grapevines growing in saline soils is that their roots must take up mineral nutrients while keeping out the toxic Na and Cl. As usual, "toxic" is a relative term; it is important to remember that plants require some Cl^- for the water-splitting reaction that produces electrons, protons, and oxygen during photosynthesis (see Section 4.1). Roots effectively "pick" the nutrient ions from the toxic ions in the soil solution or pump the toxic ions that have been taken up back out again so that >95% of Na^+ and Cl^- are prevented from entering the xylem (Munns, 2002; Munns and Tester, 2008). Nonetheless, although Na is not an essential nutrient, the ion is taken up into cells down the electrochemical gradient, competing with K for uptake (Hasegawa et al., 2000). The hypodermal and endodermal cells of salt-stressed grapevine roots may selectively accumulate K over Na and Cl. In contrast, the cortex and pericycle cells sequester large amounts of Na and Cl in their vacuoles (Storey et al., 2003a).

Grapevines take up more Cl^- than Na^+ from saline soils that have equivalent concentrations of both ions, and a small portion of each ion ends up in the xylem and is transported to the shoots with the transpiration stream (Walker et al., 1981). Consequently, Cl and, to a lesser extent, Na accumulate in the older leaves as the growing season progresses and as their availability in the soil rises (Downton, 1985; Munns, 2002; Shani and Ben-Gal, 2005; Stevens and Walker, 2002). Eventually,

FIG. 7.10

Leaf symptoms of salt injury on Merlot.

Photo by M. Keller.

salinity stress will arrest lateral shoot growth and induce necrotic leaf margins in older leaves—a symptom called marginal burn or salt burn (Fig. 7.10). The Cl^- threshold for marginal necrosis seems to be around 2.5% of the leaf dry weight (Walker et al., 1981). The necrotic symptoms gradually progress toward the petiole, whereas the main veins remain green (Williams and Matthews, 1990; Williams et al., 1994). Since ion concentrations steadily increase as the transpiration stream deposits Na^+ and/or Cl^- in the leaves, such salt injury is the result of ions accumulating in transpiring leaves to the point where the vacuoles can no longer contain them. Accumulation in the cytoplasm then leads to salt poisoning due to enzyme inhibition, whereas accumulation in the cell walls leads to cell dehydration; both outcomes result in cell death (Munns, 2002). Over time, therefore ions may accumulate to toxic concentrations; this ion-specific phase of salinity stress is associated with premature death of older leaves (Munns and Tester, 2008).

Excessive Cl^- uptake interferes with nitrogen nutrition because NO_3^- uptake responds to the concentration of $NO_3^- + Cl^-$ rather than to NO_3^- alone (Clarkson, 1985). The ensuing competition between Cl^- and NO_3^- for root uptake requires application of abundant N fertilizer to improve plant N status on saline soils. Likewise, high Na^+ availability is toxic to plants because it competes with K^+ for uptake, which reduces K-stimulated enzyme activities, metabolism, and photosynthesis. At concentrations exceeding approximately 100 mM, Na and Cl also directly inhibit many enzymes (Munns, 2002; Munns and Tester, 2008; Zhu, 2001). When high salt supply overwhelms the leaf vacuoles' capacity to sequester Cl and Na, excess Na^+ accumulates in the cytoplasm at the expense of K^+ and can even result in a loss of K^+ and Ca^{2+} from the cells. Sodium competes with Ca^{2+} and displaces it from cell walls, thus compromising cell wall integrity. Because Ca^{2+} in turn can reduce Na^+ uptake and increase Ca^{2+} uptake, Ca addition can somewhat alleviate the toxic effects of salinity (Hasegawa et al., 2000; Plieth, 2005). However, prolonged exposure to high soil Ca may itself be stressful for the plant.

Oxidative stress is another characteristic of salinity-induced injuries to plant tissues (Munns and Tester, 2008; Zhu, 2001). It is a secondary stress that results from the effects of ion imbalance and hyperosmotic stress and from the decline in photosynthesis (see Section 7.1). Oxidative stress increases the vine's light sensitivity because more photons are being absorbed than can be used by the declining photosynthesis. Salt-stressed grapevines may cope with such an energy overload by increasing the rate of photorespiration, which dissipates some of the excess energy but comes at the cost of lower photosynthetic efficiency (Cramer et al., 2007; Downton, 1977b; Downton et al., 1990; Walker et al., 1981). Another defense strategy is to boost the antioxidant systems that capture and inactivate some of the reactive oxygen species (Cramer et al., 2007). Because Mn^{2+} and Zn^{2+} act as antioxidants in plant tissues, foliar applications of Mn-chelates and Zn-chelates might alleviate effects of oxidative stress in plants subject to salinity stress (Aktas et al., 2005).

Some American *Vitis* species, especially *V. riparia*, *V. berlandieri*, and, to a lesser extent, *V. candicans* and *V. champinii*, are somewhat tolerant of salt in the rootzone (Williams et al., 1994). Rootstocks like Ramsey, 1103 Paulsen, 110 Richter, 140 Ruggeri, and 101–14 Mgt, which are derived from these species, can exclude much of the salt from root uptake and xylem transport (Antcliff et al., 1983; Sauer, 1968; Tregeagle et al., 2010; Walker et al., 2000, 2010). A similar salt exclusion also occurs in *V. vinifera* roots, although different cultivars vary somewhat in the extent of Na and Cl uptake and accumulation in the leaves (Groot Obbink and McE Alexander, 1973; Henderson et al., 2018). These rootstocks and the scions grafted to them, as well as some own-rooted *V. vinifera* vines, accumulate little salt in the leaves and fruit even at high soil salt concentrations (Downton, 1985; Stevens and Walker, 2002; Zhang et al., 2002). However, at least some rootstocks may progressively lose their salt-exclusion ability; examples include Ramsey, 1103 Paulsen, and 101–14 Mgt (Tregeagle et al., 2006). Under long-term exposure to saline conditions, which tends to lead to salt buildup in the soil over time, these rootstocks may become less salt tolerant.

The impact of salinity on grapevines seems to be more severe on heavy soils than on soils with higher sand content. Moreover, irrigation and soil management affect the extent of physical degradation of salt-affected soils. Irrigation is a common cause of agricultural land degradation because salts dissolved in the irrigation water are left in the soil following evaporation. Excessive irrigation, particularly with saline water, as well as frequent tillage and intense trafficking are a good recipe for rapid loss of soil fertility. However, salts can also build up under highly efficient drip irrigation, when ions move down the soil profile beneath the emitters and then move laterally and rise again to the soil surface with the evaporating water (Stevens and Walker, 2002). The resulting high-salt zone around the edges of the wetting zone can restrict root growth in a way similar to the restriction imposed by a pot. The "pot" size is smaller in sandy soils than in loam soils. Waterlogging due to the formation of impermeable soil layers or excessive irrigation also increases the risk of salt damage, because waterlogged roots lose the ability to avert Na and Cl uptake. Waterlogging may moreover increase the amount of Na^+ in the soil solution relative to other ions so that Na^+ uptake is often favored over Cl^- uptake (Stevens and Walker, 2002).

Even if only a portion of a vine's root system is exposed to saline conditions while other portions continue to have access to freshwater, the latter ostensibly do not compensate for the decline in water uptake by the former (Shani et al., 1993). Prolonged exposure to saline soil water ultimately results in vine death (Shani and Ben-Gal, 2005). But where soil salts can be leached out of the rootzone by using a fresh source of irrigation water, the harmful physiological effects can often be quickly reversed. Unless irreversible damage has already been caused, grapevines restore root functionality, growth, and water uptake to drain the excess ions from the leaves, and growth and gas exchange recover rapidly (Shani et al., 1993; Walker et al., 1981).

7.4 Temperature: Too hot or too cold

Grapevines growing in cool, continental climates are exposed to a large daily temperature range and often experience widely fluctuating temperatures during spring and autumn. Low temperature may limit growth by decreasing the rate of protein production or cell wall extensibility. Inhibition of respiration by low temperature leads to rapid cessation of cell wall loosening and thus of cell expansion (Cosgrove, 2016). By preventing cell expansion, restricted cell wall extensibility also inhibits cell division; the duration of cell division increases exponentially with decreasing temperature. Furthermore, cold temperatures increase the rigidity of the normally fluid cell membranes (Chinnusamy et al., 2007). Because low temperatures restrict cell division more than photosynthesis, sugar and starch tend to accumulate in the leaves during a cool episode (Wardlaw, 1990). When the temperature becomes too low, it can cause damage to grapevine tissues. The type and extent of damage depend on whether or not the temperature drops below the freezing point and on the developmental status of the vine.

By contrast, higher temperatures tend to accelerate grapevine growth and development so that phenological stages occur in more rapid succession than under cooler conditions (see Section 2.2). In other words, an increase in temperature accelerates and compresses the temporal program of plant development—up to an optimum—so long as other factors are not limiting growth. One of these other factors is water deficit (see Section 7.2), which is often associated with high temperatures. Temperature also has a strong effect on photosynthesis and respiration (see Section 5.2). Changes in the rate of photosynthetic CO_2 assimilation are reversible over the physiological range of about 10–35°C, but higher and lower temperatures can cause injury to the photosynthetic apparatus.

Many effects of temperatures on grapevine canopies are discussed in Section 5.2; on yield formation, in Section 6.1; and on fruit composition, in Section 6.3. This section is mainly concerned with extreme temperatures above and below the physiological range, which cause temporary or permanent injury to grapevine tissues. Temperatures outside the range of approximately 15–35°C progressively impair grapevine growth, yield formation, and fruit ripening (Keller, 2010). Excessive heat and cold tend to result in symptoms that resemble those caused by drought stress. This is not surprising because both heat and cold stress influence plant water relations, decreasing the RWC of tissues and inducing oxidative stress.

7.4.1 Heat acclimation and damage

Although the majority of the world's grape crop is produced in temperate climates, production, especially of table and raisin grapes, extends to some hot areas where summertime afternoon temperatures often rise above 35°C or even 40°C (Williams et al., 1994). The optimal daytime temperature for grapevine growth, photosynthesis, yield formation, and fruit ripening is below 30°C. Heat stress is usually defined as temperatures rising >5°C above those associated with optimal growing conditions. Extremely high temperatures, above approximately 40°C, cause a sharp decline in photosynthesis due to the disruption of the functional integrity of the photosynthetic machinery in the chloroplasts (Zsófi et al., 2009). Heat increases the physical distance between the light-harvesting antenna pigments and the reaction center in photosystem II (PSII; see Section 4.1). In addition, the water-splitting system of PSII is very heat sensitive; thus, heat inhibits PSII-driven electron transport (Berry and Björkman, 1980). The activity of PSI, on the other hand, is much more heat stable. The same applies to the photosynthetic enzymes involved in CO_2 assimilation, which are stable to over 50°C, although heat stress leads to protein denaturation and misfolding of newly assembled proteins (Berry and Björkman, 1980).

For instance, rubisco inactivation increases exponentially with increasing temperature, and rubisco activity strongly declines above 35 °C. Therefore, although the stomata begin to close above approximately 35 °C, this seems to be a response to the reduced photosynthesis rather than the cause of it.

Leaf temperatures exceeding 45 °C for several hours or a few days trigger severe and lasting declines in stomatal conductance (g_s) and photosynthesis, and if such extreme temperatures occur over periods of weeks they can kill grapevine leaves (Abass and Rajashekar, 1991; Gamon and Pearcy, 1989). But even a few hours at 50 °C may be lethal to leaves (Morrell et al., 1997). High temperatures are often associated with low air humidity so that the stomata close in response to the high VPD (see Section 3.2) and not because of heat. Furthermore, high light intensity exacerbates the adverse impact of high temperature on photosynthesis (Gamon and Pearcy, 1989, 1990a,b). Thus, leaves are more susceptible to heat waves under a blue sky or during periods of water deficit.

Water evaporation from the leaf surface cools the leaf tissues so that the leaf temperature is generally close to the ambient temperature. During the day the leaves of grapevines are often only 1–3 °C warmer than the surrounding air (Peña Quiñones et al., 2019). Evaporative cooling thus usually prevents overheating of sun-exposed leaves (see Section 3.2). The cooling effect of water evaporation is sometimes exploited by growers who use intermittent irrigation, for example, by microsprinklers beneath the canopy or overhead sprinklers above it, to cool leaves in the heat of the midday sun in warm climates (Gilbert et al., 1971; Greer and Weedon, 2014). Such "hydrocooling" enables the vines to maintain higher photosynthesis rates and higher rates of berry growth and sugar accumulation. Contrary to widespread opinion, water droplets on leaves do not cause sunburn, because their ellipsoidal shape focuses the solar rays far below the leaf and because they directly cool leaves through their thermal mass (Egri et al., 2010). However, the higher relative humidity in the canopy and the surface moisture on the fruit clusters associated with such water application can sometimes lead to bunch rot outbreaks (Gilbert et al., 1971; see also Section 8.2).

The temperature of transpiring, sun-exposed grapevine leaves generally tracks the diurnal changes in air temperature with a peak in the early to mid-afternoon (Peña Quiñones et al., 2019). The leaves near the top of the canopy usually experience the highest daily temperatures. However, considering vertically shoot-positioned canopies in north-south oriented vineyard rows, the leaves on the west side of the canopy experience higher temperatures than their east-side counterparts, whereas in east-west oriented rows, the leaves on the south side (Northern Hemisphere) or north side (Southern Hemisphere) experience the highest temperatures. The hottest leaves of a canopy suffer the most from heat stress and are the most likely to show symptoms of leaf yellowing and senescence (Palliotti et al., 2009).

Heat-stressed plant cells accumulate calcium ions (Ca^{2+}) in the cytosol, which helps reduce the permeability, or leakiness, of the cell membranes and maintain membrane integrity. Maintenance of membrane integrity occurs at least in part through changes in their fatty acid composition: Warm temperatures increase the amounts of saturated and monounsaturated fatty acids (Penfield, 2008; Sage and Kubien, 2007). Calcium also enhances heat tolerance by stimulating the plant's antioxidant system to protect the photosynthetic machinery from oxidative damage. Heat stress moreover appears to result in breakdown of starch to sugars and accumulation of amino acids, especially glutamine, but not the "normal" compatible solute proline (Guy et al., 2008; Rizhsky et al., 2004). Glutamine might accumulate because it is not being used for protein production or because proteins are being degraded (see Section 5.3). In addition, emission of volatile isoprenoids, such as isoprene or monoterpenes, helps the leaves to recover from brief episodes of temperatures exceeding 40 °C by stabilizing the thylakoid membranes and by scavenging reactive oxygen species that rapidly accumulate under heat stress

(Loreto and Schnitzler, 2010; Pichersky and Gershenzon, 2002; Sharkey et al., 2008). This antioxidant activity provides thermotolerance against heat spikes and strongly decreases oxidative damage in the leaves. Consequently, the heat-induced inactivation of the photosynthetic system is reversible. Leaves can recover within a few days after temperatures return to physiological levels if the heat stress was not so severe as to cause membrane leakage and necrosis (Abass and Rajashekar, 1991). Emission of isoprenoids ceases above 45 °C, perhaps because the supply of assimilates for their production becomes insufficient due to the impairment of photosynthesis (Loreto and Schnitzler, 2010).

As is the case with other stresses, the capacity of grapevines to recover from heat stress depends on the intensity and duration of a heat wave and the growth stage at which it occurs. Brief episodes of extreme heat are worse than longer periods of moderately high temperatures. On the other hand, photosynthetic acclimation to high temperatures leads to decreased performance at low temperatures and vice versa. For example, the aforementioned changes in membrane composition may heighten the chilling susceptibility of tissues (Iba, 2002). Therefore, although grapevines can adapt to seasonal changes in temperature regimes (Zsófi et al., 2009), short-term temperature fluctuations over a wide range can result in insufficient photosynthesis for growth and/or fruit ripening. Moreover, heat stress accelerates leaf aging and senescence so that the chlorophyll content often decreases along with photosynthesis (Thomas and Stoddart, 1980).

The sensitivity to heat stress varies among *Vitis* species and cultivars. Even under relatively modest light intensities, photosynthesis in *V. aestivalis* leaves at 40 °C may decline more than in *V. vinifera* leaves, and this decrease is associated with damage to PSII and subsequent chlorophyll degradation and leaf senescence (Kadir, 2006). In susceptible species, moreover, photosynthetic recovery of the leaves that survive a heat episode may be slow, extending over several weeks, and is incomplete (Gamon and Pearcy, 1989; Kadir et al., 2007). When the heat-susceptible *V. labruscana* cultivar Concord is grown in warm climates or hot growing seasons with intense solar radiation, it typically develops a condition termed "blackleaf," whereby the adaxial epidermis cells of sun-exposed leaves become discolored, turning bronze-brown, purple, or black (Fig. 7.11). The discoloration usually

FIG. 7.11

Blackleaf symptoms initially appear on south-facing ridges of Concord leaves (left) and eventually cover most of the exposed leaf surfaces (right).

Photos by M. Keller.

begins in midsummer on the leaves' sun-facing ridges and is exacerbated by water deficit. It is accompanied by chlorophyll degradation and a decline in photosynthesis and may culminate in leaf necrosis and abscission (Lang et al., 1998; Smithyman et al., 2001). These blackleaf symptoms look suspiciously like the UV damage symptoms described in other plants and may be a consequence of the polymerization of phenolic compounds in the epidermal cells that results from cell death due to oxidative stress (Kakani et al., 2003; Yamasaki et al., 1997). Therefore, blackleaf symptoms may arise from injury caused by a combination of intense UV radiation and high temperatures. But *V. vinifera* cultivars, too, differ in their vulnerability to heat stress. During an unusually hot summer in Italy, Sangiovese vines lost more than half their basal leaves, especially on the west side of the canopy that was heated by the afternoon sun, whereas Montepulciano vines did not lose any leaves (Palliotti et al., 2009).

By favoring rapid transpiration rates, high g_s lowers leaf temperature and appears to reduce the deleterious effects of heat stress. Higher g_s also increases CO_2 diffusion into the leaf and favors higher photosynthetic rates. Higher photosynthetic rates could in turn lead to more biomass production and higher crop yields. However, higher g_s may be beneficial for yield formation by a mechanism not directly related to photosynthesis, perhaps by reducing respiration. Respiration is strongly stimulated by increasing temperature and results in a net loss of assimilated carbon. Decreases in photosynthesis and/or sink activity during heat stress suppresses growth and fruit ripening. In addition, high temperature shifts carbon partitioning to favor vegetative growth at the expense of fruit growth and ripening (Greer and Weston, 2010b; Sepúlveda et al., 1986). Postveraison heat stress can markedly delay or even inhibit fruit ripening by decreasing the amount of assimilates available for export to the clusters and/or the ability of the berries to import assimilates, which may lead to uneven ripening or an overall delay of fruit maturation (Cawthon and Morris, 1982).

The upper temperature limit for maximum yield formation in grapevines seems to be about 35 °C (Keller, 2010). High yields are generally an important goal for the production of table, raisin, or juice grapes, whereas moderate to low yields are preferred in wine grape production to maximize fruit quality. Nevertheless, in areas where vegetative growth can be controlled by deficit irrigation and the crop load can be adjusted by canopy management, cultural practices aimed at favoring high g_s could be used to improve quality during hot growing seasons (see also Section 6.2). Because water deficit increases a vine's susceptibility to heat stress, drought during the bloom period is especially detrimental to fruit set if it coincides with a heat episode. But heat stress alone may also impair fruit set, probably mostly by interfering with pollen maturation and viability (De Storme and Geelen, 2014; Zinn et al., 2010; Pereira et al., 2014). Heat stress reduces auxin production in the anthers, which leads to a blockage of cell division in the pollen immediately after meiosis. In addition, the high respiration rates of maturing pollen make them susceptible to damage from reactive oxygen radicals that accumulate during heat stress (Rieu et al., 2017). The resulting premature abortion of pollen development induces male sterility and reduces the chances of successful fertilization.

Like other developmental processes, grape ripening is accelerated by moderately high temperatures (Keller, 2010). This is especially true for sugar accumulation and malate degradation. However, because grape berries are designed to minimize water loss by transpiration, especially after veraison, they cannot take advantage of the evaporative cooling mechanism that usually protects leaves from overheating (see Section 6.2). Their poor capacity for evaporative cooling makes the berries susceptible to overheating. Indeed, sun-exposed berries may heat up to 15 °C above air temperature, whereas shaded berries are usually near ambient temperature (Smart and Sinclair, 1976; Spayd et al., 2002). Too much heat can inhibit or even denature proteins in the berries. When plant organs are suddenly exposed to high temperature above about 35 °C, they produce so-called heat shock proteins that function as chaperones by helping to fold other proteins into a suitable structure, preventing accumulation of proteins that have been denatured by heat, and helping to reactivate them (Iba, 2002; Morrell et al., 1997). Sudden sun exposure is relatively common for previously shaded grape clusters following leaf removal in the fruit zone, and such clusters are vulnerable to sunburn injury. Sunburn symptoms can develop as an expression of oxidative damage resulting from a combination of high light intensity and high temperature. Heat injury usually results from the disintegration of cell membranes due to lipid phase changes and denaturation of membrane proteins. The disruption of membrane structures is irreversible and leads to leakage of cell solutes and breakdown of cell compartmentation, which is normally lethal for the affected cells.

Symptoms of sunburn or sunscald on grape clusters include initial bleaching followed by reddening or browning of the exposed portion of the skin on both red and white grapes, especially before veraison; a change in epicuticular wax structure from crystalline to amorphous, giving especially red postveraison berries a glossy appearance; and finally sinking in of the affected area or shriveling due to dehydration of the whole berry (Bondada and Keller, 2012; see also Fig. 7.12). Sunburn symptoms can become particularly acute following late-season leaf removal because, in contrast to berries that have been exposed since the beginning of the season, the previously shaded berries have not accumulated enough "sunscreen" components such as flavonols and xanthophylls (Düring and Davtyan, 2002; Kolb et al., 2003).

In addition to the influence of canopy management practices, overheating of sun-exposed grape berries also has implications for the choice of vineyard row direction, trellis design and training system, and even vineyard floor management. Because during a sunny day the air temperature in the early afternoon is generally higher than in the morning, exposed grape berries on the afternoon side of the canopy can reach considerably higher absolute temperatures than their counterparts on the morning side. Therefore, whereas cultural practices that maximize sun exposure are usually beneficial in a cool climate, they may be detrimental in a warm, sunny climate (see Section 6.2). Moreover, a bare soil surface, especially of a light-colored soil, reflects a major portion of the incident light back to the canopy, and this can aggravate heat injury on grapes during hot days.

FIG. 7.12

Delayed ripening of the sun-exposed side of a Cabernet Sauvignon cluster in a warm climate (left), symptoms of sunburn on grape berries (center), and shriveling of sunburnt berries (right).

Photos by M. Keller.

The heat-induced changes in grapevine growth and metabolism also have implications for grape production in an increasingly warmer climate. In addition to a rise in average temperatures and an associated increase in cumulative heat units and growing-season length, meteorologists and climate modelers predict that shifts will also occur in the number and extent of extreme weather events (IPCC, Intergovernmental Panel on Climate Change, 2013). This conclusion follows partly from the statistical normal distribution of temperature data around a mean, which follows a so-called bell curve, and partly from an increase in temperature variability (Schär et al., 2004). Summer heat waves are already increasing in both frequency and severity (Fischer and Knutti, 2015). An analysis for southern Australia showed that the probability of a run of 5 consecutive days with temperatures over 35 °C doubles for a 1 °C warming and increases by a factor of five for a 3 °C warming (Hennessy and Pittock, 1995). As model calculations for the United States demonstrate, such heat episodes could be detrimental to fruit quality in areas that already experience warm or hot climates (White et al., 2006). Such climates encompass many of the current grape production regions in the world.

7.4.2 Chilling stress

Damage to nondormant plant tissues caused by low but above-freezing temperatures is referred to as chilling stress. Note that the same temperature range of about 0–10 °C acting on dormant tissues is not normally viewed as being stressful but is necessary to release dormancy (see Section 2.2). The photosynthetic cell organelles, the chloroplasts, are particularly sensitive to chilling stress (Kratsch and Wise, 2000). Swelling of chloroplasts, distortion of thylakoid membranes, and starch depletion, manifested as a decrease in the number and size of starch granules, inside the chloroplasts are usually the first microscopically visible signs of chilling injury. At the same time, chilling also leads to callose-induced blocking of plasmodesmata and phloem sieve pores, which decreases phloem loading and phloem transport (Wu et al., 2018). Accumulation of sugar in the chloroplasts due to reduced export and continued starch degradation lowers their Ψ_π and may be responsible for chloroplast swelling by causing osmotic water influx. Although damage to the photosynthetic hardware usually has severe consequences for photosynthesis, the reduction in photosynthesis under chilling temperatures may be caused by feedback inhibition due to sugar accumulation. Plant growth in general seems to be more sensitive to low-temperature inhibition than is photosynthesis (Wingler, 2015). Consequently, sugar may accumulate in the leaves because cell division, and hence growth, ceases at low temperature, which decreases sink demand for assimilates and may result in an oversupply of fixed carbon (Körner, 2003; Koroleva et al., 2002).

Under prolonged chilling, the chloroplasts may disintegrate completely so that the leaves become chlorotic (Kratsch and Wise, 2000). Leaves also turn chlorotic when they develop under chilling conditions, but this is probably due to an inability to produce thylakoid proteins. Chilling-induced chlorosis is normally reversible; thus leaves may recover and regreen upon exposure to warmer temperatures if chilling injury is not too severe. However, the longer a plant is exposed to low temperatures, the more extensive and irreversible is the damage (Kratsch and Wise, 2000). Plants eventually activate an organized cell suicide process called programmed cell death that involves the systematic, energy-dependent dismantling of the cells accompanied by recovery of cell components such as proteins and DNA. The rescued breakdown products may be exported from the dying leaves to other organs to enable the plants to survive the unfavorable conditions.

The more chilling sensitive a grape species or cultivar, the sooner and the more extensive is the development of the ultrastructural changes. The most resistant plants may not suffer injury unless they are simultaneously exposed to some other stress factor, such as high light intensity or water deficit. Light greatly exacerbates chilling injury due to energy overload because light absorption decreases less than carbon fixation, which leads to oxidative stress due to the accumulation of reactive oxygen species (Kratsch and Wise, 2000). Oxygen radicals may trigger the degradation of the two photosystems, PSI and PSII, and the photosynthetic enzyme rubisco (Møller et al., 2007). Therefore, chilling stress symptoms develop sooner and are more severe during cold days than during cold nights. Nevertheless, cold nights seem to make grapevine leaves more susceptible to high light intensity (Bertamini et al., 2006). In continental climates with large diurnal temperature variations, where cold nights are often followed by bright, warm days, photosynthesis can be inhibited completely while such conditions last. The chilling-induced depression of photosynthesis appears to be similar to the photosynthetic depression due to water deficit (Bálo et al., 1986; Flexas et al., 1999a; see also Section 7.2). Unlike under water deficit, however, the closure of stomata below 15 °C is thought to be a response to the reduced photosynthesis rather than the reverse (Hendrickson et al., 2004a). Photosynthesis recovers only after several warm nights.

Photooxidative stress can trigger anthocyanin accumulation in the exterior leaves of dark-skinned grape cultivars so that the leaves may turn red, whereas leaves in the canopy interior usually do not accumulate anthocyanins. Because anthocyanins probably act as photoprotectants, their production is not activated in the dark even at high foliar sugar concentrations (Halldorson and Keller, 2018). Moreover, autumn coloration of cultivars whose leaves normally turn red often becomes more intense during cool weather, especially if the light intensity remains high (Feild et al., 2001; Hoch et al., 2003). It also appears that the amount of foliar anthocyanins fluctuates according to changes in the weather; they may be degraded or reformed depending on the temperature and light conditions (Keskitalo et al., 2005).

High humidity protects plant tissues against injury, especially under the relatively common event in spring and fall of low temperatures occurring during the night (Kratsch and Wise, 2000). Moreover, if the temperature remains low for longer periods, the leaves may acclimate to those conditions by reemitting some of the excess energy as fluorescent light (see Section 4.1) and increasing the activity of enzymes involved in sugar production (see Section 4.2). Fluorescence emission occurs especially in mature leaves that suddenly experience a cold shock, whereas enhanced sugar production may be more important in young leaves developing during cool conditions. This so-called chilling acclimation is very similar to the photoacclimation to high light intensity (Ensminger et al., 2006; Theocharis et al., 2012). The role of both processes is to limit photoinhibition—that is, to balance energy supply and demand. However, in some continental climates the temperature fluctuates too much between night and day for acclimation to protect the leaves from inhibition of photosynthesis.

If chilling conditions persist, the leaves that had grown under warm conditions will eventually be shed and will progressively be replaced by new, acclimated leaves. These adjustments lead to higher photosynthetic rates at lower temperatures and a downward shift of the optimum temperature for photosynthesis. This response also results in improved water-use efficiency (i.e., a higher photosynthesis $\div$ transpiration ratio), and hence drought tolerance, and accelerates the acquisition of freezing tolerance (Ensminger et al., 2006). Perhaps this is because the shoots and perennial parts of the vine act as alternative sinks for the surplus assimilates that are available when low temperatures inhibit growth. Therefore, repeated exposure to chilling temperatures can induce adaptive changes in a grapevine that result in subsequent tolerance of freezing temperatures. This process is referred to as cold acclimation or hardening and is discussed later.

Chilling during the period leading up to bloom impairs pollen formation in a way that is very similar to the effects of drought or heat stress (see Section 7.2). These stress factors disturb sugar metabolism in the anthers, whereas ovule fertility appears to be rather immune to low temperature (De Storme and Geelen, 2014; Oliver et al., 2005). The cold-induced inhibition of invertase activity in the anthers slows or blocks sugar unloading from the phloem even if the leaves supply abundant assimilates. This renders the pollen grains unable to accumulate starch, and they may instead temporarily accumulate sucrose. If this blockage occurs during meiosis, which occurs about 1–2 weeks before anthesis, it causes pollen sterility (De Storme and Geelen, 2014; Koblet, 1966; Oliver et al., 2005). Even brief episodes of chilling during male meiosis, such as two or three consecutive cool nights, can irreversibly inhibit pollen development. This decreases pollination and fertilization, which results in poor fruit set or, in more severe cases, abscission of inflorescences.

Chilling stress during grape ripening may lead to carbon shortage in the vine when sugar accumulates in the leaves as its export slows. Because the sink activity of grape clusters is more sensitive to temperature than is the sink activity of shoot tips, such carbon starvation can completely and temporarily inhibit berry sugar accumulation. If the stress continues for more than a few days, it may induce a syndrome termed bunch stem necrosis. This physiological disorder essentially amounts to the abandonment of some of the vine's clusters under stress and tends to be more severe on heavily cropped vines. The vascular system in the peduncle and/or rachis, sometimes only on a shoulder or the tip of a cluster, becomes dysfunctional, the berries on affected clusters stop ripening, and the cluster or its affected portion is eventually shed by the vine (Bondada and Keller 2012; Hall et al., 2011; Theiler, 1970). Thus, grapevine reproductive development is most susceptible to chilling stress during the periods leading up to anthesis and after veraison. In other words, the stress vulnerability is high before a vine has invested heavily in reproduction and after the seeds have matured. From fruit set to early post-veraison, the berries' sink strength is high enough to attract carbon from remobilized storage reserves even under severely stressful conditions (Candolfi-Vasconcelos et al., 1994; Keller et al., 2015b; Rossouw et al., 2017).

7.4.3 Cold acclimation and damage

Ice melts at $0\,°C$, but that does not imply that liquid water freezes at $0\,°C$ to form ice crystals. Freezing, or the phase transition of water from a liquid to a solid, occurs only if no heat is removed from the water. Because plant tissues lose heat to the environment as the temperature declines, their tissue water will not freeze unless some nucleation event occurs. This phenomenon is commonly referred to as supercooling, which ends with a sudden release of heat at the point where water finally freezes. At this point, the latent heat of fusion, which is the chemical energy needed to break the hydrogen bonds that hold together individual H_2O molecules in ice crystals, is converted to sensible heat, which is heat that we can measure with a thermometer. Supercooled rainwater that freezes instantly when it hits a cold ($<0\,°C$) surface causes the phenomenon known as freezing rain or ice storm.

Highly purified water can be supercooled to almost $-40\,°C$; this temperature limit is called the homogenous nucleation point. Supercooling does not occur in plants to such an extreme degree because water inside plants and on their surfaces is never absolutely pure (Wisniewski et al., 2014). Plant tissues freeze when they cannot avoid nucleation and prevent ice growth (Pearce, 2001). The vibration of tiny particles, such as bacterial or other proteins and other molecules, serves as a so-called heterogenous nucleator that brings about the formation of ice "embryos" which in turn catalyze the growth of ice crystals (Franks, 1985). However, the presence in plant tissues of osmotically active solutes, such as sugars and mineral nutrient ions, does decrease the freezing temperature below 0 °C. Like osmotic pressure, freezing-point depression (more accurately, melting-point depression) is a colligative property of solutions. As a general rule, every mole of dissolved solutes per kilogram of water lowers the melting point of the solution by 1.86 °C (Zachariassen and Kristiansen, 2000). In other words, the extent of the osmotic freezing-point depression increases as the number of solutes present in the solution or tissue rises.

During a freezing event, grapevine tissues lose heat and supercool until they reach their ice nucleation temperature. As the temperature drops, water freezes first on an organ's surface and in the intercellular spaces, cell walls, and tracheary elements. The concentration of solutes in this apoplastic space is much lower, and hence the freezing point is higher, than inside the cells (Müller-Thurgau, 1886; Wisniewski et al., 2014; Xin and Browse, 2000). Ice formation leads to exclusion of solutes, which therefore become more concentrated in the apoplast. This concentration effect lowers the apoplast water potential by $-1.16\,\text{MPa}\,°\text{C}^{-1}$, which pulls water out of the cells osmotically (Guy, 1990; Pearce, 2001; Steponkus, 1984; Thomashow, 1999). Liquid water can move freely across intact membranes, but ice crystals cannot, so that the cell interior, or symplast, is supercooled (Steponkus, 1984). The increase in the cells' solute concentration thus lowers their freezing point by another 2–3 °C. As a practical aside, the same concentration mechanism is exploited in the production of ice wine, whereby frozen grapes are harvested and pressed.

Ice formation in the apoplast, termed extracellular freezing, is usually not lethal. When the ice crystals melt, the water simply diffuses back into the cells and they resume their metabolism. However, in organs that are not cold-acclimated the cell membranes shrink irreversibly during extracellular freezing, and when the melted water moves back into the cells the resulting turgor increase can make the cells burst (Xin and Browse, 2000). While membrane shrinkage in cold-acclimated tissues is reversible, the apoplast contains only approximately 10–15% of the total tissue water. Consequently, extended exposure to freezing temperatures dehydrates the symplast and causes a water deficit, which interferes with metabolism and puts cells under physical stress due to shrinkage (Browse and Xin, 2001; Müller-Thurgau, 1886; Thomashow, 2001).

The amount of water that a cell loses during extracellular freezing depends on its solute concentration, which determines π_{cell}, and the freezing temperature, which determines $\Psi_{apoplast}$. For instance, if freezing occurs at $-10\,°\text{C}$, then $\Psi_{apoplast}=-11.6\,\text{MPa}$, which pulls 90% of the water out of a cell at $\pi_{cell}=1\,\text{MPa}$ but only 80% from an acclimated cell with $\pi_{cell}=2\,\text{MPa}$ (Xin and Browse, 2000). Very high solute concentrations can be toxic to the cells, and if the cells shrink beyond their minimum tolerated volume, osmotic stress and membrane rupture ensue (Zachariassen and Kristiansen, 2000). Thus, dehydration of the symplast by extracellular freezing is the most common cause of freezing-induced injury, and tissue survival depends on the extent of the cells' dehydration tolerance (Guy, 1990; Thomashow, 1999; Xin and Browse, 2000). Because it may take several days for the symplast to reach water potential equilibrium with the apoplast, such injury by dehydration is mainly a problem if low temperatures persist for prolonged periods (Gusta and Wisniewski, 2013). Under such conditions of chronic stress the cells may eventually die even though the temperature is not low enough to kill them instantaneously; instant death occurs during acute stress that leads to intracellular ice formation. For example, dormant grapevine buds may be able to withstand short episodes of $-15\,°\text{C}$ to $-25\,°\text{C}$, depending on the cultivar, but may be killed by dehydration when the extracellular water remains frozen below −5 °C for a week or longer (Cragin et al., 2017).

In contrast with other tissues, xylem parenchyma cells somehow seem to avoid dehydration during extracellular freezing, at least for a while. This mechanism is termed deep supercooling and enables these xylem cells to endure lower temperatures than other tissues for extended periods (Quamme, 1991). However, prolonged exposure to freezing temperatures can lead to the coalescence of small ice crystals into large crystals—a process called recrystallization (Pearce, 2001). This formation of large ice masses can deform cells and damage plant tissues by interfering with their structure, for example, by forcing cell layers apart. The explosive force of water is considerable because the volume of freezing water expands by approximately 10% (Meiering et al., 1980). Although this does not necessarily kill a grapevine, the resulting tension can cause splitting or cracking of woody organs either immediately or during the subsequent growing season, which may provide infection sites for crown gall bacteria (Paroschy et al., 1980; see also Section 8.2). Moreover, the forcing of gases out of solution during ice formation inside the xylem vessels can lead to air embolisms (Charrier et al., 2017; see also Section 3.3). The vulnerability to such embolisms increases with increasing conduit diameter and with water stress that decreases xylem pressure. Therefore, dry or frozen soil during winter generally results in more severe xylem cavitation, especially in vines that had grown vigorously during the preceding years and hence have large xylem vessels. Although this presents quite a challenge to grapevines experiencing repeated freeze–thaw cycles at higher latitudes, they are normally very effective at repairing such cavitation in spring. Such repair cannot occur, however, if unusually dry soil prevents the buildup of root pressure needed to dissolve and push out air bubbles (see Section 2.2). It is also possible that conditions such as severe drought stress or loss of functional leaf area that lead to depletion of root storage reserves render the roots unable to restore xylem function in spring, which may result in plant death (Galvez et al., 2013).

Ice formation in the apoplast is accompanied by a release of heat, called the high-temperature exotherm or freezing exotherm, which results in a rapid but transient rise in tissue temperature (Xin and Browse, 2000). This temperature is the nonlethal freezing temperature of the tissue. During freezing, the tissue temperature remains constant and above ambient until all the extracellular water is frozen. Once this has occurred, the tissue temperature returns to ambient and continues to drop. As the temperature continues to decline, a second phase of heat release occurs when the cell protoplasts freeze, and this is termed the low-temperature exotherm (Mills et al., 2006; Ristic and Ashworth, 1993). Intracellular freezing kills cells instantaneously due to the combined effects of membrane damage, symplast dehydration, and protein denaturation. The extent of intracellular ice formation also depends on the rate of tissue cooling; very rapid cooling seems to induce ice formation within the cells even at temperatures that would normally only result in extracellular freezing (Ashworth, 1982; Müller-Thurgau, 1886). Fortunately, cooling rates under field conditions are seldom fast enough to provoke intracellular freezing, except in situations in which trunks and canes facing the sun are heated during the day before the temperature plummets during a clear night (Guy, 1990).

Grapevine tissues and organs differ in their ability to tolerate freezing temperatures. Growing organs that have a high water content are very sensitive to frost, and early spring growth after budbreak is especially frost susceptible (Fuller and Telli, 1999). Ambient temperatures of −2 °C or −3 °C or lower can damage leaves, shoots, and green buds (Fennell, 2004). The lethal temperature also depends on the presence of moisture and ice nucleating agents on the surface of an organ (Luisetti et al., 1991). For instance, dry leaves can supercool to relatively low temperatures, whereas wet leaves freeze at just a

FIG. 7.13

Young Merlot cordon (left) with lethal freeze injury to the phloem (brown area), while the xylem parenchyma remains viable (green area), and frozen primary bud with surviving secondary and tertiary buds (right).

Photos by L. Mills.

few degrees below 0 °C due to the presence of nucleating agents on the leaf surface. Certain bacteria, such as *Pseudomonas syringae*, are especially potent ice nucleators. Ice forms first on the leaf surface and then penetrates through stomata, hydathodes, wounds, or cracks in the cuticle which normally acts as a barrier to ice growth, and then initiates freezing in the apoplast (Pearce, 2001; Wisniewski et al., 2014). Severe mite, insect, or pathogen infestations may create lesions through which ice can grow into the leaf.

Freezing tolerance remains low during the growing season but increases in autumn during a process known as cold acclimation and reaches a peak in midwinter. Consequently, damage to buds, canes, cordons, and trunks in winter (Fig. 7.13) occurs at much lower temperatures than during the growing season. Grapevines can survive freezing temperatures either by avoiding them, for instance via late budbreak and early shoot maturation, or by tolerating them, for instance by deep supercooling in midwinter. They also have "ice sites" or "ice sinks" to accommodate ice formation in a process termed extraorgan freezing. For example, the bud scales with their down (see Section 1.3) can cope with ice crystals that grow from water moving out of the buds' supercooling meristem tissues. Ice formation in the bud scales progressively dehydrates the interior tissues as the temperature declines (Pearce, 2001). In addition, as they enter dormancy, the buds are disconnected from the vascular tissues of the subtending bud axis on the cane (Ashworth, 1982; Xie et al., 2018; see also Fig. 2.8). Tightly packed parenchyma cells with rather impermeable cell walls form an ice diffusion barrier and enable the hydraulically isolated primordia in the buds to supercool (Fennell, 2004; Jones et al., 2000). This structural ice barrier protects the dormant buds from the ice spreading through the cane xylem, which can occur at rates of several centimeters per second (Wisniewski et al., 2014).

Although the most conspicuous sign of cold acclimation in grapevines is the abscission of leaves at the end of the growing season, acclimation is a gradual process. Leaf abscission is preceded by shoot periderm formation, acquisition of bud and cambium dormancy, and resorption of leaf nutrients to the permanent organs of the vine. The "wave" of brown periderm moving up the shoots in late summer is associated with a massive increase in storage carbohydrates in the buds and canes (Eifert et al., 1961; Koussa et al., 1998; Winkler and Williams, 1945). At the same time their water content declines, and browning correlates closely with the ability to cold acclimate when the temperature decreases (Fennell and Hoover, 1991; Howell and Shaulis, 1980; Salzman et al., 1996; Wolpert and Howell, 1985, 1986). Because the green ends of the canes are unable to acclimate, they usually die back in winter.

This "self-pruning" effect can be exploited in minimally pruned grapevines with high bud and shoot numbers (Clingeleffer, 1984; Clingeleffer and Krake, 1992; Downton and Grant, 1992; Keller and Mills, 2007).

Periderm formation, dormancy, and cold acclimation are initiated by decreasing day length in late summer, followed by cool nonfreezing temperatures around 0–5 °C (Alleweldt, 1957; Schnabel and Wample, 1987; Wake and Fennell, 2000; see also Section 2.2). The lengthening twilight periods at dawn and dusk may at least partly be responsible for the day-length effect that is detected by the leaves; the low ratio of red to far-red light in twilight may exert its influence via the vine's phytochrome system (Franklin and Whitelam, 2007; see also Section 5.2). Shorter days alone, while inducing dormancy, are insufficient for grapevines to cold acclimate fully (Fennell and Hoover, 1991; Salzman et al., 1996). Dormancy is, however, a prerequisite for the subsequent acquisition of cold hardiness (van der Schoot and Rinne, 2011): green buds cannot cold acclimate. Even brown buds need exposure to low temperatures in order to cold acclimate (Cragin et al., 2017; Ferguson et al., 2014; Salzman et al., 1996).

Chilling temperatures, or more accurately rapid rates of temperature decline, trigger a massive temporary redistribution of calcium (Ca^{2+}) from the apoplast to the symplast, and this Ca signal may activate the cold acclimation process by which plants acquire freezing tolerance (Browse and Xin, 2001; Knight and Knight, 2012; Plieth, 2005; Thomashow, 1999, 2001). If the temperature drops rapidly, the Ca signal is strong enough to induce cold acclimation even at high temperatures. The Ca signal is locally generated in the meristematic tissues and cannot be imported from other plant parts. Although it is not known whether foliar Ca applications prior to the hydraulic isolation of the buds can improve cold hardiness, high soil Ca availability, which is typical of many high-pH soils, may indeed be associated with better freezing tolerance (Percival et al., 1999).

The Ca signal is complemented by the action of plant hormones. Low temperatures favor the conversion of bioactive forms of gibberellin to inactive forms. Disappearance of active gibberellins in turn permits accumulation of growth-suppressing DELLA proteins (see Section 2.2), which increases freezing tolerance by a yet unknown mechanism (Achard et al., 2008; Wingler, 2015). Another hormone plays a key role in cold acclimation. In addition to increasing temporarily in response to low temperature, ABA reduces shoot growth within days and promotes periderm formation, bud dormancy, and freezing tolerance even at temperatures that are normally too high to induce cold acclimation (Guy, 1990; Thomashow, 1999; Zhang et al., 2011b; Zhu, 2002). This rapid response can be exploited by spraying ABA to the canopy during grape ripening or after harvest during a warm autumn (Bowen et al., 2016; Zhang and Dami, 2012). Moreover, mild to moderate water deficit, which triggers ABA production, is associated with reduced shoot growth and earlier periderm formation (Williams and Matthews, 1990). Conversely, the buds formed by overly vigorous shoots tend to be less cold hardy than those of less vigorous shoots (Todaro and Dami, 2017). Any factor that delays the cessation of shoot growth tends to delay cold acclimation as well. Examples include late planting of young vines during vineyard establishment or late field grafting. Excessive soil moisture or nitrogen availability that promote vigor, as well as overcropping, late summer pruning, defoliation, or damage caused by mechanical harvesters, also may interfere with accumulation of storage reserves and slow cold acclimation (Currle et al., 1983; Zufferey et al., 2012). Nonetheless, application of as much as 224 kg N ha^{-1} rarely compromised bud hardiness of Riesling in a study conducted in eastern Washington, an area known for its sporadically cold winters (Wample et al., 1993). Moreover, although infection with leafroll-associated viruses curtails phloem transport, it appears to have little effect on bud and cane cold acclimation (Halldorson and Keller, 2018; see also Section 8.2).

Once activated, cold acclimation accelerates following brief episodes near 0 °C, and cold hardiness reaches a maximum after a grapevine has experienced temperatures around −5 °C (Cragin et al., 2017; Fennell, 2004). Maximum freezing tolerance is thus acquired in sequential phases that are initiated by short days, chilling temperatures and, finally, freezing temperatures (van der Schoot and Rinne, 2011). The rate of cold acclimation is inversely correlated with temperature so that vines harden off faster at cooler sites or in years with cooler autumn temperatures (Ferguson et al., 2011, 2014; Keller et al., 2008; Londo and Kovaleski, 2017; Stergios and Howell, 1977). Therefore, after gradual hardening through autumn, cold acclimation accelerates in late autumn to approach maximum freezing tolerance in time for the cold winter temperatures. This acceleration usually occurs after bud dormancy has been liften by the fulfillment of the buds' chilling requirement.

Even though low temperatures lead to cold acclimation, they will not maintain vines in a very cold hardy condition. In fact, grapevines gradually lose hardiness when they are exposed to chilling temperatures for long periods. The process of cold acclimation is cumulative and hence dependent on the temperature history; it can be stopped, reversed, and restarted, depending on temperature fluctuations (Cragin et al., 2017; Ferguson et al., 2014; Gonzalez Antivilo et al., 2018; Kalberer et al., 2006; Londo and Kovaleski, 2017). Warm episodes during the acclimation period induce deacclimation, and the greater the loss of cold hardiness, the longer it takes for the vines to reacclimate. In warmer winters, moreover, grapevines convert less stored starch into sugars than in colder winters and thus may not reach their maximum level of cold hardiness; they remain less freezing tolerant and go through budbreak earlier (Gonzalez Antivilo et al., 2018). This may spell trouble when an unseasonable freeze event follows a mild autumn or generally mild winter and has implications with respect to the consequences of global climate change (Keller and Mills, 2007; Londo and Kovaleski, 2017). Future temperature increases are expected to be higher at night, at higher latitudes, and in the winter, thus continuing the trend that has already occurred since the middle of the twentieth century (IPCC, Intergovernmental Panel on Climate Change, 2013; Jones, 2005). Moreover, as the average temperature creeps upward, so will the extremes on either end of the temperature distribution curve. This statistical trend might be offset, however, because the warming trend in the Arctic slows the polar jet stream, which increases the frequency and magnitude of weather extremes in the mid-latitudes (Semenov, 2012; Screen and Simmonds, 2013). Therefore, the overall effect of climate change on cold damage in vineyards remains unknown.

Osmotic adjustment, or accumulation of so-called cryoprotectants such as sugars, is essential for proper cold acclimation because of their effects on freezing-point depression and membrane stabilization (Xin and Browse, 2000). Cryoprotectants provide cold tolerance by reducing the extent of cell dehydration and inhibiting the nucleation and growth of ice crystals. Therefore, intracellular freezing occurs at much lower temperatures in acclimated tissues than in nonacclimated tissues. Carbohydrates produced in the leaves are transported in the phloem as sucrose and accumulated in the woody organs as starch (Fig. 7.14; see also Section 5.1). Starch accumulation starts soon after bloom time but peaks only during grape ripening and, sometimes, after harvest, thereby replenishing the storage reserves that were remobilized during budbreak and initial canopy establishment (Eifert et al., 1961; Holzapfel et al., 2010; Zapata et al., 2004; Zufferey et al., 2012). The middle portion of the shoots appears to be refilled first, followed by their basal and apical portions, the cordon and trunk, and finally the roots (Winkler and Williams, 1945). Whereas the developing grape clusters dominate the hierarchy of photosynthate distribution, partitioning to the perennial organs intensifies soon after veraison, when the grape seeds are mature and ready to germinate. Thus, toward the end of the growing season, the permanent organs

348 Chapter 7 Environmental constraints and stress physiology

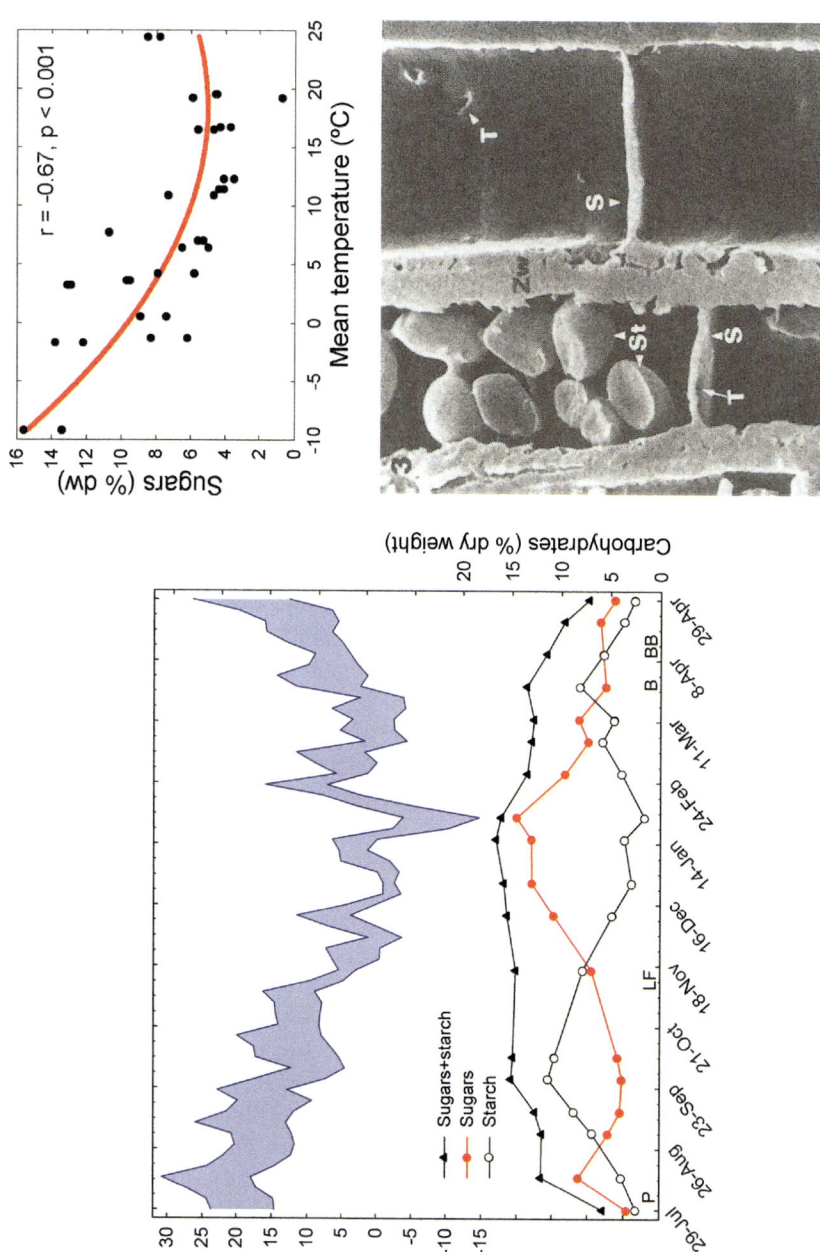

FIG. 7.14

Dynamics of grapevine cane carbohydrates with changing winter temperature (left); P, start of periderm formation; LF, leaf fall; B, start of bleeding; BB, budbreak. Association between cane sugar concentration and the mean temperature over the preceding 5 days (top right), and starch grains (St) deposited in the cane xylem (bottom right).

Left: modified with permission from Eifert, J., Pánczél, M., Eifert, A., 1961. Änderung des Stärke- und Zuckergehaltes der Rebe während der Ruheperiode. Vitis 2, 257–265; top right: calculated from data in Eifert, J., Pánczél, M., Eifert, A., 1961. Änderung des Stärke- und Zuckergehaltes der Rebe während der Ruheperiode. Vitis 2, 257–265; bottom right: reproduced with permission from Plank, S., Wolkinger, F., 1976. Holz von Vitis vinifera im Raster-Elektronenmikroskop. Vitis 15, 153–159.

become the dominant sink of the grapevine (Candolfi-Vasconcelos et al., 1994; Rossouw et al., 2017). This period is concomitant with the final stages of fruit ripening and may sometimes interfere with winemakers' quest to maximize fruit quality, especially when inclement weather limits the amount of sugar available for distribution. The change in relative sink priorities is also the reason why harvest date is usually not a significant factor in reserve carbohydrate accumulation and cold acclimation (Hamman et al., 1996; Keller et al., 2014; Koussa et al., 1998; Wample and Bary, 1992). This permits the recurring and often extreme delay in harvest of grapes destined for ice wine production near the latitudinal and altitudinal boundaries of grape growing. Delayed harvest may compromise cold acclimation only in situations of overcropping, especially under cool, cloudy conditions, when storage reserves are diverted to support fruit ripening.

Starch does not increase cold hardiness because it is osmotically inert. Acclimation requires conversion of the starch stored in phloem and xylem parenchyma cells back to osmotically active sugars—mainly sucrose, but also glucose, fructose, some galactose, and oligosaccharides such as raffinose and stachyose (Currle et al., 1983; Fennell, 2004; Guy, 1990; Hamman et al., 1996; Jones et al., 1999; Todaro and Dami, 2017). The cryoprotectant function of sucrose and other sugars may arise from their ability to bind to polar residues of membranes and replace water around these polar groups, thus stabilizing the membranes and embedded proteins and maintaining their integrity during freeze-induced dehydration (Crowe et al., 1988; Strauss and Hauser, 1986). Small amounts of raffinose may assist this process by reducing the propensity of sucrose to crystallize. The starch-to-sugar conversion seems to be initiated by short days in mid-September in the northern hemisphere (and thus *before* grapes are usually harvested in cooler climates), then accelerates as temperatures decrease below about 5°C in late autumn, and continues until midwinter (Eifert et al., 1961; Hamman et al., 1996; Koussa et al., 1998; Müller-Thurgau, 1882; Wample and Bary, 1992; Winkler and Williams, 1945; see also Fig. 7.14). A similar interconversion of starch to sugars occurs simultaneously in the buds (Grant and Dami, 2015; Hamman et al., 1996; Jones et al., 1999; Koussa et al., 1998).

Accumulation of soluble sugars also accelerates because low temperatures reduce respiration rates (see Section 5.2). Nevertheless, the concurrent decline of photosynthesis due to the cooling temperatures and shortening days requires sugars produced from stored starch to generate energy in addition to providing the substrates for the formation of cryoprotectants (Druart et al., 2007). Moreover, starch is needed to produce fatty acids because during the transition to dormancy the cambium cell vacuole splits into several smaller vacuoles, which requires synthesis of new membrane lipids (Druart et al., 2007). After reaching its maximum near the time of leaf fall, the total amount of carbohydrates in the buds and woody organs decreases only slightly and primarily due to maintenance respiration during dormancy, while the sugar $\div$ starch ratio continues to rise as temperatures decline (Eifert et al., 1961; Koussa et al., 1998; Wample et al., 1993; Winkler and Williams, 1945).

Once foliar sugar export ceases, the pores of the sieve plates separating the phloem's sieve tubes in the vines' stem portion are sealed by callose (see Section 1.3), the cork cells are waterproofed by suberization, and the perennial organs are drained of any water that is not essential to maintain cell function. Tissue dehydration reduces the water content to 42–45% and is necessary during acclimation because it is the rupture of cell membranes by ice crystals that causes freezing injury. There is a strong association between cold acclimation and declining water content in the buds and woody parts of a grapevine (Wolpert and Howell, 1985, 1986). Callose is also deposited in the plasmodesmata that connect the cells of apical meristems while they prepare for dormancy inside the buds (Rinne et al., 2016; van der Schoot and Rinne, 2011; van der Schoot et al., 2014). This process is accompanied by the impregnation of cell walls and effectively seals the meristems off from the supply of water, nutrients, and hormones from the rest of the plant until warmer temperatures induce callose degradation.

In addition to sugars and tissue water, proteins and amino acids are involved in cold acclimation too. Glycoproteins and some other soluble proteins are secreted to the apoplast, particularly in the bark during autumn; the nitrogen for their biosynthesis is provided from remobilization in the senescing leaves (Griffith and Yaish, 2004; Guy, 1990; Pearce, 2001). As well as reinforcing the cell walls and making them more elastic, the high viscosity of these cryoprotective or "antifreeze" proteins prevents the growth of ice crystals and depresses the freezing point—just as it does in polar fishes that avoid freezing in subzero saltwater temperatures (Zachariassen and Kristiansen, 2000). Antifreeze proteins, which are particularly strong inhibitors of ice recrystallization, are very similar to the so-called pathogenesis-related (PR) proteins that grapevines secrete into the apoplast upon fungal attack during the growing season (see Section 8.1). The antifreeze proteins have probably evolved from PR proteins, and they enhance the resistance of plant tissues to both cold-tolerant pathogens and freezing temperatures (Seo et al., 2010; Zachariassen and Kristiansen, 2000). Nevertheless, the PR proteins accumulated by nonacclimated tissues lack antifreeze activity so that fungal infections do not enhance cold tolerance (Griffith and Yaish, 2004). Dehydrins are another group of glycoproteins that accumulate in acclimating buds, probably in response to the osmotic stress caused by the declining water content (Salzman et al., 1996). The highly hydrophilic dehydrins can bind and thus retain water (Tompa et al., 2006). They are involved in desiccation tolerance and may protect cell membranes when water freezes in the apoplast. In addition, by acting as "metal sponges" they may support sugars in preventing protein damage that occurs when the dehydration caused by extracellular freezing increases the ion concentration in the cell (Tompa et al., 2006).

Other organic solutes, such as organic acids (particularly malate) and free amino acids (particularly proline, arginine, alanine, asparagine, and serine), may be accumulated in parallel with sugars during cold acclimation (Currle et al., 1983). The cryoprotective properties of proline and other amino acids might be linked to their function as protein and membrane stabilizers during drought stress (Thomashow, 2001; see also Section 7.2). Like sugars, such amino acids may protect proteins from being denatured during freeze-induced dehydration, but unlike sugars they become ineffective at very low water content (Crowe et al., 1988). The presence of divalent cations from the transition metal group, such as copper, cobalt, nickel, or zinc, may further enhance the cryoprotective effect of sugars and other organic solutes. Moreover, deep-supercooling xylem parenchyma cells also seem to accumulate flavonol glycosides that help them suppress ice nucleation (Kasuga et al., 2008). The extent of cold tolerance is further modified by adapting the size of individual cells, thickening of cell walls, and changes in membrane properties that reduce the membrane's sensitivity to mechanical stress due to contraction and expansion (Steponkus, 1984). Finally, activation of the antioxidant system during the cold acclimation phase enhances the protection of cell membranes from damage due to oxidative stress associated with low temperatures (Druart et al., 2007). In combination these processes can either delay the onset of freezing or diminish its adverse consequences.

During the winter, the aboveground organs of grapevines continually adjust their degree of cold hardiness in response to fluctuating temperatures (Düring, 1997; Ferguson et al., 2011, 2014; Mills et al., 2006; Fig. 7.15). After a minimum of approximately $-10\,°C$ has passed, rising temperatures induce a reversal of the starch to sugar conversion and favor starch accumulation (Currle et al., 1983; Eifert et al., 1961; Koussa et al., 1998; Winkler and Williams, 1945). This change goes along with a concomitant loss of cold hardiness as the vine deacclimates and prepares for budbreak in

7.4 Temperature: Too hot or too cold

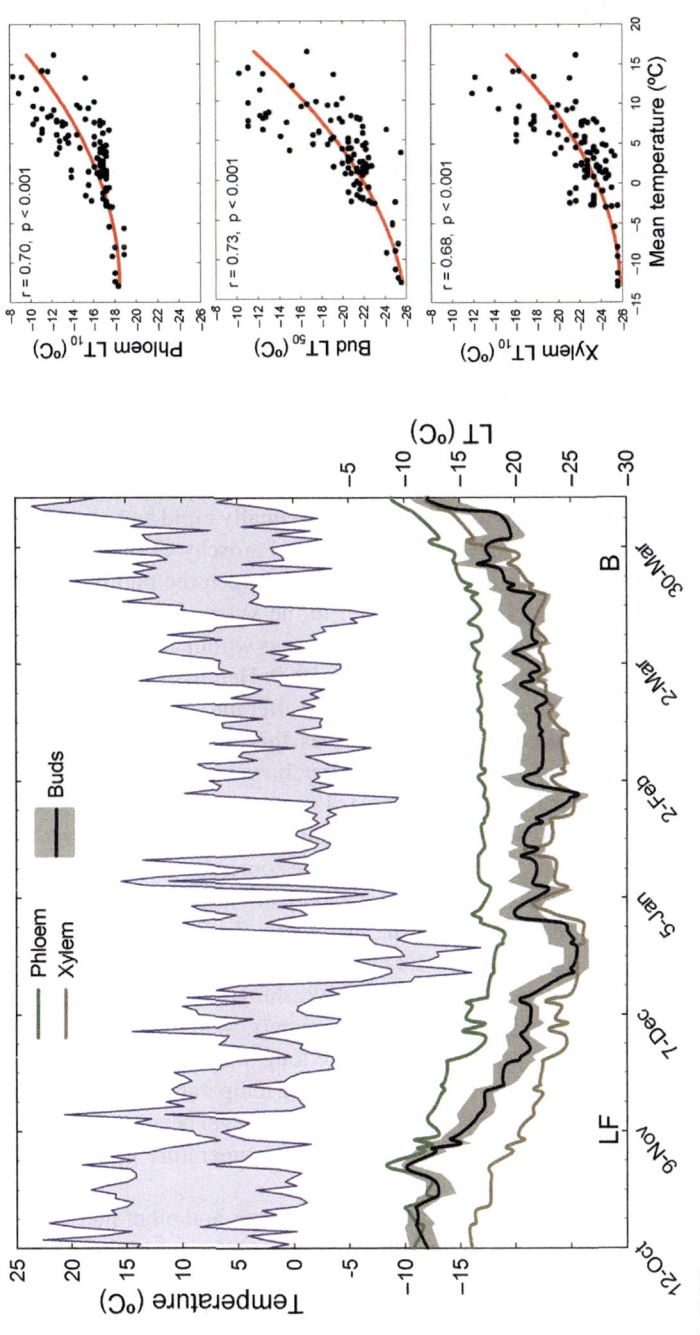

FIG. 7.15

Dynamics of Merlot cold hardiness with changing winter temperature (left): LF, leaf fall; B, start of bleeding; LT, lethal temperature; lines indicate LT for 10% of the phloem and xylem and 50% of the buds with gray area indicating the 10–90% window. Association among bud, cane phloem, and cane xylem hardiness and the mean daily temperature (right). Mills and Keller, unpublished data.

spring. Once bud dormancy has been lifted, temperature is the main driver of deacclimation; the temperature threshold that induces deacclimation is about 5 °C but varies by species and cultivar (Ferguson et al., 2011, 2014; Pagter and Arora, 2013; Stergios and Howell, 1977; Wolf and Cook, 1992).

Although the sugar ÷ starch ratio also reflects temperature fluctuations during winter, with sugars accumulating during cold episodes and starch accumulating during warm episodes, the conversion of starch back to sugar following colder temperatures gradually slows toward the end of winter (Currle et al., 1983; Grant and Dami, 2015; Wample and Bary, 1992). Moreover, the concentration of antifreeze proteins declines rapidly upon exposure to warm temperatures. Cold hardiness can be partially lost within a day during warm spells in winter (Ferguson et al., 2011, 2014; Guy, 1990; Mills et al., 2006; Pagter and Arora, 2013; Wolf and Cook, 1992). The west side of a trunk that is exposed to the afternoon sun is particularly vulnerable, especially when it warms up due to reflection of solar radiation from snow covers (Peña Quiñones et al., 2019). Although hardiness is restored when cool temperatures return, the deacclimation behavior predisposes grapevines for cold injury when the temperature drops suddenly after a warm episode. Moreover, the sensitivity to temperature fluctuations increases and deacclimation occurs more readily as spring approaches, making vines more vulnerable to sudden cold spells during late winter (Wolf and Cook, 1992). The rapid bud rehydration that occurs during budswell leading up to budbreak is associated with an equally rapid and irreversible loss of cold hardiness (Johnson and Howell, 1981; Meiering et al., 1980; Paroschy et al., 1980). It is during this budswell phase that the vascular connection is reestablished between the bud primordia and the subtending canes, thus breaking down the buds' ice barrier to the xylem (Ashworth, 1982; Wisniewski et al., 2014; see also Fig. 2.8). Swelling buds lose cold hardiness within a few days as their water content increases from <50% to almost 80% (Fuller and Telli, 1999; Hamman et al., 1990; Lavee and May 1997; Pouget, 1963). Because late winter pruning, close to or after the time of normal budbreak, delays budbreak and irreversible deacclimation in spur-pruned vines, this practice can be used in climates with a high risk of spring frosts (Friend et al., 2011). If done earlier, however, winter pruning does not impact either bud cold hardiness or budbreak timing (Wample, 1994).

The location of a vineyard site is an important determinant of the risk of freeze and frost injury. Since cold air is heavier than warm air, vineyard sites that are somewhat elevated above a valley floor and permit free air drainage down a slope are at lower risk than vineyards on the valley floor and/or where air drainage is obstructed (Widrlechner et al., 2012; Fig. 7.16). In regions with cold winters, vineyards are often located near lakes that moderate nighttime temperatures (Bowen et al., 2005). Many growers in vulnerable areas, such as New Zealand, Washington, or Ontario, use wind machines to increase the temperature around the vines during cold events. Wind machines are most effective during clear, calm nights, when heat is lost to the sky by long-wave radiation (Atwell et al., 1999; Battany, 2012; Perry, 1998; Schultz, 1961). Due to the resulting temperature inversion, the temperature 20 m above the ground can be 1–10 °C warmer than that at ground level because cold air settles near the ground after sunset. At wind speeds $>1.5\,\mathrm{m\,s^{-1}}$, however, the temperature gradient collapses so that artificial mixing of air is useless.

Although they are an important storage organ for carbohydrates and other nutrients, the roots may be less effective than the trunk, cordons, and canes in terms of the interconversion of starch and sugars during the winter season (Holzapfel et al., 2010; Winkler and Williams, 1945; Zufferey et al., 2012). In contrast to many tree species (Galvez et al., 2013), the bulk of storage carbohydrates in *V. vinifera* roots are thought to remain in the form of starch throughout the winter. Moreover, roots do not seal their phloem with callose, nor do they experience dormancy in the way the aboveground organs do

FIG. 7.16

Small gradients in temperature down a slope can mean the difference between frost damage and no damage. Note the grow tubes around replanted vines at the lower end of the slope and the wind machine on top of the hill.

Photo by M. Keller.

(Esau, 1948; Pouget, 1963; see also Section 2.2). Consequently, the roots may be unable to cold acclimate and are much more vulnerable to cold temperatures than is the stem portion of grapevines. Roots of *V. vinifera* and *V. labruscana* cultivars may be killed at −6°C, whereas *V. amurensis* roots, which have smaller cells, may withstand temperatures as low as −15°C, with many rootstocks derived from American *Vitis* species experiencing root death in the −8°C to −10°C range (Ahmedullah and Kawakami, 1986; Gale and Moyer, 2017; Guo et al., 1987). Cold injury to roots, even if not lethal, may compromise their ability to generate root pressure for xylem and bud rehydration (see Section 2.2), which may lead to uneven budbreak or canopy collapse after budbreak (Pagter and Arora, 2013). Because water has a greater heat capacity than air, temperatures fluctuate more in dry than in wet soils, and insufficient soil moisture in winter can predispose grapevine roots to cold injury. This is why growers in regions with dry and cold winters usually plant vines as deep as possible, with the trunk extending up to 50 cm below ground, and irrigate their vineyards at the end of the growing season to rewater the rootzone to field capacity. Especially dry winters, such as are common in Argentina's Mendoza region, may require some irrigation throughout the winter.

Survival of grapevines following cold injury depends on the type of tissue that is damaged, the extent of damage, and the ability to recover enough to at least partially resume proper function of the injured organs. During the cold acclimation and deacclimation phases, grapevine buds are usually slightly hardier than canes or trunks, whereas in midwinter bud damage occurs at somewhat higher temperatures than damage to the woody tissues (Mills et al., 2006). However, during calm nights the temperature near the soil surface is usually lower than at the canopy level owing to a vertical temperature gradient in the range 0.2–$2\,°C\,m^{-1}$ (Gonzalez Antivilo et al., 2017). This can sometimes result in trunk injury when canes and buds may just escape the lethal temperature. Within the compound winter bud, the primary bud is generally less hardy than the secondary bud, which in turn is less hardy than the tertiary bud (see Fig. 7.13). This is probably a reflection of the degree of bud differentiation; increasing differentiation increases the susceptibility to ice nucleation. Nonetheless, even primary buds in close proximity are not all killed at exactly the same temperature: The difference between 10% bud damage and 90% damage can be up to 7 °C during autumn acclimation and spring deacclimation, but this range is closer to 3 °C in midwinter when the buds are fully acclimated (Mills et al., 2006).

Within woody organs, the phloem is usually more vulnerable to cold injury than is the xylem (Cragin et al., 2017; Mills et al., 2006; see also Figs. 7.13 and 7.15). Even though the dilute water inside the xylem conduits freezes at much higher temperatures (above $-10\,°C$) than in the live phloem cells, the phloem cannot accommodate apoplastic ice formation in the same way that ice forms and expands rather harmlessly in the xylem vessels (Meiering et al., 1980). Phloem injury can interrupt the transport of assimilates from the permanent organs to the developing buds in spring. However, even when the phloem and vascular cambium are destroyed, starch can also be remobilized from the xylem parenchyma cells and released as sugars into the vessels along with water (Améglio et al., 2004; Sakr et al., 2003; Winkler and Williams, 1945). This generates root pressure that drives out air bubbles and enables the buds to break (see Sections 2.2 and 3.3). Moreover, although dead tissues cannot be repaired, they can be replaced by new tissues (Pratt and Pool, 1981). Phloem can be readily restored if the cambium survives. Death of the cambium is more serious, although the extent of damage depends on the position of the dead cambium. Dead cambium surrounding the trunk effectively girdles it (Fig. 7.17) and can be lethal to the vine because any surviving buds will break and begin to grow in spring. By blocking the movement of sucrose and auxin, girdling disrupts the communication between buds/leaves and roots.

If the cambium and phloem cannot be restored rapidly in spring, the shoots and inflorescences that emerge from any surviving buds may receive insufficient nutrient reserves from the trunk and the roots. Because inflorescences are poor competitors for access to the remaining reserves, they are often

FIG. 7.17

Cold injury to the phloem and cambium of a young Gewürztraminer trunk above the soil surface, with callus forming above and around the site of injury, and suckers growing from below the soil surface (left). Riesling vine with similar cold injury, with aborted clusters and leaf symptoms of sugar accumulation, chlorophyll degradation, light avoidance, and nutrient deficiency (right); note that the vines in the background experienced no cold injury.

Photos by M. Keller.

aborted to permit continued shoot growth, as long as the xylem is able to supply soil water. However, the leaves may soon begin to accumulate carbohydrates and suffer from feedback inhibition of photosynthesis, because sugar transport to the roots is inhibited—although some sugar may be used to produce callus around the injured trunk. The roots in turn will starve from photosynthate deficiency, which will decrease nutrient uptake, so that the leaves may begin to show secondary symptoms of nutrient deficiency (Fig. 7.17). In severe cases, the vine can wilt and collapse.

By producing and releasing auxin, the swelling and breaking buds in spring promote reactivation of the phloem and, subsequently, renewed cell division in the cambium for production of new vascular tissues (see Section 1.3). Therefore, at least some buds must be alive for recovery to occur, and this includes latent buds on the cordon or trunk that may break even when all the dormant buds on the canes have been killed (Keller and Mills, 2007; Pratt and Pool, 1981).

Damage to the xylem is not by itself detrimental to the vine, but it normally occurs only after the phloem and cambium have already been killed so that xylem injury is often a sign of disaster. Remarkably, however, as long as a sufficient number of immature, relatively undifferentiated xylem parenchyma cells survive, auxin may induce these cells to dedifferentiate and resume cell division to directly form new phloem and cambium cells (Pang et al., 2008; Zhang et al., 2011a). At the same time, the dedifferentiated xylem cells may also form callus on the wound surface, but the callus does not participate in the regeneration of the vascular tissues. Thus, restoration may occur via the division of surviving xylem cells, provided some bud-derived auxin can diffuse down to the injured xylem. Parenchyma cells may differentiate into vascular cells especially near surviving buds acting as auxin sources and above sections of interrupted auxin flow. However, these new vascular tissues may remain discontinuous because they cannot become properly aligned in the absence of polar auxin flow. Because the production of new cells also depends on the supply of sucrose and other nutrients stored in the parenchyma cells, a high reserve status may promote recovery from cold injury.

The maximal extent to which grapevines become freeze tolerant varies with species and cultivar (Ferguson et al., 2014; Grant and Dami, 2015; Londo and Kovaleski, 2017; Pierquet and Stushnoff, 1980; Williams et al., 1994). Hardier species or cultivars may survive with more of their extracellular water frozen than can less hardy ones. The eastern Asian species *V. amurensis* and the northeastern North American species *V. riparia* can survive midwinter temperatures as low as $-35\,°C$, whereas the southeastern North American species *M. rotundifolia* suffers severe damage at $-12\,°C$ (buds even at $-5\,°C$). Most of the other North American species are not as hardy as *V. riparia*, but they are hardier than *V. vinifera*. However, there is no conclusive evidence to suggest that rootstocks derived from American species systematically alter the hardiness of their *V. vinifera* grafting partners (Miller et al., 1988; Wolf and Pool, 1988). Most *V. vinifera* cultivars are killed at midwinter temperatures below approximately $-25\,°C$, but the vulnerability of different cultivars varies by several degrees (Davenport et al., 2008a; Ferguson et al., 2014; Mills et al., 2006). Moreover, the relative order of species and cultivar vulnerability can be modified by vineyard location, cultural practices, and timing of a cold event. For example, Concord (*V. labruscana*) is significantly winter-hardier than Riesling (*V. vinifera*), and Chardonnay (*V. vinifera*) is almost as hardy as Riesling, but the early budbreak and, consequently, early deacclimation of Concord and Chardonnay induce a substantial loss of hardiness in late winter. Species and cultivars with a low chilling requirement and hence short endodormancy period and/or a low temperature threshold for budbreak (see Section 2.2) tend to deacclimate earlier and more rapidly (by up to $3\,°C$ per day) and are more cold sensitive toward the end of winter than species and cultivars with a high chilling requirement and/or higher temperature threshold (Ferguson et al., 2014;

Londo and Johnson, 2014; Kovaleski et al., 2018). Species or cultivars that originate from higher latitudes or from regions with continental climates tend to be hardier in midwinter than are their lower-latitude or coastal counterparts (Ferguson et al., 2014; Kovács et al., 2003; Londo and Kovaleski, 2017; Wolf and Cook, 1992). When such cultivars are grown side by side, however, the former are more vulnerable to spring frost damage, because their ready response to higher temperatures leads to earlier budbreak.

CHAPTER 8

Living with other organisms

Chapter outline

8.1 Biotic stress and evolutionary arms races 357
    8.1.1 Competitors, herbivores, and pathogens 358
    8.1.2 Defense strategies 361
8.2 Pathogens: Defense and damage 366
    8.2.1 Bunch rot 367
    8.2.2 Powdery mildew 371
    8.2.3 Downy mildew 374
    8.2.4 Bacteria 376
    8.2.5 Viruses 378

8.1 Biotic stress and evolutionary arms races

Grapevines share their living quarters, both above- and belowground, with a myriad of other organisms, mainly arthropods (spiders, mites, insects) and microorganisms (fungi, oomycetes, bacteria, viruses), in addition to some nematodes, birds and mammals, as well as other plants. Some of these organisms are considered beneficial because they provide useful "services" to the vines for which growers would otherwise have to pay. Earthworms, for instance, enhance soil fertility by converting organic matter into humus, grinding up and mixing mineral particles, and creating biopores through which roots can grow easily and which aid in soil drainage and aeration. A host of microorganisms are involved in the mineralization of soil organic matter to inorganic nutrients that can then be taken up by the roots. Other microorganisms actively participate in this nutrient uptake in a symbiosis with plant roots. Most of the remaining organisms normally leave grapevines alone, do little damage, or are unable to overcome the vines' defense strategies—the latter situation is called an incompatible interaction. For example, grape berries are typically heavily colonized by various yeast species, other fungi, and bacteria, especially in the vicinity of stomata and small cracks in the cuticle that allow exudation of sap from the mesocarp (Barata et al., 2012b; Belin, 1972; Bokulich et al., 2014). Yet this is very rarely a concern in viticulture, unless berries are damaged or harvested fruit is left standing too long before processing so that uncontrolled fermentation sets in. Similarly, among the many fungal species in the order Erysiphales that cause powdery mildew in plants, only one is able to infect *Vitis* species (see Section 8.2)—and that one is unable to infect any other plant species. This is one reason that roses are planted around some vineyards: Because the two plant species are susceptible to different species of fungi that thrive under similar weather conditions, the roses may be used to indicate conditions that favor powdery mildew on grapevines. Roses offer additional benefits, too: They provide overwintering habitat for wasps that serve as biocontrol agents for leafhoppers in spring and, of course, they are aesthetically pleasing. Some organisms, however, compete with the vines for resources; if these organisms are other plants, we often call them weeds. Others make a living feeding on various grapevine structures; we usually call them herbivores if they are animals, and pathogens if they are microorganisms—in this case we speak of a compatible interaction. Grapevines perceive such organisms as biotic stress, and growers call them pests.

8.1.1 Competitors, herbivores, and pathogens

Plants sharing a vineyard with grapevines, whether they are called weeds or resident vegetation or are planted deliberately as cover crops or green floor covers, use water and tend to reduce nutrient availability for the vines (Celette et al., 2009; Celette and Gary, 2013; Guerra and Steenwerth, 2012; Morlat and Jacquet, 2003; Reeve et al., 2016; Tesic et al., 2007). The influence on grapevines of a permanent floor cover in cool and humid climates can be remarkably similar to that of deficit irrigation in warm and dry climates (see Section 7.2): Vines have diminished vigor, smaller berry size, improved fruit composition, and reduced bunch rot (Keller et al., 2001a, b; Valdés-Gómez et al., 2008). Vineyard floor vegetation is desirable for many reasons but may compete with the vines for resources if not managed carefully, especially in warm and dry climates or in dry years (Celette et al., 2009; Pool et al., 1990; Tesic et al., 2007). The impact on yield formation arising from such potential competition may be averted by judicious irrigation management that takes into account both the vines and the additional vegetation (Ingels et al., 2005; Steenwerth et al., 2013, 2016). In more humid climates, the competition may be mostly for access to soil nitrogen rather than water (Reeve et al., 2016).

Grapevine roots grow preferentially beneath emitters in drip-irrigated vineyards in dry climates (Pisciotta et al., 2018; Stevens and Douglas, 1994). Thus, even if interrow cover crops are mowed or incorporated into the soil by tilling, the mineral nutrients released during their microbial decomposition may be mostly available to cover crop regrowth rather than to the grapevines. Even in humid climates, competition from cover crops may increase vineyard fertilizer requirements if vine productivity is to be maintained (Keller et al., 2001a, b; Reeve et al., 2016). Nitrogen, at least, can also be supplied using leguminous cover crops that employ symbiotic *Rhizobium* bacteria to fix atmospheric nitrogen ($N_2 \rightarrow NH_4^+$), which mitigates the need for organic or synthetic fertilizers (Guerra and Steenwerth, 2012; Jackson et al., 2008; Patrick King and Berry, 2005). Legumes can moreover acidify their rhizosphere, which may improve P availability (Shen et al., 2011). Consequently, cover crop management should aim to synchronize nutrient availability in the soil to the seasonal demand of the vines (Keller et al., 2005; Perret et al., 1993).

Birds can be important pests of wine grapes, because grapes have evolved to have their seeds dispersed by birds (see Section 2.3). This often clashes with the need in production viticulture to mature the fruit well beyond the stage at which seeds are ready to germinate. The presence in green, immature berries of malic acid, which is an effective bird deterrent, and its degradation during ripening are probably adaptive mechanisms evolved by grapes to ensure protection of the developing seeds and their dispersal once they have reached maturity. The astringent tannins and the toxic, needle-like calcium oxalate crystals present in unripe grape berries could be another defense against fruit-eating birds as well as mammals.

A variety of insects and mites, which along with the crabs are collectively grouped as arthropods or the phylum Arthropoda (Greek *arthron* = joint, *pous* = foot), feed on grapevine organs. The most devastating insect pest of grapevines is the root-feeding, aphid-like phylloxera, *Daktulosphaira vitifoliae* Fitch (also called *Viteus vitifolii*), which can multiply in all but very sandy soils and spreads rapidly except in very dry climates. Ironically, however, the sandy soils that halt phylloxera are very conducive to the multiplication of another group of root parasites, namely the root-knot nematodes from the genus *Meloidogyne* (Huglin and Schneider, 1998). Phylloxera (Greek *phyllon* = leaf, *xeros* = dry) is indigenous to the eastern and southwestern United States and Central America. It was introduced to Europe on infested plant material in the middle of the 19th century—probably as eggs on rooted cuttings of American grape species. The subsequent epidemic across the continent wiped out most native wild vines and nearly destroyed the European wine industry, before grafting to phylloxera-tolerant American rootstocks was introduced and accepted as the only feasible remedy.

On American *Vitis* species, phylloxera has a complex life cycle, alternating between root-feeding and leaf-feeding forms (Forneck and Huber, 2009; Galet, 1996; Granett et al., 2001; Huglin and Schneider, 1998). By locally manipulating leaf development via the production of auxin, the insect induces protective leaf galls that resemble fruit carpels (Schultz et al., 2019). The leaf-feeding form usually cannot multiply on the leaves of European *V. vinifera*, but the root-feeding form is extremely damaging. Once introduced in a new region, the insect, especially its first instar larva or nymph (a.k.a. crawler), can hitchhike on people and vineyard machinery to invade new vineyards (King and Buchanan, 1986). Following infestation, the insects and secondary infections by fungal pathogens destroy the roots of susceptible *V. vinifera* cultivars. Infested vines gradually decline, often displaying symptoms of K deficiency, and eventually die as they run out of water and mineral nutrients. The decline is rapid on heavy clay soils but can be slow on sandy soils, especially in very cool or hot climates (Granett et al., 2001). Because damage to the roots of American *Vitis* and *Muscadinia* species is very limited, these phylloxera-tolerant species or their interspecific crosses are used throughout the world as rootstocks that protect vineyards from succumbing to phylloxera (Galet, 1996; Granett et al., 2001). The mechanisms of resistance or tolerance are not well understood, but many grape species may be able to form a cork layer around areas where phylloxera feeds on the roots, which may cut off the food supply to the insects. Despite this knowledge gap, the breeding and use of rootstocks, many of which continue to prevent damage after many decades of use, is an outstanding example of a highly successful biological pest control strategy. Many of these rootstocks also have at least partial resistance to or tolerance of one or more species of roundworms called nematodes (Ferris et al., 2012; see also Section 1.2).

Diverse other insect species feed on grapevine leaves or fruit. Many of these insects are able to use the mix of volatile compounds emitted by the leaves or berries to distinguish undesirable materials from edible ones, such as ripe berries (Rodríguez et al., 2013). All arthropods have natural enemies; we consider these predators and parasites beneficial organisms. Many of these are small wasps that parasitize insect eggs, and others are bugs, lady beetles, or spiders that eat other insects or mites (Mullins et al., 1992). However, one insect, the multicolored Asian lady beetle *Harmonia axyridis* Pallas, which was originally introduced to Europe and North America as a biological control agent for aphids and scale insects, may directly influence wine and grape juice aroma. When the beetles hiding in grape clusters are disturbed or crushed during juice processing, they release a yellowish fluid from their legs in a process termed reflex bleeding. The methoxypyrazine in this fluid leads to an off-flavor termed "ladybug taint" because it is extremely potent at imparting vegetal, green pepper, and herbaceous aromas to grape juice and wine (Pickering et al., 2004, 2005; see also Section 6.2). This originally beneficial lady beetle is therefore now considered an important contaminant pest in vineyards.

In addition to insects, several mite species can be potent grapevine pests; mites are related to spiders rather than insects. For instance, grape rust mites (*Calepitrimerus vitis* Nalepa) and grape bud mites (*Colomerus vitis* Pagenstecher) are responsible for a phenomenon termed restricted spring growth (Bernard et al., 2005). Symptoms include retarded budbreak, stunted shoots and distorted leaves in spring until fruit set, poor fruit set or loss of inflorescences, intense growth of lateral shoots, and bronze discoloration, called russetting, of mature leaves later in the growing season. The damage is caused by overwintering mite populations that feed on the newly emerging shoots and are dispersed mainly by movement of shoots due to wind or machinery. Rust mites are particularly destructive when cool spring temperatures slow shoot growth, whereas rapidly growing shoots during warm conditions quickly outgrow the feeding mites. In contrast, bud mites live and feed inside the developing and dormant buds, which leads to stunted shoots growing in a zigzag pattern and missing one or all of their inflorescences or the apical meristem (Bernard et al., 2005). The bud mites migrate to newly forming buds after budbreak; they are spread mainly via infested planting material. In extreme cases, mite feeding leads to bud necrosis from death of the primordia inside the buds, which can prevent budbreak altogether. In biologically diverse vineyards, both of these destructive mite species, as well as other damaging mite species, are usually kept in check by predatory mites.

Organisms also interact in other ways. For instance, social wasp species of the genera *Vespa* and *Polistes* that feed on grapes inadvertently ingest yeast cells. Because the wasps can harbor these cells in their gut during the winter and pass them on to their larvae in spring, these wasps may maintain and promote relatively site-specific yeast communities in vineyards (Stefanini et al., 2012; Vezinhet et al., 1992). Some pathogens also rely on transport agents called vectors to carry them from one plant to another. For example, some viruses are transmitted by nematodes or insects, and specialized phloem-dwelling bacteria termed phytoplasma require leafhoppers or other phloem-sucking insects for their transmission (Bovey et al., 1980; Oliver and Fuchs, 2011). Of course, vegetative propagation is also a very effective means to disperse such pathogens, which is why they are often collectively termed graft-transmissible diseases. In their native habitats the evolution of grapevines was until recently tightly linked to the evolution of other organisms that happen to like fruits or leaves. However, such coevolution has been greatly hindered since the introduction of vegetative propagation that deprived cultivated vines of the opportunity to adapt by sexual reproduction, leaving open only the much less effective avenue of somatic mutations (see Section 2.3).

Because mutations in somatic cells during cell division or mitosis (Greek *mitos* = string, loop) occur several orders of magnitude less frequently than do mutations in germ cells during meiosis, there is very little genetic variation within most modern grape cultivars compared with their wild relatives. Consequently, the genetic background of each cultivar determines its vulnerability to novel environments or newly introduced pathogens. Moreover, a population that comprises many genetically identical plants, such as is typical of the near-monocultures that make up many modern vineyards, is vulnerable to any pathogen that discovers a key to exploiting this population. This puts cultivated grapevines at great risk from organisms—especially microorganisms—that like to eat their fruit, leaves, or roots. Microorganisms such as fungi, bacteria, and viruses are extremely abundant in the environment, are exceedingly diverse, and reproduce extremely rapidly. Not surprisingly, microbes are the most copious fruit consumers (Rodríguez et al., 2013).

Large numbers, high variability, and rapid reproduction mean rapid evolution. Because random mutations happen every so often, the laws of probability dictate that in a sufficiently large or sufficiently fast-reproducing population, even unlikely changes are bound to occur. Nonrandom natural selection does the rest, quickly weeding out unfavorable mutations, ignoring irrelevant changes, and favoring genetic variants that benefit reproduction. Consequently, diverse and unlikely mutants can evolve, and any such fungal, bacterial, or viral strain with a tiny advantage over the others will soon dominate the population. These traits enable microbes to evolve extremely rapidly, under sufficient selection pressure sometimes within mere days.

In addition, bacteria have neither true biological species nor do they engage in true sexual reproduction (Mayr, 2001). They instead readily exchange DNA with one another across apparent "species" boundaries, when two not necessarily related bacteria meet and connect with each other by growing a temporary tube between them across which genes flow from one bacterial cell to its neighbor. This so-called conjugation is analogous to copying systems data files directly from one computer to another. Vegetatively propagated grapevines, by comparison, evolve only very slowly, often remaining essentially unchanged for hundreds or even thousands of years (see Section 1.2). This puts microorganisms at a tremendous competitive advantage over cultivated vines and is a major cause of disease epidemics and of the rapid emergence of pesticide-resistant strains.

Even insects can multiply at astonishing rates; a single phylloxera egg hatching in spring may result in five billion parthenogenetically generated descendants by midsummer (Battey and Simmonds, 2005). A pathogen or other pest introduced to a new area or made virulent through a mutation to which local cultivars have no defense could have devastating consequences from which wild and diverse plant communities may be somewhat protected, because they have coevolved with the pathogens. Such coevolution essentially constitutes a genetic "arms race," in which mutations occur in and natural selection acts on both the pathogens and the hosts. Moreover, mixing genes in new combinations during sexual reproduction (see Section 2.3) presents a moving and, ideally, elusive target for would-be pathogens and other pests.

Coevolution and sexual reproduction, of course, can do nothing to improve plant resistance if a pathogen and its potential host evolved in geographic isolation from one another. Both the cultivated and wild forms of the European *V. vinifera*, when confronted with phylloxera and the mildew-causing pathogens introduced from North America in the second half of the 19th century, quickly succumbed for precisely the same reason native American peoples succumbed to European diseases after the Spaniards had inadvertently introduced the pathogens to the Americas: They had not coevolved with the pathogens and therefore had not had a chance to "invent" effective genetic resistance strategies. In the early 21st century history seems to be repeating itself in China, where phylloxera is wreaking havoc among native populations of wild and cultivated grapes (Du et al., 2009).

8.1.2 Defense strategies

When leaves are damaged by arthropod feeding or other mechanical injuries, they release a blend of volatiles that is composed of ethylene, terpenoids, and other compounds derived from membrane fatty acids, such as methyl salicylate or methyl jasmonate. The emission of such volatiles is an indirect defense response and serves to attract beneficial arthropods that are predators or parasites of the herbivorous pests but do not feed on grapevines (James and Price, 2004; Pichersky and Gershenzon, 2002; Poelman et al., 2008; Takabayashi and Dicke, 1996; van den Boom et al., 2004). In addition to this indirect effect, some volatiles also appear to have a direct defensive function. Hexanals and hexenals, for instance, may be toxic to certain phloem-feeding aphids. Other volatiles deter some Lepidoptera, whose caterpillars chew on the leaves, from laying eggs on the leaves. Moreover, grapevines that have been attacked by pests may be able to use volatile emission as a signal of imminent threat to nearby vines, which can detect this signal and respond to it by activating their own defenses. There is even evidence of plants under herbivore attack releasing root exudates that then induce neighboring plants to release volatiles from their leaves (Bais et al., 2006), but it is unknown whether grapevines can do this. Such plant-to-arthropod and plant-to-plant communication by use of air-transmissible signals may be important in minimizing herbivore-induced damage because plants cannot hide or run away from pest attack.

The number and relative abundance of beneficial arthropods often increases as the diversity of plant species increases, and so does the number of indifferent or neutral arthropod species, which are neither harmful nor beneficial. Consequently, botanical diversity results in an overall increase in biodiversity and ecosystem stability and keeps pest populations in check (Altieri et al., 2005; Remund et al., 1989). Growers can enhance the biodiversity in vineyards and their vicinity by planting diverse cover crops or encouraging resident floor vegetation or adjacent brush or hedgerow vegetation. Added benefits of floor covers compared with bare soil are their ability to reduce the splash dispersal of fungal spores during rainfall and to improve vineyard water-use efficiency (Ntahimpera et al., 1998; Schultz and Stoll, 2010).

Any prospective pathogen that attempts to penetrate the epidermis must first overcome the waxy cuticle and thick outer cell walls on plant organs before it can cause disease (Yeats and Rose, 2013). Accordingly, there seems to be an inverse association between the thickness of the cuticle and outer epidermal wall of different *Vitis* cultivars and their susceptibility to powdery mildew (Heintz and Blaich, 1989). Wounding of leaves and grape berries caused by herbivore, bird, or arthropod feeding or by other mechanical damage (e.g., from wind or machinery) can provide possible "enhanced-access" points for pathogens. In addition, shedding of plant organs, such as occurs during anthesis, leads to exposed fracture surfaces that provide ideal sites for pathogen invasion (Roberts et al., 2002; Viret et al., 2004). This is why plants respond to physical damage by mechanisms that aim at healing wounds and preventing invasion, such as reinforcement of cell walls by deposition of callose, lignin glycoproteins, and phenolics, as well as production of so-called pathogenesis-related (PR) proteins such as chitinases and glucanases. Cell separation is also accompanied by accumulation of PR proteins to prevent infection; for example, these proteins accumulate at the base of anthers and calyptras at the time of shedding (Roberts et al., 2002).

If this first line of defense fails and pathogens penetrate into grapevine tissues, the vine can shed the entire infected organ to prevent the spread of the pathogen throughout the plant. However, plants have evolved an array of strategies to resist fungal infection. These strategies can be constitutive or induced. Constitutive resistance strategies are preinfection strategies; they are passive and are present regardless of infection. They include physical barriers, such as cuticles and cell walls, in addition to chemicals with antimicrobial activity called phytoanticipins, such as phenolics. Phytoanticipins are generally accumulated in the cell vacuoles—often as inactive precursors—and provide nonspecific protection against a wide range of would-be invaders. Induced resistance strategies, on the other hand, are postinfection strategies; they are actively initiated in response to pathogen invasion and specifically target pathogens that have overcome the constitutive barriers. They include the production of reactive oxygen species; fortification of cell walls by lignification, suberization, or incorporation of callose, proteins, silicon, and calcium; and production of antimicrobial compounds, such as PR proteins and the so-called phytoalexins. Active defenses are usually restricted to the site of invasion. The infected and neighboring cells accumulate the antimicrobial chemicals to high concentrations in an attempt to restrict the spreading of the invading pathogen. Cell wall reinforcement also requires the release of metabolic precursors to the apoplast before they can be incorporated in the cell walls, and this is usually accompanied by the production of reactive oxygen species that induce cross-linking of the cell walls (Field et al., 2006).

Grapevines, like other plants, have special receptor proteins that recognize invading pathogens by some of their microbial enzymes or complex carbohydrates, proteins, and lipids, especially those in the fungal cell walls, such as the polysaccharide chitin (Shibuya and Minami, 2001). In addition, plants also interpret as intruder signals the breakdown products of their own cuticle and cell walls (Claverie et al., 2018; Yeats and Rose, 2013). These compounds are collectively termed elicitors because they elicit a defense response by the plant. In fact, the defense response results from activation of various biochemical pathways by a series of signaling cascades that are triggered by the detection of a pathogen. One of the first signals is the so-called oxidative burst on the cell membrane (Apel and Hirt, 2004; Jones and Dangl, 1996; Mittler, 2002; Smith, 1996; Somssich and Hahlbrock, 1998).

Within minutes of an attempted infection by a foreign invader, reactive oxygen species, but especially superoxide (O_2^-), accumulate in the apoplast. The superoxide is rapidly converted to hydrogen peroxide (H_2O_2), which diffuses into the cells and is perceived by the vine as a signal of an imminent threat and also has antimicrobial effects. In contrast to the situation during abiotic stress (see Section 7.1), this oxidative burst is intentional: O_2^- is produced enzymatically by NADPH-oxidase, peroxidase, and amine oxidase, while the antioxidant systems remain silent to boost the amount of H_2O_2. In response to this signal, the plant cells surrounding an infection site mount structural barriers. They reinforce their cell walls by depositing callose and exploiting the stimulating effect of H_2O_2 on peroxidase for lignification, and they produce PR proteins. These proteins owe their antifungal activity to their ability to degrade chitin and glucans, which are important components of the cell walls of fungi (Selitrennikoff, 2001).

Secondary signaling molecules, including salicylic acid, jasmonic acid, and ethylene, then augment the early defense response and may even activate defenses in distant healthy tissues (Heil and Ton, 2008; Thatcher et al., 2005). In other words, the secondary signals exert a kind of remote control; that is, they act systemically. Although salicylic and jasmonic acid are not volatile, their methylated forms—methyl salicylate and methyl jasmonate—can be emitted by the infected leaves within hours of an infection and taken up by distant uninfected leaves, where they may be transformed back into the nonmethylated forms to prime or sensitize those distant leaves for defense. The soluble salicylic and jasmonic acids arriving days to weeks after an infection via the vascular pathway may then amplify the induced defense response in those leaves (Heil and Ton, 2008). The defense response may be further strengthened by the inhibitory effect of jasmonate on root and shoot growth, which may allow the plant to focus its resources on defense when required (Huang et al., 2017). The activation of defense responses by such signaling molecules may be exploited for disease control in vineyards. Foliar application of methyl jasmonate, it seems, not only triggers accumulation of defense compounds, such as PR proteins and phytoalexins, in the leaves but also reduces powdery mildew infection (Belhadj et al., 2006; Larronde et al., 2003).

In some instances, H_2O_2 in concert with secondary signals induces the infected cells and those surrounding the infection site to commit suicide in a process termed hypersensitive response. This programmed cell suicide limits food supply to the pathogen, starving and confining the invaders in a localized area (Apel and Hirt, 2004; Jones and Dangl, 1996). In addition, nearby vascular tissues may be occluded by the deposition of callose, which limits the spread of the invader or its toxins.

Induced structural barriers may fail to contain a pathogen or may work for some but not other pathogens. Several hours after an infection, the vine therefore also activates a second line of induced defense, this time of a more direct, biochemical nature. This biochemical defense includes accumulation of antimicrobial compounds, including the PR proteins and phytoalexins. Another defense strategy, especially in response to xylem-invading fungal and bacterial pathogens, is the accumulation of elemental sulfur in the vessel walls and xylem parenchyma cells (Cooper and Williams, 2004). Unfortunately, most bacteria, including the Pierce's disease–causing *Xylella fastidiosa*, are not sensitive to S. In contrast, the nonvascular pathogen *Erysiphe necator* and other fungi are highly sensitive. Application of S fertilizers may enhance the vines' resistance to *E. necator* (see Section 7.3). In addition, S dust and S–lime mixtures have been in use as fungicides for thousands of years (Rausch, 2007). Indeed, they were the world's first fungicides, and their effectiveness against powdery mildew was discovered soon after the inadvertent introduction of the fungus in Europe in the 1840s. The disadvantage of foliar S application, however, is the risk of spider mite infestations, especially when S is applied as dust: Mites like dust.

Phytoalexins (Greek *phyton*=plant, *alexein*=to ward off) are a diverse group of antimicrobial, low-molecular-weight metabolites that plants produce in response to various kinds of stresses, such as attack by pathogens or wounding (Kuć, 1995; Smith, 1996). Fungitoxic metabolites comprise more than 300 chemically diverse compounds, including phenolics, terpenoids, polyacetylenes, fatty acid derivatives, and others. These natural fungicides have been found in more than 100 plant species from over 20 families. The phytoalexins characteristic of grapevines are phenolic compounds named stilbenes and include resveratrol and its glucoside piceid, pterostilbene, and several resveratrol oligomers that are termed viniferins (Jeandet et al., 2002; see also Section 6.2). They are produced at or near sites of infection and can inhibit spore germination and mycelium growth of a variety of fungi and oomycetes. The biosynthesis of these phytoalexins is triggered by elicitors comprising high-molecular-weight microbial compounds and components of plant cell walls that are released during infection (Blaich and Bachmann, 1980; Darvill and Albersheim, 1984). In addition, stilbenes also accumulate, along with PR proteins, in response to the air pollutant ozone (O_3) that forms as part of photochemical smog and is highly phytotoxic, inducing leaf symptoms such as numerous yellowish to reddish spots ("stipple") or blotches ("mottling"), bleaching, bronzing, or, in severe cases, premature senescence (Sandermann et al., 1998; Schubert et al., 1997; Williams et al., 1994).

The rate at which phytoalexins accumulate at an infection site often determines whether or not the pathogen attack is successful (Kuć, 1995; Smith, 1996). If accumulation is too slow, the intruder has grown far beyond the infection site by the time phytoalexin concentrations are high enough to inhibit its growth. In this natural arms race, many microorganisms have found ways to degrade and thus detoxify phytoalexins. Detoxification is an important trait conferring pathogenicity, or virulence, enabling the pathogen to infect the plant tissue (Darvill and Albersheim, 1984; Jeandet et al., 2002; Sbaghi et al., 1996; Smith, 1996; VanEtten et al., 1989). For instance, by employing the all-purpose phenol-oxidizing enzyme laccase (a.k.a. stilbene oxidase), *Botrytis cinerea* can degrade resveratrol and pterostilbene (Breuil et al., 1999; Hoos and Blaich, 1990; Pezet et al., 1991). Fungi regarded as not normally pathogenic for grape berries ostensibly lack this ability. Moreover, it is possible that accumulation of the fungitoxic viniferins comes too late: They may be formed only after the infected tissues have already become necrotic (Keller et al., 2003b).

Despite their inducible nature in green plant organs, stilbenes also to be present as constitutive compounds in the vine's woody organs, including canes, trunks, and roots, as well as in peduncles. In concert with other phenolics and even terpenoids, they may contribute to the general pathogen resistance and durability of wood (Kemp and Burden, 1986; Langcake and Pryce, 1976; Pawlus et al., 2013; Pool et al., 1981).

Like the stilbenes, the PR proteins too can directly inhibit spore germination and/or hyphal growth of certain pathogens (Giannakis et al., 1998; Jacobs et al., 1999; Monteiro et al., 2003). These antimicrobial proteins are mostly accumulated in the cell walls. They include β-1,3-glucanases, which are members of the PR-2 protein family; chitinases from the PR-3 family; osmotin/thaumatin-like proteins from the PR-5 family; and perhaps some nonspecific lipid transfer proteins (nsLTPs) that belong to the PR-14 family. Some of these antifungal proteins are present in grapevine tissues at some base concentration and can thus act constitutively, and all accumulate to high concentrations in response to fungal attack of both leaves and berries. Whereas the PR-5 proteins are not produced constitutively in preveraison grape berries and other organs, they and the PR-3 (but not PR-2) proteins begin to accumulate in the berries at veraison (Negri et al., 2008; Pocock et al., 2000; Robinson et al., 1997; Salzman et al., 1998; Tattersall et al., 1997). This induction is probably a consequence of the osmotic stress caused by sugar accumulation and makes the ripening berries increasingly resistant to powdery mildew, downy mildew, and black rot. In addition, whereas moderate amounts of sugars, which are important carbon sources for microbes, can boost the colonization by some fungal pathogens, higher sugar concentrations may reverse this effect, leading to lower susceptibility. For example, *E. necator* is unable to establish new infections on grape berries whose total soluble solids concentration exceeds approximately 7°Brix, a threshold that is normally reached just before veraison (see Section 6.2). It is possible that this phenomenon, which is termed high-sugar resistance, may be at least partly due to the osmotic stress the high sugar concentration imposes on the fungal cells. Nevertheless, the rachis supporting the ripening berries remains susceptible as long as it remains green.

Some members of the PR-3, PR-5, and PR-14 families also seem to inhibit spore germination and growth of *Phomopsis viticola* Sacc., which causes cane and leaf spot disease, and *Elsinoe ampelina* (de Bary) Shear, which causes grapevine anthracnose (Monteiro et al., 2003). The mostly *V. labrusca*–derived Concord grapes accumulate very high concentrations of these or closely related PR proteins, which may contribute to their strong *B. cinerea* resistance compared with the susceptible *V. vinifera* cultivars (Salzman et al., 1998). Although the latter produce much lower amounts, their PR-3 and PR-5 proteins can make up half of the total soluble protein in mature grape berries and are major contributors to protein instability in white wines (see Section 6.2). One reason that PR proteins are only effective against *B. cinerea* at very high concentration may be that during the infection process the fungus secretes specialized proteins that bind to PR proteins and block their antifungal activity (González et al., 2017).

Because of the immense economic importance of grapevine diseases, grape growers generally impose stringent phytosanitary measures to control disease outbreaks. Cultural practices aimed at reducing disease inoculum and spread of infections include canopy management as well as nutrient and irrigation management. Judicious fertilizer and water application avoids excess plant vigor, and an open canopy facilitates rapid drying after rain, minimizes shade, and maximizes spray penetration and coverage. In addition, successful disease management often requires the intensive use of costly pesticides with alternating modes of action to optimize control and manage pesticide resistance of pathogens. Nonetheless, although such strategies usually work reasonably well for fungal and fungi-like pathogens, they fail completely for most bacteria and viruses that insert themselves within their host's cells. Once present in a victim, vegetative propagation ensures their dissemination with new planting material that is typically taken from the host in the form of cuttings or buds. Some viruses are further distributed to healthy plants in the same or neighboring vineyards by insect or nematode vectors (Oliver and Fuchs, 2011). Unlike other disease-causing agents, bacteria and viruses cannot be eradicated or contained with any curative or therapeutic control measures short of destroying infected plants (Rowhani et al., 2005).

8.2 Pathogens: Defense and damage

Pathogens can be roughly divided into necrotrophs and biotrophs. A necrotroph (Greek *nekrois*=death; *trophé*=nutrition) is defined as a pathogen that kills its host's cells and then obtains its food supply from these dead cells. The typical necrotroph attacking grapevines is *Botrytis cinerea* Pers.:Fr., which can infect all green plant organs but is best known for causing gray mold in grapes and many other plant species (Williamson et al., 2007). In a devious twist, rather than being killed by the post-infection oxidative burst inside the host tissues, such necrotrophs may actually exploit and thus subvert the plant's defense response by protecting themselves against reactive oxygen species while promoting the death of host tissues, which in turn favors disease progression (Chong et al., 2014; Elad and Evensen, 1995; Kelloniemi et al., 2015; von Tiedemann, 1997). Necrotrophs, as their name suggests, also benefit from rather than being inhibited by the cell death induced during a hypersensitive response which, by contrast, is very effective against the spread of biotrophs, again as their name suggests (Poland et al., 2009). Therefore, although killing the cells around sites of attempted infection is indeed a useful strategy to contain biotrophs, this approach comes at the cost of increased vulnerability to invasion by necrotrophs, whose mode of life evolved on a platform of thriving on dead tissue and which is only encouraged by suicidal plant cells.

A biotroph (Greek *bios*=life) is an organism that can live and multiply only on another living organism, making it an obligate parasite or symbiont, depending on whether its presence is damaging or beneficial to its host. Biotrophs, therefore, cannot be grown in culture. Examples of biotrophs include most viruses, the powdery mildew fungus *Erysiphe necator* [Schwein.] Burr. (formerly *Uncinula necator*) and the downy-mildew–causing oomycete *Plasmopara viticola* (Berk. and Curtis) Berl. and de Toni, both of which develop on green plant organs, but also the mycorrhizal fungi that enter into a symbiotic relationship with the roots of their hosts (see Section 7.3). Biotrophs typically tap into their host's epidermal cells to extract sugar (mainly glucose) and other nutrients (especially amino acids) to support their hyphae growing on the surface and act as sinks that alter carbon distribution in the host to divert nutrients for their own metabolism (Brem et al., 1986; Hall and Williams, 2000; Hayes et al., 2010; Ruan, 2014; Smith et al., 2001). In other words, biotrophs compete with their host's sinks for assimilate supply. They rely on a strategy that attempts to invade host tissues with minimal damage to their host's cells and suppress or bypass its defense system. Infections by biotrophs are usually limited to the epidermis, which they penetrate by producing a swollen "pressing organ" called appressorium from which a penetration peg emerges that uses a high-precision chemical drilling approach to digest a pore through the outer cell wall (Cantu et al., 2008b). Biotrophs may exploit the host's defense strategies to accommodate the establishment of elaborate intracellular feeding structures termed haustoria (singular haustorium) in the case of the mildews or arbuscules in the case of mycorrhizae (Hall and Williams, 2000; Heintz and Blaich, 1990; Pearson and Goheen, 1998; Smith et al., 2001). The host cell membrane becomes wrapped around the mycelium but is not penetrated by it. This maximizes the intruder's surface area inside the host for active nutrient absorption, which is a one-way street for mildews and a two-way street for mycorrhizae.

The remainder of this chapter discusses examples of some of the major pathogens, or groups of pathogens, that afflict vineyards throughout the world, their disease symptoms, and interaction with their hosts.

8.2.1 Bunch rot

The term bunch rot refers to the disintegration of ripening grape clusters due to infection by a range of different pathogens. The most important of these is the fungus *B. cinerea*, also known by the name of its sexual form *Botryotinia fuckeliana*. In fact, *B. cinerea* is one of the most ubiquitous plant pathogens in the world, with a wide range of host species. Its dry dispersal spores, called conidia, can survive temperatures as low as $-80\,°C$ for several months, can be blown by the wind over enormous distances, and can become infective by germinating over a temperature range of $1–30\,°C$, as long as the relative humidity exceeds 90% (Pearson and Goheen, 1998; Pezet et al., 2004b). Most vineyards, even in isolated areas, are under permanent pressure from wind-dispersed conidia, so *B. cinerea* can be regarded as part of a vineyard's environmental microflora. Nevertheless, its requirement for high humidity provides some protection to many vineyards in dry regions, if they are not compromised by overly dense canopies (Valdés-Gómez et al., 2008; see also Section 5.2).

The *B. cinerea* fungus is responsible for one of the worst fungal diseases of grapes, namely gray mold, which can cause serious crop losses and reductions in fruit quality. The fungus can use the otherwise stable tartrate in addition to sugars as a carbon source, converting some of the acid's degradation products into small amounts of malate and other organic acids. Enzymes, such as the polyphenol oxidase termed laccase, secreted by the fungus can readily oxidize phenolic compounds in grapes and continue to do so in the fermenting juice and wine made from these grapes (Dittrich, 1989; Dubernet et al., 1977; Ky et al., 2012; Macheix et al., 1991; Pezet et al., 2004b; Ribéreau-Gayon, 1982). When phenolics are oxidized, they turn into quinonens, which in turn can form brown polymers; this leads to discoloration of red wines and browning of white wines. The fungal intruder also degrades amino acids and aroma compounds, which shifts a wine's aroma from fruity to "phenol" or "iodine." Moreover, *B. cinerea* and other fungi that appear as secondary colonizers of infected grape clusters may produce volatiles with undesirable "mushroom" or "earthy" odors (La Guerche et al., 2006).

Ironically, however, some of the world's most highly prized dessert wines, particularly the French Sauternes, German Trockenbeerenauslesen, and Hungarian Tokaji aszús, are produced from grapes infected with *B. cinerea*. This so-called "noble rot" appears to be due to the fungus growing mainly in the epidermis of relatively mature grape berries, which leads to desiccation due to increased water permeability of the skin and to concentration of sugars and, to a lesser extent, acids in the berries (Dittrich, 1989; Pezet et al., 2004b). Additional beneficial changes induced by the fungus include the accumulation of glycerol in the berries, which contributes to the sweetness of the resulting wine; of glutathione- and cysteine-precursors of volatile thiols with an aroma of grapefruit and passion fruit;

and of fatty acid–derived lactones with a prune-like aroma (Blanco-Ulate et al., 2015; Ribéreau-Gayon et al., 2006; Robinson et al., 2014; Thibon et al., 2009). Contrary to what happens during bunch rot development, noble rot also leads to accumulation of aroma-active terpenoids such as monoterpenes and sesquiterpenes. The fungus further reprograms the metabolism of white grapes even to the extent that they produce anthocyanins at the expense of flavonols (Blanco-Ulate et al., 2015). Accumulation of cyanidin- and delphinidin-based anthocyanins leads to the characteristic color change of infected berries from green/yellow to pink and then purple/brown. Early morning fog followed by warm, dry days is thought to favor the development of noble rot during grape ripening, but rainfall can quickly turn a beneficial noble rot infection into a detrimental bunch rot epidemic (Mullins et al., 1992). Winemakers inadvertently discovered the benefits of noble rot for wine quality when a delay in the required permission to harvest in the German Rheingau in 1775 and the invasion of France by the German–Russian armies in 1815 both led to late harvests (Dittrich, 1989; Pezet et al., 2004b).

The *B. cinerea* fungus can opportunistically invade all green grapevine organs through wounds or via senescent or dead tissues. In addition, it can invade young organs directly through the cuticle and outer cell wall, both of which it degrades in the process, causing necrosis (Elad and Evensen, 1995; Martelli et al., 1986; Salinas et al., 1986). Although *B. cinerea* is commonly classified as a necrotroph, it may have a very brief biotrophic phase of no more than a few hours early on during the infection process, which enables the fungus to colonize its host's tissues before inducing their death (Veloso and van Kan, 2018). Whereas prebloom flowers are relatively well protected by the calyptra, grape flowers immediately after anthesis are particularly vulnerable to infection, especially when senescent anthers and calyptras remain stuck in the inflorescence (Bulit and Dubos, 1982; Pearson and Goheen, 1998). The fungus can penetrate flowers at the receptacle end, possibly through the scar left behind after capfall, or by breaching the cuticle, and may cause early damage on inflorescences or remain as latent mycelium inside the developing berries without causing disease symptoms until the berries begin to ripen (Cadle-Davidson, 2008; Keller et al., 2003b; Pezet et al., 2003, 2004b; Viret et al., 2004). It is possible that rapid activation of defense responses, including accumulation of stilbenes and PR proteins and lignification of cell walls, may force the fungus into this latent state (Mehari et al., 2017).

Many other fungi also seem to be able to establish latent infections in grape berries. Some of these—especially *Alternaria*, *Cladosporium*, and *Penicillium* species—may cause postharvest rot in table grapes (Dugan et al., 2002). Damage to the berries by cracking of the cuticle from pressure within or outside the berries and physical damage from insects, hail, and wind predispose clusters to berry infection during the ripening phase (Kretschmer et al., 2007). Together with wet conditions such damage leads to disease expression and fruit loss (Fig. 8.1). Cracking of berries in susceptible cultivars can ensue during rainfall or during humid nights, particularly in combination with high soil moisture (see Section 6.2). In addition, small pores in the berry surface may also provide entry sites for *B. cinerea* (Mlikota Gabler et al., 2003). Once a disease outbreak has occurred, conducive conditions will quickly lead to secondary infections via sporulating conidia. Such secondary infections spread much more readily throughout tight clusters whose berries touch each other, thus retaining surface water and rubbing off epicuticular wax, than through loose clusters whose berries are separated by gaps and dry rapidly after rain (Marois et al., 1986; Rosenquist and Morrison, 1989; Vail and Marois 1991; Vail et al., 1999).

The importance of surface water for the spread of the fungus is a major reason why removal of leaves around the fruit clusters is a critical canopy management practice in regions that are prone to rainfall during the ripening period (English et al., 1990; Zoecklein et al., 1992). The increased

FIG. 8.1

B. cinerea infection in ripening grape cluster (left), *E. necator* blotches on abaxial side of leaf (center), and *P. viticola* "oilspots" on adaxial side of leaf (right).

Photos by M. Keller.

air flow and, once the sun shines again, the higher berry surface temperature favor speedy drying of the clusters after a rainfall event. High sun exposure, in addition to loose clusters, associated with minimally pruned vines also strongly decreases *B. cinerea* infection (Intrieri et al., 2001). Where deliberate yield reduction is commonly practiced, cutting off the distal part of each cluster after bloom may more effectively reduce *B. cinerea* epidemics than does cluster thinning. This is because the reduction in berry number and compensatory rachis elongation that occurs in some cultivars following excision of a portion of the cluster decreases cluster compactness (Molitor et al., 2012).

In contrast with open flowers, young, unripe grape berries are highly resistant to *B. cinerea*, preventing the fungus to penetrate beyond the cuticle, whereas ripening berries again drop their defenses against fungal attack and become increasingly susceptible to infection (Blaich et al., 1984; Bulit and Dubos, 1982; Kelloniemi et al., 2015). This relaxation of defenses at veraison occurs as a result of the gradual modification of the berry cuticle and cell walls, decreased ability to produce H_2O_2 upon infection, and the disappearance or modification of several constitutive defensive substances. Unlike immature berries, ripening berries no longer reinforce their epidermal cell walls with lignin and extensin at sites of infection (Kelloniemi et al., 2015). Once the berries have begun to dismantle their cell walls, they are no match for the necrotroph, which secretes an arsenal of wall-degrading proteins to digest the pectins and other polysaccharides in its passing host's cell walls (Cantu et al., 2008a,b; Kelloniemi et al., 2015). Such cell wall degradation appears to be associated with the release of bound Ca^{2+}, which, due to its toxicity, may help the invader to kill the host cells (Kaile et al., 1991). Although the constitutively produced glycolate and phenolic compounds, especially tannins and their subunits but also hydroxycinnamic acids and flavonols, provide a protective barrier by inhibiting the macerating fungal enzymes as well as laccase, the increasing degree of polymerization of skin tannins toward fruit maturity (see Section 6.2) may decrease their ability to bind and denature the fungal proteins (Goetz et al., 1999; Jersch et al., 1989; Tabacchi, 1994). The continued accumulation in ripening grape berries of oxalate may also contribute to the decline of resistance because oxalate may facilitate degradation of cell walls and repress plant defenses. Although *B. cinerea* also secretes oxalate during infection, plants apparently retaliate by producing oxalate oxidase to partly disarm this fungal weapon (van Kan, 2006; Walz et al., 2008; Williamson et al., 2007). In addition to these constitutive barriers, preveraison berries produce stilbene phytoalexins after infection, which may slow the spread of the fungus, although resveratrol is a rather ineffective botryticide and is easily degraded by fungal laccase (stilbene oxidase) enzymes (Hoos and Blaich, 1990; Pezet et al., 1991). The berries' ability to produce stilbenes upon infection seems to decline over the course of berry development and, especially, during ripening (Bais et al., 2000; Jeandet et al., 1991; Kelloniemi et al., 2015).

Accumulation of stilbenes in response to infection by *B. cinerea* might occur at the cost of reduced production of flavonoids such as tannins, anthocyanins, and flavonols. After all, healthy, nonstressed leaves and berries do not produce stilbenes, which is why these compounds are also termed stress metabolites. Induction of stilbene synthase can inhibit the activity of chalcone synthase, and it has been argued that the two enzymes compete for substrates (Fischer et al., 1997; Gleitz et al., 1991). Conversely, it is possible that the accumulation of anthocyanins in red grape berries at veraison is responsible for the decrease in the berries' capacity to produce stilbenes. Like anthocyanins, resveratrol is accumulated predominantly in the berry skin, although it also seems to be present in the seeds. However, this does not explain why white grapes also produce lower amounts of stilbenes and become more susceptible to *B. cinerea* after veraison. Nor does it explain why vulnerability remains high after the cessation of pigment accumulation in red grapes and is a major cause of damage from storage rot in table grapes (Bais et al., 2000; Pearson and Goheen, 1998).

If one considers the phenomenon of loss of resistance during fruit ripening from a plant reproduction standpoint, the picture becomes clearer. Pigmentation of grapes provides visual cues that attract seed dispersers to ripe fruit, which become edible after veraison. Simultaneously, the barriers erected against microbial intruders that would be detrimental during seed development are dismantled when the seeds are viable and ready to be dispersed. Therefore, the increase in anthocyanins, the degradation of cell walls, and the decrease in stilbene-synthesizing capacity during ripening may serve the same purpose—to ensure seed dispersal. After all, when a grape berry rots away, its seeds generally remain free of infection and simply drop to the ground, where they may germinate. It seems at least plausible that this might constitute something like an insurance policy in case there are no birds or mammals present to eat the fruit. If this explanation is correct, *B. cinerea* may be regarded as a mutualistic symbiont in grape berries rather than as a pathogen, or at least as a fungus able to switch its lifestyle from pathogenic in flowers and young leaves to mutualistic in developing berries.

In contrast to grape berries, expanding leaves gradually become more resistant to *B. cinerea* infection (Langcake, 1981; Langcake and McCarthy, 1979). This may partly be caused by the accumulation of the two major phenolic biopolymers—lignin and tannin. Lignin is deposited in mature cell walls, resulting in physical strengthening similar to the role of steel rebars in ferroconcrete. Tannins are produced at variable concentrations and inhibit the cell wall–degrading enzymes secreted by *B. cinerea* (Bachmann and Blaich, 1979; Goetz et al., 1999). In addition, mature leaves contain high amounts of an array of flavonols, hydroxycinnamic acids, and phenolic acids, many of which inhibit fungal laccases, and they produce PR proteins in response to infection (Bachmann, 1978; Egger et al., 1976; Rapp and Ziegler, 1973; Renault et al., 1996; Tabacchi, 1994). Furthermore, stilbenes accumulate much more rapidly in leaves, especially older leaves, than in fruit. Leaves of the *B. cinerea*–resistant American *Vitis* species and interspecific hybrids have a greater capacity for stilbene biosynthesis than the susceptible European *V. vinifera* cultivars (Stein and Blaich, 1985). Different cultivars also vary in the amount of resveratrol they accumulate following fungal attack. For instance, Cabernet Sauvignon leaves appear to be able to produce twice the amount of stilbenes that accumulate in Pinot noir or Chardonnay leaves.

Grape cultivars, and perhaps different clones of the same cultivar, also differ in the susceptibility of their berries to *B. cinerea*. For example, Cabernet Sauvignon, Syrah, Mourvèdre, or Petit Verdot have relatively resistant berries, whereas the Pinots and Muscats, Chardonnay, Sauvignon blanc, Sémillon, and Thompson Seedless are among the most susceptible cultivars (Paňitrur-De La Fuente et al., 2018). Part of the variation in berry susceptibility among cultivars and clones can be attributed to differences in cluster compactness rather than stilbenes. More compact clusters with berries rubbing against each other are more easily infected than loose clusters, because rubbing may compromise the integrity of the cuticular wax on the berries. Pores in the berry skin, as well as the thickness of the cuticle and the amount of wax it contains also vary among cultivars. The rather resistant cultivars derived from American *Vitis* species and their interspecific crosses tend to have far fewer pores, a thicker cuticle, and more cuticular wax than do the vulnerable *V. vinifera* cultivars (Mlikota Gabler et al., 2003). The variation among cultivars in a range of developmental, morphological, and biochemical traits, each conferring partial resistance, demonstrates the existence of quantitative (or incomplete), rather than qualitative (or complete), disease resistance (Poland et al., 2009). Such variation within the genus *Vitis* and within the species *V. vinifera* results in a wide range of disease severity among cultivars under similar environmental conditions.

Even the rootstock is thought to influence the *B. cinerea* susceptibility of its grafting partner. Since plant nutrition also plays a role in disease resistance, it is possible that rootstocks exert their effect via differences in nutrient uptake and transport. However, although grapevines with high nitrogen status tend to have higher bunch rot incidence, research has failed to establish a correlation between the N content of grape berries and their susceptibility to *B. cinerea* (Delas, 1972; Delas et al., 1982; Keller et al., 2001a). On the other hand, the production of both constitutive phenolics, especially flavonols and hydroxycinnamic acids, and inducible stilbenes declines as soil N availability rises, which may render leaves and berries less able to resist fungal invasion. Of course, the unfavorable canopy microclimate that is often associated with high vine N status favors disease development as well.

In addition to gray mold, grapes can also succumb to sour rot, which is caused by a complex of various microorganisms that include bacteria, yeasts, and other fungi (Barata et al., 2012a; Hall et al., 2018). Most of these are secondary pathogens that can only penetrate grape berries through preformed wounds, but some may live as endophytes within the berries (Hall and Wilcox, 2019). Fungi of the genera *Penicillium*, *Aspergillus*, *Alternaria*, and others participate in the infection, as do yeasts (which are single-celled fungi) belonging to *Hanseniaspora*, *Candida*, *Kloeckera*, and others. These yeasts in turn attract fruit flies of the genus *Drosophila* that are vectors of *Acetobacter*. The action of these acetic acid bacteria causes the typical "vinegar" smell of grapes infected with sour rot.

8.2.2 Powdery mildew

In stark contrast to *B. cinerea*, the fungus *E. necator* that causes powdery mildew, or oidium, on grapes is unable to infect plant species that do not belong to the Vitaceae—and it is the only fungus of its family that is able to infect grapes (Dry et al., 2010a; Qiu et al., 2015). Nonetheless, it is arguably the most widespread and most consistently damaging pathogen of grapevines. This is because, unlike *B. cinerea* or *P. viticola*, *E. necator* requires only 40% relative humidity to germinate, a threshold that is easily reached on the lower surface of transpiring leaves, even if the surrounding air is much drier (Keller et al., 2003a; Pearson and Goheen, 1998). The rate of development of the fungus accelerates up to an optimum of approximately 85% humidity and 25 °C, but frequent rainfall, especially heavy rain, and temperatures outside the range 10–32 °C limit its development, with temperatures above 35 °C killing it outright (Carroll and Wilcox, 2003; Gadoury and Pearson, 1990; OEPP/EPPO, 2002).

Mild rainfall, in contrast, seems to benefit *E. necator* by enhancing spore dispersal. Spores germinate within an approximate range of 15–35 °C with a peak near 25 °C.

The *E. necator* fungus can colonize all green plant surfaces (Fig. 8.1) but thrives in shade and often develops in the interior of dense canopies (Nicholas et al., 1998). In addition to the influence of air humidity, this shade effect seems to be related to the absence of UV light, which is effectively filtered out by the leaves' epidermis (Keller et al., 2003a). Consequently, the severity of powdery mildew on both grape clusters and leaves tends to increase as their exposure to sunlight declines (Austin et al., 2011). The environmental factors that favor disease development are of particular concern for grapevines grown in greenhouses and for young vines growing inside plastic sleeves called grow tubes that are used to accelerate growth and facilitate weed control. This is because both greenhouses and grow tubes create a microclimate that is characterized by reduced light, especially UV light, and by elevated relative humidity and, in the case of grow tubes and uncooled greenhouses, elevated daytime temperature (Olmstead and Tarara, 2001). However, in climates where grow tubes tend to increase the temperature above 35 °C, they may inhibit, rather than enhance, powdery mildew development (Hall and Mahaffee, 2001). Moreover, grow tubes can also block access of wind-dispersed fungal spores.

The typical whitish "powdery" disease symptoms on leaf surfaces are due to mycelium and sporulating bodies of the fungus, whereas infected shoots become necrotic, and petioles or peduncles become brittle and break easily later in the growing season. Spores germinate on the surface of plant organs, penetrate the cuticle and outer cell walls, and rapidly establish haustoria inside the epidermis cells (Heintz and Blaich, 1990; Pearson and Goheen, 1998). Like all biotrophic pathogens, *E. necator* depends on the living host plant for assimilate supply. Consequently, it does not kill its host's cells but suppresses their defense responses in susceptible cultivars. Infected leaves generally have increased concentrations of sugars due to import of sucrose from uninfected plant parts and subsequent breakdown by invertase in the cell walls (Brem et al., 1986; Hall and Williams, 2000; Hayes et al., 2010). Since hexoses are its preferred carbohydrates, the pathogen itself induces invertase activity in the leaf by injecting cytokinin (Hayes et al., 2010; Walters and McRoberts, 2006). This is a rare example of the conversion of source leaves back to sinks and a demonstration of the pathological nature of this reversion, which involves import of both sucrose and amino acids. Nonetheless, the susceptibility to *E. necator* infection declines once a leaf has completed its transition from sink to source (Calonnec et al., 2018; Merry et al., 2013).

As a last line of attempted defense, the leaves may divert some of the sugar to produce callose that sometimes seals off the fungal haustoria in older infections (Heintz and Blaich, 1990). Infection—or the sugar accumulation resulting from it—will decrease photosynthesis and starch storage in and assimilate export from infected leaves (Brem et al., 1986; Hall and Williams, 2000; Lakso et al., 1982; Nail and Howell, 2005). Consequently, *E. necator*, being a powerful extra sink, alters assimilate partitioning in the vine at the expense of other sinks, such as fruit, roots, and storage reserves. Although the fungus-derived cytokinins can sometimes delay leaf senescence (Walters and McRoberts, 2006), the vine may shed severely infected leaves, which can further depress yield formation, fruit ripening, replenishment of storage reserves, and cold acclimation.

Infection of inflorescences early in the growing season can result in poor fruit set and consequently low yield, whereas infection after fruit set may reduce berry size, perhaps because imported sugar is diverted to and consumed by the fungus instead of being available for berry growth. Berries that are heavily infected early in development usually shrivel up or drop off, whereas later infections damage the epidermis so that the berries may split upon expansion during ripening and may attract insects that inflict further damage. Even berries that show no visible signs of infection may be sparsely covered with diffuse, nonsporulating mycelium, which results in necrotic epidermal cells at the sites of fungal appressoria formation. Necrotic cells and berry splitting open the door for secondary infections by *B. cinerea* and other bunch rot opportunists. In addition, such surface damage increases colonization by non-*Saccharomyces* yeasts and bacteria (Gadoury et al., 2007). The characteristic powdery mildew "mushroom" and "geranium-like" odor, caused by 1-octen-3-one and 1,5-octadien-3-one, respectively, renders infected grapes unpalatable and can impart off-odors in wine, even though yeasts degrade most of these compounds during fermentation (Darriet et al., 2002; Stummer et al., 2005). Moreover, even minor infections may blemish the berry surface, which can render table grapes unmarketable.

Whereas Chinese and American *Vitis* and *Muscadinia* species exhibit various degrees of resistance to *E. necator*, the European *V. vinifera* cultivars are readily infected because, unlike their American relatives, they did not coevolve with the pathogen and thus fail to mount timely and effective resistance mechanisms (Eibach, 1994; Fung et al., 2008; Qiu et al., 2015). In addition, the fungus times the release of spores from its overwintering bodies, the cleistothecia, to coincide with budbreak of the native *V. riparia* grapes. Because this occurs earlier than budbreak of *V. vinifera*, there are lots of spores around by the time the shoots emerge in most vineyards. Nevertheless, there is some variation even among *V. vinifera* cultivars in the degree of susceptibility (Doster and Schnathorst, 1985; Gaforio et al., 2011; Roy and Ramming, 1990). For instance, Chardonnay, Tempranillo, Grenache, and, to a somewhat lesser degree, Cabernet Sauvignon are among the most susceptible cultivars, whereas Pinot noir, Riesling, Malbec, and Mourvèdre seem to be much better able to resist or tolerate infection. Surprisingly, some *V. vinifera* table grape cultivars (e.g., Kishmish vatkana, a likely parent of Sultana) from central Asia are apparently as resistant as are American *Vitis* species (Coleman et al., 2009; Hoffmann et al., 2007).

Much of the resistance may be a consequence of localized programmed cell death during what is called a hypersensitive response, which refers to the rapid and deliberate suicide of infected epidermis cells that cuts off nutrient supply to the would-be invader (Qiu et al., 2015). Although some stilbene phytoalexins are effective against *E. necator*, infection of *V. vinifera* tissues does not normally trigger their production or triggers it only in the infected cells themselves, perhaps because the biotrophic fungus minimizes cell damage so as not to threaten its own survival (Keller et al., 2003a; Mendgen and Hahn, 2002; Schnee et al., 2008). In contrast, the resistant American *Vitis* species do produce stilbenes in response to infection. In addition, flavonols, which accumulate in the epidermis and cuticular wax in response to UVB radiation and provide a sunscreen for plant tissues (see Section 5.2), may be involved in partial *V. vinifera* resistance against *E. necator*. It is conceivable that enhanced biosynthesis of such constitutive phenolic compounds in sun-exposed leaves and berries contributes to an unfavorable environment for powdery mildew colonization. Flavonol production, however, is strongly reduced by high soil nitrogen availability, and high plant N status makes vines more susceptible to colonization by *E. necator* (Bavaresco and Eibach, 1987; Keller et al., 2003a). An additional resistance mechanism may be vitrification of penetrating mycelium by localized accumulation of silicates in the cell walls (Blaich and Wind, 1989).

Whereas the berries of American *Vitis* species are always resistant to *E. necator*, those of *V. vinifera* cultivars become increasingly resistant over the course of their development, which contrasts with the berries' increasing vulnerability to *B. cinerea* (Qiu et al., 2015). Although the powdery mildew fungus may continue to colonize the berry surface, it seems to be unable to penetrate berries as early as 3–5 weeks after bloom (Gadoury et al., 2003, 2007). Such age-related resistance is termed ontogenic resistance and may partly be a by-product of the accumulation of PR proteins for other reasons, such as sugar accumulation. However, the increase in PR protein content following an infection may raise the amount of protein in the juice and wine made from such grapes, which has the undesirable consequence of increased haze formation in white wines (Girbau et al., 2004). This reduced protein stability is not usually an issue in red wines whose proteins are mostly precipitated by tannins early on during the winemaking process. Unlike the berries, the cluster rachis remains susceptible to infection throughout the growing season, or at least until it forms brown periderm.

8.2.3 Downy mildew

Although often regarded as a fungus because it looks like one and produces spores, the causal agent *P. viticola* is in fact more closely related to marine algae such as kelps and diatoms. Also, in contrast to true fungi, its cell walls contain cellulose instead of chitin, and its cell nuclei are diploid, not haploid like those of fungi. It belongs to the class of Oomycetes, or water molds, and is much less related to the powdery mildew fungus than the latter is to us. Because its spores germinate at greater than 95% relative humidity, it thrives under humid, shady conditions, especially with frequent rainfall and temperatures of approximately 20–25 °C, which are common in coastal areas and other regions with summer rainfall (Galet, 1996; OEPP/EPPO, 2002; Pearson and Goheen, 1998). Although Mediterranean climates and other climates with dry summers are less favorable for the pathogen native to the southeastern United States, *P. viticola* is widespread throughout the world, except in regions with very low spring and summer rainfall, such as central California, eastern Washington, northern Chile, and Western Australia. Because the little rain that falls in Argentina's Mendoza region mostly comes in the summer, downy mildew is a major problem even in this otherwise arid region. The water mold wreaked havoc in the European wine industry when it spread from infested grape material imported in the 1870s for use in breeding programs to combat the previously introduced North American fungi *E. necator* and *G. bidwellii* and the insect phylloxera to which the susceptible *V. vinifera* cultivars were succumbing. During the event, the discovery of the so-called Bordeaux mixture, which consists of copper sulfate mixed with lime suspended in water, saved the industry. Incidentally, however, the widespread appearance of inferior wines, made by desperate producers who lacked sufficient *V. vinifera* grapes and instead used fruit from interspecific hybrids resulting from the resistance-breeding programs, indirectly spurred the advent of the rigid legislation that has since become a hallmark of the European wine industry.

Like the fungi discussed previously, *P. viticola* can infest all green plant parts to cause downy mildew, or peronospora. But unlike those fungi it ordinarily colonizes young leaves or young berries by entering through the stomata rather than by breaching the epidermis (Gindro et al., 2003; Langcake and Lovell, 1980; Martelli et al., 1986; Pearson and Goheen, 1998). Once a germtube has penetrated the substomatal cavity, it forms a primary hypha that branches into a mycelium which develops an intercellular network in the leaf mesophyll and feeds off these cells by means of cell wall–penetrating haustoria (Unger et al., 2007). Consequently, initial disease symptoms appear on the infected leaves' adaxial side as yellow or, in some cultivars, red oily spots (Fig. 8.1), which spread and later become angular necrotic patches restricted by the leaf veins.

The typical whitish "downy" mildew symptoms on the lower leaf surface arise from sporulation of the pathogen through the stomata. Even before any disease symptoms appear, however, the invading pathogen prevents the stomata from closing at night or in response to water deficit so that the unrestrained transpiration may lead to water loss and wilting of infected leaves (Allègre et al., 2007). Although unlike *E. necator*, *P. viticola* does not stimulate sugar accumulation in infected leaves, it also leads to reprogramming of sugar metabolism akin to that in a sink (Gamm et al., 2011; Hayes et al., 2010; Palmieri et al., 2012). The high sugar content suppresses photosynthesis and eventually results in shedding of severely damaged leaves. This can adversely impact yield formation, fruit ripening, replenishment of storage reserves, and cold acclimation. Moreover, infected shoot tips, petioles, tendrils, and inflorescences often become necrotic and are abscised, whereas the highly susceptible young grape berries become covered with a grayish "felt." Ripening berries that are infected with downy mildew drastically increase the production of volatile thiol precursors, probably from fatty acid degradation products derived from damaged membranes (Kobayashi et al., 2011).

Having coevolved with the pathogen, most of the North American *Vitis* species are partly or fully resistant to downy mildew. Examples of partly resistant species include *V. labrusca*, *V. rupestris*, *V. berlandieri*, and *V. aestivalis*, and the fully resistant species include *M. rotundifolia*, *V. riparia*, and *V. cinerea*. Some Asian species (e.g., *V. amurensis*) also have partial resistance, whereas *V. vinifera* is highly susceptible (Yu et al., 2012). It seems that *V. vinifera* has trouble recognizing *P. viticola* as a pathogen and consequently fails to mount effective defense strategies (Palmieri et al., 2012). Perhaps the intruder, whose own survival relies on live host tissues, actively suppresses the defense response (Milli et al., 2012). Nonetheless, even among the European grape cultivars there are various degrees of susceptibility. For instance, Riesling, the Pinot family, and especially Cabernet Sauvignon are among the less susceptible cultivars, whereas Tempranillo and Albariño are among the most susceptible (Boso and Kassemeyer, 2008; Boso et al., 2011). Some of the resistant species defend themselves against infection of their leaves by rapidly secreting callose that plugs their stomata and coats the pathogen spores (Gindro et al., 2003; Langcake and Lovell, 1980). This coating probably kills the germinating spores and stops mycelial growth while also reducing water loss from the leaves. This postinfection defense response can be activated, even in *V. vinifera* cultivars, by soil fungi belonging to the *Trichoderma* species, which may thus be used as biocontrol agents (Palmieri et al., 2012; Perazzolli et al., 2011). In addition, the leaves produce stilbenes that kill the cells surrounding infected stomata and hinder sporulation of the pathogen (Dai et al., 1995; Langcake, 1981; Langcake et al., 1979; Malacarne et al., 2011; Pezet et al., 2004a). High plant N status seems to compromise the leaves' ability to produce stilbenes and leads to higher vulnerability to infection (Bavaresco and Eibach, 1987).

Vitis species differ in their ability to accumulate stilbenes in response to *P. viticola* infection (Dai et al., 1995; Pezet et al., 2004a). For instance, stilbenes appear earlier and reach much higher concentrations in leaves of the resistant *V. riparia* than in the susceptible *V. vinifera*, whereas stilbene production is intermediate in the partly resistant *V. rupestris*. The resistant *V. cinerea* and *V. champinii*, however, are poor stilbene producers, which indicates that these phytoalexins are not the only actors on the stage of downy mildew resistance. Indeed, in addition to the constitutive chitinase in their leaves, resistant species also activate glucanase production upon challenge by *P. viticola* (Kortekamp, 2006). Accumulation of PR proteins, and perhaps peroxidase activity, in old rather than in young leaves also may partly account for the decrease in vulnerability to the pathogen as leaves age (Reuveni, 1998).

Young grape berries, even those of susceptible species and cultivars, also rapidly accumulate stilbenes around sites of infection (Keller et al., 2003a; Schmidlin et al., 2008). Although the berries appear to lose this ability around veraison, they nevertheless become increasingly resistant to infection during development (Kennelly et al., 2005). The appearance of necrotic cells near invasion sites during the localized hypersensitive response stops spreading of the pathogen to healthy tissues (Busam et al., 1997; Kortekamp, 2006; Langcake and Lovell, 1980).

As in the case of *E. necator*, the American *P. viticola* and the European *V. vinifera* met only recently by evolutionary timescales, so it is probable that the ability to produce stilbenes evolved in *V. vinifera* as a defense against intruders other than *P. viticola*. The formation of necrotic spots on the leaves of at least partly resistant *Vitis* species and of oilspots on susceptible species is, of course, a result of the way the different species respond to the attacking pathogen. Such variation demonstrates the principle that when a pathogen discovers and invades a new host species, the symptoms it provokes are often different and usually more severe than those that occur with the original species, which has honed its skills at fending off the invader through the long arms race of coevolution.

8.2.4 Bacteria

Although most bacteria live in the soil, a rare few have succeeded at becoming plant pathogens, suggesting that plants in general have evolved powerful resistance strategies (Bais et al., 2006). One of the few perpetrators is *Rhizobium vitis* (formerly known as *Agrobacterium vitis* or *Agrobacterium tumefaciens* biotype 3), the causal agent of crown gall, which is mainly confined to and spreads via xylem vessels—both up and down the trunk and roots (Tarbah and Goodman, 1987; Young et al., 2001). It may live for years within the vascular system of infected vines without any outward expression of disease, which typically develops only at sites of physical injury. Although commonly regarded as a soilborne pathogen, the bacterium may be viewed as a biotroph, as it survives only in *Vitis* tissues and is usually introduced into vineyards with contaminated, but generally symptomless, planting material that then serves as a source of inoculum (Burr and Otten, 1999; Burr et al., 1998). Infection often occurs during the propagation process in nurseries. The stage of transplanting freshly callused rooted cuttings into nursery beds, during which many of the soft callus cells may be injured due to friction at the callus–soil interface, seems to be an especially vulnerable period (Goodman et al., 1993). In established vineyards, the bacterium is thought to enter grapevines predominantly through wounds caused by cold injury but also at sites of other injuries such as those caused by machinery or grafting. Therefore, the disease thrives in climates that experience frequent freeze injury, and damage caused by galls often surpasses that of the initial injury.

The bacteria induce the tumors called crown gall when mechanical wounds rupture the xylem vessels and release the bacteria to the adjacent parenchyma cells, where they insert some of their own DNA into the genome of the infected plant (Burr and Otten, 1999; Martelli et al., 1986; Tarbah and Goodman, 1987). These bacterial DNA pieces are called tumor-inducing plasmids or Ti plasmids. It is this exceptional ability to transfer genetic material to plants and other organisms that also makes *Rhizobium* (formerly called *Agrobacterium*) species the vehicles of choice for plant genetic engineering (Nester, 2015). In this context it is important to realize that plants are by nature transgenic: During their evolution they have repeatedly acquired genes from bacteria, as well as from fungi and viruses, and even from other plants, and they continue to do so (Bock, 2010). Such asexual integration of foreign genes into the plant genome is termed horizontal gene transfer or lateral gene transfer. The fact that this process occurs naturally indicates that genetic engineering is nothing more than man-made horizontal gene transfer (Bock, 2010).

The bacterial genes induce the now genetically modified host cells to degrade tartrate, divert auxin away from its normal basipetal movement, and locally enhance auxin, cytokinin, and ethylene production in order to stimulate cell proliferation that leads to the characteristic crown gall tumors (Aloni, 2013; Burr and Otten, 1999; Sakakibara, 2006). Like human cancer, these tumors may also host other, opportunistic species of bacteria that exploit the tumor microenvironment (Faist et al., 2016).

Although no *Vitis* species have yet been found to be immune to crown gall, the species vary in their vulnerability to *R. vitis*. For instance, *V. vinifera* cultivars are highly susceptible, whereas *V. labrusca* cultivars are somewhat resistant, the rootstocks of *V. riparia* and *V. rupestris* parentage are relatively resistant, and *V. amurensis* is quite resistant (Burr and Otten, 1999; Burr et al., 1998; Stover et al., 1997; Szegedi et al., 1984). On susceptible species, galls normally develop close to the base of the trunk, from slightly above to slightly below the soil surface (Fig. 8.2), but they can also form up to approximately 1 m above the soil surface, and even on canes. However, the bacteria seem to be unable to move to the tip of growing shoots. This can be exploited to eliminate *R. vitis* from propagation material through the process of shoot-tip culturing (Burr and Otten, 1999; Burr et al., 1988, 1998). Training up suckers to form new trunks is a technique that can be used to replace injured trunks in existing vineyards, provided the suckers arise from the asymptomatic base of the trunk and not from a rootstock. On vines grafted to resistant rootstocks, galls usually form at or above the graft union (Burr and Otten, 1999).

In addition to inducing gall formation, auxin, cytokinin, and ethylene also trigger the differentiation of new vascular bundles, which are then connected to the vascular system of the vine to provide water and nutrients to the growing tumor at the expense of the shoots (Aloni, 2001; Efetova et al., 2007; Veselov et al., 2003). The lack of an epidermis around the tumor leads to water loss through its large surface area, which further promotes water flow to the tumor (Aloni, 2013). In response to this threat, the plant cells may produce osmoprotectants such as proline (see Section 7.1), as well as suberin for incorporation in the exposed cell walls as a sealing agent (Efetova et al., 2007; Veselov et al., 2003).

FIG. 8.2

Young grapevine with severe symptoms of crown gall (left) and tumor formation above the soil surface (right).

Photos by M. Keller.

Nevertheless, grapevines with crown gall symptoms often develop secondary symptoms of drought stress.

As developing galls expand, they increasingly restrict movement of water and nutrients with drastic consequences for plant vigor, longevity, and yield (Schroth et al., 1988). Because the girdling effect of the tumors inhibits phloem transport to the roots, the leaves may turn red due to the conversion of accumulating sugars into anthocyanins (Fig. 8.2). In severe cases, the portion of the vine above the gall may die, which is especially a problem during vineyard establishment. In addition to the striking disease symptoms associated with gall formation, *R. vitis* also causes black root lesions (Rodríguez-Palenzuela et al., 1991). Although infected roots do not usually form galls, the necrotic lesions compromise plant vigor even in the absence of trunk galls. Upon removal of a vineyard, the bacteria can persist in root debris for several years, which may provide a source of inoculum for newly planted vines.

The bacterium *X. fastidiosa*, the causal agent of Pierce's disease, contrasts with *R. vitis* in that it causes damage mainly in areas with mild winters that permit survival of its insect vectors (Mullins et al., 1992). In most of its many host plant species *X. fastidiosa* is a harmless endophyte, that is, a symbiotic microorganism living inside the plant. Species that have coevolved with the bacterium, such as *M. rotundifolia* and *V. arizonica*, are resistant to the bacterium, but the resistance varies according to the home range of local populations of wild vines (Fritschi et al., 2007; Ruel and Walker, 2006). In *V. vinifera*, however, which the bacterium had never encountered before European grapes were introduced to its home in Central America and southern North America, it causes severe damage and even death (Hopkins and Purcell, 2002). It lives exclusively inside xylem vessels and uses them to spread throughout its host plants after it has been deposited there by xylem-sucking insects, such as sharpshooter leafhoppers (Chatelet et al., 2006; Stevenson et al., 2004). The pathogen may facilitate its own migration through the xylem by releasing polygalacturonase, cellulase, and other cell wall–degrading enzymes that digest some of the pit membranes that get in its way (Pérez-Donoso et al., 2010; Roper et al., 2007; Sun et al., 2011a). Bacteria populations build up over time, and severe infection can lead to localized plugging of vessels. Plugging seems to result mostly from dislodged cell wall pectin fragments and from the production of tyloses and gums by the infected vine (Pérez-Donoso et al., 2010; Stevenson et al., 2004; Sun et al., 2013). Xylem occlusion could be a response to the release of ethylene by the injured vine (Sun et al., 2007). The ensuing disruption of water flow, especially in the petioles where the bacteria can induce cavitation, may lead to leaf dehydration (McElrone et al., 2008). Water stress may or may not be associated with the symptoms characteristic of Pierce's disease, which include marginal leaf necrosis that resembles salt injury; abscission of leaves but not the petioles, leaving the so-called "matchsticks" on the shoot; and green islands due to disruption of periderm formation on otherwise browning shoots (Stevenson et al., 2005; Thorne et al., 2006). The berries of severely infected plants sometimes shrivel up and dry, and the vines, too, ultimately die from infection.

8.2.5 Viruses

Grapevines are vulnerable to infection by a variety of viruses that, like some bacteria, cause graft-transmissible diseases. Grapevine viruses became widespread only in the 20th century after European growers had been forced, in response to the spread of phylloxera, to change their propagation methods from simple rooting of cuttings to grafting onto phylloxera-tolerant rootstocks. Cultivated grapes throughout the world now appear to be infected with more viruses than any other woody species. These viruses are typically introduced into vineyards by infected planting material, although some can also be spread by vectors, such as phloem-feeding insects or root-feeding nematodes (Bovey et al., 1980; Naidu et al., 2014; OEPP/EPPO, 2002; Oliver and Fuchs, 2011; Olmo, 1986). For example, leafroll-associated viruses are transmitted by aphids or mealybugs, often aided by wind, people, or machinery, whereas the fanleaf nepovirus is transmitted by the dagger nematode *Xiphinema index*. The latter, of course, also causes direct damage through feeding on the roots in addition to delivering its virus cargo. Moreover, although some grapevine genotypes, such as certain rootstocks, can tolerate very high nematode numbers, these nematodes can still feed on the roots and transmit viruses (Ferris et al., 2012; Mullins et al., 1992). Unlike most *Vitis* species (*V. arizonica* may be an exception), *Muscadinia* species are resistant to *X. index* feeding; yet even *Muscadinia* can be infected with fanleaf virus, albeit very slowly (Oliver and Fuchs, 2011). Once infected, all rootstocks will transmit the nepovirus to their grafting partners.

Viruses are classed as biotrophic pathogens because they depend on the metabolism of their hosts for multiplication, which leads to systemic infections. Almost all known viruses infecting *Vitis* species have genomes that consist of RNA, rather than DNA as in their hosts, and recruit host ribosomes to manufacture their viral proteins. This property enables such viruses to mutate readily because it dodges the host cells' ordinary "proofreading" machinery that is designed to detect and repair errors in the DNA. Consequently, novel virus genotypes evolve very rapidly, although one or a few stable genotypes typically dominate infections in plants (Coetzee et al., 2010; García-Arenal et al., 2001; Karasev, 2000; Naidu et al., 2014). Despite the virus's RNA genome, however, grapevines have incorporated fragments of leafroll-associated virus genes by reverse-transcribing them into the DNA genome of their own mitochondria (Goremykin et al., 2009). The so-called red blotch-associated virus is an exception, as its genome consists of a single DNA molecule (Sudarshana et al., 2015).

To further their own needs, many viruses suppress the expression of some of their host's genes—a phenomenon called host gene shutoff—while triggering the expression of other genes (Havelda et al., 2008). These viruses thus manipulate the host's metabolism, which ultimately leads to the development of disease symptoms. Once a virus has colonized a grapevine, it remains there for good; a cure is not possible. Cuttings and other vegetative propagation material obtained from such vines is therefore also compromised. Moreover, infections with a potpourri of different viruses are common, and the association of specific symptoms with a particular virus is often not clear-cut. For instance, many plant viruses induce localized or general yellowing of leaves.

Lower temperatures seem to strongly increase a plant's susceptibility to virus infection and may also result in more pronounced disease symptoms, possibly because cool conditions may favor the multiplication of viruses and slow the degradation of their RNA (Samach and Wigge, 2005; Wang et al., 2008). Many plant-parasitic viruses appear to induce the production by their hosts of heat shock proteins in addition to stilbenes and PR proteins, among other disease resistance mechanisms (Whitham et al., 2006). Whereas the heat shock proteins may facilitate virus infection and replication, the disease resistance responses might make virus-infected vines less susceptible to attack by fungal pathogens. The virus particles normally spread symplastically in infected tissues and organs, using special movement proteins that enable them to travel via plasmodesmata from cell to cell and via the phloem to other organs (Lough and Lucas, 2006; Oparka and Santa Cruz, 2000; Roberts and Oparka, 2003).

What we often call the leafroll virus is actually a genetically diverse bunch of ampeloviruses and closteroviruses causing similar symptoms that are collectively termed leafroll disease. The leafroll-associated viruses are arguably the most destructive of all grapevine viruses because they interfere with phloem transport and curtail root growth and shoot vigor (Karasev, 2000; Martelli et al., 2002; Naidu et al., 2014; Oliver and Fuchs, 2011). Though these viruses are mostly confined to the phloem and adjacent live vascular tissues, phloem blockage occurs not by the virus itself but due to callose deposition, which may be an attempt by the vine to limit the spread of virus particles (Castellano et al., 2000; Martelli et al., 1986). Unfortunately, this defense response restricts sugar export from the leaves and leads to sugar accumulation inside the leaves, which in turn induces feedback inhibition of photosynthesis and, often, the characteristic interveinal reddening in red cultivars or yellowing in white cultivars (Ayre, 2011; Espinoza et al., 2007; Halldorson and Keller, 2018; Naidu et al., 2014; see also Fig. 8.3). The red leaf coloration is due to accumulation of anthocyanins (Gutha et al., 2010). Given that sugar and abscisic acid act synergistically in inducing anthocyanin production (Pirie and Mullins, 1976), one would expect water deficit to enhance the expression of red leaf symptoms. Nevertheless, leafroll disease has little effect on the susceptibility of grapevines to water deficit (Halldorson and Keller, 2018). Foliar sugar accumulation not only triggers anthocyanin production but also leads to an increase in the osmotic pressure of the leaf tissues (Koroleva et al., 2002). Thus, the downward rolling of the leaves in some leafroll virus-infected cultivars might be a result of osmotically induced cell expansion, which would affect the densely packed upper palisade parenchyma more than the more loosely packed lower spongy parenchyma (Halldorson and Keller, 2018; see also Fig. 1.15).

Gradual leaf reddening in red cultivars or yellowing in white cultivars, but no rolling, also occurs in grapevines infected with the unrelated red blotch-associated virus that reduces yield and delays fruit ripening just like the leafroll virus (Sudarshana et al., 2015). Yet these plant responses and symptoms

FIG. 8.3

Symptoms of leafroll virus infection in Cabernet Sauvignon: bright red areas appear on the lamina of a mature leaf, while the veins remain green.

Photo by M. Keller.

are not specific to virus diseases. A similar sugar-induced accumulation of anthocyanins and downward rolling of leaves also occurs in red grape cultivars that are infected with the phloem-mobile phytoplasma Flavescence dorée that causes grapevine yellows disease (Margaria et al., 2014).

In contrast with *V. vinifera* and some Asian *Vitis* species, the American *Vitis* and *Muscadinia* species and their hybrids typically do not develop visual leafroll disease symptoms, but they still suffer from reduced yield and fruit quality (Naidu et al., 2014; Oliver and Fuchs, 2011). No grapevine species or cultivar is resistant to or tolerant of leafroll virus infection. The so-called graft incompatibility of certain scion–rootstock combinations is often associated with one or both grafting partners being infected with leafroll viruses. But the viruses infest own-rooted vines just as readily, and symptoms are identical to, albeit often less pronounced than, those developing on grafted vines.

In addition to causing reduced growth and plant longevity, some viruses, especially the leafroll-associated and fanleaf viruses, result in significant losses in both yield and quality of the fruit (Bovey et al., 1980; Cabaleiro et al., 1999; Goheen and Cook, 1959; Hewitt et al., 1962; Lider et al., 1975; Mannini et al., 2012; Woodham et al., 1983, 1984). This is because diminished sugar transport to the fruit clusters interferes with yield formation, grape berry development, and ripening. Clusters often appear rather loose due to poor fruit set, berry size and ripening are uneven, and grapes commonly lag behind in maturity, with poor sugar and color and high acidity. Whereas accumulation of sugars, some amino acids (e.g., proline), and anthocyanins is reduced in the fruit of leafroll virus-infected grapevines, accumulation of other amino acids (e.g., arginine), organic acids, and K^+ may be increased (Alabi et al., 2016; Kliewer and Lider, 1976; Vega et al., 2011).

The restricted sugar export from the leaves sometimes also compromises root growth and replenishment of storage reserves in the perennial parts of the vine (Rühl and Clingeleffer, 1993). This weakens the plant and leads to gradual decline and, finally, death of the plant, especially if multiple infections with different viruses occur. In addition, leafroll viruses also induce the plant to produce volatiles that attract the insect vectors of the disease, which facilitates virus dispersal from infected to healthy vines and neighboring vineyards.

Glossary

This glossary provides definitions or explanations of some chemical, anatomical, physiological, and viticultural terms that appear throughout the text. An arrow (→) in front of a word indicates that this word has its own entry in the glossary.

Abscisic acid (ABA) Plant hormone that indicates stress ("stress hormone"), induces stomatal closure and leaf fall (→abscission), inhibits growth, and controls seed dormancy. The carboxylic acid ABA is a breakdown product of →carotenoids and is produced when plant cells lose →turgor. ABA acts antagonistically to →gibberellin.

Abscission (Latin *abscissus* = to cut off) The shedding of leaves, flowers, fruits, or other plant parts, usually following the formation of an abscission zone and preceded by →senescence.

Absorption (Latin *absorbere* = to swallow) Uptake, e.g., of light photons by leaves or of nutrient ions by roots.

Acid A substance that dissociates in water, releasing hydrogen ions called protons (H^+) and thus causing a relative increase in the concentration of these ions. The most important (organic) acids in grapes are tartaric and malic acids. Acid also means having a →pH less than 7 due to the presence of more protons than hydroxyl ions (OH^-) in a solution.

Adaptation An anatomical or physiological trait that has been changed by →evolution, so that it more or less matches a particular set of environmental conditions. Such traits often appear to be "designed" for a specific purpose or environment.

Adsorption (Latin *ad* = to, at; *sorbere* = to suck) Binding of gaseous or dissolved compounds to the surface of a solid object.

Adventitious root (Latin *adventicius* = not properly belonging to) A →root that develops from the →cambium of the stem, as in grapevine cuttings.

Aglycon Organic component of a →glycoside that remains after removal of the sugar.

Allele One particular version of a →gene that has several different forms arising from sexual reproduction or mutation. Alleles can be present on the →chromosome pairs of an individual plant in single dose (→heterozygous) or double dose (→homozygous). Different alleles generally result in differences among →phenotypes in a population.

Amino acid (Greek *Ammon*, the Egyptian Sun god, near whose temple ammonium salts were first prepared from camel dung) Organic acid made up of a chain of carbon atoms to which hydrogen, oxygen, nitrogen, and, sometimes, sulfur atoms are attached; comes in 20 different forms. The units or "building blocks" from which →protein molecules are built. An important food source for yeast during fermentation.

Anther (Greek *anthos* = flower) The part of the stamen that develops and contains pollen and is usually borne on a stalk (filament).

Anthocyanin (Greek *anthos* = flower; *kyanos* = blue) A water-soluble blue or red pigment, belonging to the phenolic class of flavonoids, found in the cell vacuoles of the skin (and sometimes pulp) of dark grapes. Anthocyanins are responsible for the red-purple color of dark grape berries, leaves, and wine.

Apical control A case of correlative inhibition in →canes, whereby growing →shoots inhibit each other and prevent →buds on the same cane from breaking. Most obvious in spring when distal (apical) buds on a cane grow out prior to and inhibit the growth of the proximal (basal) buds.

Apical dominance (Latin *apex* = tip; *dominari* = to rule) A case of correlative inhibition in annual →shoots, whereby the growing shoot tip inhibits the outgrowth of →lateral shoots.

Apoplast (Greek *apo* = away, distant; *plastos* = formed) The totality of spaces within a plant that are outside its cell →membranes, including cell walls, intercellular spaces, and →xylem conduits.

Arginine An amino acid; the main nitrogen storage substance of grapevines.

Assimilate stream The flow of photosynthetic assimilates, or food materials, in the →phloem; moves from →source to →sink down a pressure gradient.

Assimilation (Latin *assimilare* = to align, to integrate) The conversion of inorganic chemicals acquired from outside the plant body into organic compounds inside the plant.

ATP Adenosine triphosphate; the almost universal energy currency of biological organisms; functions as a donor of phosphate groups. It consists of the purine base adenine linked to the sugar ribose (adenine + ribose = adenosine) and three phosphate molecules and is produced during →respiration. ATP also is a constituent of →RNA and, following removal of oxygen, of →DNA.

Auxin (Greek *auxanomai* = to grow) Plant →hormone that mainly stimulates the pattern of cell division and differentiation ("growth hormone"), and controls organ formation and →apical dominance. Very generally, auxin is transported to tissues to induce them to grow. The major auxin is indole-3-acetic acid (IAA), a carboxylic acid produced both from the →amino acid tryptophan and from one of its precursors.

Bark Tough external covering of a woody stem or root external to the →vascular cambium; composed mostly of old →phloem tissue.

Basal bud →Bud at the base of a shoot or cane, not classified as a bud at a count node. Such buds do not normally break in the same season as buds at count nodes but may remain latent on old wood for many years; if stimulated by severe pruning, they may produce shoots that may or may not be fruitful, depending on cultivar.

Blade The broad, expanded portion of a leaf; the lamina.

Bloom Flowering as indicated by shedding of the →calyptras covering the reproductive organs of grapevine flowers. Bloom also refers to the waxy coating on grape berries that gives a frosted appearance to dark-colored grape varieties.

Bleeding Sap flow. The exudation of sap from pruning wounds before and during →budbreak; serves to dissolve and push out air bubbles in the →xylem and to rehydrate the buds.

Brassinosteroids Group of plant →hormones that may induce cell expansion ("growth hormone"). Brassinosteroids are produced locally, mostly in the epidermis. The major active brassinosteroid of grapevines is castasterone, produced from a cholesterol-like compound.

Bud Rounded organ (winter bud or eye) at the node of a cane or shoot formed in the axil of the leaf, containing an undeveloped (embryonic) shoot protected by overlapping scales. Each bud appears to be single but comprises at least three "true buds": a more developed primary bud between two less prominent secondary buds.

Budbreak, budburst The stage of bud development when green tissue becomes visible; the emergence of a new shoot from a bud during spring.

Bud fruitfulness A "fruitful bud" has one or more inflorescence primordia, which give rise to clusters. One measure of fruitfulness is the number of clusters per shoot. Fruitfulness is an inherited characteristic that is also influenced by environmental factors at the time of inflorescence primordium initiation.

Bunch Australian term for grape →cluster.

Callose (Latin *callos* = hard skin) A →carbohydrate (a polysaccharide), composed of glucose units arranged in β-1,3-glucan chemical bonds with some β-1,6 branches; regulates cell-to-cell connections by variably opening or plugging →plasmodesmata.

Callus Mass of undifferentiated →somatic cells. Parenchyma tissue that grows over a wound or graft and protects it against drying or other injury, or cells that develop from plant pieces in tissue culture.

Calyptra (Greek *kalyptra* = cover, veil) The fused petals (cap) of the grape flower that fall off at bloom. Also, the root cap.

Cambium (Latin *cambiare* = to exchange) Meristematic (dividing) tissue that produces parallel rows of cells.

   Vascular cambium (Latin *vasculum* = small vessel) A thin, cylindrical sheath of undifferentiated →meristem between the bark and wood. When active, it produces secondary →xylem (to the inside) and →phloem (to the outside), resulting in growth of the diameter of stems (shoots, trunks) and roots.

   Cork cambium (=phellogen) Meristem, which is in part responsible for the development of the bark. Produces cork cells, the outer walls of which are impregnated with →suberin.

Cane A mature, woody 1-year-old →shoot from the previous growing season, which generally has more than six count nodes. A shoot becomes a cane after →periderm formation.

Canopy The aboveground portion of a grapevine, comprising the →trunk, →cordon, shoots, leaves, and fruit.

Capacity The total growth of a grapevine, including the production of crop, leaves, shoots, and roots in one growing season. Capacity can be estimated from a vine's total weight of fruit and shoots.

Carbohydrate (Latin *carbo* = ember; *hydro* = water) Organic compound made up of sugar units. Consists of a chain of carbon atoms to which hydrogen and oxygen atoms are attached in a 2÷1 ratio; examples are sugars, →starch, and →cellulose.

Carotenoid Yellow, orange, or red photosynthetic →pigment in plant cell →chloroplasts. Carotenoids become visible in leaves and grape berries when →chlorophyll breaks down and are responsible for the yellowish-orange fall colors of leaves.

Cell The structural and functional unit of all living organisms, enclosed in a plasma membrane and containing the →cytoplasm, →nucleus, and other organelles (such as →chloroplasts, →mitochondria, and →vacuoles). Plant (and animal) cells are typically 10–30 µm in diameter, but many bacteria are only 1–2 µm long.

Cellulose A linear, water-insoluble →carbohydrate (a homopolysaccharide), composed of 10,000–15,000 glucose units arranged in β-1,4-glucan chemical bonds, forming straight fibrous chains held together by hydrogen bonds; the main cell wall substance of plants and some fungi.

Chilling Low but mostly above-freezing temperature. Required to release →dormancy. Chilling stress refers to damage or injury caused by low temperatures (usually <15°C) to nondormant plant tissues.

Chimera A plant consisting of more than one genetically distinct population of →cells, usually due to a →mutation in one of the two cell layers of the shoot apical →meristem.

Chitin A linear →carbohydrate (a homopolysaccharide), composed of thousands of *N*-acetyl-d-glucosamine units arranged in β-1,4-glucan chemical bonds, forming extended rigid fibers similar to those of →cellulose. It is the main substance of the cell walls of fungi and of the hard external covering termed exoskeleton of arthropods but is not found anywhere in plants.

Chloroplast →A chlorophyll-containing photosynthetic →cell organelle. Chloroplasts descended from symbiotic microorganisms that invaded (or were "swallowed" by) larger cells and still have their own →genes.

Chlorophyll (Greek *chloros* = green, yellow; *phyllon* = leaf) Green photosynthetic pigment in plant cell →chloroplasts; the receptor of light energy in →photosynthesis.

Chlorosis Loss of →chlorophyll and yellowing of some or all the cells of a tissue or an organ.

Chromosome A segment made up of chromatin, which consists of the histone →proteins acting as spools around which a single large →DNA molecule winds that contains many →genes in the →cell →nucleus of all higher organisms (other than bacteria and archaea).

Clone (Greek *klon* = twig) A group of vines of a uniform, genetically identical type, derived by →vegetative propagation from a single original mother vine.

Cluster Fruiting (reproductive) structure of the grapevine with grape berries.

Coenzyme Organic cofactor (often containing a vitamin) required for certain →enzymes to be active.

Cold hardiness The ability of a grapevine to survive the extreme cold and drying effects of winter weather.

Competition Interaction between organs of the same organism (e.g., shoot tips and inflorescences), members of the same population (e.g., grapevines in a vineyard), or of two or more populations (e.g., grapevines and weeds) to obtain a resource that both require and that is available in limited supply.

Cordon The permanent arm arising from the trunk to form a part of the vine framework and which supports the fruiting 1-year-old wood. Cordons are usually horizontal, but may be inclined or vertical.

Correlative inhibition Broader term for the phenomenon of →apical control or →apical dominance.

Cork A secondary tissue produced by a cork →cambium in woody stems and roots; made up of nonliving cells with suberized cell walls, which resist the passage of gases and water vapor.

Crop The amount of fruit borne on a grapevine.

  Cropping The formation of crop yield on a grapevine.

  Crop level The amount of fruit (or number of clusters) per shoot or per meter of canopy length. Also used as a synonym for →crop size.

  Crop load The amount of fruit per vine in relation to vine size measured as pruning weight or leaf surface.

Crop recovery The amount of fruit produced from new growth following frost or other injury.
Crop size The amount of fruit per vine or per unit of land area. Also used as synonym for →crop level.
Cultivar A →variety of plant found only under cultivation. A group of plants that has been selected for a particular attribute or combination of attributes that is clearly distinct, uniform, and stable in these characteristics, and that retains those characteristics when propagated.
Cuticle (Latin *cutis* = skin) Waxy or fatty layer on the outer wall of epidermal cells, formed of cutin and wax; largely impervious to water vapor and carbon dioxide; prevents uncontrolled water loss.
Cutting A portion of cane usually 30–40 cm long, used for propagation.
Cytokinins (Greek *kytos* = hollow; *kinein* = to move) Group of plant hormones that stimulate cell division and formation of →meristems in shoots but reduce it in roots. Zeatin is the main active cytokinin. Cytokinins are produced from →ATP (and ADP) and isoprenoids in the root tips, shoot meristems, and immature seeds and transported throughout the plant in both the xylem and phloem.
Cytoplasm The internal, highly heterogeneous volume of a →cell, composed of the →cytosol and various insoluble, suspended particles and organelles arranged in a porous, elastic network similar to a sponge filled with liquid.
Cytosol An aqueous →cell solution with a complex composition (e.g., →enzymes, →RNA, →amino acids, metabolites, ions) and a gel-like consistency.
Differentiation Developmental process by which a relatively unspecialized cell undergoes a progressive change to a more specialized cell. The increasing specialization of cells and tissues for particular structures or functions during development.
Diffusion Net movement of molecules in the direction of lower concentration.
Diploid Containing two copies of the same set of →chromosomes in each cell, normally arising from →fertilization.
DNA Deoxyribonucleic acid; the helical chain molecule (polymer) that composes the →genes. DNA is the storage form of genetic information located in the →chromosomes. It consists of long chains of nucleotide building blocks composed of a purine or pyrimidine base, deoxyribose sugar, and phosphate. One pair of bases located on opposite strands of DNA constitutes a letter of the genetic code. Groups of three consecutive bases (triplets) specify (i.e., are the "recipe" for) the different →amino acids in addition to "start" and "stop" signals (i.e., "punctuation marks" that determine beginning and end of a protein). The sequence of triplets determines the various →proteins.
Domestication Selection of wild plants (or animals) for adaptation to cultivation and use by humans. The process usually involves selection of traits regarded as beneficial. Such traits may be present in wild plants or arise via spontaneous or induced →mutations.
Dormancy (Latin *dormire* = to sleep) Temporary suspension of visible growth of plants, buds, or seeds. Growth cannot resume without special environmental cues.
Ecosystem (Greek *oikos* = house) A major interacting system that involves both living organisms and their physical environment.
Enzyme (Greek *en* = into; *zyme* = yeast, sourdough) A →protein (or →RNA) capable of speeding up (catalyzing) specific biochemical reactions by lowering the required activation energy, but which is unaltered itself in the process; a biological catalyst, the "workhorse" of a biochemical reaction.
Ethylene $H_2C=CH_2$; The only gaseous plant hormone. Ethylene induces organ abscission, dormancy, and root hair formation. Ethylene acts by increasing the tissue's sensitivity to →gibberellin.
Evaporation The physical change of a liquid into its gas phase.
Evapotranspiration The →evaporation of water from a land surface, including →transpiration from plants and evaporation from the soil and surface water bodies.
Evolution Change in the genetic makeup of a population of organisms over time due to →mutations and subsequent unequal reproduction (natural selection) of individual organisms. →Species often evolve by giving an existing →gene a new job, rather than by "inventing" new genes.

Fatty acid Organic molecule made up of a polar (hydrophilic, i.e., water-attracting) head with two oxygen atoms and a hydrophobic (i.e., water-repellent) chain ("tail") of carbon atoms to which hydrogen atoms are bound. The hydrocarbon tail can be saturated (i.e., all carbon atoms are joined in single bonds) or unsaturated (i.e., some carbon atoms are joined in double bonds). Important constituent of cell →membranes, fats, oils, and waxes.

Feedback Regulation of the rate of a biochemical pathway by its own products; can be positive (a product speeds up the pathway) or negative (a product slows down the pathway by inhibiting an earlier step).

Fertilization Fusion of the sperm cells (male) from the pollen tube with the egg cells (female) of the ovule, resulting in the →zygote and leading to →fruit set. Also, the application of mineral plant nutrients to the soil or the leaves.

Field capacity The percentage of water a particular soil can hold against the action of gravity, i.e., the water remaining in the soil immediately after all free water has drained away and before the soil begins to dry down, which occurs at a soil water potential of about -0.01 MPa.

Fluorescence Light emission by an excited (unstable, higher energy) molecule (such as →chlorophyll upon light absorption) upon its reversion to the ground (normal, stable, lower energy) state. The emitted light is always of a longer wavelength (lower energy) than the absorbed light.

Fruit A mature ovary (berry) or cluster of mature ovaries.

Fruit set A stage of berry development 1–3 weeks after →bloom when many of the flowers have fallen and those remaining have set and developed into berries. Also called berry set stage or shatter stage; fruit set is preferred.

Gamete Female or male reproductive (sex) →cell (e.g., egg and sperm), produced by →meiosis.

Gene The fundamental hereditary unit, composed of multiple pairs of →DNA bases and usually located within a segment of a →chromosome. The "recipe" for a →protein; also, a developmental "on/off switch" in specific tissues at specific times; also a "device" to extract information from the environment. Genes can be switched on/off (i.e., "expressed"/"repressed") by acetylation/deacetylation (i.e., addition/removal of a $COCH_3$ group) of the supporting histone protein, or by demethylation/methylation (i.e., removal/addition of a CH_3 group) of the DNA itself.

Genome The complete set of →DNA of a particular organism. Grapevines, like all plants, have three distinct genomes: one in the →nucleus (where genome refers to the entire ~500 million base-pair DNA sequence of one set of →chromosomes), one in the →mitochondria, and one in the plastids (e.g., →chloroplasts). The range of genetic variation in a population of individual organisms (i.e., in a →species) forms its gene pool.

Genotype The genetic makeup (set of →genes) of an individual organism.

Gibberellins Also termed gibberellic acids. Group of over 100 structurally similar plant hormones, a few of which are biologically active and mainly induce cell elongation and cell differentiation in meristematic tissues. Gibberellins are cyclic diterpenoids and are produced in meristematic shoot and root tips and leaf vascular tissues.

Girdling The removal of a complete ring of bark (outward of the →cambium) from a shoot, cane, or trunk. Also called cincturing or ringing. Girdling temporarily interrupts the transport of assimilates in the →phloem.

Glutamine An →amino acid; the primary form in which assimilated nitrogen is transported in the →xylem and →phloem of grapevines.

Glycoside (Greek *glykys* = sweet) Compound consisting of a sugar bound to another molecule belonging to a variety of chemical classes.

Gradient Difference in concentration or other physical entity per unit distance or across a →membrane.

Grafting The (usually deliberate) fusion of parts from two or more different plants. A union of different individuals in which a portion, called the →scion, of one individual is inserted into a stem or root, called the stock or →rootstock, of the other individual.

Gravitropism (Latin *gravis* = heavy; Greek *tropos* = direction) Orientation of plants or plant organs in relation to gravity.

Guttation (Latin *gutta* = drop) The exudation of liquid water from leaves, caused by root pressure resulting from osmotic water uptake.

Haploid Containing only one copy of a →chromosome set in the cell; →gametes are haploid.

Heterozygous Having two →alleles (i.e., two distinct copies of a →gene, often one dominant and one recessive) at the same chromosome position (locus) but on different →chromosomes within a chromosome pair of a →diploid →cell. The dominant alleles determine the →phenotype, because their instructions are preferred over those of the recessive alleles.

Homozygous Having the same →allele (i.e., two identical copies of a →gene) on both →chromosomes of →diploid cells.

Hormone (Greek *hormaein* = to stimulate) Organic substance produced in minute amounts in one plant part (tissue or organ), from which it is transported to another part of the same plant on which it has a specific effect. Hormones function as highly specific chemical signals between cells.

Humus Decomposing organic matter in the soil.

Hybrid The result of a cross between differing plants or taxonomic units. Most modern grape cultivars are intraspecific hybrids (obtained by crossing vines of the same →species), but some are interspecific hybrids (obtained by crossing vines belonging to different species).

Inflorescence The flowering cluster of a grapevine.

Internode The section of a shoot or cane between two adjacent nodes.

Latent bud A bud that has remained undeveloped for a year or longer; usually the source of a →watershoot.

Lateral A branch or side shoot arising from a prompt bud in the axil of a true leaf on a main shoot. Lateral growth is promoted by topping the main shoot; lateral shoots may vary in length from a few mm to greater than 1 m. Also, a branch of the main axis of a cluster.

Leaf A flat, thin expanded organ growing from the shoot, consisting of a broad →blade, a →petiole, and two inconspicuous stipules (leaf-like structures) at the base of the petiole.

Leaf scar The scar left on the shoot after a leaf falls.

Lenticel Porous region of the →periderm with loosely packed cells; permits gas exchange.

Lipid Small, water-insoluble (i.e., hydrophobic) molecule made up of →fatty acids, sterols, or isoprenoids; important component of cell →membranes.

Macronutrient (Greek *macros* = large; Latin *nutrire* = to nourish) Inorganic chemical element that is required in large amounts for plant growth, such as nitrogen, potassium, phosphorus, calcium, magnesium, and sulfur.

Matrix The aqueous content of a →cell or cell organelle with its dissolved solutes.

Maturity Stage of fruit development when the fruit has reached satisfactory quality for its intended use.

Meiosis (Greek *meioun* = to make small) Special type of →cell division that occurs in anthers and ovaries and reduces a diploid cell to four haploid cells to become →gametes. Random halving of pairs of →chromosomes and random exchange (mixing) of maternal and paternal →genes before meiotic cell division produces genetic variation and repairs →DNA damage in sexually reproducing organisms.

Membrane (Latin *membrana* = skin) Thin, tough, hydrophobic (bi-)layer consisting of →lipids and →proteins that surrounds →cells and cell organelles, separating their contents from the surroundings. Membranes are barriers to the free passage of ions and other molecules.

Meristem (Greek *meristes* = divider) Tissue in which cell division predominates. The undifferentiated, perpetually young plant tissue from which new cells arise. See also →cambium.

    Apical meristem The meristem at the tip of a root or shoot.

    Intercalary meristem (Latin *intercalare* = to insert) Meristem that produces new primary tissues between more differentiated tissues, as in an internode of a shoot or the interveinal part of a leaf.

Micronutrient (Greek *mikros* = small; Latin *nutrire* = to nourish) Inorganic chemical element that is required only in very small, or trace, amounts for plant growth, such as iron, copper, manganese, zinc, molybdenum, nickel, and boron.

Mitochondrion, pl. mitochondria →Cell organelle responsible for →respiration; the "power plant" providing energy for metabolic processes. Mitochondria descended from symbiotic microorganisms that invaded (or were "swallowed" by) larger cells and still have their own →genes.

Mitosis (Greek *mitos*=string, loop) The process of →DNA duplication, →chromosome replication (by lengthwise splitting), and →cell division during growth. Each of two identical daughter cells receives one set of chromosomes from the mother cell.

Mutation (Latin mutatio=change) A spontaneous or engineered change in the →genotype which may alter the →phenotype. Mutations generally result from copying errors during →mitosis or →meiosis and provide the raw material for →evolution and plant breeding.

NAD, NADP Nicotinamide adenine dinucleotide, nicotinamide adenine dinucleotide phosphate; →coenzymes that serve as carriers of hydrogen atoms and electrons.

Necrosis (Greek *nekrois*=death) Uncontrolled, sudden death of some or all of the cells of a tissue or an organ due to loss of →membrane integrity and flooding of cells with water.

Node (Latin *nodus*=knot) The enlarged portion of a shoot or cane at which leaves, clusters, tendrils, and/or buds are located at regular intervals.

Nucleus The dense central organelle of the cell; contains the →chromosomes and is enclosed by a nuclear →membrane.

Osmosis (Greek *osmos*=impulse, thrust) The diffusion of water across a selectively permeable →membrane. In the absence of other forces, osmotic water movement is in the direction of higher solute concentration (lower water concentration or →water potential).

Overcropping The production of more crop than a grapevine can bring to maturity by normal harvest time.

Own-rooted A grapevine grown from a cutting that develops its own root system as opposed to a vine grafted or budded onto a →rootstock.

Pathogen A disease-causing agent or organism.

Pedicel The stalk of an individual flower or berry in a grape cluster.

Peduncle The stalk attaching a →cluster to the →shoot; the branched axis of a cluster apart from the →pedicels of individual flowers.

Perennial (Latin *per*=through; *annuus*=year) A plant, such as the grapevine, in which the vegetative structures live through multiple growing seasons.

Periderm The →cork cambium and the waterproof tissues it produces; outer bark.

Permanent wilting point The percentage of water remaining in the soil when it has dried down so much that a plant growing in that soil fails to recover from wilting even if placed in a humid chamber. Occurs at a soil water potential of about -1.5 MPa.

Pest Species, strain or biotype of any plant, animal, or pathogen harmful to plants or plant products.

Petiole The stalk attaching the →leaf blade to the shoot.

pH A symbol denoting the relative concentration of hydronium ions (H_3O^+) or protons (H^+, in mol L^{-1}) in a solution; $pH = -\log[H^+] \rightarrow [H^+] = 10^{-pH}$. pH values run from 0 to 14, and the lower the value the more acidic is a solution, i.e., the more protons it contains; pH 7 is neutral (e.g., pure water and →cytoplasm), less than 7 is acidic (e.g., grape juice), and more than 7 is alkaline (e.g., lime soil).

Phenols, phenolics A broad range of ultraviolet light–absorbing organic compounds, all of which have a hydroxyl (OH) group attached to an aromatic ring (a ring of six carbon atoms containing three double bonds); includes cinnamic acids, flavonoids, stilbenes, tannins, and lignins. Phenolics are important determinants of wine quality, mainly through →tannins and →anthocyanins, and contribute to disease resistance of grapevines.

Phenotype The whole of all the characteristics of an individual organism; the physical expression, resulting from interaction with the environment, of the →genotype during its development.

Phloem (Greek *phloos*=bark) The assimilate-conducting tissue of plants, which is composed of living cells. In grapevines old phloem comprises most of the bark.

Photosynthesis (Greek *photos* = light; *syn* = together; *tithenai* = to place) The conversion of light energy to chemical energy, and the process by which a plant converts carbon dioxide and water into carbohydrates. Solar radiation is the energy source for this process.

Phytochrome (Greek *phyton* = plant; *chroma* = color) A pigment →protein located in the →cytoplasm of plants that acts as a light sensor. Photoreceptor for red and far-red light; involved in a number of timing processes, such as leaf formation, flowering, dormancy, and seed germination, as well as shade-avoidance responses.

Phytotoxic Causing injury or death of plants or portions of plants.

Pigment A substance that absorbs light, often selectively (i.e., only particular bands of wavelengths), giving the visual appearance of a specific color; →anthocyanin, →chlorophyll, →phytochrome.

Plasmodesma (plural plasmodesmata) Minute channel that crosses plant cell walls to connect neighboring →cells that form a →symplast.

Pomace Solid material remaining after pressing of grape juice or wine.

Pollination (Latin *pollen* = fine dust) The transfer of pollen from the male part of a flower (anther) to the female part of a flower (stigma).

Primordium (Latin *primus* = first; *ordiri* = to begin to weave) A cell or an organ in its earliest stage of differentiation.

Prompt bud Axillary bud that develops into a →lateral shoot in the same growing season in which the bud is produced.

Protein (Greek *proteios* = primary) A complex organic compound, a polypeptide, composed of a string of just a few to several thousand →amino acids. Proteins are folded into a plethora of three-dimensional shapes that define their function. See also →enzyme.

Pruning The process of cutting away of a portion of the annual vegetative growth of a grapevine to maintain the desired number and spacing of →buds or →nodes per vine.

    Summer pruning Trimming or topping of growing →shoots during the growing season to regulate shoot length, reduce →cluster shading, and maintain canopy shape.

    Winter pruning Also called dormant pruning. Carried out during a grapevine's dormant period to regulate the →bud number and maintain vine form.

Q_{10} The proportional change in the rate of a (bio)chemical reaction per 10°C increase in temperature; e.g., for $Q_{10} = 2$, respiration will double for each 10°C rise in temperature.

Rachis (Greek *rachis* = backbone) The central axis of a →cluster stem.

Raisin A dried grape.

Receptacle The part of the axis of the flower or fruit stalk that bears the floral organs or grape berry.

Respiration An intracellular process in which molecules, particularly pyruvate in the Krebs cycle, are oxidized (i.e., have electrons removed) to produce energy (→ATP). The complete breakdown of sugar or other organic compounds, such as malate, to carbon dioxide (CO_2) and water.

RNA Ribonucleic acid; the chain molecule that translates the genetic code (→DNA) into →amino acids and hence →proteins. RNA is a working copy and transport form of genetic information (i.e., the link between DNA and protein). It consists of long chains of nucleotide building blocks composed of a purine or pyrimidine base, ribose sugar, and phosphate and is produced from →ATP and other nucleotide triphosphates.

Root The belowground axis of the grapevine, which serves to anchor the plant and to absorb and conduct water and nutrients; it bears only lateral roots. The true root develops from the hypocotyl of the embryo; the grapevine root is a taproot, an axis with lateral roots initiated near the tip.

Rootstock Specialized planting material to which fruiting varieties of grapes are grafted as →scions to produce a commercially acceptable vine. Grape rootstock varieties are used for their tolerance of or resistance to root parasites, such as phylloxera and nematodes, for their vigor, or for their tolerance of adverse soil conditions.

Sap The fluid contents of the →xylem or →phloem conduits. The fluid contents of the →vacuole are called cell sap.

Scion The →shoot or →cane portion that is grafted or budded onto a →rootstock.
Senescence (Latin *senescere* = to grow old) Aging. The ordered degradation of cell constituents that ends in death of the cell.
Sexual reproduction Propagation of plants by means of seeds following fertilization.
Shoot The current growing season's green stem growth. It becomes a →cane when over half of its length becomes woody with a tan or brown color.
Sink An assimilate-importing plant organ. All plant organs unable to meet their own nutritional demands may act as sinks, i.e., expanding leaves or growing fruit or storage organs when they are being replenished.
Sinus The gap in the margin between two individual lobes of a →leaf blade. The petiolar sinus is the gap at the attachment of the →petiole.
Somatic cell (Greek *soma* = body) A body →cell. All cells of a plant are somatic except the cells (spores) produced by →meiosis that develop into →gametes (although even these are derived from somatic cells, unlike in animals where somatic and germ line cells are kept separate). Somatic cells can be forced through tissue culture to form somatic embryos that can develop into plants that are →clones of their mother plant.
Source An assimilate-exporting plant organ. Photosynthesizing leaves are the primary sources, but storage organs, such as the trunk or roots, may also serve as important sources, especially during budbreak.
Species Population or groups of populations of sexually reproducing organisms that interbreed with each other but are reproductively isolated from other populations.
Spur A short fruiting unit of 1-year growth; the basal section of a →cane that is cut back to one to three →buds. These are used in spur →pruning to provide fruiting shoots for the current year and in cane pruning to provide replacement canes for the following year.
Starch A complex insoluble carbohydrate, composed of 1000 or more glucose units arranged in α-1,4- and α-1,6-glucan chemical bonds; the main food storage substance of grapevines.
Stoma, pl. stomata (Greek *stoma* = mouth) A minute opening bordered by guard cells in the epidermis of leaves, stems, and grapes through which gases pass. The stoma regulates the exchange of carbon dioxide used in photosynthesis and water used in transpiration between the plant and the atmosphere.
Suberin (Latin *suber* = cork) Fatty substance impervious to water and gases; deposited in some →cell walls, especially in the roots and leaves.
Sucker A →watershoot that emerges from the base of the established trunk of the vine, at or belowground level.
Sucrose A disaccharide (glucose plus fructose); the primary form in which sugar produced by photosynthesis is transported in the →phloem of grapevines.
Symbiosis (Greek *syn* = together, at the same time) The mutually beneficial living together of two or more species in a long-term relationship.
Symplast (Greek *plastos* = formed) The totality of protoplasts; i.e., everything inside a plant that is bound inside a membrane. The continuum of cytoplasm interconnected by →plasmodesmata and excluding the →vacuoles.
Tannin A macromolecule, belonging to the →phenolic class of flavonoids, found in the cell vacuoles of the skin and seeds of grape berries. Tannins are responsible for the astringency and bitterness of grapes and wine.
Tendril (Latin *tendere* = to extend) A part of the stem modified into a slender twining organ on a shoot opposite a leaf that can coil around an object for support.
Terpene An organic hydrocarbon or one of its derivatives, assembled from multiple isoprenoid units; responsible for certain odors or flavors of grape berries (and of flowers, fruits, and leaves of other plants).
Tissue A group of similar cells organized into a structural and functional unit.
Transpiration (Latin *trans* = over, beyond; *spirare* = to breathe) The loss of water vapor by plant parts; most transpiration occurs through →stomata. Transpiration is required for cooling of plant tissues.
Trellis A permanent structure that supports the vine framework, consisting of stakes, posts and, usually, at least one wire.
Trunk The main thickened stem of a vine between the →roots and the →head or →cordon of the vine.

Turgor Outward pressure of cell sap due to water absorption by →osmosis across a →membrane, balanced by the rigid cell wall; turgor pressure is necessary for cell expansion during growth.

Vacuole Organelle of plant →cells for storage, degradation, and recycling of a wide variety of compounds and waste products, surrounded by a membrane called tonoplast. Different types of specialized vacuoles can be present within the same cell; may represent up to 90% of the volume of a mature cell. Osmotic water absorption creates outward pressure (→turgor).

Variety A group of closely related plants of common origin and similar characteristics within a species. The so-called grape varieties are botanically →cultivars.

Vascular bundle (Latin *vasculum* = small vessel) A strand of tissue containing →xylem and →phloem; used for conducting water, nutrients, and assimilates.

Vegetative propagation Asexual propagation; the propagation of grapevines (and other organisms) without fertilization but by means of cuttings, layering, meristem or shoot-tip culture, and micropropagation.

Vein A strand of vascular tissues in a flat organ such as a leaf.

Veraison (French *vérer* = to change) Onset of ripening of a grape berry as indicated by a change in color, which is preceded by softening and the beginning of sugar accumulation.

Vigor The rate of shoot growth measured by the change in shoot length over time.

Virus Self-replicating complex of nucleic acids (either →DNA or →RNA) and →proteins that infects living organisms and requires an intact host →cell for replication.

Watershoot Rapidly growing shoot arising from a latent bud on a →trunk or →cordon. A noncount shoot that arises on wood older than 1 year and originates from a →basal bud.

Water potential A measure of water concentration. The free energy of an aqueous solution per unit molar volume of liquid water; typically the sum of the component potentials arising from the effects of →turgor pressure and dissolved molecules (→osmosis).

Weed Generally an herbaceous plant not valued for use or beauty, growing wild, and regarded as intruding on space or hindering the growth of grapevines.

Wood Secondary →xylem. The tissue produced on the inside of the vascular →cambium.

Xylem (Greek *ksúlon* = wood) The dead, woody portion of conducting tissues whose function is the transport of water and minerals and support of the roots, trunk, and shoots.

Zygote Fertilized egg resulting from the fusion of a male and a female →gamete (→fertilization) that forms a zygotic embryo which develops into a seed.

Abbreviations and symbols

A	photosynthesis rate
ABA	abscisic acid
ADP	adenosine diphosphate
ATP	adenosine triphosphate
c	solute concentration
C_a	atmospheric CO_2 concentration
CHS	chalcone synthase
C_i	substomatal CO_2 concentration
DMS	dimethyl sulfide
DNA	deoxyribonucleic acid
E	transpiration rate
ET	evapotranspiration
F	volume flow rate
FCR	ferric chelate reductase
Fd_{ox}	oxidized ferredoxin
Fd_{red}	REDUCED ferredoxin
FR	far-red light
g_b	boundary layer conductance
GDD	growing degree days
GDH	glutamate dehydrogenase
g_l	leaf conductance
GOGAT	glutamine oxoglutarate aminotransferase or glutamate synthase
GS	glutamine synthetase
g_s	stomatal conductance
J	flux or flow rate per unit area
K_c	crop factor
l_h	hydraulic conductance
MEP	methyl-erythritol 4-phosphate
NADH	nicotinamide adenine dinucleotide
NADPH	nicotinamide adenine dinucleotide phosphate
NiR	nitrite reductase
NR	nitrate reductase
P	hydrostatic pressure or turgor pressure
P_a	atmospheric pressure
PAL	phenylalanine ammonia lyase
PAR	photosynthetically active radiation
P_{fr}	active phytochrome
P_i	inorganic phosphate
P_r	inactive phytochrome
PR	pathogenesis-related
PRD	partial rootzone drying
PSI	photosystem I
PSII	photosystem II
P_x	xylem pressure

Q_{10}	temperature coefficient
R	red light
R	universal gas constant
r_b	boundary layer resistance
RDI	regulated deficit irrigation
r_h	hydraulic resistance
RH	relative humidity
RNA	ribonucleic acid
RQ	respiratory quotient
r_s	stomatal resistance
Rubisco	ribulose bisphosphate carboxylase/oxygenase
RWC	relative water content
SAR	sodium adsorption ratio
T	temperature
TCA	tricarboxylic acid, also 2,4,6-trichloroanisole
TDN	trimethyl-dihydronaphthalene
UDP	uridine diphosphate
UV	ultraviolet
VPD	vapor pressure deficit
WUE	water use efficiency
Δ	difference or gradient
π	osmotic pressure
σ	reflection coefficient
Ψ	water potential
Ψ_G	gravitational potential
Ψ_M	matrix potential
$\Psi\pi$	osmotic potential

Bibliography

Abass, M., Rajashekar, C.B., 1991. Characterization of heat injury in grapes using ^{1}H nuclear magnetic resonance methods. Plant Physiol. 96, 957–961.

Acevedo De la Cruz, A., Hilbert, G., Rivière, C., Mengin, V., Ollat, N., Bordenave, L., Decroocq, S., Delaunay, J.C., Delrot, S., Mérillon, J.M., Monti, J.P., Gomès, E., Richard, T., 2012. Anthocyanin identification and composition of wild *Vitis* spp. accessions by using LC-MS and LC-NMR. Anal. Chim. Acta 732, 145–152.

Achard, P., Genschik, P., 2009. Releasing the brakes of plant growth: how GAs shutdown DELLA proteins. J. Exp. Bot. 60, 1085–1092.

Achard, P., Gong, F., Cheminant, S., Alioua, M., Hedden, P., Genschik, P., 2008. The cold-inducible CBF1 factor-dependent signaling pathway modulates the accumulation of the growth-repressing DELLA proteins via its effect on gibberellin metabolism. Plant Cell 20, 2117–2212.

Acheampong, A.K., Hu, J., Rotman, A., Zheng, C., Halaly, T., Takebayashi, Y., Jikumaru, Y., Kamiya, Y., Lichter, A., Sun, T.P., Or, E., 2015. Functional characterization and developmental expression profiling of gibberellin signalling components in *Vitis vinifera*. J. Exp. Bot. 66, 1463–1476.

Acimovic, D., Tozzini, L., Green, A., Sivilotti, P., Sabbatini, P., 2016. Identification of a defoliation severity threshold for changing fruit-set, bunch morphology and fruit composition in Pinot Noir. Aust. J. Grape Wine Res. 22, 399–408.

Adam-Blondon, A.F., Lahogue-Esnault, F., Bouquet, A., Boursiquot, J.M., This, P., 2001. Usefulness of two SCAR markers for marker-assisted selection of seedless grapevine cultivars. Vitis 40, 147–155.

Adams, D.O., 2006. Phenolics and ripening in grape berries. Am. J. Enol. Vitic. 57, 249–256.

Adams, D.O., Liyanage, C., 1993. Glutathione increases in grape berries at the onset of ripening. Am. J. Enol. Vitic. 44, 333–338.

Adams, D.O., Franke, K.E., Christensen, L.P., 1990. Elevated putrescine levels in grapevine leaves that display symptoms of potassium deficiency. Am. J. Enol. Vitic. 41, 121–125.

Agaoglu, Y.S., 1971. A study on the differentiation and the development of floral parts in grapes (*Vitis vinifera* L. var.). Vitis 10, 20–26.

Agasse, A., Vignault, C., Kappel, C., Conde, C., Gerós, H., Delrot, S., 2009. Sugar transport and sugar sensing in grape. In: Roubelakis-Angelakis, K.A. (Ed.), Grapevine Molecular Physiology and Biotechnology. Springer, Dordrecht, The Netherlands, pp. 105–139.

Agati, G., Tattini, M., 2010. Multiple functional roles of flavonoids in photoprotection. New Phytol. 186, 786–793.

Ageorges, A., Issaly, N., Picaud, S., Delrot, S., Romieu, C., 2000. Identification and functional expression in yeast of a grape berry sucrose carrier. Plant Physiol. Biochem. 38, 177–185.

Agüero, C.B., Rodríguez, J.G., Martínez, L.E., Dangl, G.S., Meredith, C.P., 2003. Identity and parentage of Torrontés cultivars in Argentina. Am. J. Enol. Vitic. 54, 318–321.

Agusti, J., Herold, S., Schwarz, M., Sanchez, P., Ljung, K., Dun, E.A., 2011. Strigolactone signaling is required for auxin-dependent stimulation of secondary growth in plants. Proc. Natl. Acad. Sci. U. S. A. 108, 20242–20247.

Ahmedullah, M., Kawakami, A., 1986. Evaluation of laboratory tests for determining the lethal temperature of *Vitis labruscana* Bailey Concord roots exposed to subzero temperatures. Vitis 25, 142–150.

Ainsworth, E.A., Rogers, A., 2007. The response of photosynthesis and stomatal conductance to rising [CO_2]: mechanisms and environmental interactions. Plant Cell Environ. 30, 258–270.

Akiyama, K., Matsuzaki, K., Hayashi, H., 2005. Plant sesquiterpenes induce hyphal branching in arbuscular mycorrhizal fungi. Nature 435, 824–827.

Aktas, H., Karni, L., Chang, D.C., Turhan, E., Bar-Tal, A., Aloni, B., 2005. The suppression of salinity-associated oxygen radicals production, in pepper (*Capsicum annuum*) fruit, by manganese, zinc and calcium in relation to its sensitivity to blossom-end rot. Physiol. Plant. 123, 67–74.

Alabi, O.J., Casassa, L.F., Gutha, L.R., Larsen, R.C., Henick-Kling, T., Harbertson, J.F., Naidu, R.A., 2016. Impacts of grapevine leafroll disease on fruit yield and grape and wine chemistry in a wine grape (*Vitis vinifera* L.) cultivar. PLoS ONE 11, e0149666.

Alboresi, A., Gestin, C., Leydecker, M.T., Bedu, M., Meyer, C., Truong, H.N., 2005. Nitrate, a signal relieving seed dormancy in *Arabidopsis*. Plant Cell Environ. 28, 500–512.

Alcázar, R., Altabella, T., Marco, F., Bortolotti, C., Reymond, M., Koncz, C., 2010. Polyamines: molecules with regulatory functions in plant abiotic stress tolerance. Planta 231, 1237–1249.

Allègre, M., Daire, X., Héloir, M.C., Trouvelot, S., Mercier, L., Adrian, M., 2007. Stomatal deregulation in *Plasmopara viticola*-infected grapevine leaves. New Phytol. 173, 832–840.

Allen, M.S., 1982. Wine processing and its influence on pH—effects of fermentation, fortifications, water contamination and potassium bitartrate precipitation. Aust. Grapegrower Winemaker 220, 57–60.

Allen, G.J., Chu, S.P., Harrington, C.L., Schumacher, K., Hoffman, Y., Tang, Y.Y., 2001. A defined range of guard cell calcium oscillation parameters encodes stomatal movements. Nature 411, 1053–1057.

Alleweldt, G., 1957. Der Einfluß von Photoperiode und Temperatur auf Wachstum und Entwicklung von Holzpflanzen unter besonderer Berücksichtigung der Gattung *Vitis*. Vitis 1, 159–180.

Alleweldt, G., 1958. Eine Frühdiagnose zur Bestimmung der Fruchtbarkeit von Reben. Vitis 4, 230–236.

Alleweldt, G., 1959. Förderung des Infloreszenzwachstums der Reben durch Gibberellinsäure. Vitis 2, 71–78.

Alleweldt, G., 1960. Untersuchungen über den Austrieb der Winterknospen von Reben. Vitis 2, 134–152.

Alleweldt, G., 1961. Hemmung der Blütenbildung von *Vitis rupestris* durch Gibberellin. Naturwissenschaften 45, 628–629.

Alleweldt, G., 1964. Über die Nachwirkung von Umweltfaktoren auf das vegetative Wachstum von Rebensteck-lingen im Folgejahr. Z. Acker Pflanzenbau 119, 178–194.

Alleweldt, G., 1965. Über den Einfluss der Temperatur auf die Blutung von Reben. Vitis 5, 10–16.

Alleweldt, G., 1966. Die Differenzierung der Blütenorgane der Rebe. Wein-Wiss. 21, 393–402.

Alleweldt, G., Düring, H., 1972. Der Einfluss der Photoperiode auf Wachstum und Abscisinsäuregehalt der Rebe. Vitis 11, 280–288.

Alleweldt, G., Hifny, H.A.A., 1972. Zur Stiellähme der Reben II. Kausalanalytische Untersuchungen. Vitis 11, 10–28.

Alleweldt, G., Hofäcker, W., 1975. Einfluss von Umweltfaktoren auf Austrieb, Blüte, Fruchtbarkeit und Triebwachstum bei der Rebe. Vitis 14, 103–115.

Alleweldt, G., Ilter, E., 1969. Untersuchungen über die Beziehungen zwischen Blütenbildung und Triebwachstum bei Reben. Vitis 8, 286–313.

Alleweldt, G., Istar, A., 1969. Über die apikale Dominanz bei Reben. Vitis 8, 94–104.

Alleweldt, G., Merkt, N., 1992. Der Stickstoffexport der Wurzel und die Zusammensetzung des Xylemexsudats. Teil 1: Der Einfluss einer zunehmenden Stickstoffdüngung. Vitis 31, 121–130.

Alleweldt, G., Possingham, J.V., 1988. Progress in grapevine breeding. Theor. Appl. Genet. 75, 669–673.

Alleweldt, G., Düring, H., Waitz, G., 1975. Untersuchungen zum Mechanismus der Zuckereinlagerung in die wachsenden Weinbeeren. Angew. Bot. 49, 65–73.

Alleweldt, G., Engel, M., Gebbing, H., 1981. Histologische Untersuchungen an Weinbeeren. Vitis 20, 1–7.

Alleweldt, G., Düring, H., El-Sese, A.M.A., 1984a. The influence of nitrogen fertilization and water supply on photosynthesis, transpiration and dry matter production in grapevines (*Vitis vinifera*). Plant Res. Dev. 20, 45–58.

Alleweldt, G., Düring, H., Jung, K.H., 1984b. Zum Einfluß des Klimas auf Beerenentwicklung, Ertrag und Qualität bei Reben: Ergebnisse einer siebenjährigen Faktorenanalyse. Vitis 23, 127–142.

Aloni, R., 2001. Foliar and axial aspects of vascular differentiation: hypotheses and evidence. J. Plant Growth Regul. 20, 22–34.

Aloni, R., 2013. Role of hormones in controlling vascular differentiation and the mechanism of lateral root initiation. Planta 238, 819–830.

Aloni, R., Raviv, A., Peterson, C.A., 1991. The role of auxin in the removal of dormancy callose and resumption of phloem activity in *Vitis vinifera*. Can. J. Bot. 69, 1825–1832.

Aloni, R., Schwalm, K., Langhans, M., Ullrich, C.I., 2003. Gradual shifts in sites of free-auxin production during leaf-primordium development and their role in vascular differentiation and leaf morphogenesis in *Arabidopsis*. Planta 216, 841–853.

Aloni, R., Langhans, M., Aloni, E., Dreieicher, E., Ullrich, C.I., 2005. Root-synthesized cytokinin in *Arabidopsis* is distributed in the shoot by the transpiration stream. J. Exp. Bot. 56, 1535–1544.

Aloni, R., Aloni, E., Langhans, M., Ullrich, C.I., 2006a. Role of auxin in regulating *Arabidopsis* flower development. Planta 223, 315–328.

Aloni, R., Aloni, E., Langhans, M., Ullrich, C.I., 2006b. Role of cytokinin and auxin in shaping root architecture: regulating vascular differentiation, lateral root initiation, root apical dominance and root gravitropism. Ann. Bot. 97, 883–893.

Alsina, M.M., Smart, D.R., Bauerle, T., de Herralde, F., Biel, C., Stockert, C., 2011. Seasonal changes of whole root system conductance by a drought-tolerant grape root system. J. Exp. Bot. 62, 99–109.

Altieri, M.A., Ponti, L., Nicholls, C.I., 2005. Manipulating vineyard biodiversity for improved insect pest management: case studies from northern California. Int. J. Biodivers. Sci. Manag. 1, 191–203.

Amâncio, S., Tavares, S., Fernandes, J.C., Sousa, C., 2009. Grapevine and sulfur: old partners, new achievements. In: Roubelakis-Angelakis, K.A. (Ed.), Grapevine Molecular Physiology and Biotechnology. Springer, Dordrecht, The Netherlands, pp. 31–52.

Améglio, T., Ewers, F.W., Cochard, H., Martignac, M., Vandame, M., Bodet, C., 2001. Winter stem xylem pressure in walnut trees: effects of carbohydrates, cooling and freezing. Tree Physiol. 21, 387–394.

Améglio, T., Decourteix, M., Alves, G., Valentin, V., Sakr, S., Julien, J.L., 2004. Temperature effects on xylem sap osmolarity in walnut trees: evidence for a vitalistic model of winter embolism repair. Tree Physiol. 24, 785–793.

Amerine, M.A., Bailey, C.B., 1959. Carbohydrate content of various parts of the grape cluster. Am. J. Enol. Vitic. 10, 196–198.

Amerine, M.A., Root, G.A., 1960. Carbohydrate content of various parts of the grape cluster II. Am. J. Enol. Vitic. 11, 137–139.

Amerine, M.A., Winkler, A.J., 1944. Composition and quality of musts and wines of California grapes. Hilgardia 15, 493–675.

Amrani Joutei, K., Glories, Y., Mercier, M., 1994. Localisation des tanins dans la pellicule de baie de raisin. Vitis 33, 133–138.

Amthor, J.S., 2000. The McCree – de Wit – Penning de Vries – Thornley respiration paradigms: 30 years later. Ann. Bot. 86, 1–20.

Amtmann, A., Blatt, M.R., 2009. Regulation of macronutrient transport. New Phytol. 181, 35–52.

Anastasiou, E., Lenhard, M., 2007. Growing up to one's standard. Curr. Opin. Plant Biol. 10, 63–69.

Andersen, P.C., Brodbeck, B.V., 1989a. Diurnal and temporal changes in the chemical profile of xylem exudate from *Vitis rotundifolia*. Physiol. Plant. 75, 63–70.

Andersen, P.C., Brodbeck, B.V., 1989b. Temperature and temperature preconditioning on flux and chemical composition of xylem exudate from muscadine grapevines. J. Am. Soc. Hortic. Sci. 114, 440–444.

Andersen, P.C., Brodbeck, B.V., 1991. Influence of fertilization on xylem fluid chemistry of *Vitis rotundifolia* Noble and *Vitis* hybrid Suwannee. Am. J. Enol. Vitic. 42, 245–251.

Anderson, K., 2014. Changing varietal distinctiveness of the world's wine regions: evidence from a new global database. J. Wine Econ. 9, 249–272.

Anderson, L.J., Comas, L.H., Lakso, A.N., Eissenstat, D.M., 2003. Multiple risk factors in root survivorship: a 4-year study in Concord grape. New Phytol. 158, 489–501.

Angiosperm Phylogeny Group, 2016. An update of the Angiosperm Phylogeny Group classification for the orders and families of flowering plants: APG IV. Bot. J. Linn. Soc. 181, 1–20.

Antcliff, A.J., May, P., 1961. Dormancy and bud burst in Sultana vines. Vitis 3, 1–14.

Antcliff, A.J., Newman, H.P., Barrett, H.C., 1983. Variation in chloride accumulation in some American species of grapevine. Vitis 22, 357–362.

Antolín, M.C., Baigorri, H., De Luis, I., Aguirrezábal, F., Gény, L., Broquedis, M., 2003. ABA during reproductive development in non-irrigated grapevines (*Vitis vinifera* L. cv. Tempranillo). Aust. J. Grape Wine Res. 9, 169–176.

Apel, K., Hirt, H., 2004. Reactive oxygen species: metabolism, oxidative stress, and signal transduction. Annu. Rev. Plant Biol. 55, 373–399.

Aradhya, M.K., Dangl, G.S., Prins, B.H., Boursiquot, J.M., Walker, A.M., Meredith, C.P., 2003. Genetic structure and differentiation in cultivated grape, *Vitis vinifera* L. Genet. Res. 81, 179–192.

Aradhya, M., Wang, Y., Walker, M.A., Prins, B.H., Koehmstedt, A.M., Velasco, D., Gerrath, J.M., Dangl, G.S., Preece, J.E., 2013. Genetic diversity, structure, and patterns of differentiation in the genus *Vitis*. Plant Syst. Evol. 299, 317–330.

Araujo, F.J., Williams, L.E., 1988. Dry matter and nitrogen partitioning and root growth of young field-grown Thompson Seedless grapevines. Vitis 27, 21–32.

Araújo, W.L., Tohge, T., Ishizaki, K., Leaver, C.J., Fernie, A.R., 2011. Protein degradation—an alternative respiratory substrate for stressed plants. Trends Plant Sci. 16, 489–498.

Arnold, C., Gillet, F., Gobat, J.M., 1998. Situation de la vigne sauvage *Vitis vinifera* ssp. *silvestris* en Europe. Vitis 37, 159–170.

Aroca, R., Porcel, R., Ruiz-Lozano, J.M., 2012. Regulation of root water uptake under abiotic stress conditions. J. Exp. Bot. 63, 43–57.

Arrhenius, S., 1896. On the influence of carbonic acid in the air upon the temperature of the ground. Philos. Mag. J. Sci. 5 (41), 237–276.

Arrigo, N., Arnold, C., 2007. Naturalised *Vitis* rootstocks in Europe and consequences to native wild grapevine. PLoS ONE 2, e521.

Arroyo-García, R., Ruiz-García, L., Bolling, L., Ocete, R., López, M.A., Arnold, C., 2006. Multiple origins of cultivated grapevine (*Vitis vinifera* L. ssp. *sativa*) based on chloroplast DNA polymorphisms. Mol. Ecol. 15, 3707–3714.

Arvanitoyannis, I.S., Ladas, D., Mavromatis, A., 2006. Potential uses and applications of treated wine waste: a review. Int. J. Food Sci. Technol. 41, 475–487.

Aryan, A.P., Wilson, B., Strauss, C.R., Williams, P.J., 1987. The properties of glycosidases of *Vitis vinifera* and a comparison of their β-glucosidase activity with that of exogenous enzymes. An assessment of possible applications in enology. Am. J. Enol. Vitic. 38, 182–188.

Ashley, M.K., Grant, M., Grabov, A., 2006. Plant responses to potassium deficiencies: a role for potassium transport proteins. J. Exp. Bot. 57, 425–436.

Ashworth, E.N., 1982. Properties of peach flower buds which facilitate supercooling. Plant Physiol. 70, 1475–1479.

Atkin, O.K., Tjoelker, M.G., 2003. Thermal acclimation and the dynamic response of plant respiration to temperature. Trends Plant Sci. 8, 343–351.

Atkin, O.K., Bruhn, D., Hurry, V.M., Tjoelker, M.G., 2005. The hot and the cold: unravelling the variable response of plant respiration to temperature. Funct. Plant Biol. 32, 87–105.

Atwell, B., Kriedemann, P., Turnbull, C., 1999. Plants in Action. Adaptation in Nature, Performance in Cultivation. Macmillan Education, South Yarra, Australia.

Aubert, S., Gout, E., Bligny, R., Marty-Mazars, D., Barrieu, F., Alabouvette, J., 1996. Ultrastructural and biochemical characterization of autophagy in higher plant cells subjected to carbon deprivation: control by the supply of mitochondria with respiratory substrates. J. Cell Biol. 133, 1251–1263.

Aubert, S., Bligny, R., Douce, R., Gout, E., Ratcliffe, R.G., Roberts, J.K.M., 2001. Contribution of glutamate dehydrogenase to mitochondrial glutamate metabolism studied by ^{13}C and ^{31}P nuclear magnetic resonance. J. Exp. Bot. 52, 37–45.

Austin, C.N., Grove, G.G., Meyers, J.M., Wilcox, W.F., 2011. Powdery mildew severity as a function of canopy density: associated impacts on sunlight penetration and spray coverage. Am. J. Enol. Vitic. 62, 23–31.

Ayre, B.G., 2011. Membrane-transport systems for sucrose in relation to whole-plant carbon partitioning. Mol. Plant 4, 377–394.

Bachmann, O., 1978. Verbreitung von Phenolcarbonsäuren und Flavonoiden bei Vitaceen. Vitis 17, 234–257.

Bachmann, O., Blaich, R., 1979. Vorkommen und Eigenschaften kondensierter Tannine in Vitaceen. Vitis 18, 106–116.

Bacilieri, R., Lacombe, T., Le Cunff, L., Di Vecchi-Staraz, M., Laucou, V., Genna, B., 2013. Genetic structure in cultivated grapevines is linked to geography and human selection. BMC Plant Biol. 13, 25.

Bacon, M.A., Wilkinson, S., Davies, W.J., 1998. pH-Regulated leaf cell expansion in droughted plants is abscisic acid dependent. Plant Physiol. 118, 1507–1515.

Badr, G., Hoogenboom, G., Abouali, M., Moyer, M., Keller, M., 2018. Analysis of several bioclimatic indices for viticultural zoning in the Pacific Northwest. Clim. Res. 76, 203–223.

Bae, G., Choi, G., 2008. Decoding of light signals by plant phytochromes and their interacting proteins. Annu. Rev. Plant Biol. 59, 281–311.

Bailey-Serres, J., Voesenek, L.A.C.J., 2008. Flooding stress: acclimations and genetic diversity. Annu. Rev. Plant Biol. 59, 313–339.

Baillod, M., Baggiolini, M., 1993. Les stades repères de la vigne. Rev. Suisse Vitic. Arboric. Hortic. 25, 7–9.

Bair, K.E., Davenport, J.R., Stevens, R.G., 2008. Release of available nitrogen after incorporation of a legume cover crop in Concord grape. HortScience 43, 875–880.

Bais, A.J., Murphy, P.J., Dry, I.B., 2000. The molecular regulation of stilbene phytoalexin biosynthesis in *Vitis vinifera* during grape berry development. Aust. J. Plant Physiol. 27, 425–433.

Bais, H.P., Weir, T.L., Perry, L.G., Gilroy, S., Vivanco, J.M., 2006. The role of root exudates in rhizosphere interactions with plants and other organisms. Annu. Rev. Plant Biol. 57, 233–266.

Baker, N.R., 2008. Chlorophyll fluorescence: a probe of photosynthesis *in vivo*. Annu. Rev. Plant Biol. 59, 89–113.

Baker, D.A., Milburn, J.A., 1965. Lateral movement of inorganic solutes in plants. Nature 205, 306–307.

Balint, G., Reynolds, A.G., 2013. Impact of exogenous abscisic acid on vine physiology and grape composition of Cabernet Sauvignon. Am. J. Enol. Vitic. 64, 74–87.

Bálo, B., Mustárdy, L.A., Hideg, E., Faludi-Dániel, F., 1986. Studies on the effect of chilling on the photosynthesis of grapevine. Vitis 25, 1–7.

Bammi, R.K., Randhawa, G.S., 1968. Viticulture in the tropical regions of India. Vitis 7, 124–129.

Banowetz, G.M., Ammar, K., Chen, D.D., 1999. Postanthesis temperatures influence cytokinin accumulation and wheat kernel weight. Plant Cell Environ. 22, 309–316.

Barata, A., Malfeito-Ferreira, M., Loureiro, V., 2012a. Changes in sour rotten grape berry microbiota during ripening and wine fermentation. Int. J. Food Microbiol. 154, 152–161.

Barata, A., Malfeito-Ferreira, M., Loureiro, V., 2012b. The microbial ecology of wine grape berries. Int. J. Food Microbiol. 153, 243–259.

Barbier-Brygoo, H., De Angeli, A., Filleur, S., Frachisse, J.M., Gambale, F., Thomine, S., 2011. Anion channels/transporters in plants: from molecular bases to regulatory networks. Annu. Rev. Plant Biol. 62, 25–51.

Bargel, H., Neinhuis, C., 2005. Tomato (*Lycopersicon esculentum* Mill.) fruit growth and ripening as related to the biomechanical properties of fruit skin and isolated cuticle. J. Exp. Bot. 56, 1049–1060.

Bargel, H., Koch, K., Cerman, Z., Neinhuis, C., 2006. Structure–function relationships of the plant cuticle and cuticular waxes—a smart material? Funct. Plant Biol. 33, 893–910.

Barlass, M., Skene, K.G.M., Woodham, R.C., Krake, L.R., 1982. Regeneration of virus-free grapevines using *in vitro* apical culture. Ann. Appl. Biol. 101, 291–295.

Barnard, H.R., Ryan, M.G., 2003. A test of the hydraulic limitation hypothesis in fast-growing *Eucalyptus saligna*. Plant Cell Environ. 26, 1235–1245.

Barnard, H., Dooley, A.N., Areshian, G., Gasparyan, B., Faull, K.F., 2011. Chemical evidence for wine production around 4000 BCE in the Late Chalcolithic Near Eastern highlands. J. Archaeol. Sci. 38, 977–984.

Barnavon, L., Doco, T., Terrier, N., Ageorges, A., Romieu, C., Pellerin, P., 2000. Analysis of cell wall neutral sugar composition, β-galactosidase activity and a related cDNA clone throughout the development of *Vitis vinifera* grape berries. Plant Physiol. Biochem. 38, 289–300.

Barratt, D.H.P., Kölling, K., Graf, A., Pike, M., Calder, G., Findlay, K., 2011. Callose synthase GSL7 is necessary for normal phloem transport and inflorescence growth in *Arabidopsis*. Plant Physiol. 155, 328–341.

Barrios-Masias, F.H., Knipfer, T., Walker, M.A., McElrone, A.J., 2019. Differences in hydraulic traits of grapevine rootstocks are not conferred to a common *Vitis vinifera* scion. Funct. Plant Biol. 46, 228–235.

Barritt, B.H., 1970. Ovule development in seeded and seedless grapes. Vitis 9, 7–14.

Barros, J., Serk, H., Granlund, I., Pesquet, E., 2015. The cell biology of lignification in higher plants. Ann. Bot. 115, 1053–1074.

Bartlett, M.K., Klein, T., Jansen, S., Choat, B., Sack, L., 2016. The correlations and sequence of plant stomatal, hydraulic, and wilting responses to drought. Proc. Natl. Acad. Sci. U. S. A. 113, 13098–13103.

Bartley, G.E., Scolnik, P.A., 1995. Plant carotenoids: pigments for photoprotection, visual attraction, and human health. Plant Cell 7, 1027–1038.

Bass, P., Vuittenez, A., 1977. Amélioration de la thermothérapie des vignes virosées au moyen de la culture d'apex sur milieux nutritifs ou par greffage sur des vignes de semis obtenues aseptiquement *in vitro*. Ann. Phytopathol. 9, 539–540.

Bassil, E., Zhang, S., Gong, H., Tajima, H., Blumwald, E., 2019. Cation specificity of vacuolar NHX-type cation/H^+ antiporters. Plant Physiol. 179, 616–629.

Battany, M.C., 2012. Vineyard frost protection with upward-blowing wind machines. Agric. For. Meteorol. 157, 39–48.

Battany, M.C., Grismer, M.E., 2000. Rainfall runoff and erosion in Napa Valley vineyards: effects of slope, cover and surface roughness. Hydrol. Process. 14, 1289–1304.

Battey, N.H., Simmonds, P.E., 2005. Phylloxera and the grapevine: a sense of common purpose? J. Exp. Bot. 56, 3029–3031.

Bauerle, T.L., Richards, J.H., Smart, D.R., Eissenstat, D.M., 2008a. Importance of internal hydraulic redistribution for prolonging the lifespan of roots in dry soil. Plant Cell Environ. 31, 177–186.

Bauerle, T.L., Smart, D.R., Bauerle, W.L., Stockert, C., Eissenstat, D.M., 2008b. Root foraging in response to heterogeneous soil moisture in two grapevines that differ in potential growth rate. New Phytol. 179, 857–866.

Baumes, R., Wirth, J., Bureau, S., Gunata, Y., Razungles, A., 2002. Biogeneration of C_{13}-norisoprenoid compounds: experiments supportive for an apo-carotenoid pathway in grapevines. Anal. Chim. Acta 458, 3–14.

Baur, J.A., Sinclair, D.A., 2006. Therapeutic potential of resveratrol: the *in vivo* evidence. Nat. Rev. Drug Discov. 5, 493–506.

Bautista, J., Dangl, G.S., Yang, J., Reisch, B., Stover, E., 2008. Use of genetic markers to assess pedigrees of grape cultivars and breeding program selections. Am. J. Enol. Vitic. 59, 248–254.

Bauwe, H., Hagemann, M., Fernie, A.R., 2010. Photorespiration: players, partners and origin. Trends Plant Sci. 15, 330–336.

Bavaresco, L., Eibach, R., 1987. Investigations on the influence of N fertilizer on resistance to powdery mildew (*Oidium tuckeri*), downy mildew (*Plasmopara viticola*) and on phytoalexin synthesis in different grapevine varieties. Vitis 26, 192–200.

Bavaresco, L., Fregoni, M., Fraschini, P., 1991. Investigations on iron uptake and reduction by excised roots of different grapevine rootstocks and a *V. vinifera* cultivar. Plant Soil 130, 109–113.

Bavaresco, L., Giachino, E., Pezzutto, S., 2003. Grapevine rootstock effects on lime-induced chlorosis, nutrient uptake, and source–sink relationships. J. Plant Nutr. 26, 1451–1465.

Bayonove, C., Cordonnier, R., Dubois, P., 1975a. Etude d'une fraction caractéristique de l'arome du raisin de la variété Cabernet-Sauvignon; mise en évidence de la 2-methoxy-3-isobutylpyrazine. C.R. Hébd. Séances Acad. Sci. Sér. D, Sci. Nat. 281, 75–78.

Bayonove, C., Cordonnier, R., Ratier, R., 1975b. Localisation de l'arome dans la baie de raisin: variétés Muscat d'Aléxandrie et Cabernet-Sauvignon. C.R. Séances Acad. Agric. Fr. 60, 1321–1328.

Becker, H., 1976. Genetische Konstitution, züchterische Bearbeitung und Leistung der Rebsorte Müller-Thurgau. Wein-Wiss. 31, 26–37.

Becker, T., Knoche, M., 2011. Water movement through the surfaces of the grape berry and its stem. Am. J. Enol. Vitic. 62, 340–350.

Becker, T., Grimm, E., Knoche, M., 2012. Substantial water uptake into detached grape berries occurs through the stem surface. Aust. J. Grape Wine Res. 18, 109–114.

Begum, S., Nakaba, S., Oribe, Y., Kubo, T., Funada, R., 2007. Induction of cambial reactivation by localized heating in a deciduous hardwood hybrid poplar (*Populus sieboldii* x *P. grandidentata*). Ann. Bot. 100, 439–447.

Belhadj, A., Saigne, C., Telef, N., Cluzet, S., Bouscaut, J., Corio-Costet, M.F., 2006. Methyl jasmonate induces defense responses in grapevine and triggers protection against *Erysiphe necator*. J. Agric. Food Chem. 54, 9119–9125.

Belin, J.M., 1972. Recherches sur la répartition des levures à la surface de la grappe de raisin. Vitis 11, 135–145.

Bell, S.J., Henschke, P.A., 2005. Implications of nitrogen nutrition for grapes, fermentation and wine. Aust. J. Grape Wine Res. 11, 242–295.

Bell, S.J., Robson, A., 1999. Effect of nitrogen fertilization on growth, canopy density, and yield of *Vitis vinifera* L. cv. Cabernet Sauvignon. Am. J. Enol. Vitic. 50, 351–358.

Bell, A.A., Ough, C.S., Kliewer, W.M., 1979. Effects on must and wine composition, rates of fermentation, and wine quality of nitrogen fertilization of *Vitis vinifera* var. Thompson Seedless grapevines. Am. J. Enol. Vitic. 30, 124–129.

Bellini, C., Pacurar, D.I., Perrone, I., 2014. Adventitious roots and lateral roots: similarities and differences. Annu. Rev. Plant Biol. 65, 639–666.

Bellvert, J., Marsal, J., Mata, M., Girona, J., 2016. Yield, must composition, and wine quality responses to pre-veraison water deficits in sparkling base wines of Chardonnay. Am. J. Enol. Vitic. 67, 1–12.

Bengough, G.A., McKenzie, B.M., Hallett, P.D., Valentine, T.A., 2011. Root elongation, water stress, and mechanical impedance: a review of limiting stresses and beneficial root tip traits. J. Exp. Bot. 62, 59–68.

Benjak, A., Forneck, A., Casacuberta, J.M., 2008. Genome-wide analysis of the "cut-and-paste" transposons of grapevine. PLoS ONE 3, e3107.

Benjamins, R., Scheres, B., 2008. Auxin: the looping star in plant development. Annu. Rev. Plant Biol. 59, 443–465.

Bennett, J., Jarvis, P., Creasy, G.L., Trought, M.C.T., 2005. Influence of defoliation on overwintering carbohydrate reserves, return bloom, and yield of mature Chardonnay grapevines. Am. J. Enol. Vitic. 56, 386–393.

Berger, F., Grini, P.E., Schnittger, A., 2006. Endosperm: an integrator of seed growth and development. Curr. Opin. Plant Biol. 9, 664–670.

Berger, F., Hamamura, Y., Ingouff, M., Higashiyama, T., 2008. Double fertilization—caught in the act. Trends Plant Sci. 13, 437–443.

Bergqvist, J., Dokoozlian, N., Ebisuda, N., 2001. Sunlight exposure and temperature effects on berry growth and composition of Cabernet Sauvignon and Grenache in the central San Joaquin Valley of California. Am. J. Enol. Vitic. 52, 1–7.

Berleth, T., Scarpella, E., Prusinkiewicz, P., 2007. Towards the systems biology of auxin-transport-mediated patterning. Trends Plant Sci. 12, 151–159.

Berli, F., D'Angelo, J., Cavagnaro, B., Bottini, R., Wuilloud, R., Silva, M.F., 2008. Phenolic composition in grape (*Vitis vinifera* L. cv. Malbec) ripened with different solar UV-B radiation levels by capillary zone electrophoresis. J. Agric. Food Chem. 56, 2892–2898.

Berli, F.J., Fanzone, M., Piccoli, P., Bottini, R., 2011. Solar UV-B and ABA are involved in phenol metabolism of *Vitis vinifera* L. increasing biosynthesis of berry skin polyphenols. J. Agric. Food Chem. 59, 4874–4884.

Bernard, M.B., Horne, P.A., Hoffmann, A.A., 2005. Eriophyoid mite damage in *Vitis vinifera* (grapevine) in Australia: *Calepitrimerus vitis* and *Colomerus vitis* (Acari: Eriophyidae) as the common cause of the widespread 'Restricted Spring Growth' syndrome. Exp. Appl. Acarol. 35, 83–109.

Bernizzoni, F., Civardi, S., van Zeller, M., Gatti, M., Poni, S., 2011. Shoot thinning effects on seasonal whole-canopy photosynthesis and vine performance in *Vitis vinifera* L. cv. Barbera. Aust. J. Grape Wine Res. 17, 351–357.

Bernstein, Z., Lustig, I., 1985. Hydrostatic methods of measurement of firmness and turgor pressure of grape berries (*Vitis vinifera* L.). Sci. Hortic. 25, 129–136.

Berry, J., Björkman, O., 1980. Photosynthetic response and adaptation to temperature in higher plants. Annu. Rev. Plant Physiol. 31, 491–543.

Bertamini, M., Nedunchezhian, N., 2001. Decline of photosynthetic pigments, ribulose-1,5-bisphosphate carboxylase and soluble protein contents, nitrate reductase and photosynthetic activities, and changes in thylakoid membrane protein pattern in canopy shade grapevine (*Vitis vinifera* L. cv. Moscato giallo) leaves. Photosynthetica 39, 529–537.

Bertamini, M., Nedunchezhian, N., 2003. Photosynthetic fuctioning of individual grapevine leaves (*Vitis vinifera* L. cv. Pinot noir) during ontogeny in the field. Vitis 42, 13–17.

Bertamini, M., Nedunchezhian, N., 2005. Grapevine growth and physiological responses to iron deficiency. J. Plant Nutr. 28, 737–749.

Bertamini, M., Muthuchelian, K., Rubinigg, M., Zorer, R., Velasco, R., Nedunchezhian, N., 2006. Low-night temperature increased the photoinhibition of photosynthesis in grapevine (*Vitis vinifera* L. cv. Riesling) leaves. Environ. Exp. Bot. 57, 25–31.

Bertoldi, D., Larcher, R., Nicolini, G., Bertamini, M., Concheri, G., 2009. Distribution of rare earth elements in *Vitis vinifera* L. 'Chardonnay' berries. Vitis 48, 49–51.

Bertoldi, D., Larcher, R., Bertamini, M., Otto, S., Concheri, G., Nicolini, G., 2011. Accumulation and distribution pattern of macro- and microelements and trace elements in *Vitis vinifera* L. cv. Chardonnay berries. J. Agric. Food Chem. 59, 7224–7236.

Bessis, R., 1972a. Etude de l'évolution des caractères morphologiques des cires cuticulaires au cours de la vie du fruit de la vigne. C.R. Hébd. Séances Acad. Sci. Sér. D, Sci. Nat. 274, 1911–1914.

Bessis, R., 1972b. Etude de l'évolution des stomates et des tissus péristomatiques du fruit de la vigne. C.R. Hébd. Séances Acad. Sci. Sér. D, Sci. Nat. 274, 2158–2161.

Bessis, R., 2007. Evolution of the grapevine (*Vitis vinifera* L.) imprinted by natural and human factors. Can. J. Bot. 85, 679–690.

Bettner, W., 1988. Nährstoffgehalte in den Organen des grünen Rebtriebs und deren Nährstoffentzüge. Mitt. Klosterneuburg. 38, 130–137.

Beveridge, C.A., 2006. Axillary bud outgrowth: sending a message. Curr. Opin. Plant Biol. 9, 35–40.

Beyer, M., Knoche, M., 2002. Studies on water transport through the sweet cherry fruit surface: V. Conductance for water uptake. J. Am. Soc. Hortic. Sci. 127, 325–332.

Bindi, M., Fibbi, L., Gozzini, B., Orlandini, S., Miglietta, F., 1996. Modelling the impact of future climate scenarios on yield and yield variability of grapevine. Clim. Res. 7, 213–224.

Bindon, K.A., Kennedy, J.A., 2011. Ripening-induced changes in grape skin proanthocyanidins modify their interaction with cell walls. J. Agric. Food Chem. 59, 2696–2707.

Bindon, K., Dry, P., Loveys, B., 2008. Influence of partial rootzone drying on the composition and accumulation of anthocyanins in grape berries (*Vitis vinifera* cv. Cabernet Sauvignon). Aust. J. Grape Wine Res. 14, 91–103.

Bindon, K.A., Bacic, A., Kennedy, J.A., 2012. Tissue-specific and developmental modifications of grape cell walls influence the adsorption of proanthocyanidins. J. Agric. Food Chem. 60, 9249–9260.

Bindon, K.A., Varela, C., Kennedy, J., Holt, H., Herderich, M., 2013. Relationships between harvest time and wine composition in *Vitis vinifera* L. Cabernet Sauvignon 1. Grape and wine chemistry. Food Chem. 138, 1696–1705.

Bindon, K.A., Li, S., Kassara, S., Smith, P.A., 2016. Retention of proanthocyanidin in wine-like solution is conferred by a dynamic interaction between soluble and insoluble grape cell wall components. J. Agric. Food Chem. 64, 8406–8419.

Binenbaum, J., Weinstain, R., Shani, E., 2018. Gibberellin localization and transport in plants. Trends Plant Sci. 23, 410–421.

Birky, C.W., 1995. Uniparental inheritance of mitochondrial and chloroplast genes: mechanisms and evolution. Proc. Natl. Acad. Sci. U. S. A. 92, 11331–11338.

Blaich, R., Bachmann, O., 1980. Die Resveratrolsynthese bei Vitaceen. Induktion und zytologische Beobachtungen. Vitis 19, 230–240.

Blaich, R., Wind, R., 1989. Inducible silica incrusts in cell walls of *Vitis* leaves. Vitis 28, 73–80.

Blaich, R., Stein, U., Wind, R., 1984. Perforationen in der Cuticula von Weinbeeren als morphologischer Faktor der Botrytisresistenz. Vitis 23, 242–256.

Blaich, R., Konradi, J., Rühl, E., Forneck, A., 2007. Assessing genetic variation among Pinot noir (*Vitis vinifera* L.) clones with AFLP markers. Am. J. Enol. Vitic. 58, 526–529.

Blanco-Ulate, B., Amrine, K.C.H., Collins, T.S., Rivero, R.M., Vicente, A.R., Morales-Cruz, A., Doyle, C.L., Ye, Z., Allen, G., Heymann, H., Ebeler, S.E., Cantu, D., 2015. Developmental and metabolic plasticity of white-skinned grape berries in response to *Botrytis cinerea* during noble rot. Plant Physiol. 169, 2422–2443.

Blanke, M.M., 1990a. Carbon economy of the grape inflorescence. 4. Light transmission into grape berries. Wein-Wiss. 45, 21–23.

Blanke, M.M., 1990b. Carbon economy of the grape inflorescence. 5. Energy requirement of the flower bud of grape. Wein-Wiss. 45, 33–35.

Blanke, M.M., Leyhe, A., 1987. Stomatal activity of the grape berry cv. Riesling, Müller-Thurgau and Ehrenfelser. J. Plant Physiol. 127, 451–460.

Blanke, M.M., Leyhe, A., 1988. Stomatal and cuticular transpiration of the cap and berry of grape. J. Plant Physiol. 132, 250–253.

Blanke, M.M., Leyhe, A., 1989a. Carbon economy of the grape inflorescence. 1. Carbon economy in flower buds of grape. Wein-Wiss. 44, 33–36.

Blanke, M.M., Leyhe, A., 1989b. Carbon economy of the grape inflorescence. 3. Chlorophyll in the grape inflorescence. Wein-Wiss. 44, 188–191.

Blanke, M.M., Pring, R.J., Baker, E.A., 1999. Structure and elemental composition of grape berry stomata. J. Plant Physiol. 154, 477–481.

Bleby, T.M., McElrone, A.J., Jackson, R.B., 2010. Water uptake and hydraulic redistribution across large woody root systems to 20 m depth. Plant Cell Environ. 33, 2132–2148.

Bloom, A.J., Chapin, F.S., Mooney, H.A., 1985. Resource limitation in plants—an economic analogy. Annu. Rev. Ecol. Syst. 16, 363–392.

Bock, R., 2010. The give-and-take of DNA: horizontal gene transfer in plants. Trends Plant Sci. 15, 11–22.

Boerjan, W., Ralph, J., Baucher, M., 2003. Lignin biosynthesis. Annu. Rev. Plant Biol. 54, 519–546.

Bogs, J., Downey, M.O., Harvey, J.S., Ashton, A.R., Tanner, G.J., Robinson, S.P., 2005. Proanthocyanidin synthesis and expression of genes encoding leucoanthocyanidin reductase and anthocyanidin reductase in developing grape berries and grapevine leaves. Plant Physiol. 139, 652–663.

Böhlenius, H., Huang, T., Charbonnel-Campaa, L., Brunner, A.M., Jansson, S., Strauss, S.H., 2006. *CO/FT* regulatory module controls timing of flowering and seasonal growth cessation in trees. Science 312, 1040–1043.

Bohnert, H.J., Nelson, D.E., Jensen, R.G., 1995. Adaptations to environmental stresses. Plant Cell 7, 1099–1111.

Bokulich, N.A., Thorngate, J.H., Richardson, P.M., Mills, D.A., 2014. Microbial biogeography of wine grapes is conditioned by cultivar, vintage, and climate. Proc. Natl. Acad. Sci. U. S. A. 111, E139–E148.

Bonada, M., Sadras, V.O., Fuentes, S., 2013. Effect of elevated temperature on the onset and rate of mesocarp cell death in berries of Shiraz and Chardonnay and its relationship with berry shrivel. Aust. J. Grape Wine Res. 19, 87–94.

Bonanomi, G., Incerti, G., Barile, E., Capodilupo, M., Antignani, V., Mingo, A., 2011. Phytotoxicity, not nitrogen immobilization, explains plant litter inhibitory effects: evidence from solid-state ^{13}C NMR spectroscopy. New Phytol. 191, 1018–1030.

Bond, B.J., Czarnomski, N.M., Cooper, C., Day, M.E., Greenwood, M.S., 2007. Developmental decline in height growth in Douglas-fir. Tree Physiol. 27, 441–453.

Bondada, B., 2011. Micromorpho-anatomical examination of 2,4-D phytotoxicity in grapevine (*Vitis vinifera* L.) leaves. J. Plant Growth Regul. 30, 185–198.

Bondada, B.R., 2012. An array of simple, fast, and safe approaches to visualizing fine cellular structures in freehand sections of stem, leaf, and fruit using optical microscopy. Curr. Bot. 3, 11–22.

Bondada, B., 2014. Structural and compositional characterization of suppression of uniform ripening in grapevine: a paradoxical ripening disorder of grape berries with no known causative clues. J. Am. Soc. Hortic. Sci. 139, 567–581.

Bondada, B.R., Keller, M., 2012. Not all shrivels are created equal—morpho-anatomical and compositional characteristics differ among different shrivel types that develop during ripening of grape (*Vitis vinifera* L.) berries. Am. J. Plant Sci. 3, 879–898.

Bondada, B.R., Matthews, M.A., Shackel, K.A., 2005. Functional xylem in the post-veraison grape berry. J. Exp. Bot. 56, 2949–2957.

Bondada, B., Harbertson, E., Shrestha, P.M., Keller, M., 2017. Temporal extension of ripening beyond its physiological limits imposes physical and osmotic challenges perturbing metabolism in grape (*Vitis vinifera* L.) berries. Sci. Hortic. 219, 135–143.

Bongers, A.J., Hinsch, R.T., Bus, V.G., 1991. Physical and chemical characteristics of raisins from several countries. Am. J. Enol. Vitic. 42, 76–78.

Bönisch, F., Frotscher, J., Stanitzek, S., Rühl, E., Wüst, M., Bitz, O., Schwab, W., 2014a. A UDP-glucose: monoterpenol glucosyltransferase adds to the chemical diversity of the grapevine metabolome. Plant Physiol. 165, 461–581.

Bönisch, F., Frotscher, J., Stanitzek, S., Rühl, E., Wüst, M., Bitz, O., Schwab, W., 2014b. Activity-based profiling of a physiologic aglycone library reveals sugar acceptor promiscuity of family 1 UDP-glucosyltransferases from grape. Plant Physiol. 166, 23–39.

Boonman, A., Prinsen, E., Gilmer, F., Schurr, U., Peeters, A.J.M., Voesenek, L.A.C.J., 2007. Cytokinin import rate as a signal for photosynthetic acclimation to canopy light gradients. Plant Physiol. 143, 1841–1852.

Boonman, A., Prinsen, E., Voesenek, L.A.C.J., Pons, T.L., 2009. Redundant roles of photoreceptors and cytokinins in regulating photosynthetic acclimation to canopy density. J. Exp. Bot. 60, 1179–1190.

Boso, S., Kassemeyer, H.H., 2008. Different susceptibility of European grapevine cultivars for downy mildew. Vitis 47, 33–49.

Boso, S., Alonso-Villaverde, V., Gago, P., Santiago, J.L., Martínez, M.C., 2011. Susceptibility of 44 grapevine (*Vitis vinifera* L.) varieties to downy mildew in the field. Aust. J. Grape Wine Res. 17, 394–400.

Boss, P.K., Thomas, M.R., 2002. Association of dwarfism and floral induction with a grape 'green revolution' mutation. Nature 416, 847–850.

Boss, P.K., Davies, C., Robinson, S.P., 1996. Analysis of the expression of anthocyanin pathway genes in developing *Vitis vinifera* L. cv Shiraz grape berries and the implications for pathway regulation. Plant Physiol. 111, 1059–1066.

Boss, P.K., Buckeridge, E.J., Poole, A., Thomas, M.R., 2003. New insights into grapevine flowering. Funct. Plant Biol. 30, 593–606.

Boss, P.K., Davies, C., 2009. Molecular biology of anthocyanin accumulation in grape berries. In: Roubelakis-Angelakis, K.A. (Ed.), Grapevine Molecular Physiology and Biotechnology. Springer, Dordrecht, The Netherlands, pp. 263–292.

Bota, J., Medrano, H., Flexas, J., 2004. Is photosynthesis limited by decreased Rubisco activity and RuBP content under progressive water stress? New Phytol. 162, 671–681.

Böttcher, C., Boss, P.K., Davies, C., 2013. Increase in cytokinin levels during ripening in developing *Vitis vinifera* cv. Shiraz berries. Am. J. Enol. Vitic. 64, 527–531.

Böttcher, C., Burbidge, C.A., Boss, P.K., Davies, C., 2015. Changes in transcription of cytokinin metabolism and signalling genes in grape (*Vitis vinifera* L.) berries are associated with the ripening-related increase in isopentenyladenine. BMC Plant Biol. 15, 223.

Bou Nader, K., Stoll, M., Rauhut, D., Patz, C.D., Jung, R., Loehnertz, O., et al., 2019. Impact of grapevine age on water status and productivity of *Vitis vinifera* L. cv. Riesling. Eur. J. Agron. 104, 1–12.

Bouby, L., Figueiral, I., Bouchette, A., Rovira, N., Ivorra, S., Lacombe, T., Pastor, T., Picq, S., Marinval, P., Terral, J.F., 2013. Bioarchaeological insights into the process of domestication of grapevine (*Vitis vinifera* L.) during Roman times in southern France. PLoS ONE 8, e63195.

Boudaoud, A., 2010. An introduction to the mechanics of morphogenesis for plant biologists. Trends Plant Sci. 15, 353–360.

Boulton, R., 1980a. The general relationship between potassium, sodium and pH in grape juice and wine. Am. J. Enol. Vitic. 31, 182–186.

Boulton, R., 1980b. The relationships between total acidity, titratable acidity and pH in grape tissues. Vitis 19, 113–120.

Boulton, R., 2001. The copigmentation of anthocyanins and its role in the color of red wine: a critical review. Am. J. Enol. Vitic. 52, 67–87.

Boursiquot, J.M., Dessup, M., Rennes, C., 1995. Distribution des principaux caractères phénologiques, agronomiques et technologiques chez *Vitis vinifera* L. Vitis 34, 31–35.

Boursiquot, J.M., Lacombe, T., Laucou, V., Julliard, S., Perrin, F.X., Lanier, N., 2009. Parentage of Merlot and related winegrape cultivars of southwestern France: discovery of the missing link. Aust. J. Grape Wine Res. 15, 144–155.

Bovey, R., Gärtel, W., Hewitt, W.B., Martelli, G.P., Vuittenez, A., 1980. Virus and Virus-like Diseases of Grapevines. Colour Atlas of Symptoms. Editions Payot, Lausanne, Switzerland.

Bowen, P.A., Kliewer, W.M., 1990. Relationships between the yield and vegetative characteristics of individual shoots of 'Cabernet Sauvignon' grapevines. J. Am. Soc. Hortic. Sci. 115, 534–539.

Bowen, P.A., Bogdanoff, C.P., Estergaard, B.F., Marsh, S.G., Usher, K.B., Smith, C.A.S., 2005. Geology and wine 10: use of geographic information system technology to assess viticulture performance in the Okanagan and Similkameen valleys, British Columbia. Geosci. Can. 32, 161–176.

Bowen, P., Bogdanoff, C., Usher, K., Estergaard, B., Watson, M., 2011. Effects of irrigation and crop load on leaf gas exchange and fruit composition in red winegrapes grown on a loamy sand. Am. J. Enol. Vitic. 62, 9–22.

Bowen, P., Bogdanoff, C., Estergaard, B., 2012. Effects of converting from sprinkler to drip irrigation on water conservation and the performance of Merlot grown on a loamy sand. Am. J. Enol. Vitic. 63, 385–393.

Bowen, P., Shellie, K.C., Mills, L., Willwerth, J., Bogdanoff, C., Keller, M., 2016. Abscisic acid form, concentration, and application timing influence phenology and bud cold hardiness in Merlot grapevines. Can. J. Plant Sci. 96, 347–359.

Bowers, J.E., Meredith, C.P., 1997. The parentage of a classic wine grape Cabernet Sauvignon. Nat. Genet. 16, 84–87.

Bowers, J., Boursiquot, J.M., This, P., Chu, K., Johansson, H., Meredith, C., 1999. Historical genetics: the parentage of Chardonnay, Gamay, and other wine grapes of northeastern France. Science 285, 1562–1565.

Bowles, D., Lim, E.K., Poppenberger, B., Vaistij, F.E., 2006. Glycosyltransferases of lipophilic small molecules. Annu. Rev. Plant Biol. 57, 567–597.

Boyer, J.S., 1969. Measurement of the water status of plants. Annu. Rev. Plant Physiol. 20, 351–364.

Boyer, J.S., 1985. Water transport. Annu. Rev. Plant Physiol. 36, 473–516.

Boyer, J.S., 2009. Cell wall biosynthesis and the molecular mechanism of plant enlargement. Funct. Plant Biol. 36, 383–394.

Boyer, J.S., 2015. Turgor and the transport of CO_2 and water across the cuticle (epidermis) of leaves. J. Exp. Bot. 66, 2625–2633.

Boyer, J.S., Silk, W.K., 2004. Hydraulics of plant growth. Funct. Plant Biol. 31, 761–773.

Boyer, J.S., Wong, S.C., Farquhar, G.D., 1997. CO_2 and water vapor exchange across leaf cuticle (epidermis) at various water potentials. Plant Physiol. 114, 185–191.

Boyer, J.S., Silk, W.K., Watt, M., 2010. Path of water for root growth. Funct. Plant Biol. 37, 1105–1116.

Braam, J., 2005. In touch: plant responses to mechanical stimuli. New Phytol. 165, 373–389.

Bradford, K.J., 1994. Water stress and the water relations of seed development: a critical review. Crop Sci. 34, 1–11.

Braidot, E., Zancani, M., Petrussa, E., Peresson, C., Bertolini, A., Patui, S., 2008. Transport and accumulation of flavonoids in grapevine (*Vitis vinifera* L.). Plant Signal. Behav. 3, 626–632.

Braidwood, L., Breuer, C., Sugimoto, K., 2014. My body is a cage: mechanisms and modulation of plant cell growth. New Phytol. 201, 388–402.

Brancadoro, L., Rabotti, G., Scienza, A., Zocchi, G., 1995. Mechanisms of Fe-efficiency in roots of *Vitis* spp. in response to iron deficiency stress. Plant Soil 171, 229–234.

Brantjes, N.B.M., 1978. Pollinator attraction of *Vitis vinifera* subsp. *silvestris*. Vitis 17, 229–233.

Bravdo, B., Hepner, Y., Loinger, C., Cohen, S., Tabacman, H., 1984. Effect of crop level on growth, yield and wine quality of a high yielding Carignane vineyard. Am. J. Enol. Vitic. 35, 247–252.

Bravdo, B., Hepner, Y., Loinger, C., Cohen, S., Tabacman, H., 1985a. Effect of crop level and crop load on growth, yield, must and wine composition, and quality of Cabernet Sauvignon. Am. J. Enol. Vitic. 36, 125–131.

Bravdo, B., Hepner, Y., Loinger, C., Cohen, S., Tabacman, H., 1985b. Effect of irrigation and crop level on growth, yield and wine quality of Cabernet Sauvignon. Am. J. Enol. Vitic. 36, 132–139.

Bray, E.A., 1997. Plant responses to water deficit. Trends Plant Sci. 2, 48–54.

Brem, S., Rast, D.M., Ruffner, H.P., 1986. Partitioning of photosynthate in leaves of *Vitis vinifera* infected with *Uncinula necator* or *Plasmopara viticola*. Physiol. Mol. Plant Pathol. 29, 285–291.

Breuil, A.C., Jeandet, P., Adrian, M., Chopin, F., Pirio, N., Meunier, P., 1999. Characterization of a pterostilbene dehydrodimer produced by laccase of *Botrytis cinerea*. Phytopathology 89, 298–302.

Briat, J.F., Lobréaux, S., 1997. Iron transport and storage in plants. Trends Plant Sci. 2, 187–193.

Briat, J.F., Dubos, C., Gaymard, F., 2015. Iron nutrition, biomass production, and plant product quality. Trends Plant Sci. 20, 33–40.

Brickell, C.D., Alexander, C., David, J.C., Hetterscheid, W.L.A., Leslie, A.C., Malecot, V., et al., 2009. International Code of Nomenclature for Cultivated Plants, eighth ed. Scr. Hortic. 10, 184.

Brillouet, J.M., Romieu, C., Schoefs, B., Solymosi, K., Cheynier, V., Fulcrand, H., 2013. The tannosome is an organelle forming condensed tannins in the chlorophyllous organs of Tracheophyta. Ann. Bot. 112, 1003–1014.

Britto, D.T., Kronzucker, H.J., 2006. Futile cycling at the plasma membrane: a hallmark of low-affinity nutrient transport. Trends Plant Sci. 11, 529–534.

Broadley, M.R., White, P.J., Hammond, J.P., Zelko, I., Lux, A., 2007. Zinc in plants. New Phytol. 173, 677–702.

Brodersen, C.R., McElrone, A.J., Choat, B., Matthews, M.A., Shackel, K.A., 2010. The dynamics of embolism repair in xylem: *in vivo* visualizations using high-resolution computed tomography. Plant Physiol. 154, 1088–1095.

Brodersen, C.R., Lee, E.F., Choat, B., Jansen, S., Phillips, R.J., Shackel, K.A., 2011. Automated analysis of three-dimensional xylem networks using high-resolution computed tomography. New Phytol. 191, 1168–1179.

Brodersen, C.R., McElrone, A.J., Choat, B., Lee, E.F., Shackel, K.A., Matthews, M.A., 2013. *In vivo* visualization of drought-induced embolism spread in *Vitis vinifera*. Plant Physiol. 161, 1820–1829.

Brodersen, C.R., Knipfer, T., McElrone, A.J., 2018. *In vivo* visualization of the final stages of xylem vessel refilling in grapevine (*Vitis vinifera*) stems. New Phytol. 217, 117–126.

Brodribb, T.J., Holbrook, N.M., 2003. Stomatal closure during leaf dehydration, correlation with other leaf physiological traits. Plant Physiol. 132, 2166–2173.

Brodribb, T.J., Field, T.S., Jordan, G.J., 2007. Leaf maximum photosynthetic rate and venation are linked by hydraulics. Plant Physiol. 144, 1890–1898.

Brodribb, T.J., Bienaimé, D., Marmottant, P., 2016. Revealing catastrophic failure of leaf networks under stress. Proc. Natl. Acad. Sci. U. S. A. 113, 4865–4869.

Brossaud, F., Cheynier, V., Noble, A.C., 2001. Bitterness and astringency of grape and wine polyphenols. Aust. J. Grape Wine Res. 7, 33–39.

Browse, J., Xin, Z., 2001. Temperature sensing and cold acclimation. Curr. Opin. Plant Biol. 4, 241–246.

Brüggenwirth, M., Knoche, M., 2017. Cell wall swelling, fracture mode, and the mechanical properties of cherry fruit skins are closely related. Planta 245, 765–777.

Brüggenwirth, M., Fricke, H., Knoche, M., 2014. Biaxial tensile tests identify epidermis and hypodermis as the main structural elements of sweet cherry skin. AoB Plants 6, plu019.

Brummell, D.A., 2006. Cell wall disassembly in ripening fruit. Funct. Plant Biol. 33, 103–119.

Bucchetti, B., Matthews, M., Falginella, L., Peterlunger, E., Castellarin, S.D., 2011. Effect of water deficit in Merlot grape tannins and anthocyanins across four seasons. Sci. Hortic. 128, 297–305.

Bucci, S.J., Scholz, F.G., Goldstein, G., Meinzer, F.C., Franco, A.C., Campanello, P.I., 2006. Nutrient availability constrains the hydraulic architecture and water relations of savannah trees. Plant Cell Environ. 29, 2153–2167.

Buchanan, B.B., Gruissem, W., Jones, R.L., 2000. Biochemistry and Molecular Biology of Plants. ASPP, Rockville, MD.

Buchanan-Wollaston, V., 1997. The molecular biology of leaf senescence. J. Exp. Bot. 48, 181–199.

Bucher, M., 2007. Functional biology of plant phosphate uptake at root and mycorrhiza interfaces. New Phytol. 173, 11–26.

Buckley, T.N., 2015. The contributions of apoplastic, symplastic and gas phase pathways for water transport outside the bundle sheath in leaves. Plant Cell Environ. 38, 7–22.

Buckley, T.N., John, G.P., Scoffoni, C., Sack, L., 2017. The sites of evaporation within leaves. Plant Physiol. 173, 1763–1782.

Buer, C.S., Muday, G.K., Djordjevic, M.A., 2007. Flavonoids are differentially taken up and transported long distances in *Arabidopsis*. Plant Physiol. 145, 478–490.

Bulit, J., Dubos, B., 1982. Epidémiologie de la pourriture grise. Bull. OEPP 12, 37–48.

Burch-Smith, T.M., Zambryski, P.C., 2012. Plasmodesmata paradigm shift: regulation from without versus within. Annu. Rev. Plant Biol. 63, 239–260.

Burkhead, J.L., Gogolin Reynolds, K.A., Abdel-Ghany, S.E., Cohu, C.M., Pilon, M., 2009. Copper homeostasis. New Phytol. 182, 799–816.

Burr, T.J., Otten, L., 1999. Crown gall of grape: biology and disease management. Annu. Rev. Phytopathol. 37, 53–80.

Burr, T.J., Katz, B.H., Bishop, A.L., Meyers, C.A., Mittak, V.L., 1988. Effect of shoot age and tip culture propagation of grapes on systemic infestations by *Agrobacterium tumefaciens* Biovar 3. Am. J. Enol. Vitic. 39, 67–70.

Burr, T.J., Bazzi, C., Süle, S., Otten, L., 1998. Crown gall of grape: biology of *Agrobacterium vitis* and the development of disease control strategies. Plant Dis. 82, 1288–1297.

Busam, G., Kassemeyer, H.H., Matern, U., 1997. Differential expression of chitinases in *Vitis vinifera* L. responding to systemic acquired resistance activators or fungal challenge. Plant Physiol. 115, 1029–1038.

Buttrose, M.S., 1966a. The effect of reducing leaf area on the growth of roots, stems and berries of Gordo grapevines. Vitis 5, 455–464.

Buttrose, M.S., 1966b. Use of carbohydrate reserves during growth from cuttings of grape vine. Aust. J. Biol. Sci. 19, 247–256.

Buttrose, M.S., 1969a. Fruitfulness in grapevines: effects of light intensity and temperature. Bot. Gaz. 130, 166–173.

Buttrose, M.S., 1969b. The dissolution and reaccumulation of starch granules in grape vine cane. Aust. J. Biol. Sci. 22, 1297–1303.

Buttrose, M.S., 1969c. Vegetative growth of grape-vine varieties under controlled temperature and light intensity. Vitis 8, 280–285.

Buttrose, M.S., 1970a. Fruitfulness in grapevines: development of leaf primordia in buds in relation to bud fruitfulness. Bot. Gaz. 131, 78–83.

Buttrose, M.S., 1970b. Fruitfulness in grape-vines: the response of different cultivars to light, temperature and daylength. Vitis 9, 121–125.

Buttrose, M.S., 1974a. Climatic factors and fruitfulness in grapevines. Hortic. Abstr. 44, 319–325.

Buttrose, M.S., 1974b. Fruitfulness in grape-vines: effect of water stress. Vitis 12, 299–305.

Buttrose, M.S., Hale, C.R., 1973. Effect of temperature on development of the grapevine inflorescence after bud burst. Am. J. Enol. Vitic. 24, 4–16.

Buttrose, M.S., Mullins, M.G., 1968. Proportional reduction in shoot growth of grapevine with root systems maintained at constant relative volumes by repeated pruning. Aust. J. Biol. Sci. 21, 1095–1101.

Buttrose, M.S., Hale, C.R., Kliewer, W.M., 1971. Effect of temperature on the composition of 'Cabernet Sauvignon' berries. Am. J. Enol. Vitic. 22, 71–75.

Cabaleiro, C., Segura, A., García-Berrios, J.J., 1999. Effects of grapevine leafroll-associated virus 3 on the physiology and must of *Vitis vinifera* L. cv. Albariño following contamination in the field. Am. J. Enol. Vitic. 50, 40–44.

Cabezas, J.A., Cervera, M.T., Arroyo-García, R., Ibáñez, J., Rodríguez-Torres, I., Borrego, J., 2003. Garnacha and Garnacha Tintorera: genetic relationships and the origin of teinturier varieties cultivated in Spain. Am. J. Enol. Vitic. 54, 237–245.

Cabezas, J.A., Cervera, M.T., Ruiz-García, L., Carreño, J., Martínez-Zapater, J.M., 2006. A genetic analysis of seed and berry weight in grapevine. Genome 49, 1574–1585.

Cacho, J., Fernández, P., Ferreira, V., Castells, J.E., 1992. Evolution of five anthocyanidin-3-glucosides in the skin of the Tempranillo, Moristel, and Garnacha grape varieties and influence of climatological variables. Am. J. Enol. Vitic. 43, 244–248.

Cadle-Davidson, L., 2008. Monitoring pathogenesis of natural *Botrytis cinerea* infections in developing grape berries. Am. J. Enol. Vitic. 59, 387–395.

Cadle-Davidson, M.M., Owens, C.L., 2008. Genomic amplification of the *Gret1* retroelement in white-fruited accessions of wild *Vitis* and interspecific hybrids. Theor. Appl. Genet. 116, 1079–1094.

Cadot, Y., Miñana-Castelló, M.T., Chevalier, M., 2006. Anatomical, histological, and histochemical changes in grape seeds from *Vitis vinifera* L. cv Cabernet franc during fruit development. J. Agric. Food Chem. 54, 9206–9215.

Cahoon, C.A., 1986. The Concord grapes. Fruit Var. J. 40, 106–107.
Cain, D.W., Emershad, R.L., Tarailo, R.E., 1983. *In-ovulo* embryo culture and seedling development of seeded and seedless grapes (*Vitis vinifera* L.). Vitis 22, 9–14.
Cakmak, I., Kirkby, E.A., 2008. Role of magnesium in carbon partitioning and alleviating photooxidative damage. Physiol. Plant. 133, 692–704.
Cakmak, I., Marschner, H., 1986. Mechanism of phosphorus-induced zinc deficiency in cotton: I. Zinc deficiency-enhanced uptake rate of phosphorus. Physiol. Plant. 68, 483–490.
Cakmak, I., Hengeler, C., Marschner, H., 1994. Changes in phloem export of sucrose in leaves in response to phosphorus, potassium and magnesium deficiency in bean plants. J. Exp. Bot. 45, 1251–1257.
Calderón, A.A., Zapata, J.M., Ros Barceló, A., 1994. Differential expression of a cell wall-localized peroxidase isoenzyme capable of oxidizing 4-hydroxystilbenes during the cell culture of grapevine (*Vitis vinifera* cv. Airen and Monastrell). Plant Cell Tissue Organ Cult. 37, 121–127.
Calderon-Orellana, A., Mercenaro, L., Shackel, K.A., Willits, N., Matthews, M.A., 2014a. Responses of fruit uniformity to deficit irrigation and cluster thinning in commercial winegrape production. Am. J. Enol. Vitic. 65, 354–362.
Calderon-Orellana, A., Matthews, M.A., Drayton, W.M., Shackel, K.A., 2014b. Uniformity of ripeness and size in Cabernet Sauvignon berries from vineyards with contrasting crop price. Am. J. Enol. Vitic. 65, 81–88.
Caldwell, M.M., Björn, L.O., Bornman, J.F., Flint, S.D., Kulandaivelu, G., Teramura, A.H., 1998. Effects of increased solar ultraviolet radiation on terrestrial ecosystems. J. Photochem. Photobiol. B 46, 40–52.
Callen, S.T., Klein, L.L., Miller, A.J., 2016. Climatic niche characterization of 13 North American *Vitis* species. Am. J. Enol. Vitic. 67, 339–349.
Callis, J., 1995. Regulation of protein degradation. Plant Cell 7, 845–857.
Calonje, M., Cubas, P., Martínez-Zapater, J.M., Carmona, M.J., 2004. Floral meristem identity genes are expressed during tendril development in grapevine. Plant Physiol. 135, 1491–1501.
Calonnec, A., Jolivet, J., Vivin, P., Schnee, S., 2018. Pathogenicity traits correlate with the susceptible *Vitis vinifera* leaf physiology transition in the biotroph fungus *Erysiphe necator*: an adaptation to plant ontogenic resistance. Front. Plant Sci. 9, 1808.
Camacho-Cristóbal, J.J., González-Fontes, A., 1999. Boron deficiency causes a drastic decrease in nitrate content and nitrate reductase activity, and increases the content of carbohydrates in leaves from tobacco plants. Planta 209, 528–536.
Camargo Alvarez, H., Salazar-Gutiérrez, M., Zapata, D., Keller, M., Hoogenboom, G., 2018. Time-to-event analysis to evaluate dormancy status of single-bud cuttings: an example for grapevines. Plant Methods 14, 94.
Campbell, L., Turner, S., 2017. Regulation of vascular cell division. J. Exp. Bot. 68, 27–43.
Campbell-Clause, J.M., 1998. Stomatal response of grapevines to wind. Aust. J. Exp. Agric. 38, 77–82.
Candolfi-Vasconcelos, M.C., Koblet, W., 1990. Yield, fruit quality, bud fertility and starch reserves of the wood as a function of leaf removal in *Vitis vinifera*—evidence of compensation and stress recovering. Vitis 29, 199–221.
Candolfi-Vasconcelos, M.C., Koblet, W., 1991. Influence of partial defoliation on gas exchange parameters and chlorophyll content of field grown grapevines—mechanisms and limitations of the compensation capacity. Vitis 30, 129–141.
Candolfi-Vasconcelos, M.C., Candolfi, M.P., Koblet, W., 1994. Retranslocation of carbon reserves from the woody storage tissues into the fruit as a response to defoliation stress during the ripening period in *Vitis vinifera* L. Planta 192, 567–573.
Canny, M.J., 1993. The transpiration stream in the leaf apoplast: water and solutes. Philos. Trans. Biol. Sci. 341, 87–100.
Cantos, E., Espín, J.C., Tomás-Barberán, F.A., 2002. Varietal differences among the polyphenol profiles of seven table grape cultivars studied by LC-DAD-MS-MS. J. Agric. Food Chem. 50, 5691–5696.

Cantu, D., Vicente, A.R., Greve, L.C., Dewey, F.M., Bennett, A.B., Labavitch, J.M., Powell, A.L.T., 2008a. The intersection between cell wall disassembly, ripening, and fruit susceptibility to *Botrytis cinerea*. Proc. Natl. Acad. Sci. U. S. A. 105, 859–864.

Cantu, D., Vicente, A.R., Labavitch, J.M., Bennett, A.B., Powell, A.L.T., 2008b. Strangers in the matrix: plant cell walls and pathogen susceptibility. Trends Pant Sci. 13, 610–617.

Caporali, E., Spada, A., Marziani, G., Failla, O., Scienza, A., 2003. The arrest of development of abortive reproductive organs in the unisexual flower of *Vitis vinifera* ssp. *silvestris*. Sex. Plant Reprod. 15, 291–300.

Capps, E.R., Wolf, T.K., 2000. Reduction of bunch stem necrosis of Cabernet Sauvignon by increased tissue nitrogen concentration. Am. J. Enol. Vitic. 51, 319–328.

Caputi, L., Carlin, S., Ghiglieno, I., Stefanini, M., Valenti, L., Vrhovsek, U., 2011. Relationship of changes in rotundone content during grape ripening and winemaking to manipulation of the 'peppery' character of wine. J. Agric. Food Chem. 59, 5565–5571.

Caravia, L., Collins, C., Tyerman, S.D., 2015. Electrical impedance of Shiraz berries correlates with decreasing cell vitality during ripening. Aust. J. Grape Wine Res. 21, 430–438.

Carbonneau, A., de Loth, C., 1985. Influence du régime d'éclairement journalier sur la résistance stomatique et la photosynthèse brute chez Vitis vinifera L. cv. "Cabernet-Sauvignon" Agronomie 5, 631–638.

Carbonneau, A., Deloire, A., Jaillard, B., 2007. Traité de La Vigne. Physiologie, Terroir, Culture. Dunod, Paris, France.

Cardone, M.F., D'Addabbo, P., Alkan, C., Bergamini, C., Catacchio, C.R., Anaclerio, F., Chiatante, G., Marra, A., Giannuzzi, G., Perniola, R., Ventura, M., Antonacci, D., 2016. Inter-varietal structural variation in grapevine genomes. Plant J. 88, 648–661.

Carmona, M.J., Cubas, P., Martínez-Zapater, J.M., 2002. *VFL*, the grapevine *FLORICAULA/LEAFY* ortholog, is expressed in meristematic regions independently of their fate. Plant Physiol. 130, 68–77.

Carmona, M.J., Cubas, P., Calonje, M., Martínez-Zapater, J.M., 2007. Flowering transition in grapevine (*Vitis vinifera* L.). Can. J. Bot. 85, 701–711.

Carmona, M.J., Chaïb, J., Martínez-Zapater, J.M., Thomas, M.R., 2008. A molecular genetic perspective of reproductive development in grapevine. J. Exp. Bot. 59, 2579–2596.

Carpaneto, A., Geiger, D., Bamberg, E., Sauer, N., Fromm, J., Hedrich, R., 2005. Phloem-localized, proton-coupled sucrose carrier ZmSUT1 mediates sucrose efflux under the control of the sucrose gradient and the proton motive force. J. Biol. Chem. 280, 21437–21443.

Carpita, N.C., Gibeaut, D.M., 1993. Structural models of primary cell walls in flowering plants: consistency of molecular structure with the physical properties of the walls during growth. Plant J. 3, 1–30.

Carreño, J., Almela, L., Martínez, A., Fernández-López, J.A., 1997. Chemotaxonomical classification of red table grapes based on anthocyanin profile and external colour. Lebensm. Wiss. Technol. 30, 259–265.

Carrier, G., Le Cunff, L., Dereeper, A., Legrand, D., Sabot, F., Bouchez, O., 2012. Transposable elements are a major cause of somatic polymorphism in *Vitis vinifera* L. PLoS ONE 7, e32973.

Carroll, D.E., Marcy, J.E., 1982. Chemical and physical changes during maturation of Muscadine grapes (*Vitis rotundifolia*). Am. J. Enol. Vitic. 33, 168–172.

Carroll, J.E., Wilcox, W.F., 2003. Effects of humidity on the development of grapevine powdery mildew. Phytopathology 93, 1137–1144.

Carroll, D.E., Hoover, M.W., Nesbitt, W.B., 1971. Sugar and organic acid concentrations in cultivars of muscadine grapes. J. Am. Soc. Hortic. Sci. 96, 737–740.

Cartechini, A., Palliotti, A., 1995. Effect of shading on vine morphology and productivity and leaf gas exchange characteristics in grapevines in the field. Am. J. Enol. Vitic. 46, 227–234.

Carvalho, M.R., Turgeon, R., Owens, T., Niklas, K.J., 2017. The scaling of the hydraulic architecture in poplar leaves. New Phytol. 214, 145–157.

Casado, C.G., Heredia, A., 2001. Ultrastructure of the cuticle during growth of the grape berry (*Vitis vinifera*). Physiol. Plant. 111, 220–224.

Casassa, L.F., Larsen, R.C., Beaver, C.W., Mireles, M.S., Keller, M., Riley, W.R., 2013. Impact of extended maceration and regulated deficit irrigation (RDI) in Cabernet Sauvignon wines: characterization of proanthocyanidin distribution, anthocyanin extraction, and chromatic properties. J. Agric. Food Chem. 61, 6446–6457.

Casassa, L.F., Larsen, R.C., Harbertson, J.F., 2016. Effects of vineyard and winemaking practices impacting berry size on evolution of phenolics during winemaking. Am. J. Enol. Vitic. 67, 257–268.

Caspari, H.W., Lang, A., Alspach, P., 1998. Effects of girdling and leaf removal on fruit set and vegetative growth in grape. Am. J. Enol. Vitic. 49, 359–366.

Castelan-Estrada, M., Vivin, P., Gaudillère, J.P., 2002. Allometric relationships to estimate seasonal above-ground vegetative and reproductive biomass in *Vitis vinifera* L. Ann. Bot. 89, 401–408.

Castellano, M.A., Abou-Ghanem, N., Choueiri, E., Martelli, G.P., 2000. Ultrastructure of *grapevine leafroll-associated virus 2 and 7* infections. J. Plant Pathol. 82, 9–15.

Castellarin, S.D., Di Gaspero, G., 2007. Transcriptional control of anthocyanin biosynthetic genes in extreme phenotypes for berry pigmentation of naturally occurring grapevines. BMC Plant Biol. 7, 46.

Castellarin, S.D., Matthews, M.A., Di Gaspero, G., Gambetta, G.A., 2007a. Water deficits accelerate ripening and induce changes in gene expression regulating flavonoid biosynthesis in grape berries. Planta 227, 101–112.

Castellarin, S.D., Pfeiffer, A., Sivilotti, P., Degan, M., Peterlunger, E., Di Gaspero, G., 2007b. Transcriptional regulation of anthocyanin biosynthesis in ripening fruits of grapevine under seasonal water deficit. Plant Cell Environ. 30, 1381–1399.

Castellarin, S.D., Gambetta, G.A., Wada, H., Shackel, K.A., Matthews, M.A., 2011. Fruit ripening in *Vitis vinifera*: spatiotemporal relationships among turgor, sugar accumulation, and anthocyanin biosynthesis. J. Exp. Bot. 62, 4345–4354.

Castellarin, G.A., Gambetta, H., Wada, M.N., Krasnow, G.R., Cramer, E., Peterlunger, K.A., Shackel, M.A., 2016. Matthews, Characterization of major ripening events during softening in grape: turgor, sugar accumulation, abscisic acid metabolism, colour development, and their relationship with growth. J. Exp. Bot. 67, 709–722.

Cawthon, D.L., Morris, J.R., 1982. Relationship of seed number and maturity to berry development, fruit maturation, hormonal changes, and uneven ripening of 'Concord' (*Vitis labrusca* L.) grapes. J. Am. Soc. Hortic. Sci. 107, 1097–1104.

Cazzonelli, C.I., Pogson, B.J., 2010. Source to sink: regulation of carotenoid biosynthesis in plants. Trends Plant Sci. 15, 266–274.

Celette, F., Gary, C., 2013. Dynamics of water and nitrogen stress along the grapevine cycle as affected by cover cropping. Eur. J. Agron. 45, 142–152.

Celette, F., Gaudin, R., Gary, C., 2008. Spatial and temporal changes to the water regime of a Mediterranean vineyard due to the adoption of cover cropping. Eur. J. Agron. 29, 153–162.

Celette, F., Findeling, A., Gary, C., 2009. Competition for nitrogen in an unfertilized intercropping system: the case of an association of grapevine and grass cover in a Mediterranean climate. Eur. J. Agron. 30, 41–51.

Cerpa-Calderón, F.K., Kennedy, J.A., 2008. Berry integrity and extraction of skin and seed proanthocyanidins during red wine fermentation. J. Agric. Food Chem. 56, 9006–9014.

Cerreti, M., Esti, M., Benucci, I., Liburdi, K., de Simone, C., Ferranti, P., 2015. Evolution of S-cysteinylated and S-glutathionylated thiol precursors during grape ripening of *Vitis vinifera* L. cvs Grechetto, Malvasia del Lazio and Sauvignon Blanc. Aust. J. Grape Wine Res. 21, 411–416.

Chacko, E.K., Saidha, T., Swamy, R.D., Reddy, Y.N., Kohli, R.R., 1976. Studies on the cytokinins in fruits. I. Occurrence and levels of cytokinin-like substances in grape berries at different developmental stages. Vitis 15, 221–226.

Chalker-Scott, L., 1999. Environmental significance of anthocyanins in plant stress responses. Photochem. Photobiol. 70, 1–9.

Chandler, J.W., 2011. Founder cell specification. Trends Plant Sci. 16, 607–613.
Chapin, F.S., 1991. Integrated response of plants to stress. Bioscience 41, 29–36.
Chapman, D.M., Matthews, M.A., Guinard, J.X., 2004a. Sensory attributes of Cabernet Sauvignon wines made from vines with different crop yields. Am. J. Enol. Vitic. 55, 325–334.
Chapman, D.M., Thorngate, J.H., Matthews, M.A., Guinard, J.X., Ebeler, S.E., 2004b. Yield effects on 2-methoxy-3-isobutylpyrazine concentration in Cabernet Sauvignon using a solid phase microextraction gas chromatography/mass spectrometry method. J. Agric. Food Chem. 52, 5431–5435.
Chapman, N., Miller, A.J., Lindsey, K., Whalley, W.R., 2012. Roots, water, and nutrient acquisition: let's get physical. Trends Plant Sci. 17, 701–710.
Charrier, G., Torres-Ruiz, J.M., Badel, E., Burlett, R., Choat, B., Cochard, H., et al., 2016. Evidence for hydraulic vulnerability segmentation and lack of xylem refilling under tension. Plant Physiol. 172, 1657–1668.
Charrier, G., Nolf, M., Leitinger, G., Charra-Vaskou, K., Losso, A., Tappeiner, U., Améglio, T., Mayr, S., 2017. Monitoring of freezing dynamics in trees: a simple phase shift causes complexity. Plant Physiol. 173, 2196–2207.
Chatelet, D.S., Matthews, M.A., Rost, T.L., 2006. Xylem structure and connectivity in grapevine (*Vitis vinifera*) shoots provides a passive mechanism for the spread of bacteria in grape plants. Ann. Bot. 98, 483–494.
Chatelet, D.S., Rost, T.L., Matthews, M.A., Shackel, K.A., 2008a. The peripheral xylem of grapevine (*Vitis vinifera*). 2. Anatomy and development. J. Exp. Bot. 59, 1997–2007.
Chatelet, D.S., Rost, T.L., Shackel, K.A., Matthews, M.A., 2008b. The peripheral xylem of grapevine (*Vitis vinifera*). 1. Structural integrity in post-veraison berries. J. Exp. Bot. 59, 1987–1996.
Chater, C.C.C., Oliver, J., Casson, S., Gray, J.E., 2014. Putting the brakes on: abscisic acid as a central environmental regulator of stomatal development. New Phytol. 202, 376–391.
Chatonnet, P., Dubourdieu, D., 1998. Comparative study of the characteristics of American white oak (*Quercus alba*) and European oak (*Quercus petraea* and *Q. robur*) for production of barrels used in barrel aging of wines. Am. J. Enol. Vitic. 49, 79–85.
Chatonnet, P., Dubourdieu, D., Boidron, J.N., Pons, M., 1992. The origin of ethylphenols in wines. J. Sci. Food Agric. 60, 165–178.
Chaves, M.M., Harley, P.C., Tehunen, J.D., Lange, O.L., 1987. Gas exchange studies in two Portuguese grapevine cultivars. Physiol. Plant. 70, 639–647.
Chaves, M.M., Zarrouk, O., Francisco, R., Costa, J.M., Santos, T., Regalado, A.P., 2010. Grapevine under deficit irrigation: hints from physiological and molecular data. Ann. Bot. 105, 661–676.
Chen, L.S., Cheng, L., 2003a. Both xanthophyll cycle-dependent thermal dissipation and the antioxidant system are up-regulated in grape (*Vitis labrusca* L. cv. Concord) leaves in response to N limitation. J. Exp. Bot. 54, 2165–2175.
Chen, L.S., Cheng, L., 2003b. Carbon assimilation and carbohydrate metabolism of 'Concord' grape (*Vitis labrusca* L.) leaves in response to nitrogen supply. J. Am. Soc. Hortic. Sci. 128, 754–760.
Chen, L.Q., Qu, X.Q., Hou, B.H., Sosso, D., Osorio, S., Fernie, A.R., 2012. Sucrose efflux mediated by SWEET proteins as a key step for phloem transport. Science 335, 207–211.
Cheng, C.Y., Reuther, G., Gruppe, W., 1974. Untersuchungen zur Regulation der Knospenruhe verschiedener Rebsorten durch ökologische und endogene Faktoren. Vitis 13, 98–111.
Cheng, H., Qin, L., Lee, S., Fu, X., Richards, D.E., Cao, D., 2004a. Gibberellin regulates *Arabidopsis* floral development via suppression of DELLA protein function. Development 131, 1055–1064.
Cheng, L., Xia, G., Bates, T., 2004b. Growth and fruiting of young 'Concord' grapevines in relation to reserve nitrogen and carbohydrates. J. Am. Soc. Hortic. Sci. 129, 660–666.
Cheng, C., Xu, X., Singer, S.D., Li, J., Zhang, H., Gao, M., Wang, L., Song, J., Wang, X., 2013. Effect of GA_3 treatment on seed development and seed-related gene expression in grape. PLoS ONE 8, e80044.

Chervin, C., El-Kereamy, A., Roustan, J.P., Latché, A., Lamon, J., Bouzayen, M., 2004. Ethylene seems required for the berry development and ripening in grape, a non-climacteric fruit. Plant Sci. 167, 1301–1305.

Chervin, C., Tira-Umphon, A., Terrier, N., Zouine, M., Severac, D., Roustan, J.P., 2008. Stimulation of the grape berry expansion by ethylene and effects on related gene transcripts, over the ripening phase. Physiol. Plant. 134, 534–546.

Cheung, A.Y., Boavida, L.C., Aggarwal, M., Wu, H.M., Feijó, J.A., 2010. The pollen tube journey in the pistil and imaging the *in vivo* process by two-photon microscopy. J. Exp. Bot. 61, 1907–1915.

Chevalier, É., Loubert-Hudon, A., Zimmerman, E.L., Matton, D.P., 2011. Cell–cell communication and signalling pathways within the ovule: from its inception to fertilization. New Phytol. 192, 13–28.

Cheynier, V., Rigaud, J., 1986. HPLC separation and characterization of flavonols in the skins of *Vitis vinifera* var. Cinsault. Am. J. Enol. Vitic. 37, 248–252.

Cheynier, V., Dueñas-Paton, M., Salas, E., Maury, C., Souquet, J.M., Sarni-Manchado, P., 2006. Structure and properties of wine pigments and tannins. Am. J. Enol. Vitic. 57, 298–305.

Chiera, J., Thomas, J., Rufty, T., 2002. Leaf initiation and development in soybean under phosphorus stress. J. Exp. Bot. 53, 473–481.

Chinnusamy, V., Zhu, J., Zhu, J.K., 2007. Cold stress regulation of gene expression in plants. Trends Plant Sci. 12, 444–451.

Chiou, T.J., Lin, S.I., 2011. Signaling network in sensing phosphate availability in plants. Annu. Rev. Plant Biol. 62, 185–206.

Chitwood, D.H., Ranjan, A., Martinez, C.C., Headland, L.R., Thiem, T., Kumar, R., 2014. A modern ampelography: a genetic basis for leaf shape and venation patterning in grape. Plant Physiol. 164, 259–272.

Choat, B., Lahr, E.C., Melcher, P.J., Zwieniecki, M.A., Holbrook, N.M., 2005. The spatial pattern of air seeding thresholds in mature sugar maple trees. Plant Cell Environ. 28, 1082–1089.

Choat, B., Cobb, A.R., Jansen, S., 2008. Structure and function of bordered pits: new discoveries and impacts on whole-plant hydraulic function. New Phytol. 177, 608–626.

Choat, B., Drayton, W.M., Brodersen, C., Matthews, M.A., Shackel, K.A., Wada, H., 2010. Measurement of vulnerability to water stress-induced cavitation in grapevine: a comparison of four techniques applied to a long-vesseled species. Plant Cell Environ. 33, 1502–1512.

Choné, X., Lavigne-Cruège, V., Tominaga, T., Van Leeuwen, C., Castagnède, C., Saucier, C., 2006. Effect of vine nitrogen status on grape aromatic potential: flavor precursors (*S*-cysteine conjugates), glutathione and phenolic content in *Vitis vinifera* L. cv. Sauvignon blanc grape juice. J. Int. Sci. Vigne Vin 40, 1–6.

Chong, J., Piron, M.C., Meyer, S., Merdinoglu, D., Bertsch, C., Mestre, P., 2014. The SWEET family of sugar transporters in grapevine: VvSWEET4 is involved in the interaction with *Botrytis cinerea*. J. Exp. Bot. 65, 6589–6601.

Chory, J., 1997. Light modulation of vegetative development. Plant Cell 9, 1225–1234.

Christensen, P., 1975. Response of 'Thompson Seedless' grapevines to the timing of preharvest irrigation cut-off. Am. J. Enol. Vitic. 26, 188–194.

Christensen, P., 1984. Nutrient level comparisons of leaf petioles and blades in twenty-six grape cultivars over three years (1979 through 1981). Am. J. Enol. Vitic. 35, 124–133.

Christensen, L.P., Boggero, J.D., 1985. A study of mineral nutrition relationships of waterberry in Thompson Seedless. Am. J. Enol. Vitic. 36, 57–64.

Christensen, L.P., Bianchi, M.L., Peacock, W.L., Hirschfelt, D.J., 1994. Effect of nitrogen fertilizer timing and rate on inorganic nitrogen status, fruit composition, and yield of grapevines. Am. J. Enol. Vitic. 45, 377–387.

Christensen, L.P., Bianchi, M.L., Miller, M.W., Kasimatis, A.N., Lynn, C.D., 1995. The effects of harvest date on Thompson Seedless grapes and raisins. II. Relationships of fruit quality factors. Am. J. Enol. Vitic. 46, 493–498.

Christensen, N.M., Nicolaisen, M., Hansen, M., Schulz, A., 2004. Distribution of phytoplasmas in infected plants as revealed by real time PCR and bioimaging. Mol. Plant-Microbe Interact. 17, 1175–1184.

Christmann, A., Weiler, E.W., Steudle, E., Grill, E., 2007. A hydraulic signal in root-to-shoot signalling of water shortage. Plant J. 52, 167–174.

Chuine, I., Yiou, P., Viovy, N., Seguin, B., Daux, V., Le Roy Ladurie, E., 2004. Grape ripening as a past climate indicator. Nature 432, 289–290.

Cipriani, G., Spadotto, A., Jurman, I., Di Gaspero, G., Crespan, M., Meneghetti, S., 2010. The SSR-based molecular profile of 1005 grapevine (*Vitis vinifera* L.) accessions uncovers new synonymy and parentages, and reveals a large admixture amongst varieties of different geographic origin. Theor. Appl. Genet. 121, 1569–1585.

Claeys, H., De Bodt, S., Inzé, D., 2014. Gibberellins and DELLAs: central nodes in growth regulatory networks. Trends Plant Sci. 19, 231–239.

Clark, S.E., 1997. Organ formation at the vegetative shoot meristem. Plant Cell 9, 1067–1076.

Clarke, S.J., Hardie, W.J., Rogiers, S.Y., 2010. Changes in susceptibility of grape berries to splitting are related to impaired osmotic water uptake associated with losses in cell vitality. Aust. J. Grape Wine Res. 16, 469–476.

Clarke, S.J., Lamont, K.J., Pan, H.Y., Barry, L.A., Hall, A., Rogiers, S.Y., 2015. Spring root-zone temperature regulates root growth, nutrient uptake and shoot growth dynamics in grapevines. Aust. J. Grape Wine Res. 21, 479–489.

Clarkson, D.T., 1985. Factors affecting mineral nutrient acquisition by plants. Annu. Rev. Plant Physiol. 36, 77–115.

Clarkson, D.T., 1993. Roots and the delivery of solutes to the xylem. Philos. Trans. Biol. Sci. 341, 5–17.

Clarkson, D.T., Hanson, J.B., 1980. The mineral nutrition of higher plants. Annu. Rev. Plant Physiol. 31, 239–298.

Clarkson, D.T., Carvajal, M., Henzler, T., Waterhouse, R.N., Smyth, A.J., Cooke, D.T., 2000. Root hydraulic conductance: diurnal aquaporin expression and the effects of nutrient stress. J. Exp. Bot. 51, 61–70.

Claverie, J., Balacey, S., Lemaître-Guillier, C., Brulé, D., Chiltz, A., Granet, L., et al., 2018. The cell wall-derived xyloglucan is a new DAMP triggering plant immunity in *Vitis vinifera* and *Arabidopsis thaliana*. Front. Plant Sci. 9, 1725.

Cline, M.G., 1997. Concepts and terminology of apical dominance. Am. J. Bot. 84, 1064–1069.

Clingeleffer, P.R., 1984. Production and growth of minimal pruned Sultana vines. Vitis 23, 42–54.

Clingeleffer, P.R., 1989. Effect of varying node number per bearer on yield and juice composition of Cabernet Sauvignon grapevines. Aust. J. Exp. Agric. 29, 701–705.

Clingeleffer, P.R., Krake, L.R., 1992. Responses of Cabernet franc grapevines to minimal pruning and virus infection. Am. J. Enol. Vitic. 43, 31–37.

Cocucci, S., Morgutti, S., Cocucci, M., Scienza, A., 1988. A possible relationship between stalk necrosis and membrane transport in grapevine cultivars. Sci. Hortic. 34, 67–74.

Coetzee, B., Freeborough, M.J., Maree, H.J., Celton, J.M., Rees, D.J.G., Burger, J.T., 2010. Deep sequencing analysis of viruses infecting grapevines: virome of a vineyard. Virology 400, 157–163.

Cola, G., Failla, O., Mariani, L., 2009. BerryTone—a simulation model for the daily course of grape berry temperature. Agric. For. Meteorol. 149, 1215–1228.

Coleman, C., Copetti, D., Cipriani, G., Hoffmann, S., Kozma, P., Kovács, L., et al., 2009. The powdery mildew resistance gene *REN1* co-segregates with an NBS-LRR gene cluster in two Central Asian grapevines. BMC Genet. 10, 89.

Columella, L.I.M., ca. 70 ad. De Re Rustica. Available from: http://www.thelatinlibrary.com/columella.html.

Comas, L.H., Eissenstat, D.M., Lakso, A.N., 2000. Assessing root death and root system dynamics in a study of grape canopy pruning. New Phytol. 147, 171–178.

Comas, L.H., Anderson, L.J., Dunst, R.M., Lakso, A.N., Eissenstat, D.M., 2005. Canopy and environmental control of root dynamics in a long-term study of Concord grape. New Phytol. 167, 829–840.

Comas, L.H., Bauerle, T.L., Eissenstat, D.M., 2010. Biological and environmental factors controlling root dynamics and function: effects of root ageing and soil moisture. Aust. J. Grape Wine Res. 16, 131–137.

Comstock, J.P., 2002. Hydraulic and chemical signalling in the control of stomatal conductance and transpiration. J. Exp. Bot. 53, 195–200.

Conde, C., Silva, P., Fontes, N., Dias, A.C.P., Tavares, R.M., Sousa, M.J., 2007. Biochemical changes throughout grape berry development and fruit and wine quality. Food 1, 1–22.

Coniberti, A., Ferrari, V., Fariña, L., Carrau, F., Dellacassa, E., Boido, E., 2012. Role of canopy management in controlling high pH in Tannat grapes and wines. Am. J. Enol. Vitic. 63, 554–558.

Conn, S., Zhang, W., Franco, C., 2003. Anthocyanic vacuolar inclusions (AVIs) selectively bind acylated anthocyanins in *Vitis vinifera* L. (grapevine) suspension culture. Biotechnol. Lett. 25, 835–839.

Conn, S., Franco, C., Zhang, W., 2010. Characterization of anthocyanic vacuolar inclusions in *Vitis vinifera* L. cell suspension cultures. Planta 231, 1343–1360.

Conradie, W.J., 1980. Seasonal uptake of nutrients by Chenin blanc in sand culture: I. Nitrogen. S. Afr. J. Enol. Vitic. 1, 59–65.

Conradie, W.J., 1981a. Nutrient consumption by Chenin blanc grown in sand culture and seasonal changes in the chemical composition of leaf blades and petioles. S. Afr. J. Enol. Vitic. 2, 15–18.

Conradie, W.J., 1981b. Seasonal uptake of nutrients by Chenin blanc in sand culture: II. Phosphorus, potassium, calcium and magnesium. S. Afr. J. Enol. Vitic. 2, 7–13.

Conradie, W.J., 1986. Utilisation of nitrogen by the grapevine as affected by time of application and soil type. S. Afr. J. Enol. Vitic. 7, 76–83.

Conradie, W.J., Saayman, D., 1989. Effects of long term nitrogen, phosphorus, and potassium fertilization on Chenin blanc vines. II. Leaf analyses and grape composition. Am. J. Enol. Vitic. 40, 91–98.

Considine, J.A., 1981. Correlation of resistance to physical stress with fruit structure in the grape *Vitis vinifera* L. Aust. J. Bot. 29, 475–482.

Considine, J., Brown, K., 1981. Physical aspects of fruit growth: theoretical analysis of distribution of surface growth forces in fruit in relation to cracking and splitting. Plant Physiol. 68, 371–376.

Considine, J.A., Knox, R.B., 1979a. Development and histochemistry of the pistil of the grape, *Vitis vinifera*. Ann. Bot. 43, 11–22.

Considine, J.A., Knox, R.B., 1979b. Development and histochemistry of the cells, cell walls, and cuticle of the dermal system of fruit of the grape, *Vitis vinifera* L. Protoplasma 99, 347–365.

Considine, J.A., Knox, R.B., 1981. Tissue origins, cell lineages and patterns of cell division in the developing dermal system of the fruit of *Vitis vinifera* L. Planta 151, 403–412.

Considine, J.A., Kriedemann, P.E., 1972. Fruit splitting in grapes: determination of the critical turgor pressure. Aust. J. Agric. Res. 23, 17–24.

Cooke, J.E.K., Eriksson, M.E., Junttila, O., 2012. The dynamic nature of bud dormancy in trees: environmental control and molecular mechanisms. Plant Cell Environ. 35, 1707–1728.

Cookson, S.J., Clemente Moreno, M.J., Hevin, C., Nyamba Mendome, L.Z., Delrot, S., Trossat-Magnin, C., 2013. Graft union formation in grapevine induces transcriptional changes related to cell wall modification, wounding, hormone signalling, and secondary metabolism. J. Exp. Bot. 64, 2997–3008.

Cookson, S.J., Clemente Moreno, M.J., Hevin, C., Nyamba Mendome, L.Z., Delrot, S., Trossat-Magnin, C., 2014. Heterografting with nonself rootstocks induces genes involved in stress responses at the graft interface when compared with autografted controls. J. Exp. Bot. 65, 2473–2481.

Coombe, B.G., 1959. Fruit set and development in seeded grape varieties as affected by defoliation, topping, girdling, and other treatments. Am. J. Enol. Vitic. 10, 85–100.

Coombe, B.G., 1960. Relationship of growth and development to changes in sugars, auxins and gibberellins in fruit of seeded and seedless varieties of *Vitis vinifera* L. Plant Physiol. 35, 241–250.

Coombe, B.G., 1962. The effect of removing leaves, flowers and shoot tips on fruit-set in *Vitis vinifera* L. J. Hortic. Sci. 37, 1–15.

Coombe, B.G., 1976. The development of fleshy fruits. Annu. Rev. Plant Physiol. 27, 507–528.

Coombe, B.G., 1987. Distribution of solutes within the developing grape berry in relation to its morphology. Am. J. Enol. Vitic. 38, 120–127.

Coombe, B.G., 1992. Research on development and ripening of the grape berry. Am. J. Enol. Vitic. 43, 101–110.

Coombe, B.G., 1995. Adoption of a system for identifying grapevine growth stages. Aust. J. Grape Wine Res. 1, 104–110.

Coombe, B.G., 2001. Ripening berries—a critical issue. Aust. Vitic. 5, 28–34.

Coombe, B.G., Bishop, G.R., 1980. Development of the grape berry. II. Changes in diameter and deformability during veraison. Aust. J. Agric. Res. 31, 499–509.

Coombe, B.G., Dry, P.R., 2001. Viticulture. Volume 2—Practices, seventh ed. Winetitles, Adelaide, Australia.

Coombe, B.G., Hale, C.R., 1973. The hormone content of ripening grape berries and the effect of growth substance treatments. Plant Physiol. 51, 629–634.

Coombe, B.G., Monk, P.R., 1979. Proline and abscisic acid content of the juice of ripe Riesling grape berries: effect of irrigation during harvest. Am. J. Enol. Vitic. 30, 64–67.

Cooper, R.M., Williams, J.S., 2004. Elemental sulphur as an induced antifungal substance in plant defence. J. Exp. Bot. 55, 1947–1953.

Correia, M.J., Chaves, M.M., Pereira, J.S., 1990. Afternoon depression in photosynthesis in grapevine leaves—evidence for a high light stress effect. J. Exp. Bot. 41, 417–426.

Correia, M.J., Pereira, J.S., Chaves, M.M., Rodrigues, M.L., Pacheco, C.A., 1995. ABA xylem concentrations determine maximum daily leaf conductance of field-grown *Vitis vinifera* L. plants. Plant Cell Environ. 18, 511–521.

Cortell, J.M., Kennedy, J.A., 2006. Effect of shading on accumulation of flavonoid compounds in (*Vitis vinifera* L.) Pinot noir fruit and extraction in a model system. J. Agric. Food Chem. 54, 8510–8520.

Cortell, J.M., Halbleib, M., Gallagher, A.V., Righetti, T.L., Kennedy, J.A., 2005. Influence of vine vigor on grape (*Vitis vinifera* L. cv. Pinot noir) and wine proanthocyanidins. J. Agric. Food Chem. 53, 5798–5808.

Cortell, J.M., Halbleib, M., Gallagher, A.V., Righetti, T.L., Kennedy, J.A., 2007a. Influence of vine vigor on grape (*Vitis vinifera* L. cv. Pinot noir) anthocyanins. 1. Anthocyanin concentration and composition in fruit. J. Agric. Food Chem. 55, 6575–6584.

Cortell, J.M., Halbleib, M., Gallagher, A.V., Righetti, T.L., Kennedy, J.A., 2007b. Influence of vine vigor on grape (*Vitis vinifera* L. cv. Pinot noir) anthocyanins. 2. Anthocyanins and pigmented polymers in wine. J. Agric. Food Chem. 55, 6585–6595.

Coruzzi, G.M., Zhou, L., 2001. Carbon and nitrogen sensing and signaling in plants: emerging 'matrix effects'. Curr. Opin. Plant Biol. 4, 247–253.

Cosgrove, D.J., 1997. Relaxation in a high-stress environment: the molecular bases of extensible cell walls and cell enlargement. Plant Cell 9, 1031–1041.

Cosgrove, D.J., 2000. Loosening of plant cell walls by expansins. Nature 407, 321–326.

Cosgrove, D.J., 2005. Growth of the plant cell wall. Nat. Rev. Mol. Cell Biol. 6, 850–861.

Cosgrove, D.J., 2016. Plant cell wall extensibility: connecting plant cell growth with cell wall structure, mechanics, and the action of wall-modifying enzymes. J. Exp. Bot. 67, 463–476.

Cosgrove, D.J., 2018. Diffuse growth of plant cell walls. Plant Physiol. 176, 16–27.

Costantini, V., Bellincontro, A., De Santis, D., Botondi, R., Mencarelli, F., 2006. Metabolic changes of Malvasia grapes for wine production during postharvest drying. J. Agric. Food Chem. 54, 3334–3340.

Costantini, E., Landi, L., Silvestroni, O., Pandolfini, T., Spena, A., Mezzetti, B., 2007. Auxin synthesis-encoding transgene enhances grape fecundity. Plant Physiol. 143, 1689–1694.

Coupel-Ledru, A., Tyerman, S.D., Masclef, D., Lebon, E., Christophe, A., Edwards, E.J., Simonneau, T., 2017. Abscisic acid down-regulates hydraulic conductance of grapevine leaves in isohydric genotypes only. Plant Physiol. 175, 1121–1134.

Courtois-Moreau, C.L., Pesquet, E., Sjödin, A., Muñiz, L., Bollhöner, B., Kaneda, M., 2009. A unique program for cell death in xylem fibers of *Populus* stem. Plant J. 58, 260–274.

Coutand, C., 2010. Mechanosensing and thigmomorphogenesis, a physiological and biomechanical point of view. Plant J. 179, 168–182.

Cragin, J., Serpe, M., Keller, M., Shellie, K., 2017. Dormancy and cold hardiness transitions in winegrape cultivars Chardonnay and Cabernet Sauvignon. Am. J. Enol. Vitic. 68, 195–202.

Cramer, G.R., Ergül, A., Grimplet, J., Tillett, R.L., Tattersall, E.A.R., Bohlman, M.C., 2007. Water and salinity stress in grapevines: early and late changes in transcript and metabolite profiles. Funct. Integr. Genom. 7, 111–134.

Cramer, M.D., Hoffmann, V., Verboom, G.A., 2008. Nutrient availability moderates transpiration in *Ehrharta calycina*. New Phytol. 179, 1048–1057.

Crane, O., Halaly, T., Pang, X., Lavee, S., Perl, A., Vankova, R., 2012. Cytokinin-induced *VvTFL1A* expression may be involved in the control of grapevine fruitfulness. Planta 235, 181–192.

Cravero, M.C., Guidoni, S., Schneider, A., Di Stefano, R., 1994. Caractérisation variétale de cépages musqués à raisin coloré au moyen de paramètres ampélographiques descriptifs et biochimiques. Vitis 33, 75–80.

Crawford, N.M., 1995. Nitrate: nutrient and signal for plant growth. Plant Cell 7, 859–868.

Crawford, N.M., Glass, A.D.M., 1998. Molecular and physiological aspects of nitrate uptake in plants. Trends Plant Sci. 3, 389–395.

Crawford, S., Shinohara, N., Sieberer, T., Williamson, L., George, G., Hepworth, J., 2010. Strigolactones enhance competition between shoot branches by dampening auxin transport. Development 137, 2905–2913.

Creasy, G.L., Lombard, P.B., 1993. Vine water stress and peduncle girdling effects on pre- and post-veraison grape berry growth and deformability. Am. J. Enol. Vitic. 44, 193–197.

Creasy, G.L., Price, S.F., Lombard, P.B., 1993. Evidence for xylem discontinuity in Pinot noir and Merlot grapes: dye uptake and mineral composition during berry maturation. Am. J. Enol. Vitic. 44, 187–192.

Crespan, M., Milani, N., 2001. The Muscats: a molecular analysis of synonyms, homonyms and genetic relationships within a large family of grapevine cultivars. Vitis 40, 23–30.

Crespo-Martínez, S., Sobczak, M., Rózanska, E., Forneck, A., Griesser, M., 2019. The role of the secondary phloem during the development of the grapevine berry shrivel ripening disorder. Micron 116, 36–45.

Crippen, D.D., Morrison, J.C., 1986. The effects of sun exposure on the compositional development of Cabernet Sauvignon berries. Am. J. Enol. Vitic. 37, 235–242.

Crowe, J.H., Crowe, L.M., Carpenter, J.F., Rudolph, A.S., Aurell Wistrom, C., Spargo, B.J., Anchordoguy, T.J., 1988. Interactions of sugars with membranes. Biochim. Biophys. Acta 947, 367–384.

Cuéllar, T., Azeem, F., Andrianteranagna, M., Pascaud, F., Verdeil, J.L., Sentenac, H., 2013. Potassium transport in developing fleshy fruits: the grapevine inward K^+ channel VvK1.2 is activated by CIPK-CBL complexes and induced in ripening berry flesh cells. Plant J. 73, 1006–1018.

Cuevas, J.C., López-Cobollo, R., Alcázar, R., Zarza, X., Koncz, C., Altabella, T., 2008. Putrescine is involved in *Arabidopsis* freezing tolerance and cold acclimation by regulating abscisic acid levels in response to low temperature. Plant Physiol. 148, 1094–1105.

Cuneo, I.F., Knipfer, T., Brodersen, C.R., McElrone, A.J., 2016. Mechanical failure of fine root cortical cells initiates plant hydraulic decline during drought. Plant Physiol. 172, 1669–1678.

Cuneo, I.F., Knipfer, T., Mandal, P., Brodersen, C.R., McElrone, A.J., 2018. Water uptake can occur through woody portions of roots and facilitates embolism repair in grapevine. New Phytol. 218, 506–516.

Curie, C., Briat, J.F., 2003. Iron transport and signaling in plants. Annu. Rev. Plant Biol. 54, 183–206.

Curie, C., Cassin, G., Couch, D., Divol, F., Higuchi, K., Le Jean, M., 2009. Metal movement within the plant: contribution of nicotianamine and yellow stripe 1-like transporters. Ann. Bot. 103, 1–11.

Currie, H.A., Perry, C.C., 2007. Silica in plants: biological, biochemical and chemical studies. Ann. Bot. 100, 1383–1389.

Currle, O., Bauer, O., Hofäcker, W., Schumann, F., Frisch, W., 1983. Biologie der Rebe. Neustadt an der Weinstrasse, Meininger, Germany.

Dai, G.H., Andary, C., Mondolot-Cosson, L., Boubals, D., 1995. Histochemical studies on the interaction between three species of grapevine, *Vitis vinifera*, *V. rupestris* and *V. rotundifolia* and the downy mildew fungus, *Plasmopara viticola*. Physiol. Mol. Plant Pathol. 46, 177–188.

Dai, Z.W., Ollat, N., Gomès, E., Decroocq, S., Tandonnet, J.P., Bordenave, L., 2011. Ecophysiological, genetic, and molecular causes of variation in grape berry weight and composition: a review. Am. J. Enol. Vitic. 62, 413–425.

Dakora, F.D., Phillips, D.A., 2002. Root exudates as mediators of mineral acquisition in low-nutrient environments. Plant Soil 245, 35–47.

Dal Santo, S., Zenoni, S., Sandri, M., De Lorenzis, G., Magris, G., De Paoli, E., Di Gaspero, G., Del Fabbro, C., Morgante, M., Brancadoro, L., Grossi, D., Fasoli, M., Zuccolotto, P., Tornielli, G.B., Pezzotti, M., 2018. Grapevine field experiments reveal the contribution of genotype, the influence of environment and the effect of their interaction ($G \times E$) on the berry transcriptome. Plant J. 93, 1143–1159.

Daly, C., Halbleib, M., Smith, J.I., Gibson, W.P., Doggett, M.K., Taylor, G.H., 2008. Physiographically sensitive mapping of climatological temperature and precipitation across the conterminous Unites States. Int. J. Climatol. 28, 2031–2064.

Danilewicz, J.C., 2003. Review of reaction mechanisms of oxygen and proposed intermediate reduction products in wine: central role of iron and copper. Am. J. Enol. Vitic. 54, 73–85.

Darriet, P., Pons, M., Henry, R., Dumont, O., Findeling, V., Cartolaro, P., 2002. Impact odorants contributing to the fungus type aroma from grape berries contaminated by powdery mildew (*Uncinula necator*): incidence of enzymatic activities of the yeast *Saccharomyces cerevisiae*. J. Agric. Food Chem. 50, 3277–3282.

Darvill, A.G., Albersheim, P., 1984. Phytoalexins and their elicitors—a defense against microbial infection in plants. Annu. Rev. Plant Physiol. 35, 243–275.

Darwin, C., 1859. On the Origin of Species. Or the Preservation of Favoured Races in the Struggle for Life. John Murray, London, UK.

Darwin, C., 1875. The Movements and Habits of Climbing Plants. Murray, London, UK.

Davenport, J.R., Stevens, R.G., 2006. High soil moisture and low soil temperature are associated with chlorosis occurrence in Concord grape. HortScience 41, 418–422.

Davenport, J.R., Keller, M., Mills, L.J., 2008a. How cold can you go? Frost and winter protection for grape. HortScience 43, 1966–1969.

Davenport, J.R., Stevens, R.G., Whitley, K.M., 2008b. Spatial and temporal distribution of soil moisture in drip-irrigated vineyards. HortScience 43, 229–235.

Davies, C., Robinson, S.P., 2000. Differential screening indicates a dramatic change in mRNA profiles during grape berry ripening. Cloning and characterization of cDNAs encoding putative cell wall and stress response proteins. Plant Physiol. 122, 803–812.

Davies, W.J., Zhang, J., 1991. Root signals and the regulation of growth and development of plants in drying soil. Annu. Rev. Plant Physiol. Plant Mol. Biol. 42, 55–76.

Davies, C., Boss, P.K., Robinson, S.P., 1997. Treatment of grape berries, a nonclimacteric fruit with a synthetic auxin, retards ripening and alters the expression of developmentally regulated genes. Plant Physiol. 115, 1155–1161.

Davies, C., Wolf, T., Robinson, S.P., 1999. Three putative sucrose transporters are differentially expressed in grapevine tissues. Plant Sci. 147, 93–100.

Davies, W.J., Wilkinson, S., Loveys, B., 2002. Stomatal control by chemical signalling and the exploitation of this mechanism to increase water use efficiency in agriculture. New Phytol. 153, 449–460.

Davies, C., Shin, R., Liu, W., Thomas, M.R., Schachtman, D.P., 2006. Transporters expressed during grape berry (*Vitis vinifera* L.) development are associated with an increase in berry size and berry potassium accumulation. J. Exp. Bot. 57, 3209–3216.

Davies, C., Böttcher, C., 2009. Hormonal control of grape berry ripening. In: Roubelakis-Angelakis, K.A. (Ed.), Grapevine Molecular Physiology and Biotechnology. Springer, Dordrecht, The Netherlands, pp. 229–261.

Davis, J.D., Evert, R.F., 1970. Seasonal cycle of phloem development in woody vines. Bot. Gaz. 131, 128–138.

Dayer, S., Prieto, J.A., Galat, E., Perez Peña, J., 2016. Leaf carbohydrate metabolism in Malbec grapevines: combined effects of regulated deficit irrigation and crop load. Aust. J. Grape Wine Res. 22, 112–123.

De Andrés, M.T., Benito, A., Peréz-Rivera, G., Ocete, R., Lopez, A., Gaforio, L., 2012. Genetic diversity of wild grapevine populations in Spain and their genetic relationships with cultivated grapevines. Mol. Ecol. 21, 800–816.

De Boer, A.H., Volkov, V., 2003. Logistics of water and salt transport through the plant: structure and functioning of the xylem. Plant Cell Environ. 26, 87–101.

de Candolle, A., 1855. Géographie botanique raisonnée ou exposition des faits principaux et des lois concernant la distribution géographique des plants de l'époque actuelle. Librairie de Victor Masson, Paris, France.

De Freitas, V.A.P., Glories, Y., 1999. Concentration and compositional changes of procyanidins in grape seeds and skin of white *Vitis vinifera* varieties. J. Sci. Food Agric. 79, 1601–1606.

De Freitas, S.T., Handa, A.K., Wu, Q., Park, S., Mitcham, E.J., 2012. Role of pectin methylesterases in cellular calcium distribution and blossom-end rot development in tomato fruit. Plant J. 71, 824–835.

De Mora, S.J., Eschenbruch, R., Knowles, S.J., Spedding, D.J., 1986. The formation of dimethyl sulphide during fermentation using a wine yeast. Food Microbiol. 3, 27–32.

De Schepper, V., De Swaef, T., Bauweraerts, I., Steppe, K., 2013. Phloem transport: a review of mechanisms and controls. J. Exp. Bot. 64, 4839–4850.

De Smet, I., Zhang, H., Inzé, D., Beeckman, T., 2006. A novel role for abscisic acid emerges from underground. Trends Plant Sci. 11, 434–439.

De Souza, C.R., Maroco, J.P., dos Santos, T.P., Rodrigues, M.L., Lopes, C.M., Pereira, J.S., 2003. Partial rootzone drying: regulation of stomatal aperture and carbon assimilation in field-grown grapevines (*Vitis vinifera* cv. Moscatel). Funct. Plant Biol. 30, 653–662.

De Storme, N., Geelen, D., 2014. The impact of environmental stress on male reproductive development in plants: biological processes and molecular mechanisms. Plant Cell Environ. 37, 1–18.

DeBolt, S., Hardie, J., Tyerman, S., Ford, C.M., 2004. Composition and synthesis of raphide crystals and druse crystals in berries of *Vitis vinifera* L. cv. Cabernet Sauvignon: ascorbic acid as precursor for both oxalic and tartaric acids as revealed by radiolabelling studies. Aust. J. Grape Wine Res. 10, 134–142.

DeBolt, S., Cook, D.R., Ford, C.M., 2006. l-Tartaric acid synthesis from vitamin C in higher plants. Proc. Natl. Acad. Sci. U.S.A. 103, 5608–5613.

DeBolt, S., Melino, V., Ford, C.M., 2007. Ascorbate as a biosynthetic precursor in plants. Ann. Bot. 99, 3–8.

Dechorgnat, J., Nguyen, C.T., Armengaud, P., Jossier, M., Diatloff, E., Filleur, S., 2011. From the soil to the seeds: the long journey of nitrate in plants. J. Exp. Bot. 62, 1349–1359.

Dejonghe, W., Okamoto, M., Cutler, S.R., 2018. Small molecule probes of ABA biosynthesis and signaling. Plant Cell Physiol. 59, 1490–1499.

Del Bianco, M., Giustini, L., Sabatini, S., 2013. Spatiotemporal changes in the role of cytokinin during root development. New Phytol. 199, 324–338.

Del Pozo, J.C., Lopez-Matas, M.A., Ramirez-Parra, E., Gutierrez, C., 2005. Hormonal control of the plant cell cycle. Physiol. Plant. 123, 173–183.

Delas, J., 1972. Effets de la fertilisation de la vigne sur le développement de "Botrytis cinerea" Rev. Zool. Agric. Pathol. Vég. 71, 11–17.

Delas, J., Pouget, R., 1984. Action de la concentration de la solution nutritive sur quelques caractéristiques physiologiques et technologiques chez *Vitis vinifera* L. cv. Cabernet-Sauvignon. II. Composition minérale des organes végétatifs, du moût et du vin. Agronomie 4, 350–443.

Delas, J., Molot, C., Soyer, J.P., 1982. Influence d'une fertilisation azotée excessive, du porte-greffe et de la charge sur la sensibilité du cépage Merlot à *Botrytis cinerea*. Bull. OEPP 12, 177–182.

Della Penna, D., Pogson, B.J., 2006. Vitamin synthesis in plants: tocopherols and carotenoids. Annu. Rev. Plant Biol. 57, 711–738.

Deluc, L.G., Grimplet, J., Wheatley, M.D., Tillett, R.L., Quilici, D.R., Osborne, C., 2007. Transcriptomic and metabolite analysis of Cabernet Sauvignon grape berry development. BMC Genom. 8, 429.

Delwiche, J.F., 2003. Impact of color on perceived wine flavor. Foods Food Ingred. J. Jpn. 208, 349–352.

Demmig-Adams, B., Adams, W.W., 1996. The role of xanthophyll cycle carotenoids in the protection of photosynthesis. Trends Plant Sci. 1, 21–26.

Dengler, N.G., 2001. Regulation of vascular development. J. Plant Growth Regul. 20, 1–13.

Dettweiler, E., Jung, A., Zyprian, E., Töpfer, R., 2000. Grapevine cultivar Müller-Thurgau and its true to type descent. Vitis 39, 63–65.

Dewitte, W., Chiappetta, A., Azmi, A., Witters, E., Strnad, M., Rembur, J., 1999. Dynamics of cytokinins in apical shoot meristems of a day-neutral tobacco during floral transition and flower formation. Plant Physiol. 119, 111–121.

Deytieux, C., Gény, L., Lapaillerie, D., Claverol, S., Bonneu, M., Donéche, B., 2007. Proteome analysis of grape skins during ripening. J. Exp. Bot. 58, 1851–1862.

Di Filippo, M., Vila, H., 2011. Influence of different rootstocks on the vegetative and reproductive performance of *Vitis vinifera* L. Malbec under irrigated conditions. J. Int. Sci. Vigne Vin 45, 75–84.

Di Genova, A., Almeida, A.M., Muñoz-Espinoza, C., Vizoso, P., Travisany, D., Moraga, C., Pinto, M., Hinrichsen, P., Orellana, A., Maass, A., 2014. Whole genome comparison between table and wine grapes reveals a comprehensive catalog of structural variants. BMC Plant Biol. 14, 7.

Di Vecchi Staraz, M., Bandinelli, R., Boselli, M., This, P., Boursiquot, J.M., Laucou, V., 2007. Genetic structuring and parentage analysis for evolutionary studies in grapevine: kin group and origin of the cultivar Sangiovese revealed. J. Am. Soc. Hortic. Sci. 132, 514–524.

Di Vecchi-Staraz, M., Laucou, V., Bruno, G., Lacombe, T., Gerber, S., Bourse, T., 2009. Low level of pollen-mediated gene flow from cultivated to wild grapevine: consequences for the evolution of the endangered subspecies *Vitis vinifera* L. subsp. *silvestris*. J. Hered. 100, 66–75.

Diakou, P., Carde, J.P., 2001. *In situ* fixation of grape berries. Protoplasma 218, 225–235.

Díaz-Riquelme, J., Lijavetzky, D., Martínez-Zapater, J.M., Carmona, M.J., 2009. Genome-wide analysis of MIKC^C-type MADS box genes in grapevine. Plant Physiol. 149, 354–369.

Dietrich, D., 2018. Hydrotropism: how roots search for water. J. Exp. Bot. 69, 2750–2771.

Dintscheff, D., Stoeff, K., Peschakoff, G., Kolarowa, E., 1964. Untersuchungen über die Stickstoffernährung der Weinrebe unter Anwendung des stabilen Stickstoffisotops ^{15}N. Vitis 4, 347–356.

Dittrich, H.H., 1989. Die Veränderungen der Beereninhaltsstoffe und der Weinqualität durch *Botrytis cinerea*—Übersichtsreferat. Wein-Wiss. 44, 105–131.

Dixon, H.H., Joly, J., 1895. On the ascent of sap. Philos. Trans. R. Soc. B 186, 563–576.

Dixon, R.A., Xie, D.Y., Sharma, S.B., 2005. Proanthocyanidins—a final frontier in flavonoid research? New Phytol. 165, 9–28.

Do, P.T., Prudent, M., Sulpice, R., Causse, M., Fernie, A.R., 2010. The influence of fruit load on the tomato pericarp metabolome in a *Solanum chmielewskii* introgression line population. Plant Physiol. 154, 1128–1142.

Doblin, M.S., Pettolino, F., Bacic, A., 2010. Plant cell walls: the skeleton of the plant world. Funct. Plant Biol. 37, 357–381.

Doco, T., Williams, P., Pauly, M., O'Neill, M.A., Pellerin, P., 2003. Polysaccharides from grape berry cell walls. Part II. Structural characterization of the xyloglucan polysaccharides. Carbohydr. Polym. 53, 253–261.

Dodd, I.C., Egea, G., Davies, W.J., 2008. Abscisic acid signalling when soil moisture is heterogeneous: decreased photoperiod sap flow from drying roots limits abscisic acid export to the shoots. Plant Cell Environ. 31, 1263–1274.

Dokoozlian, N.K., 1999. Chilling temperature and duration interact on the budbreak of 'Perlette' grapevine cuttings. HortScience 34, 1054–1056.

Dokoozlian, N.K., Hirschfelt, D.J., 1995. The influence of cluster thinning at various stages of fruit development on Flame Seedless table grapes. Am. J. Enol. Vitic. 46, 429–436.

Dokoozlian, N.K., Kliewer, W.M., 1995a. The light environment within grapevine canopies. I. Description and seasonal changes during fruit development. Am. J. Enol. Vitic. 46, 209–218.

Dokoozlian, N.K., Kliewer, W.M., 1995b. The light environment within grapevine canopies. II. Influence of leaf area density on fruit zone light environment and some canopy assessment parameters. Am. J. Enol. Vitic. 46, 219–226.

Dokoozlian, N.K., Kliewer, W.M., 1996. Influence of light on grape berry growth and composition varies during fruit development. J. Am. Soc. Hortic. Sci. 121, 869–874.

Dokoozlian, N.K., Williams, L.E., Neja, R.A., 1995. Chilling exposure and hydrogen cyanamide interact in breaking dormancy of grape buds. HortScience 30, 1244–1247.

Domínguez, E., España, L., López-Casado, G., Cuartero, J., Heredia, A., 2009. Biomechanics of isolated tomato (*Solanum lycopersicum*) fruit cuticles during ripening: the role of flavonoids. Funct. Plant Biol. 36, 613–620.

Domínguez, E., Heredia-Guerrero, J.A., Heredia, A., 2011. The biophysical design of plant cuticles: an overview. New Phytol. 189, 938–949.

Donèche, B., Chardonnet, C., 1992. Evolution et localisation des principaux cations au cours du développement du raisin. Vitis 31, 175–181.

Dorcey, E., Urbez, C., Blázquez, M.A., Carbonell, J., Perez-Amador, M.A., 2009. Fertilization-dependent auxin response in ovules triggers fruit development through the modulation of gibberellin metabolism in *Arabidopsis*. Plant J. 58, 318–332.

Dorion, S., Lalonde, S., Saini, H.S., 1996. Induction of male sterility in wheat by meiotic-stage water deficit is preceded by a decline in invertase activity and changes in carbohydrate metabolism in anthers. Plant Physiol. 111, 137–145.

dos Santos, T.P., Lopes, C.M., Rodrigues, M.L., de Souza, C.R., Maroco, J.P., Pereira, J.S., 2003. Partial rootzone drying: effects on growth and fruit quality of field-grown grapevines (*Vitis vinifera*). Funct. Plant Biol. 30, 663–671.

Doster, M.A., Schnathorst, W.C., 1985. Comparative susceptibility of various grapevine cultivars to the powdery mildew fungus *Uncinula necator*. Am. J. Enol. Vitic. 36, 101–104.

Downey, M.O., Rochfort, S., 2008. Simultaneous separation by reversed-phase high-performance liquid chromatography and mass spectral identification of anthocyanins and flavonols in Shiraz grape skin. J. Chromatogr. A 1201, 43–47.

Downey, M.O., Harvey, J.S., Robinson, S.P., 2003a. Analysis of tannins in seeds and skins of Shiraz grapes throughout berry development. Aust. J. Grape Wine Res. 9, 15–27.

Downey, M.O., Harvey, J.S., Robinson, S.P., 2003b. Synthesis of flavonols and expression of flavonol synthase genes in the developing grape berries of Shiraz and Chardonnay (*Vitis vinifera* L.). Aust. J. Grape Wine Res. 9, 110–121.

Downey, M.O., Harvey, J.S., Robinson, S.P., 2004. The effect of bunch shading on berry development and flavonoid accumulation in Shiraz grapes. Aust. J. Grape Wine Res. 10, 55–73.

Downey, M.O., Dokoozlian, N.K., Krstic, M.P., 2006. Cultural practice and environmental impacts on the flavonoid composition of grapes and wine: a review of recent research. Am. J. Enol. Vitic. 57, 257–268.

Downton, W.J.S., 1977a. Salinity effects on the ion composition of fruiting Cabernet Sauvignon vines. Am. J. Enol. Vitic. 28, 210–214.

Downton, W.J.S., 1977b. Photosynthesis in salt-stressed grapevines. Aust. J. Plant Physiol. 4, 183–192.

Downton, W.J.S., 1985. Growth and mineral composition of the Sultana grapevine as influenced by salinity and rootstock. Aust. J. Agric. Res. 36, 425–434.

Downton, W.J.S., Grant, W.J.R., 1992. Photosynthetic physiology of spur pruned and minimal pruned grapevines. Aust. J. Plant Physiol. 19, 309–316.

Downton, W.J.S., Loveys, B.R., 1978. Compositional changes during grape berry development in relation to abscisic acid and salinity. Aust. J. Plant Physiol. 5, 415–423.

Downton, W.J.S., Loveys, B.R., 1981. Abscisic acid content and osmotic relations of salt-stressed grapevine leaves. Aust. J. Plant Physiol. 8, 443–452.

Downton, W.J.S., Grant, W.J.R., Loveys, B.R., 1987. Diurnal changes in the photosynthesis of field-grown grape vines. New Phytol. 105, 71–80.

Downton, W.J.S., Loveys, B.R., Grant, W.J.R., 1990. Salinity effects on the stomatal behaviour of grapevine. New Phytol. 116, 499–503.

Dragoni, D., Lakso, A.N., Piccioni, R.M., Tarara, J.M., 2006. Transpiration of grapevines in the humid northeastern United States. Am. J. Enol. Vitic. 57, 460–467.

Drawert, F., Steffan, H., 1966a. Biochemisch-physiologische Untersuchungen an Traubenbeeren. II. Distribution and respiration of the added ^{14}C compounds. Vitis 5, 27–34.

Drawert, F., Steffan, H., 1966b. Biochemisch-physiologische Untersuchungen an Traubenbeeren. III. Stoffwechsel von zugeführten ^{14}C-Verbindungen und die Bedeutung des Säure-Zucker-Mechanismus für die Reifung von Traubenbeeren. Vitis 5, 377–384.

Dreyer, I., Gomez-Porras, J.L., Riedelsberger, J., 2017. The potassium battery: a mobile energy source for transport processes in plant vascular tissues. New Phytol. 216, 1049–1053.

Druart, N., Johansson, A., Baba, K., Schrader, J., Sjödin, A., Bhalerao, R.R., 2007. Environmental and hormonal regulation of the activity-dormancy cycle in the cambial meristem involves stage-specific modulation of transcriptional and metabolic networks. Plant J. 50, 557–573.

Dry, P.D., 2000. Canopy management for fruitfulness. Aust. J. Grape Wine Res. 6, 109–115.

Dry, P.R., Coombe, B.G., 2004. Viticulture. Volume 1—Resources, second ed. Winetitles, Adelaide, Australia.

Dry, P.R., Loveys, B.R., 1998. Factors influencing grapevine vigour and the potential for control with partial rootzone drying. Aust. J. Grape Wine Res. 4, 140–148.

Dry, P.R., Loveys, B.R., 1999. Grapevine shoot growth and stomatal conductance are reduced when part of the root system is dried. Vitis 38, 151–156.

Dry, P.R., Loveys, B.R., Düring, H., 2000a. Partial drying of the root-zone of grape. I. Transient changes in shoot growth and gas exchange. Vitis 39, 3–8.

Dry, P.R., Loveys, B.R., Düring, H., 2000b. Partial drying of the root-zone of grape. II. Changes in the pattern of root development. Vitis 39, 9–12.

Dry, P.R., Loveys, B.R., McCarthy, M.G., Stoll, M., 2001. Strategic irrigation management in Australian vineyards. J. Int. Sci. Vigne Vin 35, 45–61.

Dry, I.B., Feechan, A., Anderson, C., Jermakow, A.M., Bouquet, A., Adam-Blondon, A.F., 2010a. Molecular strategies to enhance the genetic resistance of grapevines to powdery mildew. Aust. J. Grape Wine Res. 16, 94–105.

Dry, P.R., Longbottom, M.L., McLoughlin, S., Johnson, T.E., Collins, C., 2010b. Classification of reproductive performance of ten winegrape varieties. Aust. J. Grape Wine Res. 16, 47–55.

Du, Y.P., Zhai, H., Sun, Q.H., Wang, Z.S., 2009. Susceptibility of Chinese grapes to grape Phylloxera. Vitis 48, 57–58.

Dubernet, M., Ribereau-Gayon, P., Lerner, H.R., Harel, E., Mayer, A.M., 1977. Purification and properties of laccase from *Botrytis cinerea*. Phytochemistry 16, 191–193.

Dubourdieu, D., Tominaga, T., Masneuf, I., des Gachons, C.P., Murat, M.L., 2006. The role of yeasts in grape flavor development during fermentation: the example of Sauvignon blanc. Am. J. Enol. Vitic. 57, 81–88.

Duchêne, E., Schneider, C., 2005. Grapevine and climatic changes: a glance at the situation in Alsace. Agron. Sustain. Dev. 25, 93–99.

Duchêne, E., Huard, F., Dumas, V., Schneider, C., Merdinoglu, D., 2010. The challenge of adapting grapevine varieties to climate change. Clim. Res. 41, 193–204.

Dudareva, N., Klempien, A., Muhlemann, J.K., Kaplan, I., 2013. Biosynthesis, function and metabolic engineering of plant volatile organic compounds. New Phytol. 198, 16–32.

Dudley, R., 2004. Ethanol, fruit ripening, and the historical origins of human alcoholism in primate frugivory. Integr. Comp. Biol. 44, 315–323.

Dugan, F.M., Lupien, S.L., Grove, G.G., 2002. Incidence, aggressiveness and *in planta* interactions of *Botrytis cinerea* and other filamentous fungi quiescent in grape berries and dormant buds in central Washington state. J. Phytopathol. 150, 375–381.

Dugardeyn, J., Van Der Straeten, D., 2008. Ethylene: fine-tuning plant growth and development by stimulation and inhibition of elongation. Plant Sci. 175, 59–70.

Dun, E.A., Ferguson, B.J., Beveridge, C.A., 2006. Apical dominance and shoot branching. Divergent opinions or divergent mechanisms? Plant Physiol. 142, 812–819.

Dun, E.A., de Saint Germain, A., Rameau, C., Beveridge, C.A., 2012. Antagonistic action of strigolactone and cytokinin in bud outgrowth control. Plant Physiol. 158, 487–498.

Dundon, C.G., Smart, R.E., 1984. Effects of water relations on the potassium status of Shiraz vines. Am. J. Enol. Vitic. 35, 40–45.

Dunlevy, J.D., Kalua, C.M., Keyzers, R.A., Boss, P.K., 2009. The production of flavour and aroma compounds in grape berries. In: Roubelakis-Angelakis, K.A. (Ed.), Grapevine Molecular Physiology and Biotechnology. Springer, Dordrecht, The Netherlands, pp. 293–340.

Dunlevy, J.D., Dennis, E.G., Soole, K.L., Perkins, M.V., Davies, C., Boss, P.K., 2013a. A methyltransferase essential for the methoxypyrazine-derived flavour of wine. Plant J. 75, 606–617.

Dunlevy, J.D., Soole, K.L., Perkins, M.V., Nicholson, E.L., Maffei, S.M., Boss, P.K., 2013b. Determining the methoxypyrazine biosynthesis variables affected by light exposure and crop level in Cabernet Sauvignon. Am. J. Enol. Vitic. 64, 450–458.

Düring, H., 1980. Stomatafrequenz bei Blättern von Vitis-Arten und -Sorten. Vitis 19, 91–98.

Düring, H., 1984. Evidence for osmotic adjustment to drought in grapevines (*Vitis vinifera* L.). Vitis 23, 1–10.

Düring, H., 1987. Stomatal response to alteration of soil and air humidity in grapevines. Vitis 26, 9–18.

Düring, H., 1988. CO_2 assimilation and photorespiration of grapevine leaves: responses to light and drought. Vitis 27, 199–208.

Düring, H., 1997. Potential frost resistance of grape: kinetics of temperature-induced hardening of Riesling and Silvaner buds. Vitis 36, 213–214.

Düring, H., 1999. Photoprotection in leaves of grapevines: responses of the xanthophyll cycle to alterations of light intensity. Vitis 38, 21–24.

Düring, H., 2003. Stomatal and mesophyll conductances control CO_2 transfer to chloroplasts in leaves of grapevine (*Vitis vinifera* L.). Vitis 42, 65–68.

Düring, H., Alleweldt, G., 1973. Der Jahresgang der Abscisinsäure in vegetativen Organen von Reben. Vitis 12, 26–32.

Düring, H., Bachmann, O., 1975. Abscisic acid analysis in *Vitis vinifera* in the period of endogenous bud dormancy by high pressure liquid chromatography. Physiol. Plant. 34, 201–203.

Düring, H., Davtyan, A., 2002. Developmental changes of primary processes of photosynthesis in sun- and shade-adapted berries of two grapevine cultivars. Vitis 41, 63–67.

Düring, H., Dry, P.R., 1995. Osmoregulation in water stressed roots: responses of leaf conductance and photosynthesis. Vitis 34, 15–17.

Düring, H., Kismali, I., 1975. Die Rolle der Abscisinsäure bei der Knospenruhe di- und tetraploider Rebsorten. Vitis 14, 89–96.

Düring, H., Loveys, B.R., 1982. Diurnal changes in water relations and abscisic acid in field grown *Vitis vinifera* cvs. I. Leaf water potential components and leaf conductance under humid temperate and semiarid conditions. Vitis 21, 223–232.

Düring, H., Oggionni, F., 1986. Transpiration und Mineralstoffeinlagerung der Weinbeere. Vitis 25, 59–66.

Düring, H., Lang, A., Oggionni, F., 1987. Patterns of water flow in Riesling berries in relation to developmental changes in their xylem morphology. Vitis 26, 123–131.

Dutta, R., Robinson, K.R., 2004. Identification and characterization of stretch-activated ion channels in pollen protoplasts. Plant Physiol. 135, 1398–1406.

Eapen, D., Barroso, M.L., Ponce, G., Campos, M.E., Cassab, G.I., 2005. Hydrotropism: root growth responses to water. Trends Plant Sci. 10, 44–50.

Earles, J.M., Knipfer, T., Tixier, A., Orozco, J., Reyes, C., Zwieniecki, M.A., Brodersen, C.R., McElrone, A.J., 2018. *In vivo* quantification of plant starch reserves at micrometer resolution using X-ray microCT imaging and machine learning. New Phytol. 218, 1260–1269.

Ebadi, A., Sedgley, M., May, P., Coombe, B.G., 1996. Seed development and abortion in *Vitis vinifera* L., cv. Chardonnay. Int. J. Plant Sci. 157, 703–712.

Edson, C.E., Howell, G.S., Flore, J.A., 1993. Influence of crop load on photosynthesis and dry matter partitioning of Seyval grapevines I. Single leaf and whole vine response pre- and post-harvest. Am. J. Enol. Vitic. 44, 139–147.

Edson, C.E., Howell, G.S., Flore, J.A., 1995a. Influence of crop load on photosynthesis and dry matter partitioning of Seyval grapevines II. Seasonal changes in single leaf and whole vine photosynthesis. Am. J. Enol. Vitic. 46, 469–477.

Edson, C.E., Howell, G.S., Flore, J.A., 1995b. Influence of crop load on photosynthesis and dry matter partitioning of Seyval grapevines III. Seasonal changes in dry matter partitioning, vine morphology, yield, and fruit composition. Am. J. Enol. Vitic. 46, 478–485.

Edye, L.A., Richards, G.N., 1991. Analysis of condensates from wood smoke: components derived from polysaccharides and lignins. Environ. Sci. Technol. 25, 1133–1137.

Efetova, M., Zeier, J., Riederer, M., Lee, C.W., Stingl, N., Mueller, M., 2007. A central role of abscisic acid in drought stress protection of *Agrobacterium*-induced tumors on *Arabidopsis*. Plant Physiol. 145, 853–862.

Egerton-Warburton, L.M., Querejeta, J.I., Allen, M.F., 2007. Common mycorrhizal networks provide a potential pathway for the transfer of hydraulically lifted water between plants. New Phytol. 58, 1473–1483.

Egger, K., Reichling, J., Ammann-Schweizer, R., 1976. Flavonol-Derivate in Formen der Gattung Vitis. Vitis 15, 24–28.

Egri, Á., Horváth, Á., Kriska, G., Horváth, G., 2010. Optics of sunlit water drops on leaves: conditions under which sunburn is possible. New Phytol. 185, 979–987.

Ehret, D.L., Helmer, T., Hall, J.W., 1993. Cuticle cracking in tomato fruit. J. Hortic. Sci. 68, 195–201.

Eibach, R., 1994. Investigations about the genetic resources of grapes with regard to resistance characteristics to powdery mildew (*Oidium tuckeri*). Vitis 33, 143–150.

Eibach, R., Alleweldt, G., 1984. Einfluss der Wasserversorgung auf Wachstum, Gaswechsel und Substanzproduktion traubentragender Reben. II. Der Gaswechsel. Vitis 23, 11–20.

Eibach, R., Alleweldt, G., 1985. Einfluss der Wasserversorgung auf Wachstum, Gaswechsel und Substanzproduktion traubentragender Reben. III. Die Substanzproduktion. Vitis 24, 183–198.

Eichert, T., Goldbach, H.E., 2008. Equivalent pore radii of hydrophilic foliar uptake routes in stomatous and astomatous leaf surfaces—further evidence for a stomatal pathway. Physiol. Plant. 132, 491–502.

Eifert, J., Pánczél, M., Eifert, A., 1961. Änderung des Stärke- und Zuckergehaltes der Rebe während der Ruheperiode. Vitis 2, 257–265.

Elad, Y., Evensen, K., 1995. Physiological aspects of resistance to *Botrytis cinerea*. Phytopathology 85, 637–643.

Else, M.A., Stankiewicz-Davies, A.P., Crisp, C.M., Atkinson, C.J., 2004. The role of polar auxin transport through pedicels of *Prunus avium* L. in relation to fruit development and retention. J. Exp. Bot. 55, 2099–2109.

Else, M.A., Janowiak, F., Atkinson, C.J., Jackson, M.B., 2009. Root signals and stomatal closure in relation to photosynthesis, chlorophyll *a* fluorescence and adventitious rooting of flooded tomato plants. Ann. Bot. 103, 313–323.

Eltom, M., Winefield, C.S., Trought, M.C.T., 2014. Effect of pruning system, cane size and season on inflorescence primordia initiation and inflorescence architecture of *Vitis vinifera* L. Sauvignon blanc. Aust. J. Grape Wine Res. 20, 459–464.

Eltom, M., Trought, M.C.T., Agnew, R., Parker, A., Winefield, C.S., 2017. Pre-budburst temperature influences the inner and outer arm morphology, phenology, flower number, fruitset, TSS accumulation and variability of *Vitis vinifera* L. Sauvignon Blanc bunches. Aust. J. Grape Wine Res. 23, 280–286.

Emershad, R.L., Ramming, D.W., 1984. *In-ovulo* embryo culture of *Vitis vinifera* L. c.v. 'Thompson Seedless'. Am. J. Bot. 71, 873–877.

Emershad, R.L., Ramming, D.W., Serpe, M.D., 1989. *In ovulo* embryo development and plant formation from stenospermic genotypes of *Vitis vinifera*. Am. J. Bot. 76, 397–402.

Emes, M.J., Neuhaus, H.E., 1997. Metabolism and transport in non-photosynthetic plastids. J. Exp. Bot. 48, 1995–2005.

Endo, A., Sawada, Y., Takahashi, H., Okamoto, M., Ikegami, K., Koiwai, H., 2008. Drought induction of *Arabidopsis* 9-cis-epoxycarotenoid dioxygenase occurs in vascular parenchyma cells. Plant Physiol. 147, 1984–1993.

Engineer, C.B., Hashimoto-Sugimoto, M., Negi, J., Israelsson-Nordström, M., Azoulay-Shemer, T., Rappel, W.J., Iba, K., Schroeder, J.I., 2016. CO_2 sensing and CO_2 regulation of stomatal conductance: advances and open questions. Trends Plant Sci. 21, 16–30.

English, J.T., Bledsoe, A.M., Marois, J.J., Kliewer, W.M., 1990. Influence of grapevine canopy management on evaporative potential in the fruit zone. Am. J. Enol. Vitic. 41, 137–141.

Enquist, B.J., Niklas, K.J., 2002. Global allocation rules for patterns of biomass partitioning in seed plants. Science 295, 1517–1520.

Ensminger, I., Busch, F., Huner, N.P.A., 2006. Photostasis and cold acclimation: sensing low temperature through photosynthesis. Physiol. Plant. 126, 28–44.

Epstein, E., 1999. Silicon. Annu. Rev. Plant Physiol. Plant Mol. Biol. 50, 641–664.

Erlenwein, H., 1965a. Einfluss der Ernährung und des Pfropfpartners auf das Wurzelwachstum von Vitis-Arten und -Sorten. Vitis 5, 161–186.

Erlenwein, H., 1965b. Einfluß von Klimafaktoren auf das Wachstum von Vitis-Arten und-Sorten. Vitis 5, 94–109.

Esau, K., 1948. Phloem structure in the grapevine, and its seasonal changes. Hilgardia 18, 217–296.

Escalona, J.M., Flexas, J., Medrano, H., 1999. Stomatal and non-stomatal limitations of photosynthesis under water stress in field-grown grapevines. Aust. J. Plant Physiol. 26, 421–433.

Escalona, J.M., Fuentes, S., Tomás, M., Martorell, S., Flexas, J., Medrano, H., 2013. Responses of leaf night transpiration to drought stress in *Vitis vinifera* L. Agric. Water Manag. 118, 50–58.

Eschenbruch, R., Kleynhans, P.H., 1974. The influence of copper-containing fungicides on the copper content of grape juice and on hydrogen sulphide formation. Vitis 12, 320–324.

Escudero, A., Campo, E., Fariña, L., Cacho, J., Ferreira, V., 2007. Analytical characterization of the aroma of five premium red wines. Insights into the role of odor families and the concept of fruitiness of wines. J. Agric. Food Chem. 55, 4501–4510.

Espinoza, C., Vega, A., Medina, C., Schlauch, K., Cramer, G., Arce-Johnson, P., 2007. Gene expression associated with compatible viral diseases in grapevine cultivars. Funct. Integr. Genom. 7, 95–110.

Esteban, A., Villanueva, J., Lissarrague, J.R., 1999. Effect of irrigation on changes in berry composition of Tempranillo during maturation. Sugars, organic acids, and mineral elements. Am. J. Enol. Vitic. 50, 418–434.

Etchebarne, F., Ojeda, H., Deloire, A., 2009. Influence of water status on mineral composition of berries in 'Grenache Noir' (*Vitis vinifera* L.). Vitis 48, 63–68.

Etchebarne, F., Ojeda, H., Hunter, J.J., 2010. Leaf: fruit ratio and vine water status effects on Grenache noir (*Vitis vinfera* L.) berry composition: water, sugar, organic acids and cations. S. Afr. J. Enol. Vitic. 31, 106–115.

Etienne, A., Génard, M., Lobit, P., Mbeguié-A-Mbéguié, D., Bugaud, C., 2013. What controls fleshy fruit acidity? A review of malate and citrate accumulation in fruit cells. J. Exp. Bot. 64, 1451–1469.

Etxeberria, E., González, P., Tomlinson, P., Pozueta-Romero, J., 2005. Existence of two parallel mechanisms for glucose uptake in heterotrophic plant cells. J. Exp. Bot. 56, 1905–1912.

Evans, J.R., 1989. Photosynthesis and nitrogen relationships in leaves of C_3 plants. Oecologia 78, 9–19.

Evans, L.T., Fischer, R.A., 1999. Yield potential: Its definition, measurement and significance. Crop Sci. 39, 1544–1551.

Evans, R.G., Spayd, S.E., Wample, R.L., Kroeger, M.W., Mahan, M.O., 1993. Water use of *Vitis vinifera* grapes in Washington. Agric. Water Manag. 23, 109–124.

Evans, J.R., Kaldenhoff, R., Genty, B., Terashima, I., 2009. Resistances along the CO_2 diffusion pathway inside leaves. J. Exp. Bot. 60, 2235–2248.

Evert, R.F., 2006. Esau's Plant Anatomy. Meristems, Cells, and Tissues of the Plant Body—Their Structure, Function, and Development, third ed. John Wiley & Sons, Hoboken, NJ.

Ewart, A., Kliewer, W.M., 1977. Effects of controlled day and night temperatures and nitrogen on fruit-set, ovule fertility, and fruit composition of several wine grape cultivars. Am. J. Enol. Vitic. 28, 88–95.

Ewers, F.W., Fisher, J.B., 1991. Why vines have narrow stems: histological trends in *Bauhinia* (Fabaceae). Oecologia 88, 233–237.

Ezzahouani, A., Williams, L.E., 1995. The influence of rootstock on leaf water potential, yield, and berry composition of Ruby Seedless grapevines. Am. J. Enol. Vitic. 46, 559–563.

Failla, O., Mariani, L., Brancadoro, L., Minelli, R., Scienza, A., Murada, G., 2004. Spatial distribution of solar radiation and its effects on vine phenology and grape ripening in an alpine environment. Am. J. Enol. Vitic. 55, 128–138.

Faist, H., Keller, A., Hentschel, U., Deeken, R., 2016. Grapevine (*Vitis vinifera*) crown galls host distinct microbiota. Appl. Environ. Microbiol. 82, 5542–5552.

Fait, A., Hanhineva, K., Beleggia, R., Dai, N., Rogachev, I., Nikiforova, V.J., 2008. Reconfiguration of the achene and receptacle metabolic networks during strawberry fruit development. Plant Physiol. 148, 730–750.

Falcetti, M., Scienza, A., 1989. Influence de la densité de plantation et du porte-greffe sur la production et la qualité du moût de Pinot blanc et de Chardonnay cultivés en Italie dans le Trentin. J. Int. Sci. Vigne Vin 23, 151–164.

Falginella, L., Di Gaspero, G., Castellarin, S.D., 2012. Expression of flavonoid genes in the red grape berry of 'Alicante Bouschet' varies with the histological distribution of anthocyanins and their chemical composition. Planta 236, 1037–1051.

Famiani, F., Walker, R.P., Técsi, L., Chen, Z.H., Proietti, P., Leegood, R.C., 2000. An immunohistochemical study of the compartmentation of metabolism during the development of grape (*Vitis vinifera* L.) berries. J. Exp. Bot. 51, 675–683.

Fan, S.C., Lin, C.S., Hsu, P.K., Lin, S.H., Tsay, Y.F., 2009. The *Arabidopsis* nitrate transporter NRT1.7, expressed in phloem, is responsible for source-to-sink remobilization of nitrate. Plant Cell 21, 2750–2761.

Fanciullino, A.L., Bidel, L.P.R., Urban, L., 2014. Carotenoid responses to environmental stimuli: integrating redox and carbon controls into a fruit model. Plant Cell Environ. 37, 273–289.

Fang, Y., Qian, M.C., 2006. Quantification of selected aroma-active compounds in Pinot noir wines from different grape maturities. J. Agric. Food Chem. 54, 8567–8573.

Fanizza, G., Lamaj, F., Resta, P., Ricciardi, L., Savino, V., 2005. Grapevine cvs Primitivo, Zinfandel and Crljenak kastelanski: molecular analysis by AFLP. Vitis 44, 147–148.

Fankhauser, C., Staiger, D., 2002. Photoreceptors in *Arabidopsis thaliana*: light perception, signal transduction and entrainment of the endogenous clock. Planta 216, 1–16.

Farquhar, G.D., Wetselaar, R., Firth, P.M., 1979. Ammonia volatilization from senescing leaves of maize. Science 203, 1257–1258.

Farquhar, G.D., von Caemmerer, S., Berry, J.A., 1980. A biochemical model of photosynthetic CO_2 assimilation in leaves of C_3 species. Planta 149, 78–90.

Fasoli, M., Richter, C.L., Zenoni, S., Bertini, E., Vitulo, N., Dal Santo, S., Dokoozlian, N., Pezzotti, M., Tornielli, G.B., 2018. Timing and order of the molecular events marking the onset of berry ripening in grapevine. Plant Physiol. 178, 1187–1206.

Favero, A.C., Angelucci de Amorim, D., Vieira da Mota, R., Soares, A.M., de Souza, C.R., de Albuquerque Regina, M., 2011. Double-pruning of 'Syrah' grapevines: a management strategy to harvest wine grapes during the winter in the Brazilian Southeast. Vitis 50, 151–158.

Feild, T.S., Lee, D.W., Holbrook, N.M., 2001. Why leaves turn red in autumn. The role of anthocyanins in senescing leaves of red-osier dogwood. Plant Physiol. 127, 566–574.

Feldman, L.J., 1984. Regulation of root development. Annu. Rev. Plant Physiol. 35, 223–242.

Fennell, A., 2004. Freezing tolerance and injury in grapevines. In: Arora, R. (Ed.), Adaptations and Responses of Woody Plants to Environmental Stresses. Food Products Press, Binghamton, NY, pp. 201–235.

Fennell, A., Hoover, E., 1991. Photoperiod influences growth, bud dormancy, and cold acclimation in *Vitis labruscana* and *V. riparia*. J. Am. Soc. Hortic. Sci. 116, 270–273.

Ferguson, I.B., 1984. Calcium in plant senescence and fruit ripening. Plant Cell Environ. 7, 477–489.

Ferguson, B.J., Beveridge, C.A., 2009. Roles for auxin, cytokinin, and strigolactone in regulating shoot branching. Plant Physiol. 149, 1929–1944.

Ferguson, J.C., Tarara, J.M., Mills, L.J., Grove, G.G., Keller, M., 2011. Dynamic thermal time model of cold hardiness for dormant grapevine buds. Ann. Bot. 107, 389–396.

Ferguson, J.C., Moyer, M.M., Mills, L.J., Hoogenboom, G., Keller, M., 2014. Modeling dormant bud cold hardiness and budbreak in twenty-three *Vitis* genotypes reveals variation by region of origin. Am. J. Enol. Vitic. 65, 59–71.

Fernandez, L., Pradal, M., Lopez, G., Berud, F., Romieu, C., Torregrosa, L., 2006. Berry size variability in *Vitis vinifera* L. Vitis 45, 53–55.

Fernandez, L., Torregrosa, L., Segura, V., Bouquet, A., Martinez-Zapater, J.M., 2010. Transposon-induced gene activation as a mechanism generating cluster shape somatic variation in grapevine. Plant J. 61, 545–557.

Fernie, A.R., Carrari, F., Sweetlove, L.J., 2004. Respiratory metabolism: glycolysis, the TCA cycle and mitochondrial electron transport. Curr. Opin. Plant Biol. 7, 254–261.

Ferrandino, A., Carra, A., Rolle, L., Schneider, A., Schubert, A., 2012. Profiling of hydroxycinnamoyl tartrates and acylated anthocyanins in the skin of 34 *Vitis vinifera* genotypes. J. Agric. Food Chem. 60, 4931–4945.

Ferris, H., Zheng, L., Walker, M.A., 2012. Resistance of grape rootstocks to plant-parasitic nematodes. J. Nematol. 44, 377–386.

Feucht, W., Forche, E., Porstendörfer, J., 1975. Ermittlung der Kalium- und Calciumverteilung in Traubenachsen von *Vitis vinifera* mit Hilfe der Röntgenstrahlenmikroanalyse am Rasterelektronenmikroskop. Vitis 14, 190–197.

Fich, E.A., Segerson, N.A., Rose, J.K.C., 2016. The plant polyester cutin: biosynthesis, structure, and biological roles. Annu. Rev. Plant Biol. 67, 207–233.

Field, B., Jordán, F., Osbourn, A., 2006. First encounters—deployment of defence-related natural products by plants. New Phytol. 172, 193–207.

Field, S.K., Smith, J.P., Holzapfel, B.P., Hardie, W.J., Emery, R.J.N., 2009. Grapevine response to soil temperature: xylem cytokinins and carbohydrate reserve mobilization from budbreak to anthesis. Am. J. Enol. Vitic. 60, 164–172.

Filippetti, I., Intrieri, C., Cemtinari, M., Bucchetti, B., Pastore, C., 2005. Molecular characterization of officially registered Sangiovese clones and of other Sangiovese-like biotypes in Tuscany, Corsica and Emilia-Romagna. Vitis 44, 167–172.

Fillion, L., Ageorges, A., Picaud, S., Coutos-Thévenot, P., Lemoine, R., Romieu, C., 1999. Cloning and expression of a hexose transporter gene expressed during the ripening of grape berry. Plant Physiol. 120, 1083–1093.

Findlay, N., Oliver, K.J., Nii, N., Coombe, B.G., 1987. Solute accumulation by grape pericarp cells. IV. Perfusion of pericarp apoplast via the pedicel and evidence for xylem malfunction in ripening berries. J. Exp. Bot. 38, 668–679.

Fine, P.M., Cass, G.R., Simoneit, B.R.T., 2001. Chemical characterization of fine particle emissions from fireplace combustion of woods grown in the northeastern United States. Environ. Sci. Technol. 35, 2665–2675.

Firn, R.D., Jones, C.G., 2009. A Darwinian view of metabolism: molecular properties determine fitness. J. Exp. Bot. 60, 719–726.

Fischer, E.M., Knutti, R., 2015. Anthropogenic contribution to global occurrence of heavy-precipitation and high-temperature extremes. Nat. Clim. Change 5, 560–564.

Fischer, R., Budde, I., Hain, R., 1997. Stilbene synthase gene expression causes changes in flower colour and male sterility in tobacco. Plant J. 11, 489–498.

Fischer, W.N., André, B., Rentsch, D., Krolkiewicz, S., Tegeder, M., Breitkreuz, K., 1998. Amino acid transport in plants. Trends Plant Sci. 3, 188–195.

Fischer, T.C., Mirbeth, B., Rentsch, J., Sutter, C., Ring, L., Flachowsky, H., 2014. Premature and ectopic anthocyanin formation by silencing of anthocyanidin reductase in strawberry (*Fragaria* × *ananassa*). New Phytol. 201, 440–451.

Fishman, S., Génard, M., 1998. A biophysical model of fruit growth: simulation of seasonal and diurnal dynamics of mass. Plant Cell Environ. 21, 739–752.

Flexas, J., Escalona, J.M., Medrano, H., 1998. Down-regulation of photosynthesis by drought under field conditions in grapevine leaves. Aust. J. Plant Physiol. 25, 893–900.

Flexas, J., Badger, M., Chow, W.S., Medrano, H., Osmond, C.B., 1999a. Analysis of the relative increase in photosynthetic O_2 uptake when photosynthesis in grapevine leaves is inhibited following low night temperatures and/or water stress. Plant Physiol. 121, 675–684.

Flexas, J., Escalona, J.M., Medrano, H., 1999b. Water stress induces different levels of photosynthesis and electron transport rate regulation in grapevines. Plant Cell Environ. 22, 39–48.

Flexas, J., Bota, J., Escalona, J.M., Sampol, B., Medrano, H., 2002. Effects of drought on photosynthesis in grapevines under field conditions: an evaluation of stomatal and mesophyll limitations. Funct. Plant Biol. 29, 461–471.

Flore, J.A., Lakso, A.N., 1989. Environmental and physiological regulation of photosynthesis in fruit crops. Hortic. Rev. (Am. Soc. Hortic. Sci). 11, 111–157.

Fontes, N., Côrte-Real, M., Gerós, H., 2011a. New observations on the integrity, structure, and physiology of flesh cells from fully ripened grape berry. Am. J. Enol. Vitic. 62, 279–284.

Fontes, N., Gerós, H., Delrot, S., 2011b. Grape berry vacuole: a complex and heterogeneous membrane system specialized in the accumulation of solutes. Am. J. Enol. Vitic. 62, 270–278.

Fonti, P., Solomonoff, N., García-González, I., 2007. Earlywood vessels of *Castanea sativa* record temperature before their formation. New Phytol. 173, 562–570.

Forde, B.G., 2002. Local and long-range signaling pathways regulating plant responses to nitrate. Annu. Rev. Plant Biol. 53, 203–224.

Forde, C.G., Cox, A., Williams, E.R., Boss, P.K., 2011. Associations between the sensory attributes and volatile composition of Cabernet Sauvignon wines and the volatile composition of the grapes used for their production. J. Agric. Food Chem. 59, 2573–2583.

Forkmann, G., 1991. Flavonoids as flower pigments: the formation of the natural spectrum and its extension by genetic engineering. Plant Breed. 106, 1–26.

Forneck, A., Huber, L., 2009. (A)sexual reproduction—a review of life cycles of grape phylloxera, *Daktulosphaira vitifoliae*. Entomol. Exp. Appl. 131, 1–10.

Fortes, A.M., Teixeira, R.T., Agudelo-Romero, P., 2015. Complex interplay of hormonal signals during grape berry ripening. Molecules 20, 9326–9343.

Fougère-Rifot, M., Park, H.S., Bouard, J., 1995. Données nouvelles sur l'hypoderme et la pulpe des baies normales et des baies millerandées d'une variété de *Vitis vinifera* L., le Merlot noir. Vitis 34, 1–7.

Fournand, D., Vicens, A., Sidhoum, L., Souquet, J.M., Moutounet, M., Cheynier, V., 2006. Accumulation and extractability of grape skin tannins and anthocyanins at different advanced physiological stages. J. Agric. Food Chem. 54, 7331–7338.

Fournier-Level, A., Le Cunff, L., Gomez, C., Doligez, A., Ageorges, A., Roux, C., 2009. Quantitative genetic bases of anthocyanin variation in grape (*Vitis vinifera* L. ssp. *sativa*) berry: a quantitative trait locus to quantitative trait nucleotide integrated study. Genetics 183, 1127–1139.

Fournier-Level, A., Lacombe, T., Le Cunff, L., Boursiquot, J.M., This, P., 2010. Evolution of the *VvMybA* gene family, the major determinant of berry colour in cultivated grapevine (*Vitis vinifera* L.). Heredity 104, 351–362.

Foyer, C.H., Noctor, G., 2005. Oxidant and antioxidant signalling in plants: a re-evaluation of the concept of oxidative stress in a physiological context. Plant Cell Environ. 28, 1056–1071.

Foyer, C.H., Parry, M., Noctor, G., 2003. Markers and signals associated with nitrogen assimilation in higher plants. J. Exp. Bot. 54, 585–593.

Foyer, C.H., Bloom, A.J., Queval, G., Noctor, G., 2009. Photorespiratory metabolism: genes, mutants, energetics, and redox signaling. Annu. Rev. Plant Biol. 60, 455–484.

Foyo-Moreno, I., Alados, I., Alados-Arboledas, L., 2017. A new conventional regression model to estimate hourly photosynthetic photon flux density under all sky conditions. Int. J. Climatol. 37, 1067–1075.

Fracheboud, Y., Luquez, V., Björkén, L., Sjödin, A., Tuominen, H., Jansson, S., 2009. The control of autumn senescence in European aspen. Plant Physiol. 149, 1982–1991.

Franceschi, V.R., Nakata, P.A., 2005. Calcium oxalate in plants: formation and function. Annu. Rev. Plant Biol. 56, 41–71.

Franck, N., Morales, J.P., Arancibia-Avendaño, D., García de Cortázar, V., Perez-Quezada, J.F., Zurita-Silva, A., 2011. Seasonal fluctuations in *Vitis vinifera* root respiration in the field. New Phytol. 192, 939–951.

Frank, D., Gould, I., Millikan, M., 2004. Browning reactions during storage of low-moisture Australian sultanas: evidence for arginine-mediated Maillard reactions. Aust. J. Grape Wine Res. 10, 151–163.

Franke, R., Schreiber, L., 2007. Suberin—a biopolyester forming apoplastic plant interfaces. Curr. Opin. Plant Biol. 10, 252–259.

Franklin, K.A., 2008. Shade avoidance. New Phytol. 179, 930–944.

Franklin, K.A., Whitelam, G.C., 2007. Light-quality regulation of freezing tolerance in *Arabidopsis thaliana*. Nat. Genet. 39, 1410–1413.

Franks, F., 1985. Biophysics and Biochemistry at Low Temperatures. Cambridge University Press, Cambridge, United Kingdom.

Franks, P.J., Farquhar, G.D., 2007. The mechanical diversity of stomata and its significance in gas-exchange control. Plant Physiol. 143, 78–87.

Franks, T., Botta, R., Thomas, M.R., 2002. Chimerism in grapevines: implications for cultivar identity, ancestry and genetic improvement. Theor. Appl. Genet. 104, 192–199.

Franks, P.J., Adams, M.A., Amthor, J.S., Barbour, M.M., Berry, J.A., Ellsworth, D.S., 2013. Sensitivity of plants to changing atmospheric CO_2 concentration: from the geological past to the next century. New Phytol. 197, 1077–1094.

Fraser, P.D., Truesdale, M.R., Bird, C.R., Schuch, W., Bramley, P.M., 1994. Carotenoid biosynthesis during tomato fruit development. Plant Physiol. 105, 405–413.

Frato, K.E., 2019. Identification of hydroxypyrazine O-methyltransferase genes in *Coffea arabica*: a potential source of methoxypyrazines that cause potato taste defect. J. Agric. Food Chem. 67, 341–351.

Freeman, B.M., 1983. Effects of irrigation and pruning of Shiraz grapevines on subsequent red wine pigments. Am. J. Enol. Vitic. 34, 23–26.

Freeman, B.M., Kliewer, W.M., 1983. Effect of irrigation, crop level and potassium fertilization on Carignane vines. II. Grape and wine quality. Am. J. Enol. Vitic. 34, 197–207.

Freeman, B.M., Lee, T.H., Turkington, C.R., 1979. Interaction of irrigation and pruning level on growth and yield of Shiraz vines. Am. J. Enol. Vitic. 30, 218–223.

Freeman, B.M., Kliewer, W.M., Stern, P., 1982. Influence of windbreaks and climatic region on diurnal fluctuation of leaf water potential, stomatal conductance, and leaf temperature of grapevines. Am. J. Enol. Vitic. 33, 233–236.

Fricke, W., 2019. Night-time transpiration – favouring growth? Trends Plant Sci. 24, 311–317.

Friedel, M., Stoll, M., Patz, C.D., Will, F., Dietrich, H., 2015. Impact of light exposure on fruit composition of white 'Riesling' grape berries (*Vitis vinifera* L.). Vitis 54, 107–116.

Friend, A.P., Trought, M.C.T., Stushnoff, C., Wells, G.H., 2011. Effect of delaying budburst on shoot development and yield of *Vitis vinifera* L. Chardonnay 'Mendoza' after a spring freeze event. Aust. J. Grape Wine Res. 17, 378–382.

Frigerio, M., Alabadí, D., Pérez-Gómez, J., García-Cárcel, L., Phillips, A.L., Hedden, P., 2006. Transcriptional regulation of gibberellin metabolism genes by auxin signaling in *Arabidopsis*. Plant Physiol. 142, 553–563.

Friml, J., 2003. Auxin transport—shaping the plant. Curr. Opin. Plant Biol. 6, 7–12.

Frioni, T., Tombesi, S., Silvestroni, O., Lanari, V., Bellincontro, A., Sabbatini, P., Gatti, M., Poni, S., Palliotti, A., 2016. Postbudburst spur pruning reduces yield and delays fruit sugar accumulation in Sangiovese in central Italy. Am. J. Enol. Vitic. 67, 419–425.

Fritschi, F.B., Lin, H., Walker, M.A., 2007. *Xylella fastidiosa* population dynamics in grapevine genotypes differing in susceptibility to Pierce's disease. Am. J. Enol. Vitic. 58, 326–332.

Fritz, C., Palacios-Rojas, N., Feil, R., Stitt, M., 2006. Regulation of secondary metabolism by the carbon-nitrogen status in tobacco: nitrate inhibits large sectors of phenylpropanoid metabolism. Plant J. 46, 533–548.

Fromm, J., Lautner, S., 2007. Electrical signals and their physiological significance in plants. Plant Cell Environ. 30, 249–257.

Fu, X., Harberd, N.P., 2003. Auxin promotes *Arabidopsis* root growth by modulating gibberellin response. Nature 421, 740–743.

Fu, Q., Cheng, L., Guo, Y., Turgeon, R., 2011. Phloem loading strategies and water relations in trees and herbaceous plants. Plant Physiol. 157, 1518–1527.

Fuentes, S., Sullivan, W., Tilbrook, J., Tyerman, S., 2010. A novel analysis of grapevine berry tissue demonstrates a variety-dependent correlation between tissue vitality and berry shrivel. Aust. J. Grape Wine Res. 16, 327–336.

Fujioka, S., Yokota, T., 2003. Biosynthesis and metabolism of brassinosteroids. Annu. Rev. Plant Biol. 54, 137–164.

Fukui, K., Hayashi, K., 2018. Manipulation and sensing of auxin metabolism, transport and signaling. Plant Cell Physiol. 59, 1500–1510.

Fukui, M., Yokotsuka, K., 2003. Content and origin of protein in white and red wines: changes during fermentation and maturation. Am. J. Enol. Vitic. 54, 178–188.

Fulcrand, H., Dueñas, M., Salas, E., Cheynier, V., 2006. Phenolic reactions during winemaking and aging. Am. J. Enol. Vitic. 57, 289–297.

Fuller, M.P., Telli, G., 1999. An investigation of the frost hardiness of grapevine (*Vitis vinifera*) during bud break. Ann. Appl. Biol. 135, 589–595.

Fung, R.W.M., Gonzalo, M., Fekete, C., Kovacs, L.G., He, Y., Marsh, E., 2008. Powdery mildew induces defense-oriented reprogramming of the transcriptome in a susceptible but not in a resistant grapevine. Plant Physiol. 146, 236–249.

Furch, A.C.U., Hafke, J.B., Schulz, A., van Bel, A.J.E., 2007. Ca^{2+}-mediated remote control of reversible sieve tube occlusion in *Vicia faba*. J. Exp. Bot. 58, 2827–2838.

Furiya, T., Suzuki, S., Sueta, T., Takayanagi, T., 2009. Molecular characterization of a bud sport of Pinot gris bearing white berries. Am. J. Enol. Vitic. 60, 66–73.

Furze, M.E., Trumbore, S., Hartmann, H., 2018. Detours on the phloem sugar highway: stem carbon storage and remobilization. Curr. Opin. Plant Biol. 43, 89–95.

Gadoury, D.M., Pearson, R.C., 1990. Germination of ascospores and infection of *Vitis* by *Uncinula necator*. Phytopathology 80, 1198–1203.

Gadoury, D.M., Seem, R.C., Ficke, A., Wilcox, W.F., 2003. Ontogenic resistance to powdery mildew in grape berries. Phytopathology 93, 547–555.

Gadoury, D.M., Seem, R.C., Wilcox, W.F., Henick-Kling, T., Conterno, L., Day, A., 2007. Effects of diffuse colonization of grape berries by *Uncinula necator* on bunch rots, berry microflora, and juice and wine quality. Phytopathology 97, 1356–1365.

Gaforio, L., García-Muñoz, S., Cabello, F., Muñoz-Organero, G., 2011. Evaluation of susceptibility to powdery mildew (*Erysiphe necator*) in *Vitis vinifera* varieties. Vitis 50, 123–126.

Gagné, S., Saucier, C., Gény, L., 2006. Composition and cellular localization of tannins in Cabernet Sauvignon skins during growth. J. Agric. Food Chem. 54, 9465–9471.

Gago, P., Santiago, J.L., Boso, S., Alonso-Villaverde, V., Grando, M.S., Martínez, M.C., 2009. Biodiversity and characterization of twenty-two *Vitis vinifera* L. cultivars in the northwestern Iberian Peninsula. Am. J. Enol. Vitic. 60, 293–301.

Gajdanowicz, P., Michard, E., Sandmann, M., Rocha, M., Guedes Corrêa, L.G., Ramírez-Aguilar, S.J., 2011. Potassium (K^+) gradients serve as a mobile energy source in plant vascular tissues. Proc. Natl. Acad. Sci. U. S. A. 108, 864–869.

Gale, E.J., Moyer, M.M., 2017. Cold hardiness of *Vitis vinifera* roots. Am. J. Enol. Vitic. 68, 468–477.

Galet, P., 1985. Précis d'Ampélographie Pratique, fifth ed. Déhan, Montpellier, France.

Galet, P., 1996. Grape Diseases (J. Smith, Trans.). Oenoplurimédia, Chaintré, France.

Galet, P., 1998. Grape Varieties and Rootstock Varieties (J. Smith, Trans.). Oenoplurimédia, Chaintré, France.

Galet, P., 2000. General Viticulture (J. Smith, Trans.). Oenoplurimédia, Chaintré, France.

Gallavotti, A., 2013. The role of auxin in shaping shoot architecture. J. Exp. Bot. 64, 2593–2608.

Galmés, J., Pou, A., Alsina, M.M., Tomàs, M., Medrano, H., Flexas, J., 2007. Aquaporin expression in response to different water stress intensities and recovery in Richter-110 (*Vitis* sp.): relationship with ecophysiological status. Planta 226, 671–681.

Galvan-Ampudia, C.S., Julkowska, M.M., Darwish, E., Gandullo, J., Korver, R.A., Brunoud, G., Haring, M.A., Munnik, T., Vernoux, T., Testerink, C., 2013. Halotropism is a response of plant roots to avoid a saline environment. Curr. Biol. 23, 2044–2050.

Galvez, D.A., Landhäusser, S.M., Tyree, M.T., 2013. Low root reserve accumulation during drought may lead to winter mortality in poplar seedlings. New Phytol. 198, 139–148.

Gamalei, Y., 1989. Structure and function of leaf minor veins in trees and herbs. A taxonomic review. Trees 3, 96–110.

Gambetta, G.A., Matthews, M.A., Shaghasi, T.H., McElrone, A.J., Castellarin, S.D., 2010. Sugar and abscisic acid signaling orthologs are activated at the onset of ripening in grape. Planta 232, 219–234.

Gambetta, G.A., Fei, J., Rost, T.L., Knipfer, T., Matthews, M.A., Shackel, K.A., 2013. Water uptake along the length of grapevine fine roots: developmental anatomy, tissue-specific aquaporin expression, and pathways of water transport. Plant Physiol. 163, 1254–1265.

Gambetta, J.M., Bastian, S.E.P., Cozzolino, D., Jeffery, D.W., 2014. Factors influencing the aroma composition of Chardonnay wines. J. Agric. Food Chem. 62, 6512–6534.

Gamm, M., Heloir, M.C., Bligny, R., Vaillant-Gaveau, N., Trouvelot, S., Alcaraz, G., et al., 2011. Changes in carbohydrate metabolism in *Plasmopara viticola*-infected grapevine leaves. Mol. Plant-Microbe Interact. 24, 1061–1073.

Gamon, J.A., Pearcy, R.W., 1989. Leaf movement, stress avoidance and photosynthesis in *Vitis californica*. Oecologia 79, 475–481.

Gamon, J.A., Pearcy, R.W., 1990a. Photoinhibition in *Vitis californica*: interactive effects of sunlight, temperature and water status. Plant Cell Environ. 13, 267–275.

Gamon, J.A., Pearcy, R.W., 1990b. Photoinhibition in *Vitis californica*. The role of temperature during high-light treatment. Plant Physiol. 92, 487–494.

Gang, D.R., 2005. Evolution of flavors and scents. Annu. Rev. Plant Biol. 56, 301–325.

García-Arenal, F., Fraile, A., Malpica, J.M., 2001. Variability and genetic structure of plant virus populations. Annu. Rev. Phytopathol. 39, 157–186.

Gardea, A.A., Moreno, Y.M., Azarenko, A.N., Lombard, P.B., Daley, L.S., Criddle, R.S., 1994. Changes in metabolic properties of grape buds during development. J. Am. Soc. Hortic. Sci. 119, 756–760.

Garris, A., Clark, L., Owens, C., McKay, S., Luby, J., Mathiason, K., 2009. Mapping of photoperiod-induced growth cessation in the wild grape *Vitis riparia*. J. Am. Soc. Hortic. Sci. 134, 261–272.

Gärtel, W., 1993. Grapes. In: Bennett, W.F. (Ed.), Nutrient Deficiencies and Toxicities in Crop Plants. APS Press, St. Paul, MN, pp. 177–183.

Gastal, F., Lemaire, G., 2002. N uptake and distribution in crops: an agronomical and ecophysiological perspective. J. Exp. Bot. 53, 789–799.

Gatti, M., Bernizzoni, F., Civardi, S., Poni, S., 2012. Effects of cluster thinning and preflowering leaf removal on growth and grape composition in cv. Sangiovese. Am. J. Enol. Vitic. 63, 325–332.

Gatti, M., Pirez, F.J., Chiari, G., Tombesi, S., Palliotti, A., Merli, M.C., Poni, S., 2016. Phenology, canopy aging and seasonal carbon balance as related to delayed winter pruning of *Vitis vinifera* L. cv. Sangiovese grapevines. Front. Plant Sci. 7, 659.

Gauthier, P.P.G., Bligny, R., Gout, E., Mahé, A., Nogués, S., Hodges, M., 2010. *In folio* isotopic tracing demonstrates that nitrogen assimilation into glutamate is mostly independent from current CO_2 assimilation in illuminated leaves of *Brassica napus*. New Phytol. 185, 988–999.

Gautier, A., Cookson, S.J., Hevin, C., Vivin, P., Lauvergeat, V., Mollier, A., 2018. Phosphorus acquisition efficiency and phosphorus remobilization mediate genotype-specific differences in shoot phosphorus content in grapevine. Tree Physiol. 38, 1742–1751.

Gawel, R., 1998. Red wine astringency: a review. Aust. J. Grape Wine Res. 4, 74–95.

Gebbing, H., Schwab, A., Alleweldt, G., 1977. Mykorrhiza der Rebe. Vitis 16, 279–285.

Geiger, D., 2011. Plant sucrose transporters from a biophysical point of view. Mol. Plant 4, 395–406.

Geiger, D.R., Servaites, J.C., 1991. Carbon allocation and responses to stress. In: Mooney, H.A., Winner, W.E., Pell, E.J. (Eds.), Response of Plants to Multiple Stresses. Academic Press, San Diego, CA, pp. 103–127.

Geiger, D.R., Servaites, J.C., Fuchs, M.A., 2000. Role of starch in carbon translocation and partitioning at the plant level. Aust. J. Plant Physiol. 27, 571–582.

Geisler, G., Radler, R., 1963. Entwicklungs- und Reifevorgänge an Trauben von *Vitis*. Ber. Dtsch. Bot. Ges. 76, 112–119.

Geisler, G., Staab, J., 1958. Versuchsanstellung im Weinbau. Vitis 1, 257–281.

Geldner, N., 2013. The endodermis. Annu. Rev. Plant Biol. 64, 531–558.

Gény, L., Broquedis, M., Martin-Tanguy, J., Bouard, J., 1997a. Free, conjugated, and wall-bound polyamines in various organs of fruiting cuttings of *Vitis vinifera* L. cv. Cabernet Sauvignon. Am. J. Enol. Vitic. 48, 80–84.

Gény, L., Broquedis, M., Martin-Tanguy, J., Soyer, J.P., Bouard, J., 1997b. Effects of potassium nutrition on polyamine content of various organs of fruiting cuttings of *Vitis vinifera* L. cv. Cabernet Sauvignon. Am. J. Enol. Vitic. 48, 85–92.

Gény, L., Ollat, N., Soyer, J.P., 1998. Les boutures fructifères de vigne: validation d'un modèle d'étude de la physiologie de la vigne. II. Étude de développement de la grappe. J. Int. Sci. Vigne Vin 32, 83–90.

Gény, L., Saucier, C., Bracco, S., Daviaud, F., Glories, Y., 2003. Composition and cellular localization of tannins in grape seeds during maturation. J. Agric. Food Chem. 51, 8051–8054.

German, J.B., Walzem, R.L., 2000. The health benefits of wine. Annu. Rev. Nutr. 20, 561–593.

Gerrath, J.M., 1992. Developmental morphology and anatomy of grape flowers. Hortic. Rev. (Am. Soc. Hortic. Sci). 13, 315–337.

Gerrath, J.M., Posluszny, U., 1988a. Morphological and anatomical development in the Vitaceae. I. Vegetative development in *Vitis riparia*. Can. J. Bot. 66, 209–224.

Gerrath, J.M., Posluszny, U., 1988b. Morphological and anatomical development in the Vitaceae. II. Floral development in *Vitis riparia*. Can. J. Bot. 66, 1334–1351.

Gerrath, J.M., Posluszny, U., 2007. Shoot architecture in the Vitaceae. Can. J. Bot. 85, 691–700.

Gerrath, J.M., Posluszny, U., Dengler, N.G., 2001. Primary vascular patterns in the Vitaceae. Int. J. Plant Sci. 162, 729–745.

Gerrath, J., Posluszny, U., Melville, L., 2015. Taming the Wild Grape: Botany and Horticulture in the Vitaceae. Springer, Cham, Switzerland.

Gessler, A., Schultze, M., Schrempp, S., Rennenberg, H., 1998. Interaction of phloem-translocated amino compounds with nitrate net uptake by the roots of beech (*Fagus sylvatica*) seedlings. J. Exp. Bot. 49, 1529–1537.

Ghaffari, S., Hasnaoui, N., Zinelabidine, L.H., Ferchichi, A., Martínez-Zapater, J.M., Ibáñez, J., 2013. Genetic identification and origin of grapevine cultivars (*Vitis vinifera* L.) in Tunisia. Am. J. Enol. Vitic. 64, 538–544.

Gholami, M., Hayasaka, Y., Coombe, B.G., Jackson, J.F., Robinson, S.P., Williams, P.J., 1995. Biosynthesis of flavour compounds in Muscat Gordo Blanco grape berries. Aust. J. Grape Wine Res. 1, 19–24.

Giacomelli, L., Rota-Stabelli, O., Masuero, D., Acheampong, A.K., Moretto, M., Caputi, L., 2013. Gibberellin metabolism in *Vitis vinifera* L. during bloom and fruit-set: functional characterization and evolution of grapevine gibberellin oxidases. J. Exp. Bot. 64, 4403–4419.

Giannakis, C., Bucheli, C.S., Skene, K.G.M., Robinson, S.P., Scott, N.S., 1998. Chitinase and ß-1,3-glucanase in grapevine leaves: a possible defence against powdery mildew infection. Aust. J. Grape Wine Res. 4, 14–22.

Gifford, M.L., Dean, A., Gutierrez, R.A., Coruzzi, G.M., Birnbaum, K.D., 2008. Cell-specific nitrogen responses mediate developmental plasticity. Proc. Natl. Acad. Sci. U. S. A. 105, 803–808.

Gilbert, D.E., Meyer, J.L., Kissler, J.J., 1971. Evaporation cooling of vineyards. Trans. ASAE 14, 841–843.

Gillaspy, G., Ben-David, H., Gruissem, W., 1993. Fruits: a developmental perspective. Plant Cell 5, 1439–1451.

Gilliham, M., Dayod, M., Hocking, B.J., Xu, B., Conn, S.J., Kaiser, B.N., 2011. Calcium delivery and storage in plant leaves: exploring the link with water flow. J. Exp. Bot. 62, 2233–2250.

Gilroy, S., Jones, D.L., 2000. Through form to function: root hair development and nutrient uptake. Trends Plant Sci. 5, 56–60.

Gindro, K., Pezet, R., Viret, O., 2003. Histological study of the response of two *Vitis vinifera* cultivars (resistant and susceptible) to *Plasmopara viticola* infections. Plant Physiol. Biochem. 41, 846–853.

Giovannoni, J., Nguyen, C., Ampofo, B., Zhong, S., Fei, Z., 2017. The epigenome and transcriptional dynamics of fruit ripening. Annu. Rev. Plant Biol. 68, 61–84.

Girbau, T., Stummer, B.E., Pocock, K.F., Baldock, G.A., Scott, E.S., Waters, E.J., 2004. The effect of *Uncinula necator* (powdery mildew) and *Botrytis cinerea* infection of grapes on the levels of haze-forming pathogenesis-related proteins in grape juice and wine. Aust. J. Grape Wine Res. 10, 125–133.

Giribaldi, M., Perugini, I., Sauvage, F.X., Schubert, A., 2007. Analysis of protein changes during grape berry ripening by 2-DE and MALDI-TOF. Proteomics 7, 3154–3170.

Glad, C., Regnard, J.L., Querou, Y., Brun, O., Morot-Gaudry, J.F., 1992. Flux and chemical composition of xylem exudates from Chardonnay grapevines: temporal evolution and effect of recut. Am. J. Enol. Vitic. 43, 275–282.

Gladstone, E.A., Dokoozlian, N.K., 2003. Influence of leaf area density and trellis/training system on the light microclimate within grapevine canopies. Vitis 42, 123–131.

Gladstones, J., 1992. Viticulture and Environment. Winetitles, Adelaide, Australia.

Gleitz, J., Schnitzler, J.P., Steimle, D., Seitz, H.U., 1991. Metabolic changes in carrot cells in response to simultaneous treatment with ultraviolet light and a fungal elicitor. Planta 184, 362–367.

Gloser, V., Zwieniecki, M.A., Orians, C.M., Holbrook, N.M., 2007. Dynamic changes in root hydraulic properties in response to nitrate availability. J. Exp. Bot. 58, 2409–2415.

Godinho, S., Paulo, O.S., Morais-Cecílio, L., Rocheta, M., 2012. A new *gypsy*-like retroelement family in *Vitis vinifera*. Vitis 51, 65–72.

Goes da Silva, F., Iandolino, A., Al-Kayal, F., Bohlmann, M.C., Cushman, M.A., Lim, H., 2005. Characterizing the grape transcriptome. Analysis of expressed sequence tags from multiple *Vitis* species and development of a compendium of gene expression during berry development. Plant Physiol. 139, 574–597.

Goetz, G., Fkyerat, A., Métais, N., Kunz, M., Tabacchi, R., Pezet, R., 1999. Resistance factors to grey mould in grape berries: identification of some phenolics inhibitors of *Botrytis cinerea* stilbene oxidase. Phytochemistry 52, 759–767.

Goetz, M., Godt, D.E., Guivarc'h, A., Kahmann, U., Chriqui, D., Roitsch, T., 2001. Induction of male sterility in plants by metabolic engineering of the carbohydrate supply. Proc. Natl. Acad. Sci. U. S. A. 98, 6522–6527.

Goff, S.A., Klee, H.J., 2006. Plant volatile compounds: sensory cues for health and nutritional value? Science 311, 815–819.

Goheen, A.C., Cook, J.A., 1959. Leafroll (red-leaf or rougeau) and its effects on vine growth, fruit quality, and yields. Am. J. Enol. Vitic. 10, 173–181.

Gojon, A., Nacry, P., Davidian, J.C., 2009. Root uptake regulation: a central process for NPS homeostasis in plants. Curr. Opin. Plant Biol. 12, 328–338.

Goldberg, R., Morvan, C., Jauneau, A., Jarvis, M.C., 1996. Methyl-esterification, de-esterification and gelation of pectins in the primary cell wall. Prog. Biotechnol. 14, 151–172.

Gollop, R., Even, S., Colova-Tsolova, V., Perl, A., 2002. Expression of the grape dihydroflavonol reductase gene and analysis of its promoter region. J. Exp. Bot. 53, 1397–1409.

Gomez, C., Conejero, G., Torregrosa, L., Cheynier, V., Terrier, N., Ageorges, A., 2011. *In vivo* grapevine anthocyanin transport involves vesicle-mediated trafficking and the contribution of anthoMATE transporters and GST. Plant J. 67, 960–970.

Gomez-Roldan, V., Fermas, S., Brewer, P.B., Puech-Pagès, V., Dun, E.A., Pillot, J.P., 2008. Strigolactone inhibition of shoot branching. Nature 455, 189–194.

Gong, H., Blackmore, D.H., Walker, R.R., 2010. Organic and inorganic anions in Shiraz and Chardonnay grape berries and wine as affected by rootstock under saline conditions. Aust. J. Grape Wine Res. 16, 227–236.

Gonzalez Antivilo, F., Paz, R.C., Keller, M., Borgo, R., Tognetti, J., Roig Juñent, F., 2017. Macro- and microclimate conditions may alter grapevine deacclimation: variation in thermal amplitude in two contrasting wine regions from North and South America. Int. J. Biometeorol. 61, 2033–2045.

Gonzalez Antivilo, F., Paz, R.C., Echeverria, M., Keller, M., Tognetti, J., Borgo, R., Roig Juñent, F., 2018. Thermal history parameters drive changes in physiology and cold hardiness of young grapevine plants during winter. Agric. For. Meteorol. 262, 227–236.

Gonzalez, N., Vanhaeren, H., Inzé, D., 2012. Leaf size control: complex coordination of cell division and expansion. Trends Plant Sci. 17, 332–340.

González, M., Brito, N., González, C., 2017. The *Botrytis cinerea* elicitor protein BcIEB1 interacts with the tobacco PR5-family protein osmotin and protects the fungus against its antifungal activity. New Phytol. 215, 397–410.

González-Centeno, M.R., Jourdes, M., Femenia, A., Simal, S., Rosselló, C., Teissedre, P.L., 2013. Characterization of polyphenols and antioxidant potential of white grape pomace byproducts (*Vitis vinifera* L.). J. Agric. Food Chem. 61, 11579–11587.

Goodman, R.N., Grimm, R., Frank, M., 1993. The influence of grape rootstocks on the crown gall infection process and on tumor development. Am. J. Enol. Vitic. 44, 22–26.

Goremykin, V.V., Salamini, F., Velasco, R., Viola, R., 2009. Mitochondrial DNA of *Vitis vinifera* and the issue of rampant horizontal gene transfer. Mol. Biol. Evol. 26, 99–110.

Gorska, A., Ye, Q., Holbrook, N.M., Zwieniecki, M.A., 2008. Nitrate control of root hydraulic properties in plants: translating local information to whole plant response. Plant Physiol. 148, 1159–1167.

Gouot, J.C., Smith, J.P., Holzapfel, B.P., Walker, A.R., Barril, C., 2019. Grape berry flavonoids: a review of their biochemical responses to high and extreme high temperatures. J. Exp. Bot. 70, 397–423.

Gourieroux, A.M., McCully, M.E., Holzapfel, B.P., Scollary, G.R., Rogiers, S.Y., 2016. Flowers regulate the growth and vascular development of the inflorescence rachis in *Vitis vinifera* L. Plant Physiol. Biochem. 108, 519–529.

Gouthu, S., Deluc, L.G., 2015. Timing of ripening initiation in grape berries and its relationship to seed content and pericarp auxin levels. BMC Plant Biol. 15, 46.

Gouthu, S., O'Neill, S.T., Di, Y., Ansarolia, M., Megraw, M., Deluc, L.G., 2014. A comparative study of ripening among berries of the grape cluster reveals an altered transcriptional programme and enhanced ripening rate in delayed berries. J. Exp. Bot. 65, 5889–5902.

Grace, J., 2006. The temperature of buds may be higher than you thought. New Phytol. 170, 1–3.

Graf, A., Smith, A.M., 2011. Starch and the clock: the dark side of plant productivity. Trends Plant Sci. 16, 169–175.

Granett, J., Walker, M.A., Kocsis, L., Omer, A.D., 2001. Biology and management of grape phylloxera. Annu. Rev. Entomol. 46, 387–412.

Grant, T.N.L., Dami, I.E., 2015. Physiological and biochemical seasonal changes in *Vitis* genotypes with contrasting freezing tolerance. Am. J. Enol. Vitic. 66, 195–203.

Grant, R.S., Matthews, M.A., 1996a. The influence of phosphorus availability and rootstock on root system characteristics, phosphorus uptake, phosphorus partitioning, and growth efficiency. Am. J. Enol. Vitic. 47, 403–409.

Grant, R.S., Matthews, M.A., 1996b. The influence of phosphorus availability, scion, and rootstock on grapevine shoot growth, leaf area, and petiole phosphorus concentration. Am. J. Enol. Vitic. 47, 217–224.

Grant, O.M., Tronina, L., Jones, H.G., Chaves, M.M., 2007. Exploring thermal imaging variables for the detection of stress responses in grapevine under different irrigation regimes. J. Exp. Bot. 58, 815–825.

Grassi, F., Labra, M., Imazio, S., Spada, A., Sgorbati, S., Scienza, A., 2003. Evidence of a secondary grapevine domestication centre detected by SSR analysis. Theor. Appl. Genet. 107, 1315–1320.

Grassi, F., Labra, M., Imazio, S., Ocete Rubio, R., Failla, O., Scienza, A., 2006. Phylogeographical structure and conservation genetics of wild grapevine. Conserv. Genet. 7, 837–845.

Gray, J.D., Kolesik, P., Høj, P.B., Coombe, B.G., 1999. Confocal measurement of the three-dimensional size and shape of plant parenchyma cells in a developing fruit tissue. Plant J. 19, 229–236.

Green, B.R., 2011. Chloroplast genomes of photosynthetic eukaryotes. Plant J. 66, 34–44.

Green, M.A., Fry, S.C., 2005. Vitamin C degradation in plant cells via enzymatic hydrolysis of 4-O-oxalyl-l-threonate. Nature 433, 83–87.

Greenspan, M.D., Shackel, K.A., Matthews, M.A., 1994. Developmental changes in the diurnal water budget of the grape berry exposed to water deficits. Plant Cell Environ. 17, 811–820.

Greenspan, M.D., Schultz, H.R., Matthews, M.A., 1996. Field evaluation of water transport in grape berries during water deficits. Physiol. Plant. 97, 55–62.

Greer, D.H., Weedon, M.M., 2012. Modelling photosynthetic responses to temperature of grapevine (*Vitis vinifera* cv. Semillon) leaves on vines grown in a hot climate. Plant Cell Environ. 35, 1050–1064.

Greer, D.H., Weedon, M.M., 2014. Does the hydrocooling of *Vitis vinifera* cv. Semillon vines protect the vegetative and reproductive growth processes and vine performance against high summer temperatures? Funct. Plant Biol. 41, 620–633.

Greer, D.H., Weston, C., 2010a. Effects of fruiting on vegetative growth and development dynamics of grapevines (*Vitis vinifera* cv. Semillon) can be traced back to events at or before budbreak. Funct. Plant Biol. 37, 756–766.

Greer, D.H., Weston, C., 2010b. Heat stress affects flowering, berry growth, sugar accumulation and photosynthesis of *Vitis vinifera* cv. Semillon grapevines grown in a controlled environment. Funct. Plant Biol. 37, 206–214.

Gregan, S.M., Wargent, J.J., Liu, L., Shinkle, J., Hoffmann, R., Winefield, C., 2012. Effects of solar ultraviolet radiation and canopy manipulation on the biochemical composition of Sauvignon blanc grapes. Aust. J. Grape Wine Res. 18, 227–238.

Grenan, S., 1984. Polymorphisme foliaire consécutif à la culture *in vitro* de *Vitis vinifera* L. Vitis 23, 159–174.

Greven, M.M., Neal, S.M., Tustin, D.S., Boldingh, H., Bennett, J., Vasconcelos, M.C., 2016. Effect of postharvest defoliation on carbon and nitrogen resources of high-yielding Sauvignon blanc grapevines. Am. J. Enol. Vitic. 67, 315–326.

Griffith, M., Yaish, M.W.F., 2004. Antifreeze proteins in overwintering plants: a tale of two activities. Trends Plant Sci. 9, 399–405.

Grigg, D., Methven, D., de Bei, R., Rodríguez López, C.M., Dry, P., Collins, C., 2018. Effect of vine age on vine performance of Shiraz in the Barossa Valley, Australia. Aust. J. Grape Wine Res. 24, 75–87.

Grimplet, J., Ibañez, S., Baroja, E., Tello, J., Ibañez, J., 2019. Phenotypic, hormonal, and genomic variation among *Vitis vinifera* clones with different cluster compactness and reproductive performance. Front. Plant Sci. 9, 1917.

Grncarevic, M., Radler, F., 1971. A review of the surface lipids of grapes and their importance in the drying process. Am. J. Enol. Vitic. 22, 80–86.

Groot Obbink, J., McE Alexander, D., 1973. Response of six grapevine cultivars to a range of chloride concentrations. Am. J. Enol. Vitic. 24, 65–68.

Groot Obbink, J., McE Alexander, D., Possingham, J.V., 1973. Use of nitrogen and potassium reserves during growth of grape vine cuttings. Vitis 12, 207–213.

Groover, A., Robischon, M., 2006. Developmental mechanisms regulating secondary growth in woody plants. Curr. Opin. Plant Biol. 9, 55–58.

Grossman, A., Takahashi, H., 2001. Macronutrient utilization by photosynthetic eukaryotes and the fabric of interactions. Annu. Rev. Plant Physiol. Plant Mol. Biol. 52, 163–210.

Gruber, B., Kosegarten, H., 2002. Depressed growth of non-chlorotic vine grown in calcareous soil is an iron deficiency symptom prior to leaf chlorosis. J. Plant Nutr. Soil Sci. 165, 111–117.

Grzegorczyk, W., Walker, M.A., 1998. Evaluating resistance to grape phylloxera in *Vitis* species with an *in vitro* dual culture assay. Am. J. Enol. Vitic. 49, 17–22.

Gu, L., Hanson, P.J., Post, W.M., Kaiser, D.P., Yang, B., Nemani, R., 2008. The 2007 eastern US spring freeze: increased cold damage in a warming world? Bioscience 58, 253–262.

Gu, S., Jacobs, S., Du, G., 2011. Efficacy, rate and timing of applications of abscisic acid to enhance fruit anthocyanin contents in 'Cabernet Sauvignon' grapes. J. Hortic. Sci. Biotechnol. 86, 505–510.

Gu, S., Jacobs, S.D., McCarthy, B.S., Gohl, H.L., 2012. Forcing vine regrowth and shifting fruit ripening in a warm region to enhance fruit quality in 'Cabernet Sauvignon' grapevine (*Vitis vinifera* L.). J. Hortic. Sci. Biotechnol. 87, 287–292.

Guerra, B., Steenwerth, K., 2012. Influence of floor management technique on grapevine growth, disease pressure, and juice and wine composition: a review. Am. J. Enol. Vitic. 63, 149–164.

Guilford, J.M., Pezzuto, J.M., 2011. Wine and health: a review. Am. J. Enol. Vitic. 62, 471–486.

Guillaumie, S., Ilg, A., Réty, S., Brette, M., Trossat-Magnin, C., Decroocq, S., 2013. Genetic analysis of the biosynthesis of 2-methoxy-3-isobutylpyrazine, a major grape-derived aroma compound impacting wine quality. Plant Physiol. 162, 604–615.

Gunata, Y.Z., Bayonove, C.L., Baumes, R.L., Cordonnier, R.E., 1985. The aroma of grapes. Localisation and evolution of free and bound fractions of some grape aroma components c.v. Muscat during first development and maturation. J. Sci. Food Agric. 36, 857–862.

Guo, X.W., Fu, W.H., Wang, G.J., 1987. Studies on cold hardiness of grapevine roots. Vitis 26, 161–171.

Guo, C., Guo, R., Xu, X., Gao, M., Li, X., Song, J., 2014. Evolution and expression analysis of the grape (*Vitis vinifera* L.) *WRKY* gene family. J. Exp. Bot. 65, 1513–1528.

Gusta, L.V., Wisniewski, M., 2013. Understanding plant cold hardiness: an opinion. Physiol. Plant. 147, 4–14.

Gutermuth, T., Herbell, S., Lassig, R., Brosché, M., Romeis, T., Feijó, J.A., Hedrich, R., Konrad, K.R., 2018. Tip-localized Ca^{2+}-permeable channels control pollen tube growth via kinase-dependent R- and S-type anion channel regulation. New Phytol. 218, 1089–1105.

Guth, H., 1997a. Identification of character impact odorants of different white wine varieties. J. Agric. Food Chem. 45, 3022–3026.

Guth, H., 1997b. Quantitation and sensory studies of character impact odorants of different white wine cultivars. J. Agric. Food Chem. 45, 3027–3032.

Gutha, L.R., Casassa, L.F., Harbertson, J.F., Naidu, R.A., 2010. Modulation of flavonoid biosynthetic pathway genes and anthocyanins due to virus infection in grapevine (*Vitis vinifera* L.) leaves. BMC Plant Biol. 10, 187.

Gutierrez, L., Van Wuytswinkel, O., Castelain, M., Bellini, C., 2007. Combined networks regulating seed maturation. Trends Plant Sci. 12, 294–300.

Gutiérrez-Granda, M.J., Morrison, J.C., 1992. Solute distribution and malic enzyme activity in developing grape berries. Am. J. Enol. Vitic. 43, 323–328.

Guy, C.L., 1990. Cold acclimation and freezing stress tolerance: role of protein metabolism. Annu. Rev. Plant Physiol. Plant Mol. Biol. 41, 187–223.

Guy, C., Kaplan, F., Kopka, J., Selbig, J., Hincha, D.K., 2008. Metabolomics of temperature stress. Physiol. Plant. 132, 220–235.

Guzmán-Delgado, P., Earles, J.M., Zwieniecki, M.A., 2018. Insight into the physiological role of water absorption via the leaf surface from a rehydration kinetics perspective. Plant Cell Environ. 41, 1886–1894.

Ha, S., Vankova, R., Yamaguchi-Shinozaki, K., Shinozaki, K., Tran, L.S.P., 2012. Cytokinins: metabolism and function in plant adaptation to environmental stresses. Trends Plant Sci. 17, 172–179.

Hacke, U.G., Sperry, J.S., Wheeler, J.K., Castro, L., 2006. Scaling of angiosperm xylem structure with safety and efficiency. Tree Physiol. 26, 689–701.

Hafke, J.B., van Amerongen, J.K., Kelling, F., Furch, A.C.U., Gaupels, F., van Bel, A.J.E., 2005. Thermodynamic battle for photosynthate acquisition between sieve tubes and adjoining parenchyma in transport phloem. Plant Physiol. 138, 1527–1537.

Hale, C.R., 1962. Synthesis of organic acids in the fruit of the grape. Nature 195, 917–918.

Hale, C.R., 1977. Relation between potassium and the malate and tartrate contents of grape berries. Vitis 16, 9–19.

Hale, C.R., Buttrose, M.S., 1974. Effect of temperature on ontogeny of berries of *Vitis vinifera* L. cv. Cabernet Sauvignon. J. Am. Soc. Hortic. Sci. 99, 390–394.

Hale, C.R., Weaver, R.J., 1962. The effect of developmental stage on direction of translocation of photosynthate in *Vitis vinifera*. Hilgardia 33, 89–131.

Hales, S., 1727. Vegetable Staticks. W. and J. Innys and T, Woodward, London.

Hall, A., Jones, G.V., 2010. Spatial analysis of climate in winegrape-growing regions in Australia. Aust. J. Grape Wine Res. 16, 389–404.

Hall, T.W., Mahaffee, W.F., 2001. Impact of vine shelter use on development of grape powdery mildew. Am. J. Enol. Vitic. 52, 204–209.

Hall, M.E., Wilcox, W.F., 2019. Identification and frequencies of endophytic microbes within healthy grape berries. Am. J. Enol. Vitic. 70, 212–219.

Hall, J.L., Williams, L.E., 2000. Assimilate transport and partitioning in fungal biotrophic interactions. Aust. J. Plant Physiol. 27, 549–560.

Hall, A., Lamb, D.W., Holzapfel, B., Louis, J., 2002. Optical remote sensing applications in viticulture—a review. Aust. J. Grape Wine Res. 8, 36–47.

Hall, G.E., Bondada, B.R., Keller, M., 2011. Loss of rachis cell viability is associated with ripening disorders in grapes. J. Exp. Bot. 62, 1145–1153.

Hall, M.E., Loeb, G.M., Cadle-Davidson, L., Evans, K., Wilcox, W.F., 2018. Grape sour rot: a four-way interaction involving the host, yeast, acetic acid bacteria, and insects. Phytopathology 108, 1429–1442.

Halldorson, M.M., Keller, M., 2018. Grapevine leafroll disease alters leaf physiology but has little effect on plant cold hardiness. Planta 248, 1201–1211.

Halliwell, B., 2006. Reactive species and antioxidants. Redox biology is a fundamental theme of aerobic life. Plant Physiol. 141, 312–322.

Halliwell, B., 2007. Dietary polyphenols: good, bad, or indifferent for your health? Cardiovasc. Res. 73, 341–347.

Hamant, O., Traas, J., 2010. The mechanics behind plant development. New Phytol. 185, 369–385.

Hamman, R.A., Renquist, A.R., Hughes, H.G., 1990. Pruning effects on cold hardiness and water content during deacclimation of Merlot bud and cane tissues. Am. J. Enol. Vitic. 41, 251–260.

Hamman, R.A., Dami, I.E., Walsh, T.M., Stushnoff, C., 1996. Seasonal carbohydrate changes and cold hardiness of Chardonnay and Riesling grapevines. Am. J. Enol. Vitic. 47, 31–36.

Hammond, J.P., White, P.J., 2008. Sucrose transport in the phloem: integrating root responses to phosphorus starvation. J. Exp. Bot. 59, 93–109.

Hammond, J.P., White, P.J., 2011. Sugar signaling in root responses to low phosphorus availability. Plant Physiol. 156, 1033–1040.

Hanlin, R.L., Downey, M.O., 2009. Condensed tannin accumulation and composition in skin of Shiraz and Cabernet Sauvignon grapes during berry development. Am. J. Enol. Vitic. 60, 13–23.

Hanlin, R.L., Hrmova, M., Harbertson, J.F., Downey, M.O., 2010. Condensed tannin and grape cell wall interactions and their impact on tannin extractability into wine. Aust. J. Grape Wine Res. 16, 173–188.

Hanlin, R.L., Kelm, M.A., Wilkinson, K.L., Downey, M.O., 2011. Detailed characterization of proanthocyanidins in skin, seeds, and wine of Shiraz and Cabernet Sauvignon wine grapes (*Vitis vinifera*). J. Agric. Food Chem. 59, 13265–13276.

Hannam, K.D., Neilsen, G.H., Neilsen, D., Midwood, A.J., Millard, P., Zhang, Z., Thornton, B., Steinke, D., 2016. Amino acid composition of grape (*Vitis vinifera* L.) juice in response to applications of urea to the soil or foliage. Am. J. Enol. Vitic. 67, 47–55.

Hanson, J., Smeekens, S., 2009. Sugar perception and signaling—an update. Curr. Opin. Plant Biol. 12, 562–567.

Harbertson, J.F., Keller, M., 2012. Rootstock effects on deficit-irrigated winegrapes in a dry climate: grape and wine composition. Am. J. Enol. Vitic. 63, 40–48.

Harbertson, J.F., Kennedy, J.A., Adams, D.O., 2002. Tannin in skins and seeds of Cabernet Sauvignon, Syrah, and Pinot noir berries during ripening. Am. J. Enol. Vitic. 53, 54–59.

Harbertson, J.F., Hodgins, R.E., Thurston, L.N., Schaffer, L.J., Reid, M.S., Landon, J.L., 2008. Variability of tannin concentration in red wines. Am. J. Enol. Vitic. 59, 210–214.

Hardie, W.J., 2000. Grapevine biology and adaptation to viticulture. Aust. J. Grape Wine Res. 6, 74–81.

Hardie, W.J., Aggenbach, S.J., 1996. Effects of site, season and viticultural practices on grape seed development. Aust. J. Grape Wine Res. 2, 21–24.

Hardie, W.J., Considine, J.A., 1976. Response of grapes to water-deficit stress in particular stages of development. Am. J. Enol. Vitic. 27, 55–61.

Hardie, W.J., Aggenbach, S.J., Jaudzems, V.G., 1996a. The plastids of the grape pericarp and their significance in isoprenoid synthesis. Aust. J. Grape Wine Res. 2, 144–154.

Hardie, W.J., O'Brien, T.P., Jaudzems, V.G., 1996b. Morphology, anatomy and development of the pericarp after anthesis in grape, *Vitis vinifera* L. Aust. J. Grape Wine Res. 2, 97–142.

Hardy, P.H., 1969. Selective diffusion of basic and acidic products of CO2 fixation into the transpiration stream in grapevine. J. Exp. Bot. 20, 856–862.

Harris, J.M., Kriedemann, P.E., Possingham, J.V., 1968. Anatomical aspects of grape berry development. Vitis 7, 106–119.

Harris, J.M., Kriedemann, P.E., Possingham, J.V., 1971. Grape berry respiration: effects of metabolic inhibitors. Vitis 9, 291–298.

Hartung, W., Slovik, S., 1991. Physicochemical properties of plant growth regulators and plant tissues determine their distribution and redistribution: stomatal regulation by abscisic acid in leaves. New Phytol. 119, 361–382.

Hartweck, L.M., 2008. Gibberellin signaling. Planta 229, 1–13.

Hasegawa, P.M., Bressan, R.A., Zhu, J.K., Bohnert, H.J., 2000. Plant cellular and molecular responses to high salinity. Annu. Rev. Plant Physiol. Plant Mol. Biol. 51, 463–499.

Haselgrove, L., Botting, D., van Heeswijck, R., Høj, P.B., Dry, P.R., Ford, C., 2000. Canopy microclimate and berry composition: the effect of bunch exposure on the phenolic composition of *Vitis vinifera* L. cv. Shiraz grape berries. Aust. J. Grape Wine Res. 6, 141–149.

Hashizume, K., Samuta, T., 1997. Green odorants of grape cluster stem and their ability to cause a wine stemmy flavor. J. Agric. Food Chem. 45, 1333–1337.

Hashizume, K., Samuta, T., 1999. Grape maturity and light exposure affect berry methoxypyrazine concentration. Am. J. Enol. Vitic. 50, 194–198.

Havé, M., Marmagne, A., Chardon, F., Masclaux-Daubresse, C., 2017. Nitrogen remobilization during leaf senescence: lessons from Arabidopsis to crops. J. Exp. Bot. 68, 2513–2529.

Havelda, Z., Várallyay, E., Válóczi, A., Burgyán, J., 2008. Plant virus infection-induced persistent host gene downregulation in systemically infected leaves. Plant J. 55, 278–288.

Hawker, J.S., Walker, R.R., 1978. The effect of sodium chloride on the growth and fruiting of Cabernet Sauvignon vines. Am. J. Enol. Vitic. 29, 172–176.

Hawker, J.S., Ruffner, H.P., Walker, R.R., 1976. The sucrose content of some Australian grapes. Am. J. Enol. Vitic. 27, 125–129.

Hawker, J.S., Jenner, C.F., Niemitz, C.M., 1991. Sugar metabolism and compartmentation. Aust. J. Plant Physiol. 18, 227–237.

Hayasaka, Y., Baldock, G.A., Parker, M., Pardon, K.H., Black, C.A., Herderich, M.J., 2010. Glycosylation of smoke-derived volatile phenols in grapes as a consequence of grapevine exposure to bushfire smoke. J. Agric. Food Chem. 58, 10989–10998.

Hayes, M.A., Davies, C., Dry, I.B., 2007. Isolation, functional characterization, and expression analysis of grapevine (*Vitis vinifera* L.) hexose transporters: differential roles in sink and source tissues. J. Exp. Bot. 58, 1985–1997.

Hayes, M.A., Feechan, A., Dry, I.B., 2010. Involvement of abscisic acid in the coordinated response of a stress-inducible hexose transporter (VvHT5) and a cell wall invertase in grapevine in response to biotrophic fungal infection. Plant Physiol. 153, 211–221.

Hayward, A., Stirnberg, P., Beveridge, C., Leyser, O., 2009. Interactions between auxin and strigolactone in shoot branching control. Plant Physiol. 151, 400–412.

He, M., Dijkstra, F.A., 2014. Drought effect on plant nitrogen and phosphorus: a meta-analysis. New Phytol. 204, 924–931.

He, F., He, J.J., Pan, Q.H., Duan, C.Q., 2010a. Mass-spectrometry evidence confirming the presence of pelargonidin-3-*O*-glucoside in the berry skins of Cabernet Sauvignon and Pinot noir (*Vitis vinifera* L.). Aust. J. Grape Wine Res. 16, 464–468.

He, F., Mu, L., Yan, G.L., Liang, N.N., Pan, Q.H., Wang, J., 2010b. Biosynthesis of anthocyanins and their regulation in colored grapes. Molecules 15, 9057–9091.

He, H., Veneklaas, E.J., Kuo, J., Lambers, H., 2014. Physiological and ecological significance of biomineralization in plants. Trends Plant Sci. 19, 166–174.

Heazlewood, J.E., Wilson, S., 2004. Anthesis, pollination and fruitset in Pinot noir. Vitis 43, 65–68.

Hedrich, R., 2012. Ion channels in plants. Physiol. Rev. 92, 1777–1811.

Heil, M., Ton, J., 2008. Long-distance signalling in plant defence. Trends Plant Sci. 13, 264–272.

Heintz, C., Blaich, R., 1989. Structural characters of epidermal cell walls and resistance to powdery mildew of different grapevine cultivars. Vitis 28, 153–160.

Heintz, C., Blaich, R., 1990. Ultrastructural and histochemical studies on interactions between *Vitis vinifera* L. and *Uncinula necator* (Schw.) Burr. New Phytol. 115, 107–117.

Hellmann, H., Barker, L., Funck, D., Frommer, W.B., 2000. The regulation of assimilate allocation and transport. Aust. J. Plant Physiol. 27, 583–594.

Helwi, P., Habran, A., Guillaumie, S., Thibon, C., Hilbert, G., Gomes, E., Delrot, S., Darriet, P., van Leeuwen, C., 2015. Vine nitrogen status does not have a direct impact on 2-methoxy-3-isobutylpyrazine in grape berries and wines. J. Agric. Food Chem. 63, 9789–9802.

Henderson, S.W., Dunlevy, J.D., Wu, Y., Blackmore, D.H., Walker, R.R., Edwards, E.J., Gilliham, M., Walker, A.R., 2018. Functional differences in transport properties of natural HKT1;1 variants influence shoot Na^+ exclusion in grapevine rootstocks. New Phytol. 217, 1113–1127.

Hendrickson, L., Ball, M.C., Osmond, C.B., Furbank, R.T., Chow, W.S., 2003. Assessment of photoprotection mechanisms of grapevines at low temperature. Funct. Plant Biol. 30, 631–642.

Hendrickson, L., Ball, M.C., Wood, J.T., Chow, W.S., Furbank, R.T., 2004a. Low temperature effects on photosynthesis and growth of grapevine. Plant Cell Environ. 27, 795–809.

Hendrickson, L., Chow, W.S., Furbank, R.T., 2004b. Low temperature effects on grapevine photosynthesis: the role of inorganic phosphate. Funct. Plant Biol. 31, 789–801.

Hennessy, K.J., Pittock, A.B., 1995. The potential impact of greenhouse warming on threshold temperature events in Victoria. Int. J. Climatol. 15, 591–692.

Hepner, Y., Bravdo, B., 1985. Effect of crop level and drip irrigation scheduling on the potassium status of Cabernet Sauvignon and Carignane vines and its influence on must and wine composition and quality. Am. J. Enol. Vitic. 36, 140–147.

Hepner, Y., Bravdo, B., Loinger, C., Cohen, S., Tabacman, H., 1985. Effect of drip irrigation schedules on growth, yield, must composition and wine quality of Cabernet Sauvignon. Am. J. Enol. Vitic. 36, 77–85.

Hermans, C., Hammond, J.P., White, P.J., Verbruggen, N., 2006. How do plants respond to nutrient shortage by biomass allocation? Trends Plant Sci. 11, 610–617.

Hernández, I., Alegre, L., Van Breusegem, F., Munné-Bosch, S., 2009. How relevant are flavonoids as antioxidants in plants? Trends Plant Sci. 14, 125–132.

Hernández-Montes, E., Escalona, J.M., Tomás, M., Medrano, H., 2017. Influence of water availability and grapevine phenological stage on the spatial variation in soil respiration. Aust. J. Grape Wine Res. 23, 273–279.

Hernández-Orte, P., Guitart, A., Cacho, J., 1999. Changes in the concentration of amino acids during the ripening of *Vitis vinifera* Tempranillo variety from the Denomination d'Origine Somontano (Spain). Am. J. Enol. Vitic. 50, 144–154.

Herrera, J.C., Hochberg, U., Degu, A., Sabbatini, P., Lazarovitch, N., Castellarin, S.D., Fait, A., Alberti, G., Peterlunger, E., 2017. Grape metabolic response to postveraison water deficit is affected by interseason weather variability. J. Agric. Food Chem. 65, 5868–5878.

Hetherington, A.M., Brownlee, C., 2004. The generation of Ca^{2+} signals in plants. Annu. Rev. Plant Biol. 55, 401–427.

Hewitt, W.B., Goheen, A.C., Raski, D.J., Gooding, G.V., 1962. Studies on virus diseases of grapevines in California. Vitis 3, 57–83.

Heymann, H., Noble, A.C., 1987. Descriptive analysis of commercial Cabernet Sauvignon wines from California. Am. J. Enol. Vitic. 38, 41–44.

Hichri, I., Barrieu, F., Bogs, J., Kappel, C., Delrot, S., Lauvergeat, V., 2011. Recent advances in the transcriptional regulation of the flavonoid biosynthetic pathway. J. Exp. Bot. 62, 2465–2483.

Hifny, H.A.A., Alleweldt, G., 1972. Untersuchungen zur Stiellähme der Reben. I. Die Symptomatologie der Krankheit. Vitis 10, 298–313.

Hikosaka, K., 2005. Leaf canopy as a dynamic system: ecophysiology and optimality in leaf turnover. Ann. Bot. 95, 521–533.

Hikosaka, K., Ishikawa, K., Borjigidai, A., Muller, O., Onoda, Y., 2006. Temperature acclimation of photosynthesis: mechanisms involved in the changes in temperature dependence of photosynthetic rate. J. Exp. Bot. 57, 291–302.

Hilbert, G., Soyer, J.P., Molot, C., Giraudon, J., Milin, S., Gaudillère, J.P., 2003. Effects of nitrogen supply on must quality and anthocyanins accumulation in berries of cv. Merlot. Vitis 42, 69–76.

Hirschberg, J., 2001. Carotenoid biosynthesis in flowering plants. Curr. Opin. Plant Biol. 4, 210–218.

Hirschi, K.D., 2004. The calcium conundrum. Both versatile nutrient and specific signal. Plant Physiol. 136, 2438–2442.

Hirst, M.B., Richter, C.L., 2016. Review of aroma formation through metabolic pathways of *Saccharomyces cerevisiae* in beverage fermentations. Am. J. Enol. Vitic. 67, 361–370.

Hjelmeland, A.K., Ebeler, S.E., 2015. Glycosidically bound volatile aroma compounds in grapes and wine: a review. Am. J. Enol. Vitic. 66, 1–11.

Hoch, W.A., Singsaas, E.L., McCown, B.H., 2003. Resorption protection. Anthocyanins facilitate nutrient recovery in autumn by shielding leaves from potentially damaging light levels. Plant Physiol. 133, 1296–1305.

Hochberg, U., Albuquerque, C., Rachmilevitch, S., Cochard, H., David-Schwartz, R., Brodersen, C.R., McElrone, A., Windt, C.W., 2016a. Grapevine petioles are more sensitive to drought induced embolism than stems: evidence from *in vivo* MRI and microcomputed tomography observations of hydraulic vulnerability segmentation. Plant Cell Environ. 39, 1886–1894.

Hochberg, U., Herrera, J.C., Cochard, H., Badel, E., 2016b. Short-time xylem relaxation results in reliable quantification of embolism in grapevine petioles and sheds new light on their hydraulic strategy. Tree Physiol. 36, 748–755.

Hochberg, U., Windt, C.W., Ponomarenko, A., Zhang, Y.J., Gersony, J., Rockwell, F.E., Holbrook, N.M., 2017. Stomatal closure, basal leaf embolism, and shedding protect the hydraulic integrity of grape stems. Plant Physiol. 174, 764–775.

Hocquigny, S., Pelsy, F., Dumas, V., Kindt, S., Heloir, M.C., Merdinoglu, D., 2004. Diversification within grapevine cultivars goes through chimeric states. Genome 47, 579–589.

Hodge, A., 2006. Plastic plants and patchy soils. J. Exp. Bot. 57, 401–411.

Hodge, A., Campbell, C.D., Fitter, A.H., 2001. An arbuscular mycorrhizal fungus accelerates decomposition and acquires nitrogen directly from organic material. Nature 413, 297–299.

Hofäcker, W., 1978. Untersuchungen zur Photosynthese der Rebe. Einfluß der Entblätterung, der Dekapitierung, der Ringelung und der Entfernung der Traube. Vitis 17, 10–22.

Hofäcker, W., Alleweldt, G., Khader, S., 1976. Einfluss der Umweltfaktoren auf Beerenwachstum und Mostqualität bei der Rebe. Vitis 15, 96–112.

Hoffmann, S., Di Gaspero, G., Kovács, L., Howard, S., Kiss, E., Galbács, Z., 2007. Resistance to *Erysiphe necator* in the grapevine 'Kishmish vatkana' is controlled by a single locus through restriction of hyphal growth. Theor. Appl. Genet. 116, 427–438.

Hogewoning, S.W., Douwstra, P., Trouwborst, G., van Ieperen, W., Harbinson, J., 2010. An artificial solar spectrum substantially alters plant development compared with usual climate room irradiance spectra. J. Exp. Bot. 61, 1267–1276.

Holbrook, N.M., Ahrens, E.T., Burns, M.J., Zwieniecki, M.A., 2001. *In vivo* observation of cavitation and embolism repair using magnetic resonance imaging. Plant Physiol. 126, 27–31.

Holdaway-Clarke, T.L., Hepler, P.K., 2003. Control of pollen tube growth: role of ion gradients and fluxes. New Phytol. 159, 539–563.

Holländer, H., Amrhein, N., 1980. The site of the inhibition of the shikimate pathway by glyphosate. I. Inhibition by glyphosate of phenylpropanoid synthesis in buckwheat (*Fagopyrum esculentum* Moench). Plant Physiol. 66, 823–829.

Holmes, M.G., Smith, H., 1977a. The function of phytochrome in the natural environment. I. Characterization of daylight for studies in photomorphogenesis and photoperiodism. Photochem. Photobiol. 25, 533–538.

Holmes, M.G., Smith, H., 1977b. The function of phytochrome in the natural environment. II. The influence of vegetation canopies on the spectral energy distribution of natural daylight. Photochem. Photobiol. 25, 539–545.

Holst, T., Mayer, H., Schindler, D., 2004. Microclimate within beech stands—part II: thermal conditions. Eur. J. For. Res. 123, 13–28.

Holst, T., Rost, J., Mayer, H., 2005. Net radiation balance for two forested slopes on opposite sides of a valley. Int. J. Biometeorol. 49, 275–284.

Holt, H.E., Birchmore, W., Herderich, M.J., Iland, P.G., 2010. Berry phenolics in Cabernet Sauvignon (*Vitis vinifera* L.) during late-stage ripening. Am. J. Enol. Vitic. 61, 285–299.

Holzapfel, B.P., Treeby, M.T., 2007. Effects of timing and rate of N supply on leaf nitrogen status, grape yield and juice composition from Shiraz grapevines grafted to one of three different rootstocks. Aust. J. Grape Wine Res. 13, 14–22.

Holzapfel, B.P., Smith, J.P., Mandel, R.M., Keller, M., 2006. Manipulating the postharvest period and its impact on vine productivity of Semillon grapevines. Am. J. Enol. Vitic. 57, 148–157.

Holzapfel, B.P., Smith, J.P., Field, S.K., Hardie, W.J., 2010. Dynamics of carbohydrate reserves in cultivated grapevines. Hortic. Rev. (Am. Soc. Hortic. Sci). 37, 143–211.

Hoos, G., Blaich, R., 1990. Influence of resveratrol on germination of conidia and mycelial growth of *Botrytis cinerea* and *Phomopsis viticola*. J. Phytopathol. 129, 102–110.

Hopkins, D.L., Purcell, A.H., 2002. *Xylella fastidiosa*: cause of Pierce's disease of grapevine and other emergent diseases. Plant Dis. 86, 1056–1066.
Hörtensteiner, S., 2006. Chlorophyll degradation during senescence. Annu. Rev. Plant Biol. 57, 55–77.
Hörtensteiner, S., 2009. Stay-green regulates chlorophyll and chlorophyll-binding protein degradation during senescence. Trends Plant Sci. 14, 155–162.
Hörtensteiner, S., Feller, U., 2002. Nitrogen metabolism and remobilization during senescence. J. Exp. Bot. 53, 927–937.
Horvath, D.P., Anderson, J.V., Chao, W.S., Foley, M.E., 2003. Knowing when to grow: signals regulating bud dormancy. Trends Plant Sci. 8, 534–540.
Hotta, C.T., Gardner, M.J., Hubbard, K.E., Baek, S.J., Dalchau, N., Suhita, D., 2007. Modulation of environmental responses of plants by circadian clocks. Plant Cell Environ. 30, 333–349.
Houdaille, F., Guillon, J.M., 1895. Contribution à l'étude des pleurs de la vigne. Rev. Vitic. 67, 305–308.
Howell, G.S., Shaulis, N., 1980. Factors influencing within-vine variation in the cold resistance of cane and primary bud tissues. Am. J. Enol. Vitic. 31, 158–161.
Hrazdina, G., 1975. Anthocyanin composition of Concord grapes. Lebensm. Wiss. Technol. 8, 111–113.
Hrazdina, G., Parsons, G.F., Mattick, L.R., 1984. Physiological and biochemical events during development and maturation of grape berries. Am. J. Enol. Vitic. 35, 220–227.
Hsiao, T.C., 1973. Plant responses to water stress. Annu. Rev. Plant Physiol. 24, 519–570.
Hsiao, T.C., Xu, L.K., 2000. Sensitivity of growth of roots versus leaves to water stress: biophysical analysis and relation to water transport. J. Exp. Bot. 51, 1595–1616.
Huang, X.M., Huang, H.B., 2001. Early post-veraison growth in grapes: evidence for a two-step mode of berry enlargement. Aust. J. Grape Wine Res. 7, 132–136.
Huang, Z., Ough, C.S., 1991. Amino acid profiles of commercial grape juices and wines. Am. J. Enol. Vitic. 42, 261–267.
Huang, S., Pollack, H.N., Shen, P.Y., 2000. Temperature trends over the past five centuries reconstructed from borehole temperatures. Nature 403, 756–758.
Huang, X., Lakso, A.N., Eissenstat, D.M., 2005a. Interactive effects of soil temperature and moisture on Concord grape root respiration. J. Exp. Bot. 56, 2651–2660.
Huang, X.M., Huang, H.B., Wang, H.C., 2005b. Cell walls of loosening skin in post-veraison grape berries lose structural polysaccharides and calcium while accumulate structural proteins. Sci. Hortic. 104, 249–263.
Huang, H., Liu, B., Liu, L., Song, S., 2017. Jasmonate action in plant growth and development. J. Exp. Bot. 68, 1349–1359.
Hubbard, R.M., Ryan, M.G., Stiller, V., Sperry, J.S., 2001. Stomatal conductance and photosynthesis vary linearly with plant hydraulic conductance in ponderosa pine. Plant Cell Environ. 24, 113–121.
Huber, F., Röckel, F., Schwander, F., Maul, E., Eibach, R., Cousins, P., Töpfer, R., 2016. A view into American grapevine history: Vitis vinifera cv. 'Sémillon' is an ancestor of 'Catawba' and 'Concord'. Vitis 55, 53–56.
Hufnagel, J.C., Hofmann, T., 2008a. Orosensory-directed identification of astringent mouthfeel and bitter-tasting compounds in red wine. J. Agric. Food Chem. 56, 1376–1386.
Hufnagel, J.C., Hofmann, T., 2008b. Quantitative reconstruction of the nonvolatile sensometabolome of a red wine. J. Agric. Food Chem. 56, 9190–9199.
Hughes, N.M., Morley, C.B., Smith, W.K., 2007. Coordination of anthocyanin decline and photosynthetic maturation in juvenile leaves of three deciduous tree species. New Phytol. 175, 675–685.
Huglin, P., Balthazard, J., 1975. Variabilité et fluctuation de la composition des inflorescences et des grappes chez quelques variétés de *Vitis vinifera*. Vitis 14, 6–13.
Huglin, P., Schneider, C., 1998. Biologie et Ecologie de la Vigne, second ed. Lavoisier, Paris.
Humphreys, J.M., Chapple, C., 2002. Rewriting the lignin roadmap. Curr. Opin. Plant Biol. 5, 224–229.

Hunter, J.J., Visser, J.H., 1988. The effect of partial defoliation, leaf position and developmental stage of the vine on the photosynthetic activity of *Vitis vinifera* L. cv Cabernet Sauvignon. S. Afr. J. Enol. Vitic. 9, 9–15.

Hunter, J.J., Skrivan, R., Ruffner, H.P., 1994. Diurnal and seasonal physiological changes in leaves of *Vitis vinifera* L.: CO_2 assimilation rates, sugar levels and sucrolytic enzyme activity. Vitis 33, 189–195.

Huppe, H.C., Turpin, D.H., 1994. Integration of carbon and nitrogen metabolism in plant and algal cells. Annu. Rev. Plant Physiol. Plant Mol. Biol. 45, 577–607.

Hwang, I., Sakakibara, H., 2006. Cytokinin biosynthesis and perception. Physiol. Plant. 126, 528–538.

Hyodo, H., Terao, A., Furukawa, J., Sakamoto, N., Yurimoto, H., Satoh, S., Iwai, H., 2013. Tissue specific localization of pectin–Ca^{2+} cross-likages and pectin methyl-esterification during fruit ripening in tomato (*Solanum lycopersicum*). PLoS ONE 8, e78949.

Iacono, F., Sommer, K.J., 1996. Photoinhibition of photosynthesis and photorespiration in *Vitis vinifera* under field conditions: effects of light climate and leaf position. Aust. J. Grape Wine Res. 2, 1–11.

Iacono, F., Bertamini, M., Scienza, A., Coombe, B.G., 1995. Differential effects of canopy manipulation and shading of *Vitis vinifera* L. cv. Cabernet Sauvignon. Leaf gas exchange, photosynthetic electron transport rate and sugar accumulation in berries. Vitis 34, 201–206.

Iba, K., 2002. Acclimative response to temperature stress in higher plants: approaches of gene engineering for temperature tolerance. Annu. Rev. Plant Biol. 53, 225–245.

Ibáñez, J., Vargas, A.M., Palancar, M., Borrego, J., de Andrés, M.T., 2009. Genetic relationships among tablegrape varieties. Am. J. Enol. Vitic. 60, 35–42.

Ibáñez, J., Muñoz-Organero, G., Zinelabidine, L.H., de Andrés, M.T., Cabello, F., Martínez-Zapater, J.M., 2012. Genetic origin of the grapevine cultivar Tempranillo. Am. J. Enol. Vitic. 63, 549–553.

Ichihashi, R., Tateno, M., 2015. Biomass allocation and long-term growth patterns of temperate lianas in comparison with trees. New Phytol. 207, 604–612.

Iland, P.G., Coombe, B.G., 1988. Malate, tartrate, potassium, and sodium in flesh and skin of Shiraz grapes during ripening: concentration and compartmentation. Am. J. Enol. Vitic. 39, 71–76.

Iland, P., Dry, P., Proffitt, T., Tyerman, S., 2011. The Grapevine: From the Science to the Practice of Growing Vines for Wine. Patrick Iland Wine Promotions, Adelaide, Australia.

Inaba, A., Ishida, M., Sobajima, Y., 1976. Changes in endogenous hormone concentrations during berry development in relaton to the ripening of Delaware grapes. J. Jpn. Soc. Hortic. Sci. 45, 245–252.

Ingels, C.A., Scow, K.M., Whisson, D.A., Drenovsky, R.E., 2005. Effects of cover crops on grapevines, yield, juice composition, soil microbial ecology, and gopher activity. Am. J. Enol. Vitic. 56, 19–29.

IPCC, Intergovernmental Panel on Climate Change, 2013. Climate Change 2013: The Physical Science Basis. In: - Stocker, T.F., Qin, D., Plattner, G.K., Tignor, M., Allen, S.K., Boschung, J. (Eds.), Contribution of Working Group I to the Fifth Assessment Report of the Intergovernmental Panel on Climate Change. Cambridge University Press, Cambridge, United Kingdom.

Intrieri, C., Poni, S., Silvestroni, O., Filippetti, I., 1992. Leaf age, leaf position and photosynthesis in potted grapevines. Adv. Hortic. Sci. 6, 23–27.

Intrieri, C., Zerbi, G., Marchiol, L., Poni, S., Caiado, T., 1995. Physiological response of grapevine leaves to lightflecks. Sci. Hortic. 61, 47–59.

Intrieri, C., Poni, S., Rebucci, B., Magnanini, E., 1997. Effects of canopy manipulations on whole-vine photosynthesis: results from pot and field experiments. Vitis 36, 167–173.

Intrieri, C., Poni, S., Lia, G., Gomez del Campo, M., 2001. Vine performance and leaf physiology of conventionally and minimally pruned Sangiovese grapevines. Vitis 40, 123–130.

Intrigliolo, D.S., Castel, J.R., 2010. Response of grapevine cv. 'Tempranillo' to timing and amount of irrigation: water relations, vine growth, yield and berry and wine composition. Irrig. Sci. 28, 113–125.

Intrigliolo, D.S., Lakso, A.N., Centinari, M., 2009. Effects of the whole vine versus single shoot-crop level on fruit growth in *Vitis labruscana* 'Concord'. Vitis 48, 1–5.

Intrigliolo, D.S., Pérez, D., Risco, D., Yeves, A., Castel, J.R., 2012. Yield components and grape composition responses to seasonal water deficits in Tempranillo grapevines. Irrig. Sci. 30, 339–349.

Iriti, M., Rossoni, M., Faoro, F., 2006. Melatonin content in grape: myth or panacea? J. Sci. Food Agric. 86, 1432–1438.

Ishimaru, M., Kobayashi, S., 2002. Expression of a xyloglucan endo-transglycosylase gene is closely related to grape berry softening. Plant Sci. 162, 621–628.

Israelsson, M., Sundberg, B., Moritz, T., 2005. Tissue-specific localization of gibberellins and expression of gibberellin-biosynthetic and signaling genes in wood-forming tissues in aspen. Plant J. 44, 494–504.

Ivanchenko, M.G., Muday, G.K., Dubrovsky, J.G., 2008. Ethylene–auxin interactions regulate lateral root initiation and emergence in *Arabidopsis thaliana*. Plant J. 55, 335–347.

Iwahori, S., Weaver, R.J., Pool, R.M., 1968. Gibberellin-like activity in berries of seeded and seedless Tokay grapes. Plant Physiol. 43, 333–337.

Jaakola, L., Hohtola, A., 2010. Effect of latitude on flavonoid biosynthesis in plants. Plant Cell Environ. 33, 1239–1247.

Jackson, D.I., 1991. Environmental and hormonal effects on development of early bunch stem necrosis. Am. J. Enol. Vitic. 42, 290–294.

Jackson, R.S., 2008. Wine Science: Principles and Applications, third ed. Academic Press, San Diego, CA.

Jackson, D.I., Cherry, N.J., 1988. Prediction of a district's grape-ripening capacity using a latitude-temperature index (LTI). Am. J. Enol. Vitic. 39, 19–28.

Jackson, D.I., Coombe, B.G., 1988. Early bunchstem necrosis in grapes—a cause of poor fruit set. Vitis 27, 57–61.

Jackson, D.I., Lombard, P.B., 1993. Environmental and management practices affecting grape composition and wine quality—a review. Am. J. Enol. Vitic. 44, 409–430.

Jackson, D.I., Steans, G.F., Hemmings, P.C., 1984. Vine response to increased node numbers. Am. J. Enol. Vitic. 35, 161–163.

Jackson, L.E., Burger, M., Cavagnaro, T.R., 2008. Roots, nitrogen transformations, and ecosystem services. Annu. Rev. Plant Biol. 59, 341–363.

Jacobs, A.K., Dry, I.B., Robinson, S.P., 1999. Induction of different pathogenesis-related cDNAs in grapevine infected with powdery mildew and treated with ethephon. Plant Pathol. 48, 325–336.

Jähnl, G., 1967. Stiellähme, mikroskopisch betrachtet. Wein-Wiss. 22, 446.

Jähnl, G., 1971. Anatomische Veränderungen durch Stiellähme. Wein-Wiss. 26, 242–252.

Jaillon, O., Aury, J.M., Noel, B., Policriti, A., Clepet, C., Casagrande, A., et al., 2007. The grapevine genome sequence suggests ancestral hexaploidization in major angiosperm phyla. Nature 449, 463–468.

James, D.G., Price, T.S., 2004. Field-testing of methyl salicylate for recruitment and retention of beneficial insects in grapes and hops. J. Chem. Ecol. 30, 1613–1628.

Jansen, M.A.K., Gaba, V., Greenberg, B.M., 1998. Higher plants and UV-B radiation: balancing damage, repair and acclimation. Trends Plant Sci. 3, 131–135.

Jansen, S.C., van Dusseldorp, M., Bottema, K.C., Dubois, A.E., 2003. Intolerance to dietary biogenic amines: a review. Ann. Allergy Asthma Immunol. 91, 233–240.

Jansen, R.K., Kaittanis, C., Saski, C., Lee, S.B., Tomkins, J., Alverson, A.J., 2006. Phylogenetic analyses of *Vitis* (Vitaceae) based on complete chloroplast genome sequences: effects of taxon sampling and phylogenetic methods on resolving relationships among rosids. BMC Evol. Biol. 6, 32.

Jánváry, L., Hoffmann, T., Pfeiffer, J., Hausmann, L., Töpfer, R., Fischer, T.C., 2009. A double mutation in the anthocyanin 5-O-glucosyltransferase gene disrupts enzymatic activity in *Vitis vinifera* L. J. Agric. Food Chem. 57, 3512–3518.

Jarvis, M.C., 1984. Structure and properties of pectin gels in plant cell walls. Plant Cell Environ. 7, 153–164.

Javelle, M., Vernoud, V., Rogowsky, P.M., Ingram, G.C., 2011. Epidermis: the formation and functions of a fundamental plant tissue. New Phytol. 189, 17–39.

Jeandet, P., Bessis, R., Gautheron, B., 1991. The production of resveratrol (3,5,4′-trihydroxystilbene) by grape berries in different developmental stages. Am. J. Enol. Vitic. 42, 41–46.

Jeandet, P., Douillet-Breuil, A.C., Bessis, R., Debord, S., Sbaghi, M., Adrian, M., 2002. Phytoalexins from Vitaceae: biosynthesis, phytoalexin gene expression in transgenic plants, antifungal activity, and metabolism. J. Agric. Food Chem. 50, 2731–2741.

Jenik, P.D., Irish, V.F., 2000. Regulation of cell proliferation patterns by homeotic genes during *Arabidopsis* floral development. Development 127, 1267–1276.

Jeong, S.T., Goto-Yamamoto, N., Kobayashi, S., Esaka, M., 2004. Effects of plant hormones and shading on the accumulation of anthocyanins and the expression of anthocyanin biosynthetic genes in grape berry skins. Plant Sci. 167, 247–252.

Jersch, S., Scherer, C., Huth, G., Schlösser, E., 1989. Proanthocyanidins as basis for quiescence of *Botrytis cinerea* in immature strawberry fruits. Z. Pflanzenkr. Pflanzenschutz. 96, 365–378.

Jia, W., Davies, W.J., 2007. Modification of leaf apoplastic pH in relation to stomatal sensitivity to root-sourced abscisic acid signals. Plant Physiol. 143, 68–77.

Jia, H.F., Chai, Y.M., Li, C.L., Lu, D., Luo, J.J., Qin, L., 2011. Abscisic acid plays an important role in the regulation of strawberry fruit ripening. Plant Physiol. 157, 188–199.

Jiménez, S., Gogorcena, Y., Hévin, C., Rombolà, A.D., Ollat, N., 2007. Nitrogen nutrition influences some biochemical responses to iron deficiency in tolerant and sensitive genotypes of *Vitis*. Plant Soil 290, 343–355.

Jin, K., Shen, J., Ashton, R.W., Dodd, I.C., Parry, M.A.J., Whalley, W.R., 2013. How do roots elongate in a structured soil? J. Exp. Bot. 64, 4761–4777.

Johansson, I., Karlsson, M., Johanson, U., Larsson, C., Kjellbom, P., 2000. The role of aquaporins in cellular and whole plant water balance. Biochim. Biophys. Acta 1465, 324–342.

Johnson, D.E., Howell, G.S., 1981. Factors influencing critical temperatures for spring freeze damage to developing primary shoots on Concord grapevines. Am. J. Enol. Vitic. 32, 144–149.

Johnson, J.O., Weaver, R.J., Paige, D.F., 1982. Differences in the mobilization of assimilates of *Vitis vinifera* L. grapevines as influenced by an increased source strength. Am. J. Enol. Vitic. 33, 207–213.

Johnson, D.M., McCulloh, K.A., Woodruff, D.R., Meinzer, F.C., 2012. Evidence for xylem embolism as a primary factor in dehydration-induced declines in leaf hydraulic conductance. Plant Cell Environ. 35, 760–769.

Jones, H.G., 1998. Stomatal control of photosynthesis and transpiration. J. Exp. Bot. 49, 387–398.

Jones, G.V., 2005. Climate change in the western United States grape growing regions. Acta Hortic. 689, 41–59.

Jones, A.M., Dangl, J.L., 1996. Logjam at the Styx: programmed cell death in plants. Trends Plant Sci. 1, 114–119.

Jones, G.V., Davis, R.E., 2000. Climate influences on grapevine phenology, grape composition, and wine production and quality for Bordeaux, France. Am. J. Enol. Vitic. 51, 249–261.

Jones, R.S., Ough, C.S., 1985. Variations in the percent ethanol (v/v) per Brix conversions of wines from different climatic regions. Am. J. Enol. Vitic. 36, 268–270.

Jones, K.S., Paroschy, J., McKersie, B.D., Bowley, S.R., 1999. Carbohydrate composition and freezing tolerance of canes and buds in *Vitis vinifera*. J. Plant Physiol. 155, 101–106.

Jones, K.S., McKersie, B.D., Paroschy, J., 2000. Prevention of ice propagation by permeability barriers in bud axes of *Vitis vinifera*. Can. J. Bot. 78, 3–9.

Jones, G.V., White, M.A., Cooper, O.R., Storchmann, K., 2005. Climate change and global wine quality. Clim. Change 73, 319–343.

Jones, G.V., Duff, A.A., Hall, A., Myers, J.W., 2010. Spatial analysis of climate in winegrape growing regions in the western United States. Am. J. Enol. Vitic. 61, 313–326.

Jones, J.E., Kerslake, F.L., Dambergs, R.G., Close, D.C., 2018. Spur pruning leads to distinctly different phenolic profiles of base sparkling wines than cane pruning. Vitis 57, 103–109.

Joslyn, M.A., Dittmar, H.F.K., 1967. The proanthocyanidins of Pinot blanc grapes. Am. J. Enol. Vitic. 18, 1–10.

Joubert, C., Young, P.R., Eyéghé-Bickong, H.A., Vivier, M.A., 2016. Field-grown grapevine berries use carotenoids and the associated xanthophyll cycles to acclimate to UV exposure differentially in high and low light (shade) conditions. Front. Plant Sci. 7, 786.

Junquera, P., Lissarrague, J.R., Jiménez, L., Linares, R., Baeza, P., 2012. Long-term effects of different irrigation strategies on yield components, vine vigour, and grape composition in *cv*. Cabernet-Sauvignon (*Vitis vinifera* L.). Irrig. Sci. 30, 351–361.

Kadir, S., 2006. Thermostability of photosynthesis of *Vitis aestivalis* and *V. vinifera*. J. Am. Soc. Hortic. Sci. 131, 476–483.

Kadir, S., Von Weihe, M., Al-Khatib, K., 2007. Photochemical efficiency and recovery of photosystem II in grapes after exposure to sudden and gradual heat stress. J. Am. Soc. Hortic. Sci. 132, 764–769.

Kaile, A., Pitt, D., Kuhn, P.J., 1991. Release of calcium and other ions from various plant host tissues infected by different necrotrophic pathogens with special reference to *Botrytis cinerea* Pers. Physiol. Mol. Plant Pathol. 38, 275–291.

Kaiser, B.N., Gridley, K.L., Ngaire Brady, J., Phillips, T., Tyerman, S.D., 2005. The role of molybdenum in agricultural plant production. Ann. Bot. 96, 745–754.

Kakani, V.G., Reddy, K.R., Zhao, D., Sailaja, K., 2003. Field crop responses to ultraviolet-B radiation: a review. Agric. For. Meteorol. 120, 191–218.

Kakimoto, T., 2003. Perception and signal transduction of cytokinins. Annu. Rev. Plant Biol. 54, 605–627.

Kalberer, S.R., Wisniewski, M., Arora, R., 2006. Deacclimation and reacclimation of cold-hardy plants: current understanding and emerging concepts. Plant Sci. 171, 3–16.

Kalua, C.M., Boss, P.K., 2009. Evolution of volatile compounds during the development of Cabernet Sauvignon grapes (*Vitis vinifera* L.). J. Agric. Food Chem. 57, 3818–3830.

Kammerer, D., Claus, A., Carle, R., Schieber, A., 2004. Polyphenol screening of pomace from red and white grape varieties (*Vitis vinifera* L.) by HPLC-DAD-MS/MS. J. Agric. Food Chem. 52, 4360–4367.

Kanellis, A.K., Roubelakis-Angelakis, K.A., 1993. Grape. In: Seymour, G.B., Taylor, J.E., Tucker, G.A. (Eds.), Biochemistry of Fruit Ripening. Chapman & Hall, London, pp. 189–234.

Kang, S., Hu, T., Jerie, P., Zhang, J., 2003. The effects of partial rootzone drying on root, trunk sap flow and water balance in an irrigated pear (*Pyrus communis* L.) orchard. J. Hydrol. 280, 192–206.

Karabourniotis, G., Bornman, J.F., Liakoura, V., 1999. Different leaf surface characteristics of three grape cultivars affect leaf optical properties as measured with fibre optics: possible implication in stress tolerance. Aust. J. Plant Physiol. 26, 47–53.

Karasev, A.V., 2000. Genetic diversity and evolution of closteroviruses. Annu. Rev. Phytopathol. 38, 293–324.

Karley, A.J., White, P.J., 2009. Moving cationic minerals to edible tissues: potassium, magnesium, calcium. Curr. Opin. Plant Biol. 12, 291–298.

Kasimatis, A.N., Vilas, E.P., Swanson, F.H., Baranek, P.P., 1975. A study of the variability of 'Thompson Seedless' berries for soluble solids and weight. Am. J. Enol. Vitic. 26, 37–42.

Kassemeyer, H.H., Staudt, G., 1981. Über die Entwicklung des Embryosacks und die Befruchtung der Reben. Vitis 20, 202–210.

Kassemeyer, H.H., Staudt, G., 1982. Cytologische Untersuchungen über die Ursachen des Verrieselns bei *Vitis*. Vitis 21, 121–135.

Kassemeyer, H.H., Staudt, G., 1983. Über das Wachstum von Endosperm, Embryo und Samenanlagen von Vitis vinifera. Vitis 22, 109–119.

Kasuga, J., Hashidoko, Y., Nishioka, A., Yoshiba, M., Arakawa, K., Fujikawa, S., 2008. Deep supercooling xylem parenchyma cells of katsura tree (*Cercidiphyllum japonicum*) contain flavonol glycosides exhibiting anti-ice nucleation activity. Plant Cell Environ. 31, 1335–1348.

Kataoka, I., Sugiura, A., Utsunomiya, N., Tomana, T., 1982. Effect of abscisic acid and defoliation on anthocyanin accumulation in Kyoho grapes (*Vitis vinifera* L. × *V. labruscana* Bailey). Vitis 21, 325–332.

Katsuhara, M., Hanba, Y.T., Shiratake, K., Maeshima, M., 2008. Expanding roles of plant aquaporins in plasma membranes and cell organelles. Funct. Plant Biol. 35, 1–14.

Katul, G.G., Oren, R., Manzoni, S., Higgins, C., Parlange, M.B., 2012. Evapotranspiration: a process driving mass transport and energy exchange in the soil-plant-atmosphere-climate system. Rev. Geophys. 50, RG3002.

Kawai, Y., Benz, J., Kliewer, W.M., 1996. Effect of flooding on shoot and root growth of rooted cuttings of four grape rootstocks. J. Jpn. Soc. Hortic. Sci. 65, 455–461.

Kawamura, E., Horiguchi, G., Tsukaya, H., 2010. Mechanisms of leaf tooth formation in *Arabidopsis*. Plant J. 62, 429–441.

Kehr, J., Buhtz, A., 2008. Long distance transport and movement of RNA through the phloem. J. Exp. Bot. 59, 85–92.

Keller, M., 2005. Deficit irrigation and vine mineral nutrition. Am. J. Enol. Vitic. 56, 267–283.

Keller, M., 2010. Managing grapevines to optimise fruit development in a challenging environment: a climate change primer for viticulturists. Aust. J. Grape Wine Res. 16, 56–69.

Keller, M., Hrazdina, G., 1998. Interaction of nitrogen availability during bloom and light intensity during veraison: II. Effects on anthocyanin and phenolic development during grape ripening. Am. J. Enol. Vitic. 49, 341–349.

Keller, M., Koblet, W., 1994. Is carbon starvation rather than excessive nitrogen supply the cause of inflorescence necrosis in *Vitis vinifera* L.? Vitis 33, 81–86.

Keller, M., Koblet, W., 1995a. Dry matter and leaf area partitioning, bud fertility and second-season growth of *Vitis vinifera* L.: responses to nitrogen supply and limiting irradiance. Vitis 34, 77–83.

Keller, M., Koblet, W., 1995b. Stress-induced development of inflorescence necrosis and bunch stem necrosis in *Vitis vinifera* L. in response to environmental and nutritional effects. Vitis 34, 145–150.

Keller, M., Mills, L.J., 2007. Effect of pruning on recovery and productivity of cold-injured Merlot grapevines. Am. J. Enol. Vitic. 58, 351–357.

Keller, M., Shrestha, P.M., 2014. Solute accumulation differs in the vacuoles and apoplast of ripening grape berries. Planta 239, 633–642.

Keller, M., Tarara, J.M., 2010. Warm spring temperatures induce persistent season-long changes in shoot development in grapevines. Ann. Bot. 106, 131–141.

Keller, M., Hess, B., Schwager, H., Schärer, H., Koblet, W., 1995. Carbon and nitrogen partitioning in *Vitis vinifera* L.: responses to nitrogen supply and limiting irradiance. Vitis 34, 19–26.

Keller, M., Arnink, K.J., Hrazdina, G., 1998. Interaction of nitrogen availability during bloom and light intensity during veraison. I. Effects on grapevine growth, fruit development, and ripening. Am. J. Enol. Vitic. 49, 333–340.

Keller, M., Pool, R.M., Henick-Kling, T., 1999. Excessive nitrogen supply and shoot trimming can impair colour development in Pinot noir grapes and wine. Aust. J. Grape Wine Res. 5, 45–55.

Keller, M., Kummer, M., Vasconcelos, M.C., 2001a. Reproductive growth of grapevines in response to nitrogen supply and rootstock. Aust. J. Grape Wine Res. 7, 12–18.

Keller, M., Kummer, M., Vasconcelos, M.C., 2001b. Soil nitrogen utilisation for growth and gas exchange by grapevines in response to nitrogen supply and rootstock. Aust. J. Grape Wine Res. 7, 2–11.

Keller, M., Rogiers, S.Y., Schultz, H.R., 2003a. Nitrogen and ultraviolet radiation modify grapevines' susceptibility to powdery mildew. Vitis 42, 87–94.

Keller, M., Viret, O., Cole, F.M., 2003b. *Botrytis cinerea* infection in grape flowers: defense reaction, latency, and disease expression. Phytopathology 93, 316–322.

Keller, M., Mills, L.J., Wample, R.L., Spayd, S.E., 2004. Crop load management in Concord grapes using different pruning techniques. Am. J. Enol. Vitic. 55, 35–50.

Keller, M., Mills, L.J., Wample, R.L., Spayd, S.E., 2005. Cluster thinning effects on three deficit-irrigated *Vitis vinifera* cultivars. Am. J. Enol. Vitic. 56, 91–103.

Keller, M., Smith, J.P., Bondada, B.R., 2006. Ripening grape berries remain hydraulically connected to the shoot. J. Exp. Bot. 57, 2577–2587.

Keller, M., Mills, L.J., Smithyman, R.P., 2008. Interactive effects of deficit irrigation and crop load on Cabernet Sauvignon in an arid climate. Am. J. Enol. Vitic. 59, 221–234.

Keller, M., Tarara, J.M., Mills, L.J., 2010. Spring temperatures alter reproductive development in grapevines. Aust. J. Grape Wine Res. 16, 445–454.

Keller, M., Mills, L.J., Harbertson, J.F., 2012. Rootstock effects on deficit-irrigated winegrapes in a dry climate: vigor, yield formation, and fruit ripening. Am. J. Enol. Vitic. 63, 29–39.

Keller, M., Mills, L.J., Olmstead, M.A., 2014. Fruit ripening has little influence on grapevine cold acclimation. Am. J. Enol. Vitic. 65, 417–423.

Keller, M., Deyermond, L.S., Bondada, B.R., 2015a. Plant hydraulic conductance adapts to shoot number but limits shoot vigour in grapevines. Funct. Plant Biol. 42, 366–375.

Keller, M., Zhang, Y., Shrestha, P.M., Biondi, M., Bondada, B.R., 2015b. Sugar demand of ripening grape berries leads to recycling of surplus phloem water via the xylem. Plant Cell Environ. 38, 1048–1059.

Keller, M., Romero, P., Gohil, H., Smithyman, R.P., Riley, W.R., Casassa, L.F., Harbertson, J.F., 2016a. Deficit irrigation alters grapevine growth, physiology, and fruit microclimate. Am. J. Enol. Vitic. 67, 426–435.

Keller, M., Shrestha, P.M., Hall, G.E., Bondada, B.R., Davenport, J.R., 2016b. Arrested sugar accumulation and altered organic acid metabolism in grape berries affected by berry shrivel syndrome. Am. J. Enol. Vitic. 67, 398–406.

Kelloniemi, J., Trouvelot, S., Héloir, M.C., Simon, A., Dalmais, B., Frettinger, P., et al., 2015. Analysis of the molecular dialogue between gray mold (*Botrytis cinerea*) and grapevine (*Vitis vinifera*) reveals a clear shift in defense mechanisms during berry ripening. Mol. Plant-Microbe Interact. 28, 1167–1180.

Kemp, M.S., Burden, R.S., 1986. Phytoalexins and stress metabolites in the sapwood of trees. Phytochemistry 25, 1261–1269.

Kende, H., Zeevaart, J.A.D., 1997. The five "classical" plant hormones. Plant Cell 9, 1197–1210.

Kennedy, J.A., Troup, G.J., Pilbrow, J.R., Hutton, D.R., Hewitt, D., Hunter, C.R., 2000. Development of seed polyphenols in berries from *Vitis vinifera* L. cv. Shiraz. Aust. J. Grape Wine Res. 6, 244–254.

Kennedy, J.A., Hayasaka, Y., Vidal, S., Waters, E.J., Jones, G.P., 2001. Composition of grape skin proanthocyanidins at different stages of berry development. J. Agric. Food Chem. 49, 5348–5355.

Kennedy, J.A., Matthews, M.A., Waterhouse, A.L., 2002. Effect of maturity and vine water status on grape skin and wine flavonoids. Am. J. Enol. Vitic. 53, 268–274.

Kennelly, M.M., Gadoury, D.M., Wilcox, W.F., Magarey, P.A., Seem, R.C., 2005. Seasonal development of ontogenic resistance to downy mildew in grape berries and rachises. Phytopathology 95, 1445–1452.

Kennison, K.R., Gibberd, M.R., Pollnitz, A.P., Wilkinson, K.L., 2008. Smoke-derived taint in wine: the release of smoke-derived volatile phenols during fermentation of Merlot juice following grapevine exposure to smoke. J. Agric. Food Chem. 56, 7379–7383.

Kennison, K.R., Wilkinson, K.L., Pollnitz, A.P., Williams, H.G., Gibberd, M.R., 2009. Effect of timing and duration of grapevine exposure to smoke on the composition and sensory properties of wine. Aust. J. Grape Wine Res. 15, 228–237.

Kennison, K.R., Wilkinson, K.L., Pollnitz, A.P., Williams, H.G., Gibberd, M.R., 2011. Effect of smoke application to field-grown Merlot grapevines at key phenological growth stages on wine sensory and chemical properties. Aust. J. Grape Wine Res. 17, S5–S12.

Keppler, F., Hamilton, J.T.G., McRoberts, W.C., Vigano, I., Brass, M., Röckmann, T., 2008. Methoxyl groups of plant pectin as a precursor of atmospheric methane: evidence from deuterium labelling studies. New Phytol. 178, 808–814.

Kerstetter, R.A., Hake, S., 1997. Shoot meristem formation in vegetative development. Plant Cell 9, 1001–1010.

Kerstiens, G., 1996. Signalling across the divide: a wider perspective of cuticular structure–function relationships. Trends Plant Sci. 1, 125–129.

Keskitalo, J., Bergquist, G., Gardeström, P., Jansson, S., 2005. A cellular timetable of autumn senescence. Plant Physiol. 139, 1635–1648.

Kevan, P.G., Longair, R.W., Gadawski, R.M., 1985. Dioecy and pollen dimorphism in *Vitis riparia* (Vitaceae). Can. J. Bot. 63, 2263–2267.

Kevan, P.G., Blades, D.C.A., Posluszny, U., Ambrose, J.D., 1988. Pollen dimorphism and dioecy in *Vitis aestivalis*. Vitis 27, 143–146.

Khanal, B.P., Grimm, E., Finger, S., Blume, A., Knoche, M., 2013. Intracuticular wax fixes and restricts strain in leaf and fruit cuticles. New Phytol. 200, 134–143.

Kiba, T., Kudo, T., Kojima, M., Sakakibara, H., 2011. Hormonal control of nitrogen acquisition: roles of auxin, abscisic acid, and cytokinin. J. Exp. Bot. 62, 1399–1409.

Kidman, C.M., Olarte Mantilla, S., Dry, P.R., McCarthy, M.G., Collins, C., 2014. Effect of water stress on the reproductive performance of Shiraz (*Vitis vinifera* L.) grafted to rootstocks. Am. J. Enol. Vitic. 65, 96–108.

Kimura, P.H., Okamoto, G., Hirano, K., 1997. Flower types, pollen morphology and germination, fertilization, and berry set in *Vitis coignetiae* Pulliat. Am. J. Enol. Vitic. 48, 323–327.

King, P.D., Buchanan, G.A., 1986. The dispersal of phylloxera crawlers and spread of phylloxera infestations in New Zealand and Australian vineyards. Am. J. Enol. Vitic. 37, 26–33.

King, P.D., Smart, R.E., McClellan, D.J., 2014. Within-vineyard variability in vine vegetative growth, yield, and fruit and wine composition of Cabernet Sauvignon in Hawke's Bay, New Zealand. Aust. J. Grape Wine Res. 20, 234–246.

Klee, H.J., 2010. Improving the flavor of fresh fruits: genomics, biochemistry, and biotechnology. New Phytol. 187, 44–56.

Klenert, M., Rapp, A., Alleweldt, G., 1978. Einfluß der Traubentemperatur auf Beerenwachstum und Beerenreife der Rebsorte Silvaner. Vitis 17, 350–360.

Kliewer, W.M., 1964. Influence of environment on metabolism of organic acids and carbohydrates in *Vitis vinifera*. I. Temperature. Plant Physiol. 39, 869–880.

Kliewer, W.M., 1965a. Changes in concentration of glucose, fructose, and total soluble solids in flowers and berries of *Vitis vinifera*. Am. J. Enol. Vitic. 16, 101–110.

Kliewer, W.M., 1965b. Changes in the concentration of malates, tartrates, and total free acids in flowers and berries of *Vitis vinifera*. Am. J. Enol. Vitic. 16, 92–100.

Kliewer, W.M., 1966. Sugars and organic acids of *Vitis vinifera*. Plant Physiol. 41, 923–931.

Kliewer, W.M., 1967a. Annual cyclic changes in the concentration of free amino acids in grapevines. Am. J. Enol. Vitic. 18, 126–137.

Kliewer, W.M., 1967b. Concentration of tartrates, malates, glucose and fructose in the fruits of the genus *Vitis*. Am. J. Enol. Vitic. 18, 87–96.

Kliewer, W.M., 1967c. The glucose–fructose ratio of *Vitis vinifera* grapes. Am. J. Enol. Vitic. 18, 33–41.

Kliewer, W.M., 1968. Changes in the concentration of free amino acids in grape berries during maturation. Am. J. Enol. Vitic. 19, 166–174.

Kliewer, W.M., 1970. Free amino acids and other nitrogenous fractions in wine grapes. J. Food Sci. 35, 17–21.

Kliewer, W.M., 1973. Berry composition of *Vitis vinifera* cultivars as influenced by photo- and nycto-temperatures during maturation. J. Am. Soc. Hortic. Sci. 98, 153–159.

Kliewer, W.M., 1975. Effect of root temperature on budbreak, shoot growth, and fruit-set of 'Cabernet Sauvignon' grapevines. Am. J. Enol. Vitic. 26, 82–89.

Kliewer, W.M., 1977a. Effect of high temperatures during the bloom-set period on fruit-set, ovule fertility, and berry growth of several grape cultivars. Am. J. Enol. Vitic. 28, 215–222.

Kliewer, W.M., 1977b. Influence of temperature, solar radiation and nitrogen on coloration and composition of Emperor grapes. Am. J. Enol. Vitic. 28, 96–103.
Kliewer, W.M., Cook, J.A., 1971. Arginine and total free amino acids as indicators of the nitrogen status of grapevines. J. Am. Soc. Hortic. Sci. 96, 581–587.
Kliewer, W.M., Dokoozlian, N.K., 2005. Leaf area/crop weight ratios of grapevines: influence on fruit composition and wine quality. Am. J. Enol. Vitic. 56, 170–181.
Kliewer, W.M., Lider, L.A., 1968. Influence of cluster exposure to the sun on the composition of Thompson Seedless fruit. Am. J. Enol. Vitic. 19, 175–184.
Kliewer, W.M., Lider, L.A., 1976. Influence of leafroll virus on composition of Burger fruits. Am. J. Enol. Vitic. 27, 118–124.
Kliewer, W.M., Ough, C.S., 1970. The effect of leaf area and crop level on the concentration of amino acids and total nitrogen in 'Thompson Seedless' grapes. Vitis 9, 196–206.
Kliewer, W.M., Schultz, H.B., 1964. Influence of environment on metabolism of organic acids and carbohydrates in *Vitis vinifera*. II. Light. Am. J. Enol. Vitic. 15, 119–129.
Kliewer, W.M., Soleimani, A., 1972. Effect of chilling on budbreak in 'Thompson Seedless' and 'Carignane' grapevines. Am. J. Enol. Vitic. 23, 31–34.
Kliewer, W.M., Torres, R.E., 1972. Effect of controlled day and night temperatures on grape coloration. Am. J. Enol. Vitic. 23, 71–77.
Kliewer, W.M., Weaver, R.J., 1971. Effect of crop level and leaf area on growth, composition, and coloration of 'Tokay' grapes. Am. J. Enol. Vitic. 22, 172–177.
Kliewer, W.M., Howarth, L., Omori, M., 1967. Concentrations of tartaric acid and malic acids and their salts in *Vitis vinifera* grapes. Am. J. Enol. Vitic. 18, 42–54.
Kliewer, W.M., Lider, L.A., Ferrari, N., 1972. Effects of controlled temperature and light intensity on growth and carbohydrate levels of 'Thompson Seedless' grapevines. J. Am. Soc. Hortic. Sci. 97, 185–188.
Kliewer, W.M., Freeman, B.M., Hosssom, C., 1983. Effect of irrigation, crop level and potassium fertilization on Carignane vines. I. Degree of water stress and effect on growth and yield. Am. J. Enol. Vitic. 34, 186–196.
Kliewer, W.M., Bowen, P., Benz, M., 1989. Influence of shoot orientation on growth and yield development in Cabernet Sauvignon. Am. J. Enol. Vitic. 40, 259–264.
Klik, A., Rosner, J., Loiskandl, W., 1998. Effects of temporary and permanent soil cover on grape yield and soil chemical and physical properties. J. Soil Water Conserv. 53, 249–253.
Knight, M.R., Knight, H., 2012. Low-temperature perception leading to gene expression and cold tolerance in higher plants. New Phytol. 195, 737–751.
Knipfer, T., Eustis, A., Brodersen, C., Walker, A.M., McElrone, A.J., 2015. Grapevine species from varied native habitats exhibit differences in embolism formation/repair associated with leaf gas exchange and root pressure. Plant Cell Environ. 38, 1503–1513.
Knipfer, T., Cuneo, I.F., Brodersen, C.R., McElrone, A.J., 2016. In situ visualization of the dynamics in xylem embolism formation and removal in the absence of root pressure: a study on excised grapevine stems. Plant Physiol. 171, 1024–1036.
Knoblauch, M., Oparka, K., 2012. The structure of the phloem—still more questions than answers. Plant J. 70, 147–156.
Knoblauch, M., Peters, W.S., 2010. Münch, morphology, microfluidics—our structural problem with the phloem. Plant Cell Environ. 33, 1439–1452.
Knoblauch, J., Mullendore, D.L., Jensen, K.H., Knoblauch, M., 2014. Pico gauges for minimally invasive intracellular hydrostatic pressure measurements. Plant Physiol. 166, 1271–1279.
Knoblauch, M., Knoblauch, J., Mullendore, D.L., Savage, J.A., Babst, B.A., Beecher, S.D., et al., 2016. Testing the Münch hypothesis of long distance phloem transport in plants. eLife 5, e15341.

Kobayashi, T., Nishizawa, N.K., 2012. Iron uptake, translocation, and regulation in higher plants. Annu. Rev. Plant Biol. 63, 131–152.

Kobayashi, A., Iwasaki, K., Teranuma, T., 1964. Pollen germination and berry set, growth and quality of Delaware grapes as affected by soil oxygen concentration. J. Jpn. Soc. Hortic. Sci. 33, 265–272.

Kobayashi, S., Goto-Yamamoto, N., Hirochika, H., 2004. Retrotransposon-induced mutations in grape skin color. Science 304, 982.

Kobayashi, H., Suzuki, S., Tanzawa, F., Takayanagi, T., 2009. Low expression of flavonoid $3',5'$-hydroxylase ($F3',5'H$) associated with cyanidin-based anthocyanins in grape leaf. Am. J. Enol. Vitic. 60, 362–367.

Kobayashi, H., Takase, H., Suzuki, Y., Tanzawa, F., Takata, R., Fujita, K., 2011. Environmental stress enhances biosynthesis of flavor precursors, S-3-(hexan-1-ol)-glutathione and S-3-(hexan-1-ol)-cysteine, in grapevine through glutathione S-transferase activation. J. Exp. Bot. 62, 1325–1336.

Kobayashi, H., Matsuyama, S., Takase, H., Sasaki, K., Suzuki, S., Takata, R., 2012. Impact of harvest timing on the concentration of 3-mercaptohexan-1-ol precursors in *Vitis vinifera* berries. Am. J. Enol. Vitic. 63, 544–548.

Koblet, W., 1966. Fruchtansatz bei Reben in Abhängigkeit von Triebbehandlungen und Klimafaktoren. Wein-Wiss. 21, 297–321.

Koblet, W., 1969. Wanderung von Assimilaten in Rebtrieben und Einfluss der Blattfläche auf Ertrag und Qualität der Trauben. Wein-Wiss. 24, 277–319.

Koblet, W., 1975. Wanderung von Assimilaten aus verschiedenen Rebenblättern während der Reifephase der Trauben. Wein-Wiss. 30, 241–249.

Koblet, W., Perret, P., 1971. Kohlehydratwanderung in Geiztrieben von Reben. Wein-Wiss. 26, 202–211.

Koblet, W., Perret, P., 1972. Wanderung von Assimilaten innerhalb der Rebe. Wein-Wiss. 27, 146–154.

Koch, K., 2004. Sucrose metabolism: regulatory mechanisms and pivotal roles in sugar sensing and plant development. Curr. Opin. Plant Biol. 7, 235–246.

Koch, R., Alleweldt, G., 1978. Der Gaswechsel reifender Weinbeeren. Vitis 17, 30–44.

Koch, G.W., Sillett, S.C., Jennings, G.M., Davis, S.D., 2004. The limits to tree height. Nature 428, 851–854.

Koch, A., Doyle, C.L., Matthews, M.A., Williams, L.E., Ebeler, S.E., 2010. 2-Methoxy-3-isobutylpyrazine in grape berries and its dependence on genotype. Phytochemistry 71, 2190–2198.

Koch, A., Ebeler, S.E., Williams, L.E., Matthews, M.A., 2012. Fruit ripening in *Vitis vinifera*: light intensity before and not during ripening determines the concentration of 2-methoxy-3-isobutylpyrazine in Cabernet Sauvignon berries. Physiol. Plant. 145, 275–285.

Kochian, L.V., Hoekenga, O.A., Piñeros, M.A., 2004. How do crop plants tolerate acid soils? Mechanisms of aluminum tolerance and phosphorous efficiency. Annu. Rev. Plant Biol. 55, 459–493.

Köckenberger, W., Pope, J.M., Xia, Y., Jeffrey, K.R., Komor, E., Callaghan, P.T., 1997. A non-invasive measurement of phloem and xylem water flow in castor bean seedlings by nuclear magnetic resonance microimaging. Planta 201, 53–63.

Kohlen, W., Charnikhova, T., Liu, Q., Bours, R., Domagalska, M.A., Beguerie, S., 2011. Strigolactones are transported through the xylem and play a key role in shoot architectural response to phosphate deficiency in non-arbuscular mycorrhizal host *Arabidopsis*. Plant Physiol. 155, 974–987.

Kolb, C.A., Pfündel, E.E., 2005. Origins of non-linear and dissimilar relationships between epidermal UV absorbance and UV absorbance of extracted phenolics in leaves of grapevine and barley. Plant Cell Environ. 25, 580–590.

Kolb, C.A., Käser, M.A., Kopecký, J., Zotz, G., Riederer, M., Pfündel, E.E., 2001. Effects of natural intensities of visible and ultraviolet radiation on epidermal ultraviolet screening and photosynthesis in grape leaves. Plant Physiol. 127, 863–875.

Kolb, C.A., Kopecký, J., Riederer, M., Pfündel, E.E., 2003. UV screening by phenolics in berries of grapevine (*Vitis vinifera*). Funct. Plant Biol. 30, 1177–1186.

Kopriva, S., 2006. Regulation of sulfate assimilation in *Arabidopsis* and beyond. Ann. Bot. 97, 479–495.

Körner, C., 2003. Carbon limitation in trees. J. Ecol. 91, 4–17.

Koroleva, O.A., Tomos, A.D., Farrar, J., Pollock, C.J., 2002. Changes in osmotic and turgor pressure in response to sugar accumulation in barley source leaves. Planta 215, 210–219.

Kortekamp, A., 2006. Expression analysis of defence-related genes in grapevine leaves after inoculation wtih a host and a non-host pathogen. Plant Physiol. Biochem. 44, 58–67.

Koussa, T., Broquedis, M., Bouard, J., 1994. Importance de l'acide abscissique dans le développement des bourgeons latents de vigne (*Vitis vinifera* L. var. Merlot) et plus particulièrement dans la phase de levée de dormance. Vitis 33, 63–67.

Koussa, T., Cherrad, M., Bertrand, A., Broquedis, M., 1998. Comparaison de la teneur en amidon, en glucides solubles et en acide abscissique des bourgeons latents et des entre-noeuds au cours du cycle végétatif de la vigne. Vitis 37, 5–10.

Kovács, L.G., Byers, P.L., Kaps, M.L., Saenz, J., 2003. Dormancy, cold hardiness, and spring frost hazard in *Vitis amurensis* hybrids under continental climatic conditions. Am. J. Enol. Vitic. 54, 8–14.

Kovaleski, A.P., Reisch, B.I., Londo, J.P., 2018. Decclimation kinetics as a quantitative phenotype for delineating the dormancy transition and thermal efficiency for budbreak in *Vitis* species. AoB Plants 10, ply066.

Koyama, K., Goto-Yamamoto, N., 2008. Bunch shading during different developmental stages affects the phenolic biosynthesis in berry skins of 'Cabernet Sauvignon' grapes. J. Am. Soc. Hortic. Sci. 133, 743–753.

Koyama, K., Sadamatsu, K., Goto-Yamamoto, N., 2009. Abscisic acid stimulated ripening and gene expression in berry skins of the Cabernet Sauvignon grape. Funct. Integr. Genom. 10, 367–381.

Kramer, E.M., Bennett, M.J., 2006. Auxin transport: a field in flux. Trends Plant Sci. 11, 382–386.

Kramer, P.J., Boyer, J.S., 1995. Water Relations of Plants and Soils. Academic Press, San Diego, CA.

Kramer, E.M., Myers, D.R., 2013. Osmosis is not driven by water dilution. Trends Plant Sci. 18, 195–197.

Kramer, E.M., Frazer, N.L., Baskin, T.I., 2007. Measurement of diffusion within the cell wall in living roots of *Arabidopsis thaliana*. J. Exp. Bot. 58, 3005–3015.

Krasnow, M., Matthews, M., Shackel, K., 2008. Evidence for substantial maintenance of membrane integrity and cell viability in normally developing grape (*Vitis vinifera* L.) berries throughout development. J. Exp. Bot. 59, 849–859.

Kratsch, H.A., Wise, R.R., 2000. The ultrastructure of chilling stress. Plant Cell Environ. 23, 337–350.

Kraus, T.E.C., Yu, Z., Preston, C.M., Dahlgren, R.A., Zasoski, R.J., 2003. Linking chemical reactivity and protein precipitation to structural characteristics of foliar tannins. J. Chem. Ecol. 29, 703–730.

Kretschmer, M., Kassemeyer, H.H., Hahn, M., 2007. Age-dependent grey mould susceptibility and tissue-specific defence gene activation of grapevine berry skins after infection by *Botrytis cinerea*. J. Phytopathol. 155, 258–263.

Kreuzwieser, J., Rennenberg, H., 2014. Molecular and physiological responses of trees to waterlogging stress. Plant Cell Environ. 37, 2245–2259.

Kriedemann, P.E., 1968a. Observations on gas exchange in the developing Sultana berry. Aust. J. Biol. Sci. 21, 907–916.

Kriedemann, P.E., 1968b. Photosynthesis in vine leaves as a function of light intensity, temperature, and leaf age. Vitis 7, 213–220.

Kriedemann, P.E., Buttrose, M.S., 1971. Chlorophyll content and photosynthetic activity within woody shoots of *Vitis vinifera* (L.). Photosynthetica 5, 22–27.

Kriedemann, P.E., Goodwin, I., 2003. Regulated Deficit Irrigation and Partial Rootzone Drying. An Overview of Principles and Applications. Land and Water Australia, Canberra.

Kriedemann, P.E., Törökfalvy, E., Smart, R.E., 1973. Natural occurrence and photosynthetic utilisation of sunflecks by grapevine leaves. Photosynthetica 7, 18–27.

Kritzinger, E.C., Bauer, F.F., du Toit, W.J., 2013. Role of glutathione in winemaking: a review. J. Agric. Food Chem. 61, 269–277.

Krouk, G., Ruffel, S., Gutiérrez, R.A., Gojon, A., Crawford, N.M., Coruzzi, G.M., 2011. A framework integrating plant growth with hormones and nutrients. Trends Plant Sci. 16, 178–182.

Kruse, J., Hopmans, P., Adams, M.A., 2008. Temperature responses are a window to the physiology of dark respiration: differences between CO_2 release and O_2 reduction shed light on energy conservation. Plant Cell Environ. 31, 901–914.

Kruse, J., Rennenberg, H., Adams, M.A., 2011. Steps towards a mechanistic understanding of respiratory temperature responses. New Phytol. 189, 659–677.

Kuć, J., 1995. Phytoalexins, stress metabolism, and disease resistance in plants. Annu. Rev. Phytopathol. 33, 275–297.

Kuhlemeier, C., 2007. Phyllotaxis. Trends Plant Sci. 12, 143–150.

Kühn, C., Barker, L., Bürkle, L., Frommer, W.B., 1999. Update on sucrose transport in higher plants. J. Exp. Bot. 50, 935–953.

Kurkdjian, A., Guern, J., 1989. Intracellular pH: measurement and importance in cell activity. Annu. Rev. Plant Physiol. Plant Mol. Biol. 40, 271–303.

Kuromori, T., Seo, M., Shinozaki, K., 2018. ABA transport and plant water stress responses. Trends Plant Sci. 23, 513–522.

Kusano, T., Berberich, T., Tateda, C., Takahashi, Y., 2008. Polyamines: essential factors for growth and survival. Planta 228, 367–381.

Kutschera, U., 2008a. The growing outer epidermal wall: design and physiological role of a composite structure. Ann. Bot. 101, 615–621.

Kutschera, U., 2008b. The pacemaker of plant growth. Trends Plant Sci. 13, 105–107.

Kutschera, U., Briggs, W.R., 2012. Root phototropism: from dogma to the mechanism of blue light perception. Planta 235, 443–452.

Kwasniewski, M.T., Sacks, G.L., Wilcox, W.F., 2014. Persistence of elemental sulfur spray residue on grapes during ripening and vinification. Am. J. Enol. Vitic. 65, 453–462.

Ky, I., Lorrain, B., Jourdes, M., Pasquier, G., Fermaud, M., Gény, L., 2012. Assessment of grey mould (*Botrytis cinerea*) impact on phenolic and sensory quality of Bordeaux grapes, musts and wines for two consecutive vintages. Aust. J. Grape Wine Res. 18, 215–226.

La Guerche, S., Dauphin, B., Pons, M., Blancard, D., Darriet, P., 2006. Characterization of some mushroom and earthy off-odors microbially induced by the development of rot on grapes. J. Agric. Food Chem. 54, 9193–9200.

Lacey, M.J., Allen, M.S., Harris, R.L.N., Brown, W.V., 1991. Methoxypyrazines in Sauvignon blanc grapes and wines. Am. J. Enol. Vitic. 42, 103–108.

Lacombe, T., Boursiquot, J.M., Laucou, V., Dechesne, F., Varès, D., This, P., 2007. Relationships and genetic diversity within the accessions related to Malvasia held in the Domaine de Vassal grape germplasm repository. Am. J. Enol. Vitic. 58, 124–131.

Lacombe, T., Boursiquot, J.M., Laucou, V., Di Vecchi-Staraz, M., Péros, J.P., This, P., 2013. Large-scale parentage analysis in an extended set of grapevine cultivars (*Vitis vinifera* L.). Theor. Appl. Genet. 126, 401–414.

Lacroux, F., Tregoat, O., Van Leeuwen, C., Pons, A., Tominaga, T., Lavigne-Cruège, V., 2008. Effect of foliar nitrogen and sulphur application on aromatic expression of *Vitis vinifera* L. cv. Sauvignon blanc. J. Int. Sci. Vigne Vin 42, 125–132.

Lafon-Lafourcade, S., Guimberteau, G., 1962. Evolution des amino-acides au cours de la maturation des raisins. Vitis 3, 130–135.

Laguna, E., 2003. Sobre las formas naturalizadas de *Vitis* en la Comunidad Valenciana. I. Las especies. Flora Montiberica 23, 46–82.

Laidig, F., Piepho, H.P., Hofäcker, W., 2009. Statistical analysis of 'White Riesling' (*Vitis vinifera* ssp. *sativa* L.) clonal performance at 16 locations in the Rheinland-Pfalz region of Germany between 1971 and 2007. Vitis 48, 77–85.

Lakso, A.N., Kliewer, W.M., 1975. The influence of temperature on malic acid metabolism in rape berries. I. Enzyme responses. Plant Physiol. 56, 370–372.

Lakso, A.N., Kliewer, W.M., 1978. The influence of temperature on malic acid metabolism in grape berries. II. Temperature responses of net dark CO_2 fixation and malic acid pools. Am. J. Enol. Vitic. 29, 145–149.

Lakso, A.N., Pratt, C., Pearson, R.C., Pool, R.M., Seem, R.C., Welser, M.J., 1982. Photosynthesis, transpiration, and water use efficiency of mature grape leaves infected with *Uncinula necator* (powdery mildew). Phytopathology 72, 232–236.

Lalonde, S., Tegeder, M., Throne-Holst, M., Frommer, W.B., Patrick, J.W., 2003. Phloem loading and unloading of sugars and amino acids. Plant Cell Environ. 26, 37–56.

Lalonde, S., Wipf, D., Frommer, W.B., 2004. Transport meachnisms for organic forms of carbon and nitrogen between source and sink. Annu. Rev. Plant Biol. 55, 341–372.

Lamb, D.W., Weedon, M.M., Bramley, R.G.V., 2004. Using remote sensing to predict grape phenolics and colour at harvest in a Cabernet Sauvignon vineyard. Timing observations against vine phenology and optimising image resolution. Aust. J. Grape Wine Res. 10, 46–54.

Lane, N., 2002. Oxygen. The Molecule That Made the World. Oxford University Press, Oxford.

Lang, A., Düring, H., 1990. Grape berry splitting and some mechanical properties of the skin. Vitis 29, 61–70.

Lang, A., Thorpe, M.R., 1989. Xylem, phloem and transpiration flows in a grape: application of a technique for measuring the volume of attached fruits to high resolution using Archimedes' principle. J. Exp. Bot. 40, 1069–1078.

Lang, G.A., Early, J.D., Martin, G.C., Darnell, R.L., 1987. Endo-, para-, and eco-dormancy: physiological terminology and classification for dormancy research. HortScience 22, 371–377.

Lang, N.S., Wample, R.L., Smithyman, R., Mills, L., 1998. Photosynthesis and chlorophyll fluorescence in blackleaf-affected Concord leaves. Am. J. Enol. Vitic. 49, 367–374.

Langcake, P., 1981. Disease resistance of *Vitis* spp. and the production of the stress metabolites resveratrol, ε-viniferin, α-viniferin and pterostilbene. Physiol. Plant Pathol. 18, 213–226.

Langcake, P., Lovell, P.A., 1980. Light and electron microscopical studies of the infection of *Vitis* spp. by *Plasmopara viticola*, the downy mildew pathogen. Vitis 19, 321–337.

Langcake, P., McCarthy, W.V., 1979. The relationship of resveratrol production to infection of grapevine leaves by *Botrytis cinerea*. Vitis 18, 244–253.

Langcake, P., Pryce, R.J., 1976. The production of resveratrol by *Vitis vinifera* and other members of the Vitaceae as a response to infection or injury. Physiol. Plant Pathol. 9, 77–86.

Langcake, P., Cornford, C.A., Pryce, R.J., 1979. Identification of pterostilbene as a phytoalexin from *Vitis vinifera* leaves. Phytochemistry 18, 1025–1027.

Lara, I., Belge, B., Goulao, L.F., 2015. A focus on the biosynthesis and composition of cuticle in fruits. J. Agric. Food Chem. 63, 4005–4019.

Larronde, F., Krisa, S., Decendit, A., Chèze, C., Deffieux, G., Mérillon, J.M., 1998. Regulation of polyphenol production in *Vitis vinifera* cell suspension cultures by sugars. Plant Cell Rep. 17, 946–950.

Larronde, F., Gaudillère, J.P., Krisa, S., Decendit, A., Deffieux, G., Mérillon, J.M., 2003. Airborne methyl jasmonate induces stilbene accumulation in leaves and berries of grapevine plants. Am. J. Enol. Vitic. 54, 63–66.

Laucou, V., Lacombe, T., Dechesne, F., Siret, R., Bruno, J.P., Dessup, M., 2011. High throughput analysis of grape genetic diversity as a tool for germplasm collection management. Theor. Appl. Genet. 122, 1233–1245.

Lavee, S., May, P., 1997. Dormancy of grapevine buds—facts and speculation. Aust. J. Grape Wine Res. 3, 31–46.

Lavoie-Lamoureux, A., Sacco, D., Risse, P.A., Lovisolo, C., 2017. Factors influencing stomatal conductance in response to water availability in grapevine: a meta-analysis. Physiol. Plant. 159, 468–482.

Lawlor, D.W., 2002. Carbon and nitrogen assimilation in relation to yield: mechanisms are the key to understanding production systems. J. Exp. Bot. 53, 773–787.

Lawlor, D.W., Cornic, G., 2002. Photosynthetic carbon assimilation and associated metabolism in relation to water deficits in higher plants. Plant Cell Environ. 25, 275–294.

Lawlor, D.W., Tezara, W., 2009. Causes of decreased photosynthetic rate and metabolic capacity in water-deficient leaf cells: a critical evaluation of mechanisms and integration of processes. Ann. Bot. 103, 561–579.

Lawson, T., 2008. Guard cell photosynthesis and stomatal function. New Phytol. 181, 13–34.

Lazar, G., Goodman, H.M., 2006. *MAX1*, a regulator of the flavonoid pathway, controls vegetative axillary bud outgrowth in *Arabidopsis*. Proc. Natl. Acad. Sci. U. S. A. 103, 472–476.

Leakey, A.D.B., Xu, F., Gillespie, K.M., McGrath, J.M., Ainsworth, E.A., Ort, D.R., 2009. Genomic basis for stimulated respiration by plants growing under elevated carbon dioxide. Proc. Natl. Acad. Sci. U. S. A. 106, 3597–3602.

Lebon, E., Pellegrino, A., Tardieu, F., Lecoeur, J., 2004. Shoot development in grapevine (*Vitis vinifera*) is affected by the modular branching pattern of the stem and intra- and inter-shoot trophic competition. Ann. Bot. 93, 263–274.

Lebon, E., Pellegrino, A., Louarn, G., Lecoeur, J., 2006. Branch development controls leaf area dynamics in grapevine (*Vitis vinifera*) growing in drying soil. Ann. Bot. 98, 175–185.

Lebon, G., Wojnarowiez, G., Holzapfel, B., Fontaine, F., Vaillant-Gaveau, N., Clément, C., 2008. Sugars and flowering in the grapevine (*Vitis vinifera* L.). J. Exp. Bot. 59, 2565–2578.

Lecas, M., Brillouet, J.M., 1994. Cell wall composition of grape berry skins. Phytochemistry 35, 1241–1243.

Lecourieux, F., Kappel, C., Lecourieux, D., Serrano, A., Torres, E., Arce-Johnson, P., 2014. An update on sugar transport and signalling in grapevine. J. Exp. Bot. 65, 821–832.

Ledbetter, C.A., Ramming, D.W., 1989. Seedlessness in grapes. Hortic. Rev. (Am. Soc. Hortic. Sci). 11, 159–184.

Ledbetter, C.A., Shonnard, C.B., 1991. Berry and seed characteristics associated with stenospermocarpy in *vinifera* grapes. J. Hortic. Sci. 66, 247–252.

Lee, C.Y., Smith, N.L., Nelson, R.R., 1979. Relationship between pectin methylesterase activity and the formation of methanol in Concord grape juice and wine. Food Chem. 4, 143–148.

Lee, S.H., Seo, M.J., Riu, M., Cotta, J.P., Block, D.E., Dokoozlian, N.K., 2007. Vine microclimate and norisoprenoid concentration in Cabernet Sauvignon grapes and wines. Am. J. Enol. Vitic. 58, 291–301.

Lehnart, R., Michel, H., Löhnertz, O., Linsenmeier, A., 2008. Root dynamics and pattern of 'Riesling' on 5C rootstock using minirhizotrons. Vitis 47, 197–200.

Lei, M., Liu, Y., Zhang, B., Zhao, Y., Wang, X., Zhou, Y., 2011. Genetic and genomic evidence that sucrose is a global regulator of plant responses to phosphate starvation in *Arabidopsis*. Plant Physiol. 156, 1116–1130.

Leide, J., Hildebrandt, U., Reussing, K., Riederer, M., Vogg, G., 2007. The developmental pattern of tomato fruit wax accumulation and its impact on cuticular transpiration barrier properties: effects of a deficiency in a β-ketoacyl-coenzyme A synthase (LeCER6). Plant Physiol. 144, 1667–1679.

Leigh, J., Hodge, A., Fitter, A.H., 2009. Arbuscular mycorrhizal fungi can transfer substantial amounts of nitrogen to their host plant from organic material. New Phytol. 181, 199–207.

Lejay, L., Wirth, J., Pervent, M., Cross, J.M.F., Tillard, P., Gojon, A., 2008. Oxidative pentose phosphate pathway-dependent sugar sensing as a mechanism for regulation of root ion transporters by photosynthesis. Plant Physiol. 146, 2036–2053.

Lemaire, G., Millard, P., 1999. An ecophysiological approach to modelling resource fluxes in competing plants. J. Exp. Bot. 50, 15–28.

Lendzian, K.J., 2006. Survival strategies of plants during secondary growth: barrier properties of phellems and lenticels towards water, oxygen, and carbon dioxide. J. Exp. Bot. 57, 2535–2546.

Lenk, S., Buschmann, C., Pfündel, E.E., 2007. *In vivo* assessing flavonols in white grape berries (*Vitis vinifera* L. cv. Pinot Blanc) of different degrees of ripeness using chlorophyll fluorescence imaging. Funct. Plant Biol. 34, 1092–1104.

Lens, F., Sperry, J.S., Christman, M.A., Choat, B., Rabaey, D., Jansen, S., 2011. Testing hypotheses that link wood anatomy to cavitation resistance and hydraulic conductivity in the genus *Acer*. New Phytol. 190, 709–723.
Lesch, S.M., Suarez, D.L., 2009. A short note on calculating the adjusted SAR index. Trans. ASABE 52, 493–496.
Leustek, T., Saito, K., 1999. Sulfate transport and assimilation in plants. Plant Physiol. 120, 637–644.
Levadoux, L., 1956. Les populations sauvages et cultivées de *Vitis vinifera* L. Ann. Amélior. Plantes 6, 59–118.
Levey, D.J., 2004. The evolutionary ecology of ethanol production and alcoholism. Integr. Comp. Biol. 44, 284–289.
Lewinsohn, E., Sitrit, Y., Bar, E., Azulay, Y., Meir, A., Zamir, D., 2005. Carotenoid pigmentation affects the volatile composition of tomato and watermelon fruits, as revealed by comparative genetic analyses. J. Agric. Food Chem. 53, 3142–3148.
Leyhe, A., Blanke, M.M., 1989. Kohlenstoff-Haushalt der Infloreszenz der Rebe. 2. CO_2-Gaswechsel der Blütenstände und Weinbeere. Wein-Wiss. 44, 147–150.
Leyser, O., 2010. The power of auxin in plants. Plant Physiol. 154, 501–505.
Li, Y., Jones, L., McQueen-Mason, S., 2003. Expansins and cell growth. Curr. Opin. Plant Biol. 6, 603–610.
Li, D., Wan, Y., Wang, Y., He, P., 2008. Relatedness of resistance to anthracnose and to white rot in Chinese wild grapes. Vitis 47, 213–215.
Li, Z., Wakao, S., Fischer, B.B., Niyogi, K.K., 2009. Sensing and responding to excess light. Annu. Rev. Plant Biol. 60, 239–260.
Li, B., Feng, Z., Xie, M., Sun, M., Zhao, Y., Liang, L., 2011. Modulation of the root-sourced ABA signal along its way to the shoot in *Vitis riparia* × *Vitis labrusca* under water deficit. J. Exp. Bot. 62, 1731–1741.
Liakopoulos, G., Nikolopoulos, D., Klouvatou, A., Vekkos, K.A., Manetas, Y., Karabourniotis, G., 2006. The photoprotective role of epidermal anthocyanins and surface pubescence in young leaves of grapevine (*Vitis vinifera*). Ann. Bot. 98, 257–265.
Liakoura, V., Fotelli, M.N., Rennenberg, H., Karabourniotis, G., 2009. Should structure–function relations be considered separately for homobaric vs. heterobaric leaves? Am. J. Bot. 96, 612–619.
Liang, Y., Si, J., Römheld, V., 2005. Silicon uptake and transport is an active process in *Cucumis sativus*. New Phytol. 167, 797–804.
Liang, Z., Owens, C.L., Zhong, G.Y., Cheng, L., 2011. Polyphenolic profiles detected in the ripe berries of *Vitis vinifera* germplasm. Food Chem. 129, 940–950.
Liang, Z., Yang, Y., Cheng, L., Zhong, G.Y., 2012a. Characterization of polyphenolic metabolites in the seeds of *Vitis* germplasm. J. Agric. Food Chem. 60, 1291–1299.
Liang, Z., Yang, Y., Cheng, L., Zhong, G.Y., 2012b. Polyphenolic composition and content in the ripe berries of wild *Vitis* germplasm. Food Chem. 132, 730–738.
Liang, N.N., Pan, Q.H., He, F., Wang, J., Reeves, M.J., Duan, C.Q., 2013. Phenolic profiles of *Vitis davidii* and *Vitis quinquangularis* species native to China. J. Agric. Food Chem. 61, 6016–6027.
Lider, L.A., Goheen, A.C., Ferrari, N.L., 1975. A comparison between healthy and leafroll-affected grapevine planting stocks. Am. J. Enol. Vitic. 26, 144–147.
Lijavetzky, D., Ruiz-García, L., Cabezas, J.A., De Andrés, M.T., Bravo, G., Ibáñez, A., 2006. Molecular genetics of berry colour variation in table grape. Mol. Gen. Genom. 276, 427–435.
Lijavetzky, D., Carbonell-Bejerano, P., Grimplet, J., Bravo, G., Flores, P., Fenoll, J., 2012. Berry flesh and skin ripening features in *Vitis vinifera* as assessed by transcriptional profiling. PLoS ONE 7, e39547.
Lillo, C., Lea, U.S., Ruoff, P., 2008. Nutrient depletion as a key factor for manipulating gene expression and product formation in different branches of the flavonoid pathway. Plant Cell Environ. 31, 587–601.
Lilov, D., Andonova, T., 1976. Cytokinins, growth, flower and fruit formation in *Vitis vinifera*. Vitis 15, 160–170.
Lim, P.O., Kim, H.J., Nam, H.G., 2007. Leaf senescence. Annu. Rev. Plant Biol. 58, 115–136.

Lin, C.H., Lin, J.H., Chang, L.R., Lin, H.S., 1985. The regulation of the Golden Muscat grape production season in Taiwan. Am. J. Enol. Vitic. 36, 114–117.

Linsenmeier, A., Rauhut, D., Kürbel, H., Löhnertz, O., Schubert, S., 2007. Untypical ageing off-flavour and masking effects due to long-term nitrogen fertilization. Vitis 46, 33–38.

Linsenmeier, A.W., Loos, U., Löhnertz, O., 2008. Must composition and nitrogen uptake in a long-term trial as affected by timing of nitrogen fertilization in a cool-climate Riesling vineyard. Am. J. Enol. Vitic. 59, 255–264.

Lisch, D., 2009. Epigenetic regulation of transposable elements in plants. Annu. Rev. Plant Biol. 60, 43–66.

Liu, W.T., Pool, R., Wenkert, W., Kriedemann, P.E., 1978. Changes in photosynthesis, stomatal resistance and abscisic acid of *Vitis labruscana* through drought and irrigation cycles. Am. J. Enol. Vitic. 29, 239–246.

Liu, H.F., Wu, B.H., Fan, P.G., Li, S.H., Li, L.S., 2006. Sugar and acid concentrations in 98 grape cultivars analyzed by principal component analysis. J. Sci. Food Agric. 86, 1526–1536.

Liu, L., Gregan, S., Winefield, C., Jordan, B., 2015. From UVR8 to flavonol synthase: UV-B-induced gene expression in Sauvignon blanc grape berry. Plant Cell Environ. 38, 905–919.

Llorens, N., Arola, L., Blade, C., Mas, A., 2002. Nitrogen metabolism in a grapevine *in vitro* system. J. Int. Sci. Vigne Vin 36, 157–159.

Loescher, W.H., McCamant, T., Keller, J.D., 1990. Carbohydrate reserves, translocation, and storage in woody plant roots. HortScience 25, 274–281.

Loewus, F.A., 1999. Biosynthesis and metabolism of ascorbic acid in plants and of analogs of ascorbic acid in fungi. Phytochemistry 52, 193–210.

Logan, D.C., 2006. The mitochondrial compartment. J. Exp. Bot. 57, 1225–1243.

Löhnertz, O., Schaller, K., Mengel, K., 1989. Nährstoffdynamik in Reben. III. Mitteilung: Stickstoffkonzentration und Verlauf der Aufnahme in der Vegetation. Wein-Wiss. 44, 192–204.

Londo, J.P., Johnson, L.M., 2014. Variation in the chilling requirement and budburst rate of wild *Vitis* species. Environ. Exp. Bot. 106, 138–147.

Londo, J.P., Kovaleski, A.P., 2017. Characterization of wild North American grapevine cold hardiness using differential thermal analysis. Am. J. Enol. Vitic. 68, 203–212.

Long, S.P., Ainsworth, E.A., Rogers, A., Ort, D.R., 2004. Rising atmospheric carbon dioxide: plants FACE the future. Annu. Rev. Plant Biol. 55, 591–628.

Longbottom, M.L., Dry, P.R., Sedgley, M., 2008. Observations on the morphology and development of star flowers of *Vitis vinifera* L. cvs Chardonnay and Shiraz. Aust. J. Grape Wine Res. 14, 203–210.

Longbottom, M.L., Dry, P.R., Sedgley, M., 2010. Effects of sodium molybdate foliar sprays on molybdenum concentration in the vegetative and reproductive structures and on yield components of *Vitis vinifera* cv. Merlot. Aust. J. Grape Wine Res. 16, 477–490.

Longo, E., Cansado, J., Agrelo, D., Villa, T.G., 1991. Effect of climatic conditions on yeast diversity in grape musts from northwest Spain. Am. J. Enol. Vitic. 42, 141–144.

López-Bucio, J., Cruz-Ramírez, A., Herrera-Estrella, L., 2003. The role of nutrient availability in regulating root architecture. Curr. Opin. Plant Biol. 6, 280–287.

Lopez-Casado, G., Salamanca, A., Heredia, A., 2010. Viscoelastic nature of isolated tomato (*Solanum lycopersicum*) fruit cuticles: a mathematical model. Physiol. Plant. 140, 79–88.

Loreto, F., Schnitzler, J.P., 2010. Abiotic stresses and induced BVOCs. Trends Plant Sci. 15, 154–166.

Lott, R.V., Barrett, H.C., 1967. The dextrose, levulose, sucrose, and acid content of the juice from 39 grape clones. Vitis 6, 257–268.

Louarn, G., Guedon, Y., Lecoeur, J., Lebon, E., 2007. Quantitative analysis of the phenotypic variability of shoot architecture in two grapevine (*Vitis vinifera*) cultivars. Ann. Bot. 99, 425–437.

Lough, T.J., Lucas, W.J., 2006. Integrative plant biology: role of phloem long-distance macromolecular trafficking. Annu. Rev. Plant Biol. 57, 203–232.

Loulakakis, C.A., Roubelakis-Angelakis, K.A., 1990. Intracellular localization and properties of NADH-glutamate dehydrogenase from *Vitis vinifera* L.: Purification and characterization of the major leaf isoenzyme. J. Exp. Bot. 41, 1223–1230.

Loulakakis, K.A., Roubelakis-Angelakis, K.A., 2001. Nitrogen assimilation in grapevine. In: Roubelakis-Angelakis, K.A. (Ed.), Molecular Biology and Biotechnology of the Grapevine. Kluwer, Dordrecht, The Netherlands, pp. 59–85.

Loulakakis, K.A., Primikirios, N.I., Nikolantonakis, M.A., Roubelakis-Angelakis, K.A., 2002. Immunocharacterization of *Vitis vinifera* L. ferredoxin-dependent glutamate synthase, and its spatial and temporal changes during leaf development. Planta 215, 630–638.

Loulakakis, K.A., Morot-Gaudry, J.F., Velanis, C.N., Skopelitis, D.S., Moschou, P.N., Hirel, B., Roubelakis-Angelakis, K.A., 2009. Advancements in nitrogen metabolism in grapevine. In: Roubelakis-Angelakis, K.A. (Ed.), Grapevine Molecular Physiology and Biotechnology. Springer, Dordrecht, The Netherlands, pp. 161–205.

Loveys, B.R., 1984. Diurnal changes in water relations and abscisic acid in field-grown *Vitis vinifera* cultivars. III. The influence of xylem-derived abscisic acid on leaf gas exchange. New Phytol. 98, 563–573.

Loveys, B.R., Düring, H., 1984. Diurnal changes in water relations and abscisic acid in field-grown *Vitis vinifera* cultivars. II. Abscisic acid changes under semi-arid conditions. New Phytol. 97, 37–47.

Loveys, B.R., Scheurwater, I., Pons, T.L., Fitter, A.H., Atkin, O.K., 2002. Growth temperature influences the underlying components of relative growth rate: an investigation using inherently fast- and slow-growing plant species. Plant Cell Environ. 25, 975–987.

Lovisetto, A., Guzzo, F., Tadiello, A., Toffali, K., Favretto, A., Casadoro, G., 2012. Molecular analyses of MADS-box genes trace back to Gymnosperms the invention of fleshy fruits. Mol. Biol. Evol. 409–419.

Lovisolo, C., Schubert, A., 1998. Effects of water stress on vessel size and xylem hydraulic conductivity in *Vitis vinifera* L. J. Exp. Bot. 49, 693–700.

Lovisolo, C., Schubert, A., 2000. Downward shoot positioning affects water transport in field-grown grapevines. Vitis 39, 49–53.

Lovisolo, C., Schubert, A., 2006. Mercury hinders recovery of shoot hydraulic conductivity during grapevine rehydration: evidence from a whole-plant approach. New Phytol. 172, 469–478.

Lovisolo, C., Hartung, W., Schubert, A., 2002a. Whole-plant hydraulic conductance and root-to-shoot flow of abscisic acid are independently affected by water stress in grapevines. Funct. Plant Biol. 29, 1349–1356.

Lovisolo, C., Schubert, A., Sorce, C., 2002b. Are xylem radial development and hydraulic conductivity in downwardly-growing grapevine shoots influenced by perturbed auxin metabolism? New Phytol. 156, 65–74.

Lovisolo, C., Perrone, I., Hartung, W., Schubert, A., 2008. An abscisic acid-related reduced transpiration promotes gradual embolism repair when grapevines are rehydrated after drought. New Phytol. 180, 642–651.

Lovisolo, C., Perrone, I., Carra, A., Ferrandino, A., Flexas, J., Medrano, H., 2010. Drought-induced changes in development and function of grapevine (*Vitis* spp.) organs and in their hydraulic and non-hydraulic interactions at the whole-plant level: a physiological and molecular update. Funct. Plant Biol. 37, 98 116.

Lu, Y., Foo, L.Y., 1999. The polyphenol constituents of grape pomace. Food Chem. 65, 1–8.

Luan, F., Wüst, M., 2002. Differential incorporation of 1-deoxy-d-xylulose into (3*S*)-linalool and geraniol in grape berry exocap and mesocarp. Phytochemistry 60, 451–459.

Luan, F., Mosandl, A., Münch, A., Wüst, M., 2005. Metabolism of geraniol in grape berry mesocarp of *Vitis vinifera* L. cv. Scheurebe: demonstration of stereoselective reduction, E/Z isomerization, oxidation and glycosylation. Phytochemistry 66, 295–303.

Luan, F., Mosandl, A., Gubesch, M., Wüst, M., 2006. Enantioselective analysis of monoterpenes in different grape varieties during berry ripening using stir bar sorptive extraction- and solid phase extraction-enantioselective-multidimensional gas chromatography-mass spectrometry. J. Chromatogr. A 1112, 369–374.

Lucas, R.E., Davis, J.F., 1961. Relationships between pH values of organic soils and availabilities of 12 plant nutrients. Soil Sci. 92, 177–182.

Luisetti, J., Gaignard, J.L., Devaux, M., 1991. *Pseudomonas syringae* pv. *syringae* as one of the factors affecting the ice nucleation of grapevine buds in controlled conditions. J. Phytopathol. 133, 334–344.

Lund, S.T., Peng, F.Y., Nayar, T., Reid, K.E., Schlosser, J., 2008. Gene expression analyses in individual grape (*Vitis vinifera* L.) berries during ripening initiation reveal that pigmentation intensity is a valid indicator of developmental staging within the cluster. Plant Mol. Biol. 68, 301–315.

Lundgren, M.R., Osborne, C.P., Christin, P.A., 2014. Deconstructing Kranz anatomy to understand C_4 evolution. J. Exp. Bot. 65, 3357–3369.

Lushai, G., Loxdale, H.D., 2002. The biological improbability of a clone. Genet. Res. 79, 1–9.

Lustig, I., Bernstein, Z., 1985. Determination of the mechanical properties of the grape berry skin by hydraulic measurements. Sci. Hortic. 25, 279–285.

Luu, D.T., Maurel, C., 2005. Aquaporins in a challenging environment: molecular gears for adjusting plant water status. Plant Cell Environ. 28, 85–96.

Ma, J.F., Yamaji, N., 2006. Silicon uptake and accumulation in higher plants. Trends Plant Sci. 11, 392–397.

Ma, J.F., Yamaji, N., 2015. A cooperative system of silicon transport in plants. Trends Plant Sci. 20, 435–442.

Ma, Z.Y., Wen, J., Ickert-Bond, S.M., Chen, L.Q., Liu, X.Q., 2016. Morphology, structure, and ontogeny of trichomes of the grape genus (*Vitis*, Vitaceae). Front. Plant Sci. 7, 704.

Macheix, J.J., Fleuriet, A., Billot, J., 1990. Fruit Phenolics. CRC Press, Boca Raton, FL.

Macheix, J.J., Sapis, J.C., Fleuriet, A., 1991. Phenolic compounds and polyphenoloxidase in relation to browning in grapes and wines. Crit. Rev. Food Sci. Nutr. 30, 441–486.

MacNeill, G.J., Mehrpouyan, S., Minow, M.A.A., Patterson, J.A., Tetlow, I.J., Emes, M.J., 2017. Starch as a source, starch as a sink: the bifunctional role of starch in carbon allocation. J. Exp. Bot. 68, 4433–4453.

Madronich, S., McKenzie, R.L., Björn, L.O., Caldwell, M.M., 1998. Changes in biologically active ultraviolet radiation reaching the Earth's surface. J. Photochem. Photobiol. B 46, 5–19.

Maga, J.A., 1989. The contribution of wood to the flavor of alcoholic beverages. Food Rev. Int. 5, 39–99.

Maghradze, D., Rustioni, L., Turok, J., Scienza, A., Failla, O., 2012. Caucasus and Northern Black Sea Region Ampelography. Vitis Special Issue, Maierdruck, Lingenfeld, Germany.

Malacarne, G., Vrhovsek, U., Zulini, L., Cestaro, A., Stefanini, M., Mattivi, F., 2011. Resistance to *Plasmopara viticola* in a grapevine segregating population is associated with stilbenoid accumulation and with specific host transcriptional responses. BMC Plant Biol. 11, 114.

Malamy, J.E., 2005. Intrinsic and environmental response pathways that regulate root system architecture. Plant Cell Environ. 28, 67–77.

Malamy, J.E., Benfey, P.N., 1997. Down and out in *Arabidopsis*: the formation of lateral roots. Trends Plant Sci. 2, 390–396.

Maletić, E., Pejić, I., Karoglan Kontić, J., Piljac, J., Dangl, G.S., Vokurka, A., 2004. Zinfandel, Dobričić, and Plavac mali: the genetic relationship among three cultivars of the Dalmatian coast of Croatia. Am. J. Enol. Vitic. 55, 174–180.

Mancuso, S., 1999. Hydraulic and electrical transmission of wound-induced signals in *Vitis vinifera*. Aust. J. Plant Physiol. 26, 55–61.

Mancuso, S., Boselli, M., 2002. Characterisation of the oxygen fluxes in the division, elongation and mature zones of *Vitis* roots: influence of oxygen availability. Planta 214, 767–774.

Mancuso, S., Marras, A.M., 2006. Adaptive response of *Vitis* root to anoxia. Plant Cell Physiol. 47, 401–409.

Mannini, F., Mollo, A., Credi, R., 2012. Field performance and wine quality modification in a clone of *Vitis vinifera* cv. Dolcetto after GLRaV-3 elimination. Am. J. Enol. Vitic. 63, 144–147.

Maoz, I., Kaplunov, T., Beno-Mualem, D., Lewinsohn, E., Lichter, A., 2018. Variability in volatile composition of Crimson Seedless (*Vitis vinifera*) in association with maturity at harvest. Am. J. Enol. Vitic. 63, 125–132.

Mapfumo, E., Aspinall, D., Hancock, T.W., 1994a. Growth and development of roots of grapevine (*Vitis vinifera* L.) in relation to water uptake from soil. Ann. Bot. 74, 75–85.

Mapfumo, E., Aspinall, D., Hancock, T.W., 1994b. Vessel-diameter distribution in roots of grapevines (*Vitis vinifera* L. cv. Shiraz). Plant Soil 160, 49–55.

Marais, J., Hunter, J.J., Haasbroek, P.D., 1999. Effect of canopy microclimate, season and region on Sauvignon blanc grape composition and wine quality. S. Afr. J. Enol. Vitic. 20, 19–30.

Marais, J., Calitz, F., Haasbroek, P.D., 2001. Relationship between microclimatic data, aroma component concentrations and wine quality parameters in the prediction of Sauvignon blanc wine quality. S. Afr. J. Enol. Vitic. 22, 22–26.

Marangoni, B., Vitagliano, C., Peterlunger, E., 1986. The effect of defoliation on the composition of xylem sap from Cabernet franc grapevines. Am. J. Enol. Vitic. 37, 259–262.

Marga, F., Pesacreta, T.C., Hasenstein, K.H., 2001. Biochemical analysis of elastic and rigid cuticles of *Cirsium horridulum* Michx. Planta 213, 841–848.

Margaria, P., Ferrandino, A., Caciagli, P., Kedrina, O., Schubert, A., Palmano, S., 2014. Metabolic and transcript analysis of the flavonoid pathway in diseased and recovered Nebbiolo and Barbera grapevines (*Vitis vinifera* L.) following infection by Flavescence dorée phytoplasma. Plant Cell Environ. 37, 2183–2200.

Margna, U., Vainjsärv, T., Laanest, L., 1989. Different l-phenylalanine pools available for the biosynthesis of phenolics in buckwheat seedling tissues. Phytochemistry 28, 469–475.

Markham, K.R., Gould, K.S., Winefield, C.S., Mitchell, K.A., Bloor, S.J., Boase, M.R., 2000. Anthocyanic vacuolar inclusions—their nature and significance in flower colouration. Phytochemistry 55, 327–336.

Marois, J.J., Nelson, J.K., Morrison, J.C., Lile, L.S., Bledsoe, A.M., 1986. The influence of berry contact within grape clusters on the development of *Botrytis cinerea* and epicuticular wax. Am. J. Enol. Vitic. 37, 293–296.

Marschner, H., 1995. Mineral Nutrition of Higher Plants. Academic Press, London.

Marschner, H., Cakmak, I., 1989. High light intensity enhances chlorosis and necrosis in leaves of zinc, potassium, and magnesium deficient bean (*Phaseolus vulgaris* L.) plants. J. Plant Physiol. 134, 308–315.

Marshall, D.A., Spiers, J.M., Stringer, S.J., Curry, K.J., 2007. Laboratory method to estimate rain-induced splitting in cultivated blueberries. HortScience 42, 1551–1553.

Martelli, G.P., Graniti, A., Ercolani, G.L., 1986. Nature and physiological effects of grapevine diseases. Experientia 42, 933–942.

Martelli, G.P., Agranovsky, A.A., Bar-Joseph, M., Boscia, D., Candresse, T., Coutts, R.H.A., 2002. The family *Closteroviridae* revised. Arch. Virol. 147, 2039–2044.

Marten, I., Hoth, S., Deeken, R., Ache, P., Ketchum, K.A., Hoshi, T., 1999. AKT3, a phloem-localized K^+ channel, is blocked by protons. Proc. Natl. Acad. Sci. U. S. A. 96, 7581–7586.

Martens, H.J., Roberts, A.G., Oparka, K.J., Schulz, A., 2006. Quantification of plasmodesmatal endoplasmic reticulum coupling between sieve elements and companion cells using fluorescence redistribution after photobleaching. Plant Physiol. 142, 471–480.

Martin, C., Smith, A.M., 1995. Starch biosynthesis. Plant Cell 7, 971–985.

Martin, D.M., Toub, O., Chiang, A., Lo, B.C., Ohse, S., Lund, S.T., 2009. The bouquet of grapevine (*Vitis vinifera* L. cv. Cabernet Sauvignon) flowers arises from the biosynthesis of sesquiterpene volatiles in pollen grains. Proc. Natl. Acad. Sci. U. S. A. 106, 7245–7250.

Martinelli, L., Gribaudo, I., 2001. Somatic embryogenesis in grapevine. In: Roubelakis-Angelakis, K.A. (Ed.), Molecular Biology and Biotechnology of the Grapevine. Kluwer, Dordrecht, The Netherlands, pp. 327–351.

Martínez, L., Cavagnaro, P., Boursiquot, J.M., Agüero, C., 2008. Molecular characterization of Bonarda-type grapevine (*Vitis vinifera* L.) cultivars from Argentina, Italy, and France. Am. J. Enol. Vitic. 59, 287–291.

Martínez-Esteso, M.J., Sélles-Marchart, S., Lijavetzky, D., Pedreño, M.A., Bru-Martínez, R., 2011. A DIGE-based quantitative proteomic analysis of grape berry flesh development and ripening reveals key events in sugar and organic acid metabolism. J. Exp. Bot. 62, 2521–2569.

Martínez-Lüscher, J., Sánchez-Díaz, M., Delrot, S., Aguirreolea, J., Pascual, I., Gomès, E., 2014. Ultraviolet-B radiation and water deficit interact to alter flavonol and anthocyanin profiles in grapevine berries through transcriptomic regulation. Plant Cell Physiol. 55, 1925–1936.

Martinoia, E., Meyer, S., De Angeli, A., Nagy, R., 2012. Vacuolar transporters in their physiological context. Annu. Rev. Plant Biol. 63, 183–213.

Martins, V., Bassil, E., Hanana, M., Blumwald, E., Gerós, H., 2014a. Copper homeostasis in grapevine: functional characterization of the *Vitis vinifera* copper transporter. Planta 240, 91–101.

Martins, V., Teixeira, A., Bassil, E., Hanana, M., Blumwald, E., Gerós, H., 2014b. Copper-based fungicide Bordeaux mixture regulates the expression of *Vitis vinifera* copper transporters. Aust. J. Grape Wine Res. 20, 451–458.

Marty, F., 1999. Plant vacuoles. Plant Cell 11, 587–599.

Masclaux, C., Valadier, M.H., Brugière, N., Morot-Gaudry, J.F., Hirel, B., 2000. Characterization of the sink/source transition in tobacco (*Nicotiana tabacum* L.) shoots in relation to nitrogen management and leaf senescence. Planta 211, 510–518.

Masclaux-Daubresse, C., Daniel-Vedele, F., Dechorgnat, J., Chardon, F., Gaufichon, L., Suzuki, A., 2010. Nitrogen uptake, assimilation and remobilization in plants: challenges for sustainable and productive agriculture. Ann. Bot. 105, 1141–1157.

Mason, J.R., Clark, L., Shah, P.S., 1991. Ortho-aminoacetophenone repellency to birds: similarities to methyl anthranilate. J. Wildl. Manag. 55, 334–340.

Mason, M.G., Ross, J.J., Babst, B.A., Wienclaw, B.N., Beveridge, C.A., 2014. Sugar demand, not auxin, is the initial regulator of apical dominance. Proc. Natl. Acad. Sci. U. S. A. 111, 6092–6097.

Massonnet, M., Fasoli, M., Tornielli, G.B., Altieri, M., Sandri, M., Zuccolotto, P., Paci, P., Gardiman, M., Zenoni, S., Pezzotti, M., 2017. Ripening transcriptomic program in red and white grapevine varieties correlates with berry skin anthocyanin accumulation. Plant Physiol. 174, 2376–2396.

Mathiason, K., He, D., Grimplet, J., Vekateswari, J., Galbraith, D.W., Or, E., 2009. Transcript profiling in *Vitis riparia* during chilling requirement fulfillment reveals coordination of gene expression patterns with optimized bud break. Funct. Integr. Genom. 9, 81–96.

Mathieu, S., Terrier, N., Procureur, J., Bigey, F., Günata, Z., 2005. A carotenoid cleavage dioxygenase from *Vitis vinifera* L.: functional characterization and expression during grape berry development in relation to C_{13}-norisoprenoid accumulation. J. Exp. Bot. 56, 2721–2731.

Matsui, S., Ryugo, K., Kliewer, W.M., 1986. Growth inhibition of Thompson Seedless and Napa Gamay berries by heat stress and its partial reversibility by applications of growth regulators. Am. J. Enol. Vitic. 37, 67–71.

Matsumoto-Kitano, M., Kusumoto, T., Tarkowski, P., Kinoshita-Tsujimura, K., Václavíková, K., Miyawaki, K., 2008. Cytokinins are central regulators of cambial activity. Proc. Natl. Acad. Sci. U. S. A. 105, 20027–20031.

Matthews, M.A., 2015. Terroir and Other Myths of Winegrowing. University of California Press, Oakland, CA.

Matthews, M.A., Anderson, M.M., 1988. Fruit ripening in *Vitis vinifera* L.: responses to seasonal water deficits. Am. J. Enol. Vitic. 39, 313–320.

Matthews, M.A., Anderson, M.M., 1989. Reproductive development in grape (*Vitis vinifera* L.): responses to seasonal water deficits. Am. J. Enol. Vitic. 40, 52–60.

Matthews, M.A., Cheng, G., Weinbaum, S.A., 1987. Changes in water potential and dermal extensibility during grape berry development. J. Am. Soc. Hortic. Sci. 112, 314–319.

Matthews, M.A., Shackel, K.A., 2005. Growth and water transport in fleshy fruit. In: Holbrook, N.M., Zwieniecki, M.A. (Eds.), Vascular Transport in Plants. Elsevier, Boston, MA, pp. 181–195.

Matthews, J.S.A., Vialet-Chabrand, S.R.M., Lawson, T., 2017. Diurnal variation in gas exchange: the balance between carbon fixation and water loss. Plant Physiol. 174, 614–623.

Mattivi, F., Guzzon, R., Vrhovsek, U., Stefanini, M., Velasco, R., 2006. Metabolite profiling of grape: flavonols and anthocyanins. J. Agric. Food Chem. 54, 7692–7702.

Mattivi, F., Vrhovsek, U., Masuero, D., Trainotti, D., 2009. Differences in the amount and structure of extractable skin and seed tannins amongst red grape varieties. Aust. J. Grape Wine Res. 15, 27–35.

Matus, J.T., Cavallini, E., Loyola, R., Höll, J., Finezzo, L., Dal Santo, S., et al., 2017. A group of grapevine MYBA transcription factors located in chromosome 14 control anthocyanin synthesis in vegetative organs with different specificities compared with the berry color locus. Plant J. 91, 220–236.

Maurel, C., Verdoucq, L., Luu, D.T., Santoni, V., 2008. Plant aquaporins: membrane channels with multiple integrated functions. Annu. Rev. Plant Biol. 59, 595–624.

Maurel, C., Verdoucq, L., Rodrigues, O., 2016. Aquaporins and plant transpiration. Plant Cell Environ. 39, 2580–2587.

May, P., 1960. Effect of direction of growth on Sultana canes. Nature 185, 394–395.

May, P., 1994. Using Grapevine Rootstocks. The Australian Perspective. Winetitles, Adelaide, Australia.

May, P., 2000. From bud to berry, with special reference to inflorescence and bunch morphology in *Vitis vinifera* L. Aust. J. Grape Wine Res. 6, 82–98.

May, P., 2004. Flowering and Fruitset in Grapevines. Lythrum Press, Adelaide, Australia.

May, P., Antcliff, A.J., 1963. The effect of shading on fruitfulness and yield in the Sultana. J. Hortic. Sci. 38, 85–94.

May, B., Wüst, M., 2012. Temporal development of sesquiterpene hydrocarbon profiles of different grape varieties during ripening. Flav. Fragr. J. 27, 280–285.

May, P., Shaulis, N.J., Antcliff, A.J., 1969. The effect of controlled defoliation in the Sultana vine. Am. J. Enol. Vitic. 20, 237–250.

May, P., Clingeleffer, P.R., Brien, C.J., 1976. Sultana (*Vitis vinifera* L.) canes and their exposure to light. Vitis 14, 278–288.

Mayer, H., Holst, T., Schindler, D., 2002. Mikroklima in Buchenbeständen—Teil I: Photosynthetisch aktive Strahlung. Forstwiss. Centralbl. 121, 301–321.

Mayr, E., 2001. What Evolution Is. Basic Books, New York, NY.

Mazza, G., Miniati, E., 1993. Anthocyanins in Fruits, Vegetables, and Grains. CRC Press, Boca Raton, FL.

Mc Intyre, G.N., Kliewer, W.M., Lider, L.A., 1987. Some limitations of the degree day system as used in viticulture in California. Am. J. Enol. Vitic. 38, 128–132.

McAdam, S.A.M., Sussmilch, F.C., Brodribb, T.J., 2016. Stomatal responses to vapour pressure deficit are regulated by high speed gene expression in angiosperms. Plant Cell Environ. 39, 485–491.

McAinsh, M.R., Pittman, J.K., 2009. Shaping the calcium signature. New Phytol. 181, 275–294.

McCann, M.C., Wells, B., Roberts, K., 1990. Direct visualization of cross-links in the primary plant cell wall. J. Cell Sci. 96, 323–334.

McCarthy, M.G., 1997. The effect of transient water deficit on berry development of cv. Shiraz (*Vitis vinifera* L.). Aust. J. Grape Wine Res. 3, 102–108.

McCarthy, M.G., Cirami, R.M., 1990. The effect of rootstocks on the performance of Chardonnay from a nematode-infested Barossa Valley vineyard. Am. J. Enol. Vitic. 41, 126–130.

McClung, C.R., 2001. Circadian rhythms in plants. Annu. Rev. Plant Physiol. Plant Mol. Biol. 52, 139–162.

McClung, C.R., 2008. Comes a time. Curr. Opin. Plant Biol. 11, 514–520.

McDowell, N.G., 2011. Mechanisms linking drought, hydraulics, carbon metabolism, and vegetation mortality. Plant Physiol. 155, 1051–1059.

McElrone, A.J., Jackson, S., Habdas, P., 2008. Hydraulic disruption and passive migration by a bacterial pathogen in oak tree xylem. J. Exp. Bot. 59, 2649–2657.

McGarvey, D.J., Croteau, R., 1995. Terpenoid metabolism. Plant Cell 7, 1015–1026.

McGovern, P.E., 2003. Ancient Wine. The Search for the Origins of Viniculture. Princeton University Press, Princeton, NJ.

McGovern, P.E., Glusker, D.L., Exner, L.J., Voigt, M.M., 1996. Neolithic resinated wine. Nature 381, 480–481.

McGovern, P.E., Zhang, J., Tang, J., Zhang, Z., Hall, G.R., Moreau, R.A., 2004. Fermented beverages of pre- and proto-historic China. Proc. Natl. Acad. Sci. U. S. A. 101, 17593–17598.

McGovern, P., Jalabadze, M., Batiuk, S., Callahan, M.P., Smith, K.E., Hall, G.R., et al., 2017. Early Neolithic wine of Georgia in the South Caucasus. Proc. Natl. Acad. Sci. U. S. A. 114, E10309–E10318.

McKenry, M.V., 1984. Grape root phenology relative to control of parasitic nematodes. Am. J. Enol. Vitic. 35, 206–211.

McKenzie, R., Connor, B., Bodeker, G., 1999. Increased summertime UV radiation in New Zealand in response to ozone loss. Science 285, 1709–1711.

McKey, D., Elias, M., Pujol, B., Duputié, A., 2010. The evolutionary ecology of clonally propagated domesticated plants. New Phytol. 186, 318–332.

McLeod, A.R., Fry, S.C., Loake, G.J., Messenger, D.J., Reay, D.S., Smith, K.A., 2008. Ultraviolet radiation drives methane emissions from terrestrial plant pectins. New Phytol. 180, 124–132.

McQueen-Mason, S., 2005. Cell walls: the boundaries of plant development. New Phytol. 166, 717–722.

McSteen, P., Leyser, O., 2005. Shoot branching. Annu. Rev. Plant Biol. 56, 353–374.

Mehari, Z.H., Pilati, S., Sonego, P., Malacarne, G., Vrhovsek, U., Engelen, K., Tudzynski, P., Zottini, M., Baraldi, E., Moser, C., 2017. Molecular analysis of the early interaction between the grapevine flower and *Botrytis cinerea* reveals that prompt activation of specific host pathways leads to fungus quiescence. Plant Cell Environ. 40, 1409–1428.

Meiering, A.G., Paroschy, J.H., Peterson, R.L., Hostetter, G., Neff, A., 1980. Mechanical freezing injury in grapevine trunks. Am. J. Enol. Vitic. 31, 81–89.

Meitha, K., Konnerup, D., Colmer, T.D., Considine, J.A., Foyer, C.H., Considine, M.J., 2015. Spatio-temporal relief from hypoxia and production of reactive oxygen species during bud burst in grapevine (*Vitis vinifera*). Ann. Bot. 116, 703–711.

Melino, V.J., Soole, K.L., Ford, C.M., 2009. Ascorbate metabolism and the developmental demand for tartaric and oxalic acids in ripening grape berries. BMC Plant Biol. 9, 145.

Mendgen, K., Hahn, M., 2002. Plant infection and the establishment of fungal biotrophy. Trends Plant Sci. 7, 1–5.

Meneghetti, S., Gardiman, M., Calò, A., 2006. Flower biology of grapevine: a review. Adv. Hortic. Sci. 20, 317–325.

Menge, J.A., Raski, D.J., Lider, L.A., Johnson, E.L.V., Jones, N.O., Kissler, J.J., 1983. Interactions between mycorrhizal fungi, soil fumigation, and growth of grapes in California. Am. J. Enol. Vitic. 34, 117–121.

Mengel, K., 1994. Iron availability in plant tissues: iron chlorosis on calcareous soils. Plant Soil 165, 275–283.

Mengel, K., Bübl, W., 1983. Verteilung von Eisen in Blättern von Weinreben mit HCO_3^--induzierter Fe-Chlorose. Z. Pflanzenern. Bodenk. 146, 560–571.

Mengel, K., Malissiovas, N., 1982. Light dependent proton excretion by roots of entire vine plants (*Vitis vinifera* L.). Z. Pflanzenernähr. Bodenkd. 145, 261–267.

Mengel, K., Breininge, M.T., Bübl, W., 1984. Bicarbonate, the most important factor inducing iron chlorosis in vine grapes on calcareous soils. Plant Soil 81, 333–344.

Mengin, V., Pyl, E.T., Moraes, T.A., Sulpice, R., Krohn, N., Encke, B., Stitt, M., 2017. Photosynthate partitioning to starch in *Arabidopsis thaliana* is insensitive to light intensity but sensitive to photoperiod due to a restriction on growth in the light in short photoperiods. Plant Cell Environ. 40, 2608–2627.

Mercurio, M.D., Dambergs, R.G., Cozzolino, D., Herderich, M.J., Smith, P.A., 2010. Relationship between red wine grades and phenolics. 1. Tannin and total phenolics concentrations. J. Agric. Food Chem. 58, 12313–12319.

Meredith, C.P., Bowers, J.E., Riaz, S., Handley, V., Bandman, E.B., Dangl, G.S., 1999. The identity and parentage of the variety known in California as Petite Sirah. Am. J. Enol. Vitic. 50, 236–242.

Merry, A.M., Evans, K.J., Corkrey, R., Wilson, S.J., 2013. Coincidence of maximum severity of powdery mildew on grape leaves and the carbohydrate sink-to-source transition. Plant Pathol. 62, 842–850.

Merzlyak, M.N., Solovchenko, A.E., Chivkunova, O.B., 2002. Patterns of pigment changes in apple fruits during adaptation to high sunlight and sunscald development. Plant Physiol. Biochem. 40, 679–684.

Messenger, D.J., McLeod, A.R., Fry, S.C., 2009. The role of ultraviolet radiation, photosensitizers, reactive oxygen species and ester groups in mechanisms of methane formation from pectin. Plant Cell Environ. 32, 1–9.

Metzner, R., Schneider, H.U., Breuer, U., Thorpe, M.R., Schurr, U., Schroeder, W.H., 2010a. Tracing cationic nutrients from xylem into stem tissue of French bean by stable isotope tracers and cryo-secondary ion mass spectrometry. Plant Physiol. 152, 1030–1043.

Metzner, R., Thorpe, M.R., Breuer, U., Blümler, P., Schurr, U., Schneider, H.U., 2010b. Contrasting dynamics of water and mineral nutrients in stems shown by stable isotope tracers and cryo-SIMS. Plant Cell Environ. 33, 1393–1407.

Meyer, R.C., Steinfath, M., Lisec, J., Becher, M., Witucka-Wall, H., Törjék, O., 2007. The metabolic signature related to high plant growth rate in *Arabidopsis thaliana*. Proc. Natl. Acad. Sci. U. S. A. 104, 4759–4764.

Meyer, S., De Angeli, A., Fernie, A.R., Martinoia, E., 2010. Intra- and extra-cellular excretion of carboxylates. Trends Plant Sci. 15, 40–47.

Meynhardt, J.T., 1964. Some studies on berry splitting of Queen of the Vineyard grape. S. Afr. J. Agric. Sci. 7, 179–186.

Meynhardt, J.T., Malan, A.H., 1963. Translocation of sugars in double-stem grape vines. S. Afr. J. Agric. Sci. 6, 337–338.

Miele, A., Bouard, J., Bertrand, A., 1993. Fatty acids from lipid fractions of leaves and different tissues of Cabernet Sauvignon grapes. Am. J. Enol. Vitic. 44, 180–186.

Miguel, C., Mesias, J.L., Maynar, J.I., 1985. Évolution des acides aminés pendant la maturation des raisins des variétés *Cayetana* et *Macabeo* (*Vitis vinifera*). Sci. Aliment. 5, 599–605.

Millar, A.A., 1972. Thermal regime of grapevines. Am. J. Enol. Vitic. 23, 173–176.

Millar, A.H., Whelan, J., Soole, K.L., Day, D.A., 2011. Organization and regulation of mitochondrial respiration in plants. Annu. Rev. Plant Biol. 62, 79–104.

Miller, D.P., Howell, G.S., 1998. Influence of vine capacity and crop load on canopy development, morphology, and dry matter partitioning in Concord grapevines. Am. J. Enol. Vitic. 49, 183–190.

Miller, D.P., Howell, G.S., Striegler, R.K., 1988. Cane and bud hardiness of own-rooted White Riesling and scions of White Riesling and Chardonnay grafted to selected rootstocks. Am. J. Enol. Vitic. 39, 60–66.

Miller, A.J., Fan, X., Shen, Q., Smith, S.J., 2008. Amino acids and nitrate as signals for the regulation of nitrogen acquisition. J. Exp. Bot. 59, 111–119.

Miller, A.J., Matasci, N., Schwaninger, H., Aradhya, M.K., Prins, B., Zhong, G., Simon, C., Buckler, E.S., Myles, S., 2013. *Vitis* phylogenomics: hybridization intensities from a SNP array outperform genotype calls. PLoS ONE 8, e78680.

Milli, A., Cecconi, D., Bortesi, L., Persi, A., Rinalducci, S., Zamboni, A., 2012. Proteomic analysis of the compatible interaction between *Vitis vinifera* and *Plasmopara viticola*. J. Proteome 75, 1284–1302.

Mills, L.J., Ferguson, J.C., Keller, M., 2006. Cold hardiness evaluation of grapevine buds and cane tissues. Am. J. Enol. Vitic. 57, 194–200.

Minchin, P.E.H., Lacointe, A., 2005. New understanding on phloem physiology and possible consequences for modelling long-distance carbon transport. New Phytol. 166, 771–779.

Mirabet, V., Das, P., Boudaoud, A., Hamant, O., 2011. The role of mechanical forces in plant morphogenesis. Annu. Rev. Plant Biol. 62, 365–385.

Mirás-Avalos, J.M., Buesa, I., Llacer, E., Jiménez-Bello, M.A., Risco, D., Castel, J.R., Intrigliolo, D.S., 2017. Water versus source–sink relationships in a semiarid Tempranillo vineyard: vine performance and fruit composition. Am. J. Enol. Vitic. 68, 11–22.

Mitani, N., Azuma, A., Fukai, E., Hirochika, H., Kobayashi, S., 2009. A retrotransposon-inserted *VvmybA1a* allele has been spread among cultivars of *Vitis vinifera* but not North American of East Asian *Vitis* species. Vitis 48, 55–56.

Mittler, R., 2002. Oxidative stress, antioxidants and stress tolerance. Trends Plant Sci. 7, 405–410.

Miyashita, Y., Good, A.G., 2008. NAD(H)-dependent glutamate dehydrogenase is essential for the survival of *Arabidopsis thaliana* during dark-induced carbon starvation. J. Exp. Bot. 59, 667–680.

Mlikota Gabler, F., Smilanick, J.L., Mansour, M., Ramming, D.W., Mackey, B.E., 2003. Correlations of morphological, anatomical, and chemical features of grape berries with resistance to *Botrytis cinerea*. Phytopathology 93, 1263–1273.

Mohr, H.D., 1996. Periodicity of root tip growth of vines in the Moselle valley. Wein-Wiss. 51, 83–90.

Mok, D.W.S., Mok, M.C., 2001. Cytokinin metabolism and action. Annu. Rev. Plant Physiol. Plant Mol. Biol. 52, 89–118.

Molitor, D., Keller, M., 2016. Yield of Müller-Thurgau and Riesling grapevines is altered by meteorological conditions in the current and previous growing seasons. OENO One 50, 245–258.

Molitor, D., Behr, M., Hoffmann, L., Evers, D., 2012. Impact of grape cluster division on cluster morphology and bunch rot epidemic. Am. J. Enol. Vitic. 63, 508–514.

Møller, I.M., Jensen, P.E., Hansson, A., 2007. Oxidative modifications to cellular components in plants. Annu. Rev. Plant Biol. 58, 459–481.

Monagas, M., Bartolomé, B., Gómez-Cordovés, C., 2005. Updated knowledge about the presence of phenolic compounds in wine. Crit. Rev. Food Sci. Nutr. 45, 85–118.

Monagas, M., Garrido, I., Bartolomé, B., Gómez-Cordovés, C., 2006. Chemical characterization of commercial dietary ingredients from *Vitis vinifera* L. Anal. Chim. Acta 563, 401–410.

Moncada, X., Pelsy, F., Merdinoglu, D., Hinrichsen, P., 2006. Genetic diversity and geographical dispersal in grapevine clones revealed by microsatellite markers. Genome 49, 1459–1472.

Mongélard, G., Seemann, M., Boisson, A.M., Rohmer, M., Bligny, R., Rivasseau, C., 2011. Measurement of carbon flux through the MEP pathway for isoprenoid synthesis by ^{31}P-NMR spectroscopy after specific inhibition of 2-*C*-methyl-d-erythritol 2,4-cyclodiphosphate reductase. Effect of light and temperature. Plant Cell Environ. 34, 1241–1247.

Montagu, K.D., Conroy, J.P., Atwell, B.J., 2001. The position of localized soil compaction determines root and subsequent shoot growth responses. J. Exp. Bot. 52, 2127–2133.

Monteiro, S., Barakat, M., Piçarra-Pereira, M.A., Teixeira, A.R., Ferreira, R.B., 2003. Osmotin and thaumatin from grape: a putative general defense mechanism against pathogenic fungi. Phytopathology 93, 1505–1512.

Monteiro, S., Piçarra-Pereira, M.A., Loureiro, V.B., Teixeira, A.R., Ferreira, R.B., 2007. The diversity of pathogenesis-related proteins decreases during grape maturation. Phytochemistry 68, 416–425.

Moran, M.A., Sadras, V.O., Petrie, P.R., 2017. Late pruning and carry-over effects on phenology, yield components and berry traits in Shiraz. Aust. J. Grape Wine Res. 23, 390–398.

Morano, L.D., Walker, M.A., 1995. Soils and plant communities associated with three *Vitis* species. Am. Midl. Nat. 134, 254–263.

Moreau, L., Vinet, E., 1923. Sur la composition des pleurs de vignes. C.R. Acad. Agric. France 9, 554–557.

Morelli, G., Ruberti, I., 2002. Light and shade in the photocontrol of *Arabidopsis* growth. Trends Plant Sci. 7, 399–404.

Morgan, J.M., 1984. Osmoregulation and water stress in higher plants. Annu. Rev. Plant Physiol. 35, 299–319.

Morgan, D.C., Stanley, C.J., Warrington, I.J., 1985. The effects of simulated daylight and shadelight on vegetative and reproductive growth in kiwifruit and grapevine. J. Hortic. Sci. 60, 473–484.

Mori, K., Saito, H., Goto-Yamamoto, N., Kitayama, M., Kobayashi, S., Sugaya, S., 2005. Effects of abscisic acid treatment and night temperatures on anthocyanin composition in Pinot noir grapes. Vitis 44, 161–165.

Mori, K., Goto-Yamamoto, N., Kitayama, M., Hashizume, K., 2007. Loss of anthocyanins in red-wine grape under high temperature. J. Exp. Bot. 58, 1935–1945.

Morinaga, K., Imai, S., Yakushiji, H., Koshita, Y., 2003. Effects of fruit load on partitioning of ^{15}N and ^{13}C, respiration, and growth of grapevine roots at different fruit stages. Sci. Hortic. 97, 239–253.

Morita, M.T., Tasaka, M., 2004. Gravity sensing and signaling. Curr. Opin. Plant Biol. 7, 712–718.

Morlat, R., Jacquet, A., 1993. The soil effects on the grapevine root system in several vineyards of the Loire Valley, France. Vitis 32, 35–42.

Morlat, R., Jacquet, A., 2003. Grapevine root system and soil characteristics in a vineyard maintained long-term with or without interrow sward. Am. J. Enol. Vitic. 54, 1–7.

Morlat, R., Symoneaux, R., 2008. Long-term additions of organic amendments in a Loire Valley vineyard on a calcareous sandy soil. III. Effects on fruit composition and chemical and sensory characteristics of Cabernet franc wine. Am. J. Enol. Vitic. 59, 375–386.

Morrell, A.M., Wample, R.L., Mink, G.I., Ku, M.S.B., 1997. Heat shock protein expression in leaves of Cabernet Sauvignon. Am. J. Enol. Vitic. 48, 459–464.

Morris, D.A., 2000. Transmembrane auxin carrier systems—dynamic regulators of polar auxin transport. Plant Growth Regul. 32, 161–172.

Morris, J.R., Cawthon, D.L., Fleming, J.W., 1980. Effects of high rates of potassium fertilization on raw product quality and changes in pH and acidity during storage of Concord grape juice. Am. J. Enol. Vitic. 31, 323–328.

Morris, J.R., Sims, C.A., Cawthon, D.L., 1983. Effects of excessive potassium levels on pH, acidity and color of fresh and stored grape juice. Am. J. Enol. Vitic. 34, 35–39.

Morris, J.R., Sims, C.A., Striegler, R.K., Cackler, S.D., Donley, R.A., 1987. Effects of cultivar, maturity, cluster thinning, and excessive potassium fertilization on yield and quality of Arkansas wine grapes. Am. J. Enol. Vitic. 38, 260–264.

Morrison, J.C., 1991. Bud development in *Vitis vinifera* L. Bot. Gaz. 152, 304–315.

Morrison, J.C., Iodi, M., 1990. The influence of waterberry on the development and composition of Thompson Seedless grapes. Am. J. Enol. Vitic. 41, 301–305.

Morrison, J.C., Noble, A.C., 1990. The effects of leaf and cluster shading on the composition of Cabernet Sauvignon grapes and fruit and wine sensory properties. Am. J. Enol. Vitic. 41, 193–200.

Morrot, G., Brochet, F., Dubourdieu, D., 2001. The color of odors. Brain Lang. 79, 309–320.

Moschou, P.N., Paschalidis, K.A., Roubelakis-Angelakis, K.A., 2008. Plant polyamine catabolism. The state of the art. Plant Signal. Behav. 3, 1061–1066.

Moskowitz, A.H., Hrazdina, G., 1981. Vacuolar contents of fruit subepidermal cells from *Vitis* species. Plant Physiol. 68, 686–692.

Motomura, Y., 1990. Distribution of ^{14}C-assimilates from individual leaves on clusters in grape shoots. Am. J. Enol. Vitic. 41, 306–312.

Motomura, Y., 1993. ^{14}C-Assimilate partitioning in grapevine shoots: effects of shoot pinching, girdling of shoot, and leaf-halving on assimilates partitioning from leaves into clusters. Am. J. Enol. Vitic. 44, 1–7.

Mpelasoka, B.S., Schachtmann, D.P., Treeby, M.T., Thomas, M.R., 2003. A review of potassium nutrition in grapevines with special emphasis on berry accumulation. Aust. J. Grape Wine Res. 9, 154–168.

Muday, G.K., Rahman, A., Binder, B.M., 2012. Auxin and ethylene: collaborators or competitors? Trends Plant Sci. 17, 181–195.

Muganu, M., Bellincontro, A., Barnaba, F.E., Paolocci, M., Bignami, C., Gambellini, G., 2011. Influence of bunch position in the canopy on berry epicuticular wax during ripening and on weight loss during postharvest dehydration. Am. J. Enol. Vitic. 62, 91–98.

Müller, K., 1985. Abhängigkeit des pflanzenverfügbaren Stickstoffs (N_{min}) von den Humusgehalten und der mineralischen N-Düngung auf einem fränkischen Weinbaustandort. Z. Pflanzenernähr. Bodenkd. 148, 169–178.

Müller, D., Leyser, O., 2011. Auxin, cytokinin and the control of shoot branching. Ann. Bot. 107, 1203–1212.

Müller, T., Ulrich, M., Ongania, K.H., Kräutler, B., 2007. In reifen Früchten gefundene farblose tetrapyrrolische Chlorophyll-Kataboliten sind wirksame Antioxidantien. Angew. Chem. 119, 8854–8857.

Muller, B., Pantin, F., Génard, M., Turc, O., Freixes, S., Piques, M., 2011. Water deficits uncouple growth from photosynthesis, increase C content, and modify the relationships between C and growth in sink organs. J. Exp. Bot. 62, 1715–1729.

Müller-Stoll, W.R., 1950. Mutative Färbungsänderungen bei Weintrauben. Züchter 20, 288–291.

Müller-Thurgau, H., 1882. Über Zuckeranhäufung in Pflanzenteilen infolge niederer Temperatur. Landwirtsch. Jahrb. 11, 751–828.

Müller-Thurgau, H., 1883a. Über das Abfallen der Rebenblüten und die Entstehung kernloser Traubenbeeren. Der Weinbau. 9, 87–89.

Müller-Thurgau, H., 1883b. Über die Fruchtbarkeit der aus den älteren Theilen der Weinstöcke hervorgehenden Triebe, sowie der sog. Nebentriebe. Botanisches Centralblatt 4, 85–86.

Müller-Thurgau, H., 1886. Über das Gefrieren und Erfrieren der Pflanzen. Landwirtsch. Jahrb. 15, 453–610.

Mullins, M.G., Srinivasan, C., 1976. Somatic embryos and plantlets from an ancient clone of the grapevine (cv. Cabernet-Sauvignon) by apomixis *in vitro*. J. Exp. Bot. 27, 1022–1030.

Mullins, M.G., Bouquet, A., Williams, L.E., 1992. Biology of the Grapevine. Cambridge University Press, Cambridge, UK.

Münch, E., 1930. Die Stoffbewegungen in der Pflanze. Gustav Fischer, Jena, Germany.

Munné-Bosch, S., 2008. Do perennials really senesce? Trends Plant Sci. 13, 216–220.

Munns, R., 2002. Comparative physiology of salt and water stress. Plant Cell Environ. 25, 239–250.

Munns, R., Tester, M., 2008. Mechanisms of salinity tolerance. Annu. Rev. Plant Biol. 59, 651–681.

Muñoz, C., Gomez-Talquenca, S., Chialva, C., Ibáñez, J., Martinez-Zapater, J.M., Peña-Neira, Á., 2014. Relationships among gene expression and anthocyanin composition of Malbec grapevine clones. J. Agric. Food Chem. 62, 6716–6725.

Müntz, K., 2007. Protein dynamics and proteolysis in plant vacuoles. J. Exp. Bot. 58, 2391–2407.

Murphey, J.M., Spayd, S.E., Powers, J.R., 1989. Effect of grape maturation on soluble protein characteristics of Gewürztraminer and White Riesling juice and wine. Am. J. Enol. Vitic. 40, 199–207.

Mwange, K.N.K., Hou, H.W., Wang, Y.Q., He, X.Q., Cui, K.M., 2005. Opposite patterns in the annual distribution and time-course of endogenous abscisic acid and indole-3-acetic acid in relation to the periodicity of cambial activity in *Eucommia ulmoides* Oliv. J. Exp. Bot. 56, 1017–1028.

Myles, S., Chia, J.M., Hurwitz, B., Simon, C., Zhong, G.Y., Buckler, E., 2010. Rapid genomic characterization of the genus *Vitis*. PLoS ONE 5, e8218.

Myles, S., Boyko, A.R., Owens, C.L., Brown, P.J., Grassi, F., Aradhya, M.K., 2011. Genetic structure and domestication history of the grape. Proc. Natl. Acad. Sci. U. S. A. 108, 3530–3535.

Naidu, R., Rowhani, A., Fuchs, M., Golino, D., Martelli, G.P., 2014. Grapevine leafroll: a complex viral disease affecting a high-value fruit crop. Plant Dis. 98, 1172–1185.

Nail, W.R., Howell, G.S., 2005. Effects of timing of powdery mildew infection on carbon assimilation and subsequent seasonal growth of potted Chardonnay grapevines. Am. J. Enol. Vitic. 56, 220–227.

Nakagawa, S., Nanjo, Y., 1965. A morphological study of Delaware grape berries. J. Jpn. Soc. Hortic. Sci. 34, 85–95.

Nambara, E., Marion-Poll, A., 2005. Abscisic acid biosynthesis and catabolism. Annu. Rev. Plant Biol. 56, 165–185.

Naor, A., Gal, Y., Bravdo, B., 1997. Crop load affects assimilation rate, stomatal conductance, stem water potential and water relations of field-grown Sauvignon blanc grapevines. J. Exp. Bot. 48, 1675–1680.

Nardini, A., Tyree, M.T., Salleo, S., 2001. Xylem cavitation in the leaf of *Prunus laurocerasus* L. and its impact on leaf hydraulics. Plant Physiol. 125, 1700–1709.

Nardini, A., Salleo, S., Jansen, S., 2011. More than just a vulnerable pipeline: xylem physiology in the light of ion-mediated regulation of plant water transport. J. Exp. Bot. 62, 4701–4718.

Negi, S., Ivanchenko, M.G., Muday, G.K., 2008. Ethylene regulates lateral root formation and auxin transport in *Arabidopsis thaliana*. Plant J. 55, 175–187.

Negri, A.S., Prinsi, B., Rossoni, M., Failla, O., Scienza, A., Cocucci, M., 2008. Proteome changes in the skin of the grape cultivar Barbera among different stages of ripening. BMC Genom. 9, 388.

Negrul, A.M., 1936. Variabilität und Vererbung des Geschlechts bei der Rebe. Gartenbauwissenschaft 10, 215–231.

Nelson, N., Yocum, C.F., 2006. Structure and function of photosystems I and II. Annu. Rev. Plant Biol. 57, 521–565.

Nelson, C.C., Kennedy, J.A., Zhang, Y., Kurtural, S.K., 2016. Applied water and rootstock affect productivity and anthocyanin composition of Zinfandel in central California. Am. J. Enol. Vitic. 67, 18–28.

Nester, E.W., 2015. *Agrobacterium*: nature's genetic engineer. Front. Plant Sci. 5, 730.

Newsham, K.K., Low, M.N.R., McLeod, A.R., Greenslade, P.D., Emmett, B.A., 1997. Ultraviolet-B radiation influences the abundance and distribution of phylloplane fungi on pedunculate oak (*Quercus robur*). New Phytol. 136, 287–297.

Nibau, C., Gibbs, D.J., Coates, J.C., 2008. Branching out in new directions: the control of root architecture by lateral root formation. New Phytol. 179, 595–614.

Nicholas, P., Magarey, P., Wachtel, M., 1998. Diseases and Pests: Grape Production Series Number 1, third ed. Winetitles, Adelaide, Australia.

Nieminen, K., Immanen, J., Laxell, M., Kauppinen, L., Tarkowski, P., Dolezal, K., 2008. Cytokinin signaling regulates cambial development in poplar. Proc. Natl. Acad. Sci. U. S. A. 105, 20032–20037.

Nieves-Cordones, M., Andrianteranagna, M., Cuéllar, T., Chérel, I., Gibrat, R., Boeglin, M., et al., 2019. Characterization of the grapevine Shaker K^+ channel VvK3.1 supports its function in massive potassium fluxes necessary for berry potassium loading and pulvinus-actuated leaf movements. New Phytol. 222, 286–300.

Niimi, Y., Torikata, H., 1979. Changes in photosynthesis and respiration during berry development in relation to the ripening of Delaware grapes. J. Jpn. Soc. Hortic. Sci. 47, 448–453.

Niimi, J., Boss, P.K., Jeffery, D.W., Bastian, S.E.P., 2018. Linking the sensory properties of Chardonnay grape *Vitis vinifera* cv. berries to wine characteristics. Am. J. Enol. Vitic. 113–124.

Niklas, K.J., 2006. Plant allometry, leaf nitrogen and phosphorus stoichiometry, and interspecific trends in annual growth rates. Ann. Bot. 97, 155–163.

Niklas, K.J., Cobb, E.D., 2006. Biomass partitioning and leaf N,P—stoichiometry: comparisons between tree and herbaceous current-year shoots. Plant Cell Environ. 29, 2030–2042.

Nikolic, M., Römheld, V., Merkt, N., 2000. Effect of bicarbonate on uptake and translocation of ^{59}Fe in two grapevine rootstocks differing in their resistance to Fe deficiency chlorosis. Vitis 39, 145–149.

Nitsch, J.P., Pratt, C., Nitsch, C., Shaulis, N.J., 1960. Natural growth substances in Concord and Concord Seedless grapes in relation to berry development. Am. J. Bot. 47, 566–576.

Niyogi, K.K., 2000. Safety valves for photosynthesis. Curr. Opin. Plant Biol. 3, 455–460.

Nobel, P.S., 2009. Physicochemical and Environmental Plant Physiology, fourth ed. Academic Press, San Diego, CA.

Noble, A.C., 1979. Evaluation of Chardonnay wines obtained from sites with different soil compositions. Am. J. Enol. Vitic. 30, 214–217.

Noble, A.C., Bursick, G.F., 1984. The contribution of glycerol to perceived viscosity and sweetness in white wine. Am. J. Enol. Vitic. 35, 110–112.

Noble, A.C., Ebeler, S.E., 2002. Use of multivariate statistics in understanding wine flavor. Food Rev. Int. 18, 1–21.

Noble, A.C., Strauss, C.R., Williams, P.J., Wilson, B., 1988. Contribution of terpene glycosides to bitterness in Muscat wines. Am. J. Enol. Vitic. 39, 129–131.

Noctor, G., Foyer, C.H., 1998. Ascorbate and glutathione: keeping active oxygen under control. Annu. Rev. Plant Physiol. Plant Mol. Biol. 49, 249–279.

Nogales, A., Santos, E.S., Abreu, M.M., Arán, D., Victorino, G., Pereira, H.S., et al., 2019. Mycorrhizal inoculation differentially affects grapevine's performance in copper contaminated and non-contaminated soils. Front. Plant Sci. 9, 1906.

Nonami, H., Boyer, J.S., 1987. Origin of growth-induced water potential. Solute concentration is low in apoplast of enlarging tissues. Plant Physiol. 83, 596–601.

Nonami, H., Wu, Y., Boyer, J.S., 1997. Decreased growth-induced water potential. A primary cause of growth inhibition at low water potentials. Plant Physiol. 114, 501–509.

Nord, E.A., Lynch, J.P., 2009. Plant phenology: a critical controller of soil resource acquisition. J. Exp. Bot. 60, 1927–1937.

Nordström, A., Tarkowski, P., Tarkowska, D., Norbaek, R., Åstot, C., Dolezal, K., 2004. Auxin regulation of cytokinin biosynthesis in *Arabidopsis thaliana*: a factor of potential importance for auxin-cytokinin-regulated development. Proc. Natl. Acad. Sci. U. S. A. 101, 8039–8044.

Noyce, P.W., Harper, J.D.I., Steel, C.C., Wood, R.M., 2016. A new description and the rate of development of inflorescence primordia over a full season in *Vitis vinifera* L. cv. Chardonnay. Am. J. Enol. Vitic. 67, 86–93.

Ntahimpera, N., Ellis, M.A., Wilson, L.L., Madden, L.V., 1998. Effects of a cover crop on splash dispersal of *Colletotrichum acutatum* conidia. Phytopathology 88, 536–543.

Nunan, K.J., Sims, I.M., Bacic, A., Robinson, S.P., Fincher, G.B., 1997. Isolation and characterization of cell walls from the mesocarp of mature grape berries. Planta 203, 93–100.

Nunan, K.J., Sims, I.M., Bacic, A., Robinson, S.P., Fincher, G.B., 1998. Changes in cell wall composition during ripening of grape berries. Plant Physiol. 118, 783–792.

Nunan, K.J., Davies, C., Robinson, S.P., Fincher, G.B., 2001. Expression patterns of cell wall-modifying enzymes during grape berry development. Planta 214, 257–264.

Nuzzo, V., Matthews, M.A., 2006. Response of fruit growth and ripening to crop level in dry-farmed Cabernet Sauvignon on four rootstocks. Am. J. Enol. Vitic. 57, 314–324.

O'Brien, E.E., Gersani, M., Brown, J.S., 2005. Root proliferation and seed yield in response to spatial heterogeneity of below-ground competition. New Phytol. 168, 401–412.

O'Neill, S.D.O., 1997. Pollination regulation of flower development. Annu. Rev. Plant Physiol. Plant Mol. Biol. 48, 547–574.

O'Neill, M.A., Ishii, T., Albersheim, P., Darvill, A.G., 2004. Rhamnogalacturonan II: structure and function of a borate cross-linked cell wall pectic polysaccharide. Annu. Rev. Plant Biol. 55, 109–139.

Oaks, A., Hirel, B., 1985. Nitrogen metabolism in roots. Annu. Rev. Plant Physiol. 36, 345–365.

OEPP/EPPO, 2002. Good plant protection practice. Grapevine. EPPO Bull. 32, 371–392.

Ogren, W.L., 1984. Photorespiration: pathways, regulation, and modification. Annu. Rev. Plant Physiol. 35, 415–442.

Ojeda, H., Deloire, A., Carbonneau, A., Ageorges, A., Romieu, C., 1999. Berry development of grapevines: relations between the growth of berries and their DNA content indicate cell multiplication and enlargement. Vitis 38, 145–150.

Ojeda, H., Deloire, A., Carbonneau, A., 2001. Influence of water deficits on grape berry growth. Vitis 40, 141–145.

Ojeda, H., Andary, C., Kraeva, E., Carbonneau, A., Deloire, A., 2002. Influence of pre- and postveraison water deficit on synthesis and concentration of skin phenolic compounds during berry growth of *Vitis vinifera* cv. Shiraz. Am. J. Enol. Vitic. 53, 261–267.

Okuda, T., Yokotsuka, K., 1999. Levels of glutathione and activities of related enzymes during ripening of Koshu and Cabernet Sauvignon grapes and during winemaking. Am. J. Enol. Vitic. 50, 264–270.

Olarte Mantilla, S.M., Collins, C., Iland, P.G., Kidman, C.M., Ristic, R., Boss, P.K., Jordans, C., Bastian, S.E.P., 2018. Shiraz (*Vitis vinifera* L.) berry and wine sensory profiles and composition are modulated by rootstocks. Am. J. Enol. Vitic. 69, 32–44.

Olien, W.C., 1990. The muscadine grape: botany, viticulture, history, and current industry. HortScience 25, 732–739.

Oliveira, C., Ferreira, A.C., Costa, P., Guerra, J., Guedes de Pinho, P., 2004. Effect of some viticultural parameters on the grape carotenoid profile. J. Agric. Food Chem. 52, 4178–4184.

Oliver, J.E., Fuchs, M., 2011. Tolerance and resistance to viruses and their vectors in *Vitis* sp.: a virologist's perspective of the literature. Am. J. Enol. Vitic. 62, 438–451.

Oliver, S.N., Van Dongen, J.T., Alfred, A.C., Mamun, E.A., Zhao, X., Saini, H.S., 2005. Cold-induced repression of the rice anther-specific cell wall invertase gene *OSINV4* is correlated with sucrose accumulation and pollen sterility. Plant Cell Environ. 28, 1534–1551.

Ollat, N., Gaudillere, J.P., 1998. The effect of limiting leaf area during stage I of berry growth on development and composition of berries of *Vitis vinifera* L. cv. Cabernet Sauvignon. Am. J. Enol. Vitic. 49, 251–258.

Ollat, N., Diakou-Verdin, P., Carde, J.P., Barrieu, F., Gaudillère, J.P., Moing, A., 2002. Grape berry development: a review. J. Int. Sci. Vigne Vin 36, 109–131.

Ollat, N., Laborde, B., Neveux, M., Diakou-Verdin, P., Renaud, C., Moing, A., 2003. Organic acid metabolism in roots of various grapevine (*Vitis*) rootstocks submitted to iron deficiency and bicarbonate nutrition. J. Plant Nutr. 26, 2165–2176.

Olmo, H.P., 1936. Pollination and the setting of fruit in the black Corinth grape. Proc. Am. Soc. Hortic. Sci. 34, 402–404.

Olmo, H.P., 1946. Correlations between seed and berry development in some seeded varieties of *Vitis vinifera*. Proc. Am. Soc. Hortic. Sci. 48, 291–297.

Olmo, H.P., 1952. Wine grape varieties of the future. Am. J. Enol. Vitic. 3, 45–51.

Olmo, H.P., 1986. The potential role of (*vinifera* × *rotundifolia*) hybrids in grape variety improvement. Experientia 42, 921–926.

Olmstead, M.A., Tarara, J.M., 2001. Physical principles of row covers and grow tubes with application to small fruit crops. Small Fruits Rev. 1, 29–46.

Olsen, K.M., Slimestad, R., Lea, U.S., Brede, C., Løvdal, T., Ruoff, P., 2009. Temperature and nitrogen effects on regulators and products of the flavonoid pathway: experimental and kinetic model studies. Plant Cell Environ. 32, 286–299.

Olszewski, N., Sun, T., Gubler, F., 2002. Gibberellin signaling: biosynthesis, catabolism, and response pathways. Plant Cell 14, S61–S80.

Ong, B.Y., Nagel, C.W., 1978. High pressure liquid chromatographic analysis of hydroxycinnamic acid-tartaric acid esters and their glucose esters in *Vitis vinifera*. J. Chromatogr. 157, 345–355.

Ongaro, V., Leyser, O., 2008. Hormonal control of shoot branching. J. Exp. Bot. 59, 67–74.

Oparka, K.J., Prior, D.A.M., 1992. Direct evidence for pressure-generated closure of plasmodesmata. Plant J. 2, 741–750.

Oparka, K.J., Santa Cruz, S., 2000. The great escape: phloem transport and unloading of macromolecules. Annu. Rev. Plant Physiol. Plant Mol. Biol. 51, 323–347.

Orrù, M., Mattana, E., Pritchard, H.W., Bacchetta, G., 2012. Thermal thresholds as predictors of seed dormancy release and germination timing: altitude-related risks from climate warming for the wild grapevine *Vitis vinifera* subsp. *sylvestris*. Ann. Bot. 110, 1651–1660.

Ortega-Regules, A., Romero-Cascales, I., López-Roca, J.M., Ros-García, J.M., Gómez-Plaza, E., 2006a. Anthocyanin fingerprint of grapes: environmental and genetic variations. J. Sci. Food Agric. 86, 1460–1467.

Ortega-Regules, A., Romero-Cascales, I., Ros-García, J.M., López-Roca, J.M., Gómez-Plaza, E., 2006b. A first approach towards the relationship between grape skin cell-wall composition and anthocyanin extractability. Anal. Chim. Acta 563, 26–32.

Ortega-Regules, A., Ros-García, J.M., Bautista-Ortín, A.B., López-Roca, J.M., Gómez-Plaza, E., 2008. Changes in skin cell wall composition during the maturation of four premium wine grape varieties. J. Sci. Food Agric. 88, 420–428.

Ortoidze, T., Düring, H., 2001. Light utilisation and thermal dissipation in light- and shade-adapted leaves of *Vitis* genotypes. Vitis 40, 131–136.

Osmont, K.S., Sibout, R., Hardtke, C.S., 2007. Hidden branches: developments in root system architecture. Annu. Rev. Plant Biol. 58, 93–113.

Ou, C., Du, X., Shellie, K., Ross, C., Qian, M.C., 2010. Volatile compounds and sensory attributes of wine from cv. Merlot (*Vitis vinifera* L.) grown under differential levels of water deficit with or without kaolin-based, foliar reflectant particle film. J. Agric. Food Chem. 58, 12890–12898.

Ough, C.S., Amerine, M.A., 1963. Regional, varietal, and type influences on the degree Brix and alcohol relationship of grape musts and wines. Hilgardia 34, 585–600.

Ough, C.S., Nagaoka, R., 1984. Effect of cluster thinning and vineyard yields on grape and wine composition and wine quality of Cabernet Sauvignon. Am. J. Enol. Vitic. 35, 30–34.

Owen, S.J., Lafond, M.D., Bowen, P., Bogdanoff, C., Usher, K., Abrams, S.R., 2009. Profiles of abscisic acid and its catabolites in developing Merlot grape (*Vitis vinifera*) berries. Am. J. Enol. Vitic. 60, 277–284.

Ozga, J.A., Reinecke, D.M., 2003. Hormonal interactions in fruit development. J. Plant Growth Regul. 22, 73–81.

Padmalatha, K., Weksler, H., Mugzach, A., Acheampong, A.K., Zheng, C., Halaly-Basha, T., Or, E., 2017. ABA application during flowering and fruit set reduces berry number and improves cluster uniformity. Am. J. Enol. Vitic. 61, 275–282.

Pagay, V., Cheng, L., 2010. Variability in berry maturation of Concord and Cabernet franc in a cool climate. Am. J. Enol. Vitic. 61, 61–67.

Pagay, V., Zufferey, V., Lakso, A.N., 2016. The influence of water stress on grapevine (*Vitis vinifera* L.) shoots in a cool, humid climate: growth, gas exchange and hydraulics. Funct. Plant Biol. 43, 827–837.

Pagter, M., Arora, R., 2013. Winter survival and deacclimation of perennials under warming climate: physiological perspectives. Physiol. Plant. 147, 75–87.

Paiva, E.A.S., Buono, R.A., Lombardi, J.A., 2009. Food bodies in *Cissus verticillata* (Vitaceae): ontogenesis, structure and functional aspects. Ann. Bot. 103, 517–524.

Palanivelu, R., Brass, L., Edlund, A.F., Preuss, D., 2003. Pollen tube growth and guidance is regulated by *POP2*, an *Arabidopsis* gene that controls GABA levels. Cell 114, 47–59.

Palejwala, V.A., Parikh, H.R., Modi, V.V., 1985. The role of abscisic acid in the ripening of grapes. Physiol. Plant. 65, 498–502.

Pallas, B., Christophe, A., 2015. Relationships between biomass allocation, axis organogenesis and organ expansion under shading and water deficit conditions in grapevine. Funct. Plant Biol. 42, 1116–1128.

Pallas, B., Louarn, G., Christophe, A., Lebon, E., Lecoeur, J., 2008. Influence of intra-shoot trophic competition on shoot development in two grapevine cultivars (*Vitis vinifera*). Physiol. Plant. 134, 49–63.

Palliotti, A., Cartechini, A., 2001. Developmental changes in gas exchange activity in flowers, berries, and tendrils of field-grown Cabernet Sauvignon. Am. J. Enol. Vitic. 52, 317–323.

Palliotti, A., Cartechini, A., Ferranti, F., 2000. Morpho-anatomical and physiological characteristics of primary and lateral shoot leaves of Cabernet franc and Trebbiano toscano grapevines under two irradiance regimes. Am. J. Enol. Vitic. 51, 122–130.

Palliotti, A., Silvestroni, O., Petoumenou, D., 2009. Photosynthetic and photoinhibition behavior of two field-grown grapevine cultivars under multiple summer stresses. Am. J. Enol. Vitic. 60, 189–198.

Palliotti, A., Silvestroni, O., Petoumenou, D., 2010. Seasonal patterns of growth rate and morphophysiological features in green organs of Cabernet Sauvignon grapevines. Am. J. Enol. Vitic. 61, 74–82.

Palliotti, A., Gatti, M., Poni, S., 2011. Early leaf removal to improve vineyard efficiency: gas exchange, source-to-sink balance, and reserve storage responses. Am. J. Enol. Vitic. 62, 219–228.

Palliotti, A., Tombesi, S., Frioni, T., Famiani, F., Silvestroni, O., Zamboni, M., 2014. Morpho-structural and physiological response of container-grown Sangiovese and Montepulciano cvv. (*Vitis vinifera*) to re-watering after pre-veraison limiting water deficit. Funct. Plant Biol. 41, 634–647.

Palmieri, M.C., Perazzolli, M., Matafora, V., Moretto, M., Bacchi, A., Pertot, I., 2012. Proteomic analysis of grapevine resistance induced by *Trichoderma harzianum* T39 reveals specific defence pathways activated against downy mildew. J. Exp. Bot. 63, 6237–6251.

Pan, Q.H., Li, M.J., Peng, C.C., Zhang, N., Zou, X., Zou, K.Q., 2005. Abscisic acid activates acid invertases in developing grape berries. Physiol. Plant. 125, 157–170.

Pandolfini, T., Molesini, B., Spena, A., 2007. Molecular dissection of the role of auxin in fruit initiation. Trends Plant Sci. 12, 327–329.

Pang, X., Halaly, T., Crane, O., Keilin, T., Keren-Keiserman, A., Ogrodovitch, A., 2007. Involvement of calcium signalling in dormancy release of grape buds. J. Exp. Bot. 58, 3249–3262.

Pang, Y., Zhang, J., Cao, J., Yin, S.Y., He, X.Q., Cui, K.M., 2008. Phloem transdifferentiation from immature xylem cells during bark regeneration after girdling in *Eucommia ulmoides* Oliv. J. Exp. Bot. 59, 1341–1351.

Pañitrur-De La Fuente, C., Valdés-Gómez, H., Roudet, J., Acevedo-Opazo, C., Verdugo-Vásquez, N., Araya-Alman, M., et al., 2018. Classification of winegrape cultivars in Chile and France according to their susceptibility to *Botrytis cinerea* related to fruit maturity. Aust. J. Grape Wine Res. 24, 145–157.

Pantin, F., Simonneau, T., Muller, B., 2012. Coming of leaf age: control of growth by hydraulics and metabolics during leaf ontogeny. New Phytol. 196, 349–366.

Pantin, F., Monnet, F., Jannaud, D., Costa, J.M., Renaud, J., Muller, B., 2013. The dual effect of abscisic acid on stomata. New Phytol. 197, 65–72.

Parent, B., Millet, E.J., Tardieu, F., 2019. The use of thermal time in plant studies has a sound theoretical basis provided that confounding effects are avoided. J. Exp. Bot. 70, 2359–2370.

Park, S.K., Morrison, J.C., Adams, D.O., Noble, A.C., 1991. Distribution of free and glycosidically bound monoterpenes in the skin and mesocarp of Muscat of Alexandria grapes during development. J. Agric. Food Chem. 39, 514–518.

Parker, A.K., de Cortázar-Atauri, I.G., van Leeuwen, C., Chuine, I., 2011. General phenological model to characterise the timing of flowering and veraison of *Vitis vinifera* L. Aust. J. Grape Wine Res. 17, 206–216.

Parker, A., Garcia de Cortázar-Atauri, I., Chuine, I., Barbeau, G., Bois, B., Boursiquot, J.M., et al., 2013. Classification of varieties for their timing of flowering and veraison using a modelling approach: a case study for the grapevine species *Vitis vinifera* L. Agric. For. Meteorol. 180, 249–264.

Parker, M., Capone, D.L., Francis, I.L., Herderich, M.J., 2018. Aroma precursors in grapes and wine: flavor release during wine production and consumption. J. Agric. Food Chem. 66, 2281–2286.

Paroschy, J.H., Meiering, A.G., Peterson, R.L., Hostetter, G., Neff, A., 1980. Mechanical winter injury in grapevine trunks. Am. J. Enol. Vitic. 31, 227–232.

Parpinello, G.P., Heymann, H., Vasquez, S., Cathline, K.A., Fidelibus, M.W., 2012. Grape maturity, yield, quality, sensory properties, and consumer acceptance of Fiesta and Selma Pete dry-on-vine raisins. Am. J. Enol. Vitic. 63, 212–219.

Passioura, J.B., 2002. Soil conditions and plant growth. Plant Cell Environ. 25, 311–318.

Passioura, J.B., 2006. The perils of pot experiments. Funct. Plant Biol. 33, 1075–1079.

Pastor del Rio, J.L., Kennedy, J.A., 2006. Development of proanthocyanidins in *Vitis vinifera* L. cv. Pinot noir grapes and extraction into wine. Am. J. Enol. Vitic. 57, 125–132.

Pastore, C., Dal Santo, S., Zenoni, S., Movahed, N., Allegro, G., Valentini, G., Filippetti, I., Tornielli, G.B., 2017. Whole plant temperature manipulation affects flavonoid metabolism and the transcriptome of grapevine berries. Front. Plant Sci. 8, 929.

Patakas, A., Noitsakis, B., Stavrakas, D., 1997. Adaptation of leaves of *Vitis vinifera* L. to seasonal drought as affected by leaf age. Vitis 36, 11–14.

Patakas, A., Nikolaou, N., Zioziou, E., Radoglou, K., Noitsakis, B., 2002. The role of organic solute and ion accumulation in osmotic adjustment in drought-stressed grapevines. Plant Sci. 163, 361–367.

Patakas, A.A., Zotos, A., Beis, A.S., 2010. Production, localisation and possible roles of nitric oxide in drought-stressed grapevines. Aust. J. Grape Wine Res. 16, 203–209.

Pate, J.S., 1980. Transport and partitioning of nitrogenous solutes. Annu. Rev. Plant Physiol. 31, 313–340.

Patrick, J.W., 1997. Phloem unloading: sieve element unloading and post-sieve element transport. Annu. Rev. Plant Physiol. Plant Mol. Biol. 48, 191–222.

Patrick King, A., Berry, A.M., 2005. Vineyard δ^{15}N, nitrogen and water status in perennial clover and bunch grass cover crop systems of California's central valley. Agric. Ecosyst. Environ. 109, 262–272.

Patrick, J.W., Zhang, W., Tyerman, S.D., Offler, C.E., Walker, N.A., 2001. Role of membrane transport in phloem translocation of assimilates and water. Aust. J. Plant Physiol. 28, 695–707.

Pattison, R.J., Catalá, C., 2012. Evaluating auxin distribution in tomato (*Solanum lycopersicum*) through an analysis of the *PIN* and *AUX/LAX* gene families. Plant J. 70, 585–598.

Pattison, R.J., Csukasi, F., Catalá, C., 2014. Mechanisms regulating auxin action during fruit development. Physiol. Plant. 151, 62–72.

Paul, M.J., Foyer, C.H., 2001. Sink regulation of photosynthesis. J. Exp. Bot. 52, 1383–1400.

Pawlus, A.D., Sahli, R., Bisson, J., Rivière, C., Delaunay, J.C., Richard, T., 2013. Stilbenoid profiles of canes from *Vitis* and *Muscadinia* species. J. Agric. Food Chem. 61, 501–511.

Peacock, W.L., Christensen, L.P., Broadbent, F.E., 1989. Uptake, storage, and utilization of soil-applied nitrogen by Thompson Seedless as affected by time of application. Am. J. Enol. Vitic. 40, 16–20.

Pearce, R.S., 2001. Plant freezing and damage. Ann. Bot. 87, 417–424.

Pearcy, R.W., 1990. Sunflecks and photosynthesis in plant canopies. Annu. Rev. Plant Physiol. Plant Mol. Biol. 41, 421–453.

Pearsall, K.R., Williams, L.E., Castorani, S., Bleby, T.M., McElrone, A.J., 2014. Evaluating the potential of a novel dual heat-pulse sensor to measure volumetric water use in grapevines under a range of flow conditions. Funct. Plant Biol. 41, 874–883.

Pearson, H.M., 1932. Parthenocarpy and seed abortion in *Vitis vinifera*. Proc. Am. Soc. Hortic. Sci. 29, 169–175.

Pearson, R.C., Goheen, A.C., 1998. Compendium of Grape Diseases. APS Press, St. Paul, MN.

Pecinka, A., Abdelsamad, A., Vu, G.T.H., 2013. Hidden genetic nature of epigenetic natural variation in plants. Trends Plant Sci. 18, 625–632.

Peer, W.A., Murphy, A.S., 2007. Flavonoids and auxin transport: modulators or regulators? Trends Plant Sci. 12, 556–563.

Pelsy, F., 2010. Molecular and cellular mechanisms of diversity within grapevine varieties. Heredity 104, 331–340.

Pelsy, F., Hocquigny, S., Moncada, X., 2010. An extensive study of the genetic diversity within seven French wine grape variety collections. Theor. Appl. Genet. 120, 1219–1231.

Peña Quiñones, A.J., Keller, M., Salazar Gutierrez, M.R., Khot, L., Hoogenboom, G., 2019. Comparison between grapevine tissue temperature and air temperature. Sci. Hortic. 247, 407–420.

Peña-Gallego, A., Hernández-Orte, P., Cacho, J., Ferreira, V., 2012. S-cysteinylated and S-glutathionylated thiol precursors in grapes. A review. Food Chem. 131, 1–13.

Penfield, S., 2008. Temperature perception and signal transduction in plants. New Phytol. 179, 615–628.
Penning de Vries, F.W.T., 1975. The cost of maintenance processes in plant cells. Ann. Bot. 39, 77–92.
Penning de Vries, F.W.T., Brunsting, A.H.M., Van Laar, H.H., 1974. Products, requirements and efficiency of biosynthesis: a quantitative approach. J. Theor. Biol. 45, 339–377.
Pensec, F., Pączkowski, C., Grabarczyk, M., Woźniak, A., Bénard-Gellon, M., Bertsch, C., Chong, J., Szakiel, A., 2014. Changes in triterpenoid content of cuticular waxes during fruit ripening of eight grape (*Vitis vinifera*) cultivars grown in the upper Rhine Valley. J. Agric. Food Chem. 62, 7998–8007.
Peppi, M.C., Walker, M.A., Fidelibus, M.W., 2008. Application of abscisic acid rapidly upregulated UFGT gene expression and improved color of grape berries. Vitis 47, 11–14.
Perazzolli, M., Roatti, B., Bozza, E., Pertot, I., 2011. *Trichoderma harzianum* T39 induces resistance against downy mildew by priming for defense without costs for grapevine. Biol. Control 58, 74–82.
Percival, G.C., Boyle, C., Baird, L., 1999. The influence of calcium supplementation on the freezing tolerance of woody plants. J. Arboric. 25, 285–291.
Pereira, G.E., Gaudillère, J.P., Pieri, P., Hilbert, G., Maucourt, M., Deborde, C., 2006a. Microclimate influence on mineral and metabolic profiles of grape berries. J. Agric. Food Chem. 54, 6765–6775.
Pereira, G.E., Gaudillere, J.P., van Leeuwen, C., Hilbert, G., Maucourt, M., Deborde, C., 2006b. ^{1}H NMR metabolite fingerprints of grape berry: comparison of vintage and soil effects in Bordeaux grapevine growing areas. Anal. Chim. Acta 563, 346–352.
Pereira, H.S., Delgado, M., Avó, A.P., Barão, A., Serrano, I., Viegas, W., 2014. Pollen grain development is highly sensitive to temperature stress in *Vitis vinifera*. Aust. J. Grape Wine Res. 20, 474–484.
Pérez Harvey, J., Gaete, L., 1986. Efecto del microclima luminoso sobre la calidad de la uva Sultanina en sistema de parronal español. II. Desgrane, palo negro y pudrición gris. Cienc. Invest. Agrar. 13, 113–120.
Perez, J.R., Kliewer, W.M., 1982. Influence of light regime and nitrate fertilization on nitrate reductase activity and concentrations of nitrate and arginine in tissues of three cultivars of grapevines. Am. J. Enol. Vitic. 33, 86–93.
Perez, J., Kliewer, W.M., 1990. Effect of shading on bud necrosis and bud fruitfulness of Thompson Seedless grapevines. Am. J. Enol. Vitic. 41, 168–175.
Perez Peña, J., Tarara, J., 2004. A portable whole canopy gas exchange system for several mature field-grown grapevines. Vitis 43, 7–14.
Pérez, F.J., Viani, C., Retamales, J., 2000. Bioactive gibberellins in seeded and seedless grapes: identification and changes in content during berry development. Am. J. Enol. Vitic. 51, 315–318.
Pérez, F.J., Villegas, D., Mejia, N., 2002. Ascorbic acid and flavonoid-peroxidase reaction as a detoxifying system of H_2O_2 in grapevine leaves. Phytochemistry 60, 573–580.
Pérez, F.J., Rubio, S., Ormeño-Núñez, J., 2007. Is erratic bud-break in grapevines grown in warm winter areas related to disturbances in mitochondrial respiratory capacity and oxidative metabolism? Funct. Plant Biol. 34, 624–632.
Pérez, F.J., Vergara, R., Or, E., 2009. On the mechanism of dormancy release in grapevine buds: a comparative study between hydrogen cyanamide and sodium azide. Plant Growth Regul. 59, 145–152.
Pérez-Donoso, A.G., Sun, Q., Roper, M.C., Greve, C., Kirkpatrick, B., Labavitch, J.M., 2010. Cell wall-degrading enzymes enlarge the pore size of intervessel pit membranes in healthy and *Xylella fastidiosa*-infected grapevines. Plant Physiol. 152, 1748–1759.
Péros, J.P., Berger, G., Portemont, A., Boursiquot, J.M., Lacombe, T., 2011. Genetic variation and biogeography of the disjunct *Vitis* subg. *Vitis* (Vitaceae). J. Biogeogr. 38, 471–486.
Perret, P., Roth, I., 1996. Nährstoffverluste durch Versickerung in ostschweizerischen Rebbergen. Schweiz. Z. Obst-Weinbau 132, 396–399.
Perret, P., Weissenbach, P., Schwager, H., Heller, W., Koblet, W., 1993. "Adaptive nitrogen-management"—a tool for the optimisation of N-fertilisation in vineyards. Wein-Wiss. 48, 124–126.

Perrone, I., Gambino, G., Chitarra, W., Vitali, M., Pagliarani, C., Riccomagno, N., 2012. The grapevine root-specific aquaporin *VvPIP2;4N* controls root hydraulic conductance and leaf gas exchange under well-watered conditions but not under water stress. Plant Physiol. 160, 965–977.

Perry, K.B., 1998. Basics of frost and freeze protection for horticultural crops. HortTechnology 8, 10–15.

Petit, J.R., Jouzel, J., Raynaud, D., Barkov, N.I., Barnola, J.M., Basile, I., 1999. Climate and atmospheric history of the past 420,000 years from the Vostok ice core, Antarctica. Nature 399, 429–436.

Petrášek, J., Friml, J., 2009. Auxin transport routes in plant development. Development 136, 2675–2688.

Petrie, P.R., Clingeleffer, P.R., 2005. Effects of temperature and light (before and after budburst) on inflorescence morphology and flower number of Chardonnay grapevines (*Vitis vinifera* L.). Aust. J. Grape Wine Res. 11, 59–65.

Petrie, P.R., Trought, M.C.T., Howell, G.S., 2000a. Fruit composition and ripening of Pinot noir (*Vitis vinifera* L.) in relation to leaf area. Aust. J. Grape Wine Res. 6, 46–51.

Petrie, P.R., Trought, M.C.T., Howell, G.S., 2000b. Growth and dry matter partitioning of Pinot noir (*Vitis vinifera* L.) in relation to leaf area and crop load. Aust. J. Grape Wine Res. 6, 40–45.

Petrie, P.R., Trought, M.C.T., Howell, G.S., Buchan, G.D., Palmer, J.W., 2009. Whole-canopy gas exchange and light interception of vertically trained *Vitis vinifera* L. under direct and diffuse light. Am. J. Enol. Vitic. 60, 173–182.

Peuke, A.D., 2010. Correlations in concentrations, xylem and phloem flows, and partitioning of elements and ions in intact plants. A summary and statistical re-evaluation of modelling experiments in Ricinus communis. J. Exp. Bot. 61, 635–655.

Peuke, A.D., Rokitta, M., Zimmermann, U., Schreiber, L., Haase, A., 2001. Simultaneous measurement of water flow velocity and solute transport in xylem and phloem of adult plants of *Ricinus communis* over a daily time course by nuclear magnetic resonance spectrometry. Plant Cell Environ. 24, 491–503.

Peyrot des Gachons, C., Tominaga, T., Dubourdieu, D., 2002a. Localisation in the berry of *S*-cysteine conjugates, aroma precursors of *Vitis vinifera* L. cv. Sauvignon blanc grapes. The effect of skin contact on the aromatic potential of the musts. Am. J. Enol. Vitic. 53, 144–146.

Peyrot des Gachons, C., Tominaga, T., Dubourdieu, D., 2002b. Sulfur aroma precursor present in *S*-glutathione conjugate form: identification of *S*-3-(hexan-1-ol)-glutathione in must from *Vitis vinifera* L. cv. Sauvignon blanc. J. Agric. Food Chem. 50, 4076–4079.

Peyrot des Gachons, C., Van Leeuwen, C., Tominaga, T., Soyer, J.P., Gaudillère, J.P., Dubourdieu, D., 2005. Influence of water and nitrogen deficit on fruit ripening and aroma potential of *Vitis vinifera* L cv Sauvignon blanc in field conditions. J. Sci. Food Agric. 85, 73–85.

Pezet, R., Pont, V., Hoang-Van, K., 1991. Evidence for oxidative detoxication of pterostilbene and resveratrol by a laccase-like stilbene oxidase produced by *Botrytis cinerea*. Physiol. Mol. Plant Pathol. 39, 441–450.

Pezet, R., Viret, O., Perret, C., Tabacchi, R., 2003. Latency of *Botrytis cinerea* Pers.: Fr. and biochemical studies during growth and ripening of two grape berry cultivars, respectively susceptible and resistant to grey mould. J. Phytopathol. 151, 208–214.

Pezet, R., Gindro, K., Viret, O., Richter, H., 2004a. Effects of resveratrol, viniferins and pterostilbene on *Plasmopara viticola* zoospore mobility and disease development. Vitis 43, 145–148.

Pezet, R., Viret, O., Gindro, K., 2004b. Plant–Microbe Interaction: The *Botrytis* Grey Mould of Grapes—Biology, Biochemistry, Epidemiology and Control Management. Scientific Publishers, Jodhpur, India, pp. 71–116.

Pezzuto, J.M., 2008. Grapes and human health: a perspective. J. Agric. Food Chem. 56, 6777–6784.

Pfanz, H., 2008. Bark photosynthesis. Trees 22, 137–138.

Pfautsch, S., Renard, J., Tjoelker, M.G., Salih, A., 2015. Phloem as capacitor: radial transfer of water into xylem of tree stems occurs via symplastic transport in ray parenchyma. Plant Physiol. 167, 963–971.

Phillips, J.G., Riha, S.J., 1994. Root growth, water uptake and canopy development in *Eucalyptus viminalis* seedlings. Aust. J. Plant Physiol. 21, 69–78.

Piazza, P., Jasinski, S., Tsiantis, M., 2005. Evolution of leaf developmental mechanisms. New Phytol. 167, 693–710.

Pichersky, E., Gershenzon, J., 2002. The formation and function of plant volatiles: perfumes for pollinator attraction and defense. Curr. Opin. Plant Biol. 5, 237–243.

Pickering, G., Lin, J., Riesen, R., Reynolds, A., Brindle, I., Soleas, G., 2004. Influence of *Harmonia axyridis* on the sensory properties of white and red wine. Am. J. Enol. Vitic. 55, 153–159.

Pickering, G.J., Lin, Y., Reynolds, A., Soleas, G., Riesen, R., Brindle, I., 2005. The influence of *Harmonia axyridis* on wine composition and aging. J. Food Sci. 70, 128–135.

Pien, S., Wyrzykowska, J., McQueen-Mason, S., Smart, C., Fleming, A., 2001. Local expression of expansin induces the entire process of leaf development and modifies leaf shape. Proc. Natl. Acad. Sci. U. S. A. 98, 11812–11817.

Pierik, R., Cuppens, M.L.C., Voesenek, L.A.C.J., Visser, E.J.W., 2004. Interactions between ethylene and gibberellins in phytochrome-mediated shade avoidance responses in tobacco. Plant Physiol. 136, 2928–2936.

Pierquet, P., Stushnoff, C., 1980. Relationship of low temperature exotherms to cold injury in *Vitis riparia* Michx. Am. J. Enol. Vitic. 31, 1–6.

Pilati, S., Perazzolli, M., Malossini, A., Cestaro, A., Dematte, L., Fontana, P., 2007. Genome-wide transcriptional analysis of grapevine berry ripening reveals a set of genes similarly modulated during three seasons and the occurrence of an oxidative burst at veraison. BMC Genom. 8, 428.

Pilati, S., Brazzale, D., Guella, G., Milli, A., Ruberti, C., Biasioli, F., Zottini, M., Moser, C., 2014. The onset of grapevine berry ripening is characterized by ROS accumulation and lipoxygenase-mediated membrane peroxidation in the skin. BMC Plant Biol. 14, 87.

Pineau, B., Barbe, J.C., Van Leeuwen, C., Dubourdieu, D., 2007. Which impact for β-damascenone on red wines aroma? J. Agric. Food Chem. 55, 4103–4108.

Pinelo, M., Arnous, A., Meyer, A.S., 2006. Upgrading of grape skins: significance of plant cell-wall structural components and extraction techniques for phenol release. Trends Food Sci. Technol. 17, 579–590.

Pirie, A., Mullins, M.G., 1976. Changes in anthocyanin and phenolics content of grapevine leaf and fruit tissue treated with sucrose, nitrate, and abscisic acid. Plant Physiol. 58, 468–472.

Pirie, A., Mullins, M.G., 1977. Interrelationships of sugars, anthocyanins, total phenols and dry weight in the skin of grape berries during ripening. Am. J. Enol. Vitic. 28, 204–209.

Pirie, A.J.G., Mullins, M.G., 1980. Concentration of phenolics in the skin of grape berries during fruit development and ripening. Am. J. Enol. Vitic. 31, 34–36.

Pisciotta, A., Di Lorenzo, R., Santalucia, G., Barbagallo, M.G., 2018. Response of grapevine (Cabernet Sauvignon cv) to above ground and subsurface drip irrigation under arid conditions. Agric. Water Manag. 197, 122–131.

Pittman, J.K., 2005. Managing the manganese: molecular mechanisms of manganese transport and homeostasis. New Phytol. 167, 733–742.

Piva, C.R., Lopez Garcia, J.L., Morgan, W., 2006. The ideal table grapes for the Spanish market. Rev. Bras. Frutic. 28, 258–261.

Plank, S., Wolkinger, F., 1976. Holz von *Vitis vinifera* im Raster-Elektronenmikroskop. Vitis 15, 153–159.

Plavcová, L., Hacke, U.G., Almeida-Rodriguez, A.M., Li, E., Douglas, C.J., 2012. Gene expression patterns underlying changes in xylem structure and function in response to increased nitrogen availability in hybrid poplar. Plant Cell Environ. 36, 186–199.

Plaxton, W.C., Tran, H.T., 2011. Metabolic adaptations of phosphate-starved plants. Plant Physiol. 156, 1006–1015.

Plieth, C., 2005. Calcium: just another regulator in the machinery of life? Ann. Bot. 96, 1–8.

Pliny the Elder, G.P.S., ca. 70 ad. Historia Naturalis. Available from: http://www.perseus.tufts.edu/hopper/text;jsessionid=19A3C191276E730051F35B6E88D5084A?doc=Perseus%3atext%3a1999.02.0138.

Plomion, C., Leprovost, G., Stokes, A., 2001. Wood formation in trees. Plant Physiol. 127, 1513–1523.

Pocock, K.F., Hayasaka, Y., Peng, Z., Williams, P.J., Waters, E.J., 1998. The effect of mechanical harvesting and long-distance transport on the concentration of haze-forming proteins in grape juice. Aust. J. Grape Wine Res. 4, 23–29.

Pocock, K.F., Hayasaka, Y., McCarthy, M.G., Waters, E.J., 2000. Thaumatin-like proteins and chitinases, the haze-forming proteins of wine, accumulate during ripening of grape (*Vitis vinifera*) berries and drought stress does not affect the final levels per berry at maturity. J. Agric. Food Chem. 48, 1637–1643.

Poelman, E.H., van Loon, J.J.A., Dicke, M., 2008. Consequences of variation in plant defense for biodiversity at higher trophic levels. Trends Plant Sci. 13, 534–541.

Poland, J.A., Balint-Kurti, P.J., Wisser, R.J., Pratt, R.C., Nelson, R.J., 2009. Shades of gray: the world of quantitative disease resistance. Trends Plant Sci. 14, 21–29.

Pomar, F., Novo, M., Masa, A., 2005. Varietal differences among the anthocyanin profiles of 50 red table grape cultivars studied by high performance liquid chromatography. J. Chromatogr. A 1094, 34–41.

Pongrácz, D.P., 1983. Rootstocks for Grape-Vines. David Philip, Cape Town, South Africa.

Poni, S., Intrieri, C., Silvestroni, O., 1994a. Interactions of leaf age, fruiting, and exogenous cytokinins in Sangiovese grapevines under non-irrigated conditions. I. Gas exchange. Am. J. Enol. Vitic. 45, 71–78.

Poni, S., Intrieri, C., Silvestroni, O., 1994b. Interactions of leaf age, fruiting, and exogenous cytokinins in Sangiovese grapevines under non-irrigated conditions. II. Chlorophyll and nitrogen content. Am. J. Enol. Vitic. 45, 278–284.

Poni, S., Magnanini, E., Bernizzoni, F., 2003. Degree of correlation between total light interception and whole-canopy net CO_2 exchange rate in two grapevine growth systems. Aust. J. Grape Wine Res. 9, 2–11.

Poni, S., Casalini, L., Bernizzoni, F., Civardi, S., Intrieri, C., 2006. Effects of early defoliation on shoot photosynthesis, yield components, and grape composition. Am. J. Enol. Vitic. 57, 397–407.

Poni, S., Bernizzoni, F., Civardi, S., Gatti, M., Porro, D., Camin, F., 2008. Performance and water-use efficiency (single-leaf vs. whole-canopy) of well-watered and half-stressed split-root Lambrusco grapevines grown in Po Valley (Italy). Agric. Ecosyst. Environ. 129, 97–106.

Poni, S., Bernizzoni, F., Civardi, S., Libelli, N., 2009. Effects of pre-bloom leaf removal on growth of berry tissues and must composition in two red *Vitis vinifera* L. cultivars. Aust. J. Grape Wine Res. 15, 185–193.

Poni, S., Tombesi, S., Palliotti, A., Ughini, V., Gatti, M., 2016. Mechanical winter pruning of grapevine: physiological bases and applications. Sci. Hortic. 204, 88–98.

Pons, T.L., Jordi, W., Kuiper, D., 2001. Acclimation of plants to light gradients in leaf canopies: evidence for a possible role for cytokinins transported in the transpiration stream. J. Exp. Bot. 52, 1563–1574.

Pool, R.M., Pratt, C., Hubbard, H.D., 1978. Structure of base buds in relation to yield of grapes. Am. J. Enol. Vitic. 29, 36–41.

Pool, R.M., Creasy, L.L., Frackelton, A.S., 1981. Resveratrol and the viniferins, their application to screening for disease resistance in grape breeding programs. Vitis 20, 136–145.

Pool, R., Dunst, R., Lakso, A., 1990. Comparison of sod, mulch, cultivation, and herbicide floor management practices for grape production in non-irrigated vineyards. J. Am. Soc. Hortic. Sci. 115, 872–877.

Poorter, H., Nagel, O., 2000. The role of biomass allocation in the growth response of plants to different levels of light, nutrients and water: a quantitative review. Aust. J. Plant Physiol. 27, 595–607.

Poorter, H., Pepin, S., Rijkers, T., de Jong, Y., Evans, J.R., Körner, C., 2006. Construction costs, chemical composition and payback time of high- and low-irradiance leaves. J. Exp. Bot. 57, 355–371.

Poorter, H., Bühler, J., van Dusschoten, D., Climent, J., Postma, J.A., 2012. Pot size matters: a meta-analysis of the effects of rooting volume on plant growth. Funct. Plant Biol. 39, 839–850.

Poovaiah, B.W., Leopold, A.C., 1973. Inhibition of abscission by calcium. Plant Physiol. 51, 848–851.

Possingham, J.V., 1994. New concepts in pruning grapevines. Hortic. Rev. 16, 235–254.

Possingham, J.V., Groot Obbink, J., 1971. Endotrophic mycorrhiza and the nutrition of grape vines. Vitis 10, 120–130.
Possingham, J.V., Chambers, T.C., Radler, F., Grncarevic, M., 1967. Cuticular transpiration and wax structure and composition of leaves and fruit of *Vitis vinifera*. Aust. J. Biol. Sci. 20, 1149–1153.
Possner, D.R.E., Kliewer, W.M., 1985. The localisation of acids, sugars, potassium and calcium in developing grape berries. Vitis 24, 229–240.
Potters, G., Pasternak, T.P., Guisez, Y., Palme, K.J., Jansen, M.A.K., 2007. Stress-induced morphogenic responses: growing out of trouble? Trends Plant Sci. 12, 98–105.
Potters, G., Pasternak, T.P., Guisez, Y., Jansen, M.A.K., 2009. Different stresses, similar morphogenic responses: integrating a plethora of pathways. Plant Cell Environ. 32, 158–169.
Pouget, R., 1963. Recherches physiologiques sur le repos végétatif de la vigne (*Vitis vinifera* L.): La dormance des bourgeons et le mécanisme de sa disparition. Ann. Amélior. Plantes 13, 1–247.
Pouget, R., 1972. Considérations générales sur le rythme végétatif et la dormance des bourgeons de la vigne. Vitis 11, 198–217.
Pouget, R., 1981. Action de la température sur la différenciation des inflorescences et des fleurs durant les phases de pré-débourrement et de post-débourrement des bourgeons latents de la vigne. Connaiss. Vigne Vin 15, 65–79.
Pouget, R., 1984. Action de la concentration de la solution nutritive sur quelques caractéristiques physiologiques et technologiques chez *Vitis vinifera* L. cv. Cabernet-Sauvignon. I. Vigueur, rendement, qualité du moût et du vin. Agronomie 4, 437–442.
Pouget, R., 1988. Le débourrement des bourgeons de la vigne: méthode de prévision et principes d'établissement d'une échelle de précocité de débourrement. Connaiss. Vigne Vin 22, 105–123.
Pourcel, L., Routaboul, J.M., Cheynier, V., Lepiniec, L., Debeaujon, I., 2007. Flavonoid oxidation in plants: from biochemical properties to physiological functions. Trends Plant Sci. 12, 29–36.
Pourtchev, P., 2003. To the problem of depth of root system penetration of grapevine. Soil Sci. Agrochem. Ecol. 2, 47–52.
Powles, S.B., Yu, Q., 2010. Evolution in action: plants resistant to herbicides. Annu. Rev. Plant Biol. 61, 317–347.
Pradubsuk, S., Davenport, J.R., 2010. Seasonal uptake and partitioning of macronutrients in mature 'Concord' grape. J. Am. Soc. Hortic. Sci. 135, 474–483.
Pradubsuk, S., Davenport, J.R., 2011. Seasonal distribution of micronutrients in mature 'Concord' grape: boron, iron, manganese, copper, and zinc. J. Am. Soc. Hortic. Sci. 136, 69–77.
Pratt, C., 1971. Reproductive anatomy in cultivated grapes—a review. Am. J. Enol. Vitic. 22, 92–109.
Pratt, C., 1973. Reproductive system of "Concord" and two sports (*Vitis labruscana* Bailey). J. Am. Soc. Hortic. Sci. 98, 489–496.
Pratt, C., 1974. Vegetative anatomy of cultivated grapes—a review. Am. J. Enol. Vitic. 25, 131–150.
Pratt, C., 1978. Shoot nodes of *Vitis labruscana* Bailey cv. Concord. Vitis 17, 329–334.
Pratt, C., 1979. Shoot and bud development during the prebloom period of *Vitis*. Vitis 18, 1–5.
Pratt, C., Coombe, B.G., 1978. Shoot growth and anthesis in *Vitis*. Vitis 17, 125–133.
Pratt, C., Pool, R.M., 1981. Anatomy of recovery of canes of *Vitis vinifera* L. from simulated freezing injury. Am. J. Enol. Vitic. 32, 223–227.
Price, S.F., Breen, P.J., Valladao, M., Watson, B.T., 1995. Cluster sun exposure and quercetin in Pinot noir grapes and wine. Am. J. Enol. Vitic. 46, 187–194.
Price, C.A., Gilooly, J.F., Allen, A.P., Weitz, J.S., Niklas, K.J., 2010. The metabolic theory of ecology: prospects and challenges for plant biology. New Phytol. 188, 696–710.
Price, N.P.J., Vermillion, K.E., Eller, F.J., Vaughn, S.F., 2015. Frost grape polysaccharide (FGP), an emulsion-forming arabinogalactan gum from the stems of native North American grape species *Vitis riparia* Michx. J. Agric. Food Chem. 63, 7286–7293.

Priestley, J.H., Wormall, A., 1925. On the solutes exuded by root pressure from vines. New Phytol. 24, 24–38.

Prieto, J.A., Louarn, G., Perez Peña, J., Ojeda, H., Simonneau, T., Lebon, E., 2012. A leaf gas exchange model that accounts for intra-canopy variability by considering leaf nitrogen content and local acclimation to radiation in grapevine (*Vitis vinifera* L.). Plant Cell Environ. 35, 1313–1328.

Prieur, C., Rigaud, J., Cheynier, V., Moutounet, M., 1994. Oligomeric and polymeric procyanidins from grape seeds. Phytochemistry 36, 781–784.

Prusinkiewicz, P., Crawford, S., Smith, R.S., Ljung, K., Bennett, T., Ongaro, V., 2009. Control of bud activation by an auxin transport switch. Proc. Natl. Acad. Sci. U. S. A. 106, 17431–17436.

Puech, J.L., Feuillat, F., Mosedale, J.R., 1999. The tannins of oak heartwood: structure, properties, and their influence on wine flavor. Am. J. Enol. Vitic. 50, 469–478.

Qiu, W., Feechan, A., Dry, I., 2015. Current understanding of grapevine defense mechanisms against the biotrophic fungus (*Erysiphe necator*), the causal agent of powdery mildew disease. Hortic. Res. 2, 15020.

Quamme, H.A., 1991. Application of thermal analysis to breeding fruit crops for increased cold hardiness. HortScience 26, 513–517.

Querejeta, J.I., Egerton-Warburton, L.M., Allen, M.F., 2003. Direct nocturnal water transfer from oaks to their mycorrhizal symbionts during severe soil drying. Oecologia 134, 55–64.

Quick, W.P., Chaves, M.M., Wendler, R., David, M., Rodrigues, M.L., Passaharinho, J.A., 1992. The effect of water stress on photosynthetic carbon metabolism in four species grown under field conditions. Plant Cell Environ. 15, 25–35.

Quijada-Morín, N., Regueiro, J., Simal-Gándara, J., Tomás, E., Rivas-Gonzalo, J.C., Escribano-Bailón, M.T., 2012. Relationship between the sensory-determined astringency and the flavanolic composition of red wines. J. Agric. Food Chem. 60, 12355–12361.

Quinlan, J.D., Weaver, R.J., 1969. Influence of benzyladenine, leaf darkening, and ringing on movement of ^{14}C-labeled assimilates into expanding leaves of *Vitis vinifera* L. Plant Physiol. 44, 1247–1252.

Quinlan, J.D., Weaver, R.J., 1970. Modification of pattern of the photosynthate movement within and between shoots of *Vitis vinifera* L. Plant Physiol. 46, 527–530.

Rabe, E., 1990. Stress physiology: the functional significance of the accumulation of nitrogen-containing compounds. J. Hortic. Sci. 65, 231–243.

Radler, F., 1965a. Reduction of loss of moisture by the cuticle wax components of grapes. Nature 207, 1002–1003.

Radler, F., 1965b. The main constituents of the surface waxes of varieties and species of the genus *Vitis*. Am. J. Enol. Vitic. 16, 159–167.

Radler, F., 1970. Untersuchungen über das Kutikularwachs von *Vitis vinifera* L. ssp. *sylvestris* (Gmel.) Beger und *Vitis vinifera* L. ssp. *vinifera*. Angew. Bot. 44, 187–195.

Rafei, M.S., 1941. Anatomical studies in *Vitis* and allied genera. I. Development of the fruit. II. Floral anatomy. PhD thesis, Oregon State College.

Raghavan, V., 2003. Some reflections on double fertilization, from its discovery to the present. New Phytol. 159, 565–583.

Raghothama, K.G., 1999. Phosphate acquisition. Annu. Rev. Plant Physiol. Plant Mol. Biol. 50, 665–693.

Ramming, D.W., 2009. Water loss from fresh berries of raisin cultivars under controlled drying conditions. Am. J. Enol. Vitic. 60, 208–214.

Rankine, B.C., Cellier, K.M., Boehm, E.W., 1962. Studies on grape variability and field sampling. Am. J. Enol. Vitic. 13, 58–72.

Rapp, A., Klenert, M., 1974. Einfluß der Samen auf die Beerenreife bei *Vitis vinifera* L. Vitis 13, 222–232.

Rapp, A., Versini, G., 1996. Influence of nitrogen compounds in grapes on aroma compounds of wines. Wein-Wiss. 51, 193–203.

Rapp, A., Ziegler, A., 1973. Bestimmung der Phenolcarbonsäuren (Hydroxybenzoesäuren und Hydroxyzimtsäuren) in Rebblättern, Weintrauben und Wein mittels Mikro-Polyamid-Dünnschichtchromatographie. Vitis 12, 226–236.

Rascher, U., Nedbal, L., 2006. Dynamics of photosynthesis in fluctuating light. Curr. Opin. Plant Biol. 9, 671–678.

Raschke, K., 1975. Stomatal action. Annu. Rev. Plant Physiol. 26, 309–340.

Rasmussen, A., Mason, M.G., De Cuyper, C., Brewer, P.B., Herold, S., Agusti, J., 2012. Strigolactones suppress adventitious rooting in *Arabidopsis* and pea. Plant Physiol. 158, 1976–1987.

Rasmussen, A., Depuydt, S., Goormachtig, S., Geelen, D., 2013. Strigolactones fine-tune the root system. Planta 238, 615–626.

Rathgeber, C.B.K., Cuny, H.E., Fonti, P., 2016. Biological basis of tree-ring formation: a crash course. Front. Plant Sci. 7, 734.

Rausch, T., 2007. When plant life gets tough sulfur gets going. Plant Biol. 9, 551–555.

Rayle, D.L., Cleland, R.E., 1992. The Acid Growth Theory of auxin-induced cell elongation is alive and well. Plant Physiol. 99, 1271–1274.

Razungles, A., Bayonove, C.L., Cordonnier, R.E., Sapis, J.C., 1988. Grape carotenoids: changes during the maturation period and localization in mature berries. Am. J. Enol. Vitic. 39, 44–48.

Razungles, A.J., Babic, I., Sapis, J.C., Bayonove, C.L., 1996. Particular behavior of epoxy xanthophylls during veraison and maturation of grape. J. Agric. Food Chem. 44, 3821–3825.

Razungles, A.J., Baumes, R.L., Dufour, C., Sznaper, C.N., Bayonove, C.L., 1998. Effect of sun exposure on carotenoids and C-13-norisoprenoid glycosides in Syrah berries (*Vitis vinifera* L.). Sci. Aliment. 18, 361–373.

Rebucci, B., Poni, S., Intrieri, C., Magnanini, E., Lakso, A.N., 1997. Effects of manipulated grape berry transpiration and post-veraison sugar accumulation. Aust. J. Grape Wine Res. 3, 57–65.

Redl, H., 1984. Die Entblätterung der Traubenzone als vorbeugende Massnahme gegen die Stiellähme. Wein-Wiss. 39, 75–82.

Reeve, A.L., Skinkis, P.A., Vance, A.J., Lee, J., Tarara, J.M., 2016. Vineyard floor management influences 'Pinot noir' vine growth and productivity more than cluster thinning. HortScience 51, 1233–1244.

Regner, F., Stadlbauer, A., Eisenheld, C., Kaserer, H., 2000. Genetic relationships among Pinots and related cultivars. Am. J. Enol. Vitic. 51, 7–14.

Reich, P.B., Falster, D.S., Ellsworth, D.S., Wright, I.J., Westoby, M., Oleksyn, J., 2009. Controls on declining carbon balance with leaf age among 10 woody species in Australian woodland: do leaves have zero daily net carbon balances when they die? New Phytol. 183, 153–166.

Reid, R.J., 2001. Mechanisms of micronutrient uptake in plants. Aust. J. Plant Physiol. 28, 659–666.

Reimers, H., Steinberg, B., Kiefer, W., 1994. Ergebnisse von Wurzeluntersuchungen an Reben bei offenem und begrüntem Boden. Wein-Wiss. 49, 136–145.

Reinhardt, D., Mandel, T., Kuhlemeier, C., 2000. Auxin regulates the initiation and radial position of plant lateral organs. Plant Cell 12, 507–518.

Reisch, B.I., Goodman, R.N., Martens, M.H., Weeden, N.F., 1993. The relationship between Norton and Cynthiana, red wine cultivars derived from *Vitis aestivalis*. Am. J. Enol. Vitic. 44, 441–444.

Reiter, W.D., 2002. Biosynthesis and properties of the plant cell wall. Curr. Opin. Plant Biol. 5, 536–542.

Rellán-Álvarez, R., Lobet, G., Dinneny, J.R., 2016. Environmental control of root system biology. Annu. Rev. Plant Biol. 67, 619–642.

Remund, U., Niggli, U., Boller, E.F., 1989. Faunistische und botanische Erhebungen in einem Rebberg der Ostschweiz. Einfluss der Unterwuchsbewirtschaftung auf das Oekosystem Rebberg. Landwirtsch. Schweiz. 2, 393–408.

Renault, A.S., Deloire, A., Bierne, J., 1996. Pathogenesis-related proteins in grapevines induced by salicylic acid and *Botrytis cinerea*. Vitis 35, 49–52.

Rennenberg, H., Herschbach, C., 2014. A detailed view on sulphur metabolism at the cellular and whole-plant level illustrates challenges in metabolite flux analyses. J. Exp. Bot. 65, 5711–5724.

Rennie, E.A., Turgeon, R., 2009. A comprehensive picture of phloem loading strategies. Proc. Natl. Acad. Sci. U. S. A. 106, 14162–14167.

Renouf, V., Tregoat, O., Roby, J.P., Van Leeuwen, C., 2010. Soils, rootstocks and grapevine varieties in prestigious Bordeaux vineyards and their impact on yield and quality. J. Int. Sci. Vigne Vin 44, 1–8.

Reuther, G., Reichardt, A., 1963. Temperatureinflüsse auf Blutung und Stoffwechsel bei *Vitis vinifera*. Planta 59, 391–410.

Reuveni, M., 1998. Relationships between leaf age, peroxidase and β-1,3-glucanase activity, and resistance to downy mildew in grapevines. J. Phytopathol. 146, 525–530.

Revilla, E., García-Beneytiz, E., Cabello, F., 2009. Anthocyanin fingerprint of clones of Tempranillo grapes and wines made with them. Aust. J. Grape Wine Res. 15, 70–78.

Reynolds, A.G., Vanden Heuvel, J.E., 2009. Influence of grapevine training systems on vine growth and fruit composition: a review. Am. J. Enol. Vitic. 60, 251–268.

Reynolds, A.G., Wardle, D.A., 1989a. Impact of various canopy manipulation techniques on growth, yield, fruit composition, and wine quality of Gewürztraminer. Am. J. Enol. Vitic. 40, 121–129.

Reynolds, A.G., Wardle, D.A., 1989b. Influence of fruit microclimate on monoterpene levels of Gewürztraminer. Am. J. Enol. Vitic. 40, 149–154.

Reynolds, A.G., Pool, R.M., Mattick, L.R., 1986. Influence of cluster exposure on fruit composition and wine quality of Seyval blanc grapes. Vitis 25, 85–95.

Reynolds, A.G., Edwards, C.G., Wardle, D.A., Webster, D.R., Dever, M., 1994a. Shoot density affects 'Riesling' grapevines. I. Vine performance. J. Am. Soc. Hortic. Sci. 119, 874–880.

Reynolds, A.G., Edwards, C.G., Wardle, D.A., Webster, D.R., Dever, M., 1994b. Shoot density affects 'Riesling' grapevines. II. Wine composition and sensory response. J. Am. Soc. Hortic. Sci. 119, 881–892.

Reynolds, A.G., Wardle, D.A., Dever, M.J., 1996. Vine performance, fruit composition, and wine sensory attributes of Gewürztraminer in response to vineyard location and canopy manipulation. Am. J. Enol. Vitic. 47, 77–92.

Reynolds, A.G., Molek, T., De Savigny, C., 2005. Timing of shoot thinning in *Vitis vinifera*: impacts on yield and fruit composition variables. Am. J. Enol. Vitic. 56, 343–356.

Reynolds, A.G., Senchuk, I.V., van der Reest, C., de Savigny, C., 2007. Use of GPS and GIS for elucidation of the basis for terroir: spatial variation in an Ontario Riesling vineyard. Am. J. Enol. Vitic. 58, 145–162.

Reynolds, A.G., Taylor, G., de Savigny, C., 2013. Defining Niagara terroir by chemical and sensory analysis of Chardonnay wines from various soil textures and vine sizes. Am. J. Enol. Vitic. 64, 180–194.

Riaz, S., Garrison, K.E., Dangl, G.S., Boursiquot, J.M., Meredith, C.P., 2002. Genetic divergence and chimerism within ancient asexually propagated winegrape cultivars. J. Am. Soc. Hortic. Sci. 127, 508–514.

Ribéreau-Gayon, P., 1964. Les flavonosides de la baie dans le genre *Vitis*. C.R. Hébd. Séances Acad. Sci. (Paris) 258, 1335–1337.

Ribéreau-Gayon, P., 1973. Interprétation chimique de la couleur des vins rouges. Vitis 12, 119–142.

Ribéreau-Gayon, P., 1982. Incidences oenologiques de la pourriture du raisin. Bull. OEPP 12, 201–214.

Ribéreau-Gayon, J., Ribéreau-Gayon, P., 1958. The anthocyans and leucoanthocyans of grapes and wines. Am. J. Enol. Vitic. 9, 1–9.

Ribéreau-Gayon, P., Boidron, J.N., Terrier, A., 1975. Aroma of Muscat grape varieties. J. Agric. Food Chem. 23, 1042–1047.

Ribéreau-Gayon, P., Dubourdieu, D., Donéche, B., Lonvaud, A., 2006. Handbook of Enology. Volume 1: The Microbiology of Wine and Vinifications, second ed. Wiley & Sons, Chichester, UK. Trans. J.M. Branco.

Rice-Evans, C.A., Miller, N.J., Paganga, G., 1997. Antioxidant properties of phenolic compounds. Trends Plant Sci. 2, 152–159.

Richards, D., 1983. The grape root system. Hortic. Rev. 5, 127–168.

Richardson, A.C., Marsh, K.B., Boldingh, H.L., Pickering, A.H., Bulley, S.M., Frearson, N.J., 2004. High growing temperatures reduce fruit carbohydrate and vitamin C in kiwifruit. Plant Cell Environ. 27, 423–435.

Riederer, M., Schreiber, L., 2001. Protecting against water loss: analysis of the barrier properties of plant cuticles. J. Exp. Bot. 52, 2023–2032.

Rieu, I., Twell, D., Firon, N., 2017. Pollen development at high temperature: from acclimation to collapse. Plant Physiol. 173, 1967–1976.

Rinaldo, A.R., Cavallini, E., Jia, Y., Moss, S.M.A., McDavid, D.A.J., Hooper, L.C., Robinson, S.P., Tornielli, G.B., Zenoni, S., Ford, C.M., Boss, P.K., Walker, A.R., 2015. A grapevine anthocyanin acyltransferase, transcriptionally regulated by VvMYBA, can produce most acylated anthocyanins present in grape skins. Plant Physiol. 169, 1897–1916.

Rinne, P.L.H., Paul, L.K., Vahala, J., Kangasjärvi, J., van der Schoot, C., 2016. Axillary buds are dwarfed shoots that tightly regulate GA pathway and GA-inducible 1,3-β-glucanase genes during branching in hybrid aspen. J. Exp. Bot. 67, 5975–5991.

Ristic, Z., Ashworth, E.N., 1993. Ultrastructural evidence that intracellular ice formation and possibly cavitation are the sources of freezing injury in supercooling wood tissue of *Cornus florida* L. Plant Physiol. 103, 753–761.

Ristic, R., Iland, P.G., 2005. Relationships between seed and berry development of *Vitis vinifera* L. cv Shiraz: developmental changes in seed morphology and phenolic composition. Aust. J. Grape Wine Res. 11, 43–58.

Ristic, R., Bindon, K., Francis, L.I., Herderich, M.J., Iland, P.G., 2010. Flavonoids and C_{13}-norisoprenoids in *Vitis vinifera* L. cv. Shiraz: relationships between grape and wine composition, wine colour and wine sensory properties. Aust. J. Grape Wine Res. 16, 369–388.

Rivera Núñez, D., Walker, M.J., 1989. A review of palaeobotanical findings of early *Vitis* in the Mediterranean and of the origins of cultivated grape-vines, with special reference to new pointers to prehistoric exploitation in the western Mediterranean. Rev. Palaeobot. Palynol. 61, 205–237.

Rizhsky, L., Liang, H., Shuman, J., Shulaev, V., Davletova, S., Mittler, R., 2004. When defense pathways collide. The response of *Arabidopsis* to a combination of drought and heat stress. Plant Physiol. 134, 1683–1696.

Roach, M.J., Johnson, D.L., Bohlmann, J., van Vuuren, H.J.J., Jones, S.J.M., Pretorius, I.S., et al., 2018. Population sequencing reveals clonal diversity and ancestral inbreeding in the grapevine cultivar Chardonnay. PLoS Genet. 14, e1007807.

Roberts, A.G., Oparka, K.J., 2003. Plasmodesmata and the control of symplastic transport. Plant Cell Environ. 26, 103–124.

Roberts, S.K., Snowman, B.N., 2000. The effects of ABA on channel-mediated K^+ transport across higher plant roots. J. Exp. Bot. 51, 1585–1594.

Roberts, J.K.M., Callis, J., Jardetzky, O., Walbot, V., Freeling, M., 1984. Cytoplasmic acidosis as a determinant of flooding intolerance in plants. Proc. Natl. Acad. Sci. U. S. A. 81, 6029–6033.

Roberts, J.A., Elliott, K.A., Gonzalez-Carranza, Z.H., 2002. Abscission, dehiscence, and other cell separation processes. Annu. Rev. Plant Biol. 53, 131–158.

Robertson, G.L., Eschenbruch, R., Cresswell, K.J., 1980. Seasonal changes in the pectic substances of grapes and their implication in juice extraction. Am. J. Enol. Vitic. 31, 162–164.

Robichaud, J.L., Noble, A.C., 1990. Astringency and bitterness of selected phenolics in wine. J. Sci. Food Agric. 53, 343–353.

Robinson, D., 1994. The response of plants to non-uniform supplies of nutrients. New Phytol. 127, 635–674.

Robinson, S.P., Davies, C., 2000. Molecular biology of grape berry ripening. Aust. J. Grape Wine Res. 6, 175–188.

Robinson, S.A., Slade, A.P., Fox, G.G., Phillips, R., Ratcliffe, R.G., Stewart, G.R., 1991. The role of glutamate dehydrogenase in plant nitrogen metabolism. Plant Physiol. 95, 509–516.

Robinson, S.P., Jacobs, A.K., Dry, I.B., 1997. A class IV chitinase is highly expressed in grape berries during ripening. Plant Physiol. 114, 771–778.

Robinson, J., Harding, J., Vouillamoz, J., 2012. Wine Grapes. Harper Collins, New York, NY.

Robinson, A.L., Boss, P.K., Solomon, P.S., Trengove, R.D., Heymann, H., Ebeler, S.E., 2014. Origins of grape and wine aroma. Part 1. Chemical components and viticultural impacts. Am. J. Enol. Vitic. 65, 1–24.

Roby, G., Matthews, M.A., 2004. Relative proportions of seed, skin and flesh, in ripe berries from Cabernet Sauvignon grapevines grown in a vineyard either well irrigated or under water deficit. Aust. J. Grape Wine Res. 10, 74–82.

Rockwell, N.C., Su, Y.S., Lagarias, J.C., 2006. Phytochrome structure and signaling mechanisms. Annu. Rev. Plant Biol. 57, 837–858.

Rodríguez Montealegre, R., Romero Peces, R., Chacón Vozmediano, J.L., Martínez Gascueña, J., García Romero, E., 2006. Phenolic compounds in skins and seeds of ten grape *Vitis vinifera* varieties grown in a warm climate. J. Food Compos. Anal. 19, 687–693.

Rodríguez, A., Alquézar, B., Peña, L., 2013. Fruit aromas in mature fleshy fruits as signals of readiness for predation and seed dispersal. New Phytol. 197, 36–48.

Rodriguez-Dominguez, C.M., Buckley, T.N., Egea, G., de Cires, A., Hernandez-Santana, V., Martorell, S., Diaz-Espejo, A., 2016. Most stomatal closure in woody species under moderate drought can be explained by stomatal responses to leaf turgor. Plant Cell Environ. 39, 2014–2026.

Rodriguez-Lovelle, B., Gaudillére, J.P., 2002. Carbon and nitrogen partitioning in either fruiting or non-fruiting grapevines: effects of nitrogen limitation before and after veraison. Aust. J. Grape Wine Res. 8, 86–94.

Rodríguez-Palenzuela, P., Burr, T.J., Collmer, A., 1991. Polygalacturonase is a virulence factor in *Agrobacterium tumefaciens* biovar 3. J. Bacteriol. 173, 6547–6552.

Roelfsema, M.R.G., Hedrich, R., 2005. In the light of stomatal opening: new insights into 'the Watergate'. New Phytol. 167, 665–691.

Rogiers, S.Y., Clarke, S.J., 2013. Nocturnal and daytime stomatal conductance respond to root-zone temperature in 'Shiraz' grapevines. Ann. Bot. 111, 433–444.

Rogiers, S.Y., Smith, J.A., White, R., Keller, M., Holzapfel, B.P., Virgona, J.M., 2001. Vascular function in berries of *Vitis vinifera* (L) cv. Shiraz. Aust. J. Grape Wine Res. 7, 47–51.

Rogiers, S.Y., Hatfield, J.M., Jaudzems, V.G., White, R.G., Keller, M., 2004a. Grape berry cv. Shiraz epicuticular wax and transpiration during ripening and preharvest weight loss. Am. J. Enol. Vitic. 55, 121–127.

Rogiers, S.Y., Hatfield, J.M., Keller, M., 2004b. Irrigation, nitrogen, and rootstock effects on volume loss of berries from potted Shiraz vines. Vitis 43, 1–6.

Rogiers, S.Y., Greer, D.H., Hatfield, J.M., Orchard, B.A., Keller, M., 2006. Mineral sinks within ripening grape berries (*Vitis vinifera* L.). Vitis 45, 115–123.

Rogiers, S.Y., Greer, D.H., Hutton, R.J., Landsberg, J.J., 2009. Does night-time transpiration contribute to anisohydric behaviour in a *Vitis vinifera* cultivar? J. Exp. Bot. 60, 3751–3763.

Rogiers, S.Y., Smith, J.P., Holzapfel, B.P., Hardie, W.J., 2011. Soil temperature moderates grapevine carbohydrate reserves after bud break and conditions fruit set responses to photoassimilatory stress. Funct. Plant Biol. 38, 899–909.

Rogiers, S.Y., Greer, D.H., Hatfield, J.M., Hutton, R.J., Clarke, S.J., Hutchinson, P.A., Somers, A., 2012. Stomatal response of an anisohydric grapevine cultivar to evaporative demand, available soil moisture and abscisic acid. Tree Physiol. 32, 249–261.

Rogiers, S.Y., Clarke, S.J., Schmidtke, L.M., 2014. Elevated root-zone temperature hastens vegetative and reproductive development in Shiraz grapevines. Aust. J. Grape Wine Res. 20, 123–133.

Rohde, A., Bhalerao, R.P., 2007. Plant dormancy in the perennial context. Trends Plant Sci. 12, 217–223.

Roitsch, T., 1999. Source-sink regulation by sugar and stress. Curr. Opin. Plant Biol. 2, 198–206.

Roitsch, T., González, M.C., 2004. Function and regulation of plant invertases: sweet sensations. Trends Plant Sci. 9, 606–613.

Rojas-Lara, B.A., Morrison, J.C., 1989. Differential effects of shading fruit or foliage on the development and composition of grape berries. Vitis 28, 199–208.

Rolland, F., Baena-Gonzalez, E., Sheen, J., 2006. Sugar sensing and signaling in plants: conserved and novel mechanisms. Annu. Rev. Plant Biol. 57, 675–709.

Rolle, L., Siret, R., Segade, S.R., Maury, C., Gerbi, V., Jourjon, F., 2012. Instrumental texture analysis parameters as markers of table-grape and winegrape quality: a review. Am. J. Enol. Vitic. 63, 11–28.

Romero, P., Martinez-Cutillas, A., 2012. The effects of partial root-zone irrigation and regulated deficit irrigation on the vegetative and reproductive development of field-grown Monastrell grapevines. Irrig. Sci. 30, 377–396.

Romero, P., Fernández-Fernández, J.I., Martinez-Cutillas, A., 2010. Physiological thresholds for efficient regulated deficit-irrigation management in winegrapes grown under semiarid conditions. Am. J. Enol. Vitic. 61, 300–312.

Romero, P., Gil-Muñoz, R., del Amor, F.M., Valdés, E., Fernández, J.I., Martinez-Cutillas, A., 2013. Regulated deficit irrigation based upon optimum water status improves phenolic composition in Monastrell grapes and wines. Agric. Water Manag. 121, 85–101.

Romero, P., Botía, P., Keller, M., 2017. Hydraulics and gas exchange recover more rapidly from severe drought stress in small pot-grown grapevines than in field-grown plants. J. Plant Physiol. 216, 58–73.

Romero-Pérez, A.I., Ibern-Gómez, M., Lamuela-Raventós, R.M., de la Torre-Boronat, M.C., 1999. Piceid, the major resveratrol derivative in grape juices. J. Agric. Food Chem. 47, 1533–1536.

Roper, T.R., Williams, L.E., 1989. Net CO_2 assimilation and carbohydrate partitioning of grapevine leaves in response to trunk girdling and gibberellic acid application. Plant Physiol. 89, 1136–1140.

Roper, M.C., Greve, L.C., Warren, J.G., Labavitch, J.M., Kirkpatrick, B.C., 2007. *Xylella fastidiosa* requires polygalacturonase for colonization and pathogenicity in *Vitis vinifera* grapevines. Mol. Plant-Microbe Interact. 20, 411–419.

Ros Barceló, A., Pomar, F., López-Serrano, M., Pedreño, M.A., 2003. Peroxidase: a multifunctional enzyme in grapevines. Funct. Plant Biol. 30, 577–591.

Rose, J.K.C., Bennett, A.B., 1999. Cooperative disassembly of the cellulose-xyloglucan network of plant cell walls: parallels between cell expansion and fruit ripening. Trends Plant Sci. 4, 176–183.

Rose, J.K.C., Lee, H.H., Bennett, A.B., 1997. Expression of a divergent expansin gene is fruit-specific and ripening-regulated. Proc. Natl. Acad. Sci. U. S. A. 94, 5955–5960.

Rose, J.K.C., Hadfield, K.A., Labavitch, J.M., Bennett, A.B., 1998. Temporal sequence of cell wall disassembly in rapidly ripening melon fruit. Plant Physiol. 117, 345–361.

Rose, J.K.C., Saladié, M., Catalá, C., 2004. The plot thickens: new perspectives of primary cell wall modification. Curr. Opin. Plant Biol. 7, 296–301.

Rosenheim, O., 1920. Observations on anthocyanins. I. The anthocyanins of the young leaves of the grape vine. Biochem. J. 14, 178–188.

Rosenquist, J.K., Morrison, J.C., 1988. The development of the cuticle and epicuticular wax of the grape berry. Vitis 27, 63–70.

Rosenquist, J.K., Morrison, J.C., 1989. Some factors affecting cuticle and wax accumulation on grape berries. Am. J. Enol. Vitic. 40, 241–244.

Rosner, N., Cook, J.A., 1983. Effects of differential pruning on Cabernet Sauvignon grapevines. Am. J. Enol. Vitic. 34, 243–248.

Ross, J.J., Weston, D.E., Davidson, S.E., Reid, J.B., 2011. Plant hormone interactions: how complex are they? Physiol. Plant. 141, 299–309.

Rossouw, G.C., Smith, J.P., Barril, C., Deloire, A., Holzapfel, B.P., 2017. Carbohydrate distribution during berry ripening of potted grapevines: impact of water availability and leaf-to-fruit ratio. Sci. Hortic. 216, 215–225.

Roubelakis, K.A., Kliewer, W.M., 1976. Influence of light intensity and growth regulators on fruit-set and ovule fertilization in grape cultivars under low temperature conditions. Am. J. Enol. Vitic. 27, 163–167.

Roubelakis-Angelakis, K.A., Kliewer, W.M., 1979. The composition of bleeding sap from Thompson Seedless grapevines as affected by nitrogen fertilization. Am. J. Enol. Vitic. 30, 14–18.

Roubelakis-Angelakis, K.A., Kliewer, W.M., 1983. Ammonia assimilation in *Vitis vinifera* L.: II. Leaf and root glutamine synthetase. Vitis 22, 299–305.

Roubelakis-Angelakis, K.A., Kliewer, W.M., 1992. Nitrogen metabolism in grapevine. Hortic. Rev. 14, 407–452.

Roufet, M., Bayonove, C.L., Cordonnier, R.E., 1987. Etude de la composition lipidique du raisin, *Vitis vinifera* L.: evolution au cours de la maturation et localisation dans la baie. Vitis 26, 85–97.

Roujou de Boubée, D., Van Leeuwen, C., Dubourdieu, D., 2000. Organoleptic impact of 2-methoxy-3-isobutylpyrazine on red Bordeaux and Loire wines. Effect of environmental conditions on concentrations in grapes during ripening. J. Agric. Food Chem. 48, 4830–4834.

Roujou de Boubée, D., Cumsille, A.M., Pons, M., Dubourdieu, D., 2002. Location of 2-methoxy-3-isobutylpyrazine in Cabernet Sauvignon grape bunches and its extractability during vinification. Am. J. Enol. Vitic. 53, 1–5.

Rousseaux, M.C., Hall, A.J., Sánchez, R.A., 2000. Basal leaf senescence in a sunflower (*Helianthus annuus*) canopy: responses to increased R/FR ratio. Physiol. Plant. 110, 477–482.

Rowhani, A., Uyemoto, J.K., Golino, D.A., Martelli, G.P., 2005. Pathogen testing and certification of *Vitis* and *Prunus* species. Annu. Rev. Phytopathol. 43, 261–278.

Roy, R.R., Ramming, D.W., 1990. Varietal resistance of grape to the powdery mildew fungus, *Uncinula necator*. Fruit Var. J. 44, 149–155.

Royo, C., Carbonell-Bejerano, P., Torres-Pérez, R., Nebish, A., Martínez, Ó., Rey, M., Aroutiounian, R., Ibáñez, J., Martínez-Zapater, J.M., 2016. Developmental, transcriptome, and genetic alterations associated with parthenocarpy in the grapevine seedless somatic variant Corinto bianco. J. Exp. Bot. 67, 259–273.

Royo, C., Torres-Pérez, R., Mauri, N., Diestro, N., Cabezas, J.A., Marchal, C., Lacombe, T., Ibáñez, J., Tornel, M., Carreño, J., Martínez-Zapater, J.M., Carbonell-Bejerano, P., 2018. The major origin of seedless grapes is associated with a missense mutation in the MADS-box gene *VviAGL11*. Plant Physiol. 177, 1234–1253.

Roytchev, V.R., 1998. Microsporogenesis and male gametophyte development in seedless grapevine (*Vitis vinifera* L.) cultivars. Wein-Wiss. 53, 162–167.

Roytchev, V.R., 2000. Megasporogenesis and female gametophyte development in seedless grapevine (*Vitis vinifera* L.) cultivars. Wein-Wiss. 55, 21–28.

Rozema, J., van de Staaij, J., Björn, L.O., Caldwell, M., 1997. UV-B as an environmental factor in plant life: stress and regulation. Trees 12, 22–28.

Ruan, Y.L., 2014. Sucrose metabolism: gateway to diverse carbon use and sugar signaling. Annu. Rev. Plant Biol. 65, 33–67.

Ruan, Y.L., Patrick, J.W., Bouzayen, M., Osorio, S., Fernie, A.R., 2012. Molecular regulation of seed and fruit set. Trends Plant Sci. 17, 656–665.

Rubio, M., Alvarez-Ortí, M., Alvarruiz, A., Fernández, E., Pardo, J.E., 2009. Characterization of oil obtained from grape seeds collected during berry development. J. Agric. Food Chem. 57, 2812–2815.

Ruel, J.J., Walker, M.A., 2006. Resistance to Pierce's disease in *Muscadinia rotundifolia* and other native grape species. Am. J. Enol. Vitic. 57, 158–165.

Ruffner, H.P., 1982a. Metabolism of tartaric and malic acids in *Vitis*: a review—Part A. Vitis 21, 247–259.

Ruffner, H.P., 1982b. Metabolism of tartaric and malic acids in *Vitis*: a review—Part B. Vitis 21, 346–358.

Ruffner, H.P., Hawker, J.S., 1977. Control of glycolysis in ripening berries of *Vitis vinifera*. Phytochemistry 16, 1171–1175.

Ruffner, H.P., Koblet, W., Rast, D., 1975. Gluconeogenese in reifenden Beeren von *Vitis vinifera*. Vitis 13, 319–328.

Rufty, T.W., MacKown, C.T., Volk, R., 1989. Effects of altered carbohydrate availability on whole-plant assimilation of $^{15}NO_3^-$. Plant Physiol. 89, 457–463.

Rüger, S., Netzer, Y., Westhoff, M., Zimmermann, D., Reuss, R., Ovadiya, S., 2010. Remote monitoring of leaf turgor pressure of grapevines subjected to different irrigation treatments using the leaf patch clamp pressure probe. Aust. J. Grape Wine Res. 16, 405–412.

Rühl, E.H., Clingeleffer, P.R., 1993. Effect of minimal pruning and virus inoculation on the carbohydrate and nitrogen accumulation in Cabernet franc vines. Am. J. Enol. Vitic. 44, 81–85.

Rühl, E., Imgraben, H.J., 1985. Einfluß der Stickstoffversorgung auf die Stomatazahl und den Blattaufbau von Weinreben (*Vitis vinifera* L.). Wein-Wiss. 40, 160–171.

Rühl, E.H., Alleweldt, G., Hofäcker, W., 1981. Untersuchungen zum Lichtkompensationspunkt der Rebe. Vitis 20, 122–129.

Ruhl, E.H., Clingeleffer, P.R., Nicholas, P.R., Cirami, R.M., McCarthy, M.G., Whiting, J.R., 1988. Effect of rootstocks on berry weight and pH, mineral content and organic acid concentrations of grape juice of some wine varieties. Aust. J. Exp. Agric. 28, 119–125.

Ruhl, E.H., Fuda, A.P., Treeby, M.T., 1992. Effect of potassium, magnesium and nitrogen supply on grape juice composition of Riesling, Chardonnay and Cabernet Sauvignon vines. Aust. J. Exp. Agric. 32, 645–649.

Ruperti, B., Botton, A., Populin, F., Eccher, G., Brilli, M., Quaggiotti, S., et al., 2019. Flooding responses on grapevine: a physiological, transcriptional, and metabolic perspective. Front. Plant Sci. 10, 339.

Ruyter-Spira, C., Kohlen, W., Charnikhova, T., van Zeijl, A., van Bezouwen, L., de Ruijter, N., 2011. Physiological effects of the synthetic strigolactone analog GR24 on root system architecture in *Arabidopsis*: another belowground role for strigolactones? Plant Physiol. 155, 721–734.

Ruyter-Spira, C., Al-Babili, S., van der Krol, S., Bouwmeester, H., 2013. The biology of strigolactones. Trends Plant Sci. 18, 72–83.

Ryan, M.G., Yoder, B.J., 1997. Hydraulic limits to tree height and growth. Bioscience 47, 235–242.

Ryan, M.G., Phillips, N., Bond, B.J., 2006. The hydraulic limitation hypothesis revisited. Plant Cell Environ. 29, 367–381.

Ryona, I., Pan, B.S., Intrigliolo, D.S., Lakso, A.N., Sacks, G.L., 2008. Effects of cluster light exposure on 3-isobutyl-2-methoxypyrazine accumulation and degradation patterns in red wine grapes (*Vitis vinifera* L. cv. Cabernet franc). J. Agric. Food Chem. 56, 10838–10846.

Ryona, I., Pan, B.S., Sacks, G.L., 2009. Rapid measurement of 3-alkyl-2-methoxypyrazine content of winegrapes to predict levels in resultant wines. J. Agric. Food Chem. 57, 8250–8257.

Ryona, I., Leclerc, S., Sacks, G.L., 2010. Correlation of 3-isobutyl-2-methoxypyrazine to 3-isobutyl-2-hydroxypyrazine during maturation of bell pepper (*Capsicum annuum*) and wine grapes (*Vitis vinifera*). J. Agric. Food Chem. 58, 9723–9730.

Sack, L., Holbrook, N.M., 2006. Leaf hydraulics. Annu. Rev. Plant Biol. 57, 361–381.

Sack, L., Cowan, P.D., Jaikumar, N., Holbrook, N.M., 2003. The 'hydrology' of leaves: co-ordination of structure and function in temperate woody species. Plant Cell Environ. 26, 1343–1356.

Sack, L., John, G.P., Buckley, T.N., 2018. ABA accumulation in dehydrating leaves is associated with decline in cell volume, not turgor pressure. Plant Physiol. 176, 489–493.

Sadras, V.O., Moran, M.A., 2012. Elevated temperature decouples anthocyanins and sugars in berries of Shiraz and Cabernet franc. Aust. J. Grape Wine Res. 18, 115–122.

Sadras, V.A., Moran, M.A., 2013. Nonlinear effects of elevated temperature on grapevine phenology. Agric. For. Meteorol. 173, 107–115.

Sáenz-Navajas, M.P., Tao, Y.S., Dizy, M., Ferreira, V., Fernández-Zurbano, P., 2010. Relationship between nonvolatile composition and sensory properties of premium Spanish red wines and their correlation to quality perception. J. Agric. Food Chem. 58, 12407–12416.

Sage, R.F., Kubien, D.S., 2007. The temperature response of C_3 and C_4 photosynthesis. Plant Cell Environ. 30, 1086–1106.

Sage, R.F., Sage, T.L., Kocacinar, F., 2012. Photorespiration and the evolution of C_4 photosynthesis. Annu. Rev. Plant Biol. 63, 19–47.

Saini, H.S., 1997. Effects of water stress on male gametophyte development in plants. Sex. Plant Reprod. 10, 67–73.

Saito, K., Ohmoto, J., Kuriha, N., 1997. Incorporation of ^{18}O into oxalic, l-threonic and l-tartaric acids during cleavage of l-ascorbic and 5-keto-d-gluconic acids in plants. Phytochemistry 44, 805–809.

Sakakibara, H., 2006. Cytokinins: activity, biosynthesis, and translocation. Annu. Rev. Plant Biol. 57, 431–449.

Sakakibara, H., Takei, K., Hirose, N., 2006. Interactions between nitrogen and cytokinin in the regulation of metabolism and development. Trends Plant Sci. 11, 440–448.

Sakr, S., Alves, G., Morillon, R., Maurel, K., Decourteix, M., Guilliot, A., 2003. Plasma membrane aquaporins are involved in winter embolism recovery in walnut tree. Plant Physiol. 133, 630–641.

Sala, A., Hoch, G., 2009. Height-related growth declines in ponderosa pine are not due to carbon limitation. Plant Cell Environ. 32, 22–30.

Sala, C., Busto, O., Guasch, J., Zamora, F., 2005. Contents of 3-alkyl-2-methoxypyrazines in musts and wines from *Vitis vinifera* variety Cabernet Sauvignon: influence of irrigation and plantation density. J. Sci. Food Agric. 85, 1131–1136.

Sala, A., Piper, F., Hoch, G., 2010. Physiological mechanisms of drought-induced tree mortality are far from being resolved. New Phytol. 186, 274–281.

Salazar-Parra, C., Aranjuelo, I., Pascual, I., Erice, G., Sanz-Sáez, Á., Aguirreolea, J., Sánchez-Díaz, M., Irigoyen, J.J., Araus, J.L., Morales, F., 2015. Carbon balance, partitioning and photosynthetic acclimation in fruit-bearing grapevine (*Vitis vinifera* L. cv. Tempranillo) grown under simulated climate change (elevated CO_2, elevated temperature and moderate drought) scenarios in temperature gradient greenhouses. J. Plant Physiol. 174, 97–109.

Salinas, J., Warnaar, F., Verhoeff, K., 1986. Production of cutin hydrolyzing enzymes by *Botrytis cinerea in vitro*. J. Phytopathol. 116, 299–307.

Salisbury, F.B., Ross, C.W., 1992. Plant Physiology, fourth ed. Wadsworth Publishing, Belmont, CA.

Salmasco, M., Dalla Valle, R., Lucchin, M., 2008. Gene pool variation and phylogenetic relationships of an indigenous northeast Italian grapevine collection revealed by nuclear and chloroplast SSRs. Genome 51, 838–855.

Salomé, P.A., McClung, C.R., 2005. What makes the *Arabidopsis* clock tick on time? A review on entrainment. Plant Cell Environ. 28, 21–38.

Salón, J.L., Chirivella, C., Castel, J.R., 2005. Response of cv. Bobal to timing of deficit irrigation in Requena, Spain: water relations, yield, and wine quality. Am. J. Enol. Vitic. 56, 1–8.

Salzman, R.A., Bressan, R.A., Hasegawa, P.M., Ashworth, E.N., Bordelon, B.P., 1996. Programmed accumulation of LEA-like proteins during desiccation and cold acclimation of overwintering grape buds. Plant Cell Environ. 19, 713–720.

Salzman, R.A., Tikhonova, I., Bordelon, B.P., Hasegawa, P.M., Bressan, R.A., 1998. Coordinate accumulation of antifungal proteins and hexoses constitutes a developmentally controlled defense response during fruit ripening in grape. Plant Physiol. 117, 465–472.

Samach, A., Wigge, P.A., 2005. Ambient temperature perception in plants. Curr. Opin. Plant Biol. 8, 483–486.

Sampson, B., Noffsinger, S., Gupton, C., Magee, J., 2001. Pollination biology of the muscadine grape. HortScience 36, 120–124.

Samuels, L., Kunst, L., Jetter, R., 2008. Sealing plant surfaces: cuticular wax formation by epidermal cells. Annu. Rev. Plant Biol. 59, 683–707.

Sánchez, L.A., Dokoozlian, N.K., 2005. Bud microclimate and fruitfulness in *Vitis vinifera* L. Am. J. Enol. Vitic. 56, 319–329.

Sanchez, P., Nehlin, L., Greb, T., 2012. From thin to thick: major transitions during stem development. Trends Plant Sci. 17, 113–121.
Sandermann, H., Ernst, D., Heller, W., Langebartels, C., 1998. Ozone: an abiotic elicitor of plant defence reactions. Trends Plant Sci. 3, 47–50.
Santana, J.C., Heuertz, M., Arranz, C., Rubio, J.A., Martínez-Zapater, J.M., Hidalgo, E., 2010. Genetic structure, origins, and relationships of grapevine cultivars from the Castilian Plateau of Spain. Am. J. Enol. Vitic. 61, 214–224.
Santesteban, L.G., Royo, J.B., 2006. Water status, leaf area and fruit load influence on berry weight and sugar accumulation of cv. 'Tempranillo' under semiarid conditions. Sci. Hortic. 109, 60–65.
Santner, A., Calderon-Villalobos, L.I.A., Estelle, M., 2009. Plant hormones are versatile chemical regulators of plant growth. Nat. Chem. Biol. 5, 301–307.
Santos, T.P., Lopes, C.M., Rodrigues, M.L., de Souza, C.R., Ricardo-Da-Silva, J.M., Maroco, J.P., 2005. Effects of partial root-zone drying irrigation on cluster microclimate and fruit composition of field-grown Castelão grapevines. Vitis 44, 117–125.
Sapis, J.C., Macheix, J.J., Cordonnier, R.E., 1983. The browning capacity of grapes. II. Browning potential and polyphenol oxidase activities in different mature grape varieties. Am. J. Enol. Vitic. 34, 157–162.
Saqib, M., Zörb, C., Schubert, S., 2008. Silicon-mediated improvement in the salt resistance of wheat (*Triticum aestivum*) results from increased sodium exclusion and resistance to oxidative stress. Funct. Plant Biol. 35, 633–639.
Sarry, J.E., Sommerer, N., Sauvage, F.X., Bergoin, A., Rossignol, M., Albagnac, G., 2004. Grape berry biochemistry revisited upon proteomic analysis of the mesocarp. Proteomics 4, 201–215.
Sartorius, O., 1926. Zur Entwicklung und Physiologie der Rebblüte. Angew. Bot. 8, 29–62.
Sartorius, O., 1968. Die Blütenknospen der Rebe. Wein-Wiss. 23, 309–338.
Sartorius, O., 1973. Holzreserven und Traubenreife. Wein-Wiss. 28, 100–107.
Sassi, M., Vernoux, T., 2013. Auxin and self-regulation at the shoot apical meristem. J. Exp. Bot. 64, 2579–2592.
Sattelmacher, B., 2001. The apoplast and its significance for plant mineral nutrition. New Phytol. 149, 167–192.
Sauer, M.R., 1968. Effects of vine rootstocks on chloride concentration in Sultana scions. Vitis 7, 223–226.
Saulnier, L., Brillouet, J.M., 1989. An arabinogalactan-protein from the pulp of grape berries. Carbohydr. Res. 188, 137–144.
Savaldi-Goldstein, S., Peto, C., Chory, J., 2007. The epidermis both drives and restricts plant shoot growth. Nature 446, 199–202.
Saveyn, A., Steppe, K., Ubierna, N., Dawson, T.E., 2010. Woody tissue photosynthesis and its contribution to trunk growth and bud development in young plants. Plant Cell Environ. 33, 1949–1958.
Savoi, S., Wong, D.C.J., Arapitsas, P., Miculan, M., Bucchetti, B., Peterlunger, E., Fait, A., Mattivi, F., Castellarin, S.D., 2016. Transcriptome and metabolite profiling reveals that prolonged drought modulates the phenylpropanoid and terpenoid pathway in white grapes (*Vitis vinifera* L.). BMC Plant Biol. 16, 67.
Savoi, S., Wong, D.C.J., Degu, A., Herrera, J.C., Bucchetti, B., Peterlunger, E., Fait, A., Mattivi, F., Castellarin, S.D., 2017. Multi-omics and integrated network analyses reveal new insights into the systems relationships between metabolites, structural genes, and transcriptional regulators in developing grape berries (*Vitis vinifera* L.) exposed to water deficit. Front. Plant Sci. 8, 1124.
Sawler, J., Reisch, B., Aradhya, M.K., Prins, B., Zhong, G.Y., Schwaninger, H., 2013. Genomics assisted ancestry deconvolution in grape. PLoS ONE 8, e80791.
Sbaghi, M., Jeandet, P., Bessis, R., Leroux, P., 1996. Degradation of stilbene-type phytoalexins in relation to the pathogenicity of *Botrytis cinerea* to grapevines. Plant Pathol. 45, 139–144.
Schachtman, D.P., Shin, R., 2007. Nutrient sensing and signaling: NPKS. Annu. Rev. Plant Biol. 58, 47–69.

Schaefer, H., 1981. Der jahreszeitliche Verlauf des Stoffwechsels der löslichen und unlöslichen Proteine in den verholzten Reborganen. Wein-Wiss. 36, 3–20.

Schaller, K., Löhnertz, O., 1991. N-Stoffwechsel von Reben. 3. Mitteilung: Arginin- und Prolindynamik in vegetativen und generativen Teilen der Müller-Thurgau-Rebe im Verlaufe einer Vegetationsperiode. Wein-Wiss. 46, 151–160.

Schaller, K., Löhnertz, O., Oswald, D., Sprengart, B., 1985. Nitratanreicherung in Reben. 3. Mitteilung: Nitratdynamik in Rappen und Beeren während einer Vegetationsperiode in verschiedenen Rebsorten. Wein-Wiss. 40, 147–159.

Schaller, K., Löhnertz, O., Geiben, R., Breit, N., 1989. N-Stoffwechsel von Reben. 1. Mitteilung: N- und Arginindynamik im Holzkörper der Sorte Müller-Thurgau im Verlaufe einer Vegetationsperiode. Wein-Wiss. 44, 91–101.

Schaller, K., Löhnertz, O., Geiben, R., 1990. N-Stoffwechsel von Reben. 2. Mitteilung: N- und Aminosäuredynamik in vegetativen und generativen Teilen der Müller-Thurgau-Rebe im Verlaufe einer Vegetationsperiode. Wein-Wiss. 45, 160–166.

Schär, C., Vidale, P.L., Lüthi, D., Frei, C., Häberli, C., Liniger, M.A., Appenzeller, C., 2004. The role of increasing temperature variability in European summer heatwaves. Nature 427, 332–336.

Scharwies, J.D., Tyerman, S.D., 2017. Comparison of isohydric and anisohydric *Vitis vinifera* L. cultivars reveals a fine balance between hydraulic resistances, driving forces and transpiration in ripening berries. Funct. Plant Biol. 44, 324–338.

Scheenen, T.W.J., Vergeldt, F.J., Heemskerk, A.M., Van As, H., 2007. Intact plant magnetic resonance imaging to study dynamics in long-distance sap flow and flow-conducting surface area. Plant Physiol. 144, 1157–1165.

Scheffeldt, P., Hrazdina, G., 1978. Co-pigmentation of anthocyanins under physiological conditions. J. Food Sci. 43, 517–520.

Scheible, W.R., Lauerer, M., Schulze, E.D., Caboche, M., Stitt, M., 1997. Accumulation of nitrate in the shoot acts as a signal to regulate shoot-root allocation in tobacco. Plant J. 11, 671–691.

Scheible, W.R., Morcuende, R., Czechowski, T., Fritz, C., Osuna, D., Palacios-Rojas, N., 2004. Genome-wide reprogramming of primary and secondary metabolism, protein synthesis, cellular growth processes, and the regulatory infrastructure of *Arabidopsis* in response to nitrogen. Plant Physiol. 136, 2483–2499.

Scheiner, J.J., Sacks, G.L., Pan, B., Ennahli, S., Tarlton, L., Wise, A., 2010. Impact of severity and timing of basal leaf removal on 3-isobutyl-2-methoxypyrazine concentrations in red winegrapes. Am. J. Enol. Vitic. 61, 358–364.

Scheiner, J.J., Vanden Heuvel, J.E., Pan, B., Sacks, G.L., 2012. Modeling impacts of viticultural and environmental factors on 3-isobutyl-2-methoxypyrazine in Cabernet franc grapes. Am. J. Enol. Vitic. 63, 94–105.

Scherz, W., 1939. Sind selbstfertile hermaphrodite Weinreben obligat autogam? Züchter 11, 244–249.

Schippers, J.H.M., Schmidt, R., Wagstaff, C., Jing, H.C., 2015. Living to die and dying to live: the survival strategy behind leaf senescence. Plant Physiol. 169, 914–930.

Schlosser, J., Olsson, N., Weis, M., Reid, K., Peng, F., Lund, S., 2008. Cellular expansion and gene expression in the developing grape (*Vitis vinifera* L.). Protoplasma 232, 255–265.

Schmidlin, L., Poutaraud, A., Claudel, P., Mestre, P., Prado, E., Santos-Rosa, M., 2008. A stress-inducible resveratrol *O*-methyltransferase involved in the biosynthesis of pterostilbene in grapevine. Plant Physiol. 148, 1630–1639.

Schmidt, W., 2003. Iron solutions: acquisition strategies and signaling pathways in plants. Trends Plant Sci. 8, 188–193.

Schmitt, B., Stadler, R., Sauer, N., 2008. Immunolocalization of solanaceous SUT1 proteins in companion cells and xylem parenchyma: new perspectives for phloem loading and transport. Plant Physiol. 148, 187–199.

Schnabel, B.J., Wample, R.L., 1987. Dormancy and cold hardiness of *Vitis vinifera* L. cv. White Riesling as influenced by photoperiod and temperature. Am. J. Enol. Vitic. 38, 265–272.

Schnee, S., Viret, O., Gindro, K., 2008. Role of stilbenes in the resistance of grapevine to powdery mildew. Physiol. Mol. Plant Pathol. 72, 128–133.

Schneider, W., Staudt, G., 1978. Zur Abhängigkeit des Verrieselns von Umwelt und Genom bei *Vitis vinifera*. Vitis 17, 45–53.

Scholander, P.F., Love, W.E., Kanwisher, J.W., 1955. The rise of sap in tall grapevines. Plant Physiol. 30, 93–104.

Scholefield, P.B., Ward, R.C., 1975. Scanning electron microscopy of the developmental stages of the Sultana inflorescence. Vitis 14, 14–19.

Scholefield, P.B., May, P., Neales, T.F., 1977a. Harvest-pruning and trellising of 'Sultana' vines. 1. Effects on yield and vegetative growth. Sci. Hortic. 7, 115–122.

Scholefield, P.B., May, P., Neales, T.F., 1977b. Harvest-pruning and trellising of Sultana vines. 2. Effects on early spring development. Sci. Hortic. 7, 123–132.

Scholefield, P.B., Neales, T.F., May, P., 1978. Carbon balance of the Sultana vine (*Vitis vinifera* L.) and the effects of autumn defoliation by harvest-pruning. Aust. J. Plant Physiol. 5, 561–570.

Schönherr, J., 2000. Calcium chloride penetrates plant cuticles via aqueous pores. Planta 212, 112–118.

Schönherr, J., 2006. Characterization of aqueous pores in plant cuticles and permeation of ionic solutes. J. Exp. Bot. 57, 2471–2491.

Schönherr, J., Eckl, K., Gruler, H., 1979. Water permeability of plant cuticles: the effect of temperature on diffusion of water. Planta 147, 21–26.

Schopfer, P., 2006. Biomechanics of plant growth. Am. J. Bot. 93, 1415–1425.

Schopp, L.M., Lee, J., Osborne, J.P., Chescheir, S.C., Edwards, C.G., 2013. Metabolism of nonesterified and esterified hydroxycinnamic acids in red wines by *Brettanomyces bruxellensis*. J. Agric. Food Chem. 61, 11610–11617.

Schreiber, L., 2005. Polar paths of diffusion across plant cuticles: new evidence for an old hypothesis. Ann. Bot. 95, 1069–1073.

Schreiber, L., 2010. Transport barriers made of cutin, suberin and associated waxes. Trends Plant Sci. 15, 546–553.

Schreiber, L., Skrabs, M., Hartmann, K.D., Diamantopoulos, P., Simanova, E., Santrucek, J., 2001. Effect of humidity on culticular water permeability of isolated culticular membranes and leaf disks. Planta 214, 274–282.

Schreiner, R.P., 2005. Spatial and temporal variation of roots, arbuscular mycorrhizal fungi, and plant and soil nutrients in a mature Pinot noir (*Vitis vinifera* L.) vineyard in Oregon, USA. Plant Soil 276, 219–234.

Schreiner, R.P., 2016. Nutrient uptake and distribution in young Pinot noir grapevines over two seasons. Am. J. Enol. Vitic. 67, 436–448.

Schreiner, R.P., Osborne, J., 2018. Defining phosphorus requirements for Pinot noir grapevines. Am. J. Enol. Vitic. 69, 351–359.

Schreiner, R.P., Scagel, C.F., Baham, J., 2006. Nutrient uptake and distribution in a mature 'Pinot noir' vineyard. HortScience 41, 336–345.

Schreiner, R.P., Tarara, J.M., Smithyman, R.P., 2007. Deficit irrigation promotes arbuscular colonization of fine roots by mycorrhizal fungi in grapevines (*Vitis vinifera* L.) in an arid climate. Mycorrhiza 17, 551–562.

Schreiner, R.P., Lee, J., Skinkis, P.A., 2013. N, P, and K supply to Pinot noir grapevines: impact on vine nutrient status, growth, physiology, and yield. Am. J. Enol. Vitic. 64, 26–38.

Schreiner, R.P., Scagel, C.F., Lee, J., 2014. N, P, and K supply to Pinot noir grapevines: impact on berry phenolics and free amino acids. Am. J. Enol. Vitic. 65, 43–49.

Schreiner, R.P., Osborne, J., Skinkis, P.A., 2018. Nitrogen requirements of Pinot noir based on growth parameters, must composition, and fermentation behavior. Am. J. Enol. Vitic. 69, 45–58.

Schröder, J., 1997. A family of plant-specific polyketide synthases: facts and predictions. Trends Plant Sci. 2, 373–378.

Schroth, M.N., McCain, A.H., Foott, J.H., Huisman, O.C., 1988. Reduction in yield and vigor of grapevine caused by crown gall disease. Plant Dis. 72, 241–246.

Schubert, A., Restagno, M., Novello, V., Peterlunger, E., 1995. Effects of shoot orientation on growth, net photosynthesis, and hydraulic conductivity of *Vitis vinifera* L. cv. Cortese. Am. J. Enol. Vitic. 46, 324–328.

Schubert, R., Fischer, R., Hain, R., Schreier, P.H., Bahnweg, G., Ernst, D., 1997. An ozone-responsive region of the grapevine resveratrol synthase promoter differs from the basal pathogen-responsive sequence. Plant Mol. Biol. 34, 417–426.

Schubert, A., Lovisolo, C., Peterlunger, E., 1999. Shoot orientation affects vessel size, shoot hydraulic conductivity and shoot growth rate in *Vitis vinifera* L. Plant Cell Environ. 22, 197–204.

Schuetz, M., Smith, R., Ellis, B., 2013. Xylem tissue specification, patterning, and differentiation mechanisms. J. Exp. Bot. 64, 11–31.

Schultz, H.B., 1961. Microclimates on spring frost nights in Napa Valley vineyards. Am. J. Enol. Vitic. 12, 81–87.

Schultz, H.R., 1991. Seasonal and nocturnal changes in leaf age dependent dark respiration of grapevine sun and shade leaves. Modelling the temperature effect. Wein-Wiss. 46, 129–141.

Schultz, H.R., 2000. Climate change and viticulture: a European perspective on climatology, carbon dioxide and UV-B effects. Aust. J. Grape Wine Res. 6, 2–12.

Schultz, H.R., 2003. Differences in hydraulic architecture account for near-isohydric and anisohydric behaviour of two field-grown *Vitis vinifera* L. cultivars during drought. Plant Cell Environ. 26, 1393–1405.

Schultz, H.R., Matthews, M.A., 1988a. Resistance to water transport in shoots of *Vitis vinifera* L. relation to growth at low water potential. Plant Physiol. 88, 718–724.

Schultz, H.R., Matthews, M.A., 1988b. Vegetative growth distribution during water deficits in *Vitis vinifera* L. Aust. J. Plant Physiol. 15, 641–656.

Schultz, H.R., Matthews, M.A., 1993a. Growth, osmotic adjustment, and cell-wall mechanics of expanding grape leaves during water deficits. Crop Sci. 33, 287–294.

Schultz, H.R., Matthews, M.A., 1993b. Xylem development and hydraulic conductance in sun and shade shoots of grapevine (*Vitis vinifera* L.): evidence that low light uncouples water transport capacity from leaf area. Planta 190, 393–406.

Schultz, H.R., Stoll, M., 2010. Some critical issues in environmental physiology of grapevines: future challenges and current limitations. Aust. J. Grape Wine Res. 16, 4–24.

Schultz, H.R., Kiefer, W., Gruppe, W., 1996. Photosynthetic duration, carboxylation efficiency and stomatal limitation of sun and shade leaves of different ages in field-grown grapevine (*Vitis vinifera* L.). Vitis 35, 169–176.

Schultz, J.C., Edger, P.P., Body, M.J.A., Appel, H.M., 2019. A galling insect activates plant reproductive programs during gall development. Sci. Rep. 9, 1833.

Schumann, F., 1974. Beziehungen zwischen Edelreis und Unterlagen—langjährige Ergebnisse aus Adaptationsversuchen. Wein-Wiss. 29, 216–229.

Schumann, F., Frieß, H., 1976. Beziehung zwischen Edelreis und Unterlagen. 2. Teil: Unterlagen für kalkreiche Böden. Wein-Wiss 31, 94–120.

Schütz, M., Kunkee, R.E., 1977. Formation of hydrogen sulfide from elemental sulfur during fermentation by wine yeast. Am. J. Enol. Vitic. 28, 137–144.

Schwab, W., Wüst, M., 2015. Understanding the constitutive and induced biosynthesis of mono- and sesquiterpenes in grapes (*Vitis vinifera*): a key to unlocking the biochemical secrets of unique grape aroma profiles. J. Agric. Food Chem. 63, 10591–10603.

Schwab, W., Davidovich-Rikanati, R., Lewinsohn, E., 2008. Biosynthesis of plant-derived flavor compounds. Plant J. 54, 712–732.

Schwarz, G., Mendel, R.R., 2006. Molybdenum cofactor biosynthesis and molybdenum enzymes. Annu. Rev. Plant Biol. 57, 623–647.

Scienza, A., Boselli, M., 1981. Fréquence et caractéristiques biométriques des stomates de certains porte-greffes de vigne. Vitis 20, 281–292.

Scienza, A., Düring, H., 1980. Stickstoffernährung und Wasserhaushalt bei Reben. Vitis 19, 301–307.
Scienza, A., Miravalle, R., Visai, C., Fregoni, M., 1978. Relationships between seed number, gibberellin and abscisic acid levels and ripening in Cabernet Sauvignon grape berries. Vitis 17, 361–368.
Scienza, A., Failla, O., Romano, F., 1986. Untersuchungen zur sortenspezifischen Mineralstoffaufnahme bei Reben. Vitis 25, 160–168.
Scoffoni, C., Albuquerque, C., Brodersen, C.R., Townes, S.V., John, G.P., Bartlett, M.K., Buckley, T.N., McElrone, A.J., Sack, L., 2017. Outside-xylem vulnerability, not xylem embolism, controls leaf hydraulic decline during dehydration. Plant Physiol. 173, 1197–1210.
Scrase-Field, S.A.M.G., Knight, M.R., 2003. Calcium: just a chemical switch? Curr. Opin. Plant Biol. 6, 500–506.
Screen, J.A., Simmonds, I., 2013. The central role of diminishing sea ice in recent Artic temperature amplification. Nature 464, 1334–1337.
Secchi, F., Zwieniecki, M.A., 2010. Patterns of PIP gene expression in *Populus trichocarpa* during recovery from xylem embolism suggest a major role for the PIP1 aquaporin subfamily as moderators of refilling process. Plant Cell Environ. 33, 1285–1297.
Secchi, F., Zwieniecki, M.A., 2012. Analysis of xylem sap from functional (nonembolized) and nonfunctional (embolized) vessels of *Populus nigra*: chemistry of refilling. Plant Physiol. 160, 955–964.
Seddon, T.J., Downey, M.O., 2008. Comparison of analytical methods for the determination of condensed tannins in grape skin. Aust. J. Grape Wine Res. 14, 54–61.
Seemann, J.R., Sharkey, T.D., Wang, J., Osmond, C.B., 1987. Environmental effects on photosynthesis, nitrogen-use efficiency, and metabolite pools in leaves of sun and shade plants. Plant Physiol. 84, 796–802.
Sefc, K.M., Steinkellner, H., Glössl, J., Kampfer, S., Regner, F., 1998. Reconstruction of a grapevine pedigree by microsatellite analysis. Theor. Appl. Genet. 97, 227–231.
Sefc, K.M., Steinkellner, H., Lefort, F., Botta, R., da Câmara Machado, A., Borrego, J., 2003. Evaluation of the genetic contribution of local wild vines to European grapevine cultivars. Am. J. Enol. Vitic. 54, 15–21.
Sefton, M.A., Francis, I.L., Williams, P.J., 1990. Volatile norisoprenoid compounds as constituents of oak woods used in wine and spirit maturation. J. Agric. Food Chem. 38, 2045–2049.
Sefton, M.A., Francis, I.L., Williams, P.J., 1993. The volatile composition of Chardonnay juices: a study by flavor precursor analysis. Am. J. Enol. Vitic. 44, 359–370.
Sefton, M.A., Skouroumounis, G.K., Elsey, G.M., Taylor, D.K., 2011. Occurrence, sensory impact, formation, and fate of damascenone in grapes, wines, and other foods and beverages. J. Agric. Food Chem. 59, 9717–9746.
Segurel, M.A., Razungles, A.J., Riou, C., Salles, M., Baumes, R.L., 2004. Contribution of dimethyl sulfide to the aroma of Syrah and Grenache noir wines and estimation of its potential in grapes of these varieties. J. Agric. Food Chem. 52, 7084–7093.
Sekse, L., 1995. Fruit cracking in sweet cherries (*Prunus avium* L.). Some physiological aspects—a mini review. Sci. Hortic. 63, 135–141.
Selitrennikoff, C.P., 2001. Antifungal proteins. Appl. Environ. Microbiol. 67, 2883–2894.
Semenov, V.A., 2012. Meteorology: Arctic warming favours extremes. Nat. Clim. Change 2, 315–316.
Seo, P.J., Kim, M.J., Park, J.Y., Kim, S.Y., Jeon, J., Lee, Y.H., 2010. Cold activation of a plasma membrane-tethered NAC transcription factor induces a pathogen resistance response in *Arabidopsis*. Plant J. 61, 661–671.
Sepúlveda, G., Kliewer, W.M., 1986. Effect of high temperature on grapevines (*Vitis vinifera* L.). II. Distribution of soluble sugars. Am. J. Enol. Vitic. 37, 20–25.
Sepúlveda, G., Kliewer, W.M., Ryugo, K., 1986. Effect of high temperature on grapevines (*Vitis vinifera* L.). I. Translocation of ^{14}C-photosynthates. Am. J. Enol. Vitic. 37, 13–19.
Serrani, J.C., Fos, M., Atarés, A., García-Martínez, J.L., 2007. Effect of gibberellin and auxin on parthenocarpic fruit growth induction in the cv. Micro-Tom of tomato. J. Plant Growth Regul. 26, 211–221.
Serrani, J.C., Ruiz-Rivero, O., Fos, M., García-Martínez, J.L., 2008. Auxin-induced fruit-set in tomato is mediated in part by gibberellins. Plant J. 56, 922–934.

Serratosa, M.P., Lopez-Toledano, A., Merida, J., Medina, M., 2008. Changes in color and phenolic compounds during the raisining of grape cv. Pedro Ximenez. J. Agric. Food Chem. 56, 2810–2816.

Seymour, G.B., Østergaard, L., Chapman, N.H., Knapp, S., Martin, C., 2013. Fruit development and ripening. Annu. Rev. Plant Biol. 64, 219–241.

Shackel, K.A., Matthews, M.A., Morrison, J.C., 1987. Dynamic relation between expansion and cellular turgor in growing grape (*Vitis vinifera* L.) leaves. Plant Physiol. 84, 1166–1171.

Shani, U., Ben-Gal, A., 2005. Long-term response of grapevines to salinity: osmotic effects and ion toxicity. Am. J. Enol. Vitic. 56, 148–154.

Shani, U., Waisel, Y., Eshel, A., Xue, S., Ziv, G., 1993. Responses to salinity of grapevine plants with split root systems. New Phytol. 124, 695–701.

Sharkey, T.D., Wiberley, A.E., Donohue, A.R., 2008. Isoprene emission from plants: why and how. Ann. Bot. 101, 5–18.

Sharp, R.E., LeNoble, M.E., 2002. ABA, ethylene and the control of shoot and root growth under water stress. J. Exp. Bot. 53, 33–37.

Sharp, R.E., Poroyko, V., Hejlek, L.G., Spollen, W.G., Springer, G.K., Bohnert, H.J., 2004. Root growth maintenance during water deficits: physiology to functional genomics. J. Exp. Bot. 55, 2343–2351.

Shatil-Cohen, A., Attia, Z., Moshelion, M., 2011. Bundle-sheath cell regulation of xylem-mesophyll water transport via aquaporins under drought stress: a target of xylem-borne ABA? Plant J. 67, 72–80.

Shaul, O., 2002. Magnesium transport and function in plants: the tip of the iceberg. Biometals 15, 309–323.

Shaulis, N., Amberg, H., Crowe, D., 1966. Response of Concord grapes to light exposure and Geneva double courtain training. Proc. Am. Soc. Hortic. Sci. 89, 268–280.

Shellie, K.C., 2011. Interactive effects of deficit irrigation and berry exposure aspect on Merlot and Cabernet Sauvignon in an arid climate. Am. J. Enol. Vitic. 62, 462–470.

Shellie, K.C., 2014. Water productivity, yield, and berry composition in sustained versus regulated deficit irrigation of Merlot grapevines. Am. J. Enol. Vitic. 65, 197–205.

Shellie, K.C., Bowen, P., 2014. Isohydrodynamic behavior in deficit-irrigated Cabernet Sauvignon and Malbec and its relationship between yield and berry composition. Irrig. Sci. 32, 87–97.

Shellie, K.C., King, B.A., 2013. Kaolin particle film and water deficit influence red winegrape color under high solar radiation in an arid climate. Am. J. Enol. Vitic. 64, 214–222.

Shen, J., Yuan, L., Zhang, J., Li, H., Bai, Z., Chen, X., 2011. Phosphorus dynamics: from soil to plant. Plant Physiol. 156, 997–1005.

Shepherd, T., Griffiths, D.W., 2006. The effects of stress on plant cuticular waxes. New Phytol. 171, 469–499.

Sheppard, A.E., Ayliffe, M.A., Blatch, L., Day, A., Delaney, S.K., Khairul-Fahmy, N., 2008. Transfer of plastid DNA to the nucleus is elevated during male gametogenesis in tobacco. Plant Physiol. 148, 328–336.

Shibuya, N., Minami, E., 2001. Oligosaccharide signaling for defense responses in plants. Physiol. Mol. Plant Pathol. 59, 223–233.

Shimazaki, K.I., Doi, M., Assmann, S.M., Kinoshita, T., 2007. Light regulation of stomatal movement. Annu. Rev. Plant Biol. 58, 219–247.

Shinozaki, Y., Nicolas, P., Fernandez-Pozo, N., Ma, Q., Evanich, D.J., Shi, Y., et al., 2018. High-resolution spatiotemporal transcriptome mapping of tomato fruit development and ripening. Nat. Commun. 9, 364.

Shiratake, K., Martinoia, E., 2007. Transporters in fruit vacuoles. Plant Biotechnol. 24, 127–133.

Shirke, P.A., Pathre, U.V., 2004. Influence of leaf-to-air vapour pressure deficit (VPD) on the biochemistry and physiology of photosynthesis in *Prosopis juliflora*. J. Exp. Bot. 55, 2111–2120.

Shure, K.B., Acree, T.E., 1994. Changes in the odor-active compounds in *Vitis labruscana* cv. Concord during growth and development. J. Agric. Food Chem. 42, 350–353.

Siddall, E.C., Marples, N.M., 2011. The effect of pyrazine odor on avoidance learning and memory in wild robins. Curr. Zool. 57, 208–214.

Sidlowski, J.J., Phillips, W.S., Kuykendall, J.R., 1971. Phloem regeneration across girdles of grape vines. J. Am. Soc. Hortic. Sci. 96, 97–102.

Silacci, M.W., Morrison, J.C., 1990. Changes in pectin content of Cabernet Sauvignon grape berries during maturation. Am. J. Enol. Vitic. 41, 111–115.

Simonin, K.A., Burns, E., Choat, B., Barbour, M.M., Dawson, T.E., Franks, P.J., 2015. Increasing leaf hydraulic conductance with transpiration rate minimizes the water potential drawdown from stem to leaf. J. Exp. Bot. 66, 1303–1315.

Simpson, R.F., Miller, G.C., 1983. Aroma composition of aged Riesling wine. Vitis 22, 51–63.

Simpson, R.F., Miller, G.C., 1984. Aroma composition of Chardonnay wine. Vitis 23, 143–158.

Singleton, V.L., Trousdale, E., 1983. White wine phenolics: varietal and processing differences as shown by HPLC. Am. J. Enol. Vitic. 34, 27–34.

Singleton, V.L., Ough, C.S., Nelson, K.E., 1966. Density separations of wine grape berries and ripeness distribution. Am. J. Enol. Vitic. 17, 95–105.

Singleton, V.L., Trousdale, E., Zaya, J., 1985. One reason sun-dried raisins brown so much. Am. J. Enol. Vitic. 36, 111–113.

Singleton, V.L., Zaya, J., Trousdale, E., 1986. Compositional changes in ripening grapes: caftaric and coutaric acids. Vitis 25, 107–117.

Singleton, V.L., 1992. Tannins and the qualities of wines. In: Hemingway, R.W., Laks, P.E. (Eds.), Plant Polyphenols. Plenum Press, New York, NY, pp. 859–880.

Sinilal, B., Ovadia, R., Nissim-Levi, A., Perl, A., Carmeli-Weissberg, M., Oren-Shamir, M., 2011. Increased accumulation and decreased catabolism of anthocyanins in red grape cell suspension culture following magnesium treatment. Planta 234, 61–71.

Skene, K.G.M., Antcliff, A.J., 1972. A comparative study of cytokinin levels in bleeding sap of *Vitis vinifera* (L.) and the two grapevine rootstocks, Salt Creek and 1613. J. Exp. Bot. 23, 283–293.

Skene, K.G.M., Kerridge, G.H., 1967. Effect of root temperature on cytokinin activity in root exudate of *Vitis vinifera* L. Plant Physiol. 42, 1131–1139.

Skinkis, P.A., Gregory, K.M., 2017. Spur pruning may be a viable option for Oregon Pinot noir producers despite industry fears of lower productivity. Catalyst (2), 62–72.

Skinkis, P.A., Bordelon, B.P., Wood, K.V., 2008. Comparison of monoterpene constituents in Traminette, Gewürztraminer, and Riesling winegrapes. Am. J. Enol. Vitic. 59, 440–445.

Skinner, P.W., Matthews, M.A., 1990. A novel interaction of magnesium translocation with the supply of phosphorus to roots of grapevine (*Vitis vinifera* L.). Plant Cell Environ. 13, 821–826.

Skinner, P.W., Cook, J.A., Matthews, M.A., 1988. Responses of grapevine cvs. Chenin blanc and Chardonnay to phosphorus fertilizer applications under phosphorus-limited conditions. Vitis 27, 95–109.

Skopelitis, D.S., Paranychianakis, N.V., Paschalidis, K.A., Pliakonis, E.D., Delis, I.D., Yakoumakis, D.I., et al., 2006. Abiotic stress generates ROS that signal expression of anionic glutamate dehydrogenases to form glutamate for proline synthesis in tobacco and grapevine. Plant Cell 18, 2767–2781.

Skypala, I.J., Williams, M., Reeves, L., Meyer, R., Venter, C., 2015. Sensitivity to food additives, vaso-active amines and salicylates: a review of the evidence. Clin. Transl. Allergy 5, 34.

Smart, R.E., 1974. Aspects of water relations of the grapevine (*Vitis vinifera*). Am. J. Enol. Vitic. 25, 84–91.

Smart, R.E., 1985. Principles of grapevine canopy microclimate manipulation with implications for yield and quality. A review. Am. J. Enol. Vitic. 36, 230–239.

Smart, R.E., 1988. Shoot spacing and canopy light microclimate. Am. J. Enol. Vitic. 39, 325–333.

Smart, R.E., Sinclair, T.R., 1976. Solar heating of grape berries and other spherical fruits. Agric. Meteorol. 17, 241–259.

Smart, R.E., Shaulis, N.J., Lemon, E.R., 1982. The effect of Concord vineyard microclimate on yield. I. The effects of pruning, training, and shoot positioning on radiation microclimate. Am. J. Enol. Vitic. 33, 99–108.

Smart, R.E., Coombe, B.G., 1983. Water relations in grapevine. In: Kozlowski, T.T. (Ed.), Water Deficits and Plant Growth. Vol. VII. Additional Woody Crop Plants. Academic Press, New York, NY, pp. 137–196.

Smart, R.E., Robinson, J.B., Due, G.R., Brien, C.J., 1985a. Canopy microclimate modification for the cultivar Shiraz. I. Definition of canopy microclimate. Vitis 24, 17–31.

Smart, R.E., Robinson, J.B., Due, G.R., Brien, C.J., 1985b. Canopy microclimate modification for the cultivar Shiraz. II. Effects on must and wine composition. Vitis 24, 119–128.

Smart, R.E., Smith, S.M., Winchester, R.V., 1988. Light quality and quantity effects on fruit ripening for Cabernet Sauvignon. Am. J. Enol. Vitic. 39, 250–258.

Smart, R.E., Dick, J.K., Gravett, I.M., Fisher, B.M., 1990. Canopy management to improve grape yield and wine quality—principles and practices. S. Afr. J. Enol. Vitic. 11, 3–17.

Smart, D.R., Kocsis, L., Walker, M.A., Stockert, C., 2003. Dormant buds and adventitious root formation by *Vitis* and other woody plants. J. Plant Growth Regul. 21, 296–314.

Smart, D.R., Carlisle, E., Goebel, M., Núñez, B.A., 2005. Transverse hydraulic redistribution by a grapevine. Plant Cell Environ. 28, 157–166.

Smart, D.R., Schwass, E., Lakso, A., Morano, L., 2006. Grapevine rooting patterns: a comprehensive analysis and a review. Am. J. Enol. Vitic. 57, 89–104.

Smith, C.J., 1996. Accumulation of phytoalexins: defence mechanism and stimulus response system. New Phytol. 132, 1–45.

Smith, H., 2000. Phytochromes and light signal perception by plants—an emerging synthesis. Nature 407, 585–591.

Smith, A.M., 2012. Starch in the *Arabidopsis* plant. Starch-Stärke 64, 421–434.

Smith, F.A., Raven, J.A., 1979. Intracellular pH and its regulation. Annu. Rev. Plant Physiol. 30, 289–311.

Smith, S.E., Smith, F.A., 2011. Roles of arbuscular mycorrhizas in plant nutrition and growth: new paradigms from cellular to ecosystem scales. Annu. Rev. Plant Biol. 62, 227–250.

Smith, H., Whitelam, G.C., 1997. The shade avoidance syndrome: multiple responses mediated by multiple phytochromes. Plant Cell Environ. 20, 840–844.

Smith, S.E., Dickson, S., Smith, F.A., 2001. Nutrient transfer in arbuscular mycorrhizas: how are fungal and plant processes integrated? Aust. J. Plant Physiol. 28, 683–694.

Smith, A.M., Zeeman, S.C., Smith, S.M., 2005. Starch degradation. Annu. Rev. Plant Biol. 56, 73–98.

Smithyman, R.P., Howell, G.S., Miller, D.P., 1998. The use of competition for carbohydrates among vegetative and reproductive sinks to reduce fruit set and *Botrytis* bunch rot in Seyval blanc grapevines. Am. J. Enol. Vitic. 49, 163–170.

Smithyman, R.P., Wample, R.L., Lang, N.S., 2001. Water deficit and crop level influences on photosynthetic strain and blackleaf symptom development in Concord grapevines. Am. J. Enol. Vitic. 52, 364–375.

Snedden, W.A., Fromm, H., 2001. Calmodulin as a versatile calcium signal transducer in plants. New Phytol. 151, 35–66.

Snyder, J.C., 1933. Flower bud formation in the Concord grape. Bot. Gaz. 94, 771–779.

Soar, C.J., Loveys, B.R., 2007. The effect of changing patterns in soil-moisture availability on grapevine root distribution, and viticultural implications for converting full-cover irrigation into a point-source irrigation system. Aust. J. Grape Wine Res. 13, 2–13.

Soar, C.J., Speirs, J., Maffei, S.M., Loveys, B.R., 2004. Gradients in stomatal conductance, xylem sap ABA and bulk leaf ABA along canes of *Vitis vinifera* cv. Shiraz: molecular and physiological studies investigating their source. Funct. Plant Biol. 31, 659–669.

Soar, C.J., Dry, P.R., Loveys, B.R., 2006a. Scion photosynthesis and leaf gas exchange in *Vitis vinifera* L. cv. Shiraz: mediation of rootstock effects via xylem sap ABA. Aust. J. Grape Wine Res. 12, 82–96.

Soar, C.J., Speirs, J., Maffei, S.M., Penrose, A.B., McCarthy, M.G., Loveys, B.R., 2006b. Grape vine varieties Shiraz and Grenache differ in their stomatal response to VPD: apparent links with ABA physiology and gene expression in leaf tissue. Aust. J. Grape Wine Res. 12, 2–12.

Soares, S., Mateus, N., de Freitas, V., 2012. Carbohydrates inhibit salivary proteins precipitation by condensed tannins. J. Agric. Food Chem. 60, 3966–3972.

Soltis, P.S., Soltis, D.E., 2009. The role of hybridization in plant speciation. Annu. Rev. Plant Biol. 60, 561–588.

Somers, T.C., 1968. Pigment profiles of grapes and of wines. Vitis 7, 303–320.

Somers, T.C., 1971. The polymeric nature of wine pigments. Phytochemistry 10, 2175–2186.

Sommer, K.J., Clingeleffer, P.R., Shulman, Y., 1995. Comparative study of vine morphology, growth, and canopy development in cane-pruned and minimal-pruned Sultana. Aust. J. Exp. Agric. 35, 265–273.

Sommer, K.J., Islam, M.T., Clingeleffer, P.R., 2000. Light and temperature effects on shoot fruitfulness in *Vitis vinifera* L. cv. Sultana: influence of trellis type and grafting. Aust. J. Grape Wine Res. 6, 99–108.

Somssich, I.E., Hahlbrock, K., 1998. Pathogen defence in plants—a paradigm of biological complexity. Trends Plant Sci. 3, 86–90.

Sondergaard, T.E., Schulz, A., Palmgren, M.G., 2004. Energization of transport processes in plants. Roles of the plasma membrane H^+-ATPase. Plant Physiol. 136, 2475–2482.

Soubeyrand, E., Basteau, C., Hilbert, G., van Leeuwen, C., Delrot, S., Gomès, E., 2014. Nitrogen supply affects anthocyanin biosynthetic and regulatory genes in grapevine cv. Cabernet-Sauvignon berries. Phytochemistry 103, 38–49.

Souquet, J.M., Cheynier, V., Brossaud, F., Moutounet, M., 1996. Polymeric proanthocyanidins from grape skins. Phytochemistry 43, 509–512.

Souquet, J.M., Labarbe, B., Le Guerneve, C., Cheynier, V., Moutounet, M., 2000. Phenolic composition of grape stems. J. Agric. Food Chem. 48, 1076–1080.

Spalding, E.P., Folta, K.M., 2005. Illuminating topics in plant photobiology. Plant Cell Environ. 28, 39–53.

Spayd, S.E., Nagel, C.W., Hayrynen, L.D., Ahmedullah, M., 1987. Effect of freezing fruit on the composition of musts and wines. Am. J. Enol. Vitic. 38, 243–245.

Spayd, S.E., Wample, R.L., Stevens, R.G., Evans, R.G., Kawakami, A.K., 1993. Nitrogen fertilization of White Riesling in Washington: effects on petiole nutrient concentration, yield, yield components, and vegetative growth. Am. J. Enol. Vitic. 44, 378–386.

Spayd, S.E., Wample, R.L., Evans, R.G., Stevens, R.G., Seymour, B.J., Nagel, C.W., 1994. Nitrogen fertilization of White Riesling grapes in Washington. Must and wine composition. Am. J. Enol. Vitic. 45, 34–42.

Spayd, S.E., Nagel, C.W., Edwards, C.G., 1995. Yeast growth in Riesling juice as affected by vineyard nitrogen fertilization. Am. J. Enol. Vitic. 46, 49–55.

Spayd, S.E., Tarara, J.M., Mee, D.L., Ferguson, J.C., 2002. Separation of sunlight and temperature effects on the composition of *Vitis vinifera* cv. Merlot berries. Am. J. Enol. Vitic. 53, 171–182.

Speirs, J., Binney, A., Collins, M., Edwards, E., Loveys, B., 2013. Expression of ABA synthesis and metabolism genes under different irrigation strategies and atmospheric VPDs is associated with stomatal conductance in grapevine (*Vitis vinifera* L. cv Cabernet Sauvignon). J. Exp. Bot. 64, 1907–1916.

Speiser, A., Silbermann, M., Dong, Y., Haberland, S., Vural Uslu, V., Wang, S., et al., 2018. Sulfur partitioning between glutathione and protein synthesis determines plant growth. Plant Physiol. 177, 927–937.

Sperry, J.S., 2004. Coordinating stomatal and xylem functioning—an evolutionary perspective. New Phytol. 162, 568–570.

Sperry, J.S., Holbrook, N.M., Zimmermann, M.H., Tyree, M.T., 1987. Spring filling of xylem vessels in wild grapevine. Plant Physiol. 83, 414–417.

Sperry, J.S., Adler, F.R., Campbell, G.S., Comstock, J.P., 1998. Limitation of plant water use by rhizosphere and xylem conductance: results from a model. Plant Cell Environ. 21, 347–359.

Sperry, J.S., Hacke, U.G., Oren, R., Comstock, J.P., 2002. Water deficits and hydraulic limits to leaf water supply. Plant Cell Environ. 25, 251–263.

Sperry, J.S., Stiller, V., Hacke, U.G., 2003. Xylem hydraulics and the soil-plant-atmosphere continuum: opportunities and unresolved issues. Agron. J. 95, 1362–1370.

Spreitzer, R.J., Salvucci, M.E., 2002. RUBISCO: structure, regulatory interactions, and possibilities for a better enzyme. Annu. Rev. Plant Biol. 53, 449–475.

Srinivasan, C., Mullins, M.G., 1978. Control of flowering in the grapevine (*Vitis vinifera* L.). Plant Physiol. 61, 127–130.

Srinivasan, C., Mullins, M.G., 1980. Effects of temperature and growth regulators on formation of anlagen, tendrils and inflorescences in *Vitis vinifera* L. Ann. Bot. 45, 439–446.

Srinivasan, C., Mullins, M.G., 1981. Physiology of flowering in the grapevine—a review. Am. J. Enol. Vitic. 32, 47–63.

Stafford, H.A., 1988. Proanthocyanidins and the lignin connection. Phytochemistry 27, 1–6.

Stahl, E., 1900. Der Sinn der Mycorrhizenbildung. Jahrb. wiss. Bot. 34, 539–668.

Starkenmann, C., Le Calvé, B., Niclass, Y., Cayeux, I., Beccucci, S., Troccaz, M., 2008. Olfactory perception of cysteine–*S*-conjugates from fruits and vegetables. J. Agric. Food Chem. 56, 9575–9580.

Staudt, G., 1982. Pollenkeimung und Pollenschlauchwachstum *in vivo* bei *Vitis* und die Abhängigkeit von der Temperatur. Vitis 21, 205–216.

Staudt, G., 1999. Opening of flowers and time of anthesis in grapevines, *Vitis vinifera* L. Vitis 38, 15–20.

Staudt, G., Kassemeyer, H.H., 1984. Entstehen kleine Beeren bei *Vitis vinifera* durch Parthenocarpie? Vitis 23, 205–213.

Staudt, G., Kassrawi, M., 1973. Untersuchungen über das Rieseln di- und tetraploider Reben. Vitis 12, 1–15.

Staudt, G., Schneider, W., Leidel, J., 1986. Phases of berry growth in *Vitis vinifera*. Ann. Bot. 58, 789–800.

Steel, C.C., Keller, M., 2000. Influence of UV-B irradiation on the carotenoid content of *Vitis vinifera* tissues. Biochem. Soc. Trans. 28, 883–885.

Steenwerth, K.L., McElrone, A.J., Calderon-Orellana, A., Hanifin, R.C., Storm, C., Collatz, W., 2013. Cover crops and tillage in a mature Merlot vineyard show few effects on grapevines. Am. J. Enol. Vitic. 64, 515–521.

Steenwerth, K.L., Calderón-Orellana, A., Hanifin, R.C., Storm, C., McElrone, A.J., 2016. Effects of various vineyard floor management techniques on weed community shifts and grapevine water relations. Am. J. Enol. Vitic. 67, 153–162.

Stefanini, I., Dapporto, L., Legras, J.L., Calabretta, A., Di Paola, M., De Filippo, C., 2012. Role of social wasps in *Saccharomyces cerevisiae* ecology and evolution. Proc. Natl. Acad. Sci. U. S. A. 109, 13398–13403.

Steffan, H., Rapp, A., 1979. Ein Beitrag zum Nachweis unterschiedlicher Malatpools in Beeren der Rebe. Vitis 18, 100–105.

Stegemann, S., Bock, R., 2009. Exchange of genetic material between cells in plant tissue grafts. Science 324, 649–651.

Stein, U., Blaich, R., 1985. Untersuchungen über Stilbenproduktion und Botrytisanfälligkeit bei *Vitis*-Arten. Vitis 24, 75–87.

Steponkus, P.L., 1984. Role of the plasma membrane in freezing injury and cold acclimation. Annu. Rev. Plant Physiol. 35, 543–584.

Steppe, K., Lemeur, R., 2004. An experimental system for analysis of the dynamic sap-flow characteristics in young trees: results of a beech tree. Funct. Plant Biol. 31, 83–92.

Stergios, B.G., Howell, G.S., 1977. Effect of site on cold acclimation and deacclimation of Concord grapevines. Am. J. Enol. Vitic. 28, 43–48.

Steudle, E., 2000. Water uptake by roots: effects of water deficit. J. Exp. Bot. 51, 1531–1542.

Steudle, E., 2001. The cohesion-tension mechanism and the acquisition of water by plant roots. Annu. Rev. Plant Physiol. Plant Mol. Biol. 52, 847–875.

Steudle, E., Peterson, C.A., 1998. How does water get through roots? J. Exp. Bot. 49, 775–788.
Stevens, R.M., Douglas, T., 1994. Distribution of grapevine roots and salt under drip and full-ground cover micro-jet irrigation systems. Irrig. Sci. 15, 147–152.
Stevens, R.M., Prior, L.D., 1994. The effect of transient waterlogging on the growth, leaf gas exchange, and mineral composition of potted Sultana grapevines. Am. J. Enol. Vitic. 45, 285–290.
Stevens, R.M., Walker, R.R., 2002. Response of grapevines to irrigation-induced saline–sodic soil conditions. Aust. J. Exp. Agric. 42, 323–331.
Stevens, R.M., Harvey, G., Aspinall, D., 1995. Grapevine growth of shoots and fruit linearly correlate with water stress indices based on root-weighted soil matric potential. Aust. J. Grape Wine Res. 1, 58–66.
Stevens, R.M., Harvey, G., Johns, R.E., 1999. Waterlogging reduces shoot growth and bud fruitfulness in pot-grown grapevines with a split-root system. Aust. J. Grape Wine Res. 5, 99–103.
Stevens, R.M., Pech, J.M., Gibberd, M.R., Jones, J.A., Taylor, J., Nicholas, P.R., 2008. Effect of reduced irrigation on growth, yield, ripening rates and water relations of Chardonnay vines grafted to five rootstocks. Aust. J. Grape Wine Res. 14, 177–190.
Stevens, R.M., Pech, J.M., Gibberd, M.R., Walker, R.R., Nicholas, P.R., 2010. Reduced irrigation and rootstock effects on vegetative growth, yield and its components, and leaf physiological responses of Shiraz. Aust. J. Grape Wine Res. 16, 413–425.
Stevens, R.M., Pech, J.M., Taylor, J., Clingeleffer, P., Walker, R.R., Nicholas, P.R., 2016. Effects of irrigation and rootstock on *Vitis vinifera* (L.) cv. Shiraz berry composition and shrivel, and wine composition and wine score. Aust. J. Grape Wine Res. 22, 124–136.
Stevenson, J.F., Matthews, M.A., Greve, L.C., Labavitch, J.M., Rost, T.L., 2004. Grapevine susceptibility to Pierce's disease II: progression of anatomical symptoms. Am. J. Enol. Vitic. 55, 238–245.
Stevenson, J.F., Matthews, M.A., Rost, T.L., 2005. The developmental anatomy of Pierce's disease symptoms in grapevines: green islands and matchsticks. Plant Dis. 89, 543–548.
Stiles, K.A., Van Volkenburgh, E., 2004. Role of K^+ in leaf growth: K^+ uptake is required for light-stimulated H^+ efflux but not solute accumulation. Plant Cell Environ. 27, 315–325.
Stines, A.P., Grubb, J., Gockowiak, H., Henschke, P.A., Høj, P.B., van Heeswijck, R., 2000. Proline and arginine accumulation in developing berries of *Vitis vinifera* L. in Australian vineyards: influence of vine cultivar, berry maturity and tissue type. Aust. J. Grape Wine Res. 6, 150–158.
Stitt, M., 1999. Nitrate regulation of metabolism and growth. Curr. Opin. Plant Biol. 2, 178–186.
Stitt, M., Müller, C., Matt, P., Gibon, Y., Carillo, P., Morcuende, R., 2002. Steps towards an integrated view of nitrogen metabolism. J. Exp. Bot. 53, 959–970.
Stitt, M., Gibon, Y., Lunn, J.E., Piques, M., 2007. Multilevel genomics analysis of carbon signalling during low carbon availability: coordinating the supply and utilisation of carbon in a fluctuating environment. Funct. Plant Biol. 34, 526–549.
Stoev, K.D., Dobreva, S.I., Wosteninez, G., 1966. Über die Synthese von Aminosäuren im Wurzelsystem der Rebe. Vitis 5, 265–287.
Stoll, M., Loveys, B., Dry, P., 2000. Hormonal changes induced by partial rootzone drying of irrigated grapevine. J. Exp. Bot. 51, 1627–1634.
Storey, R., 1987. Potassium localization in the grape berry pericarp by energy-dispersive X-ray microanalysis. Am. J. Enol. Vitic. 38, 301–309.
Storey, R., Schachtman, D.P., Thomas, M.R., 2003a. Root structure and cellular chloride, sodium and potassium distribution in salinized grapevines. Plant Cell Environ. 26, 789–800.
Storey, R., Wyn Jones, R.G., Schachtman, D.P., Treeby, M.T., 2003b. Calcium-accumulating cells in the meristematic region of grapevine root apices. Funct. Plant Biol. 30, 719–727.
Stover, E.W., Swartz, H.J., Burr, T.J., 1997. Crown gall formation in a diverse collection of *Vitis* genotypes inoculated with *Agrobacterium vitis*. Am. J. Enol. Vitic. 48, 26–32.

Strader, L.C., Chen, G.L., Bartel, B., 2010. Ethylene directs auxin to control root cell expansion. Plant J. 64, 874–884.

Stradwick, L., Inglis, D., Kelly, J., Pickering, G., 2017. Development and application of assay for determining β-glucosidase activity in human saliva. Flavour 6, 1.

Strauss, G., Hauser, H., 1986. Stabilization of lipid bilayer vesicles by sucrose during freezing. Proc. Natl. Acad. Sci. U. S. A. 83, 2422–2426.

Strauss, C.R., Wilson, B., Anderson, R., Williams, P.J., 1987. Development of precursors of C_{13} nor-isoprenoid flavorants in Riesling grapes. Am. J. Enol. Vitic. 38, 23–27.

Strefeler, M.S., Weeden, N.F., Reisch, B.I., 1992. Inheritance of chloroplast DNA in two full-sib *Vitis* populations. Vitis 31, 183–187.

Striegler, R.K., Howell, G.S., Flore, J.A., 1993. Influence of rootstock on the response of Seyval grapevines to flooding stress. Am. J. Enol. Vitic. 44, 313–319.

Stummer, B.E., Francis, I.L., Zanker, T., Lattey, K.A., Scott, E.S., 2005. Effects of powdery mildew on the sensory properties and composition of Chardonnay juice and wine when grape sugar ripeness is standardized. Aust. J. Grape Wine Res. 11, 66–76.

Suárez, N., 2010. Leaf lifetime photosynthetic rate and leaf demography in whole plants of *Ipomoea pes-caprae* growing with a low supply of calcium, a 'non-mobile' nutrient. J. Exp. Bot. 61, 843–855.

Sudarshana, M.R., Perry, K.L., Fuchs, M.F., 2015. Grapevine red blotch-associated virus, an emerging threat to the grapevine industry. Phytopathology 105, 1026–1032.

Sulpice, R., Pyl, E.T., Ishihara, H., Trenkamp, S., Steinfarth, M., Witucka-Wall, H., Gibon, Y., Usadel, B., Poree, F., Conceição Piques, M., Von Korff, M., Steinhauser, M.C., Keurentjes, J.J.B., Guenther, M., Hoehne, M., Selbig, J., Fernie, A.R., Altmann, T., Stitt, M., 2009. Starch as a major integrator in the regulation of plant growth. Proc. Natl. Acad. Sci. U. S. A. 106, 10348–10353.

Sun, B.S., Pinto, T., Leandro, M.C., Ricardo-da-Silva, J.M., Spranger, M.I., 1999. Transfer of catechins and proanthocyanidins from solid parts of the grape cluster into wine. Am. J. Enol. Vitic. 50, 179–184.

Sun, Q., Rost, T.L., Matthews, M.A., 2006. Pruning-induced tylose development in stems of current-year shoots of *Vitis vinifera* (Vitaceae). Am. J. Bot. 93, 1567–1576.

Sun, Q., Rost, T.L., Reid, M.S., Matthews, M.A., 2007. Ethylene and not embolism is required for wound-induced tylose development in stems of grapevines. Plant Physiol. 145, 1629–1636.

Sun, Q., Rost, T.L., Matthews, M.A., 2008. Wound-induced vascular occlusions in *Vitis vinifera* (Vitaceae): tyloses in summer and gels in winter. Am. J. Bot. 95, 1498–1505.

Sun, L., Zhang, M., Ren, J., Qi, J., Zhang, G., Leng, P., 2010. Reciprocity between abscisic acid and ethylene at the onset of berry ripening and after harvest. BMC Plant Biol. 10, 257.

Sun, Q., Greve, L.C., Labavitch, J.M., 2011a. Polysaccharide compositions of intervessel pit membranes contribute to Pierce's disease resistance of grapevines. Plant Physiol. 155, 1976–1987.

Sun, Q., Gates, M.J., Lavin, E.H., Acree, T.E., Sacks, G.L., 2011b. Comparison of odor-active compounds in grapes and wines from *Vitis vinifera* and non-foxy American grape species. J. Agric. Food Chem. 59, 10657–10664.

Sun, Q., Sun, Y., Walker, M.A., Labavitch, J.M., 2013. Vascular occlusions in grapevines with Pierce's disease make disease symptom development worse. Plant Physiol. 161, 1529–1541.

Swanepoel, J.J., Archer, E., 1988. The ontogeny and development of *Vitis vinifera* L. cv. Chenin blanc inflorescence in relation to phenological stages. Vitis 27, 133–141.

Swanepoel, J.J., de la Harpe, A.C., Orffer, C.J., 1984. A comparative anatomical study of the grapevine shoot: I Epidermis. S. Afr. J. Enol. Vitic. 5, 51–57.

Swanson, C.A., El-Shishiny, E.D.H., 1958. Translocation of sugars in the Concord grape. Plant Physiol. 33, 33–37.

Sweet, W.J., Morrison, J.C., Labavitch, J.M., Matthews, M.A., 1990. Altered synthesis and composition of cell wall of grape (*Vitis vinifera* L.) leaves during expansion and growth-inhibiting water deficits. Plant Cell Physiol. 31, 407–414.

Sweetlove, L.J., Beard, K.F.M., Nunes-Nesi, A., Fernie, A.R., Ratcliffe, R.G., 2010. Not just a circle: flux modes in the plant TCA cycle. Trends Plant Sci. 15, 462–470.

Sweetman, C., Deluc, L.G., Cramer, G.R., Ford, C.M., Soole, K.L., 2009. Regulation of malate metabolism in grape berry and other developing fruits. Phytochemistry 70, 1329–1344.

Sweetman, C., Sadras, V.O., Hancock, R.D., Soole, K.L., Ford, C.M., 2014. Metabolic effects of elevated temperature on organic acid degradation in ripening *Vitis vinifera* fruit. J. Exp. Bot. 65, 5975–5988.

Swiegers, J.H., Bartowsky, E.J., Henschke, P.A., Pretorius, I.S., 2005. Yeast and bacterial modulation of wine aroma and flavour. Aust. J. Grape Wine Res. 11, 139–173.

Swift, J.G., Buttrose, M.S., Possingham, J.V., 1973. Stomata and starch in grape berries. Vitis 12, 38–45.

Symons, G.M., Davies, C., Shavrukov, Y., Dry, I.B., Reid, J.B., Thomas, M.R., 2006. Grapes on steroids. Brassinosteroids are involved in grape berry ripening. Plant Physiol. 140, 150–158.

Szegedi, E., Korbuly, J., Koleda, I., 1984. Crown gall resistance in East-Asian *Vitis* species and in their *V. vinifera* hybrids. Vitis 23, 21–26.

Szyjewicz, E., Rosner, N., Kliewer, W.M., 1984. Ethephon ((2-chloroethyl) phosphonic acid, Ethrel, CEPA) in viticulture—a review. Am. J. Enol. Vitic. 35, 117–123.

Tabacchi, R., 1994. Secondary phytotoxic metabolites from pathogenic fungi: structure, synthesis and activity. Pure Appl. Chem. 66, 2299–2302.

Taiz, L., 1984. Plant cell expansion: regulation of cell wall mechanical properties. Annu. Rev. Plant Physiol. 35, 585–657.

Taiz, L., Zeiger, E., 2006. Plant Physiology, fourth ed. Sinauer, Sunderland, MA.

Takabayashi, J., Dicke, M., 1996. Plant-carnivore mutualism through herbivore-induced carnivore attractants. Trends Plant Sci. 1, 109–113.

Takahashi, S., Badger, M.R., 2011. Photoprotection in plants: a new light on photosystem II damage. Trends Plant Sci. 16, 53–60.

Takahashi, T., Kakehi, J.I., 2010. Polyamines: ubiquitous polycations with unique roles in growth and stress responses. Ann. Bot. 105, 1–6.

Takahashi, S., Murata, N., 2008. How do environmental stresses accelerate photoinhibition? Trends Plant Sci. 13, 178–182.

Takahashi, N., Yamazaki, Y., Kobayashi, A., Higashitani, A., Takahaski, H., 2003. Hydrotropism interacts with gravitropism by degrading amyloplasts in seedling roots of *Arabidopsis* and radish. Plant Physiol. 132, 805–810.

Takahashi, H., Kopriva, S., Giordano, M., Saito, K., Hell, R., 2011. Sulfur assimilation in photosynthetic organisms: molecular functions and regulations of transporters and assimilatory enzymes. Annu. Rev. Plant Biol. 62, 157–184.

Takahashi, K., Hayashi, K., Kinoshita, T., 2012. Auxin activates the plasma membrane H^+-ATPase by phosphorylation during hypocotyl elongation in *Arabidopsis*. Plant Physiol. 159, 632–641.

Takano, J., Miwa, K., Fujiwara, T., 2008. Boron transport mechanisms: collaboration of channels and transporters. Trends Plant Sci. 13, 451–457.

Takei, K., Takahashi, T., Sugiyama, T., Yamaya, T., Sakakibara, H., 2002. Multiple routes communicating nitrogen availability from roots to shoots: a signal transduction pathway mediated by cytokinin. J. Exp. Bot. 53, 971–977.

Takimoto, K., Saito, K., Kasai, Z., 1976. Diurnal change of tartrate dissimilation during the ripening of grapes. Phytochemistry 15, 927–930.

Talbott, L.D., Zeiger, E., 1996. Central roles for potassium and sucrose in guard-cell osmoregulation. Plant Physiol. 111, 1051–1057.

Tallman, G., 2004. Are diurnal patterns of stomatal movement the result of alternating metabolism of endogenous guard cell ABA and accumulation of ABA delivered to the apoplast around guard cells by transpiration? J. Exp. Bot. 55, 1963–1976.

Tanaka, M., Takei, K., Kojima, M., Sakakibara, H., Mori, H., 2006. Auxin controls local cytokinin biosynthesis in the nodal stem in apical dominance. Plant J. 45, 1028–1036.

Tanaka, Y., Sasaki, N., Ohmiya, A., 2008. Biosynthesis of plant pigments: anthocyanins, betalains and carotenoids. Plant J. 54, 733–749.

Tang, D., Wang, Y., Cai, J., Zhao, R., 2009. Effects of exogenous application of plant growth regulators on the development of ovule and subsequent embryo rescue of stenospermic grape (*Vitis vinifera* L.). Sci. Hortic. 120, 51–57.

Tanner, W., Beevers, H., 2001. Transpiration, a prerequisite for long-distance transport of minerals in plants? Proc. Natl. Acad. Sci. U. S. A. 98, 9443–9447.

Tarancón, C., González-Grandío, E., Oliveros, J.C., Nicolas, M., Cubas, P., 2017. A conserved carbon starvation response underlies bud dormancy in woody and herbaceous species. Front. Plant Sci. 8, 788.

Tarara, J.M., Ferguson, J.C., 2006. Two algorithms for variable power control of heat-balance sap flow gauges under high flow rates. Agron. J. 98, 830–838.

Tarara, J.M., Hellman, E.W., 1991. 'Norton' and 'Cynthiana'—premium native wine grapes. Fruit Var. J. 45, 66–69.

Tarara, J.M., Perez Peña, J.E., 2015. Moderate water stress from regulated deficit irrigation decreases transpiration similarly to net carbon exchange in grapevine canopies. J. Am. Soc. Hortic. Sci. 140, 413–426.

Tarara, J.M., Ferguson, J.C., Hoheisel, G.A., Perez Peña, J.E., 2005. Asymmetrical canopy architecture due to prevailing wind direction and row orientation creates an imbalance in irradiance at the fruiting zone of grapevines. Agric. For. Meteorol. 135, 144–155.

Tarara, J.M., Lee, J., Spayd, S.E., Scagel, C.F., 2008. Berry temperature and solar radiation alter acylation, proportion, and concentration of anthocyanin in Merlot grapes. Am. J. Enol. Vitic. 59, 235–247.

Tarara, J.M., Perez Peña, J.E., Keller, M., Schreiner, R.P., Smithyman, R.P., 2011. Net carbon exchange in grapevine canopies responds rapidly to timing and extent of regulated deficit irrigation. Funct. Plant Biol. 38, 386–400.

Tarbah, F., Goodman, R.N., 1987. Systemic spread of *Agrobacterium tumefaciens* biovar 3 in the vascular system of grapes. Phytopathology 77, 915–920.

Tardaguila, J., Martinez de Toda, F., Poni, S., Diago, M.P., 2010. Impact of early leaf removal on yield and fruit and wine composition of *Vitis vinifera* L. Graciano and Carignan. Am. J. Enol. Vitic. 61, 372–381.

Tardieu, F., Simonneau, T., 1998. Variability among species of stomatal control under fluctuating soil water status and evaporative demand: modelling isohydric and anisohydric behaviours. J. Exp. Bot. 49, 419–432.

Tardieu, F., Simonneau, T., Muller, B., 2018. The physiological basis of drought tolerance in crop plants: a scenario-dependent probabilistic approach. Annu. Rev. Plant Biol. 69, 733–759.

Tattersall, D.B., van Heeswijck, R., Høj, P.B., 1997. Identification and characterization of a fruit-specific, thaumatin-like protein that accumulates at very high levels in conjunction with the onset of sugar accumulation and berry softening in grapes. Plant Physiol. 114, 759–769.

Tattersall, D.B., Pocock, K.F., Hayasaka, Y., Adams, K., van Heeswijck, R., Waters, E.J., 2001. Pathogenesis-related proteins—their accumulation in grapes during berry growth and their involvement in white wine heat instability. In: Roubelakis-Angelakis, K.A. (Ed.), Molecular Biology and Biotechnology of the Grapevine. Kluwer, Dordrecht, The Netherlands, pp. 183–196.

Taylor, J.E., Whitelaw, C.A., 2001. Signals in abscission. New Phytol. 151, 323–339.

Tcherkez, G., Gauthier, P., Buckley, T.N., Busch, F.A., Barbour, M.M., Bruhn, D., et al., 2017. Leaf day respiration: low CO_2 flux but high significance for metabolism and carbon balance. New Phytol. 216, 986–1001.

Tegeder, M., Masclaux-Daubresse, C., 2018. Source and sink mechanisms of nitrogen transport and use. New Phytol. 217, 35–53.

Tello, J., Ibáñez, J., 2018. What do we know about grapevine bunch compactness? A state-of-the-art review. Aust. J. Grape Wine Res. 24, 6–23.

Temple, S.J., Vance, C.P., Gantt, J.S., 1998. Glutamate synthase and nitrogen assimilation. Trends Plant Sci. 3, 51–56.

Terashima, I., Hanba, Y.T., Tholen, D., Niinemets, Ü., 2011. Leaf functional anatomy in relation to photosynthesis. Plant Physiol. 155, 108–116.

Terral, J.F., Tabard, E., Bouby, L., Ivorra, S., Pastor, T., Figueiral, I., 2010. Evolution and history of grapevine (*Vitis vinifera*) under domestication: new morphometric perspectives to understand seed domestication syndrome and reveal origins of ancient European cultivars. Ann. Bot. 105, 443–455.

Terrier, N., Deguilloux, C., Sauvage, F.X., Martinoia, E., Romieu, C., 1998. Proton pumps and anion transport in *Vitis vinifera*: the inorganic pyrophosphatase plays a predominant role in the energization of the tonoplast. Plant Physiol. Biochem. 36, 367–377.

Terrier, N., Glissant, D., Grimplet, J., Barrieu, F., Abbal, P., Couture, C., 2005. Isogene specific oligo arrays reveal multifaceted changes in gene expression during grape berry (*Vitis vinifera* L.) development. Planta 222, 832–847.

Terrier, N., Ollé, D., Verriés, C., Cheynier, V., 2009. Biochemical and molecular aspects of flavan-3-ol synthesis during berry development. In: Roubelakis-Angelakis, K.A. (Ed.), Grapevine Molecular Physiology and Biotechnology. Springer, Dordrecht, The Netherlands, pp. 365–388.

Tesic, D., Keller, M., Hutton, R., 2007. Influence of vineyard floor management practices on grapevine growth, yield, and fruit composition. Am. J. Enol. Vitic. 58, 1–11.

Tesniere, C., Davies, C., Sreekantan, L., Bogs, J., Thomas, M., Torregrosa, L., 2006. Analysis of the transcript levels of *VvAdh1*, *VvAdh2* and *VvGrip4*, three genes highly expressed during *Vitis vinifera* L. berry development. Vitis 45, 75–79.

Tester, M., Leigh, R.A., 2001. Partitioning of nutrient transport processes in roots. J. Exp. Bot. 52, 445–457.

Tetlow, I.J., Morell, M.K., Emes, M.J., 2004. Recent developments in understanding the regulation of starch metabolism in higher plants. J. Exp. Bot. 55, 2131–2145.

Tezara, W., Mitchell, V.J., Driscoll, S.D., Lawlor, D.W., 1999. Water stress inhibits plant photosynthesis by decreasing coupling factor and ATP. Nature 401, 914–917.

Thatcher, L.F., Anderson, J.P., Singh, K.B., 2005. Plant defence responses: what have we learnt from *Arabidopsis*? Funct. Plant Biol. 32, 1–19.

Theiler, R., 1970. Anatomische Untersuchungen an Traubenstielen im Zusammenhang mit der Stiellähme. Wein-Wiss. 25, 381–417.

Theiler, R., Coombe, B.G., 1985. Influence of berry growth and growth regulators on the development of grape peduncles in *Vitis vinifera* L. Vitis 24, 1–11.

Theiler, R., Müller, H., 1986. Beziehungen zwischen Klimafaktoren und dem Stiellähmebefall bei Riesling × Sylvaner. Vitis 25, 8–20.

Theiler, R., Müller, H., 1987. Beziehung zwischen mittlerer Tagesmaximumtemperatur während der Blüteperiode und Stiellähmebefall für verschiedene Rebsorten und Standorte. Mitt. Klosterneuburg 37, 102–108.

Theocharis, A., Clément, C., Ait Barka, E., 2012. Physiological and molecular changes in plants grown at low temperatures. Planta 235, 1091–1105.

Thibon, C., Dubourdieu, D., Darriet, P., Tominaga, T., 2009. Impact of noble rot on the aroma precursor of 3-sulfanylhexanol content in *Vitis vinifera* L. cv Sauvignon blanc and Semillon grape juice. Food Chem. 114, 1359–1364.

This, P., Lacombe, T., Thomas, M.R., 2006. Historical origins and genetic diversity of wine grapes. Trends Genet. 22, 511–519.

This, P., Lacombe, T., Cadle-Davidson, M., Owens, C.L., 2007. Wine grape (*Vitis vinifera* L.) color associates with allelic variation in the domestication gene *VvmybA1*. Theor. Appl. Genet. 114, 723–730.

Thomas, H., 2013. Senescence, ageing and death of the whole plant. New Phytol. 197, 696–711.

Thomas, M.R., Scott, N.S., 1993. Microsatellite repeats in grapevine reveal DNA polymorphisms when analysed as sequence-tagged sites (STSs). Theor. Appl. Genet. 86, 985–990.

Thomas, H., Stoddart, J.L., 1980. Leaf senescence. Annu. Rev. Plant Physiol. 31, 83–111.

Thomas, C.S., Gubler, W.D., Silacci, M.W., Miller, R., 1993. Changes in elemental sulfur residues on Pinot noir and Cabernet Sauvignon grape berries during the growing season. Am. J. Enol. Vitic. 44, 205–210.

Thomas, T.R., Matthews, M.A., Shackel, K.A., 2006. Direct *in situ* measurement of cell turgor in grape (*Vitis vinifera* L.) berries during development and in response to plant water deficits. Plant Cell Environ. 29, 993–1001.

Thomas, T.R., Shackel, K.A., Matthews, M.A., 2008. Mesocarp cell turgor in *Vitis vinifera* L. berries throughout development and its relation to firmness, growth, and the onset of ripening. Planta 228, 1067–1076.

Thomashow, M.F., 1999. Plant cold acclimation: freezing tolerance genes and regulatory mechanisms. Annu. Rev. Plant Physiol. Plant Mol. Biol. 50, 571–599.

Thomashow, M.F., 2001. So what's new in the field of plant cold acclimation? Lots!. Plant Physiol. 125, 89–93.

Thompson, M.V., 2006. Phloem: the long and the short of it. Trends Plant Sci. 11, 26–32.

Thompson, M.V., Holbrook, N.M., 2003. Scaling phloem transport: water potential equilibrium and osmoregulatory flow. Plant Cell Environ. 26, 1561–1577.

Thompson, M.M., Olmo, H.P., 1963. Cytohistological studies of cytochimeric and tetraploid grapes. Am. J. Bot. 50, 901–906.

Thompson, D.S., Davies, W.J., Ho, L.C., 1998. Regulation of tomato fruit growth by epidermal cell wall enzymes. Plant Cell Environ. 21, 589–599.

Thorne, E.T., Stevenson, J.F., Rost, T.L., Labavitch, J.M., Matthews, M.A., 2006. Pierce's disease symptoms: comparison with symptoms of water deficit and the impact of water deficits. Am. J. Enol. Vitic. 57, 1–11.

Thorngate, J.H., Singleton, V.L., 1994. Localization of procyanidins in grape seeds. Am. J. Enol. Vitic. 45, 259–262.

Tibbetts, T.J., Ewers, F.W., 2000. Root pressure and specific conductivity in temperate lianas: exotic *Celastrus orbiculatus* (Celastraceae) vs. native *Vitis riparia* (Vitaceae). Am. J. Bot. 87, 1272–1278.

Tilbrook, J., Tyerman, S.D., 2008. Cell death in grape berries: varietal differences linked to xylem pressure and berry weight loss. Funct. Plant Biol. 35, 173–184.

Tilbrook, J., Tyerman, S.D., 2009. Hydraulic connection of grape berries to the vine: varietal differences in water conductance into and out of berries, and potential for backflow. Funct. Plant Biol. 36, 541–550.

Timmis, J.N., Ayliffe, M.A., Huang, C.Y., Martin, W., 2004. Endosymbiotic gene transfer: organelle genomes forge eukaryotic chromosomes. Nat. Rev. Genet. 5, 123–135.

Todaro, T.M., Dami, I.E., 2017. Cane morphology and anatomy influence freezing tolerance in *Vitis vinifera* Cabernet franc. Intl. J. Fruit Sci. 17, 391–406.

Tognetti, R., Raschi, A., Longobucco, A., Lanini, M., Bindi, M., 2005. Hydraulic properties and water relations of *Vitis vinifera* L. exposed to elevated CO_2 concentrations in a free air CO_2 enrichment (FACE). Phyton 45, 243–256.

Tominaga, T., des Gachons, C.P., Dubourdieu, D., 1998. A new type of flavor precursors in *Vitis vinifera* L. cv. Sauvignon blanc: *S*-cysteine conjugates. J. Agric. Food Chem. 46, 5215–5219.

Tominaga, T., Baltenweck-Guyot, R., des Gachons, C.P., Dubourdieu, D., 2000. Contribution of volatile thiols to the aromas of white wines made from several *Vitis vinifera* grape varieties. Am. J. Enol. Vitic. 51, 178–181.

Tompa, P., Bánki, P., Bokor, M., Kamasa, P., Kovács, D., Lasanda, G., Tompa, K., 2006. Protein-water and protein-buffer interactions in the aqueous solution of an intrinsically unstructured plant dehydrin: NMR intensity and DSC aspects. Biophys. J. 91, 2243–2249.

Tonietto, J., Carbonneau, A., 2004. A multicriteria climatic classification system for grape-growing regions worldwide. Agric. For. Meteorol. 124, 81–97.

Torabinejad, J., Caldwell, M.M., Flint, S.D., Durham, S., 1998. Susceptibility of pollen to UV-B radiation: an assay of 34 taxa. Am. J. Bot. 85, 360–369.

Törnroth-Horsefield, S., Wang, Y., Hedfalk, K., Johanson, U., Karlsson, M., Tajkhorshid, E., et al., 2006. Structural mechanism of plant aquaporin gating. Nature 439, 688–694.

Toselli, M., Baldi, E., Marcolini, G., Malaguti, D., Quartieri, M., Sorrenti, G., 2009. Response of potted grapevines to increasing soil copper concentration. Aust. J. Grape Wine Res. 15, 85–92.

Tournaire-Roux, C., Sutka, M., Javot, H., Gout, E., Gerbeau, P., Luu, D.T., 2003. Cytosolic pH regulates root water transport during anoxic stress through gating of aquaporins. Nature 425, 393–397.

Tränkner, M., Tavakol, E., Jákli, B., 2018. Functioning of potassium and magnesium in photosynthesis, photosynthate translocation and photoprotection. Physiol. Plant. 163, 414–431.

Treeby, M.T., Wheatley, D.M., 2006. Effect of nitrogen fertiliser on nitrogen partitioning and pool sizes in irrigated Sultana grapevines. Aust. J. Exp. Agric. 46, 1207–1215.

Treeby, M.T., Holzapfel, B.P., Walker, R.R., Nicholas, P.R., 1998. Profiles of free amino acids in grapes of grafted Chardonnay grapevines. Aust. J. Grape Wine Res. 4, 121–126.

Tregeagle, J.M., Tisdall, J.M., Blackmore, D.H., Walker, R.R., 2006. A diminished capacity for chloride exclusion by grapevine rootstocks following long-term saline irrigation in an inland versus a coastal region of Australia. Aust. J. Grape Wine Res. 12, 178–191.

Tregeagle, J.M., Tisdall, J.M., Tester, M., Walker, R.R., 2010. Cl^- uptake, transport and accumulation in grapevine rootstocks of differing capacity for Cl^- exlusion. Funct. Plant Biol. 37, 665–673.

Triantaphylidès, C., Krischke, M., Hoeberichts, F.A., Ksas, B., Gresser, G., Havaux, M., 2008. Singlet oxygen is the major reactive oxygen species involved in photooxidative damage to plants. Plant Physiol. 148, 960–968.

Trieb, G., Becker, H., 1969. Untersuchungen über den Einfluß verschiedener Unterlagen auf die mineralische Ernährung des Edelreises. Wein-Wiss. 24, 258–266.

Trifilò, P., Gascò, A., Raimondo, F., Nardini, A., Salleo, S., 2003. Kinetics of recovery of leaf hydraulic conductance and vein functionality from cavitation-induced embolism in sunflower. J. Exp. Bot. 54, 2323–2330.

Tromp, A., De Klerk, C.A., 1988. Effect of copperoxychloride on the fermentation of must and on wine quality. S. Afr. J. Enol. Vitic. 9, 31–36.

Tropf, S., Lanz, T., Rensing, S., Schröder, J., Schröder, G., 1994. Evidence that stilbene synthases have developed from chalcone synthases several times in the course of evolution. J. Mol. Evol. 38, 610–618.

Trought, M.C.T., Bramley, R.G.V., 2011. Vineyard variability in Marlborough, New Zealand: characterising spatial and temporal changes in fruit composition and juice quality in the vineyard. Aust. J. Grape Wine Res. 17, 79–89.

Tsay, Y.F., Ho, C.H., Chen, H.Y., Lin, S.H., 2011. Integration of nitrogen and potassium signaling. Annu. Rev. Plant Biol. 62, 207–226.

Tsukaya, H., 2006. Mechanism of leaf-shape determination. Annu. Rev. Plant Biol. 57, 477–496.

Tucker, S.C., Hoefert, L.L., 1968. Ontogeny of the tendril in *Vitis vinifera*. Am. J. Bot. 55, 1110–1119.

Turgeon, R., 1984. Efflux of sucrose from minor veins of tobacco leaves. Planta 161, 120–128.

Turgeon, R., 2000. Plasmodesmata and solute exchange in the phloem. Aust. J. Plant Physiol. 27, 521–529.

Turgeon, R., 2010. The role of phloem loading reconsidered. Plant Physiol. 152, 1817–1823.

Turgeon, R., Wolf, S., 2009. Phloem transport: cellular pathways and molecular trafficking. Annu. Rev. Plant Biol. 60, 207–221.

Turnbull, C.G.N., Lopez-Cobollo, R.M., 2013. Heavy traffic in the fast lane: long-distance signalling by macromolecules. New Phytol. 198, 33–51.

Turner, S., Gallois, P., Brown, D., 2007. Tracheary element differentiation. Annu. Rev. Plant Biol. 58, 407–433.

Tyerman, S.D., Bohnert, H.J., Maurel, C., Steudle, E., Smith, J.A.C., 1999. Plant aquaporins: their molecular biology, biophysics and significance for plant water relations. J. Exp. Bot. 50, 1055–1071.

Tyerman, S.D., Tilbrook, J., Pardo, C., Kotula, L., Sullivan, W., Steudle, E., 2004. Direct measurement of hydraulic properties in developing berries of *Vitis vinifera* L. cv Shiraz and Chardonnay. Aust. J. Grape Wine Res. 10, 170–181.

Tyerman, S.D., Vandeleur, R.K., Shelden, M.C., Tilbrook, J., Mayo, G., Gilliham, M., 2009. Water transport and aquaporins in grapevine. In: Roubelakis-Angelakis, K.A. (Ed.), Grapevine Molecular Physiology and Biotechnology, second ed. Springer, Dordrecht, The Netherlands, pp. 73–104.

Tyree, M.T., Ewers, F.W., 1991. The hydraulic architecture of trees and other woody plants. New Phytol. 119, 345–360.

Tyree, M.T., Zimmermann, M.H., 2002. Xylem Structure and the Ascent of Sap, second ed. Springer, Berlin, Germany.

Tyree, M.T., Christy, L., Ferrier, J.M., 1974. A simpler iterative steady state solution of Münch pressure-flow systems applied to long and short translocation paths. Plant Physiol. 54, 589–600.

Uhlig, B.A., Clingeleffer, P.R., 1998. Ripening characteristics of the fruit from *Vitis vinifera* L. drying cultivars Sultana and Merbein seedless under furrow irrigation. Am. J. Enol. Vitic. 49, 375–382.

Umehara, M., Hanada, A., Yoshida, S., Akiyama, K., Arite, T., Takeda-Kamiya, N., 2008. Inhibition of shoot branching by new terpenoid plant hormones. Nature 455, 195–200.

Unger, S., Büche, C., Boso, S., Kassemeyer, H.H., 2007. The course of colonization of two different *Vitis* genotypes by *Plasmopara viticola* indicates compatible and incompatible host–pathogen interactions. Phytopathology 97, 780–786.

Ureta, F., Boidron, J.N., Bouard, J., 1981. Influence of dessèchement de la rafle on grape quality. Am. J. Enol. Vitic. 32, 90–92.

Vail, M.E., Marois, J.J., 1991. Grape cluster architecture and the susceptibility of berries to *Botrytis cinerea*. Phytopathology 81, 188–191.

Vail, M.E., Gubler, W.D., Rademacher, M.R., 1999. Effect of cluster tightness on *Botrytis* bunch rot in six Chardonnay clones. Plant Dis. 82, 107–109.

Valdés-Gómez, H., Fermaud, M., Roudet, J., Calonnec, A., Gary, C., 2008. Grey mould incidence is reduced on grapevines with lower vegetative and reproductive growth. Crop Prot. 27, 1174–1186.

van Bel, A.J.E., 2003. The phloem, a miracle of ingenuity. Plant Cell Environ. 26, 125–149.

van Bel, A.J.E., Ehlers, K., Knoblauch, M., 2002. Sieve elements caught in the act. Trends Plant Sci. 7, 126–132.

Van de Wal, B.A.E., Leroux, O., Steppe, K., 2018. Post-veraison irreversible stem shrinkage in grapevine (*Vitis vinifera*) is caused by periderm formation. Tree Physiol. 38, 745–754.

van den Boom, C.E.M., Van Beek, T.A., Posthumus, M.A., De Groot, A., Dicke, M., 2004. Qualitative and quantitative variation among volatile profiles induced by *Tetranychus urticae* feeding on plants from various families. J. Chem. Ecol. 30, 69–89.

van der Schoot, C., Rinne, P.L.H., 2011. Dormancy cycling at the shoot apical meristem: transitioning between self-organization and self-arrest. Plant Sci. 180, 120–131.

van der Schoot, C., Paul, L.K., Rinne, P.L.H., 2014. The embryonic shoot: a lifeline through winter. J. Exp. Bot. 65, 1699–1712.

van Dongen, J.T., Schurr, U., Pfister, M., Geigenberger, P., 2003. Phloem metabolism and function have to cope with low internal oxygen. Plant Physiol. 131, 1529–1543.

van Doorn, W.G., Woltering, E.J., 2004. Senescence and programmed cell death: substance or semantics? J. Exp. Bot. 55, 2147–2153.

van Doorn, W.G., Hiemstra, T., Fanourakis, D., 2011. Hydrogel regulation of xylem water flow: an alternative hypothesis. Plant Physiol. 157, 1642–1649.

van Heeswijck, R., Stines, A.P., Grubb, J., Skrumsager Møller, I., Høj, P.B., 2001. Molecular biology and biochemistry of proline accumulation in developing grape berries. In: Roubelakis-Angelakis, K.A. (Ed.), Molecular Biology and Biotechnology of the Grapevine. Kluwer, Dordrecht, The Netherlands, pp. 87–108.

van Ieperen, W., van Meeteren, U., van Gelder, H., 2000. Fluid ionic composition influences hydraulic conductance of xylem conduits. J. Exp. Bot. 51, 769–776.

van Kan, J.A.L., 2006. Licensed to kill: the lifestyle of a necrotrophic plant pathogen. Trends Plant Sci. 11, 247–253.

van Leeuwen, C., Friant, P., Choné, X., Tregoat, O., Koundouras, S., Dubourdieu, D., 2004. Influence of climate, soil, and cultivar on terroir. Am. J. Enol. Vitic. 55, 207–217.

van Leeuwen, C., Tregoat, O., Choné, X., Bois, B., Pernet, D., Gaudillère, J.P., 2009. Vine water status is a key factor in grape ripening and vintage quality for red Bordeaux wine. How can it be assessed for vineyard management purposes? J. Int. Sci. Vigne Vin 43, 121–134.

van Leeuwen, C., Roby, J.P., Alonso-Villaverde, V., Gindro, K., 2013. Impact of clonal variability in *Vitis vinifera* Cabernet franc on grape composition, wine quality, leaf blade stilbene content, and downy mildew resistance. J. Agric. Food Chem. 61, 19–24.

van Volkenburgh, E., 1999. Leaf expansion—an integrating plant behaviour. Plant Cell Environ. 22, 1463–1473.

van't Hoff, J.H., 1887. Die Rolle des osmotischen Druckes in der Analogie zwischen Lösungen und Gasen. Z. Phys. Chem. 1, 481–508.

Vandeleur, R.K., Mayo, G., Shelden, M.C., Gilliham, M., Kaiser, B.N., Tyerman, S.D., 2009. The role of plasma membrane intrinsic protein aquaporins in water transport through roots: diurnal and drought stress responses reveal different strategies between isohydric and anisohydric cultivars of grapevine. Plant Physiol. 149, 445–460.

Vandenbussche, F., Suslov, D., De Grauwe, L., Leroux, O., Vissenberg, K., Van der Straeten, D., 2011. The role of brassinosteroids in shoot gravitropism. Plant Physiol. 156, 1331–1336.

VanEtten, H.D., Matthews, D.E., Matthews, P.S., 1989. Phytoalexin detoxification: importance for pathogenicity and practical implications. Annu. Rev. Phytopathol. 27, 143–164.

Varanini, Z., Maggioni, A., 1982. Iron reduction and uptake by grapevine roots. J. Plant Nutr. 5, 521–529.

Vasconcelos, M.C., Castagnoli, S., 2000. Leaf canopy structure and vine performance. Am. J. Enol. Vitic. 51, 390–396.

Vasconcelos, M.C., Greven, M., Winefield, C.S., Trought, M.C.T., Raw, V., 2009. The flowering process of *Vitis vinifera*: a review. Am. J. Enol. Vitic. 60, 411–434.

Vasudevan, L., Wolf, T.K., Welbaum, G.G., Wisniewski, M.E., 1998. Anatomical developments and effects of artificial shade on bud necrosis of Riesling grapevines. Am. J. Enol. Vitic. 49, 429–439.

Vega, A., Gutiérrez, R.A., Peña-Neira, A., Cramer, G.R., Arce-Johnson, P., 2011. Compatible GLRaV-3 viral infections affect berry ripening decreasing sugar accumulation and anthocyanin biosynthesis in *Vitis vinifera*. Plant Mol. Biol. 77, 261–274.

Velasco, R., Zharkikh, A., Troggio, M., Cartwright, D.A., Cestaro, A., Pruss, D., et al., 2007. A high quality draft consensus sequence of the genome of a heterozygous grapevine variety. PLoS ONE 2, e1326.

Veloso, J., van Kan, J.A.L., 2018. Many shades of grey in *Botrytis*-host plant interactions. Trends Plant Sci. 23, 613–622.

Veluthambi, K., Poovaiah, B.W., 1984. Calcium-promoted protein phosphorylation in plants. Science 223, 167–169.

Vergara, R., Noriega, X., Aravena, K., Prieto, H., Pérez, F.J., 2017. ABA represses the expression of cell cycle genes and may modulate the development of endodormancy in grapevine buds. Front. Plant Sci. 8, 812.

Versini, G., Rapp, A., Dalla Serra, A., Pichler, U., Ramponi, M., 1994. Methyl *trans* geranate and farnesoate as markers for Gewürztraminer grape skins and related distillates. Vitis 33, 139–142.

Véry, A.A., Davies, J.M., 2000. Hyperpolarization-activated calcium channels at the tip of *Arabidopsis* root hairs. Proc. Natl. Acad. Sci. U. S. A. 97, 9801–9806.

Véry, A.A., Sentenac, H., 2003. Molecular mechanisms and regulation of K$^+$ transport in higher plants. Annu. Rev. Plant Biol. 54, 575–603.

Veselov, D., Langhans, M., Hartung, W., Aloni, R., Feussner, I., Götz, C., 2003. Development of *Agrobacterium tumefaciens* C58-induced plant tumors and impact on host shoots are controlled by a cascade of jasmonic acid, auxin, cytokinin, ethylene and abscisic acid. Planta 216, 512–522.

Vezinhet, F., Hallet, J.N., Valade, M., Poulard, A., 1992. Ecological survey of wine yeast strains by molecular methods of identification. Am. J. Enol. Vitic. 43, 83–86.

Vezzulli, S., Leonardelli, L., Malossini, U., Stefanini, M., Velasco, R., Moser, C., 2012. Pinot blanc and Pinot gris arose as independent somatic mutations of Pinot noir. J. Exp. Bot. 63, 6359–6369.

Viala, P., Vermorel, V., 1901–1909. Ampélographie. Tomes I-VII. Masson, Paris.

Vicens, A., Fournand, D., Williams, P., Sidhoum, L., Moutounet, M., Doco, T., 2009. Changes in polysaccharide and protein composition of cell walls in grape berry skin (cv. Shiraz) during ripening and over-ripening. J. Agric. Food Chem. 57, 2955–2960.

Vidal, S., Williams, P., O'Neill, M.A., Pellerin, P., 2001. Polysaccharides from grape berry cell walls. Part I: tissue distribution and structural characterization of the pectic polysaccharides. Carbohydr. Polym. 45, 315–323.

Vidal, S., Francis, L., Guyot, S., Marnet, N., Kwiatkowski, M., Gawel, R., 2003. The mouth-feel properties of grape and apple proanthocyanins in a wine-like medium. J. Sci. Food Agric. 83, 564–573.

Vidal, S., Meudec, E., Cheynier, V., Skouroumounis, G., Hayasaka, Y., 2004. Mass spectrometric evidence for the existence of oligomeric anthocyanins in grape skins. J. Agric. Food Chem. 52, 7144–7151.

Vignault, C., Vachaud, M., Cakir, B., Glissant, D., Dédaldéchamp, F., Büttner, M., 2005. *VvHT1* encodes a monosaccharide transporter expressed in the conducting complex of the grape berry phloem. J. Exp. Bot. 56, 1409–1418.

Viret, O., Keller, M., Jaudzems, V.G., Cole, F.M., 2004. *Botrytis cinerea* infection of grape flowers: light and electron microscopical studies of infection sites. Phytopathology 94, 850–857.

Vitrac, X., Larronde, F., Krisa, S., Decendit, A., Deffieux, G., Mérillon, J.M., 2000. Sugar sensing and Ca^{2+}–calmodulin requirement in *Vitis vinifera* cells producing anthocyanins. Phytochemistry 53, 659–665.

Vivin, P., Castelan-Estrada, M., Gaudillére, J.P., 2003. Seasonal changes in chemical composition and construction costs of grapevine tissues. Vitis 42, 5–12.

Vogel, S., 2009. Leaves in the lowest and highest winds: temperature, force and shape. New Phytol. 183, 13–26.

Volder, A., Smart, D.R., Bloom, A.J., Eissenstat, D.M., 2005. Rapid decline in nitrate uptake and respiration with age in fine lateral roots of grape: implications for root efficiency and competitive effectiveness. New Phytol. 165, 493–502.

von Bassermann-Jordan, F., 1923. Geschichte des Weinbaus. Bände I-III, second ed. Frankfurter Verlags-Anstalt, Frankfurt, Germany.

von Goethe, J.W., 1790. Versuch, die Metamorphose der Pflanzen zu erklären. Ettinger, Gotha, Germany.

von Tiedemann, A., 1997. Evidence for a primary role of active oxygen species in induction of host cell death during infection of bean leaves with *Botrytis cinerea*. Physiol. Mol. Plant Pathol. 50, 151–166.

Vondras, A.M., Gouthu, S., Schmidt, J.A., Petersen, A.R., Deluc, L.G., 2016. The contribution of flowering time and seed content to uneven ripening initiation among fruits within *Vitis vinifera* L. cv. Pinot noir clusters. Planta 243, 1191–1202.

Vouillamoz, J.F., Grando, M.S., 2006. Genealogy of wine grape cultivars: 'Pinot' is related to 'Syrah'. Heredity 97, 102–110.

Vouillamoz, J.F., Schneider, A., Grando, M.S., 2007. Microsatellite analysis of Alpine grape cultivars (*Vitis vinifera* L.): alleged descendants of Pliny the Elder's Raetica are genetically related. Genet. Resour. Crop Evol. 54, 1095–1104.

Vršič, A., Ivančič, A., Šušek, A., Zagradišnik, B., Valdhuber, J., Šiško, M., 2011. The World's oldest living grapevine specimen and its genetic relationships. Vitis 50, 167–171.

Wada, H., Shackel, K.A., Matthews, M.A., 2008. Fruit ripening in *Vitis vinifera*: apoplastic solute accumulation accounts for pre-veraison turgor loss in berries. Planta 227, 1351–1361.

Wada, H., Matthews, M.A., Shackel, K.A., 2009. Seasonal pattern of apoplastic solute accumulation and loss of cell turgor during ripening of *Vitis vinifera* fruit under field conditions. J. Exp. Bot. 60, 1773–1781.

Wake, C.M.F., Fennell, A., 2000. Morphological, physiological and dormancy responses of three *Vitis* genotypes to short photoperiod. Physiol. Plant. 109, 203–210.

Walbot, V., Evans, M.M.S., 2003. Unique features of the plant life cycle and their consequences. Nat. Rev. Genet. 4, 369–379.

Walker, R.R., Blackmore, D.H., 2012. Potassium concentration and pH inter-relationships in grape juice and wine of Chardonnay and Shiraz from a range of rootstocks in different environments. Aust. J. Grape Wine Res. 18, 183–193.

Walker, R.R., Törökfalvy, E., Scott, N.S., Kriedemann, P.E., 1981. An analysis of photosynthetic response to salt treatment in *Vitis vinifera*. Aust. J. Plant Physiol. 8, 359–374.

Walker, R.R., Clingeleffer, P.R., Kerridge, G.H., Ruhl, E.H., Nicholas, P.R., Blackmore, D.H., 1998. Effects of the rootstock Ramsey (*Vitis champini*) on ion and organic acid composition of grapes and wine, and on wine spectral characteristics. Aust. J. Grape Wine Res. 4, 100–110.

Walker, R.P., Chen, Z.H., Técsi, L.I., Famiani, F., Lea, P.J., Leegood, R.C., 1999. Phosphoenolpyruvate carboxykinase plays a role in interactions of carbon and nitrogen metabolism during grape seed development. Planta 210, 9–18.

Walker, R.R., Read, P.E., Blackmore, D.H., 2000. Rootstock and salinity effects on rates of berry maturation, ion accumulation and colour development in Shiraz grapes. Aust. J. Grape Wine Res. 6, 227–239.

Walker, A.R., Lee, E., Bogs, J., McDavid, D.A.J., Thomas, M.R., Robinson, S.P., 2007. White grapes arose through the mutation of two similar and adjacent regulatory genes. Plant J. 49, 772–785.

Walker, R.R., Blackmore, D.H., Clingeleffer, P.R., 2010. Impact of rootstock on yield and ion concentrations in petioles, juice and wine of Shiraz and Chardonnay in different viticultural environments with different irrigation water salinity. Aust. J. Grape Wine Res. 16, 243–257.

Walters, D.R., McRoberts, N., 2006. Plants and biotrophs: a pivotal role for cytokinins? Trends Plant Sci. 11, 581–586.

Walz, A., Zingen-Sell, I., Loeffler, M., Sauer, M., 2008. Expression of an oxalate oxidase gene in tomato and severity of disease caused by *Botrytis cinerea* and *Sclerotinia sclerotiorum*. Plant Pathol. 57, 453–458.

Wample, R.L., 1994. A comparison of short- and long-term effects of mid-winter pruning on cold hardiness of Cabernet Sauvignon and Chardonnay buds. Am. J. Enol. Vitic. 45, 388–392.

Wample, R.L., Bary, A., 1992. Harvest date as a factor in carbohydrate storage and cold hardiness of Cabernet Sauvignon grapevines. J. Am. Soc. Hortic. Sci. 117, 32–36.

Wample, R.L., Spayd, S.E., Evans, R.G., Stevens, R.G., 1993. Nitrogen fertilization of White Riesling grapes in Washington: nitrogen seasonal effects on bud cold hardiness and carbohydrate reserves. Am. J. Enol. Vitic. 44, 159–167.

Wan, Y., Schwaninger, H., He, P., Wang, Y., 2007. Comparison of resistance to powdery mildew and downy mildew in Chinese wild grapes. Vitis 46, 132–136.

Wan, Y., Schwaninger, H., Li, D., Simon, C.J., Wang, Y., He, P., 2008a. The eco-geographic distribution of wild grape germplasm in China. Vitis 47, 77–80.

Wan, Y., Schwaninger, H., Li, D., Simon, C.J., Wang, Y., Zhang, C., 2008b. A review of taxonomic research on Chinese wild grapes. Vitis 47, 81–88.

Wan, Y., Wang, Y., Li, D., He, P., 2008c. Evaluation of agronomic traits in Chinese wild grapes and screening superior accessions for use in a breeding program. Vitis 47, 153–158.

Wan, Y., Schwaninger, H.R., Baldo, A.M., Labate, J.A., Zhong, G.Y., Simon, C.J., 2013. A phylogenetic analysis of the grape genus (*Vitis* L.) reveals broad reticulation and concurrent diversification during neogene and quaternary climate change. BMC Evol. Biol. 13, 141.

Wang, J., De Luca, V., 2005. The biosynthesis and regulation of biosynthesis of Concord grape fruit esters, including 'foxy' methylanthranilate. Plant J. 44, 606–619.

Wang, Y., Jiao, Y., 2018. Auxin and above-ground meristems. J. Exp. Bot. 69, 147–154.

Wang, Y., Li, J., 2008. Molecular basis of plant architecture. Annu. Rev. Plant Biol. 59, 253–279.

Wang, L., Ruan, Y.L., 2016. Shoot–root carbon allocation, sugar signalling and their coupling with nitrogen uptake and assimilation. Funct. Plant Biol. 43, 105–113.

Wang, Y., Wu, W.H., 2013. Potassium transport and signaling in higher plants. Annu. Rev. Plant Biol. 64, 451–476.

Wang, S., Okamoto, G., Hirano, K., Lu, J., Zhang, C., 2001. Effects of restricted rooting volume on vine growth and berry development of Kyoho grapevines. Am. J. Enol. Vitic. 52, 248–253.

Wang, Q., Cuellar, W.J., Rajamäki, M.L., Hirata, Y., Valkonen, J.P.T., 2008. Combined thermotherapy and cryotherapy for efficient virus eradication: relation of virus distribution, subcellular changes, cell survival and viral RNA degradation in shoot tips. Mol. Plant Pathol. 9, 237–250.

Wang, W.Q., Song, S.Q., Li, S.H., Gan, Y.Y., Wu, J.H., Cheng, H.Y., 2011. Seed dormancy and germination in *Vitis amurensis* and its variation. Seed Sci. Res. 21, 255–265.

Wang, Y.Y., Hsu, P.K., Tsay, Y.F., 2012. Uptake, allocation and signaling of nitrate. Trends Plant Sci. 17, 458–467.

Wang, D., Yeats, T.H., Uluisik, S., Rose, J.K.C., Seymour, G.B., 2018. Fruit softening: revisiting the role of pectin. Trends Plant Sci. 23, 302–310.

Wardlaw, I.F., 1990. The control of carbon partitioning in plants. New Phytol. 116, 341–381.

Wareing, P.F., Nasr, T.A.A., 1961. Gravimorphism in trees. I. Effects of gravity on growth and apical dominance in fruit trees. Ann. Bot. 25, 321–340.

Waterhouse, A.L., Laurie, V.F., 2006. Oxidation of wine phenolics: a critical evaluation and hypotheses. Am. J. Enol. Vitic. 57, 306–313.

Watt, M., Silk, W.K., Passioura, J.B., 2006. Rates of root and organism growth, soil conditions, and temporal and spatial development of the rhizosphere. Ann. Bot. 97, 839–855.

Weaver, R.J., 1958. Effect of gibberellic acid on fruit set and berry enlargement in seedless grapes of *Vitis vinifera*. Nature 181, 851–852.

Weaver, R.J., 1960. Toxicity of gibberellin to seedless and seeded varieties of *Vitis vinifera*. Nature 187, 1135–1136.

Weaver, L.M., Herrmann, K.M., 1997. Dynamics of the shikimate pathway in plants. Trends Plant Sci. 2, 346–351.

Weaver, R.J., Ibrahim, I.M., 1968. Effect of thinning and seededness on maturation of *Vitis vinifera* L. grapes. Proc. Am. Soc. Hortic. Sci. 92, 311–318.

Weaver, R.J., Pool, R.M., 1968. Effect of various levels of cropping on *Vitis vinifera* grapevines. Am. J. Enol. Vitic. 19, 185–193.

Weaver, R.J., Pool, R.M., 1971. Chemical thinning of grape clusters (*Vitis vinifera* L.). Vitis 10, 201–209.

Weaver, R.J., Amerine, M.A., Winkler, A.J., 1957. Preliminary report on effect of level of crop on development of color in certain red wine grapes. Am. J. Enol. Vitic. 8, 157–166.

Weaver, R.J., Shindy, W., Kliewer, W.M., 1969. Growth regulator induced movement of photosynthetic products into fruits of 'Black Corinth' grapes. Plant Physiol. 44, 183–188.

Webb, L.B., Whetton, P.H., Barlow, E.W.R., 2007. Modelled impact of future climate change on the phenology of winegrapes in Australia. Aust. J. Grape Wine Res. 13, 165–175.

Weber, B., Hoesch, L., Rast, D.M., 1995. Protocatechualdehyde and other phenols as cell wall components of grapevine leaves. Phytochemistry 40, 433–437.
Wegner, L.H., Zimmermann, U., 2009. Hydraulic conductance and K^+ transport into the xylem depend on radial volume flow, rather than on xylem pressure, in roots of intact, transpiring maize seedlings. New Phytol. 181, 361–373.
Wegscheider, E., Benjak, A., Forneck, A., 2009. Clonal variation in Pinot noir revealed by S-SAP involving universal retrotransposon-based sequences. Am. J. Enol. Vitic. 60, 104–109.
Weir, T.L., Park, S.W., Vivanco, J.M., 2004. Biochemical and physiological mechanisms mediated by allelochemicals. Curr. Opin. Plant Biol. 7, 472–479.
Weiss, A., Allen, L.H., 1976. Air-flow patterns in vineyard rows. Agric. Meteorol. 16, 329–432.
Weiss, D., Ori, N., 2007. Mechanisms of cross talk between gibberellin and other hormones. Plant Physiol. 144, 1240–1246.
Went, F.W., 1944. Plant growth under controlled conditions. II. Thermoperiodicity in growth and fruiting of the tomato. Am. J. Bot. 31, 135–150.
Wenter, A., Zanotelli, D., Montagnani, L., Tagliavini, M., Andreotti, C., 2018. Effect of different timings and intensities of water stress on yield and berry composition of grapevine (cv. Sauvignon blanc) in a mountain environment. Sci. Hortic. 236, 137–145.
Wenzel, K., Dittrich, H.H., Heimfarth, M., 1987. Die Zusammensetzung der Anthocyane in den Beeren verschiedener Rebsorten. Vitis 26, 65–78.
Wermelinger, B., Koblet, W., 1990. Seasonal growth and nitrogen distribution in grapevine leaves, shoots and grapes. Vitis 29, 15–26.
Werner, T., Motyka, V., Laucou, V., Smets, R., Van Onckelen, H., Schmülling, T., 2003. Cytokinin-deficient transgenic *Arabidopsis* plants show multiple developmental alterations indicating opposite functions of cytokinins in the regulation of shoot and root meristem activity. Plant Cell 15, 2532–2550.
Weyand, K.M., Schultz, H.R., 2006a. Light interception, gas exchange and carbon balance of different canopy zones of minimally and cane-pruned field grown Riesling grapevines. Vitis 45, 105–114.
Weyand, K.M., Schultz, H.R., 2006b. Long-term dynamics of nitrogen and carbohydrate reserves in woody parts of minimally and severely pruned Riesling vines in a cool climate. Am. J. Enol. Vitic. 57, 172–182.
Weyand, K.M., Schultz, H.R., 2006c. Regulating yield and wine quality of minimal pruning systems through the application of gibberellic acid. J. Int. Sci. Vigne Vin 40, 151–163.
White, M.A., Diffenbaugh, N.S., Jones, G.V., Pal, J.S., Giorgi, F., 2006. Extreme heat reduces and shifts United States premium wine production in the 21st century. Proc. Natl. Acad. Sci. U. S. A. 103, 11217–11222.
White, A.C., Rogers, A., Rees, M., Osborne, C.P., 2016. How can we make plants grow faster? A source-sink perspective on growth rate. J. Exp. Bot. 67, 31–45.
Whitelam, G.C., Devlin, P.F., 1998. Light signalling in *Arabidopsis*. Plant Physiol. Biochem. 36, 125–133.
Whitham, S.A., Yang, C., Goodin, M.M., 2006. Global impact: elucidating plant responses to viral infection. Mol. Plant-Microbe Interact. 19, 1207–1215.
Widrlechner, M.P., Daly, C., Keller, M., Kaplan, K., 2012. Horticultural applications of a newly revised USDA Plant Hardiness Zone Map. HortTechnology 22, 6–19.
Wiedemann, P., Neinhuis, C., 1998. Biomechanics of isolated plant cuticles. Bot. Acta 111, 28–34.
Wiley, E., Helliker, B., 2012. A re-evaluation of carbon storage in trees lends greater support for carbon limitation to growth. New Phytol. 195, 285–289.
Wilkinson, S., Davies, W.J., 2002. ABA-based chemical signalling: the co-ordination of responses to stress in plants. Plant Cell Environ. 25, 195–210.
Wille, A.C., Lucas, W.J., 1984. Ultrastructural and histochemical studies on guard cells. Planta 160, 129–142.
Williams, L.E., 1987. Growth of 'Thompson Seedless' grapevines: II. Nitrogen distribution. J. Am. Soc. Hortic. Sci. 112, 330–333.

Williams, L.E., 2010. Interaction of rootstock and applied water amounts at various fractions of estimated evapotranspiration (Et$_c$) on productivity of Cabernet Sauvignon. Aust. J. Grape Wine Res. 16, 434–444.

Williams, L.E., 2014. Determination of evapotranspiration and crop coefficients for a Chardonnay vineyard located in a cool climate. Am. J. Enol. Vitic. 65, 159–169.

Williams, L.E., Ayars, J.E., 2005. Grapevine water use and the crop coefficient are linear functions of the shaded area measured beneath the canopy. Agric. For. Meteorol. 132, 201–211.

Williams, L.E., Baeza, P., 2007. Relationships among ambient temperature and vapor pressure deficit and leaf and stem water potentials of fully irrigated, field-grown grapevines. Am. J. Enol. Vitic. 58, 173–181.

Williams, L.E., Biscay, P.J., 1991. Partitioning of dry weight, nitrogen, and potassium in Cabernet Sauvignon grapevines from anthesis until harvest. Am. J. Enol. Vitic. 42, 113–117.

Williams, L.E., Smith, R.J., 1985. Net CO_2 assimilation rate and nitrogen content of grape leaves subsequent to fruit harvest. J. Am. Soc. Hortic. Sci. 110, 846–850.

Williams, L.E., Matthews, M.A., 1990. Grapevine. In: Stewart, B.A., Nielson, N.R. (Eds.), Irrigation of Agricultural Crops, Agronomy Monograph No. 30. ASA-CSSA-SSSA, Madison, WI, pp. 1019–1055.

Williams, L.E., Dokoozlian, N.K., Wample, R., 1994. Grape. In: Schaffer, B., Andersen, P.C. (Eds.), Handbook of Environmental Physiology of Fruit Crops, vol. I. Temperate Crops. CRC Press, Boca Raton, FL, pp. 85–133.

Williams, L.E., 1996. Grape. In: Zamski, E., Schaffer, A. (Eds.), Photoassimilate Distribution in Plants and Crops: Source-Sink Relationships. Dekker, New York, NY, pp. 851–881.

Williams, L.E., Retzlaff, W.A., Yang, W., Biscay, P.J., Ebisuda, N., 2000. Effect of girdling on leaf gas exchange, water status, and non-structural carbohydrates of field-grown *Vitis vinifera* L. (cv. Flame Seedless). Am. J. Enol. Vitic. 51, 49–54.

Williams, L.E., Phene, C.J., Grimes, D.W., Trout, T.J., 2003. Water use of mature Thompson Seedless grapevines in California. Irrig. Sci. 22, 11–18.

Williams, C.M.J., Maier, N.A., Bartlett, L., 2004. Effect of molybdenum foliar sprays on yield, berry size, seed formation, and petiolar nutrient composition of "Merlot" grapevines. J. Plant Nutr. 27, 1891–1916.

Williams, L.E., Grimes, D.W., Phene, C.J., 2010. The effects of applied water at various fractions of measured evapotranspiration on water relations and vegetative growth of Thompson Seedless grapevines. Irrig. Sci. 28, 221–232.

Williamson, B., Tudzynsk, B., Tudzynski, P., van Kan, J.A.L., 2007. *Botrytis cinerea*: the cause of grey mould disease. Mol. Plant Pathol. 8, 561–580.

Wilson, E.O., Brown, W.L., 1953. The subspecies concept and its taxonomic application. Syst. Zool. 2, 97–111.

Wilson, B., Strauss, C.R., Williams, P.J., 1984. Changes in free and glycosidically bound monoterpenes in developing Muscat grapes. J. Agric. Food Chem. 32, 919–924.

Wilson, B., Strauss, C.R., Williams, P.J., 1986. The distribution of free and glycosidically-bound monoterpenes among skin, juice, and pulp fractions of some white grape varieties. Am. J. Enol. Vitic. 37, 107–111.

Windt, C.W., Vergeldt, F.J., de Jager, P.A., Van As, H., 2006. MRI of long-distance water transport: a comparison of the phloem and xylem flow characteristics and dynamics in poplar, castor bean, tomato and tobacco. Plant Cell Environ. 29, 1715–1729.

Wingler, A., 2015. Comparison of signaling interactions determining annual and perennial plant growth in response to low temperature. Front. Plant Sci. 5, 794.

Wingler, A., Masclaux-Daubresse, C., Fischer, A.M., 2009. Sugars, senescence, and ageing in plants and heterotrophic organisms. J. Exp. Bot. 60, 1063–1066.

Winkel, T., Rambal, S., 1993. Influence of water stress on grapevines growing in the field: from leaf to whole-plant response. Aust. J. Plant Physiol. 20, 143–157.

Winkel-Shirley, B., 2002. Biosynthesis of flavonoids and effects of stress. Curr. Opin. Plant Biol. 5, 218–223.

Winkler, A.J., 1929. The effect of dormant pruning on the carbohydrate metabolism of *Vitis vinifera*. Hilgardia 4, 153–173.

Winkler, A.J., 1954. Effects of overcropping. Am. J. Enol. Vitic. 5, 4–12.
Winkler, A.J., 1958. The relation of leaf area and climate to vine performance and grape quality. Am. J. Enol. Vitic. 9, 10–23.
Winkler, A.J., 1959. The effect of vine spacing at Oakville on yields, fruit composition, and wine quality. Am. J. Enol. Vitic. 10, 39–43.
Winkler, A.J., 1969. Effect of vine spacing in an unirrigated vineyard on vine physiology, production and wine quality. Am. J. Enol. Vitic. 20, 7–15.
Winkler, A.J., Williams, W.O., 1935. Effect of seed development on the growth of grapes. Proc. Am. Soc. Hortic. Sci. 33, 430–434.
Winkler, A.J., Williams, W.O., 1945. Starch and sugars of *Vitis vinifera*. Plant Physiol. 20, 412–432.
Winkler, A.J., Cook, J.A., Kliewer, W.M., Lider, L.A., 1974. General Viticulture. University of California Press, Berkeley, CA.
Winship, L.J., Obermeyer, G., Geitmann, A., Hepler, P.K., 2010. Under pressure, cell walls set the pace. Trends Plant Sci. 15, 363–369.
Winterhalter, P., Schreier, P., 1994. C_{13}-Norisoprenoid glycosides in plant tissues: an overview on their occurrence, composition and role as flavour precursors. Flavour Fragr. J. 9, 281–287.
Wisniewski, M., Gusta, L., Neuner, G., 2014. Adaptive mechanisms of freeze avoidance in plants: a brief update. Environ. Exp. Bot. 99, 133–140.
Witte, C.P., Rosso, M.G., Romeis, T., 2005. Identification of three urease accessory proteins that are required for urease activation in *Arabidopsis*. Plant Physiol. 139, 1155–1162.
Wittmann, C., Pfanz, H., 2014. Bark and woody tissue photosynthesis: a means to avoid hypoxia or anoxia in developing stem tissues. Funct. Plant Biol. 41, 940–953.
Wobus, U., Weber, H., 1999. Seed maturation: genetic programmes and control signals. Curr. Opin. Plant Biol. 2, 33–38.
Wolf, T.K., Cook, M.K., 1992. Seasonal deacclimation patterns of three grape cultivars at constant, warm temperature. Am. J. Enol. Vitic. 43, 171–179.
Wolf, T.K., Pool, R.M., 1988. Nitrogen fertilization and rootstock effects on wood maturation and dormant bud cold hardiness of cv. Chardonnay grapevines. Am. J. Enol. Vitic. 39, 308–312.
Wolf, S., Hématy, K., Höfte, H., 2012. Growth control and cell wall signaling in plants. Annu. Rev. Plant Biol. 63, 381–407.
Wolfe, J.A., 1992. An analysis of present-day terrestrial lapse rates in the western conterminous United States and their significance to paleoaltitudinal estimates. U.S. Geol. Surv. Bull. 1964.
Wolfe, D.W., Schwartz, M.D., Lakso, A.N., Otsuki, Y., Pool, R.M., Shaulis, N.J., 2005. Climate change and shifts in spring phenology of three horticultural woody perennials in northeastern USA. Int. J. Biometeorol. 49, 303–309.
Wollmann, N., Hofmann, T., 2013. Compositional and sensory characterization of red wine polymers. J. Agric. Food Chem. 61, 2045–2061.
Wolpert, J.A., Howell, G.S., 1985. Cold acclimation of Concord grapevines. II. Natural acclimation pattern and tissue moisture decline in canes and primary buds of bearing vines. Am. J. Enol. Vitic. 36, 189–194.
Wolpert, J.A., Howell, G.S., 1986. Cold acclimation of Concord grapevines. III. Relationship between cold hardiness, tissue water content, and shoot maturation. Vitis 25, 151–159.
Wood, W.F., Sollers, B.G., Dragoo, G.A., Dragoo, J.W., 2002. Volatile components in defensive spray of the hooded skunk, *Mephitis macroura*. J. Chem. Ecol. 28, 1865–1870.
Wood, C., Siebert, T.E., Parker, M., Capone, D.L., Elsey, G.M., Pollnitz, A.P., 2008. From wine to pepper: rotundone, an obscure sesquiterpene, is a potent spicy aroma compound. J. Agric. Food Chem. 56, 3738–3744.
Woodham, R.C., Alexander, D.M., 1966. The effect of root temperature on development of small fruiting Sultana vines. Vitis 5, 345–350.

Woodham, R.C., Krake, L.R., Cellier, K.M., 1983. The effect of grapevine leafroll plus yellow speckle disease on annual growth, yield and quality of grapes from Cabernet Franc under two pruning systems. Vitis 22, 324–330.

Woodham, R.C., Antcliff, A.J., Krake, L.R., Taylor, R.H., 1984. Yield differences between Sultana clones related to virus status and genetic factors. Vitis 23, 73–83.

Woodrow, I.E., Berry, J.A., 1988. Enzymatic regulation of photosynthetic CO_2 fixation in C_3 plants. Annu. Rev. Plant Physiol. Plant Mol. Biol. 39, 533–594.

Woodward, F.I., 1987. Stomatal numbers are sensitive to CO_2 increases from pre-industrial levels. Nature 327, 617–618.

Woodward, A.W., Bartel, B., 2005. Auxin: regulation, action, and interaction. Ann. Bot. 95, 707–735.

Wu, Y., Cosgrove, D.J., 2000. Adaptation of roots to low water potentials by changes in cell wall extensibility and cell wall proteins. J. Exp. Bot. 51, 1543–1553.

Wu, S.W., Kumar, R., Iswanto, A.B.B., Kim, J.Y., 2018. Callose balancing at plasmodesmata. J. Exp. Bot. 69, 5325–5339.

Wyka, T.P., Oleksyn, J., Karolewski, P., Schnitzer, S.A., 2013. Phenotypic correlates of the lianescent growth form: a review. Ann. Bot. 112, 1667–1681.

Xia, G., Cheng, L., 2004. Foliar urea application in the fall affects both nitrogen and carbon storage in young 'Concord' grapevines grown under a wide range of nitrogen supply. J. Am. Soc. Hortic. Sci. 129, 653–659.

Xiao, Z., Rogiers, S.Y., Sadras, V.O., Tyerman, S.D., 2018. Hypoxia in grape berries: the role of seed respiration and lenticels on the berry pedicel and the possible link to cell death. J. Exp. Bot. 69, 2071–2083.

Xie, D.Y., Sharma, S.B., Paiva, N.L., Ferreira, D., Dixon, R.A., 2003. Role of anthocyanidin reductase, encoded by *BANYULS* in plant flavonoid biosynthesis. Science 299, 396–399.

Xie, Z., Kapteyn, J., Gang, D.R., 2008. A systems biology investigation of the MEP/terpenoid and shikimate/phenylpropanoid pathways points to multiple levels of metabolic control in sweet basil glandular trichomes. Plant J. 54, 349–361.

Xie, Z., Forney, C.F., Bondada, B., 2018. Renewal of vascular connections between grapevine buds and canes during bud break. Sci. Hortic. 233, 331–338.

Xin, Z., Browse, J., 2000. Cold comfort farm: the acclimation of plants to freezing temperatures. Plant Cell Environ. 23, 893–902.

Xing, R.R., Li, S.Y., He, F., Yang, Z., Duan, C.Q., Li, Z., Wang, J., Pan, Q.H., 2015. Mass spectrometric and enzymatic evidence confirm the existence of anthocyanidin 3,5-*O*-diglucosides in Cabernet Sauvignon (*Vitis vinifera* L.) grape berries. J. Agric. Food Chem. 63, 3251–3260.

Xiong, J., Bauer, C.E., 2002. Complex evolution of photosynthesis. Annu. Rev. Plant Biol. 53, 503–521.

Xu, G., Fan, X., Miller, A.J., 2012. Plant nitrogen assimilation and use efficiency. Annu. Rev. Plant Biol. 63, 153–182.

Yamaguchi, S., 2008. Gibberellin metabolism and its regulation. Annu. Rev. Plant Biol. 59, 225–251.

Yamane, T., Jeong, S.T., Goto-Yamamoto, N., Koshita, Y., Kobayashi, S., 2006. Effects of temperature on anthocyanin biosynthesis in grape berry skins. Am. J. Enol. Vitic. 57, 54–59.

Yamasaki, H., Sakihama, Y., Ikehara, N., 1997. Flavonoid-peroxidase reaction as a detoxification mechanism of plant cells against H_2O_2. Plant Physiol. 115, 1405–1412.

Yang, Y.S., Hori, Y., 1979. Studies on retranslocation of accumulated assimilates in 'Delaware' grapevines. I. Retranslocation of ^{14}C-assimilates in the following spring after ^{14}C feeding in summer and autumn. Tohoku J. Agric. Res. 30, 43–56.

Yang, Y.S., Hori, Y., 1980. Studies on retranslocation of accumulated assimilates in 'Delaware' grapevines. III. Early growth of new shoots as dependent on accumulated and current year assimilates. Tohoku J. Agric. Res. 31, 120–129.

Yang, T., Poovaiah, B.W., 2002. Hydrogen peroxide homeostasis: activation of plant catalase by calcium/calmodulin. Proc. Natl. Acad. Sci. U. S. A. 99, 4097–4102.

Yang, J., Zhang, J., Wang, Z., Zhu, Q., Liu, L., 2002. Abscisic acid and cytokinins in the root exudates and leaves and their relationship to senescence and remobilization of carbon reserves in rice subjected to water stress during grain filling. Planta 215, 645–652.

Yang, Y., Duan, C., Du, H., Tian, J., Pan, Q., 2010. Trace element and rare earth element profiles in berry tissues of three grape cultivars. Am. J. Enol. Vitic. 61, 401–407.

Yang, Y., Labate, J.A., Liang, Z., Cousins, P., Prins, B., Preece, J.E., Aradhya, M., Zhong, G.Y., 2014. Multiple loss-of-function 5-*O-glucosyltransferase* alleles revealed in *Vitis vinifera*, but not in other *Vitis* species. Theor. Appl. Genet. 127, 2433–2451.

Ye, Z.H., 2002. Vascular tissue differentiation and pattern formation in plants. Annu. Rev. Plant Biol. 53, 183–202.

Ye, Q., Wiera, B., Steudle, E., 2004. A cohesion/tension mechanism explains the gating of water channels (aquaporins) in *Chara* internodes by high concentration. J. Exp. Bot. 55, 449–461.

Yeats, T.H., Rose, J.K.C., 2013. The formation and function of plant cuticles. Plant Physiol. 163, 5–20.

Yin, R., Messner, B., Faus-Kessler, T., Hoffman, T., Schwab, W., Hajirezaei, M.R., 2012. Feedback inhibition of the general phenylpropanoid and flavonol biosynthetic pathways upon a compromised flavonol-3-*O*-glycosylation. J. Exp. Bot. 63, 2465–2478.

Yokotsuka, K., Fukui, M., 2002. Changes in nitrogen compounds in berries of six grape cultivars during ripening over two years. Am. J. Enol. Vitic. 53, 69–77.

Young, J.M., Kuykendall, L.D., Martínez-Romero, E., Kerr, A., Sawada, H., 2001. A revision of *Rhizobium* Frank 1889, with an amended description of the genus, and the inclusion of all species of *Agrobacterium* Conn 1942 and *Allorhizobium undicola* de Lajudie *et al.* 1998 as new combinations: *Rhizobium radiobacter, R. rhizogenes, R. rubi, R. undicola* and *R. vitis*. Int. J. Syst. Evol. Microbiol. 51, 89–103.

Young, P.R., Lashbrooke, J.G., Alexandersson, E., Jacobson, D., Moser, C., Velasco, R., Vivier, M.A., 2012. The genes and enzymes of the carotenoid metabolic pathway in *Vitis vinifera* L. BMC Genom. 13, 243.

Young, P.R., Eyeghe-Bickong, H.A., du Plessis, K., Alexanderson, E., Jacobson, D.A., Coetze, Z., Deloire, A., Vivier, M.A., 2016. Grapevine plasticity in response to an altered microclimate: Sauvignon blanc modulates specific metabolites in response to increased berry exposure. Plant Physiol. 170, 1235–1254.

Yruela, I., 2009. Copper in plants: acquisition, transport and interactions. Funct. Plant Biol. 36, 409–430.

Yu, O., Jez, J.M., 2008. Nature's assembly line: biosynthesis of simple phenylpropanoids and polyketides. Plant J. 54, 750–762.

Yu, Y., Zhang, Y., Yin, L., Lu, J., 2012. The mode of host resistance to *Plasmopara viticola* infection of grapevines. Phytopathology 102, 1094–1101.

Yu, S.M., Lo, S.F., Ho, T.H.D., 2015. Source-sink communication: regulated by hormone, nutrient, and stress cross-signaling. Trends Plant Sci. 20, 844–857.

Yuan, F., Schreiner, R.P., Osborne, J., Qian, M.C., 2018a. Effects of soil NPK supply on Pinot noir wine phenolics and aroma composition. Am. J. Enol. Vitic. 69, 371–385.

Yuan, F., Schreiner, R.P., Qian, M.C., 2018b. Soil nitrogen, phosphorus, and potassium alter β-damascenone and other volatiles in Pinot noir berries. Am. J. Enol. Vitic. 69, 157–166.

Zabalza, A., van Dongen, J.T., Froehlich, A., Oliver, S.N., Faix, B., Jagadis Gupta, K., 2009. Regulation of respiration and fermentation to control the plant internal oxygen concentration. Plant Physiol. 149, 1087–1098.

Zachariassen, K.E., Kristiansen, E., 2000. Ice nucleation and antinucleation in nature. Cryobiology 41, 257–279.

Zamboni, A., Minoia, L., Ferrarini, A., Tornielli, G.B., Zago, E., Delledonne, M., 2008. Molecular analysis of postharvest withering in grape by AFLP transcriptional profiling. J. Exp. Bot. 59, 4145–4159.

Zapata, C., Deléens, E., Chaillou, S., Magné, C., 2004. Partitioning and mobilization of starch and N reserves in grapevine (*Vitis vinifera* L.). J. Plant Physiol. 161, 1031–1040.

Zapata, D., Salazar, M., Chaves, B., Keller, M., Hoogenboom, G., 2015. Estimation of the base temperature and growth phase duration in terms of thermal time for four grapevine cultivars. Int. J. Biometeorol. 59, 1771–1781.

Zapata, D., Salazar-Gutierrez, M., Chaves, B., Keller, M., Hoogenboom, G., 2017. Predicting key phenological stages for 17 grapevine cultivars (*Vitis vinifera* L.). Am. J. Enol. Vitic. 68, 60–72.

Zarrouk, O., Brunetti, C., Egipto, R., Pinheiro, C., Genebra, T., Gori, A., Lopes, C.M., Tattini, M., Chaves, M.M., 2016. Grape ripening is regulated by deficit irrigation/elevated temperatures according to cluster position in the canopy. Front. Plant Sci. 7, 1640.

Zecca, G., Abbott, J.R., Sun, W.B., Spada, A., Sala, F., Grassi, F., 2012. The timing and the mode of evolution of wild grapes (*Vitis*). Mol. Phylogenet. Evol. 62, 736–747.

Zeeman, S.C., Kossmann, J., Smith, A.M., 2010. Starch: its metabolism, evolution, and biotechnological modification in plants. Annu. Rev. Plant Biol. 61, 209–234.

Zell, M.B., Fahnenstich, H., Maier, A., Saigo, M., Voznesenskaya, E.V., Edwards, G.E., 2010. Analysis of *Arabidopsis* with highly reduced levels of malate and fumarate sheds light on the role of these organic acids as storage carbon molecules. Plant Physiol. 152, 1251–1262.

Zelleke, A., Düring, H., 1994. Evidence for tertiary buds within latent buds of Müller-Thurgau grapevines. Vitis 33, 96.

Zelleke, A., Kliewer, W.M., 1979. Influence of root temperature and rootstock on budbreak, shoot growth, and fruit composition of Cabernet Sauvignon grapevines grown under controlled conditions. Am. J. Enol. Vitic. 30, 312–317.

Zelleke, A., Kliewer, W.M., 1981. Factors affecting the qualitative and quantitative levels of cytokinins in xylem sap of grapevines. Vitis 20, 93–104.

Zerihun, A., Treeby, M.T., 2002. Biomass distribution and nitrate assimilation in response to N supply for *Vitis vinifera* L. cv. Cabernet Sauvignon on five *Vitis* rootstock genotypes. Aust. J. Grape Wine Res. 8, 157–162.

Zhang, Y., Dami, I.E., 2012. Foliar application of abscisic acid increases freezing tolerance of field-grown *Vitis vinifera* Cabernet franc grapevines. Am. J. Enol. Vitic. 63, 377–384.

Zhang, Y., Keller, M., 2015. Grape berry transpiration is determined by vapor pressure deficit, cuticular conductance, and berry size. Am. J. Enol. Vitic. 66, 454–462.

Zhang, Y., Keller, M., 2017. Discharge of surplus phloem water may be required for normal grape ripening. J. Exp. Bot. 68, 585–595.

Zhang, X., Walker, R.R., Stevens, R.M., Prior, L.D., 2002. Yield–salinity relationships of different grapevine (*Vitis vinifera* L.) scion-rootstock combinations. Aust. J. Grape Wine Res. 8, 150–156.

Zhang, X.Y., Wang, X.L., Wang, X.F., Xia, G.H., Pan, Q.H., Fan, R.C., 2006. A shift of phloem unloading from symplasmic to apoplasmic pathway is involved in developmental onset of ripening in grape berry. Plant Physiol. 142, 220–232.

Zhang, J., Ma, H., Feng, J., Zeng, L., Wang, Z., Chen, S., 2008. Grape berry plasma membrane proteome analysis and its differential expression during ripening. J. Exp. Bot. 59, 2979–2990.

Zhang, C., Tateishi, N., Tanabe, K., 2010a. Pollen density on the stigma affects endogenous gibberellin metabolism, seed and fruit set, and fruit quality in *Pyrus pyrifolia*. J. Exp. Bot. 61, 4291–4302.

Zhang, H., Han, W., De Smet, I., Talboys, P., Loya, R., Hassan, A., 2010b. ABA promotes quiescence of the quiescent centre and suppresses stem cell differentiation in the *Arabidopsis* primary root meristem. Plant J. 64, 764–774.

Zhang, J., Gao, G., Chen, J.J., Taylor, G., Cui, K.M., He, X.Q., 2011a. Molecular features of secondary vascular tissue regeneration after bark girdling in *Populus*. New Phytol. 192, 869–884.

Zhang, Y., Mechlin, T., Dami, I., 2011b. Foliar application of abscisic acid induces dormancy responses in greenhouse-grown grapevines. HortScience 46, 1271–1277.

Zhang, Y., Oren, R., Kang, S., 2012. Spatiotemporal variation of crown-scale stomatal conductance in an arid *Vitis vinifera* L. cv. Merlot vineyard: direct effects of hydraulic properties and indirect effects of canopy leaf area. Tree Physiol. 32, 262–279.

Zhang, P., Fuentes, S., Wang, Y., Deng, R., Krstic, M., Herderich, M., Barlow, E.W.R., Howell, K., 2016. Distribution of rotundone and possible translocation of related compounds amongst grapevine tissues in *Vitis vinifera* L. cv. Shiraz. Front. Plant Sci. 7, 859.

Zhao, Y., 2018. Essential roles of local auxin biosynthesis in plant development and in adaptation to environmental changes. Annu. Rev. Plant Biol. 69, 417–435.

Zhao, J., Pang, Y., Dixon, R.A., 2010. The mysteries of proanthocyanidin transport and polymerization. Plant Physiol. 153, 437–443.

Zheng, C., Halaly, T., Acheampong, A.K., Takebayashi, Y., Jikumaru, Y., Kamiya, Y., Or, E., 2015. Abscisic acid (ABA) regulates grape bud dormancy, and dormancy release stimuli may act through modification of ABA metabolism. J. Exp. Bot. 66, 1527–1542.

Zheng, C., Acheampong, A.K., Shi, Z., Halaly, T., Kamiya, Y., Ophir, R., Galbraith, D.W., Or, E., 2018. Distinct gibberellin functions during and after grapevine bud dormancy release. J. Exp. Bot. 69, 1635–1648.

Zhou, Y., Massonnet, M., Sanjak, J.S., Cantu, D., Gaut, B.S., 2017. Evolutionary genomics of grape (*Vitis vinifera* ssp. *vinifera*) domestication. Proc. Natl. Acad. Sci. U. S. A. 114, 11715–11720.

Zhu, J.K., 2001. Plant salt tolerance. Trends Plant Sci. 6, 66–71.

Zhu, J.K., 2002. Salt and drought stress signal transduction in plants. Annu. Rev. Plant Biol. 53, 247–273.

Zimman, A., Waterhouse, A.L., 2004. Incorporation of malvidin-3-glucoside into high molecular weight polyphenols during fermentation and wine aging. Am. J. Enol. Vitic. 55, 139–146.

Zinn, K.E., Tunc-Ozdemir, M., Harper, J.F., 2010. Temperature stress and plant sexual reproduction: uncovering the weakest links. J. Exp. Bot. 61, 1959–1968.

Zoecklein, B.W., Wolf, T.K., Duncan, N.W., Judge, J.M., Cook, M.K., 1992. Effects of fruit zone leaf removal on yield, fruit composition, and fruit rot incidence of Chardonnay and White Riesling (*Vitis vinifera* L.) grapes. Am. J. Enol. Vitic. 43, 139–148.

Zohary, D., Hopf, M., 2001. Domestication of Plants in the Old World—The Origin and Spread of Cultivated Plants in West Asia, Europe, and the Nile Valley, third ed. Oxford University Press, Oxford, UK.

Zohary, D., Spiegel-Roy, P., 1975. Beginnings of fruit growing in the old world. Science 187, 319–327.

Zonia, L., Munnik, T., 2007. Life under pressure: hydrostatic pressure in cell growth and function. Trends Plant Sci. 12, 90–97.

Zsófi, Z., Váradi, G., Bálo, B., Marschall, M., Nagy, Z., Dulai, S., 2009. Heat acclimation of grapevine leaf photosynthesis: mezo- and macroclimatic aspects. Funct. Plant Biol. 36, 310–322.

Zufferey, V., 2016. Leaf respiration in grapevine (*Vitis vinifera* 'Chasselas') in relation to environmental and plant factors. Vitis 55, 65–72.

Zufferey, V., Murisier, F., Schultz, H.R., 2000. A model analysis of the photosynthetic response of *Vitis vinifera* L. cvs Riesling and Chasselas leaves in the field: I. Interaction of age, light and temperature. Vitis 39, 19–26.

Zufferey, V., Cochard, H., Ameglio, T., Spring, J.L., Viret, O., 2011. Diurnal cycles of embolism formation and repair in petioles of grapevine (*Vitis vinifera* cv. Chasselas). J. Exp. Bot. 62, 3885–3894.

Zufferey, V., Murisier, F., Vivin, P., Belcher, S., Lorenzini, F., Spring, J.L., 2012. Carbohydrate reserves in grapevine (*Vitis vinifera* L. 'Chasselas'): the influence of the leaf to fruit ratio. Vitis 51, 103–110.

Zufferey, V., Murisier, F., Belcher, S., Lorenzini, F., Vivin, P., Spring, J.L., Viret, O., 2015. Nitrogen and carbohydrate reserves in the grapevine (*Vitis vinifera* L. 'Chasselas'): the influence of the leaf to fruit ratio. Vitis 54, 183–188.

Zufferey, V., Spring, J.L., Verdenal, T., Dienes, A., Belcher, S., Lorenzini, F., Koestel, C., Rösti, J., Gindro, K., Spangenberg, J., Viret, O., 2017. Influence of water stress on plant hydraulics, gas exchange, berry composition and quality of Pinot noir wines in Switzerland. OENO One 51, 17–27.

Zwieniecki, M.A., Melcher, P.J., Holbrook, N.M., 2001. Hydrogel control of xylem hydraulic resistance in plants. Science 291, 1059–1062.

Internet resources

Angiosperm Phylogeny Website: http://www.mobot.org/MOBOT/research/APweb

American Society for Enology and Viticulture (including *American Journal of Enology and Viticulture* and *Catalyst: Discovery into Practice*): http://www.asev.org

American Society of Plant Biologists (including various journals): https://aspb.org

Australian Society of Viticulture and Oenology (including *Australian Journal of Grape and Wine Research*): http://www.asvo.com.au

European *Vitis* Database: http://www.eu-vitis.de

French Network of Grapevine Repositories: https://bioweb.supagro.inra.fr/collections_vigne/Home.php

Grape Genome Browser: http://www.genoscope.cns.fr/externe/GenomeBrowser/Vitis

International Grape Genome Program: https://www6.inra.fr/iggp

International Society for Horticultural Science (including *Acta Horticulturae*): http://www.ishs.org

OENO One, Vine and Wine Open Access Journal: http://oeno-one.eu

PlantStress: http://www.plantstress.com

The Virtual Vine Doctor (in German): https://www.rebendoktor.de

USDA Natural Resources Conservation Service Plant Profile: https://plants.usda.gov/core/profile?symbol=VITIS

Vitis International Variety Catalogue: http://www.vivc.de

Vitis Journal of Grapevine Research: https://ojs.openagrar.de/index.php/VITIS

Vitis-VEA, Viticulture and Enology Abstracts: https://www.vitis-vea.de

Vitisvinum, The Multilingual Wine Dictionary: http://www.vitisvinum.info

Index

Note: Page numbers followed by *f* indicate figures and *t* indicate tables.

A

Abortion of flowers and ovaries, 96–97
Abscisic acid (ABA), 28, 63–65, 145, 159
    ATPases and, 159
    grape ripening and, 211–212
    in inducing anthocyanin production, 379–380
    organ's sensitivity to, 280–281
    reaction of stomata to, 111
    senescence and abscission process, 283
Accessory buds, 46
Accessory pigments, 130–131
Acetyl coenzyme A (acetyl-CoA), 144
Acid growth theory, 108
Acidification, 330
Acids. *See also* Organic acids
    grape composition and fruit quality, 223–226, 223*f*
Active (or reactive) oxygen species, 133, 281
Adventitious roots, 29
Aerenchyma, 296
Agrobacterium species, 376
    A. tumefaciens, 376
    A. vitis, 3, 376
Alcohols, fusel, 271
Allocation, photosynthate, 161–167
    assimilate surplus in, 166–167
    assimilation rate determined by source in, 165
    categories, 161
    patterns in, 162–164
    phosphates availability in regulation of, 161
    source-sink relation with, 161, 162*f*
    storage in, 161
    transport in, 161
    utilization, 161
Allorhizic root system, 28–29
Alphonse Lavallée, 19
American group, 8 10
    Vitis aestivalis Michaux, 9
    Vitis berlandieri Planchon, 9
    Vitis candicans Engelmann, 10
    Vitis cinerea Engelmann, 10
    Vitis labrusca L., 8
    Vitis riparia Michaux, 9
    Vitis rupestris Scheele, 9
Amino acids, 1–2, 15, 35, 120–121, 145, 151, 173, 216, 226, 251, 256, 262, 266, 268–274, 276, 289, 304, 327, 350, 367, 381
    ammonium assimilation into, 188, 190–192
    arginine, 315–316
    aromatic, 228, 233
    assembly of, 191
    in berries, 270–272
    cysteine, 282
    free, 226–227, 350
    glutamine and glutamate, 194
    in leaf, 270, 273–274
    mesophyll cell produced, 157
    methionine, 285
    N:C ratio of, 197
    nitrate and sulfate assimilation, 147, 194, 196, 270–271
    nitrogen-containing methoxypyrazines, 250
    nonprotein, 322–323
    nutrient deficiency and, 304–305
    phenylalanine, 140
    during photorespiratory carbon oxidation cycle, 140–141
    production, 157, 191, 268–269, 289
    profile of a wine, 228
    protein breakdown during ripening and, 227
    proteolysis production of, 196
    root-delivered, 157
    root pressure and, 65–66
    rootstocks and, 276
    S-containing, 251, 270–271
    in shaded berries, 256
    during springtime remobilization of storage, 124
    in sucrose concentration, 152–153
    sulfur in, 316
    TCA cycle and, 143*f*, 145
    temperature influencing concentration of, 262
    tryptophan, 39
    xylem/phloem sap with, 152*t*
Ammonium assimilation, 188, 190–192
    alternative pathway for, 192
    glutamate dehydrogenase (GDH) catalyzation, 192
    glutamine produced by GS, 191
    GS/GOGAT cycle, 191
    structure and reactions of compounds in, 191*f*
Ampelography, 11–12, 47–48
Amylopectin, 137–138
Amylose, 137–138
Anastomosis, 47
Androecium, 56
Anemic cultivars, 20
Angiospermae (Magnoliophyta), 4

Angiosperms, 4, 89–90
  fertilization in, 93
Anisohydric behavior, water deficit, 292–294, 293t
Antenna pigments, 131–132
Antheraxanthin, 133
Anthesis, 63–65, 89–94
  abortion of flowers and ovaries following, 96–97
  berry cytokinin concentration decline following, 97–98
  effect of arrested ovule development, 95
  environmental conditions, effect on, 207
  flower caps during, 362
  grape flowers immediately before, 90f
  grape flowers, vulnerability of, 368
  during meiosis, 297–298, 342
  proportion of flowers that develop into berries following, 96
  skin cell division and, 97–98
  stomatal density and, 209
  volume of mesocarp cells and, 97–98
  weather conditions, effect on, 206
Anthocyanidin reductase, 236–237
Anthocyanins, 15–16, 40–41, 145, 177, 212, 214, 222–223, 231–232t, 236–237, 240–243, 253, 267–269, 310, 320, 367–368
  accumulation of, 186, 240, 254–256, 341, 379–380
    in exocarp, 100–101
  cloudy or overcast conditions, impact of, 256–257
  color of, 240
  composition of, 242
  conversion of accumulating sugars into, 378
  cyanidin-based, 270
  disappearance of, 70
  free, 240–243
  malvidin-based, 243–244, 256–257, 263, 267, 270
  molecular forms, 243
  peonidin-based, 270
  petunidin-based, 267
  photoprotective, 242–243, 310, 320–321
  production of, 76–77, 263, 367–368
    by crushing of grapes, 243
    glucose, role in, 238–239, 241–242
  protection of, 241–243
  reaction with acetaldehyde or pyruvate, 244
  red, 76–77
  reddish or bluish berry skin, role in, 242
  self-association, 244
  specific metabolic pathway, 242
  sun exposure, impact on, 256
  transport proteins and, 214
  in wine, 244
Anthocyanoplasts, 241
Apianic cultivars, 20
Apical meristem, 30, 37

Apoplast, 107–108, 120–122, 121f, 156, 189, 213–215, 304, 343–346, 350, 354, 362–363
  boron concentration in, 328
  Ca^{2+} concentration in, 317–318
  Fe concentration in, 324
  K^+ concentration in, 317–318
  nonxylem, 124
  pH, 295, 322–323
Apoplastic ice formation, 354
Apoplastic pathway, 121–122, 156
Apoplastic phloem unloading, 215–216
Apoplastic (extracellular) sap, 295
Apoplastic unloading, 156
Appressorium, 366–367
Aquaporins, 105–106, 113, 117, 122, 124, 134, 153, 217, 285, 290–291, 305, 311–312, 331–332
Arginine, 197
Aromatic amino acid phenylalanine, 140
Ascorbate–glutathione cycle, 282
Assimilation of carbon, 134–140
Assimilation rate of a grapevine canopy, 165
ATP to ADP, 135–137, 136f
Axillary buds, 38–39
Axillary meristems, 39

B

Bacteria, 376–378, 377f
Bark, 43
Basal buds, 47
Basal inflorescence of a shoot, 89
Basic function of lamina, 47–48
Bell curve, 340
Berry
  development, 97–103
    pericarp cells, 159–160, 159f, 209
    pigment mutants, 18
Bicarbonate-induced chlorosis, 324
Biologically effective day degrees, 75
Biopores, 84–85
Biotic stress, 357–366
Biotrophs, 366–367
Bleeding sap, 65, 65f, 307
Bleeding, vegetative cycle, 65–66
Bloom, 59–60, 63–64f
  application of gibberellin sprays at, 260
  beginning of, 89, 90f
  bloom-time temperature effects, 207–208
  inflorescences with, 89–90, 90f
  yield potential in, 206
    cell divisions before and after bloom, 208
    conditions, 204, 206–207
    nitrogen (N) availability, 205
Bonarda, 19

Index

Bordeaux mixture, 326, 374
Boron, 188–189, 230, 306, 328–329
   deficiency, 328–329, 329f
   toxicity, 328–329, 329f
Botanical classification, 1–11
   American group, 8–10
   class Dicotyledoneae, 4
   division Angiospermae, 4
   domain Eukaryota, 3
   Eurasian group, 10–11
   family Vitaceae, 5
   genus *Muscadinia*, 5–6
   genus *Vitis*, 6–7
   kingdom Plantae, 3–4
   order Rhamnales, 4–5
   order Vitales, 4–5
Botrytis cinerea, 4, 97, 364–374
   infection in ripening grape cluster, 369f
   stilbenes in response to infection, 370
Box cut, 36
Bracts, 39, 45, 54
Brassinosteroids, 31–32, 34–35
Brettanomyces, 235
Browning of shoots, 43
Budbreak, 66–69
Buds, 44–47
   accessory, 46
   apical meristem, 46
   basal, 47
   bracts with, 44–47
   cell division and auxin production in, 66–67
   development, 78–82
   dormant, 46
   duration of chilling and lower temperatures and budbreak, 67
   fruitfulness, 203–205, 208, 269
   grapevine, 44–47
   higher-order, 46
   latent, 46–47
   lateral, 44–47
   necrosis, 67–68
   secondary, 46
   during shoot's subsequent free-growth phase, 46
   tertiary, 46
   water contentduring budswell phase, 66
   winter, 46
Bunch rot, 367–371
Bunch stem necrosis (BSN), 321

C

Cabernet franc, 18
Cabernet Sauvignon, 201, 370, 373, 375, 380f
   berry growth, 99f
   rate of sap flow, 118
   tannin concentrations, 238
Calcium, 317–319
Calvin cycle, 135–136
   ATP to ADP in, 136f, 137
   carbon into starch production with, 138–139
   carboxylation step, 136f
   first stage: carboxylation, 135–136
   glucose molecules in, 137–140
   NADPH to NADP in, 137
   3-phosphoglycerate formation, 137
   rubisco in, 136
   second stage: reduction, 137
   sucrose in, 136f
   three stages of, 136f
Calvin, Melvin, 135–136
Calyptra, 56–57
Canalization, 71–72
Cane pruning, 36
Canopy, defined, 168
Canopy–Environment Interactions, 168–187
   assimilation rate of a grapevine canopy, 165
   climatic conditions for resource availability in
      macroclimate, 168
      mesoclimate, 169
      microclimate, 169
   effect of altitude, 169
   grapevines productivity, influence of, 168
   humidity in, 184–185
      leaf growth with, 184
      transpiration with, 184
      vine water status with, 185
   ideal, 185–187
      canopy height/row width ratio in, 185
      canopy surface area in, 185
      fruit exposure in, 186
      fruit zone in, 186
      lateral shoot growth limited in, 186
      north/south oriented rows in, 186
      productivity upper limit with, 185–187
      pruning weight in, 186
      renewal zone in, 186
      shoots 15 nodes long in, 186
      15 shoots per meter of canopy in, 186
      vertically trained foliage in, 185
      yield to pruning weight ratio in, 186
   light in, 170–179
      compensation point, 171–172
      in the fruiting zone, 172
      grapevine adjustment to, 173
      importance for plant development of, 170
      leaf evolution for, 170
      leaf hairs in scatter of UV, 177–178
      leaf layers alter quantity/quality of, 175

Canopy–Environment Interactions *(Continued)*
    leaf's net CO_2 assimilation with zero, 171–172
    limited, 172
    photosynthetically active radiation (PAR), 170, 172
    range for vine vigor of, 172
    red/far red leaf absorption of, 173
    red/far red variations for day's, 175–176
    UV range with, 177–178
  in regions at higher latitude, 169
  south-facing slopes, 169–170
  temperature in, 179–182
    clouds alter, 181
    CO_2 assimilation influenced by, 179, 181
    global climate change with increase in, 181
    leaf growth with, 182
    leaf temperature changes, 179, 180f
    photosynthesis stimulated by, 180–181
  wind in, 183–184
    damage from, 183
    down row *vs.* across row turbulence, 184
    leaf's boundary layer resistance decreased with, 183–184
Canopy photosynthesis, 168
Capfall, 89–90
Carbohydrates
  construction costs of, 145, 146t
  production, 137–140
Carbon metabolism, 188–198
Carbon uptake, 134–140
  Calvin cycle, 135–136
    ATP to ADP in, 136f, 137
    carbon into starch production with, 138–139
    carboxylation step, 136f
    first stage: carboxylation, 135–136
    glucose molecules in, 137–140
    NADPH to NADP in, 137
    3-phosphoglycerate formation, 137
    rubisco in, 136
    second stage: reduction, 137
    sucrose in, 136f
    three stages of, 136f
  chloroplast with, 140
    starch production in, 137
    sugar production in, 137, 138f
  H_2O loss with, 134, 135f
  sucrose production with, 137
  uptake of CO_2 by leaves with, 134
Carboxylic acids, 223–224
Cardinal, 19
Carotenoids, 245–246
Catechin, 236–238
Cavitation, 115
"Cell-expansion" protein expansin, 50
Cell membrane, oxidative burst on, 363

Cellulose, 139–140, 145
Cell walls
  loosening in pulp, 212–213
  and membranes, 211–215, 213f
Chalcone synthase (CHS), 233–235
Channels, and transporters, 122–123
Chelates, 322–323
Chicks, 96–97
Chilling acclimation, 341
Chilling requirement, 67, 80
Chilling stress, 340–342
Chlorophyll *a* and *b*, 130–131, 131f, 133
Chloroplast, 140
Cinnamic acid, 233–237
Clones, 22–24
  Gewürztraminer, 17
Clusters, 53–56
Cofactors, 244
Cold acclimation, 342–356
Collection phloem, 156
Columella, 30
Competitors, 358–361
Concord, 19
  rate of sap flow, 118
Condensed tannins, 236–237
Conidia, 367
Convection, 114
Copigmentation, 244
Copper (Cu), 326
  deficiency, 326
  toxicity, 326
Cordon-trained grapevines, 36
Corolla, 88–89
Coulure, 96–97
4-coumaryl-CoA, 233–235
C_3 plants, 135–136
C_4 plants, 142
Crop load, 199–200, 272–275
Cropping, 199–200
Crown buds, 47
Crown gall, 376
Cryoprotectants, 347–349
Cultivar classification, 20–22
  anemic, 20
  apianic, 20
  based on use of grapes, 24
    brandy (distillation) grapes, 22
    juice grapes, 21
    raisin grapes, 21
    table grapes, 21
    wine grapes, 22
  based on viticultural characteristics, 21, 22t
  on the basis of winemaking characteristics, 22

grape composition, 22
production costs, 22
varietal aroma, 22
European wine grapes, 21
Vinum francicum, 21
Vinum hunicum, 21
grape varieties *vs.* grape cultivars, 11–20
by local soil, climatic conditions and cultural practices, 21
mutation variation, 17
natural hybridization, 18–19
Negrul's grouping, 21
nomentanic, 20
proles, 21
Proles occidentalis, 21
Proles orientalis, 21
Proles pontica, 21
root growth and distribution is influenced by, 85–86
V. sylvestris, 21, 29
Cultivated grapevines, 12. *See also* Cultivar classification
code of nomenclature, 12
commercial, 13
historical evidence, 11–12
by hybridization, 13–15
with "muscat" aroma, 18
mutations, 15
berry-pigment mutants, 18
genetic influence, 15–16
somatic, 13–15
propagation by cuttings, 20
for winemaking, 12–13
Cytokinins, 28, 42, 83, 85, 124, 145, 298, 308, 312–313, 376–378
action in root tips, 31
in buds, 71
callose deposition, role in, 43
fruit development and, 208–209
fungus-derived, 372
isopentenyladenine-type, 42
meristem activity and, 30
production of, 89, 99, 295, 305
production of pollen stimulated by, 90–91
ratio of auxin to, 280–281
root-derived, 42, 66–67, 71, 173, 280–281, 295
root growth and, 29
in stimulating cell division and cell expansion, 38–39, 49–50, 53, 66–67, 97–98, 308–309
support of cambial activity, 42
synthesis in the shoot, 71–72
thermosensitivity of, 208–209
zeatin-type, 42
Cytosol, 61–62, 108, 112, 159–160, 159f, 214, 226, 241, 247–248, 285, 312, 316–317
cytosolic Ca^{2+} concentration, 61–62, 318
cytosolic Fe concentrations, 323–324
cytosolic glutamine synthetase (GS1), 191, 290
of root epidermis and cortex cells, 189
sucrose production with, 137, 138f

D

2,4-D (2,4-dichlorophenoxyacetic acid), 49
Daktulosphaira vitifoliae Fitch, 4, 359
Day length, 61–65
at higher latitude, 62–63
typical responses to, 62
Dehydrins, 350
DELLA proteins, 72
Depolymerization, 222
Developmental physiology
yield formation, 196, 199–211
components of grapevine yield, 199–203
interactions between genotype and environment, 200–201
internal and external factors, 200
optimal balance between vine's vegetative and reproductive growth, 201
production of fruit, 202
pruning and other cultural practices, 202
reproductive performance, 201
shoot vigor for, 202
vine capacity, 201–202
vine self-regulation for, 201–202
vine's vegetative and reproductive growth, 201
vineyard yield, 200
yield potential, 203–211
berry and seed photosynthesis, effect of, 209
bloom occurrence with, 203, 206
bud development in dense canopies, 204
bud fruitfulness, 203–204
conditions during bloom, 204, 206–207
effect of light intensity on fruitfulness, 204
effect of seed number, 208
final fruit yield, 206
functional leaf area for inflorescence development, 205
of grapevine bud, 203
"heat wave" during the bloom–fruit set period, 209
inflorescence primordia, 204
limitation in assimilate supply, effect of, 210
maximum number of potentially fruitful shoots, 203
need to supply carbon and other nutrients, 205–206
nitrogen (N) availability during bloom, 205
period of inflorescence differentiation, 205
reserve status, 205–206
temperature effects, 204–209
time of day and, 210

Developmental physiology *(Continued)*
  timing and rate of water or nutrient application, 210–211
  in viticulture, 203–204
  water deficit effects, 205
Dicotyledoneae (Magnoliopsida), 4
Dimethyl sulfide, 251
Dormancy, 78–82
Dormant buds, 37, 37–38*f*, 46, 355
  adventitious root formation, 83
  chilling requirement to resume growth, 67, 80
  rehydration of, 66
  structural ice barrier protection, 345
Dormant-bud side (ventral side), 46
Downy mildew, 326, 366–367, 374–376
Drained upper limit, 284
Drought stress, 118
Druses, 52–53
Drying on the vine, 218–219

E

Ecodormancy, 81–82
Electromagnetic spectrum, 129, 130*f*, 170, 175
Electron transport chain, 132, 142–145, 181–182, 190
  photosynthetic, 190
  standard, 144–145
  transfer of electrons from NADH to oxygen, 144–145
Elicitors, 363–364
Embolisms in xylem network, 115
Embryonic root, 28–29
Endocarp, 59–60, 225
Endocytosis, 159–160
Endodormancy, 79
Endodormant buds, 80*f*
Endosperm nucleus, 94–95
Energy capture, 129–133
Epicatechin, 236–238
Epicuticular wax, 59–60, 59*f*
Erysiphe necator, 158–159, 364–367, 369*f*, 371–374, 376
Ethanol production, 141
Eugenol, 235–236, 251
Eukaryota, 3
Eurasian group, 10–11
  Vitis amurensis Ruprecht, 11
  Vitis coignetiae Pulliat, 11
  Vitis davidii Foëx, 11
  Vitis romanetii Romanet du Caillaud, 11
  Vitis sylvestris Hegi, 10
  Vitis vinifera L., 10
Euvitis, 5
Evapotranspiration, 110
Exocarp, 59–60, 95, 98, 220
Extracellular freezing, 343–344

F

Feedback inhibition, 138–139
Ferredoxin, 132, 316
Fertilization, 18, 57–58, 91, 95. *See also* Pollination
  in angiosperms, 93
  boron toxicity and, 328–329, 329*f*
  Ca deficiency, impact of, 319
  division of endosperm nucleus, 94–95
  double, 93
  formation of outer and inner integuments, 57–58
  fuel status and, 297–298
  K deficiency, impact of, 315
  ovary to develop into a berry, 96–97
  pollen tubes, growth of, 92
  pollination and, 89–94
  rate of pollen tube growth, 93
  reproductive cycle, 93
  in seeded grapes, 97–98
  temperature effects and, 338
Ferulic acid, 235
Flame Tokay, 19
Flavanols, 236–242, 262
Flavonoids, 233–236
Flavonols, 212, 231–232*t*, 236, 242–244, 263, 268–269, 275, 367–368, 370–371, 373
  absorption of UV light, 177–178
  accumulation of, 256, 350
  aglycones, 244
  in dark-skinned cultivars, 242
  hydrolyzation of, 244
  inhibition of auxin transport, 177–178
  as "sunscreen", 90–91, 177–178, 243, 339
  useful markers of light exposure, 256
  UV-protective, 257–258
Floaters, 95
Floral organogenesis, 88–89
Flowers, 161–162
  androecium of, 56
  anthesis and, 89–90
  from basal inflorescence of a shoot, 89
  calyptra of, 56–57
  flower initiation, 88–89
  formation, 86–89
  growth stages, 63–64*f*
  gynoecium of, 56–57
  infection of, 372–373
  morphology/anatomy of, 56–60
  ovule of, 56–58
  pistil of, 56–60
  proportion developing into berries, 96
  stamens of, 56
  star, 89–90

vascular bundles, 58–59
Foliage leaves, 47
Free energy per unit volume, 107
Freezing exotherm, 344
Freezing rain, 342
Fruitfulness, 202
　of basal buds, 47
　buds, 203–205, 208, 269
　effect of light intensity on, 204
Fruit maturity, 254–255
Fruit set, 94–97
Fusel alcohols, 271

G

Galloylation, 237–238
Gamma-amino butyrate (GABA), 92–93
Gating, 105–106
Geneva double-curtain, 73–74
Geographical distribution
　American group, 8–10
　　Vitis aestivalis Michaux, 9
　　Vitis berlandieri Planchon, 9
　　Vitis candicans Engelmann, 10
　　Vitis cinerea Engelmann, 10
　　Vitis labrusca L., 8
　　Vitis riparia Michaux, 9
　　Vitis rupestris Scheele, 9
　Eurasian group, 10–11
　　Vitis amurensis Ruprecht, 11
　　Vitis coignetiae Pulliat, 11
　　Vitis davidii Foëx, 11
　　Vitis romanetii Romanet du Caillaud, 11
　　Vitis sylvestris Hegi, 10
　　Vitis vinifera L., 10
Gewürztraminer clones, 17
Gibberellins, 72, 208
Girdling, 207–208
Global climate change
　consequences, 62, 347
　　adaptation to a conversion, 102
　　increase in photosynthesis, 147
　　rise in temperature, 75, 181, 259
　photorespiration with, 141–142
　reasons rising atmospheric CO_2 concentration, 141–142
Glucose phosphate molecule, 137
Glutamate dehydrogenase (GDH), 195–196
　ability to oxidize glutamate of, 196
　catalyzation, 192
Glutathione, 251
Glycolate pathway, 140–141

Glycolysis, 142–145, 143f, 181–182, 225–226, 233, 289
　ATP production via, 285
Glycosyltransferases, 236
Goethe, Johann Wolfgang von, 38–39
Gouais blanc, 19–20
Graft incompatibility, 381
Grafting, 24–26, 253, 275–276, 324, 355–356, 376, 378–379, 381
Graft-transmissible diseases, 360
Grape, 12–13
　brandy, 22
　composition, 22
　hybridization, 18–19
　juice, 21
　mutations in, 15
　production costs, 22
　raisin, 21
　table, 21
　varietal aroma, 22
　wine, 22
Grape berries, 56–60, 218–219
Grapevines. *See also* Morphology/anatomy
　buds of, 44–47
　clones, 22–24
　cooling tactics of, 113–114
　　convection, 114
　　radiation, 113
　　transpiration, 114
　cultivar classification, 20–22, 22t
　diffuse-porous, 42
　family, 5
　flowers, formation and development of, 56–60
　flow rate in field-grown vines, 118
　genus *Vitis*, 6–7
　meristems, 167
　monitoring of seasons, 62
　net assimilation rate of a grapevine canopy, 165
　order Rhamnales, 4–5
　phytoalexins characteristic of, 364
　production of phenolics (A) and terpenoids (B) in, 234f
　resistances to water flow through, 125
　rootstocks, 24–26, 25t
　root xylem of, 115
　shoot change and source-sink relations, 161, 162f
　tendrils and fruiting clusters of, 53–56
　variety *vs.* cultivar, 11–20
　water balance of, 287
Gravimorphism, 73–74
Gravity-sensing columella cells, 85
Green leaf volatiles, 245
Ground state, 130–131
Growing degree days (GDD), 74–75

Growth cycle
   day length, 61–65
   reproductive cycle, 86–103
   seasons, 61–65
   vegetative cycle, 65–86
Guaiacol, 235–236, 254–255
Guttation, 49
Gynoecium, 56–57

H

Harvest pruning, 218–219
Haustoria, 366–367
Heat acclimation, 335–340
Heat dissipation, 113–114
   convection, 114
   radiation, 113
   transpiration, 114
Heat shock proteins, 338–339
Heliothermal index, 75
Hemicellulose, 139–140
Hens, 96–97
Herbivores, 358–361
Heritability, 11–12
Higher-order buds, 46
High-sugar resistance, 365
High-temperature exotherm, 344
Homogenous nucleation point, 342–343
Horizontal gene transfer, 376
Host gene shutoff, 379
Humidity, 184–185
   leaf growth with, 184
   transpiration with, 184
   vine water status with, 185
Hydathodes, 49
Hydraulic capacitance, 125
Hydraulic conductance, 116
Hydraulic effect of stomatal opening or closing, 112
Hydraulic lift, 120
Hydraulic redistribution, 120
Hydrolysable tannins, 236–237
Hydrotropism, 86
Hydroxybenzoic acids, 231–232t
Hydroxycinnamic acids, 231–232t, 235–236
Hypersensitive response, 364, 373
Hypoxia, 215

I

Ice storm, 342
Idioblasts, 60, 225, 318
Imbibition, 28–29
Inflorescence necrosis, 196–197, 207

Inflorescences, 5, 38–39, 46, 55–56, 96, 196–197, 200, 285–286, 297, 323, 342, 354–355, 360, 368, 372–375
   abortion of, 206–207, 210
   basal, 89
   beginning of bloom with, 89–90
   buds with, 88–89
   flower differentiation with, 88–89, 88f
   and flower formation, 86–89
   on lateral shoots, 89
   nitrogen deficiency and excess, effect of, 205
   number of flowers per, 89, 205
   practice of cutting through, 260
   temperature effect on, 204
Interfascicular cambium, 42
Intrafascicular cambium, 42
Ion channels, 105–106, 122–123, 296
Iron, 322–325
Isohydric behavior, water deficit, 292–294, 293t
Isoprenoids, 145

J

Jumping genes, 15–16, 93, 250

K

α-ketoglutarate, 191
Kingdom Plantae, 3–4
Königin der Weingärten, 19

L

Laccase, 367
Ladybug taint, 359–360
Lamina, 47–53
   basic function of, 47–48
   growth of, 50
   petiole and, 47
   vein network of, 48–49
Latent buds, 46–47
Lateral buds, 44–47
Lateral gene transfer, 376
Lateral meristems, 37
Lateral roots, 28–29, 31–32
Lateral shoots, 44–47
Latitude-temperature index, 75
Leafroll disease, 379–380
Leaves, 47–53, 48f
   auxin production in, 49
   basic function of lamina, 47–48
   basis of ampelography, 47–48
   bracts, 47
   bundle sheaths, 48–49
   cell division and expansion in formation of, 50

Chardonnay, 48f
cotyledons, 47
cross section of, 52f
differentiation of, 47
epidermis, 50–51
foliage, 47
grapevine, 47–48, 51
hydathodes, 49
margins of, 49
Merlot, 48f
mesophyll, 52
  cells of, 47–48
N concentration in, 195–196
nitrogen assimilation in, 188–189, 193
number and length of trichomes in, 50
palisade parenchyma, 52–53
petiole of, 47
photosynthetic efficiency of, 52–53
primordia, 47
scales, 47
shape and basic architecture of its vascular system, 49
stomatal density of, 51
vascular bundles of, 47
vein network, 48–49
of *Vitis*, 49
V. vinifera, cross section of, 48f
Leeaceae, 5
Lenticels, 43
Leucoanthocyanidin reductase, 236–237
Light, 170–179
  anthocyanin accumulation in, 255–256
  compensation point, 171–172
  cultivar variation and fruit's susceptibility to shade, 256–257
  DOXP/MEP pathway, role of, 258
  in the fruiting zone, 172
  grape norisoprenoids and, 258
  grapevine adjustment to, 173
  importance for plant development of, 170
  leaf evolution for, 170
  leaf hairs in scatter of UV, 177–178
  leaf layers alter quantity/quality of, 175
  leaf's net CO_2 assimilation with zero, 171–172
  limited, 172
  methoxypyrazine accumulation and, 258
  photosynthetically active radiation (PAR), 170, 172
  range for vine vigor of, 172
  red/far red leaf absorption of, 173
  red/far red variations for day's, 175–176
  shaded fruit *vs.* sun-exposed fruit, 260
  tannin accumulation and, 257
  tartrate and, 256
  UV range with, 177–178
  variation in fruit composition, 255–259
  within-canopy microclimate, 255–256
  xanthophyll carotenoid production and, 257–258
  on yeast microflora, effect of, 258–259
Light absorption, 129–133
  conversion of light energy into chemical energy, 130
  electromagnetic spectrum with, 129, 130f
  green chlorophyll, role of, 130–131
  maximizing, 47–48
  oxidative stress, 320–321
  visible light with, 129, 257
Light compensation point, 171–172
Lignin, 139–140, 145
Linoleic acid, 245
Lipids, 50–51, 77–78, 100–101, 145, 281–282, 318, 349, 363
  construction costs of, 145, 146t
  grape berries composition/quality influenced by, 245–251
  mutations or cell death by oxidizing, 132–133

M

Macroclimate, 168
Macronutrients, 307–322
  calcium, 317–319
  magnesium, 320–322, 320f
  nitrogen, 307–311
  phosphorus, 311–314
  potassium, 314–316, 315f
  sulfur, 316–317
Magnesium (Mg), 229–230, 302–303, 320–322, 320f
Magnoliophyta, 4
Magnoliopsida, 4
Maillard reaction, 227–228
Malate, 65–66, 99, 106, 112, 143–144, 143f, 152–153, 191, 193, 214, 223–226, 247–248, 250, 252, 254, 256, 260–262, 268–272, 295–296, 313, 321, 324, 338–339, 350, 367
Malvidin-based pigment, 242
Malvidin-3-glucoside, 242
Manganese (Mn), 131–133, 230, 306, 313, 327
Marginal burn, 332–333
Marginal leaf necrosis, 378
Matchsticks on shoot, 378
Matrix potential, 107
Melanoidins, 227–228
Merlot, 201
  tannin concentrations, 238
Mesocarp, 95–98, 100–101, 159–160, 210, 212–213, 215, 217, 226, 229, 239, 241, 245, 251, 265, 271–272
  cell vacuoles, 213–216
  cell walls, 212–213, 215
  expansion, 214
  exudation of sap from, 357–358
  fatty acids, 245

Mesocarp *(Continued)*
  of a mature berry, 222
  nitrogen, 226–227
  pH, 226
  turgor pressure, 217
Mesoclimate, 169
Mesophyll cells, 47–48, 52, 70, 77–78, 110, 125, 134–135, 156, 160, 170, 174, 189, 283, 294–295, 324, 374
Methyl-erythritol 4-phosphate (MEP) pathway, 245–246
Microclimate, 169
Micronutrients, 322–330
  boron, 328–329, 329f
  copper, 326
  iron, 322–325
  manganese, 327
  molybdenum, 327–328
  nickel, 329
  silicon, 330
  zinc, 325
Micropyle, 57–58
Mild plant water stress, 287–288
Millerandage, 96–97
Mineral nutrients
  deficiency and excess, 301–334
  grape berries composition/quality with, 226–230
  macronutrients, 307–322
    calcium, 317–319
    magnesium, 320–322, 320f
    nitrogen, 307–311
    phosphorus, 311–314
    potassium, 314–316, 315f
    sulfur, 316–317
  transition metals/micronutrients, 322–330
    boron, 328–329, 329f
    copper, 326
    iron, 322–325
    manganese, 327
    molybdenum, 327–328
    nickel, 329
    silicon, 330
    zinc, 325
Mitochondria, 3
Mitochondrial electron transfer chain, 144–145
Molybdenum (Mo), 306, 327–328
  cofactor synthesis, 326
Molybdoenzymes, 327
Morphology/anatomy, 26–60
  buds, 44–47
    apical meristem, 46
    axillary, 38–39
    basal, 44f, 47
    basal-bud fruitfulness, 47
    crown, 47
    dormant, 46
    dormant Cabernet Sauvignon, 45f
    higher-order, 46
    initiation of bracts, 44–47
    latent, 44–47
    lateral, 44–47
    prompt, 44–47
    secondary, 46
  callose deposition, 43
  cambial activity, 42
  cell differentiation
    cortex, 40–41
    endodermis, 40–41
    epidermis, 40–41
    pearls, pearl glands and pearl bodies, 40
    phloem, 40–41
    procambium, 40
    specialized protein storage vacuoles, 40–41
    starch sheath, 40–41
    vascular bundles, 41
    vascular tissues, 41
    xylem, 41
  clusters, 53–56
    architecture, 55–56
    auxin and, 55–56
    bracts in, 54
    hypoclade, 55
    pedicel, 54–55
    peduncle, 55
    rachis, 54
    vascular bundles of, 55
  flowers, 56–60
    epicotyl, 58
    female, 56–57
    gynoecium of, 56–57
    male, 56–57
    pistil of, 56–60
    sepals and petals, 56
    stamens of, 56
  grape berries, 56–60
    endocarp of, 59–60
    epicuticular wax, 59–60
    epidermis of, 57–58
    exocarp, 59–60
    locules, 57
    mesocarp of, 59–60
    pericarp, 58–59
    structure of a ripe grape berry, 58f
    sun-exposed berries, 59–60
    tracheary elements, 58–59

lateral organs, development of, 38–39
    accumulation of auxin, 39
    cytokinins, role of, 39
leaf-opposed tendrils and clusters, production of, 39
leaves, 47–53, 48f
    auxin production in, 49
    basic function of lamina, 47–48
    basis of ampelography, 47–48
    bracts, 47
    bundle sheaths, 48–49
    cell division and expansion in formation of, 50
    Chardonnay, 48f
    cotyledons, 47
    cross section of, 52f
    differentiation of, 47
    epidermis, 50–51
    foliage, 47
    grapevine, 47–48, 51
    hydathodes, 49
    margins of, 49
    Merlot, 48f
    mesophyll, 52
    mesophyll cells of, 47–48
    number and length of trichomes in, 50
    palisade parenchyma, 52–53
    petiole of, 47
    photosynthetic efficiency of, 52–53
    primordia, 47
    scales, 47
    shape and basic architecture of its vascular system, 49
    stomatal density of, 51
    vascular bundles of, 47
    vein network, 48–49
    of *Vitis*, 49
    V. vinifera, cross section of, 48f
nodes, 44–47
phloem reactivation, 44
phyllotactic positioning of lateral organs, 39
rings of xylem, 42
root system, 28–35
    absorption zone, 32
    cell differentiation in formation of, 30f, 31
    cell division in, 30
    elongation zone of, 31
    endodermis, 31
    of established grapevines, 29–30
    horizontal roots, 29–30
    hypocotyl development of, 28–29
    initiation and elongation of root hairs, 32
    lateral roots, 28–29, 31–32
    lignification and, 34–35
    meristem cells of, 30
    Merlot, 30f
    parenchyma cells, 35
    perforation plates, 35
    pericycle, 31–32
    phloem, 32
    programmed cell death, 34–35
    root cap, 30
    root tip or apex, 30
    surface area of, 29–30
    tracheary elements, 34–35
    tracheids, 34–35
    uptake of nutrients and water, 30
    vascular cambium, 32
    vessel elements, 35
    V. vinifera, 38f
    xylem cells, 32–33
shoots, 36–44
    apical meristem, 37
    browning of, 43
    central zone, 37
    Concord shoot tip, 37f
    epicotyl (embryonic shoot), 36
    "fixed" growth and "free" growth of, 37
    internodes, 39
    juvenile phase of growth, 36
    lateral, 38–39, 44–47
    lateral meristems, 37
    pearls of, 40
    peripheral (or morphogenetic) zone, 37
    phyllotaxy of, 36
    pruning of, 36
    rib zone, 37
    tendrils, 36
    tertiary, 45–46
    three-node pattern of Syrah, 37f
    V. vinifera, 48f
tendrils, 53–56, 53f
    intercalary growth of, 53–54
    thigmotropic movement, 53–54
    tips, 53–54
trunks, 36–44
    sucker, 38
    tyloses, development of, 42–43
Mottle leaf, 325
Mucigel or mucilage, 30
Müller–Thurgau, 207f, 227, 309–310f, 320f
Muscadinia species, 5–6
    M. munsoniana Small, 6
    M. popenoei Fennell, 6
    M. rotundifolia, 242, 324, 355–356
    M. rotundifolia Small, 6, 56
    resistance to fungus, 373
Muscat of Alexandria, 18, 47

N

NADPH to NADP in Calvin cycle, 137
Necrotrophs, 366
Nectaries, 56–57
Nematodes, 359
Nickel (Ni), 306, 329
Nicotianamine, 322–323
Nitrite reductase (NiR), 189–190
Nitrogen, 307–311
Nitrogen assimilation, 145, 188–198
   ammonium assimilation, 190–192
   from cells to plants, 192–198
      application of N fertilizer, 197–198
      energy availability, 192–193
      environmental factors, 194–196
      leaves in assimilation for, 193–195
      nitrogen as nucleic acid component for, 192–193
      nitrogen distribution in soil for, 193
      photosynthesis effect on, 193, 196
      response of roots, 193
      roots/shoots nitrate assimilation for, 193
      sucrose transportation and, 194–195
      sufficient photosynthate energy for, 194
   definition, 188
   in grapevines, 188–198
      energy-dependent, 193
      with high carbohydrate status, 193
   nitrate uptake/reduction with, 188–190
      activity and forms of NR, 189
      boron (B) deficiency and, 188–189
      grapevines store/distribute, 189–190
      nitrate reductase forms for, 189
      protein ferredoxin in, 190
      in roots, 188–189
      vacuolar storage pool for, 189–190
   nitrogen-assimilating GS/GOGAT cycle, 233
   process, 188–198
Nitrogen deficiency, 193, 197, 307–308
Nitrogenous compounds, grape berries composition/quality with, 226–230
Noble rot, 367–368
Nodes
   ideal canopy with shoots of, 186
   morphology/anatomy of, 37, 44–47
   three-node pattern of a Syrah shoot, 180f
Nomentanic cultivars, 20
Nonflavonoids, 233–236
Nonspecific lipid transfer proteins (nsLTPs), 365
Nucleotides, 145
Nutrients, 124
   assimilation, photosynthesis with, 149–150, 196
   availability in soil, 302f
   deficiency and excess, 301–334
   macronutrients, 307–322
      calcium, 317–319
      magnesium, 320–322, 320f
      nitrogen, 307–311
      phosphorus, 311–314
      potassium, 314–316, 315f
      sulfur, 316–317
   remobilization, 196
   shoot/root growth influenced by, 85–86
   transition metals and micronutrients, 322–330
      boron, 328–329, 329f
      copper, 326
      iron, 322–325
      manganese, 327
      molybdenum, 327–328
      nickel, 329
      silicon, 330
      zinc, 325
   variation in fruit composition, 268–272
Nutrient uptake and transport, 114–127, 304
   active transport of, 123
   ATP hydrolysis, 122–123
   carriers of macronutrients, 123
   Casparian bands with, 121–122, 121f
   cell-to-cell movement of solutes, 120–121
   driving forces and resistances, 114–118
   hydraulic lift or hydraulic redistribution, 120
   inside xylem conduits, 124
   ion channels for, 122–123
   organic acids, 124
   passive, 123
   in phloem, 126
   root pressure and, 122
   semipermeability of cell membranes, 122
   soil–root–shoot pathway, 120
   symplastic and apoplastic pathways, 120–122, 121f
   transpiration stream, 124
   transport proteins, role of, 121–122
   via endodermis cell membranes, 120–121
   via roots, 120, 121f
   water flow dependent, 120

O

Ohm's law, 110, 116
Old vines, 253–254
Ontogenic resistance, 373–374
Organic acids, 60, 65–66, 73, 99, 106, 124, 143–145, 146t, 152–153, 157, 188, 191, 220, 223–226, 239–240, 243, 253, 262, 268–269, 273, 289, 295–296, 313, 324, 350, 367
Osmosis, 105–110

aquaporins with, 105–106
cell expansion and, 108
definition, 106
diffusion with, 107f
osmotic potential, 107
osmotic pressure, defined, 106
osmotic solutes (osmolytes), 106, 108
turgor pressure, 106
Osmotic pressure, 106
Overcropping, 199–200
    cluster thinning, 275
    effects of, 274
Ovules, 56–58, 95
Oxidative burst, on cell membrane, 363
Oxidative phosphorylation, 144
Oxidative stress, 281–283, 290, 310, 313–314, 320–321, 323–325, 327, 329, 334, 337–338, 341, 350
Oxidized tannins, 238
Oxygenation, 140

P

Palisade parenchyma, 52–53
Paradormant buds, 80f
Parthenocarpic fruit development, 95
Partial rootzone drying (PRD), 300
Partitioning, 161–167
    communication with, 163
    competition with, 164
    connection with, 163
    definition, 161–167
    development with, 164
    environmental variables with, 164
    interference with, 163
    patterns of, 162–164
    proximity with, 162
Pathogenesis-related (PR) proteins, 350, 362–365, 368, 370, 373–374
Pathogens, 358–361
    bacteria, 376–378, 377f
    bunch rot, 367–371
    downy mildew, 366–367, 374–376
    powdery mildew, 371–374
    viruses, 378–381, 380f
Peonidin-based pigment, 242
Pericarp, 58–59, 95, 97–101, 158–159, 215, 230
Peroxidases, 236, 256–257
Petiole, 47
Petunidin-3-glucoside, 242
Phenolics, 50–51, 143f, 146t, 211–212, 214–215, 230–244, 362–363
    amino acids with, 230–233

anthocyanins, 240–243
bonding with cell wall polysaccharides and polymers, 212–213
class of phenolic compounds, 231–232t
in cutin matrix, 214
enzymatic oxidation of, 230
flavonoids, 233–236
flavonols, 243–244
glycosylated, 235–236
incorporation into leaf cells, 177–178
nonflavonoids, 233–236
perception of astringency from, 237–238
production of, 230–244
    biosynthetic pathways, 234f
reactions in winemaking practices, 243
in seed coat, 225
smoke-derived, 235–236
synthesis pathways
    malonate, 233
    phenylpropanoid, 233
    shikimate, 233
tannins, 236–240
ultraviolet (UV)-protecting, 62, 177–178
Phenology, 61
Phenylalanine ammonia lyase (PAL), 233
Phloem
    bulk flow rate of solutes, 154
    cell differentiation, 40–41
    collection, 156
    minor-vein, 156
    nutrient uptake and transport in, 126
    pathways for sugar to vacuole, 159f
    pressure flow mechanism and, 160
    reactivation, 44
    release, 156–158
    root system, 32
    sequential sectors of, 156
    short pieces of RNA, 153
    sucrose concentration in, 158
    sucrose loading into, 157
    in sugar metabolism, 153
    symplastic to apoplastic, 160
    transport, 157
        rate in, 154
    water movement into and out of, 154–155
Phloem sap, 124, 152t, 153–154, 216
    along transport path, 154
    composition of, 152t, 153–154, 304–305
    flow velocities of, 154, 160
    pressure flow theory of, 154
    sugar concentration in, 288–289, 304–305
Phosphate deficiency, 138–139

Phosphate translocator, 137
3-phosphoglycerate, 135–136
Phosphorus, 311–314
   deficiency, effects of, 205
Photoinhibition, 133, 141
Photophosphorylation, 132
Photorespiration, 140–142
Photorespiratory carbon oxidation cycle, 140–141
Photosynthate, 147–148
   allocation, 161–167
      assimilate surplus in, 166–167, 166f
      assimilation rate determined by source in, 165
      categories, 161
      patterns in, 162–164
      phosphates availability in regulation of, 161
      source-sink relation with, 161, 162f
      storage in, 161
      transport in, 161
      utilization, 161
   photosynthetic machinery, 150, 174–175
   production of, 149–150
   sufficient energy for, nitrogen assimilation with, 194
   transport and distribution, 149–167
      amino acids in, 152t, 157
      flexibility with, 152
      nutrient assimilation with, 149–150, 196
      phloem sap in, 152–153, 152t
      sink with, 149–151, 158
      solute unloading in, 158
      sucrose in, 152–153, 159–160
      xylem sap in, 152t, 153–154
   in woody organs of the vine, 151
Photosynthesis, 62, 142
   capacity per unit leaf area, 173
   carbon uptake/assimilation, Calvin cycle, 135–136
      ATP to ADP in, 136f, 137
      carbon into starch production with, 138–139
      carboxylation step, 136f
      first stage: carboxylation, 135–136
      glucose molecules in, 137–140
      NADPH to NADP in, 137
      3-phosphoglycerate formation, 137
      rubisco in, 136
      second stage: reduction, 137
      sucrose in, 136f
      three stages of, 136f
   from cells to plants, 145–148
   chlorophyll content and, 151
   chloroplast with, 140
      starch production in, 137
      sugar production in, 137, 138f
   construction costs of compounds with, 145, 146t
   in C_4 plants, 142
   electromagnetic spectrum, 129, 130f, 170, 175
   energy content (E) of photons and, 129
   energy overload with, 133
   energy transfer with, 132–133
   and export rate of canopy, 165
   light absorption/energy capture with, 129–133
   light limited, 181
   light-saturated, 171–172
   nutrient assimilation with, 196
   organic compounds produced during, 149–150
   photochemical capture of solar energy, 130
   photosynthate $(CH_2O)_n$, 137
   processes
      chlorophyll excitation, 130–131, 131f
      conversion of light energy into chemical energy, 132–133
      electron transport chain, 132
      energy-releasing mechanisms, 131–132
      multiprotein complex, formation of, 131–132
      photochemical quenching, 132
      photoinhibition, 133, 141
      photophosphorylation, 132
      production of ATP from ADP, 132–133
      water-splitting reaction, 131–132
   product of, 141
   quantum yield or quantum efficiency, 132–133
   shade on the leaves and, 256
   sucrose export with, 138–139
   temperature change and, 179, 180f
   visible light with, 129
   whole-canopy, 172
Photosynthetic carbon reduction cycle, 135–136
Photosystem II (PSII), 131–132
Phyllotaxy or phyllotaxis, 36
Phytoalexins, 362–365, 369–370
   detoxification of, 364–365
   production of, 364
   rate of accumulation at the site of infection, 364–365
   stilbene, 369–370, 373, 375
Phytoanticipins, 362–363
Phytoplasma, 360
Phytosterols, 245
Pierce's disease, 364, 378
Pinot family, 17f, 18
   Pinot blanc, 16, 242–243
   Pinot gris, 16, 242
   Pinot noir, 16, 242–243
   tannin concentrations, 238
Pistil, 56–57, 88–89, 92–93, 95
Plantae kingdom, 3–4
Plasmodesmata, 34, 120–121
Plasmopara viticol, 365–367, 369f, 374–376

Plastochron, 47
Plastoglobuli, 246
Plugging of vessels, 378
Pollen, 89–94
    production of, 89–94
    tube growth, 90–92
Pollination, 57–58, 89–94, 206, 297–298, 342. *See also* Fertilization
    chilling stress and, 342
    cross, 13–15, 18, 91
    insect, 91
    K deficiency, impact of, 315
    self, 13–15
Polymerization of tannins, 238–239
Polyphenoloxidases, 236
Porphyrins, 145
Potassium, 314–316, 315f
Potato-taste defect, 250
Powdery mildew, 186, 271, 315, 357–358, 362–364, 366–367, 371–374
    fungus *E. necator*, 183–184, 330, 373
        resistance against, 373
    infection, 330, 371–372
Preprimordium, 38–39
Primitivo, 19
Proanthocyanidins, 236–237
Procambium, 40
Programmed cell death, 340
Prokaryotes, 3
Proles occidentalis, 21
Proles orientalis, 21
Proles pontica, 21
Proline, 197
Prompt buds, 45–47
Prompt-bud side (dorsal side), 46
Proteolysis, 103
Pruning
    cane, 36
    of shoots, 36
    spur, 36
    strategies in different seasons, 62–63
    vine capacity depressed with, 202
    weight in canopy–environment interactions, 186
    yield formation and, 202
Pseudomonas syringae, 344–345
Pyruvate, 144

Q

Quanta, 129
Quantum leap, 130–131
Quercus suber L., 235

R

Radiation, 113
Radicle, 28–29
Rain shadow effect, 109–110
Raphides, 52–53, 60, 225
Reaction center, 131–132
Reactive oxygen species, 281
Recrystallization process, 344
Red blotch-associated virus, 379
Red/far red light, 173, 346
Reductive pentose phosphate pathway, 135–136
Reference crops, 299
Reflex bleeding, 359–360
Regeneration, 137
Release phloem, 156, 158, 160
Reproductive cycle, 86–103. *See also* Vegetative cycle
    abscission of stamens, 90–91
    anthesis and, 89–90, 90f
    beginning of bloom with, 89–90
    cell division in berry, rate and duration of, 97–98
    cell expansion during cluster development, 97
    chromosome mixing, 94
    cluster initiation/differentiation in, 88f
    cytokinins and auxins, production of, 89, 97–98
    endosperm development, 93
    endosperm nucleus division, 94–95
    extent of branching before and after dormancy period, 89
    fertilization, 93, 97–98
    floral organogenesis, 88–89
    flower formation, 86–89
    flower initiation, 88–89
    formation of lateral meristems, 86–89
    fruit development and growth, 95
        fertilized berries, 96–97
        lag phase, 99–100
        parthenocarpic grapes, 95
        rapid increase in size of seeds and pericarp, 99–100
        ripening period, 100–101
        stenospermocarpic grapes, 95
    inflorescence and flower differentiation, 88–89, 88f
    inflorescences, formation of, 86–89
    inflorescences per bud with, 87–88
    meiosis process, 93–94, 96
    millerandage, 96–97
    pollen production, 89–94
    pollination, 89–94
    primordia
        branching, 89
        development of, 87
    seed development, 94–95
    seed germination, 102
    seedless grape berries, 101

Reproductive cycle *(Continued)*
  seed maturation, 101
  shoot organogenesis, 93
  somatic embryogenesis, 93
  stratification of seeds, 102
Resident vegetation, 358
Resorption, 196
Respiration, 142–145
  ADP to ATP, 144–145
  carbon intermediates from, 145
  citric acid cycle, 144
  efficiency, 143–144
  electron transport chain, 144–145
  glycolysis, 142–145, 143f
  mitochondria and, 144
  oxygen intermediates from, 144–145
  rate of meristems, 147
  of a vineyard soil, total annual, 147
Restricted spring growth, 360
Resveratrol, 369–370
Rhamnales, 4–5
Rhizobia, 188
Rhizobium vitis, 376
Rhizosphere, 115, 123
Ripe grape berry, structure of, 58f
  exocarp, 59–60
  mesocarp, 60
Roots, 118, 147
  ABA derived, ability of, 112
  ability to find water, 86
  absorption zone, 32
  adventitious, 29, 83
  allorhizic root system, 28–29
  biomass of grapevine, 29–30
  cell differentiation in formation of, 30f, 31
  cell division in, 30
  cytokinin action, 28
  delivered amino acids, 157
  derived cytokinins, 42, 66–67, 71, 173, 280–281, 295
  elongation zone of, 31
  endodermis, 31
  of established grapevines, 29–30
  flow of water from the soil toward, 118
  gravity response, 86
  growth, 82–86
    soil temperature, influence of, 65
    of tips, 82–86
  horizontal, 29–30
  hypocotyl development of, 28–29
  information exchange between shoots and, 28
  initiation and elongation of root hairs, 32
  interception, 301
  lateral, 28–29, 31–32
  lignification and, 34–35
  of mature vines, growth pattern of, 86
  meristem cells of, 30
  Merlot, 30f
  nitrate uptake/reduction with in, 188–189
  nutrient uptake and transport in, 120, 121f
  osmotic pressure of xylem and osmotic water uptake by, 65–66
  parenchyma cells, 35
  perforation plates, 35
  pericycle, 31–32
  phloem, 32
  programmed cell death, 34–35
  response in nitrogen assimilation, 193
  root cap, 30
  root tip or apex, 30
  surface area of, 29–30
  tracheary elements, 34–35
  tracheids, 34–35
  uptake of nutrients and water, 30
  vascular cambium, 32
  vessel elements, 35
  Vitaceae, 5
  V. vinifera, 38f
  water uptake, 119–120
  water uptake and transport, 115, 118
  whole-plant respiration and growth via, 147
  woody, 29–30
  xylem cells, 32–33
  xylem of grapevines, 115
Rootstocks, 24–26, 275–277
  agronomic characteristics of, 25t
  amino acids and, 276
  commonly used, 24–25
  grafting of, 24
  influence on composition of the fruit, 26
  in management of vine vigor, 25–26
  root growth and distribution is influenced by, 85–86
Rubisco, 131–132, 140–142, 195–196, 309, 313, 341
  Calvin cycle with, 136
  carboxylation function of, 141–142
  in C_4 plants, 142
  during drought stress, 289
  effect on photosynthesis, 283
  N content and, 195–196
  oxygenation rate, 181–182
  temperature changes and, 181–182
Russetting, 360

S

Saccharomyces cerevisiae, 4
Salinity, 330–334

Salt burn, 332–333
Sauvignon blanc berries, 211*f*
Savagnin, 18–19
Scavenging cells, 49
Scion cultivars, 24–26, 85–86, 334
Scott–Henry systems, 73–74
Seasons, 61–65
   endogenous "clock", 61–62
   pruning strategies and, 62–63
   reproductive cycle, 86–103
   time of active growth, 62–63
   vegetative cycle, 65–86
Secondary meristems, 38–39
Second crop, 45–46
Seedless table grapes, 95. *See also* Thompson Seedless
Seeds
   development, 97–103
   germination, 102
   maturation, 101
   stratification of, 102
Senescence, 144–145
Shatter, 96–97
Shedding, 96–97
Shoot growth, 69–76, 126
   accumulation of auxin in, 73–74
   in correlative inhibition, 69
   cycle, 76–77
   following budbreak, 69–70
   internodes development, 72
   leaf unfolding, 74
   secondary, 78–79
   shoot apex, sensitivity of, 76–77
   shoot "maturation", 72–73
   slowing down of, 79
   soil compaction, impact of, 84–85
   temperature effect on, 74
   time of rapid, 78–79
   upward and downward, 69–70
   vigor, 73
Shoot organogenesis, 93
Shoots, 36–44, 138–139
   browning of, 43
   central zone, 37
   Concord shoot tip, 37*f*
   epicotyl (embryonic shoot), 36
   "fixed" growth and "free" growth of, 37
   internodes, 39
   juvenile phase of growth, 36
   lateral, 38–39, 44–47
   leaf-opposed tendril production with, 39
   meristems, 147
      apical, 37
      lateral, 37
   pearls of, 40
   peripheral (or morphogenetic) zone, 37
   phyllotaxy of, 36
   pruning of, 36
   rib zone, 37
   tendrils, 36
   tertiary, 45–46
   three-node pattern of Syrah, 37*f*
   V. vinifera, 48*f*
Shot berries, 96–97, 207
Silicon, 330
Sink, 149–151, 158
   sink-to-source transition, 150–151
   solutes unloaded in, 158
   source-sink relation with, 161, 162*f*
      shoot change, 162*f*
   strength, 161–162
   sucrose imported into tissues of, 162–164
Sinkers, 95
Sodicity, 331
Sodium adsorption ratio (SAR), 331
Soil–plant–air continuum, 115
Soil solution, 301
Somatic embryogenesis, 93
Specific conductivity, 116
Spring fever, 316
Spur pruning, 36
Stamens, 56, 88–89
Starch accumulation, 139
Starch "assembly line", 137–138
Stenospermocarpic fruit, 95
Stenospermocarpy, 95
Steroids, 245
Stilbenes, 231–232*t*
   accumulation of, 370
Stomatal action, 110–114
   guard cells of closed stomata, 112
   protection of xylem conduits, 111
   reaction of stomata to ABA, 112
   role in regulating diffusional water loss, 111
   stomatal aperture, 111
   stomatal conductance, 111
   stomatal opening and closure, 111
Stress metabolites, 370
Stress physiology
   abiotic stress, 279–283
   biotic stress, 357–366
   carbon depletion, 283, 304
   flooding, 285
   mild plant water stress, 287–288
   nutrients: deficiency or excess in, 301–334
      macronutrients, 307–322
      transition metals and micronutrients, 322–330

Stress physiology *(Continued)*
  oxidative stress and damages, 290
  pathogens: defense and damage, 358–361
    bacteria, 376–378, 377f
    bunch rot, 367–371
    downy mildew, 366–367, 374–376
    powdery mildew, 371–374
    viruses, 378–381, 380f
  responses to abiotic stress, 279–283
    antioxidant defense systems, 290
    auxin and cytokinin concentrations, change in ratio of, 280–281
    compatible solutes, role of, 281
    optimum resource allocation, 280
    organ sacrifice, 283
    reduction of leaves' photosynthetic rate, 282–283
    rubisco, role of, 283
    senescence and abscission process, 283
    shoot and root apical meristems, adaptation of, 279–280
  salinity, 330–334
  temperature: too hot or too cold in, 335–356
    chilling stress, 340–342
    cold acclimation/damage, 342–356
    heat acclimation/damage, 335–340
  water deficit, 287–300
    drought signals and growth, 294–296
    irrigation, 299–300
    isohydric to anisohydric behavior, 292–294, 293t
    root growth and nutrient uptake, 296–297
    stomatal limitation and metabolic limitation, 288–290, 288f
    xylem cavitation and repair, 290–292
    yield formation, 297–299
  waterlogging, 285–287
    Vitis species, susceptibility of, 286
  water stress, 287–288
  water surplus, 285–287, 286f
Stress–strain ratio, 214
Strigolactones, 42, 71, 83
Sucrose, 137
  in amino acids, 152–153
  in Calvin cycle, 136f
  carbon uptake and production of, 137
  in cytosol, 137, 138f
  loading into phloem, 157
  in phloem, 158
  phosphate groups (P_i), release of, 137
  photosynthesis and export of, 138–139
  synthesis, 138–139
  transport and distribution of photosynthate, 152–153, 159–160
  transportation, 194–195
  transporter (ST) proteins, 159

Sugars, grape composition and fruit quality, 219–223, 221f
Sulfur (S), 316–317
Supercooled tissue, 342
Symplast, 120–122, 121f, 156–160, 343–344, 346
Sympodial bundles, 41–42
Syrah, 18
  sugar concentration, 221f
  tannin concentrations, 238
Systemic N acquisition response, 308

T

Tannins, 231–232t, 236–240
  classification of, 236–240
  colorless, 237
  concentration in red wines, 239
  condensed, 236–237
  extractability during ripening, 239
  functional properties, 239
  genes responsible for biosynthesis of, 238
  hydrolysable, 236–237
  polymerization of, 238–239
  production of flavanols, 238–239
  in seeds, 238
  subunits in the skin, 238–240
  wine's astringency and, 237–238
Tartrate, 235, 256
Temperature, 179–182
  application of gibberellin sprays at bloom, 260
  clouds alter, 181
  CO_2 assimilation influenced by, 179, 181
  global climate change with increase in, 181
  influencing concentration of amino acids, 262
  inversions, 169
  leaf growth with, 182
  leaf temperature changes, 179, 180f
  malvidin-based anthocyanins, 263
  photosynthesis stimulated by, 180–181
  stress physiology, 335–356
    chilling stress, 340–342
    cold acclimation/damage, 342–356
    heat acclimation/damage, 335–340
  variation in fruit composition, 259–264
Tendrils, 53–56, 53f
  intercalary growth of, 53–54
  thigmotropic movement, 53–54
  tips, 53–54
Terpenes, 245–246
Terpenoids, 245–246
Thigmomorphogenetic response, 183
Thigmotropic movement, 53–54
Thompson Seedless, 21, 47, 73, 95, 370–371

fruitfulness of buds, 87
inflorescence initiation and development, 204
maintenance of water flow, 118
rate of sap flow, 118
vigor, 73
Thylakoid lumen, 132
Tonoplast, 214
Torus, 56
Tracheobionta, 3
Traminer, 18, 20
Transaminases, 197
Transition metals and micronutrients, 322–330
  boron, 328–329, 329f
  copper, 326
  iron, 322–325
  manganese, 327
  molybdenum, 327–328
  nickel, 329
  silicon, 330
  zinc, 325
Transpiration, 110–114
  boundary-layer resistance and, 110
  of canopies, 110
  definition, 110
  driving water flow up xylem, 113
  equation, 111
  factors influencing, 110
  in grapevine leaf, 111
  negative pressure of xylem and, 113
  rate per unit leaf area, 110
  stomatal pore resistance and, 110
  stream and growth, 124–127
  temperature as determinant of, 109–110
  turnover of water in transpiring leaf, 113
  in a typical grapevine leaf, 111
  water uptake via, 115
Transpiration–adhesion–cohesion–tension theory of xylem sap, 115
Transporters, channels and, 122–123
Trichoblasts, 32
Triose phosphate transporter (TPT), 137, 138f
Trunks, 36–44
  sucker, 38
Tumor-inducing (Ti) plasmids, 376
Turgor pressure, 106
Tyloses, 42–43

U
UDP-glucose, 236
Uridine triphosphate (UTP), 137
UV light, 130f, 177–178
  leaf hairs in scatter of, 177–178

V
Vapor pressure deficit (VPD), 109
Vascular cambium, 42–43
Vegetative cycle, 65–86. *See also* Reproductive cycle
  anthocyanins, action of, 70, 76–77
  auxin production, 66–67, 74, 87–88
  bleeding, initiation of, 65–66
    soil moisture and, 65
  budbreak phenomenon, 66–69
  bud dormancy, 62, 79–80
  canalization, 71–72
  cell division and auxin production in buds, 66–67
  chilling requirement, 67
  cytokinin production, 72
  DELLA proteins, development of, 72
  early canopy development, 83–84
  ecodormancy, 81–82
  endodormancy, 79
  evolution of tendrils, 69
  gibberellins, action of, 72
  grapevine growth stages, 63–64f
  growth cessation and senescence, 76–78
  leaf senescence, 76–77
  root growth and distribution, 85–86
    auxin, role of, 83
    root density, 84
    root tips, 83
  root pressure, 65–66
  shoot growth, 69–76
    internodes development, 72
    leaf unfolding, 74
    rapid, 78–79
    secondary, 78–79
    shoot apex, sensitivity of, 76
    shoot "maturation", 72–73
    soil compaction, impact of, 84–85
    temperature effect on, 74
    upward and downward, 69–70
    vigor, 66–67
  strigolactones production, 71
  suppression of axillary bud outgrowth, 71
  uptake of water and nutrients, 61–62
Vine balance, 199–200
Vine capacity, 201–202
Vineyard floor vegetation, 358
  root density and, 85
Viniferins, 364
Vinum francicum, 21
Vinum hunicum, 21
Viognier, 20
Violaxanthin, 133
Viruses, 378–381, 380f

Visible light, 129, 170, 177, 256–257, 260
Vitaceae family, 5
Vitaceae roots, 5
Vitales, 4–5
Viteus vitifolii, 359
Vitis species, 6–7, 49, 375, 377
    leaves of, 49
    pith and cortex, 44–45
    reproductive growth in, 87
    root pressure of, 115
    susceptibility of waterlogging, 286
    tannin subunits in, 237–238
    V. aestivalis, 9
    V. amurensis, 11, 375, 377
    V. berlandieri, 9, 56
    V. candicans, 10
    V. champinii, 375
    V. cinerea, 10, 235–236
    V. coignetiae, 11
    V. davidii, 11
    V. labrusca, 8, 39, 58–59, 365, 375, 377
    V. labruscana, 87–90
        shoots of, 89–90
    V. riparia, 9, 87–88, 286, 373, 375, 377
    V. romanetii, 11
    V. rupestris, 9, 87–88, 286, 375, 377
    V. sylvestris, 10
    V. vinifera, 10, 38f, 58–59, 62, 89–90, 204, 287, 359, 361, 365, 370–371, 373–378, 381
        cross section of, 48f
        grafting partners, 355–356
        primordia in, 87–88
        resistance to fungus, 373
        salinity effects on, 331
        shoots of, 89–90
Volatiles, grape berries composition/quality with, 245–251
Vugava bijela, 20

W

Water
    grape composition and fruit quality, 215–219
    variation in fruit composition, 264–268
Water deficit, 287–300
    drought signals and growth, 294–296
    irrigation, 299–300
    isohydric to anisohydric behavior, 292–294, 293t
    root growth and nutrient uptake, 296–297
    stomatal limitation and metabolic limitation, 288–290, 288f
    xylem cavitation and repair, 290–292
    yield formation, 297–299
Water potential, 105–110
    equation, 119

Water relations
    aquaporins, 105–106
    cell expansion, 105–110
    matrix potential, 107
    nutrient uptake and transport, 114–127
    osmosis, 105–110
    stomatal action, 110–114
    transpiration, 110–114
    water potential, 105–110
    water uptake/transport, 114–127
Water surplus, 285–287, 286f
Water uptake, 119–120
Water uptake and transport, 114–127
    aquaporin activity of root cell membranes, 118
    canopy size and, 118
    cavitation and, 115, 118
    direction of movement, 114
        transpiration-driven water flow, 115, 119, 122–123
        xylem sap, 115
    driving forces and resistances, 114–118
    drought stress and imbalance between, 118
    friction between water and conduit walls, effect of, 117
    gas blockages and, 115
    hydraulic architecture for, 117
    long-distance transport, 124
    pit permeability, 117–118
    resistances to water flow through a grapevine, 125
    root pressure for, 114, 119
    surface area of root system and, 120
    surface tension, role of, 115
    symplastic and apoplastic pathways, 120–122, 121f
    transpiration rate and soil-to-leaf flow rate, 118
    within xylem, 116, 125
Water vapor concentration, 110
Water vapor pressure, 110
Weeds, 358
White grapes, 242–243
White wines, 266
Wind, 183–184
    damage from, 183
    down row *vs.* across row turbulence, 184
    leaf's boundary layer resistance decreased with, 183–184
    physical damages from strong, 183
    shoot growth and, 183
    thigmomorphogenetic response, 183
Winkler regions, 75
Winter buds, 46
Woody roots, 29–30

X

Xanthophyll cycle, 133
Xylella fastidiosa, 3, 378

Xylem
  ABA derived in, 113
  cell differentiation, 41, 126
  cells of roots, 32–33
  cells of wood, 140
  embolisms in, 115
  negative pressure of, and transpiration, 113
  nutrient uptake and transport in, 124
  osmotic pressure of, and osmotic water uptake by, 65–66
  protection of, 111
  rings of, 42
  transpiration driving water flow up, 113
  water uptake and transport within, 115–116, 125
Xylem sap, 216
  sustaining of surface tension, 115
  transpiration–adhesion–cohesion–tension theory of, 115
  transport and distribution of photosynthate, 152t, 153–154
Xylem vessel, 41–42, 49, 113, 376, 378
  air embolisms, 344
  bacterial infection, 378
  cavitation of, 115, 284, 290–292
  ice formation within, 354
  nitrogen (N) status and, 297, 308, 311
  scavenging cells of, 49
  shoot growth and, 73–74
  water and solute uptake, 125

Y

Yeast-assimilable nitrogen (YAN), 226
Yield component compensation principle, 167, 203, 210–211
Yield formation, 196, 199–211
  components of grapevine yield, 199–203
  of grapevine bud, 203
  interactions between genotype and environment, 200–201
  internal and external factors, 200
  optimal balance between vine's vegetative and reproductive growth, 201
  production of fruit, 202
  pruning and other cultural practices, 202
  reproductive performance, 201
  shoot vigor for, 202
  vine capacity, 201–202
  vine self-regulation for, 201–202
  vine's vegetative and reproductive growth, 201
  vineyard yield, 200
  water deficit, 297–299
Yield potential, 203–211
  berry and seed photosynthesis, effect of, 209
  bloom occurrence with, 203, 206
  bud development in dense canopies, 204
  bud fruitfulness, 203–204
  conditions during bloom, 204, 206–207
  effect of light intensity on fruitfulness, 204
  effect of seed number, 208
  final fruit yield, 206
  functional leaf area for inflorescence development, 205
  of grapevine bud, 203
  "heat wave" during the bloom–fruit set period, 209
  inflorescence primordia, 204
  limitation in assimilate supply, effect of, 210
  maximum number of potentially fruitful shoots, 203
  need to supply carbon and other nutrients, 205–206
  nitrogen (N) availability during bloom, 205
  period of inflorescence differentiation, 205
  reserve status, 205–206
  temperature effects, 204–209
  time of day and, 210
  timing and rate of water or nutrient application, 210–211
  in viticulture, 203–204
  water deficit effects, 205

Z

Zeatin-type cytokinins, 42
Zeaxanthin, 133
Zinc, 325
Zygotic embryo, 93

9780128163658